SECOND EDITION

Introduction to Modern Physics

John D. McGervey

Case Western Reserve University

ACADEMIC PRESS, INC.
Harcourt Brace Jovanovich, Publishers
San Diego   New York   Berkeley   Boston
London   Sydney   Tokyo   Toronto

ACADEMIC PRESS, INC.
1250 Sixth Avenue, San Diego, California 92101

United Kingdom Edition Published by
Academic Press, Inc. (London) Ltd.
24/28 Oval Road, London NW1 7DX

ISBN: 0-12-483560-0
LCCCN: 82-73936

PRINTED IN THE UNITED STATES OF AMERICA

88 89   9 8 7 6 5 4

Contents

Preface   vii

Preface to the First Edition   ix

Acknowledgments   xi

1   The Atomic Nature of Matter and Electricity

1.1	Kinetic Theory of Gases	2
1.2	The Electron	12
1.3	Determination of Avogadro's Number	16
	Problems	23

2   The Theory of Relativity

2.1	Experiments Preceding the Theory of Relativity	29
2.2	Explanations	37
2.3	The Lorentz Transformation	40
2.4	Relativistic Dynamics; Four-Vectors	50
	Problems	59

3   The Old Quantum Theory

3.1	Black-Body Radiation	64
3.2	The Photoelectric Effect	73
3.3	Line Spectra and the Bohr Atom	75
3.4	The Correspondence Principle	82
3.5	The Franck-Hertz Experiment	92
	Problems	92

4 Waves and Particles

4.1	X-Rays	96
4.2	Matter Waves	106
4.3	Wave Packets	112
4.4	The Uncertainty Principle	124
	Problems	131

5 Schrödinger Equation I: One Dimension

5.1	Construction of the Schrödinger Equation	138
5.2	Boundary Conditions	141
5.3	Probability Current	146
5.4	One-Dimensional Square Well Potential	150
5.5	Barrier Penetration	164
5.6	Expectation Values and Operators	170
5.7	The Harmonic Oscillator	176
	Problems	182

6 Schrödinger Equation II: Three Dimensions

6.1	Extension of the Schrödinger Equation to Three Dimensions	188
6.2	Spherically Symmetric Potentials and Angular Momentum	190
6.3	Measurement of Angular Momentum	199
*6.4	An Example: The Three-Dimensional Harmonic Oscillator	211
6.5	The Radial Equation	214
*6.6	The Three-Dimensional Square Well	217
*6.7	Scattering of Particles from a Spherically Symmetric Potential	223
	Problems	234

7 The Hydrogen Atom

7.1	Wavefunctions for More than One Particle	240
7.2	Energy Levels of the Hydrogen Atom	242
7.3	Fine Structure in the Hydrogen Spectrum	247
7.4	Spin and Relativity	266
	Problems	273

8 Further Applications of Quantum Theory

8.1	Time-Independent Perturbation Theory	280
8.2	Identical Particles	288
*8.3	The Helium Atom	296
*8.4	The Periodic Table of the Elements	304
	Problems	311

9 Atomic and Molecular Spectra

9.1	Atomic Spectroscopy	314
9.2	The Zeeman Effect Revisited	326
9.3	Molecular Structure	331
9.4	Molecular Spectra	335
	Problems	342

10 Atomic Radiation

10.1	Time-Dependent Perturbation Theory; Transition Rates	346
10.2	Spontaneous Transitions	355
10.3	Selection Rules	359
*10.4	Amplification by Stimulated Emission of Radiation — The Maser and the Laser	367
	Problems	376

11 Quantum Statistics

11.1	The Three Kinds of Statistics; An Example	380
11.2	Derivation of the General Form for Each Distribution Function	384
11.3	Applications of Bose-Einstein Statistics	394
11.4	Application of Fermi-Dirac Statistics: Free Electron Theory of Metals	410
	Summary	425
	Problems	425

12 The Electronic Structure of Solids

12.1	Energy Levels for a System of N Atoms	430
12.2	Traveling Electron Waves in a Solid	432
12.3	Solutions of Schrödinger's Equation for a Periodic Potential	438
12.4	Superconductivity	446
*12.5	Brillouin Zones and the Fermi Surface	449
*12.6	Insulators and Semiconductors	462
	Problems	467

13 Nuclear Radiation

13.1	Early Work with Radioactivity	472
*13.2	Passage of Radiation through Matter	480
*13.3	Positron Annihilation	490
*13.4	Recoilless Resonant Absorption of Gamma Rays (Mössbauer Effect)	501
13.5	Biological Effects of Radiation	514
	Problems	520

14 Properties of the Nucleus

14.1	Charge	526
14.2	Radius	526
14.3	Angular Momentum and Magnetic Dipole Moment	531
14.4	Electric Quadrupole Moment	533
14.5	Mass and Binding Energy	537
14.6	Parity	542
14.7	Stability	543
14.8	Shell Structure	551
14.9	Collective Motions	555
14.10	Properties of the Deuteron	556
	Problems	560

15 Nuclear Transformations

*15.1	Theory of Alpha Decay	566
*15.2	Theory of Beta Decay	571
*15.3	Gamma Decay	594
15.4	Nuclear Reactions	600
	Problems	618

16 Particles and Interactions

16.1	Meson Theory of the Nuclear Force	623
16.2	Properties of the Pion	626
16.3	Leptons and the Weak Interaction	630
16.4	Discovery of "Strange Particles"	633
16.5	Conservation Laws	637
*16.6	Properties of the K Mesons	641
16.7	Resonances	652
*16.8	Classification of Particles	655
16.9	Unified Theories	663
	Problems	669

Appendix A: Probability and Statistics	677
Appendix B: Derivation of the Bragg Scattering Law	686
Appendix C: Solution of the Radial Equation for the Hydrogen Atom or Hydrogenlike Ion	690
Appendix D: Table of Atomic Species	695
Appendix E: "Stable" Particles	738
Appendix F: Table of Physical Constants	740
Answers to Selected Problems	742
Index	747

Preface

I am pleased that the response to the first edition has justified this second edition. The second edition maintains the purpose and the level of the first edition, while improving it in the light of current knowledge. Like the first, this edition can be used for a one-semester undergraduate quantum mechanics course, using chapters three through ten; the other chapters then provide a good selection of illustrative examples, and should be useful to students as supplementary reading for subsequent more specialized courses.

New discoveries, as well as better ways to present the older material, have been incorporated into this edition wherever possible. Some older material has been abbreviated to make room for these improvements. Many figures have been improved, and new figures have been added. Newer references have been inserted. Problem solutions are now included, and there are seventy-six additional problems, including some simpler "confidence-building" problems that were relatively scarce in the first edition.

The scope of the book has been broadened by inclusion of material on molecular spectra, superconductivity, and the biological effects of radiation. Most of this new material meets the original aim of illustrating fundamental quantum theory. For example, the discussion of molecular spectra draws upon the quantum theory of angular momentum and the harmonic oscillator.

The final chapter has been revised extensively to convey the recent discoveries in particle physics, which have deepened our understanding of quarks and leptons. I am indebted to R. W. Brown for his many helpful suggestions and corrections for this chapter.

I would like to thank B. S. Chandrasekhar, T. G. Eck, W. J. Fickinger (who provided Figure 10, Chapter 16), L. L. Foldy, L. S. Kisslinger, and W. Tobocman for helpful suggestions, and to thank the numerous students who made constructive comments. Special thanks are due to Franklin Miller, Jr., who used this text for many years and who volunteered many detailed and useful comments concerning all aspects of the book.

Preface to the First Edition

This book evolved from lecture notes originally used in a course for advanced undergraduates at Western Reserve University and (after the federation of Case Institute of Technology and Western Reserve University) at Case Western Reserve. The notes, which at first supplemented a number of books used in conjunction with the modern physics course, were revised and expanded during the years in which the author lectured on this subject. Thus the material of this text has been extensively class-tested with students of various backgrounds and interests.

The purpose of the book is to bridge the gap between the mainly descriptive treatment of phenomena given in elementary texts and the extremely formal, theoretical accounts which one finds in graduate-level texts. It is not encyclopedic in its coverage; it aims to give the student a feeling for the principles of modern physics through the selection of topics which fit together and draw upon each other for reinforcement. For example, only those parts of solid-state physics which build directly upon the fundamentals of quantum theory are included, and such topics as crystallography, which require a digression from the main thrust of the book, are excluded.

Any book of this type inevitably reflects the personal preferences of the author in the final selection of the many interesting topics that can justifiably be covered. I have chosen to give the details of some classic experiments (for example, Millikan oil drop, gravitational red-shift measurement, detection of the antineutrino) in order to convey some appreciation of the care which must be taken in experimental physics, and to convince the student that these discoveries really can be believed. In addition, some recently developed areas of physics are rather expansively treated; there are lengthy sections on the laser and maser, on the Mössbauer effect and its applications, on the applications of positron annihilation, and on the properties of K mesons.

To save space for these topics some special material is introduced in the problem sections rather than in the body of the text, and some of the more complicated formal mathematical manipulations are either omitted or relegated to an appendix. Students who go on to graduate school will have ample opportunity later to manipulate differential equations, and those who do not go on will find that some knowledge of experimental research areas is more valuable to them. This is not to say that mathematics should be ignored, and in fact it is not; there is unusually extensive treatment of such things as raising and lowering operators and angular momentum operators and their eigenfunctions.

The often-maligned "historical" approach is used in introducing many of the topics. But an attempt was made to avoid the *pseudo*historical treatment which glosses over the difficulties and gives the impression that one discovery followed another in regular progression. There are many references to original papers so that students can discover some history for themselves. And in some cases wrong approaches are discussed, so that students can see why certain ideas are rejected.

I have found that the total amount here is more than can be covered in a two-semester course. Individual sections which may be omitted without affecting the continuity are indicated by an asterisk in the table of contents. Many of these sections, however, contain the more interesting special topics, so it would be a pity to omit too many. Since much of the material of the early chapters is now taught in freshman and sophomore courses, one could save some time by starting a course with Chapter 3 or even Chapter 4; one could then "flash back" to the earlier chapters when background material is needed. One might then have enough time to expand certain topics which are of special interest to the instructor; ample references are given to make this possible.

Acknowledgments

I would like to acknowledge the assistance of countless students in my classes, who have caught errors, ambiguities, and cloudy phrasing. In addition, I am grateful to colleagues and graduate students who have read over various chapters and suggested changes or additions, in particular, to Virginia Walters, who read every chapter (sometimes three or four versions of each) and made numerous suggestions; also to Peter Bond, Arnold Dahm, Bill Fickinger, Bill Frisken, C.-Y. Huang, Leonard Kisslinger, Stefan Machlup, and Brian Murray, who read various chapters and suggested improvements. I have also benefited greatly from discussions with Les Foldy, Ben Green, Berol Robinson, Joe Weinberg, and Paul Zilsel. Of course responsibility for any remaining errors is solely mine.

CHAPTER 1

The Atomic Nature of Matter and Electricity

All of modern physics is based upon the analysis of matter by reference to its elementary constituents— molecules, atoms, and "elementary particles." But the atomicity of matter is not at all obvious, and the student may well wonder how we can place so much confidence in our analysis when our observations of these particles are necessarily so indirect.

Therefore, it is the aim of this chapter to show how it is possible to obtain a surprising amount of information about the atomic structure of matter even with relatively simple apparatus. This information leads us into an investigation of the electrical properties of atoms and a study of the "atom of electricity"—the electronic charge. We shall find that the atomic nature of electricity provides us with a great deal of our evidence for the atomic nature of matter.

1.1 KINETIC THEORY OF GASES

Observations of gases provided the earliest clues to the atomicity of matter. Around 60 B.C. Lucretius observed that the random motion of dust particles in the air could be explained by assuming that it was caused by the impact of smaller, invisible particles of which the air itself was composed. He said:

> "It is meet that you should give greater heed to those bodies which are seen to tumble about in the sun's rays, because such tumblings imply that motions also of matter latent and unseen are at the bottom. Motion moves up from the first-beginnings (atoms) and step by step issues forth to our senses, so that those bodies also move, which we can discern in the sunlight, though it is not clearly seen by what blows they so act."

Similar movements of microscopic particles suspended in a liquid were observed eighteen hundred years later by the botanist Brown, and the motion became known as Brownian motion.

The idea of Lucretius was not developed further until Bernoulli, in 1739, made the first attempt to explain a natural law by reference to atoms. By assuming that the molecules of a gas all move with the same speed, he worked out a relation between this speed and the pressure and density of the gas, showing that it led to Boyle's law: pressure × volume = constant at a given temperature.

To understand Bernoulli's result, suppose that an atom of mass m is bouncing back and forth inside a cube of side l. When the atom strikes the top of the cube, its vertical velocity component v_z is reversed, and momentum equal to $2mv_z$ is transferred to the top of the cube. The atom then bounces off the bottom of the cube and rebounds to the top, taking a time equal to $2l/v_z$ between successive collisions with the top of the cube. (If v_z is large, we may neglect the change in v_z caused by gravitational acceleration of the atom.) The average force exerted on the top of the cube is just the rate at which momentum is transferred to it, or $2mv_z$ divided by $2l/v_z$, which is mv_z^2/l.

The total force exerted on the top by all the atoms in the cube is therefore $\overline{Mv_z^2}/l$, where M is the total mass of the atoms in the cube and $\overline{v_z^2}$ is the average value of v_z^2 for the various atoms. The pressure on the top is simply the total force divided by the area l^2, so $P = \overline{Mv_z^2}/l^3 = \overline{Mv_z^2}/V$, where V is the volume of the cube. Since the x, y, and z directions should be equivalent, $\overline{v_x^2} = \overline{v_y^2} = \overline{v_z^2}$; thus $\frac{1}{3}(\overline{v_x^2} + \overline{v_y^2} + \overline{v_z^2}) = \overline{v_z^2} = \frac{1}{3}\overline{v^2}$, and we have

$$PV = \tfrac{1}{3} M\overline{v^2} \tag{1}$$

where $\overline{v^2}$, the average value of v^2, is determined by the temperature of the gas and is constant for a fixed temperature. By substituting typical values

$P = 10^5$ N/m^2 and $M/V = 1$ kg/m^3, we find that $\sqrt{\overline{v^2}}$ is of the order of 10^3 m/s at standard temperature and pressure.

The advantages of the atomic description of matter became more apparent when Dalton formulated the law of multiple proportions. The law states:

> If a given amount of substance A combines with one amount of B to form C, and with another amount of B to form D, the ratio of the two amounts of B involved is a small integer.

The connection with the atomic picture is obvious. Closely related to this law is the law of combining volumes (Gay-Lussac's Law) which states that volumes of gases combining and the volume of gas produced are always in a small integer ratio.

These two laws taken together seem to imply that equal volumes of gases contain equal numbers of atoms (at a given temperature and pressure). However, there remains a problem, illustrated by the case of NO; one volume of N combines with one volume of O to form *two* volumes of NO, but if each atom of N were combining with one atom of O to form one molecule of NO, only *one* volume of NO should result. Such examples led some scientists, including Dalton, to disbelieve the law of combining volumes. The solution was provided by Avogadro, who suggested (in 1811) that many gases, including nitrogen and oxygen, consist of diatomic molecules, so that the reaction mentioned above is

$$N_2 + O_2 \longrightarrow 2NO$$

Avogadro did not explain his ideas very well, and many persons still wondered why precisely two atoms of nitrogen should combine to form a molecule. One might ask why three or more nitrogen atoms could not combine. It was not until 1860 that Avogadro's idea became generally accepted, and the question about the number of atoms in a molecule was not answered satisfactorily until 1926, when the modern quantum theory was born. It was also many years after Avogadro's time before "Avogadro's number," the number of atoms or molecules in a standard quantity of matter, was actually determined. At present, there are many ways of measuring this number, and several of them will be discussed in this chapter.

Mean Free Path. Toward the last half of the nineteenth century, as more people began to think in terms of atoms, refinements were developed in the atomic theory. One refinement overcame an objection based on the great molecular speeds involved: If molecules move so rapidly, why are gases observed to diffuse so slowly? The answer was found by considering the size of molecules: the process of diffusion is impeded by collisions between molecules.

Let us analyze the effect of these collisions. We begin by defining the *mean free path* λ as the mean distance traveled by a molecule between collisions with other molecules, and we try to relate this to the "diameter" d of a molecule. At first thought, it is not even clear that a molecule possesses a definable diameter, for a molecule is not a hard sphere like a billiard ball; but we can make an operational definition of a molecular diameter simply by saying that two identical molecules "collide" if their centers come within a distance d of each other. A collision, in turn, may be defined as an event which causes a molecule to change its direction of motion.

To see where these ideas lead, imagine that a stream of molecules issues from a hole and is then collimated so that the molecules are moving in parallel paths. If a plate is then placed perpendicular to the stream at any point, the amount of matter deposited on the plate per second should be proportional to the number of molecules per second which would normally have passed that point. If the plate is moved further from the hole, we should expect matter to be deposited on it at a slower rate, because molecules have been removed from the stream by collisions with molecules in the air through which the stream passes. (Even a good laboratory "vacuum" may contain 10^{10} molecules/cm^3.)

We may find the rate at which molecules are removed from the stream by considering the effect of a thin slab of air, with area A and thickness dx, containing n molecules/cm^3. Each air molecule in the slab covers an area of πd^2, in the sense that an incident molecule is scattered out of the stream if its center strikes that area; and the number of molecules per unit area in the slab is $n\, dx$. If dx is sufficiently small, no two molecules cover the same area, and the fraction of the total area which is covered is simply the product of the area covered by a molecule and the number of molecules per unit area in the slab, or $\pi d^2 \cdot n\, dx$. We define $\alpha = n\pi d^2$. The fraction of molecules removed from the stream as it traverses the slab of air is then $\alpha\, dx$.

Now suppose N_0 molecules start out in the stream at $x = 0$. We wish to find an expression for N, the number of particles left in the stream after traveling a distance x. The number removed from the stream in the distance dx is $N\alpha\, dx$, so we may write

$$dN = -N\alpha\, dx$$

This may be integrated to give

$$N = N_0\, e^{-\alpha x} \tag{2}$$

In other words, the probability that a given molecule goes a distance x without a collision is $e^{-\alpha x}$.

The mean free path may now be found by straightforward application of the definition. We multiply each possible value of distance x by the number of molecules traveling that distance but not further; we integrate this over x

to obtain the *total* free path, and we divide by the total number N_0 to find the *mean* free path. The probability that a molecule travels a distance x but no farther is equal to the probability $e^{-\alpha x}$ that it travels a distance x *without* a collision, multiplied by the probability $\alpha\,dx$ that it *does* make a collision in the next small distance interval dx. The number of molecules scattered at the distance x is $N_0\,e^{-\alpha x}\alpha\,dx$; the total free path traveled by these molecules is $N_0\,e^{-\alpha x}\alpha x\,dx$; and the total free path traveled by all N_0 molecules is $\int_0^\infty N_0\,e^{-\alpha x}\alpha x\,dx$. The mean free path λ is given by

$$\lambda = \frac{\int_0^\infty N_0\,e^{-\alpha x}\alpha x\,dx}{N_0} = -xe^{-\alpha x}\Big]_0^\infty + \int_0^\infty e^{-\alpha x}\,dx$$

$$= \frac{1}{\alpha} \tag{3}$$

In terms of the molecular diameter d, the mean free path is

$$\lambda = \frac{1}{n\pi d^2} \tag{4}$$

This connection between λ and α can also be seen from Poisson statistics (see Appendix A). If λ is the mean free path, then the mean number of collisions made by a molecule in traveling a distance x must be x/λ. Therefore, since Poisson statistics gives e^{-m} as the probability of zero events when the mean is m, the probability that a molecule goes a distance x without a collision must be $e^{-x/\lambda}$. Comparison with Eq. (2) shows that $1/\lambda = \alpha$.

An accurate experiment to check these ideas was not performed until 1925, when Bielz measured the amount of silver deposited on a plate by a beam of silver atoms issuing from an oven. By repeating the experiment for equal times at different pressures and applying Eq. (2), he showed that λ is inversely proportional to the pressure, as expected from Eq. (4), since n, the number of molecules per cubic centimeter, is of course proportional to the pressure.

The alert reader can see a flaw in the derivation of λ. We treated the target particles as if they were stationary, but of course they are also moving. When this motion is taken into account, the formula becomes $\lambda = 1/n\pi d^2\,\sqrt{2}$, differing from the previous result by the factor $\sqrt{2}$ in the denominator. This factor can be made plausible by the following argument.

One can ignore components of velocity perpendicular to the plane of slab; these do not change the area covered by the molecules in the slab, and the velocity of the normally incident molecule did not enter the calculation. But suppose that a molecule in the slab moves sideways (in the plane of the slab), with the same speed as that of the incident molecule. Then the *relative* velocity of the incident molecule with respect to the target molecule makes an angle of 45° with the normal. But if the incident molecule were actually approaching at a 45° angle, the total area presented by the slab to the incident molecule would be smaller by a factor $\sqrt{2}$, while the area covered by each target molecule would be unchanged, so the chance of a hit would be increased by a factor $\sqrt{2}$. This argument should not be considered a proof, because, for one thing, it does not take into account the fact that the molecules all move with different speeds; but it should make the $\sqrt{2}$ factor seem reasonable.

EXAMPLE PROBLEM 1. Estimate the mean free path of an air molecule at 0°C and atmospheric pressure, and compute the probability that a molecule could travel 1 cm before colliding with another molecule.

Solution. We need to know n and d to insert into Eq. (4). We know that under the given conditions, 2.24×10^4 cm^3 of air contains 6×10^{23} molecules, so $n = 2.7 \times 10^{19}$/cm^3. Let us find d by estimating the size of a water molecule; that should be close enough in size to an air molecule, and it is easy to estimate. Since water is relatively incompressible, there must be very little empty space between the molecules. Eighteen grams of water occupies 18 cm^3 and contains 6×10^{23} molecules, so the volume occupied by each molecule is roughly 3×10^{-23} cm^3. The diameter of a molecule should therefore be about $(30 \times 10^{-24})^{1/3}$ cm, or 3×10^{-8} cm. The mean free path is then

$$\lambda = 1/[2.7 \times 10^{19} \ \ \text{cm}^{-3} \times 3 \times (3 \times 10^{-8} \ \ \text{cm})^2] = 10^{-5} \ \ \text{cm}$$

The probability that a molecule could travel 1 cm before colliding with another molecule is then $e^{-x/\lambda} = e^{-10^5} = e^{-100,000}$ or $10^{-43,000}$, a number so small that it defies description in other terms.

The Maxwell Velocity Distribution. A very important contribution to our understanding of atomic theory was made by Maxwell, who derived the "Maxwell velocity distribution," a formula giving the distribution of velocities among the various atoms or molecules in a gas. To be precise, Maxwell derived a function $f(v)$ such that $\int_{v_1}^{v_2} f(v) \, dv$ gives the probability that a molecule's velocity lies between v_1 and v_2.

It is instructive to derive the function $f(v)$ in a way suggested later by Boltzmann. Consider a column of gas, acted on by gravity, so that the pressure and density vary with height but the temperature is uniform throughout (see Fig. 1). Let ρ and P be the density and pressure, respectively, at height z. Then at height $z + dz$ the pressure is $P + dP$, where

$$dP = -\rho g \, dz \tag{5}$$

because $\rho g \, dz$ is the weight per unit area of a column of gas of height dz. We also have the relation $PV = NRT$, the ideal gas law, which may be written

$$P = \frac{\rho RT}{M} \tag{6}$$

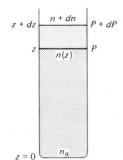

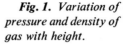

Fig. 1. *Variation of pressure and density of gas with height.*

where M is the mass of one mole of gas, and $\rho/M = N/V$ is the number of moles of gas per unit volume. Dividing Eq. (5) by Eq. (6), we obtain

$$\frac{dP}{P} = -Mg\,\frac{dz}{RT}$$

which may be integrated to give the dependence of P on z:

$$P = P_0\,e^{-Mgz/RT}$$

where P_0 is, of course, the pressure at $z = 0$. Since P is proportional to n, the number of molecules per unit volume, n must vary with z the same way as P, so

$$n = n_0\,e^{-Mgz/RT} = n_0\,e^{-mgz/kT} \tag{7}$$

where k, the Boltzmann constant, is simply the gas constant R divided by Avogadro's number, N_A, and m is the mass of one molecule. (By definition, $M = N_A \times m$.)

Before we proceed with the derivation of the velocity distribution, notice that the exponential factor in Eq. (7) may be written $e^{-E/kT}$, since $E = mgz$ is the potential energy of a molecule of mass at height z. The factor $e^{-E/kT}$, known as the Boltzmann factor, is one factor determining the distribution of particles among the various possible energies in all systems which obey classical statistics.[1] In deriving the velocity distribution, we seek the distribution of the particles among the possible values of *kinetic* energy, and we shall see that the Boltzmann factor appears again, in the form $e^{-mv^2/2kT}$.

To convert the distribution of heights into a velocity distribution, we first define $h(v_z)$ as the one-dimensional velocity distribution analogous to $f(v)$, so that $h(v_z)\,dv_z$ is the probability that a molecule's z component of velocity is between v_z and $v_z + dv_z$.

---

[1] Later, in the chapter on quantum statistics, you will see a more general derivation of the Boltzmann factor as well as the analogous factors for nonclassical statistics.

From conservation of energy, and assuming no collisions on the way up, we can say that each molecule leaving $z = 0$ with upward velocity between v_z and $v_z + dv_z$ will reach a maximum height between z and $z + dz$, where

$$v_z^2 = 2gz \quad \text{and } v_z \, dv_z = g \, dz \tag{8}$$

These molecules are the ones that produce the difference between the density at height z and that at height $z + dz$, because they reach height z but they fail to reach height $z + dz$. But this density difference is already known (implicitly) from Eq. (7); we need only take the differential of both sides of (7), obtaining

$$dn = -(n_0 mg/kT) \, e^{-mgz/kT} dz \tag{9}$$

Therefore, the magnitude of this differential must equal $h(v_z)dv_z$, and it only remains to change the variable in (9) from z to v_z, using Eq. (8):

$$|dn| = h(v_z)dv_z = (n_0 mg/kT) \, e^{-mv_z^2/2kT} (v_z dv_z/g) \tag{10}$$

In terms of average velocity, defined by

$$\bar{v} \equiv \int_{v_z=0}^{v_z=\infty} v_z \, h(v_z) \, dv_z$$

you can show that Eq. (10) may be written

$$h(v_z) = \frac{m\bar{v}_z}{kT} \, e^{-mv_z^2/2kT} = \text{constant} \times e^{-mv_z^2/2kT} \tag{11}$$

Now that we have one-dimensional distribution function, we make the reasonable assumptions that the x and y components of v are distributed in the same way as the z component, and that the three components are independent of one another. Since the probability of simultaneous occurrence of independent events is equal to the product of the probabilities of occurrence of the individual events,[2] the product

$$Ce^{-mv_x^2/2kT} e^{-mv_y^2/2kT} e^{-mv_z^2/2kT} dv_x \, dv_y \, dv_z \tag{12}$$

with the proper choice of the constant C, gives the probability that a molecule simultaneously has an x component of v between v_x and $v_x + dv_x$, a y component of v between v_y and $v_y + dv_y$, and z component of v between v_z and $v_z + dv_z$.

If we draw a set of axes for v_x, v_y, and v_z, then Eq. (12) gives the probability that the velocity may be represented by a vector drawn from the origin to a point in the volume element $dv_x \, dv_y \, dv_z$. In order to introduce the magnitude v, we may write this volume element in spherical coordinates as $v^2 dv \sin\theta \, d\theta \, d\phi$. If we then integrate over all angles, expression (12) becomes

$$4\pi C v^2 e^{-mv^2/2kT} dv = f(v)dv$$

[2] See Appendix A.

where $f(v)$ is the function we set out to find. The factor $4\pi v^2\,dv$ is simply the volume of a spherical shell of radius v and thickness dv. The constant factor C is evaluated by using the fact that the probability that a molecule has *some* velocity between zero and infinity must be unity, so that

$$\int_0^\infty f(v)\,dv = 1 \tag{13}$$

Use of condition (13) shows that

$$f(v) = \left(\frac{2}{\pi}\right)^{1/2}\left(\frac{m}{kT}\right)^{3/2} v^2 e^{-mv^2/2kT} \tag{14}$$

Notice that the essential difference between the distribution of speeds in three dimensions and the one-dimensional distribution Eq. (11) appears only in the weighting factor v^2. In both cases the Boltzmann factor is present, but in three dimensions the weighting factor enters because of the velocity dependence of the volume which is available for the velocity vector in velocity space. We shall later find that the energy distribution of the particles in every classical system depends on the product of the Boltzmann factor, which is common to all systems, and a weighting factor characteristic of the particular system.

Unfortunately, it was not possible to make a real test of the Maxwell distribution until 70 years after it was derived. By that time the existence of atoms was not in doubt, but a test of the formula was still important, and one was made by I. F. Zartman in 1931. He used an evacuated system (Fig. 2) containing an oven and a rotating drum with a slit in it. When the slit was aligned with a collimator over the oven, a pulse of atoms was admitted to the drum. A glass plate opposite the slit collected the atoms, and the variation in the density of the deposit on the glass could be used to deduce the velocity distribution, the fast atoms being deposited at one end of the plate and the slow atoms at the other end. The agreement with Maxwell was good.

Another test was made by Rainwater and Havens in 1946,[3] using neutrons which had reached thermal equilibrium in a block of paraffin. A pulse of neutrons from a cyclotron entered the paraffin and reached thermal equilibrium with the paraffin in a very short time. The neutrons which emerged from the paraffin were collimated and detected by a pulsed neutron counter located at a distance d from the paraffin. If the counter was activated for a brief time interval at a time t after the cyclotron pulse, the neutrons counted were only those whose velocity was equal to d/t. Varying the time t enabled the experimenters to obtain the distribution of neutron velocities. The result was in excellent agreement with the Maxwell velocity distribution, and it demonstrated the generality of the distribution by showing that it governed

---

[3] J. Rainwater and W. Havens, *Phys. Rev.* **70**, 136 (1946).

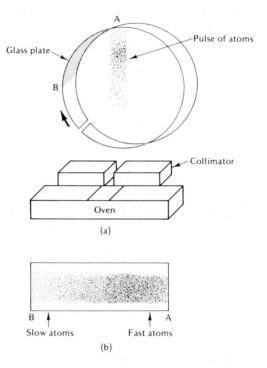

Fig. 2. *Velocity distribution measurement.*
(a) *Arrangement of apparatus;* (b) *Appearance of glass plate.*

the behavior of particles whose existence was not even suspected when Maxwell originally derived the law.

EXAMPLE PROBLEM 2. Show that the rate of escape of molecules from a container through a small aperture of area A is given by $\frac{1}{4}n\bar{v}A$, where n is the number of molecules per cubic centimeter, and $\bar{v} = \int_0^\infty vf(v)\,dv$ is the mean speed of the molecules in the container.

Solution. We must take into account the angle of impact of the molecules on the area A, so we consider those molecules which occupy a ring such that each point in the ring is at a distance between r and $r + dr$ from the area A. The area A lies on the axis of the ring. A molecule traveling from the ring to A strikes A at an angle of incidence between θ and $\theta + d\theta$ (see Fig. 3). The radius of the ring is $r\sin\theta$ and its cross-sectional area is $r\,dr\,d\theta$, so its volume is $2\pi r^2\sin\theta\,dr\,d\theta$, and the number of molecules in the ring is $n2\pi r^2\sin\theta\,dr\,d\theta$. The probability that a given molecule in the ring will strike A is $(A\cos\theta)/4\pi r^2$,

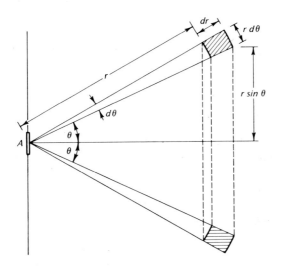

Fig. 3. *Coordinates for calculation of escape rate from a container. Shaded area shows intersection of a ring with plane of drawing.*

since $A \cos \theta / r^2$ is the solid angle subtended by A at the ring. The number of molecules from this ring which strike A with speed between v and $v + dv$ is therefore

$$n 2\pi r^2 \sin \theta \, dr \, d\theta \frac{A \cos \theta}{4\pi r^2} f(v) \, dv$$

or

$$\frac{nA \sin \theta \cos \theta f(v) \, dv \, d\theta \, dr}{2} \tag{15}$$

In a time t, the molecules which strike A with speed v come from rings such that $r \leqslant vt$. Therefore to find the total number of molecules striking A in time t, we integrate expression (15) on r between limits 0 and vt, and we then integrate over all values of v and over all values of θ, obtaining

(number of molecules striking area A in time t)

$$= \int_0^\infty f(v) \, dv \int_0^{\pi/2} d\theta \int_0^{vt} nA \frac{\sin \theta \cos \theta \, dr}{2}$$

$$= \tfrac{1}{4} nAt \int_0^\infty v f(v) \, dv$$

$$= \tfrac{1}{4} nAt\bar{v} \qquad\qquad \text{Q.E.D.}$$

The approach used here may also be used to verify that the pressure is $\frac{1}{3}M\overline{v^2}$, in agreement with Eq. (1) (see Problem 6). We shall later use this same approach to calculate the rate of escape of electrons from the surface of a metal, even though the distribution of speeds is not the Maxwell distribution in that case.

1.2   THE ELECTRON

We have seen that the ideas mentioned above were not subjected to experimental test until well into the twentieth century, yet by that time almost everyone had already been convinced of the correctness of the atomic theory. This came about through observations of the "atom of electricity," the electron. It had been very difficult to observe individual neutral atoms, but effects of individual charges, or electrons, were relatively easy to see.

The first atomic model of electricity was proposed by Faraday in 1833, after his electrolysis experiments. Faraday found that the flow of a total of 96,500 C always deposited at an electrode an amount of matter equal to one mole divided by the valence of the element deposited. Faraday cautiously reasoned that *if* matter were atomic in nature, then a definite quantity of electricity is associated with each atom. The total charge of one mole of electrons is, of course, 96,500 C, a quantity which has been named the faraday.

Cathode Rays.   Faraday also made observations of electrical discharges which eventually led to the identification of electrons in "cathode rays." Unfortunately, vacuum techniques available then did not permit him to advance his experimental results very far. After the invention of an improved vacuum pump by Geissler in 1855, detailed studies of these "rays" were possible. The pace of discovery was a bit slower in those days, but by 1870 Plücker and Hittorf had observed many properties of the rays. They showed that the rays emanate from the cathode in a vacuum tube, that they travel in straight lines, and that they cause glass to fluoresce. All three properties were demonstrated in one experiment, by placing a solid object between the cathode and the glass wall of the tube, and observing that the glass in the "shadow" does not fluoresce while the rest of the glass does. Plücker and Hittorf also showed that the rays could be deflected by a magnetic field.

Other experiments were performed by Goldstein and Hertz in Germany and by Crookes in England. The former believed that the rays were some new form of electromagnetic wave, while Crookes felt that they were molecules which had hit the cathode and had picked up a negative charge from it. Since both sides were wrong, a bitter argument raged; each side saw the flaws in the other's arguments. The argument against molecules was that the rays traveled much farther than the mean free path of a molecule in the tube; the argument against electromagnetic waves was that the rays were deflected by a magnetic field.

The aspect which confused all of the investigators was that it did not seem possible to deflect the rays in an *electric* field. Hertz calculated the speed that charged particles must have in order that they could be deflected by the earth's magnetic field of less than 1 G (as was observed), yet could not be deflected by the electric field of 1 V/mm which Hertz had applied. Hertz concluded that, if the rays were charged particles, " The requisite speed would exceed eleven earth quadrants per second [about 10^8 m/sec], . . . a speed which will scarcely be regarded as probable."[4]

But in 1895, J. P. Perrin, by collecting the rays in an insulated cup, "showed" that they were negatively charged,[5] and in 1897, J. J. Thomson succeeded in deflecting the rays in an electric field as well as a magnetic field. Thomson was thereby able to determine the charge-to-mass ratio e/m for the rays.

Thomson's cathode-ray tube is sketched in Fig. 4. External coils, not shown, produced the magnetic field, perpendicular to the plane of the drawing. The parallel plates produced the electric deflecting field. The first collimating slit was in the anode.

If the speed of the particles were known, e/m could be determined simply by measuring the deflection of the rays in the magnetic field alone. In the magnetic field B, the rays follow an arc of a circle of radius R such that

$$evB = \frac{mv^2}{R} \quad \text{or} \quad \frac{e}{m} = \frac{v}{RB} \tag{16}$$

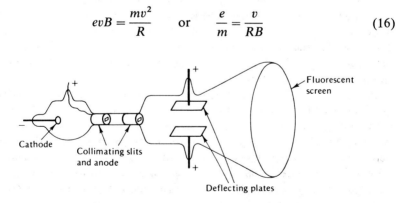

Fig. 4. *J. J. Thomson's apparatus.*

[4] H. Hertz, " Miscellaneous Papers." Macmillan, London, 1896. Quoted in D. L. Anderson, "The Discovery of the Electron," p. 37, Van Nostrand, Princeton, New Jersey, 1964.
[5] Perrin's demonstration was a useful clue at the time, but it was mere luck that it worked, for the electrons impinging on his cup could easily have knocked off additional electrons and produced the opposite effect.

R can be determined from geometry by observing the deflection of the rays when the magnetic field is turned on. Since v was not known, the electric field was used to determine it. Thomson adjusted the electric field so that the deflection of the beam in the combined electric and magnetic fields was zero. Under these conditions the electric force was equal in magnitude to the magnetic force, so

$$eE = evB \quad \text{or} \quad v = \frac{E}{B} \qquad (17)$$

Substitution of Eq. (17) into Eq. (16) yields $e/m = E/RB^2$.

The speed of the particles did turn out to be high, although not quite so high as the value Hertz had calculated (see Problem 10). Hertz's failure to see a deflection in an electric field was explained (by Thomson) as an effect of gas in his tube. The residual gas in Hertz's tube had apparently been ionized by the rays, and the resulting conduction of current between the deflecting plates had reduced the field between the plates.

Thomson's measurement of e/m showed that the particles in the cathode rays had either a much greater charge or a much smaller mass than the hydrogen atom. The mass of the hydrogen atom was not very well known at the time, but the charge-to-mass ratio was known. The charge on one gram of hydrogen ions is one faraday, 96,500 C, so e/m is 96,500 C/gm for hydrogen ions. For the cathode rays, on the other hand, Thomson found $e/m \cong 10^8$ C/gm. (This was, however, somewhat less than the correct value.)

Thomson now believed that the cathode ray particles were constituents of all atoms. He said, "Thus on this view we have in the cathode rays matter in a new state, a state in which the subdivision of matter is carried very much farther than in the ordinary gaseous state: a state in which all matter ... is of one and the same kind, this matter being the substance from which all the chemical elements are built up."[6]

This is close to the truth, except of course that matter contains other constituents as well as electrons. But the idea of "atoms of electricity" was hard for Thomson's contemporaries to accept. Maxwell's equations had fostered the idea that electricity is some sort of continuous fluid, or perhaps simply a manifestation of displacements of the "ether." As late as 1897, Wiechert said, "Electricity denotes something imaginary—an entity which exists not actually but only in thought."[7] Thomson found that some of his colleagues thought he was "pulling their legs" when he suggested his ideas.[8]

[6] J. J. Thomson, *Phil. Mag.* **44**, 312 (1897).

[7] P. Lenard, "Über Kathodenstrahlen." Berlin-Leipzig, 1920. Quoted in D. L. Anderson, "The Discovery of the Electron," p. 75. Van Nostrand, Princeton, New Jersey, 1964.

[8] For a number of references on this point, see D. L. Anderson, "The Discovery of the Electron," pp. 74–75. Van Nostrand, Princeton, New Jersey, 1964.

The Zeeman Effect. However, at this same time, Dutch physicist Pieter Zeeman was finding evidence that the electron exists inside the atom. Zeeman, repeating an experiment also attempted (unsuccessfully) by Faraday, placed a light source between the poles of a magnet and found that the spectral lines of the source were broadened. Upon using a stronger magnet, he found that each line split into three equally spaced lines when viewed along a line of sight perpendicular to the field. The middle line was of the same frequency as the original line (seen in the absence of a field), and it was polarized in the direction of the field. The other two lines were polarized perpendicular to the field. Zeeman then drilled a hole through one pole-piece of the magnet, in order to view the source along a line parallel to the field. There he found that the middle line was missing and the other two lines were circularly polarized in opposite directions.

In search of an explanation of his observations, Zeeman consulted the great Dutch theoretician, H. A. Lorentz. Lorentz was then engaged in developing an "electron theory" of matter in order to explain the failure of various attempts to detect motion through the "ether." (We shall see what this was all about in Chapter 2.) Lorentz was delighted to explain Zeeman's results on the basis of his theory.

Lorentz assumed that the light from the source is produced by accelerated charges—electrons—inside the atom of the source. If the motion of one of these charges is in general an ellipse, it can be resolved into three linear oscillations, along the three coordinate axes. Let the direction of the field be the z axis. When a magnetic field is applied, the motion in the z direction is unchanged. Since the emitted light is polarized in the same direction as the motion of the charge, this z component of the motion accounts for the middle line and its polarization. This line is missing when the source is viewed from the z direction, because z-polarized light cannot be propagated in the z direction. The two linear oscillations in the xy plane can be considered to be the sum of two opposite circular motions. Each of these circular motions changes frequency when the field is applied, thereby accounting for the other two lines and their polarizations.

To compute the magnitude of the effect, let $\delta\omega$ be the change in frequency, and let ω be the original frequency of the circular motions. The original circular motion is caused by a centripetal force $F_0 = mr\omega^2$. (The cause of F_0 is unspecified.) When the field is applied, it exerts an additional force $\mathbf{F} = q\mathbf{v} \times \mathbf{B}$, which is also directed toward the center if the charge q is positive and $\mathbf{v}$ and $\mathbf{B}$ are in the directions shown in Fig. 5. The frequency of the motion therefore changes to $\omega + \delta\omega$, and

$$F + F_0 = mr(\omega + \delta\omega)^2$$
$$\approx mr\omega^2 + 2mr\omega\,\delta\omega$$

if $\delta\omega$ is small compared to ω. Subtraction of $F_0 = mr\omega^2$ yields $F = qvB = 2mr\omega\,\delta\omega$. Since $v = \omega r$, we have $q/m = 2\delta\omega/B$.

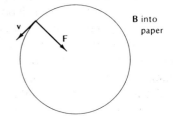

Fig. 5. *Possible set of directions for* **v**, **B**, *and* **F** *in circular motion.*

Observation of the direction of circular polarization of the lines showed that q is negative. Furthermore, q/m was found to be nearly equal to e/m for the cathode rays. Such agreement could hardly be accidental. Thus the field of electron physics was opened.

1.3 DETERMINATION OF AVOGADRO'S NUMBER

Although the charge-to-mass ratio of the electron had been measured in two quite different ways, the electronic charge e itself had not been determined in these experiments. However, it now seemed fairly clear that e was identical to the charge on a monovalent ion in electrolysis, so that the product of e and Avogadro's number N_A should equal 1 faraday, a well-determined quantity. Therefore determination of Avogadro's number should yield the value of e, and conversely, a measurement of e would reveal the value of N_A. Determination of these quantities was therefore approached from both directions, and the agreement of the various results justified the assumption that the charge of the electron is equal (in magnitude) to the charge on a monovalent ion.

Some clues to the values of N_A had been obtained before 1900. As early as 1871 Lord Rayleigh had pointed out that the blueness of the sky is an indication of the value of N_A. This blueness is caused by the inhomogeneity of the air and the fact that blue light is scattered more than light of longer wavelength in such a medium. The amount of scattering depends on the magnitude of the statistical fluctuations in the density of the air, and the magnitude of these fluctuations in turn depends on the number of molecules per cubic centimeter in the air. By comparing the spectral distribution of skylight with that of direct sunlight, Rayleigh was able to determine that N_A is between 3 and 15 times 10^{23}.

Rayleigh's method is only one of many clever ways of estimating or measuring N_A. In 1909, Perrin[9] published a summary of 14 distinct methods for determining N_A. You will be able to appreciate some of these methods after studying later chapters. One method, based on analysis of the black-body radiation spectrum, will be discussed in Chapter 3. Three other methods are based on radioactivity; for example, one can measure the amount of helium produced as a result of a known number of alpha-particle emissions by a radioactive element. (Alpha particles are helium nuclei; see Section 12.1.)

Perrin's own methods are quite instructive. One is based on Eq. (7). If we had a gas of particles of known mass, and the particles were large enough to see, we could observe the dependence of $n(z)$ on z by counting the particles directly, and we could then use Eq. (7) to determine k. Once we know k, we also know N_A, since $N_A = R/k$.

Perrin recognized that a collection of small particles in a fluid behaves like a gas in thermal equilibrium with the fluid, the particles distributing themselves according to Eq. (7) almost as if the molecules of the fluid were not present. Of course if the particles are too large (bowling balls, for example), then the m in Eq. (7) becomes too large, and the exponential is virtually zero for values of z which are smaller than the radius of the "particle." But Perrin was able to make a suspension of almost identical particles of very convenient size, about 10^{-4} to 10^{-5} cm in diameter, and he was able to count the numbers suspended at successive heights differing by about 3×10^{-3} cm. A plot of $\ln n$ versus z then gave a straight line with a slope of $-N_A \mu g/RT$, where μ was the "effective mass" of the suspended particles. Since the suspended particles were so much larger than the fluid molecules, the *collective* effect of the molecules could not be ignored, so a correction for the buoyant effect of the fluid was introduced by using the effective mass $\mu = \frac{4}{3}\pi a^3(\rho - \rho')$, where ρ is the density of a particle, ρ' is the fluid density, and a is the radius of the particle (assumed to be spherical).

One way to find ρ was to suspend the particles in a salt solution whose density could be varied until centrifuging failed to affect the particles. To find a, Perrin watched the particles descend and observed their terminal velocity (a few millimeters per day), which according to Stokes's law is

$$v_g = \frac{\frac{2}{9}ga^2(\rho - \rho')}{\eta} \tag{18}$$

where η is the viscosity of the fluid. Since μ, g, T, and R were now known, the value of N_A could then be determined from the slope of the straight-line graph of $\log n$ versus z. Perrin obtained a value of 7.05×10^{23} for N_A by this method.

[9] J. P. Perrin, "Mouvement Brownien et le Realité Moléculaire," *Ann. Chim. Phys.* (Ser. 8) **18**, 1 (1909). Quoted by D. L. Anderson, "The Discovery of the Electron," p. 83. Van Nostrand, Princeton, New Jersey, 1964.

Millikan's Oil-Drop Experiment.   The most famous of the measurements of N_A or e was that of R. A. Millikan, who began in 1906 to measure the charge e directly. Millikan's experiment is notable for its precision as well as for the directness of its approach. Perhaps for these reasons, this experiment also has the distinction of having since been performed by many thousands of physics students, and you will probably perform it yourself, if you have not done so already. Therefore, this experiment will be described in some detail.

Millikan's method, a refinement of a technique used earlier by H. A. Wilson, consists of attaching the charge e to a particle of known mass, so that the unknown mass of the electron itself does not enter the problem. The particles used are oil drops, which are allowed to fall through a small hole into a region between two flat plates. Voltage applied to the plates produces an electric field which acts on the charged drops, and the motion of individual drops is observed through a telescope. X-rays are used to ionize the air in the region, so that the charge on a drop could change in a collision with an ion.

The great precision of the experiment is achieved by following a single drop for a long time. First the drop is timed as it falls under gravity; then the field is turned on and the drop is timed as it rises (if its charge has the right sign); then the drop falls again with the field off and the cycle is repeated. Observations of the *falling* drop are used to find the drop's mass, as follows: The drop is always timed after it reaches its terminal velocity. When the terminal velocity is reached, the viscous force of the air just equals the resultant of the other forces on the drop. For a falling drop, this resultant equals μg, where μ, the effective mass, includes, as in Perrin's experiment, a correction for the buoyancy of the medium. Since the viscous force is always proportional to the speed, and the viscous force is μg when the speed is the terminal velocity v_g, we see that

$$v_g = K\mu g \tag{19}$$

The constant K may be evaluated by using Stokes's law, Eq. (18), where this time ρ is the density of the oil drop and a its radius, while ρ' and η are the density and viscosity, respectively, of the air between the plates. Equating the right-hand sides of (18) and (19), with $\mu = \frac{4}{3}\pi a^3(\rho - \rho')$, shows that $K = 1/6\pi a\eta$. The radius, and hence the mass, of the drop may be found by solving Eq. (18) for a.

The charge on the drop may now be deduced from observations of the drop as it rises in the electric field E, and the electronic charge e may be deduced from *changes* in the drop's charge. When the drop rises, the terminal velocity v_e must, by the same reasoning used above, be proportional to the resultant of the gravitational and electrical forces on the drop, with the same proportionality constant K:

$$v_e = K(Eq - \mu g) \tag{20}$$

If the charge on the drop changes, the terminal velocity changes by an amount proportional to the change in charge. If the drop rises once with charge q_1 and another time with charge q_2, the respective terminal velocities v_{e1} and v_{e2} are

$$v_{e1} = K(Eq_1 - \mu g)$$
$$v_{e2} = K(Eq_2 - \mu g)$$

so that

$$v_{e1} - v_{e2} = K(Eq_1 - Eq_2)$$

or

$$q_2 - q_1 = \left(\frac{v_{e2} - v_{e1}}{KE} \right)$$

Millikan's data for one drop are shown in Table 1. These are preliminary measurements which he made with a stopwatch; they are not so precise as his later measurements, but they illustrate the technique nicely. Notice that $v_{e2} - v_{e1}$ is always very close to an integer multiple of 0.0090. This means that $q_2 - q_1$ was always an integer multiple of the same value of charge, which we assume to be e. The actual calculation of e from these data is left as a problem. The data also show that the total charge which the drop had originally was an integer multiple of the same e. Since $v_g = K\mu g$, we may write Eq. (20) as

$$v_e + v_g = KEq \qquad \text{or} \qquad q = \frac{v_e + v_g}{KE} \tag{21}$$

Thus the total charge is found from $v_e + v_g$ in the same way that the change in charge is found from $v_{e2} - v_{e1}$. Using the first set of values in the table, plus the fact that the distance traveled was 0.5222 cm, we find that $v_g = 0.5222$ cm/13.6 sec $= 0.0384$ cm/sec, and $v_e + v_g = 0.0802$ cm/sec. This value is very close to an integer multiple of the same common divisor 0.0090 found for the changes in v_e. The integer is nine, so the original charge on the drop was nine electron charges. The significance of this point is that the original charge came not from the ions in the air, but from the friction of blowing in the oil spray, so that there is strong evidence for the universality of this unit of charge.

Millikan also made careful observations of the way in which the charge changed on the oil drop, and his results gave a vivid picture of the behavior of air molecules. He noticed several points:

1. The charge changed only when the electric field either was off or had just been turned on.
2. If a drop was near the positive plate when the field was turned on, its charge became more negative (or less positive); if it was near the negative

Table 1

Millikan's Data for One Drop[a]

Fall time,[b] t_g (sec)	Rise time,[b] t_e (sec)	v_e (cm/sec)	Change in v_e (cm/sec)
13.6	12.5	0.0418	
			+0.0003
13.8	12.4	0.0421	
			−0.0182
13.4	21.8	0.0239	
			−0.0089
13.4	34.8	0.0150	
			−0.0088
13.6	84.5	0.0062	
			−0.0001
13.6	85.5	0.0061	
			+0.0090
13.7	34.6	0.0151	
			−0.0001
13.5	34.8	0.0150	
			+0.0176
13.5	16.0	0.0326	
			−0.0176
13.8	34.8	0.0150	
			+0.0001
13.7	34.6	0.0151	
			+0.0087
13.8	21.9	0.0238	

[a] Data from R. A. Millikan, "The Electron," p. 67. Univ. of Chicago Press, Chicago, Illinois, 1917.
[b] Distance of rise=distance of fall = 0.5222 cm.

plate, the opposite occurred. Thus the charge on a drop could be increased or decreased at will by positioning it near one plate or the other and then turning on the field.

3. The frequency with which the charge changed was directly proportional to the air pressure.

4. When a drop held a charge of ten or fewer electron charges, it was just about as likely to increase as to decrease its net charge when the field was off. But when a drop was heavily charged (100 electron charges or more), it almost never acquired an additional charge of the same sign.

The explanation for each of these observations is left to the reader.

During this experiment, Millikan made another discovery about the atomicity of matter. Since the charge on a drop can be deduced from the values of v_e and v_g, Millikan knew that one should be able to find a proportionality factor between speed and charge so that the fundamental charge e came out to be a universal constant which did not depend on the size of the drop. But Millikan reported, "The attempt to do this . . . by the assumption of Stokes's law, heretofore taken for granted by all observers, led to the interesting discovery that this law is not valid."[10]

In order to follow Millikan's reasoning, notice that Eq. (21) may be written in terms of measured quantities by replacing K by $1/6\pi\eta a$ and using for a the expression obtained by solving for a in Eq. (18). The result is

$$q = \frac{v_e + v_g}{E} 6\pi\eta a$$

$$= \frac{v_e + v_g}{E} 6\pi\eta \left[\frac{9v_g \eta}{2g(\rho - \rho')}\right]^{1/2}$$

or

$$q = \frac{18\pi\eta^{3/2}}{E[2g(\rho - \rho')]^{1/2}} (v_e + v_g)v_g^{1/2} \tag{22}$$

If for $v_e + v_g$ we use the greatest common divisor of the series of values of $v_e + v_g$ and $v_{e1} - v_{e2}$ for a single drop, the value of q should equal e, the electronic charge. But, quoting Millikan again:

"When drops of different speeds [v_g], and, therefore, of different sizes, were used, the values of e obtained were consistently larger the smaller the velocity under gravity. For example, e for one drop for which $v_g = 0.01085$ cm/sec came out 5.49×10^{-10} [statcoulombs], while for another of almost the same speed, namely $v_g = 0.01176$, it came out 5.482; but for two drops whose speeds were five times as large, namely 0.0536 and 0.0553, e came out 5.143 and 5.145, respectively. This could mean nothing save that Stokes's law did not hold for drops of the order of magnitude here used, and it was surmised that the reason for its failure lay in the fact that the drops were so small that they could no longer be thought of as moving through the air as they would through a continuous, homogeneous medium . . . The law ought to begin to fail as soon as the inhomogeneities in the medium—i.e.,

[10] R. A. Millikan, "The Electron," p. 88. Univ. of Chicago Press, Chicago, Illinois, 1917.

the distances between molecules—began to be at all comparable with the size of the drop. Furthermore, it is easy to see that as soon as the holes in the medium begin to be comparable with the size of the drop, the latter begins to increase its speed, for it may then be thought of as beginning to reach the stage in which it can fall freely through the holes in the medium. This would mean that the observed speed of fall would be more and more in excess of that given by Stokes's law the smaller the drop became. But the apparent value of the electronic charge . . . is seen . . . to vary directly with the speed $[v_e + v_g]$ imparted by a given force. Hence e should come out larger and larger the smaller the radius of the drop, that is, the smaller its velocity under gravity. Now this was exactly the behavior shown consistently by all the oil drops studied. Hence it looked as though we had discovered, not merely the failure of Stokes's law, but also the line of approach by means of which it might be corrected."[11]

First Millikan plotted the apparent value of e (which we shall now call e'), as calculated from Eq. (22), against v_g, obtaining the curve shown in Fig. 6.

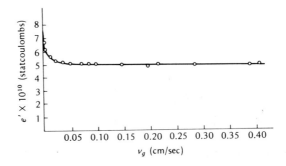

Fig. 6. *Variation of e', the apparent value of electronic charge, with terminal velocity v_g of oil drop in Millikan's experiment.*

It is clear from this curve that e' does vary systematically with v_g, and that drops for which $v_g \geqslant 0.2$ cm/sec are large enough to obey Stokes's law, for there is little variation in e' beyond $v_g = 0.2$ cm/sec. Smaller drops than these have radii which are comparable to the mean free path of an air molecule. Millikan therefore, using the mean free path λ as a measure of the air's inhomogeneity, worked out a corrected form of Stoke's Law. Using this corrected form of the law, he found that the smaller drops gave the same value of e as the larger ones.

[11] R. A. Millikan, "The Electron," pp. 90–91. Univ. of Chicago Press, Chicago, Illinois, 1917.

Millikan ran a series of measurements in which 58 drops were used, on 60 consecutive days, embracing a twelvefold variation in a (from 4.69×10^{-5} cm to 58.56×10^{-5} cm) and a seventeen-fold variation in air pressure, obtaining a thirtyfold variation in λ/a (from 0.016 to 0.444). The value of e' departed from the mean by as much as 0.5 percent for only one drop. The resulting value of N_A was $(6.062 \pm 0.006) \times 10^{23}$.

Millikan's work stood as the most accurate determination of N_A until 1931. In that year J. A. Bearden used X-rays to measure the distance between rows of atoms in calcite crystals.[12] The X-ray wavelength was precisely measured by diffraction from a ruled grating. Bearden was therefore able to find N_A by computing the number of atoms in a given volume of calcite, and he found that $N_A = (6.019 \pm 0.003) \times 10^{23}$. This value was so far from Millikan's value that at first Bearden did not believe it himself, thinking that perhaps some inhomogeneity in the crystal was responsible. However, it was then found that the value used by Millikan for the viscosity of air had been too low, by about 0.4 percent; correction of this would increase Millikan's value of e by 0.6 percent, bringing his value of N_A into agreement with Bearden's. The currently accepted value of N_A is $(6.022045 \pm 0.000031) \times 10^{23}$, obtained by a least-squares fit of all the fundamental constants to the results of all relevant experiments.[13]

PROBLEMS

1. Calculate the approximate diameter of an argon atom, given the density of solid argon (1.65 g/cm³) and the atomic mass of argon (40). (See Example Problem 1.)

2. An atomic beam is directed toward a plate which is large enough to stop all of the atoms in the beam. When the plate is 5 mm from the source of the beam, 10.00 mg is deposited from the beam onto the plate in a given time interval. When the plate is 10 mm from the source, 0.50 mg is deposited in the same time interval. Find the mean free path of the atoms in the beam.

3. After the measurements described in Problem 2 were made, the pressure in the chamber was doubled. Assuming that the density of atoms in the beam was unchanged at the source, find the respective amounts deposited on the plate in the same time interval at the same two positions.

[12] J. A. Bearden, *Phys. Rev.* **37,** 1210 (1931).
[13] Particle Data Group, *Rev. Mod. Phys.* **52,** 2, Part II (1980).

4. At what height in the earth's atmosphere would the density be 1 molecule/cm³, if the temperature were 273 K?

5. Show from expression (11) for $h(v_z)$ that the average kinetic energy in *one-dimensional* motion is $kT/2$. (*Hint*: You must use $h(v_z) \, dv_z$ to find $n(E) \, dE$ such that $\int_{E_1}^{E_2} n(E) \, dE$ is the probability that the energy is between E_1 and E_2. The average kinetic energy may then be found by means of the appropriate integral.) Also use the Maxwell distribution $f(v)$ to show that the average kinetic energy in *three-dimensional* motion is $3kT/2$.

6. Use the method of Example Problem 2 to show that the pressure on the wall of the container is $M\overline{v^2}/3V$ [Eq. (1)].

7. Use the Maxwell distribution and any convenient numerical approximation to find the number of air molecules per cubic centimeter with energy greater than 1 eV, at 0°C and atmospheric pressure.

8. Find the value of B required to observe the Zeeman splitting of a spectral line whose wavelength is 5000 Å, if the spectrometer has a resolving power $\lambda/\delta\lambda = 10^4$.

9. In one of Zartman's measurements (Sec. 1.1), the drum of radius 5.0 cm revolved at 241 revolutions/sec. The average position of the deposited metal was shifted 1.8 cm from the position directly opposite the slit. What atomic speed is required for this shift?

10. J. J. Thomson reported the following data for one measurement of e/m: The beam was deflected through an angle of 0.12 rad in passing through a magnetic field of 6.6×10^{-4} T for a distance of 5 cm. An electric field of 1.5×10^4 V/m, applied over the same distance, returned the deflection to zero. Find the speed of the particles and the value of e/m.

11. In an experiment similar to Perrin's, $a = 2 \times 10^{-5}$ cm, $\rho = 1.2$ gm/cm³ and the particles are suspended in water at 300°K. Particles are counted at intervals of 3×10^{-3} cm in height, and the numbers of particles seen at successive levels are: 201, 130, 74, 49, 18, 16, 12. Deduce the value of N_A by making a suitable graph of these numbers. Estimate the error introduced by possible statistical fluctuations in the numbers, and compare with the actual error in your value of N_A.

12. Find the value of a for a Millikan oil drop that fell at a speed of $v_g = 0.01$ cm/sec, using Eq. (18) and data from Problem 13. Would the corrected form of Stokes's Law yield a larger or a smaller value of a?

13. Compute the electronic charge from the data in Table 1. Use Eq. (22), with $\rho = 0.920$ gm/cm³, $\rho' = 0.001$ gm/cm³, $\eta = 1.83 \times 10^{-4}$ gm/sec-cm, and

$E = 3.16 \times 10^5$ V/m. Be sure to use consistent units and to indicate the units in your answer.

14. Explain observation 4 of Millikan's oil-drop experiment, p. 21, as follows. Calculate the potential energy of a singly charged negative ion at the surface of a drop of radius 0.0002 cm, if the drop carries 10 electron charges (negative), and if it carries 100 electron charges (also negative). Then compare this energy in each case with the mean kinetic energy of air molecules at room temperature, and draw the appropriate conclusions.

CHAPTER

The Theory of Relativity

The concept of motion is fundamental in physics, yet it is not immediately obvious what we mean when we speak of motion. Newton defined his laws of motion in terms of "absolute" motion, by which he meant change of position with respect to some fixed, absolute frame of reference. However, this idea is not central to Newton's laws of motion, which are valid in any inertial frame of reference. An inertial frame is defined as one which is moving with constant velocity with respect to the fixed, "absolute" frame, but if one is given a class of inertial frames of reference, there is no need, on the basis of Newton's laws, to single out one frame and call it the fixed frame.

In mechanics, what one always observes is the change in position of one body *relative to another body*; there is no way to directly observe motion relative to an incorporeal "absolute" reference frame. But the study of light seems to suggest a possible reference frame. Nineteenth century physicists, discovering that light behaves like a wave, and knowing that all other waves are disturbances of some material medium on which the wave travels, began to inquire into the nature of the medium on which light travels through apparently empty space. The medium became known as the luminiferous ether, and it obviously had to have remarkable properties. It had no mass, and it offered no resistance to the movement of material bodies through it. If, in addition, the ether itself were not affected by the passage of bodies through it, then the ether would have all the properties of Newton's absolute reference frame.

The velocity of a wave with respect to a given medium depends only on the properties of the medium and is independent of the motion of the source. This was assumed to be true for a light wave as well, and the velocity of light in free space was assumed to be a constant c *with respect to the ether*. If the speed of light were then measured with respect to a body, such as the earth, which is moving through the ether, the result should differ from c. By measuring the velocity, relative to the earth, of light rays moving in different directions, one could then, in principle, find the "true" velocity of the earth.

Since the velocity of light can be derived from Maxwell's equations, the statement that this velocity is constant with respect to some universal ether is equivalent to saying that Maxwell's equations are valid only in the frame of reference which is fixed in the ether. (If you are unfamiliar with Maxwell's equations, do not be alarmed; it is not necessary to know the form of these equations in order to appreciate the point.) If this is true, then one might detect absolute motion of the earth by observing electromagnetic forces. Since a moving charge constitutes a current, the force between two charges would depend on the absolute velocity of each charge as it moves with the earth through the ether.

Many attempts have been made to detect variations either in the speed of light or in the force between charged bodies as the earth's motion through the ether changes. We shall briefly discuss some of these attempts, and examine various theories which have been devised to explain their failure. After we see some of these theories, we should be ready to accept Einstein's theory as the simplest explanation of the observations.

We will then see that this theory makes additional predictions which are startling but nevertheless verifiable. Our confidence in the theory of relativity rests upon the success of these predictions. They will form the basis for much of the discussion in later chapters of this book, particularly in the chapters on nuclear physics.

2.1 EXPERIMENTS PRECEDING THE THEORY OF RELATIVITY

The first well-verified observation concerning the connection between the earth's motion and the velocity of light was made by the astronomer James Bradley in 1727. He observed a seasonal change in the apparent position of the star γ Draconis, relative to other stars. The effect, now known as stellar *aberration,* is not to be confused with the stellar *parallax,* a much smaller effect which was first observed a century later. Stellar aberration is caused by the change in direction of the velocity of the earth in its orbit; it can be distinguished from parallax, because it depends on the *position of the star in the sky,* and not on the distance of the star from the earth (see Fig. 1).

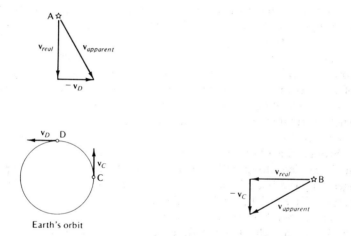

Fig. 1. *Aberration of starlight. Stars are located at A and B; light travels in direction* v_{real} *to reach earth. When earth is at C, light from B moves in direction* $v_{apparent}$ *relative to earth; when earth is at D, light from A moves in direction* $v_{apparent}$ *relative to earth.*

Stars A and B lie in the plane of the earth's orbit; in order to reach the solar system from each star, rays of light must travel in the direction of the vector v_{real}, shown at A and B, respectively. When the earth is at point C, moving with velocity v_C, the ray from B appears to an observer on earth to be moving in the direction of the velocity of the ray *relative to the earth,* or $v_{real} - v_C$ labeled $v_{apparent}$. The ray from A is still seen in the direction of

$\mathbf{v_{real}}$, because $\mathbf{v_{real}}$ for this ray is parallel to $\mathbf{v_C}$. The angle between $\mathbf{v_{real}}$ at A and $\mathbf{v_{apparent}}$ at B is less than 90°, so the two stars *are less than* 90° *apart in the sky* when the earth is at C. Similar analysis shows that the *same two stars are more than* 90° *apart in the sky* when the earth is at D, three months later.

The angle between $\mathbf{v_{real}}$ and $\mathbf{v_{apparent}}$ should equal the ratio of the earth's orbital velocity to the velocity of light, or $(30\,\text{km/sec})/(30 \times 10^5\,\text{km/sec}) = 10^{-4}$. More recent observations of aberration have confirmed this figure to a high degree of accuracy. Notice that any additional *constant* aberration, caused by motion of the *whole solar system* through space, would be undetected; it is only the *change* in the aberration angle which is seen, as the earth changes direction in its orbit.

After the wave nature of light was established (in the early nineteenth century), people began to worry more about problems related to Bradley's observation. It was suggested that the focal length of a lens should depend on its state of motion through the ether, because such motion would affect the velocity of light *relative to the lens*. But when this idea was tested experimentally, it was found that starlight was always refracted *as it would be if the earth were at rest* and the *apparent* direction of the ray could be considered its "true" direction. Augustin Fresnel, whose calculations had led to general acceptance of the wave theory of light, then sought to derive such a result from a theory of the velocity of light in a moving material medium.

Fresnel postulated that in a medium whose index of refraction is n, the ether "density" is proportional to n^2. He then assumed that when a body moves through the ether, the *excess* ether that it contains over the free space amount is carried along with the body. The fraction of the ether which is "dragged" by the body is then $(n^2 - 1)/n^2$; this fraction is called the Fresnel dragging coefficient. The "center of mass" of the ether in a body moving with velocity v then has the velocity $v(n^2 - 1)/n^2$. A light wave traveling within the body has the velocity c/n with respect to this "center of mass."

To see how this idea leads to the desired result, suppose that a light ray is normally incident on a plane surface of a medium of index of refraction n (Fig. 2). If the medium is at rest, the ray continues into the medium in the same direction, with velocity c/n. If the medium is moving with velocity $\mathbf{v}$ parallel to the plane surface (a), the only effect is that the ray acquires a horizontal component of velocity equal to $v(n^2 - 1)/n^2$ when it enters the medium. Now look at the ray from the point of view (b) of an observer moving with the medium. Since he is moving with velocity $\mathbf{v}$, the relative velocity of the ray with respect to him is found by adding the vector $-\mathbf{v}$ to the "true" velocity of the ray, both inside and outside the medium. This gives the ray velocity a horizontal component $-\mathbf{v}$ outside and $\mathbf{v}(n^2 - 1)/n^2 - \mathbf{v} = -\mathbf{v}/n^2$ inside the medium. The ray then makes angles α and β, respectively, with the normal to the surface. Neglecting terms of order v^2/c^2, we find that $\alpha = v/c$ and $\beta = v/nc$, so that $\alpha/\beta = n$. This is, of course, the correct law of refraction for small angles. Therefore light appears to be refracted as it would be if observer

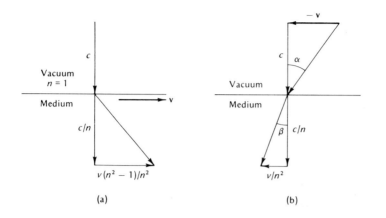

(a)   (b)

Fig. 2. *"Refraction" of a light ray in accordance with Fresnel dragging coefficient. (a) Path in "absolute" reference frame. (b) Path in reference frame moving with medium.*

and medium were at rest and the apparent direction of the ray were its "true" direction. *An observer cannot, by observing refraction of light, determine that he is moving through the ether.*

Fresnel showed that his dragging coefficient produced the same result in more general cases of refraction, provided always that the speed of the refracting medium is much less than c. He inferred from this that the aberration of starlight should not change when it was viewed through a water-filled telescope. Many years later (in 1871), Sir George Airy verified this experimentally.

Fresnel's dragging coefficient also predicts the absence of another possible first-order effect, caused by motion of an observer and his apparatus *parallel* to the direction of the ray. This effect, involving the transit time of a light ray, is best explained by describing the experiment performed by Hoek (1868) to search for it. The experimental arrangement is shown in Fig. 3. Light from a source S shines on a partially silvered plate so that one ray passes through and traverses the rectangle $PM_1M_2M_3$ in counterclockwise fashion, while another ray is reflected by P and travels clockwise via $M_3M_2M_1$. A refracting medium of length l and refractive index n is located between M_2 and M_3. When the rays meet again and enter the telescope T, an interference pattern results from the differences in the optical paths traveled by the two rays meeting at any given point. We now demonstrate that, if the apparatus were moving through a *stationary* ether, the interference pattern would show this.

Let us suppose that the apparatus is moving toward the right with velocity v through the ether, and let us compute the difference between the times taken

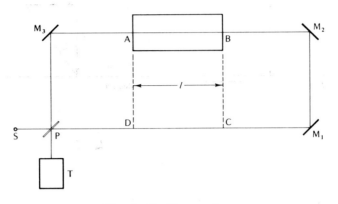

Fig. 3. *Hoek's experiment.*

by the two rays to circumnavigate the rectangle. We obviously need only consider the times taken to cross the refracting medium and to travel the same distance in the parallel side of the rectangle. With respect to the apparatus, the clockwise-moving ray moves with velocity $c/n - v$ from A to B and velocity $c + v$ from C to D; the other ray moves with velocity $c - v$ from D to C and velocity $c/n + v$ from B to A. The time difference is then

$$\delta t = l\left[\frac{1}{\dfrac{c}{n} - v} + \frac{1}{c + v} - \left(\frac{1}{\dfrac{c}{n} + v} + \frac{1}{c - v}\right)\right] \tag{1}$$

$$= l\left[\frac{-2v}{c^2 - v^2} + \frac{2v}{\dfrac{c^2}{n^2} - v^2}\right] \tag{2}$$

If we neglect v^2 in comparison to c^2 or c^2/n^2,

$$\delta t = \frac{2vl}{c^2}(n^2 - 1) \tag{3}$$

Rotating the apparatus through $180°$ has the effect of reversing its velocity through the ether, and thus reversing this time difference. This means that the interference pattern should shift when the rotation is performed, but no shift was observed when this was done.

The result is easily explained by the Fresnel dragging coefficient. Inside the medium, the fraction of ether *not* dragged is $1/n^2$, so the relative velocity of ether and medium is effectively v/n^2 rather than v. A light ray, moving with speed *c/n relative to the ether* in the medium, has the speed $c/n - v/n^2$ when going from A to B, and the speed $c/n + v/n^2$ in going from B to A. Substitution of these speeds into the first and third terms in Eq. (1) makes the numerator of the second term in Eq. (2) $2v/n^2$ rather than $2v$, and this neatly nullifies the whole expression, when terms of order v^2/c^2 are neglected. Therefore, to terms of order v/c, the Fresnel dragging coefficient renders all effects of the earth's motion through the ether undetectable.

In the discussion so far we have assumed that the ether near the earth forms a fixed reference frame which is unaffected by the motion of the earth. The ether drag of Fresnel is confined to the ether *inside* a refracting body. The results can be explained in another way. Stokes suggested that the earth drags with it *all* of the ether *near* its surface, so that any optical instrument fixed on the earth is surrounded by a blanket of ether which is stationary with respect to the instrument. This hypothesis explains all of the null results, but it does not explain aberration of starlight. Starlight, in passing through the ether surrounding the earth, would be deflected from its original path, and one could no longer assume that the ray follows a straight line in "absolute space." Stokes attempted to derive a velocity field for the ether surrounding the earth so that the correct aberration angle would result, but this effort was not convincing. It was later observed (1892) that the speed of light passing by the rim of a massive spinning wheel was changed by less than 1/200 of the speed of the rim. This observation killed Stokes's version of ether drag.

The Fresnel dragging coefficient was verified in a more positive manner by Fizeau in 1851. He observed how the motion of water affects the speed of light passing through the water. The setup is illustrated in Fig. 4. As in the

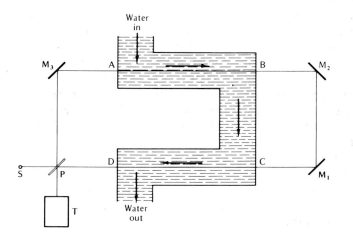

Fig. 4. *Fizeau's experiment.*

later experiment of Hoek, interference is observed between rays traversing the rectangle in opposite directions. We know already that there is no transit-time difference between these rays when the water is not moving, so let us assume that the entire apparatus is at rest in the ether. When the water is moving, one ray travels with the flow of water, while the other ray travels against the flow, and the interference pattern is shifted, relative to the pattern observed when the water is motionless.

We compute the time difference between rays as before, using the Fresnel dragging coefficient. The speeds *relative to the medium* (water) are $c/n \pm v/n^2$ as before, but now the medium is moving with velocity v relative to the observer. The speed of the clockwise ray, relative to the observer, is $c/n + v(1 - 1/n^2)$ for the entire time that it spends in the water; the corresponding speed for the counterclockwise ray is $c/n - v(1 - 1/n^2)$. The time difference between the rays is then

$$\delta t = l \left[\frac{1}{\dfrac{c}{n} - v\left(1 - \dfrac{1}{n^2}\right)} - \frac{1}{\dfrac{c}{n} + v\left(1 - \dfrac{1}{n^2}\right)} \right]$$

$$\delta t \approx \frac{2lv}{c^2}(n^2 - 1) \qquad (v \ll c) \tag{4}$$

where l is the entire distance traveled by each ray in the water. When Fizeau turned on the water, he observed a shift in the interference fringes in agreement with Eq. (4).

Although the Fresnel dragging coefficient neatly disposes of *first-order* effects of the earth's motion through the ether, one might still expect a careful experiment to reveal the presence of *second-order* effects. In a beautifully executed experiment, A. A. Michelson and E. W. Morley searched for such an effect; their failure, reported in 1887,[1] ultimately led to the downfall of the entire concept of the ether. Figure 5 shows the essentials of their apparatus, which is based on interference of two rays, as in the experiments of Hoek and Fizeau. One ray passes through P to M_2, then back, reflecting off P to T. The path of the other ray is perpendicular to that of the first ray, as it is reflected from P to M_2 and then back through P to T. Suppose that $PM_1 = PM_2 = l$, and suppose that the entire apparatus is moving with velocity **v** parallel to PM_1. Then the first ray has speed $c - v$, with respect to the apparatus, as it moves toward M_1, and speed $c + v$ on its return. The other ray, whose motion *relative to the apparatus* is perpendicular to **v**, has speed $(c^2 - v^2)^{1/2}$ relative to the apparatus (see Fig. 6). The time difference between the two rays is then

[1] A. A. Michelson and E. W. Morley, *Amer. J. Sci.* **34**, 333 (1887).

$$\delta t = \frac{l}{c-v} + \frac{l}{c+v} - \frac{2l}{(c^2-v^2)^{1/2}} \tag{5}$$

$$= \frac{2l}{c^2-v^2}\left[c - (c^2-v^2)^{1/2}\right] \tag{6}$$

$$= \frac{2l}{c^2-v^2}\left[c - c\left(1 - \frac{v^2}{c^2}\right)^{1/2}\right] \tag{7}$$

$$\approx \frac{2l}{c^2-v^2}\left(\frac{1}{2}\frac{v^2}{c}\right)$$

$$\approx \frac{lv^2}{c^3} \tag{8}$$

For light of frequency v, the phase difference between the two rays is then, in cycles, equal to $v\,\delta t$ or $(v/c)^2/(\lambda/l)$. The apparatus was mounted on a massive stone disk, floating on mercury, so that it could easily be rotated; a rotation through 90° should reverse the phase difference, so that the interference pattern should shift by a number of fringes equal to twice the above phase difference. By using additional mirrors to reflect each ray several times, Michelson and Morley obtained an effective length of 11 m for each arm. With v equal to the earth's orbital speed of 30 km/sec, the fringe shift should have been about 0.4 fringes. Although Michelson estimated that a shift of 1/100 of a fringe should have been detectable, no shift was seen. The experiment was repeated at different seasons, to allow for the possibility that the

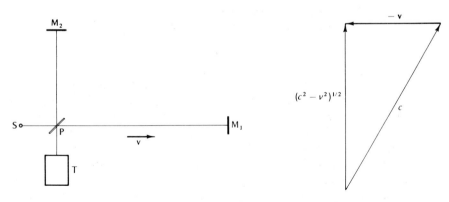

Fig. 5. *Michelson–Morley experiment.*

Fig. 6. *Adding the vector* $-\mathbf{v}$ *to the "absolute" ray velocity of magnitude c gives ray velocity of magnitude $(c^2 - v^2)^{1/2}$ with respect to apparatus.*

earth happened to be stationary in the ether when the experiment was first done, but the null result persisted.

It seemed that motion through the ether simply could not be detected by observation of light rays. Michelson himself believed that his result confirmed the Stokes version of ether drag. But when he tested his idea by repeating his experiment on a mountain, he again saw no shift. As we have seen, the Stokes hypothesis is deficient in other ways, so another explanation was needed.

Before examining the various explanations which were advanced, let us look at one more experiment which yielded a null result of a different kind. In 1904 Trouton and Noble[2] reported an attempt to detect a torque on a charged capacitor. Figure 7 shows the idea behind the attempt. Suppose

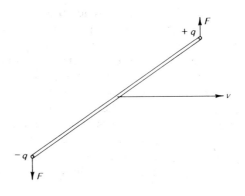

Fig. 7. Trouton–Noble experiment; magnetic torque on charges moving through ether.

charges $+q$ and $-q$ are connected by an insulated rod. If the charges are at rest, the electrostatic force between them is directed along the line joining them, producing no torque. But if Maxwell's equations are valid only in a frame of reference fixed in the ether, and if the earth is moving through the ether, each charge, moving with the earth, constitutes a current. Therefore one expects a magnetic force on each charge, in the directions shown. Trouton and Noble therefore hung a charged capacitor from a delicate suspension, and attempted to observe the torque by means of the rotation of a mirror mounted on the suspension. Although they made observations at different times, when the plane of the capacitor made different angles with the earth's orbital motion, they never observed a deflection of more than 5 percent of

[2] F. T. Trouton and H. R. Noble, *Proc. Roy. Soc.* **72**, 132 (1904).

their calculated value. They concluded, "The deflection observed was a purely capricious action and could in no way be attributed to the relative motion of the earth and the ether."

2.2 *EXPLANATIONS*

We have seen that the Fresnel dragging coefficient explains the lack of a first-order effect of the earth's motion through the ether. By the end of the nineteenth century it became apparent that this was not enough; it seemed that it would never be possible to detect such motion.

One way out of the dilemma is simply to deny that light always has the velocity c (or c/n) relative to the ether. Ritz proposed that light has the velocity c relative to the source of the light. This explains the Michelson–Morley result, but the proposal runs into difficulty in other ways. The biggest objection to the theory is based on observations of binary stars, which show that the two members of the pair follow regular orbits about the common center of mass. If the speed of light emitted by one of these were c relative to the star, and the star's orbital speed is v, then to an observer at rest with respect to the center of mass of the two stars, the light would have a speed varying from $c - v$ to $c + v$. Light emitted in the x direction (Fig. 8) when the

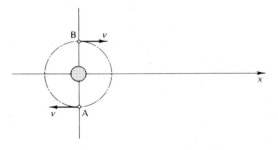

Fig. 8. *Orbit of one star of a binary star system.*

star is at point A would have the speed $c - v$, while light emitted in this direction when the star is at B would have the speed $c + v$. Even though the orbital speed is typically only $10^{-4}c$, light from B would eventually overtake the light emitted earlier from A. To an observer at the right distance from the system, the star would sometimes appear at A and B simultaneously, while at other times the star would not be visible at all! Effects of this sort are never observed, either with visible light or with X-rays emitted by such stars.[3]

---

[3] One might argue that visible light is absorbed and reradiated by the interstellar medium, so any initial dependence on the star's speed would be lost. But this objection does not apply to X-rays, which have also been carefully studied for a number of binary star systems. See P. E. Hodgson, "Speed of Light and Relativity", Nature **271**, 13 (1978).

Fitzgerald and, independently, H. A. Lorentz, came up with another hypothesis which is, in a sense, the correct explanation of the Michelson-Morley result. They pointed out, as is clear from Eq. (5), that no fringe shift would be observed if the arm which is parallel to the earth's motion is shortened by the factor $(1 - v^2/c^2)^{1/2}$, so that its length became $l(1 - v^2/c^2)^{1/2}$. Whichever arm is turned parallel to the earth's motion would have to be shortened in this amount, so they postulated that any body is contracted by this factor in the direction of its motion through the ether. The effect is always unobservable except insofar as it produces the null result in this experiment, because any measuring rod used to measure the contraction would itself be contracted. The absence of a fringe shift in the Michelson-Morley experiment can be taken as a proof of the contraction, just as the observation of a shift when one arm is heated can be taken as proof that the arm has expanded.

Lorentz went on to develop an electron theory of matter which asserted that the motion of any body through the ether produces changes in the electromagnetic forces which hold the body together. Such changes always alter the dimensions of the body just enough to prevent the detection of the motion by the methods discussed above. For example, concerning the Trouton and Noble experiment, Lorentz concludes[4]:

"The only effect of the translation must have been a contraction of the whole system of electrons and other particles constituting the charged condenser and the beam and thread of the torsion balance. Such a contraction does not give rise to a sensible change of direction."

Lorentz based his conclusions on a new *transformation of coordinates* from the coordinate system at rest in the ether to the moving system. This transformation expressed the Lorentz contraction of the moving coordinate system, and it provided for a *different time scale*, which he called "local time," in the moving system. Lorentz also assumed that the electric and magnetic fields seen by an observer in a moving system differ from the "true" fields seen in the rest system. Then when the coordinates $xyzt$ of the rest system were transformed into the moving coordinates $x'y'z't'$, by the "Lorentz transformation," the relationships of these "apparent" fields to each other, involving derivatives with respect to x', y', z', and t', were as given by Maxwell's equations. A pair of charges which were at rest relative to a moving observer would therefore appear *to that observer* to attract or repel each other just as they would if the charges *and the observer* were "truly" at rest.

In developing his theory, Lorentz made several assumptions about the nature of matter and of electromagnetism. For example, he assumed that

[4] H. A. Lorentz, *Proc. Acad. Sci. Amsterdam* 6 (1904). Translated in "The Principle of Relativity," Dover, New York, 1952.

matter is composed of elementary electric charges whose inertia, or mass, is solely produced by their internal electromagnetic energy, and that these charges are spherical when at rest but are contracted in the direction of their motion when they move. While this approach explains the observations, it does little to simplify our view of nature.

The Principle of Relativity. In 1905, the year after Lorentz published his work, Albert Einstein published the paper "On the Electrodynamics of Moving Bodies,"[5] in which he sets forth the special theory of relativity. The results of this theory are basically the same as those of Lorentz, but the underlying philosophy is quite different. Einstein decided that there was no use referring to the ether or to an absolute frame of reference if no experiment could detect its existence. He adopted as a *postulate* the *principle of relativity*:

The same laws of electrodynamics and optics (as well as of mechanics) are valid in all inertial frames of reference.

He further *postulated* that

light is always propagated in empty space with a definite velocity c with respect to any inertial frame of reference, regardless of the state of motion of the emitting body.

In these ideas Einstein was somewhat anticipated by Poincaré, but the credit for developing a complete theory goes rightfully to Einstein.[6] Starting with the idea that absolute motion is undetectable, Einstein *derived* a number of things—Lorentz contraction, change of time scales—which Lorentz had adopted as postulates.

Einstein's approach has the advantage of generality; *it is not restricted to electromagnetic phenomena*, although it was originally stated in those terms. It is not necessary to assume anything about the structure of matter or the nature of electrons, for the principle of relativity must apply to all phenomena. The "speed of light" is then *a universal constant which governs all phenomena* rather than a property peculiar to light alone.

[5] A. Einstein, *Ann. Phys.* (*Leipzig*) **17**, 891 (1905). Translated in "The Principle of Relativity," Dover, New York, 1952.
[6] Poincaré as early as 1899 had suggested the principle of relativity. He also guessed that new laws of dynamics were needed, in which the speed of light would be an upper limit on all speeds. These remarks so impressed Sir Edmund Whittaker that in his book ("A History of the Theories of Aether and Electricity," Vol. II, "The Modern Theories, 1900–1926." Philosophical Library, New York, 1954), he entitles his relativity chapter, The Relativity Theory of Poincaré and Lorentz, and he mentions Einstein's paper only as one which set forth the theory of Poincaré and Lorentz "with some amplification." However, a more balanced viewpoint which recognizes each man's contribution is presented by G. Holton in an article, On the Origins of the Special Theory of Relativity, *Amer. J. Phys.* **28**, 627 (1960). It is clear that Poincaré did not develop a theory, but he made some remarkably good guesses as to the form a fully developed theory should take.

2.3 *THE LORENTZ TRANSFORMATION*

We now wish to see how the Lorentz transformation follows from Einstein's second postulate. First let us see why it is necessary to transform the time coordinate as well as the space coordinates. Here, as in most other investigations, one does not find the right answers until one asks the right questions. The questions in this case are: What is time? How does one measure time? Any child will answer the second of these; you use a clock. But I would not venture to predict what the child would answer to the first question. (It is possible that he would say that time is what you measure with a clock, and it is hard to find a better answer.)

As the child grows up, he acquires the prejudice that time is something absolute which flows of itself, independent of any events which occur. But it is impossible to define time without introducing a sequence of events which serves as a sort of "coordinate axis." One arbitrarily assumes that a certain sequence of events is equally spaced in time, and one then uses this sequence to define time intervals between other events. Historically, the sequence of events was the occupation of successive positions by the sun and stars in the sky; when clocks were invented, they were adjusted to agree with astronomical observations. Nowadays one uses an atomic clock, because one believes that the frequency of vibration of radiation from a free atom should not vary, and one indeed finds that the frequency of one atom does not vary when compared with another atom. But if the frequencies of all atoms and planets were to "change" in the same ratio, we would have no way of knowing it; thus absolute time is as unobservable as absolute space.

The Relativity of Simultaneity. After this discussion, you may be willing to accept the idea that a moving system could have a different time scale from that of a stationary system; the frequencies of all natural vibrations could be altered by the motion. This was Lorentz's point of view. But the principle of relativity demands more than this, for it says that, although the time scales may differ for two systems in relative motion, *either* system may be considered to be the stationary one, for only the relative motion has physical meaning. For example, we will later find that a moving clock runs slowly, when compared with an identically constructed clock at rest. This means that, if observers A and B are in relative motion, *each* finds the *other* one's clock to be slower than his own. A may say, "When my clock reads two o'clock, B's clock reads one o'clock", while B may say, "When my clock reads two o'clock, A's reads one o'clock." Many persons have rebelled when asked to accept such a seemingly paradoxical statement, but the paradox is resolved when one considers that two events which are simultaneous to A may not be simultaneous to B.

Before Einstein, everyone assumed, without even thinking about it, that the simultaneity of two events was determined by absolute time, and was independent of the observer. Having discarded the concept of absolute time, we must find another way to define simultaneity. We may use a single clock to determine time intervals between events occurring at the same place; a time interval of zero denotes simultaneity. All observers will agree on such events. But to compare times of events occurring at different places, we must send a signal from one place to another. If we know the speed of the signal, we can deduce the time of any event in any place. If we now adopt Einstein's postulate, and use light signals to determine times of events, we find that different observers may disagree about the simultaneity of two events. For example, a man (observer A) at rest in the center of a room sends out a light signal. The wavefront goes out in all directions with speed c, so A says that the wavefront strikes opposite walls of the room simultaneously. But another observer, B, is moving across the room; he sees the room and observer A moving past him. He says that A's light signal travels *in all directions* with speed c relative to *himself* (B). Relative to B, one wall moves away from the place where the source emitted the light, while the other wall approaches this place, so the wavefronts *do not strike the opposite walls simultaneously.*

One who stubbornly clings to the idea that simultaneity is absolute will say that this example proves that Einstein's postulate is "wrong." It does not. The example merely shows that the postulate that simultaneity is absolute is incompatible with the postulate of Einstein. When two postulates conflict, one must be discarded. The observations which we have discussed all indicate that Einstein's postulate is reasonable, while there is no evidence for the postulate that simultaneity is absolute; the latter is merely a preconceived notion, which we now reject.

Of course, since the observers A and B are not located at the walls, they must each use remote measurements to decide on the simultaneity of events which take place there. Each must know how far it is from his own position to each wall position so that each can determine when the light struck the wall, and this he may do by determining the time when the reflection reached him. But the observers, because of the Lorentz contraction, do not agree on the size of the room, so they must disagree on the time of the events.

Derivation of the Lorentz Transformation.   We shall now use Einstein's postulates to derive the Lorentz transformation between the coordinates $xyzt$ of a system S and the coordinates $x'y'z't'$ of another system S' which is moving with velocity v relative to S (see Fig. 9). To simplify the mathematics, we choose axes so that corresponding axes are parallel and the z and z' axes are in the direction of v. We let the origins of the two systems coincide at $t = 0$, and let a light wave originate at this common origin. The equation of the wavefront of this wave is

$$x^2 + y^2 + z^2 = c^2 t^2 \qquad \text{in} \quad S \tag{9}$$

and

$$x'^2 + y'^2 + z'^2 = c^2 t'^2 \qquad \text{in} \quad S' \tag{10}$$

because the speed of light is c in each system (Einstein's postulate).

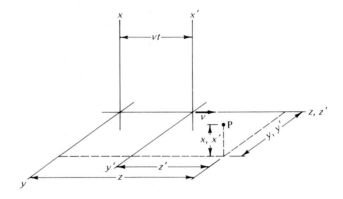

Fig. 9. *Coordinates of a point P in coordinate systems in relative motion.*

To simplify the form of the equations, let us substitute $x_1 = x$, $x_2 = y$, $x_3 = z$, and $x_4 = ct\sqrt{-1} = ict$ in Eq. (9), and make a similar substitution in Eq. (10). A set of values of x_1, x_2, x_3, and x_4 then locates an event in "space–time." The use of an imaginary coordinate to replace the time coordinate has no physical significance; it is merely a mathematical trick to help us to see how the transformation should behave. Because $x_4^2 = -c^2 t^2$, Eqs. (9) and (10) may now be written

$$\sum_{\mu=1}^{4} x_\mu^2 = 0 \qquad \text{and} \qquad \sum_{\mu=1}^{4} x_\mu'^2 = 0 \tag{11}$$

for the coordinates of those events on the wavefront of the light wave. For other events, $\sum_{\mu=1}^{4} x_\mu^2$ and $\sum_{\mu=1}^{4} x_\mu'^2$ are not zero, but $\sum_{\mu=1}^{4} x_\mu'^2$ must be proportional to $\sum_{\mu=1}^{4} x_\mu^2$, because of Eq. (11). Furthermore, systems S and S' are identical, except for their relative motion, and either system may be considered to be "stationary," so that the symmetry of the situation demands that the constant of proportionality be unity, and

$$\sum_{\mu=1}^{4} x_\mu'^2 = \sum_{\mu=1}^{4} x_\mu^2 \tag{12}$$

for all events.

Equation (12) tells us that the Lorentz transformation must be mathematically equivalent to a rotation, which preserves the lengths of vectors, because $\sum_{\mu=1}^{4} x_\mu^2$ is the four-dimensional analog of the length of a vector, and this "length" is unchanged by the transformation from S to S', according to Eq. (12). Only the x_3 and x_4 coordinates are affected by this "rotation," so that the transformation equations must have the form

$$
\begin{aligned}
x_1' &= x_1 \\
x_2' &= x_2 \\
x_3' &= x_3 \cos \alpha + x_4 \sin \alpha \\
x_4' &= -x_3 \sin \alpha + x_4 \cos \alpha
\end{aligned}
\tag{13}
$$

If the above reasoning seems unconvincing, you should verify by direct substitution that Eqs. (13) satisfy condition (12). The equations for x_3' and x_4' are the standard equations for a rotation in two dimensions. But we must remember that this transformation is a rotation *only in a formal sense*; the "angle" α *is not a real angle*. From Eq. (13) we see that $\sin \alpha$ is imaginary, because x_4', being equal to ict', is imaginary, making $x_3 \sin \alpha$ imaginary. The expressions $\cos \alpha$ and $\sin \alpha$ were used for the coefficients in Eqs. (13) simply because the condition $\cos^2 \alpha + \sin^2 \alpha = 1$, valid for imaginary as well as real angles, automatically causes Eqs. (13) to satisfy the requirement Eq. (12).

All that remains is to find the connection between α and the relative velocity v of the two coordinate systems. This follows from the definition of relative velocity; an observer in the S system sees the origin of the S' system, where $x_3' = 0$, moving along the x_3 axis with velocity v. Therefore, when

$$
x_3' = 0, \qquad \text{then} \qquad x_3 = vt
\tag{14}
$$

We can substitute $x_3' = 0$ in the third of Eqs. (13) to obtain

$$
0 = x_3 \cos \alpha + x_4 \sin \alpha
$$

or

$$
\begin{aligned}
x_3 &= -x_4 \tan \alpha \\
&= -ict \tan \alpha
\end{aligned}
\tag{15}
$$

Comparison of Eqs. (14) and (15) shows that $\tan \alpha = iv/c$. Then from the definition of $\tan \alpha$ we find

$$
\cos \alpha = \frac{1}{(1-\beta^2)^{1/2}}, \qquad \sin \alpha = \frac{i\beta}{(1-\beta^2)^{1/2}}
\tag{16}
$$

where β is defined as v/c. The reader should verify that Eq. (16) satisfies the condition $\cos^2 \alpha + \sin^2 \alpha = 1$. Equations (13), the Lorentz transformation, may now be written

$$x'_1 = x_1$$

$$x'_2 = x_2$$

$$x'_3 = \frac{x_3}{(1 - \beta^2)^{1/2}} + \frac{i\beta x_4}{(1 - \beta^2)^{1/2}}$$

$$x'_4 = \frac{- i\beta x_3}{(1 - \beta^2)^{1/2}} + \frac{x_4}{(1 - \beta^2)^{1/2}}$$

(17a)

Instead of saying that S' is moving with velocity v with respect to S, we could equally well have said that S is moving with velocity $-v$ relative to S'. Therefore one can find the transformation which gives the coordinates of S in terms of the coordinates of S' simply by interchanging the primed and unprimed coordinates and replacing β by $-\beta$ in Eqs. (17a). It is left as an exercise for the reader to solve Eqs. (17a) for x_3 and x_4 as functions of x'_3 and x'_4 in order to demonstrate this.

Converting to the real coordinates t and t', we have

$$x_3' = (x_3 - vt)\gamma$$

$$t' = (t - vx_3/c^2)\gamma$$

(17b)

where we have introduced the symbol γ to represent $(1 - \beta^2)^{-1/2}$.

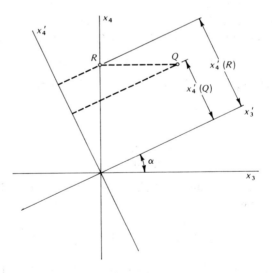

Fig. 10. *Graphical illustration of relativity of simultaneity.*

Lorentz Contraction and Time Dilation. We can represent the Lorentz transformation graphically, as long as we remember that α is an imaginary angle. We plot the $x_3 x_4$ plane, representing an event as a point in this plane. The x_3' and x_4' axes are rotated through the angle α relative to the x_3 and x_4 axes, respectively (see Fig. 10). The coordinates of an event are different in the two systems; for example, events Q and R are simultaneous in S, but in S', the event Q occurs earlier. (Its x_4' coordinate is smaller.) The graph contains all of the information in Eqs. (17), as long as we remember that α is an imaginary angle; it is very convenient to use the graph to determine distances and time intervals between events as seen in the two coordinate systems.

The Lorentz contraction can be seen directly from such a graph (Fig. 11). Suppose a rod of length l_0 is at rest in system S', with one end at $x_3' = 0$ and

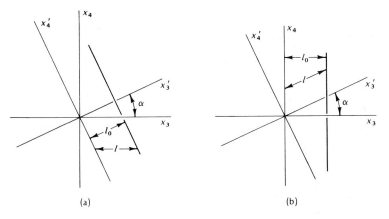

(a)   (b)

Fig. 11. *Lorentz contraction (a) of a rod at rest in system S', and (b) of a rod at rest in system S.*

the other end at $x_3' = l_0$. As time passes, x_4' increases; the ends of the rod trace out two lines parallel to the x_4' axis. Now suppose an observer in S measures the length of the rod as it passes by him with velocity v. He does this by measuring the distance between the respective positions occupied by the two ends of the rod *at the same time*, that is, when they have the same x_4 coordinate. He therefore measures the length l. But $l = l_0/\cos \alpha$, and $\cos \alpha = 1/(1 - \beta^2)^{1/2}$, so that $l = l_0(1 - \beta^2)^{1/2}$. The length l is shorter than l_0; the rod undergoes the Lorentz contraction because of its motion relative to the observer. Notice that l appears longer than l_0 in this diagram, because α is represented as a real angle; it is rather difficult to draw an imaginary angle. For real angles, $\cos \alpha \leq 1$, but here, of course, $\cos \alpha = 1/(1 - \beta^2)^{1/2} \geq 1$.

The same rod, if it were at rest in system S, would appear shorter to an observer in S'. The observer in S' would then measure the length $l_0(1 - \beta^2)^{1/2}$. Each observer sees the other's rod shortened.

The reader who is uneasy about the representation of imaginary angles on a real graph can derive the same results by substitution of the appropriate values into Eq. (17). For example, one end of the rod may be placed at the origin in both systems. If the rod is stationary in S', its other end is always at $x_3' = l_0$. The length measured in system S is then the value of x_3 when $x_4 = 0$.

The behavior of moving clocks may also be seen graphically. A clock at rest in S marks off equally spaced events along a line parallel to the x_4 axis; a clock at rest in S' does the same on a line parallel to the x_4' axis. If the time interval between successive ticks of the clock is t_0 in the system in which that clock is at rest, then Fig. 12 shows that the time interval between these events

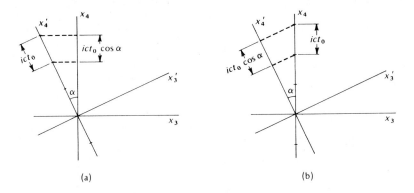

(a)   (b)

Fig. 12. *Time dilation. A clock at rest in each system has period t_0 in that system, but (a) clock at rest in S' has period of $t_0 \cos \alpha$ when observed in S; (b) clock at rest in S has period of $t_0 \cos \alpha$ when observed in S'.*

is $t_0 \cos \alpha = t_0/(1 - \beta^2)^{1/2}$ in the system in which the clock is moving. Since the interval between ticks is longer, the clock appears to be running slowly, relative to an identical clock at rest. In Fig. 12a, the observer is in system S, so simultaneous events lie on a line parallel to x_3, while in Fig. 12b, the observer is in S', and simultaneous events lie on a line parallel to x_3'. Thus *each* observer deduces that the *other's* clock is slow.

This effect, called time dilation, has been well verified by observation of elementary particles. For example, the mu meson (muon) has a mean lifetime of 2.2×10^{-6} sec; we may regard it as having an internal clock which tells it how long to live. But muons traveling at high speed have been observed to have mean lifetimes which are longer, by precisely the time dilation factor for their observed speed, than the mean lifetime of a resting muon.

EXAMPLE PROBLEM 1. A train zips past a platform at velocity $v = c/5$. A man standing at point A on the train (Fig. 13) and a man standing at point B shoot each other. The distance between A and B, *as measured by someone on the train*, is 15 m. The conductor on the train says that the two men fired simultaneously. What does an observer on the platform say? Which man fired first, and *how much* sooner did he fire, according to the observer on the platform?

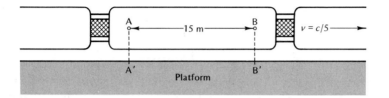

Fig. 13. *Schematic representation of a train speeding alongside a platform. Points A, A', B, and B' are shown as they appear to an observer on the train.*

Solution. Both conductor and man on platform *must agree* that A fired when he was alongside point A' and B fired when he was alongside point B'; this locates both events in space–time. The conductor says that points A' and B' are therefore 15 m apart, because A and B are 15 m apart, and they simultaneously passed A' and B', respectively. But the conductor saw the platform shortened by the Lorentz contraction. To the man on the platform, the distance between A' and B' is $15 \text{ m}/(1 - \beta^2)^{1/2} = 15 \text{ m}/0.98 = 15.3 \text{ m}$. Furthermore, the man on the platform says that A and B are only $15 \text{ m} \times 0.98 = 14.7 \text{ m}$ apart. Therefore when point A reaches A', and A fires, B has not reached B', according to the man on the platform; *points B and B' are still 0.6 m apart* (Fig. 14). At a speed of $c/5 = 0.6 \times 10^8$ m/sec, B travels 0.6 m in 10^{-8} sec, so the man on the platform says that *B fired 10^{-8} sec after A fired.*

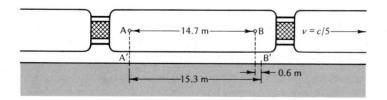

Fig. 14. *The train of Fig. 13, with points A, A', B, and B' shown as they appear to an observer on the platform.*

We may also solve the problem by more explicit use of the Lorentz transformation. Let the train be system S and the platform be S'. Give the event "A fires" the coordinates $x_3 = x_4 = x'_3 = x'_4 = 0$. Then "B fires" has coordinates $x_3 = 15$ m and $x_4 = 0$. We wish to find x'_4, which gives the time of "B fires" in system S'. We insert the values of x_3 and x_4 into Eq. (17), using $\beta = -c/5$ (because S' is moving in the *negative* x_3 direction relative to S), and we find

$$t' = \frac{x'_4}{ic} = \frac{-15i\beta}{ic(1 - \beta^2)^{1/2}}$$
$$= 10^{-8} \quad \text{sec}$$

Since A's gun fired at $t' = 0$, again we find that B's gun fired 10^{-8} seconds later than A's in the system S'. It should be clear from this example that relativistic effects are unobservable in ordinary experience. That is why relativity seems so strange to us.

Addition of Velocities.   Consider a system S'' moving with a velocity $\mathbf{v}'$ relative to S'. If S' has a velocity $\mathbf{v}$ relative to S, what is the velocity $\mathbf{v}''$ of S'' relative to S? Prerelativity physics says that $\mathbf{v}''$ is simply equal to the vector sum $\mathbf{v}' + \mathbf{v}$. This result follows from the so-called Galilean transformation, which may be written

$$\begin{aligned} x'_1 &= x_1 \\ x'_2 &= x_2 \\ x'_3 &= x_3 - vt \\ t' &= t \end{aligned} \tag{18}$$

when $\mathbf{v}$ is parallel to the x_3 axis. The Galilean transformation and the old velocity addition rule were, of course, used in the analyses of all the prerelativity experiments; they appeared to be so obvious that nobody questioned them. (That is what caused all the trouble.) It is easy to see that Eqs. (17a) and (17b) reduce to (18) in the limit $c \to \infty$.

In relativity theory one must deduce the velocity of S'' relative to S by finding the Lorentz transformation which transforms system S to system S''. It is clear that the old rule cannot work, for it would yield relative velocities greater than c (even though v_1 and v_2 were each smaller than c), and there is no allowable Lorentz transformation for such a relative velocity. (A relative velocity greater than c makes $(1 - \beta^2)^{1/2}$ imaginary, and this leads to an imaginary space coordinate in the transformed system.)

The relativistic solution is a bit complicated when $\mathbf{v}_1$ and $\mathbf{v}_2$ are not parallel, for the Lorentz transformation from S to S'' then involves a spatial rotation, and we can no longer leave the x_1 and x_2 axes out of the problem. But we can easily handle the case when $\mathbf{v}$ and $\mathbf{v}'$ are parallel. The transformation from S to S'' is then simply another rotation in the x_3x_4 plane (Fig. 15). The system S'' is rotated through the angle $\alpha'' = \alpha + \alpha'$ relative to S. We know that

$$\tan \alpha'' = i\beta'' = iv''/c,$$

where v' is the speed of S'' relative to S. But

$$\tan \alpha'' = \tan(\alpha + \alpha')$$
$$= \frac{\tan \alpha + \tan \alpha'}{1 - \tan \alpha \tan \alpha'}$$

from trigonometry. Therefore

$$\beta'' = \frac{\beta + \beta'}{1 + \beta\beta'} \qquad \text{or} \qquad v'' = \frac{v + v'}{1 + vv'/c^2} \tag{19}$$

This is the Einstein law for addition of parallel velocities. It reduces to the familiar $v'' = v + v'$ when all $v \ll c$. Notice too that if either v or v' equals c, v'' is also equal to c. Equation (19) automatically prevents v'' from becoming greater than c (unless we put v or v' greater than c to begin with).

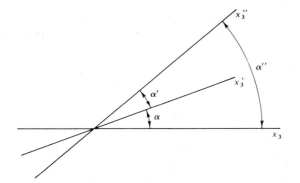

Fig. 15. *Addition of velocities by adding " angles "*
in the Lorentz transformation.

EXAMPLE PROBLEM 2.   Suppose that system S' is moving toward the *left* with speed $c/4$ relative to S, and S'' is moving toward the *right* with speed $c/2$ relative to S'. Find the velocity of S'' relative to S, and check the result by combining this velocity with the velocity of S relative to S' to obtain the given velocity of S'' relative to S'.

Solution.   We simply substitute $v = -c/4$ and $v' = +c/2$ into Eq. (19). This gives

$$v'' = \frac{\frac{1}{4}c}{1 - \frac{1}{8}} = \frac{2c}{7}$$

therefore S'' is moving with velocity $2c/7$ toward the right, relative to S. To check the result, let v now be the speed of S relative to S', v' be the speed of S'' relative to S, and v'' be the speed of S'' relative to S'. Then Eq. (19) can be used again, with v equal to $+c/4$ and v' equal to $+2c/7$. The result is:

$$v = \frac{\dfrac{2c}{7} + \dfrac{c}{4}}{1 + \dfrac{1}{14}} = \frac{\dfrac{15c}{28}}{\dfrac{15}{14}} = \frac{c}{2}$$

in agreement with the value originally given. Thus we see that Eq. (19) may be used for combining relative velocities in either direction, simply by using the proper sign for each v in the equation.

2.4 RELATIVISTIC DYNAMICS; FOUR-VECTORS

The Einstein velocity addition law implies that it is impossible to accelerate any body to a speed greater than or equal to c, for it says that addition of two velocities smaller than c cannot produce a relative velocity greater than or equal to c. We therefore should not be surprised to find that, as the speed of a body approaches c, an increasing force is required to produce a given acceleration. If mass is *defined* as force divided by acceleration, a body moving at high speed has a greater mass than the same body has when it is at rest. There is nothing mysterious about this increase in mass; it simply means that the " law" $\mathbf{F} = m\mathbf{a}$, *with m constant*, is really an approximation valid only at low speeds. Before proceeding to find the laws of dynamics which are valid at all speeds, we must pause to define our terms carefully.

Many laws of classical mechanics ($\mathbf{F} = m\mathbf{a}$, conservation of momentum, etc.) are expressed as three-dimensional vector equations. In any given coordinate system, such a law may be expressed as three independent equations, one for each component of the vectors. But the vector equation has a meaning beyond that of the component equations, for if one were to rotate the coordinate axes, each component of each vector would change, but the vector equation would remain valid. We say that the vectors are covariant with respect to rotation of axes, which means that when the axes are rotated, the components of each vector are transformed in exactly the same way as the coordinates of a point. For example, consider two three-dimensional coordinate systems, S and S', which differ only by a rotation about their common x_1 axis through an angle θ. As we have seen before in discussing the analogous Lorentz transformation, the coordinates of a point are expressed as $x_1 x_2 x_3$ in S and $x_1' x_2' x_3'$ in S', where

$$x_1' = x_1$$
$$x_2' = x_2 \cos \theta + x_3 \sin \theta$$
$$x_3' = -x_2 \sin \theta + x_3 \cos \theta$$

The covariant vector components $A'_1 A'_2 A'_3$ in S' and $A_1 A_2 A_3$ in S *are related in exactly the same way*, namely,

$$A'_1 = A_1$$
$$A'_2 = A_2 \cos\theta + A_3 \sin\theta \qquad\qquad (20)$$
$$A'_3 = -A_2 \sin\theta + A_3 \cos\theta$$

A vector equation of the form $\mathbf{A} = \mathbf{B}$ is equally valid in both frames of reference if the components of both $\mathbf{A}$ and $\mathbf{B}$ are covariant, for in S' the component equations

$$A'_2 = B'_2 \qquad \text{and} \qquad A'_3 = B'_3$$

may be written, from Eq. (20)

$$A_2 \cos\theta + A_3 \sin\theta = B_2 \cos\theta + B_3 \sin\theta$$
$$-A_2 \sin\theta + A_3 \cos\theta = -B_2 \sin\theta + B_3 \cos\theta$$

which are clearly true if $A_2 = B_2$ and $A_3 = B_3$.

To put it in more general and concise form, the components A'_i of a vector $\mathbf{A}$ in system S' are related to the components A_j in system S by the equations

$$A'_i = \sum_{j=1}^{3} a_{ij} A_j$$

where the coefficients a_{ij} *are exactly the same coefficients which relate the x'_i coordinates of a point to the x_j coordinates.*

Since the choice of the directions of the coordinate axes is completely arbitrary, it is clear that a legitimate physical law must behave in the above manner. It would not do, for example, to have a law of nature which relates only the x components of vectors.

It is not just luck that the vectors we use have the property of covariance. They are constructed in such a way that they must have this property. Each component of a velocity vector, for example, is found by taking the difference between the corresponding coordinates of two points and dividing by a time interval. Since the time interval is unchanged by a rotation of axes, the velocity components must be transformed in the same way as the coordinates of the points which were used to construct it. The same reasoning may be applied to acceleration vectors, which are constructed from velocity vectors, or to momentum and force vectors, because the mass factor in them is unchanged by the rotation.

Guided by the above principles, we may now search for the relativistically correct laws of dynamics. The laws we seek must reduce to the older laws at low speeds, and they must be valid in all inertial frames of reference. The old laws were valid in all inertial frames as long as we used the Galilean transformation; the new laws are dictated by the requirement that we use the Lorentz transformation to change inertial frames. The vectors in these new laws must therefore be vectors with *four* components which are transformed in the same way as the $x_1 x_2 x_3 x_4$ coordinates of an *event*.

Just as we constructed the velocity vector in three dimensions by dividing coordinate differences by a time interval, we must now construct a "four-velocity" by dividing corresponding coordinate differences by a generalized time interval. This generalized time interval must be a quantity which approaches the ordinary time interval when the velocity is small, and it must have the property that it be *invariant*—that it have the same value in all frames of reference—so that it does not affect the transformation properties of the vector in which it is used. The "proper time" interval $d\tau$, which is defined as the time interval between two events on the path of the particle, as measured in the system in which the particle is at rest, meets these requirements. Although observers in different frames may disagree about the time interval between two events, because each is using his own time axis, *all agree on the value of the time interval which would be observed in the system moving with the particle*. This may be clarified by the example of the muon. Although different observers, who see a muon traveling at different speeds, disagree about the lifetime of the muon, all agree on the lifetime which the resting muon has. In other words, regardless of the speed of the muon beam, one always finds the same intrinsic mean lifetime for the muons, by application of the appropriate time dilation factor.

The components of the "four-velocity" of a particle are therefore defined as

$$u_\mu = \frac{dx_\mu}{d\tau} \tag{21}$$

where dx_μ is the differential of the coordinate x_μ of the particle in a frame of reference S. The relation between the time interval dt in the frame S and the proper time interval $d\tau$ may be deduced from Fig. 12a by letting $t_0 = d\tau$. The particle, at rest is S', acts as the moving clock, marking off intervals $dx'_4 = ic\, d\tau$. Then from the figure, $ic\, dt = ic\, d\tau \cos \alpha$, or

$$d\tau = dt(1 - \beta^2)^{1/2} \tag{22}$$

We may now rewrite Eq. (21) completely in terms of quantities observed in system S as

$$u_\mu = \frac{1}{(1 - \beta^2)^{1/2}} \frac{dx_\mu}{dt} = \gamma \frac{dx_\mu}{dt} \tag{23}$$

In terms of the ordinary velocity components $v_1 v_2 v_3$,

$$\beta c = (v_1^2 + v_2^2 + v_3^2)^{1/2}$$

and

$$u_1 = \frac{v_1}{(1 - \beta^2)^{1/2}}, \qquad u_2 = \frac{v_2}{(1 - \beta^2)^{1/2}}$$

$$u_3 = \frac{v_3}{(1 - \beta^2)^{1/2}}, \qquad u_4 = \frac{ic}{(1 - \beta^2)^{1/2}}$$

(24)

The four-velocity has now been constructed so that the components are covariant under Lorentz transformation, and its length must therefore be invariant, just as the length of any three-dimensional vector is invariant under rotation. We can easily demonstrate that this is so, for

$$\sum_{\mu=1}^{4} u_\mu^2 = \frac{v^2}{1 - \beta^2} - \frac{c^2}{1 - \beta^2}$$
$$= -c^2$$

(25)

In this view, all particles move at the same rate through space–time, along lines called world lines. Acceleration of a particle changes the direction of the world line by changing the space components u_μ, but the magnitude of the four-velocity remains unchanged.

We can now construct the momentum four-vector for a particle simply by multiplying the four-velocity by the mass of the particle, provided that the mass is invariant:

$$p_\mu = mu_\mu$$

(26)

We ensure that the mass is invariant by specifying that it is the rest mass—the mass observed in the frame in which the particle is at rest. *When we say that this mass is invariant we mean that it is a quantity whose value is the same for all observers, not that it cannot change as a result of an interaction with other particles.* We shall see that the mass *can* change in an interaction (i.e., an inelastic collision), in which case all observers agree on the value of the new mass as well as that of the old mass.[7]

At last we are ready to state some laws of relativistic dynamics. We begin with the law of conservation of momentum, which applies to the momentum four-vector. We form the *total* momentum four-vector by adding corresponding components of the four-momentum of each particle: $P_\mu = p_{A\mu} + p_{B\mu} + \cdots$.

---

[7] Here and in the remainder of this book, when the word "mass" is used without qualification, it will refer to the rest mass. The so-called "relativistic mass," defined simply as force/acceleration, is not invariant but depends on the speed of the particle; use of this relativistic mass is confusing and should be avoided.

The relativistic form of the law of conservation of momentum states that each component of the total momentum P_μ is conserved in an interaction between particles.[8] This law has the properties that we demand of a relativistically correct law: It reduced to the classical law at low speed, and it is true in any reference frame, because of the covariance of the four-vectors.

This law of conservation of momentum involves four components instead of the three components to which we are accustomed. In order for the law to hold in all reference frames, the fourth component must be conserved as well as the other three, because when we transform from one frame to another, the first three components in the new frame may depend on the fourth component in the old frame. We shall see that the conservation of the fourth component is a law with which we have long been familiar—the law of conservation of energy. The separate laws of conservation of momentum and conservation of energy are incorporated in a single law in relativity theory.

In order to see how this works, we now examine the definition of energy, and this requires that we define force. We can easily construct a force four-vector by again using the invariant time interval $d\tau$ and the four-momentum in a definition analogous to Newton's second law. The components of the four-force are

$$K_\mu = \frac{dp_\mu}{d\tau}$$

This four-force is *not* the force observed in a laboratory, which is the force given by the various force laws of physics—for example, the Lorentz force $\mathbf{F} = q(\mathbf{E} + \mathbf{v} \times \mathbf{B})$. However the laboratory force is related to the four-force in the same way as the laboratory velocity is related to the four-velocity. In each case, one obtains the quantity measured in the laboratory simply by differentiating with respect to the laboratory time t rather than the invariant time τ. The first three components of the four-force are therefore related to the components of F by

$$K_i = F_i \frac{dt}{d\tau}$$

or, using Eq. (22),

$$K_i = \frac{F_i}{(1 - \beta^2)^{1/2}} \tag{27}$$

The energy E is now deduced by means of the classical equation

$$\mathbf{F} \cdot \mathbf{v} = \frac{dE}{dt} \tag{28}$$

---

[8] We restrict ourselves to interactions at a point, in order to avoid problems of simultaneity in determining which values of p_μ to use for each particle.

which expresses the fact that the rate at which work is done on a body is equal
to the rate of change of the body's energy. To evaluate $\mathbf{F} \cdot \mathbf{v}$ we construct the
sum

$$\sum_{\mu=1}^{4} K_\mu u_\mu = \frac{\mathbf{F} \cdot \mathbf{v}}{1 - \beta^2} + K_4 u_4 \tag{29}$$

from Eqs. (24) and (27), and the definition of the dot product. But for a body of
mass m,

$$\sum_{\mu=1}^{4} K_\mu u_\mu = \sum_{\mu=1}^{4} u_\mu \frac{d}{d\tau} m u_\mu = \frac{d}{d\tau} \sum_{\mu=1}^{4} \frac{m}{2} u_\mu u_\mu = 0 \tag{30}$$

because $\sum_{\mu=1}^{4} u_\mu u_\mu$ is a constant, $-c^2$ [see Eq. (25)]. Therefore

$$\mathbf{F} \cdot \mathbf{v} = -(1 - \beta^2) K_4 u_4 \tag{31}$$

But

$$K_4 = \frac{1}{(1 - \beta^2)^{1/2}} \frac{dp_4}{dt} = \frac{1}{(1 - \beta^2)^{1/2}} \frac{d}{dt} \frac{imc}{(1 - \beta^2)^{1/2}}$$

and

$$u_4 = \frac{ic}{(1 - \beta^2)^{1/2}}$$

so substitution in (31) gives

$$\mathbf{F} \cdot \mathbf{v} = \frac{d}{dt} \left(\frac{mc^2}{(1 - \beta^2)^{1/2}} \right) \tag{32}$$

Comparison with (28) shows that, to within a constant of integration,

$$\boxed{E = \frac{mc^2}{(1 - \beta^2)^{1/2}}} \tag{33}$$

This gives the familiar kinetic energy for small v, for we may write

$$E = mc^2 (1 - \beta^2)^{-1/2}$$
$$= mc^2 \left(1 + \frac{1}{2} \frac{v^2}{c^2} + \cdots \right)$$
$$= mc^2 + \tfrac{1}{2} mv^2 + \cdots \tag{34}$$

Since the zero of energy is arbitrary, and mc^2 is independent of v, it would
seem that we could dispense with the mc^2 term in Eq. (34). But if we leave it
as it is, the fourth component of momentum p_4 is then equal to iE/c, and the
conservation of energy becomes one part of the law of conservation of

momentum. If energy is defined in this way, it must be conserved in all collisions, elastic or inelastic. We may define *kinetic* energy as $E - mc^2$, in order to distinguish the velocity-dependent part of the energy. Kinetic energy is *not* conserved in an inelastic collision, and we are accustomed to saying that in such a case the energy has been converted to heat, or internal energy. But the energy E is always conserved, so when *kinetic* energy *disappears* in an interaction, there must be a corresponding *increase* in the *total rest mass of the particles involved.* Similarly, *if rest mass is decreased in a reaction, kinetic energy increases,* as in an "atomic bomb." *The rest mass of any body is therefore a direct measure of its internal energy.* As we shall see later, accurate measurements of the rest masses of atomic nuclei have been combined with measurements of kinetic energy in nuclear reactions, and the results have fully confirmed Eq. (33) and the interpretation given here.

As the argument leading up to Eq. (33) is rather long, it should be pointed out that the definitions of force were used only to show that the quantity E in Eq. (33) satisfies the conventional definition of energy. If we simply use Eq. (33) to *define* a quantity E, it is perfectly clear that

1. E is conserved in all interactions, because $p_4 = iE/c$; and
2. $E - mc^2$ is equal to the classical kinetic energy for small v.

The conclusions about conversion of rest mass into kinetic energy follow immediately from these two statements.

Let us illustrate the foregoing by considering the collision of two mud balls, each with rest mass m. We suppose that after the collision the balls stick together, and the combined ball has a rest mass of m_f. We wish to find the value of m_f in terms of m, and to demonstrate that m_f has the same value in two different frames

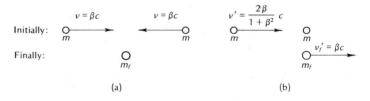

(a)   (b)

Fig. 16. *Perfectly inelastic collision of two equal masses* (a) *as seen in system in which center of mass is at rest, and* (b) *as seen in system in which one mass is initially at rest.*

of reference. (Thus the new mass is also an invariant mass, according to the definition of that expression; it has the same value in all frames of reference.)

We first choose a frame of reference in which the balls have equal and opposite velocities. (See Fig. 16a.) It is easy to find m_f in this, the center-of-mass frame, because the combined ball must be at rest after the collision, with energy $m_f c^2$.

The original energy of each ball was $mc^2(1 - \beta^2)^{-1/2}$, and conservation of energy demands that

$$m_f c^2 = 2mc^2(1 - \beta^2)^{-1/2} \quad,$$

so that

$$m_f = 2m(1 - \beta^2)^{-1/2} \quad, \tag{35}$$

a quantity greater than the original total rest mass of $2m$. The collision has transformed kinetic energy into rest energy.

Now we consider the same collision in a frame in which one ball is originally at rest. (Fig. 16b) This second frame is moving at a speed v relative to the first frame, so the ball that is moving in this frame has a speed of

$$v' = \frac{2v}{1 + \beta^2} \quad, \tag{36}$$

according to Eq. (19). In this frame, we may use the conservation of momentum p to determine the final mass m_f. Before the collision, a ball of mass m is moving at speed v', so

$$p = mv'(1 - \beta'^2)^{-1/2} \quad, \tag{37}$$

where, according to Eq. (36), $\beta' = v'/c = 2\beta/(1 + \beta^2)$. After the collision, a ball of mass m_f is moving at speed v, so

$$p = m_f v(1 - \beta^2)^{-1/2} \quad, \tag{38}$$

Equating the two expressions for p allows us to solve for m_f.

We leave it to the reader to carry out the algebra and thus verify that m_f has the same value as in Eq. (35). You may also verify that momentum in the form mv is *not* conserved in this frame of reference.

In addition to the increase in rest mass, there is also, in both frames of reference, an increase in the temperature of the mud when the balls collide. The internal energy acquired by the mud as a result of the collision may in principle be detected by measuring either the change in the temperature or the change in the total rest mass of the mud. If the same change in temperature had been produced by other means, such as by building a fire under the mud, the rest mass of the mud would have increased by the same amount.

An increase in mass when a body is heated may be understood in the light of our previous reasoning by considering a box containing an ideal gas. When the gas is heated, the atoms in the gas move about at higher speeds. Although the box as a whole remains at rest, it is more difficult to accelerate the box because to do so one must accelerate the high-speed atoms inside; therefore the box is observed to have greater mass. Of course this effect is undetectable on a macroscopic scale, because the speeds involved are so much smaller than c that the change in mass is a tiny fraction of the total mass.

The equivalence of rest mass and internal energy applies to any kind of internal energy, potential as well as kinetic. If you could measure to extreme accuracy the mass of a hydrogen atom, you would find that its mass is smaller than the total mass of its constituent parts, the proton and the electron, when they are measured separately. The mass difference is just $13.6 \text{ eV}/c^2$, because it would take an energy of 13.6 eV to separate the electron from the proton. This "mass defect" effect is much more pronounced in atomic nuclei, where many MeV may be needed to separate the particles, and we shall see later that nuclear physics has provided such strong experimental proof of the mass–energy relation that there is no longer any doubt about its correctness.

In experiments involving elementary particles, energy and momentum are observed directly, but the speed of each particle is not usually measured. Therefore, it is convenient to write explicity the relation between E and the space components of momentum. We know from Eqs. (25) and (26) that $\sum_{\mu=1}^{4} p_\mu^2 = -m^2 c^2$, so that we can write

$$p_1^2 + p_2^2 + p_3^2 - \frac{E^2}{c^2} = -m^2 c^2$$

or

$$E^2 = p^2 c^2 + m^2 c^4 \tag{39}$$

where $p^2 \equiv p_1^2 + p_2^2 + p_3^2$. Notice that when p is small relative to mc, E becomes approximately equal to $p^2/2m + mc^2$, in agreement with the classical expression $p^2/2m$ for kinetic energy.

EXAMPLE PROBLEM 3. A pi meson (pion) of mass π comes to rest and then disintegrates into a muon of rest mass μ and a neutrino of zero rest mass. Show that the kinetic energy of the muon is $c^2(\pi - \mu)^2/2\pi$.

Solution. The space components of momentum were initially zero, so the muon and neutrino fly off in opposite directions with the same magnitude of p. The initial energy was πc^2, the neutrino energy is pc, and the muon energy is $(p^2 c^2 + \mu^2 c^4)^{1/2}$, so conservation of energy gives

$$\pi c^2 = pc + (p^2 c^2 + \mu^2 c^4)^{1/2} \tag{40}$$

We solve Eq. (40) for p, obtaining $p = (\pi^2 - \mu^2)c/2\pi$. From (40), the muon's total energy is $\pi c^2 - pc$, so its kinetic energy is

$$\pi c^2 - pc - \mu c^2 = \pi c^2 - \frac{(\pi^2 - \mu^2)c^2}{2\pi} - \mu c^2$$

$$= \frac{(\pi - \mu)^2 c^2}{2\pi}$$

Q.E.D.

PROBLEMS

1. Criticize the following statements, which claim to "refute" the Einstein theory:

   (a) "It is clear that the motion of a spectroscope toward or away from a star, caused by the orbital and rotational motions of the earth, cannot in any way affect the *wavelength* of the light of the star. This means that when these motions of the earth cause the spectroscope to approach a star, the shift of the spectrum of the star toward the violet clearly indicates an increase in the *frequency* of the star light as received by the spectroscope . . . Now since velocity equals wavelength multiplied by frequency, it follows that . . . the velocity of the light relative to the spectroscope changes . . . contrary to Einstein's postulate of the 'constant velocity of light.'"

   (b) "Let us say we have a pendulum clock that ticks off exact seconds at sea level. Let us say we are able to take it to the top of a mountain sufficiently high that, because of the reduced force of gravity, the clock runs only half as fast as it did at sea level. Would we say that time at the top of the mountain flowed only half as fast as at sea level? Would we not consider that time flowed on at the same rate at the top of the mountain as at sea level, regardless of the varying of clock rates at different altitudes? Why then must we accept Einstein's concept that whatever happens to clock rates also happens to the flow of time itself?"

2. An astronaut travels at constant speed from the solar system to the star Sirius in one year, according to his clock. The distance he traveled is eight light years in our frame of reference. What is the distance from the solar system to Sirius in the astronaut's reference frame? How long did the trip take, according to our clock? If the astronaut and his ship have a combined rest mass of 10^4 kg, what is their kinetic energy, in joules?

3. In our frame of reference, event *B* occurs on the moon one second after event *A* occurs on the earth. Find a frame of reference in which the two events are simultaneous. Show that, if event *B* had occurred two seconds after event *A*, it would be impossible to find a frame in which the events are simultaneous.

4. Two electrons, 1 cm apart, are moving with velocity $0.9\,c$ in the direction shown. Find the acceleration of each electron.

$$v = 0.9\,c$$
$$v = 0.9\,c$$

5. Use the Einstein velocity addition formula, Eq. (19), to derive the speed

of each light ray in the moving water in Fizeau's experiment. Compare your result with that given by Fresnel's formula.

6. Using Eqs. (17a), derive the velocity addition law (for parallel velocities) by evaluating dx'_3/dt' in terms of dx_3/dt and v. ($v = c$.) Set v'' equal to dx_3/dt and v' equal to dx_3'/dt' to obtain an equation like Eq. (19). Why does this work?

7. Show that as $\beta \to 1$, $1 - \beta \to \frac{1}{2}(mc^2/E)^2$, where E is the total energy and m the rest mass. Find the value of $1 - \beta$, to two significant figures, for protons of 500 GeV kinetic energy. (1 GeV $= 10^9$ eV.)

8. Momentum is often quoted in units of MeV/c. Find the kinetic energy in MeV of (a) an electron, (b) a proton, and (c) a body of mass one gram, if the momentum of each is 1000 MeV/c.

9. The pion's rest mass is 140 MeV/c^2, and in its rest frame its half-life is 1.8×10^{-8} sec. How far will a beam of pions travel before one-half of them decay, if their *kinetic* energy is 140 MeV?

10. In a collision of two protons, some of the energy may be used to create a proton–antiproton pair. The total mass created is $2M_p$. (M_p = proton mass.) Show that when one of the colliding protons is initially at rest, the other proton's kinetic energy must be at least $6\,M_p c^2$ in order for this process to occur.

11. A gamma ray (high-frequency electromagnetic wave) behaves in collisions like a particle with zero rest mass. A gamma ray may strike an electron and disappear, its energy being used to create an electron and a positron, each of rest mass m_e. Show that if the original electron is at rest, this process requires a gamma-ray energy of at least $4\,m_e c^2$.

12. Carry out the algebra needed to show that Eqs. (37) and (38) require $m_f = 2m(1 - \beta^2)^{-1/2}$.

13. If a neutron at rest decays symmetrically into a proton p, an electron e, and an antineutrino $\bar{\nu}$ (see diagram), find the kinetic energy of each. (See Appendix E; ν and $\bar{\nu}$ have equal mass.)

14. If a source vibrates with frequency ν, emitting waves of speed c, an observer will detect waves of frequency ν' which is given by the classical Doppler formula as

$$\nu' = \nu\left(\frac{c + v}{c}\right)$$

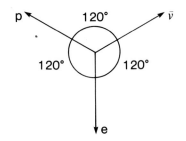

if the observer is moving toward a stationary source, or

$$v' = v\left(\frac{c}{c-v}\right)$$

if the source is moving toward a stationary observer.

(a) Show that both of these formulas give approximately the same result when $v/c \ll 1$.

(b) Show, by considering time dilation for the moving source, that the correct formula for light waves is

$$v' = v\left(\frac{c+v}{c-v}\right)^{1/2}$$

(c) Show that this formula holds whenever the relative speed of approach of source and observer is v, regardless of which one is considered to be moving. (Notice that this formula gives a result for v' which lies between the two classical results.)

15. A particle of mass M moves in the $+x$ direction with speed $0.6\,c$. A particle of mass $2M$ moves in the $-x$ direction with speed $0.8\,c$. If these two particles were the result of a single particle's splitting into two parts, what were the energy, momentum, and mass of that single particle?

The Old Quantum Theory

We have seen that Newtonian mechanics is really an approximation which is useful when one is dealing with slowly moving objects, but which must be replaced by relativistic mechanics for objects moving with speeds close to the speed of light. It should not be a surprise to find that Newtonian mechanics also fails when the range of our experience is extended in other ways. In particular, many phenomena involving the motion of very small bodies are inexplicable if one holds to the idea that dynamical quantities such as energy are continuously variable.

We find that, on a small scale, energy and other variables *cannot* take on a continuous range of values; instead, radiation of frequency v contains indivisible units of energy, or quanta, of magnitude hv, where h is a constant, called Planck's constant. Other variables are similarly quantized in a way which involves this same constant. We shall see later (Chapter 5) that this quantization follows naturally from a "wave" theory of matter.

63

For the present, however, we can introduce the experimental evidence in the context of the "old quantum theory," in which it is simply *postulated* that only certain discrete values of dynamical variables are permitted in some systems. By merely adding these postulates on as appendages to Newtonian mechanics, we are able to explain a wide variety of phenomena in atomic and solid state physics. Although the old quantum theory has been superseded by the more fundamental theories developed since 1926, it is still useful in treating certain problems, just as Newtonian mechanics is still useful for designing bridges or for calculating satellite orbits.

3.1 BLACK-BODY RADIATION

It is remarkable that the radical idea of quantization of energy resulted from the study of such a commonplace phenomenon as the radiation of heat. There seems to be nothing in the observations to stir the imagination. A "black body" is observed to emit electromagnetic waves over a continuous range of frequencies; the spectrum of radiated frequencies is peaked at a frequency v which is directly proportional to the absolute temperature of the body, and the radiated intensity goes smoothly to zero as $v \to 0$ or as $v \to \infty$. It seemed that one should be able to explain at least these simple features of the radiation by application of classical theories, but some of the best minds of the 19th century had only partial success in their attempts to do this. In the closing months of the 19th century, Max Planck finally found the formula which fit the observations perfectly, but his explanation, involving the quantization of energy, was so bold and original that very few persons were willing to accept it at first. Even Planck himself spent the next fifteen years trying to derive his formula in a "classical" way. It was not until quantization of energy was observed more directly in other phenomena that Planck's achievement received the recognition that it deserved.

Because of its unique historical importance, the Planck theory deserves careful study. To follow the analysis, we must begin with a theoretical model of a black body. A black body is defined as one which absorbs all radiation incident upon it, regardless of frequency. There is no real body which completely satisfies this ideal definition, but a close approximation to a black body consists of a small hole in a large enclosure, or cavity. If the hole is sufficiently small, entering radiation has a negligible chance of getting back out through the hole (by means of multiple reflections) before being absorbed. Thus the hole is "black." Because this is an ideal model to use in analyzing black-body radiation, such radiation is sometimes called "cavity radiation."

It is not obvious from the definition that there should be a unique spectrum of radiation which is emitted by all black bodies at a given temperature. Let us first try to understand this fact, by analyzing what is going on inside our model cavity.

Thermal agitation of electrons in the walls of the cavity causes, according to classical electromagnetic theory, the emission of electromagnetic waves. The thermal agitation may be considered to be a superposition of simple harmonic oscillations of various frequencies, and the waves emitted are of these same frequencies. Since the observed black-body spectrum is a continuous distribution, one must assume that a continuous range of frequencies of oscillation is possible for these electrons. The electrons are therefore also capable of absorbing electromagnetic waves of all frequencies. An electromagnetic wave emitted by one electron is reflected back and forth inside the cavity until it is finally absorbed by another electron. Frequencies which are not the resonant frequencies of the cavity are absorbed much more quickly than the resonant frequencies, and a standing-wave pattern, consisting of a superposition of waves of the various resonant frequencies, is set up inside the cavity. It happens that most of the energy in the standing-wave pattern is contained in waves of such high frequency that the wavelength is very short compared to the dimensions of any macroscopic cavity. Under these conditions, the successive resonant frequencies are very close together; the frequency spectrum of the energy contained in the cavity appears to be continuous, and the shape of the cavity has no observable effect on the spectrum.

The problem of calculating a theoretical black-body radiation spectrum now reduces to one of finding, as a function of frequency, the energy density of the standing waves inside a cavity at a temperature T. In other words, we must find the function $u(v)$ such that $\int_{v_1}^{v_2} u(v)\,dv$ is the energy per unit volume in a cavity of temperature T, of the radiation between frequencies v_1 and v_2. After finding $u(v)$, one then makes the reasonable assumption that the black-body spectrum, consisting of the radiant energy escaping from the cavity through a hole, is identical to the spectrum of the radiation inside the cavity. This is certainly true as long as the hole is large compared to the wavelengths involved.

From a thermodynamic analysis of adiabatic expansion of the walls of the cavity, Wien showed that $u(v)$ had to be of the form

$$u(v) = v^3 F\left(\frac{v}{T}\right) \tag{1}$$

where $F(v/T)$ is any differentiable function which depends only on the *ratio* of v and T, and not on the individual values of these variables.

From this expression one can easily show that the frequency at which $u(v)$ is maximum is directly proportional to T. We differentiate with respect to v, obtaining

$$\frac{du(v)}{dv} = 3v^2 F\left(\frac{v}{T}\right) + \frac{v^3}{T} F'\left(\frac{v}{T}\right)$$

$$= v^2 \left[3F\left(\frac{v}{T}\right) + \frac{v}{T} F'\left(\frac{v}{T}\right)\right]$$

where F' means the derivative of F with respect to its *argument*, v/T. The quantity in brackets can now be seen to be simply another function of v/T, $G(v/T)$ so that

$$\frac{du(v)}{dv} = v^2 G\left(\frac{v}{T}\right)$$

and at temperature T_1 the maximum value of $u(v)$ occurs at frequency v_1 such that $G(v_1/T_1) = 0$. Whatever its form, the function G must be zero when its argument equals v_1/T_1, so when the temperature is T_2, the function G is zero [and hence $u(v)$ is maximum] at a frequency v_2 such that $v_2/T_2 = v_1/T_1$. (Remember that *experiment* shows that $u(v)$ has one, and only one, maximum.) The proportionality between the frequency of maximum intensity and the temperature is known as Wien's displacement law.

Equipartition of Energy.   The Wien law is about as far as one can go with classical theory. Difficulty arises when one tries to derive the form of the function $F(v/T)$. Lord Rayleigh and Sir James Jeans attacked the problem by applying the principle of equipartition of energy to the standing waves in the cavity. According to this principle, for a system in thermal equilibrium at temperature T, the average energy of each degree of freedom is $kT/2$. We have seen (Chapter 1, Problem 3) that this result holds for gas molecules, whose average translational energy is $3kT/2$ for motion in three dimensions (three degrees of freedom per molecule) or $kT/2$ for one-dimensional motion. Rayleigh and Jeans counted the degrees of freedom of electromagnetic radiation in the cavity by considering the electromagnetic field to be a superposition of waves belonging to various modes of oscillation, each mode being characterized by a particular frequency and by a particular set of nodal lines or surfaces. Each mode contains two degrees of freedom[1] and thus should have average energy kT. It is shown below that the number of modes whose frequency is less than or equal to v, in a cavity of volume V, is $8\pi Vv^3/3c^3$. The total energy of the modes with frequency less than or equal to v should therefore be $(8\pi Vv^3/3c^3)(kT)$. The energy in the frequency range from v to $v + dv$ is simply the differential of this expression, and the energy per unit volume is therefore

---

[1] Each mode contains the same number of degrees of freedom as a one-dimensional simple harmonic oscillator. The latter's energy may be written as the sum of two terms—kinetic energy and potential energy—each having average energy $kT/2$ and representing one degree of freedom. To put it another way, the amplitude and the phase are independent quantities, and so there are two degrees of freedom.

$$u(v)\, dv = \frac{8\pi v^2}{c^3} (kT)\, dv \qquad (2)$$

Notice that this is of the form required by Eq. (1); $F(v/T)$ is $(8\pi k/c^3)(v/T)^{-1}$.

To verify the statement about the number of modes, one can consider the cavity to be a cube of side a. Electric field components which satisfy the boundary conditions (that the tangential component of $\mathbf{E}$ vanish at each wall) may be written

$$E_x = A \cos \frac{l\pi x}{a} \sin \frac{m\pi y}{a} \sin \frac{n\pi z}{a} \cos(2\pi vt)$$

$$E_y = B \sin \frac{l\pi x}{a} \cos \frac{m\pi y}{a} \sin \frac{n\pi z}{a} \cos(2\pi vt) \qquad (3)$$

$$E_z = C \sin \frac{l\pi x}{a} \sin \frac{m\pi y}{a} \cos \frac{n\pi z}{a} \cos(2\pi vt)$$

where l, m, and n are positive integers. The constants A, B, and C are related by the condition $\mathbf{V} \cdot \mathbf{E} = 0$, so that $nC = -Al - Bm$. Two independent constants remain, so that there are two linearly independent modes of oscillation for each choice of the set of integers l, m, and n; in other words, two independent directions of polarization are possible.

Application of the wave equation $\nabla^2 \mathbf{E} = (1/c^2)(\partial^2 \mathbf{E}/\partial t^2)$ to Eqs. (3) shows that $v^2 = (c^2/4a^2)(l^2 + m^2 + n^2)$. The number of modes whose frequency is *less* than any *given* frequency v_0 is therefore twice the number of sets of positive integers l, m, n such that

$$\frac{c^2}{4a^2} (l^2 + m^2 + n^2) < v_0^2 \qquad \text{or} \qquad l^2 + m^2 + n^2 < \frac{4a^2 v_0^2}{c^2} \qquad (4)$$

It is not too difficult to count the number of such sets. If we consider l, m, and n to be the coordinates of a point in three dimensions, then each distinct set of these numbers occupies one unit of volume in this "space," and every set which satisfies Eq. (4) lies within a *sphere* whose radius is $(l^2 + m^2 + n^2)^{1/2} = 2av_0/c$. Since we are restricted to positive integers, the number of *sets* is very nearly equal to $\frac{1}{8}$ of the volume of such a sphere, and the number of *modes* is twice this number, or

$$2 \cdot \left(\frac{1}{8}\right) \cdot \left(\frac{4\pi}{3}\right) \cdot \left(\frac{2av_0}{c}\right)^3 = \frac{8\pi V v_0^3}{3c^3}$$

which is just the result stated above.

Unfortunately, as Rayleigh and Jeans were well aware, their result cannot possibly be correct, for $u(v)$ as given by Eq. (2) increases without limit as $v \to \infty$. An infinite energy density is clearly impossible. The implications of this failure were so serious that the behavior of the theory at high frequencies was called the "ultraviolet catastrophe." It was clear that something was radically wrong with classical thinking, for there were no apparent errors in the derivation of $u(v)$. One could not question the result of the simple exercise of counting the modes of oscillation. The only remaining possibility was that the principle of equipartition of energy is incorrect, or at least inapplicable to this case. Since the number of modes per unit frequency interval increases as v increases, the *average energy per mode must decrease*, so that $u(v)$ may approach zero rather than infinity as $v \to \infty$.

But why should the high frequency modes not have acquired their share of energy? Jeans attempted to answer this question by assuming that true thermal equilibrium was never established in the cavity, and that the higher frequency modes were much slower in reaching the equilibrium value of energy.[2] But Jeans was never able to work out the correct form of $u(v)$ on the basis of this assumption. Indeed, on his assumption there is no unique spectrum, for $u(v)$ varies with time; if the walls of the cavity were held at a fixed temperature for a long time, the spectrum would continually shift toward higher frequencies.

The Planck Distribution. While Jeans was wrestling with these ideas, Planck had already solved the problem in every detail. The frequency dependence of the energy in the cavity conforms precisely to Planck's formula:

$$u(v) = \frac{8\pi v^2}{c^3} \left(\frac{hv}{e^{hv/kT} - 1} \right) \tag{5}$$

Quantum physics began with this formula. It is the first formula whose derivation requires the assumption that energy is quantized.

Planck first presented Eq. (5) as an empirical formula which he hoped could be fitted to the black-body spectrum. He was then told that his formula fit the observations better than any merely empirical formula had a right to do. This convinced Planck that there must be some fundamental significance to the formula, and he set out to find what it was. Two months later, in December, 1900, he presented his theory which introduced the concept of the quantum.[3]

To see how the quantum explains Planck's formula, let us focus our attention on the expression $hv/(e^{hv/kT} - 1)$. Insertion of this expression in place of kT converts the Rayleigh–Jeans formula to the Planck formula; this is therefore the average energy of a mode of oscillation in the Planck theory. (Planck himself considered the energy of the "oscillators"—electrons—in the walls of the cavity, rather than the energy of the standing waves in the cavity, but the result is the same.) According to both theories, the energy of the oscillators should be determined by Boltzmann statistics; that is, the number of oscillators with energy between ε and $\varepsilon + d\varepsilon$ should be proportional to the Boltzmann factor $e^{-\varepsilon/kT}$. If the energy is a continuous variable, the total

[2] J. H. Jeans, *Phil. Mag.* **10** (Ser. 6), 91–97 (1905).
[3] The sequence of events is described in a fascinating article by E. U. Condon, 60 Years of Quantum Physics, *Phys. Today*, 37–48 (Oct. 1962).

energy of the oscillators in this energy range is proportional to $\varepsilon e^{-\varepsilon/kT}\, d\varepsilon$, and the total energy of *all* the oscillators is found by integrating over all values of ε. Similarly, the total number of oscillators is proportional to $\int_0^\infty e^{-\varepsilon/kT}\, d\varepsilon$, and the average energy of an oscillator is thus

$$\bar{\varepsilon} = \frac{\displaystyle\int_0^\infty \varepsilon e^{-\varepsilon/kT}\, d\varepsilon}{\displaystyle\int_0^\infty e^{-\varepsilon/kT}\, d\varepsilon}$$

By means of the substitution $x = \varepsilon/kT$, we can write the numerator as $(kT)^2 \int_0^\infty x\, e^{-x}\, dx = (kT)^2 \Gamma(2) = (kT)^2$, where $\Gamma(n)$, the gamma function, is defined as $\Gamma(n) \equiv \int_0^\infty x^{n-1} e^{-x}\, dx$, and is equal to $(n-1)!$ when n is an integer. The denominator equals $kT\, \Gamma(1)$, or simply kT, so we have $\bar{\varepsilon} = kT$, which is, of course, the value assumed by Rayleigh and Jeans.

We have arrived at the crucial point. Planck's departure from Rayleigh and Jeans was in assuming that ε is not a continuous variable, but that the energy of an oscillator whose frequency is v could only be an integral multiple of hv, where h is a small constant whose value may be determined by fitting the observations. Let us see the effect of this assumption on the computation of the average energy. We fix our attention on *one group* of oscillators, all of which oscillate with the *single* frequency v, and we find the average energy of *these* oscillators *only*. For these oscillators, the allowed values of energy are 0, hv, $2hv$, etc. The Boltzmann factor determines the relative number of oscillators with each possible energy: if the number with zero energy is A, then the number with energy hv is $Ae^{-hv/kT}$, the number with energy $2hv$ is $Ae^{-2hv/kT}$, etc. The total number of oscillators with frequency v is then

$$N = A + Ae^{-hv/kT} + Ae^{-2hv/kT} + \cdots$$
$$= A[1 + e^{-hv/kT} + (e^{-hv/kT})^2 + \cdots]$$

Since $e^{-hv/kT}$ is always less than 1, the infinite series converges to $1/(1 - e^{-hv/kT})$. The total energy of this group of oscillators is found by multiplying each term in the above series by the appropriate energy, obtaining

$$E = A \cdot 0 + hv \cdot Ae^{-hv/kT} + 2hv \cdot Ae^{-2hv/kT} + \cdots$$
$$= A \cdot hv \cdot e^{-hv/kT}[1 + 2e^{-hv/kT} + 3(e^{-hv/kT})^2 + \cdots]$$

Again the series in brackets converges, this time equaling $1/(1 - e^{-hv/kT})^2$. The average energy per oscillator is then

$$\frac{E}{N} = \frac{\dfrac{Ahve^{-hv/kT}}{(1 - e^{-hv/kT})^2}}{\dfrac{A}{(1 - e^{-hv/kT})}}$$

$$= \frac{hve^{-hv/kT}}{1 - e^{-hv/kT}} = \frac{hv}{e^{hv/kT} - 1}$$

precisely the factor which appears in Planck's formula.

Notice that Planck's factor reduces to kT in the limit as $v \to 0$:

$$\frac{hv}{(e^{hv/kT} - 1)} \to \frac{hv}{\left(1 + \dfrac{hv}{kT} - 1\right)} = kT$$

Thus when hv is small compared to kT, the discrete nature of the energy is unimportant, and the result of the summation is identical with the result of integration on the continuous variable ε. As v increases, the average energy per oscillator decreases, as anticipated by our previous reasoning. It is instructive to depict this behavior graphically. If h were zero, ε would be continuous, and the situation would be as shown in Fig. 1. The number of

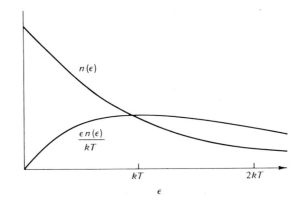

Fig. 1. *Variation of $n(\varepsilon)$ and $\varepsilon n(\varepsilon)/kT$ with energy ε,*
for continuous ε. The ratio of the areas under the two
curves is $\bar{\varepsilon}/kT$, which equals 1.

oscillators equals the area under the curve $n(\varepsilon)$, and the total energy of these is the area under the curve $\varepsilon n(\varepsilon)$. In this case there are, of course, no oscillators with *exactly* zero energy, and a great many have energy between zero and $2kT$, contributing the bulk of the total energy. When the energy is discrete, the situation is only slightly different as long as $hv \ll kT$ (Fig. 2). But when hv is large, for example when $hv = 2kT$ as in Fig. 3, almost all of the oscillators have zero energy, so that the average energy per oscillator is drastically reduced. Those oscillators which would (if hv were small) have had energies between zero and $2kT$ must now have energy of either exactly zero or exactly $2kT$, and most of them have zero. Thus the higher frequency modes have lower average energy, and the average energy goes to zero as v approaches infinity.

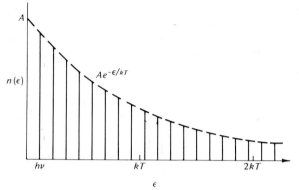

Fig. 2. *Variation of $n(\varepsilon)$ with ε when only discrete values of ε are allowed. Vertical lines indicate numbers of particles with various values of ε.*

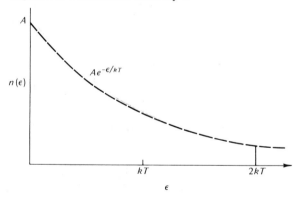

Fig. 3. *Variation of $n(\varepsilon)$ with $\varepsilon = 0, 2kT, \ldots,$ only.*

It is interesting to read what Jeans had to say about this. In comparing his theory to that of Planck, he says[4]:

" The methods of both are in effect the methods of statistical mechanics and of the theorem of equipartition of energy, but I carry the method further than Planck, since Planck stops short of the step of putting $h = 0$. I venture to express the opinion that it is not legitimate to stop short at this point,

[4] J. H. Jeans, *Nature* **72**, 294 (1905).

as the hypotheses upon which Planck has worked lead to the relation $h = 0$ as a necessary consequence.

"Of course, I am aware that Planck's law is in good agreement with experiment if h is given a value different from zero, while my own law, obtained by putting $h = 0$, cannot possibly agree with experiment. This does not alter my belief that the value $h = 0$ is the only value which it is possible to take, my view being that the supposition that the energy of the ether is in equilibrium with that of matter is utterly erroneous in the case of ether vibrations of short wavelength under experimental conditions."

EXAMPLE PROBLEM 1.   Use Planck's formula to show that the total intensity (total energy crossing unit area in unit time) of the emitted radiation is proportional to T^4, and find the constant of proportionality.

Solution.   First we express $I(v)$, the emitted *intensity*, in terms of $u(v)$, the *energy density* in the cavity. We may do this by using the result of Example Problem 2, Chapter 1. In that case the "intensity" of molecular escape from an aperture was found as a function of the density of molecules inside a container. The nature of the molecules—indeed, even the fact that the molecules are discrete particles—was not an essential part of that problem, so we may carry over the result by substituting $u(v)$ for molecular density and c for v, obtaining an expression which must equal $I(v)At$, the energy crossing area A in time t. Therefore

$$I(v) = \tfrac{1}{4} c u(v)$$

To find the total emitted intensity we integrate over all frequencies, obtaining

$$\int_0^\infty I(v)\, dv = \frac{c}{4} \int_0^\infty \frac{8\pi v^2}{c^3} \frac{hv}{e^{hv/kT} - 1}\, dv$$

$$= \frac{2\pi k^4 T^4}{c^2 h^3} \int_0^\infty \frac{\left(\dfrac{hv}{kT}\right)^3 d\left(\dfrac{hv}{kT}\right)}{e^{hv/kT} - 1}$$

$$= \frac{2\pi k^4}{c^2 h^3} \left[\int_0^\infty \frac{x^3\, dx}{e^x - 1} \right] T^4$$

A table of integrals shows that

$$\int_0^\infty \frac{x^3\, dx}{e^x - 1} = \frac{\pi^4}{15}$$

so that the constant of proportionality is

$$\sigma = \frac{2\pi^5 k^4}{15 c^2 h^3} \tag{6}$$

The fact that the integrated intensity equals σT^4 is known as the Stefan–Boltzmann law, formulated some time before Planck's work. Planck's distribution, however, gives the theoretical formula for σ in terms of h and k. In order to determine h and k, one needs one more equation which relates these constants to a measurable quantity. One such equation is the expression for the frequency ν_m at which $u(\nu)$ is a maximum; this can be found from Planck's formula by numerical methods, and it is

$$\nu_m = 2.8214 \frac{kT}{h} \tag{7}$$

Equations (6) and (7) may be solved for h and k, giving the values of these two fundamental constants in terms of measurable quantities. When k is determined, it can be used to find Avogadro's number (since $k = R/N_A$), and, as we have seen, this tells us the value of e, the electronic charge. In this way Planck, in 1900, found a value of e only 2.3 per cent below the currently accepted value!

In summary, Planck achieved the following:

1. He invented the radical idea of quantization, which is now a fundamental part of all atomic theory.
2. He used this idea to explain in detail the shape of the black-body radiation spectrum.
3. He invented the fundamental constant h (now known as Planck's constant, of course), and he determined its value.
4. He determined the value of e, another fundamental constant which many physicists were then attempting to measure, and his determination of it was by far the most accurate which had been made up to that time.

Planck's work did not create nearly the sensation that one might expect from this impressive list of achievements. Fortunately, however, his work was not unnoticed by Albert Einstein, as we shall now see.

3.2   THE PHOTOELECTRIC EFFECT

Planck had assumed that the energy of oscillators in the walls of a cavity is quantized, but he said nothing about the energy in the electromagnetic field in the cavity. However, if the oscillators can possess energy only in multiples of $h\nu$, they can only radiate energy in multiples of $h\nu$ also, and therefore, at least in this case, the field energy must also be quantized. In 1905 (when most people either had not heard of Planck's theory, or, having heard of it, did not believe it), Einstein made a logical generalization. He wrote[5]:

[5] A. Einstein, *Ann. Phys.* (*Leipzig*) **17**, 132–148 (1905). Translated by J. McGervey.

"According to the assumption considered here, the spreading of a light beam emanating from a point source does not cause the energy to be distributed continuously over larger and larger volumes, but rather the energy consists of a finite number of energy quanta, localized at space points, which move without breaking up and which can be absorbed or emitted only as wholes."

The energy of each quantum was of course assumed to be $h\nu$, equal to the quanta of energy possessed by Planck's oscillators.

Einstein was not engaging in mere idle speculation; he went on to suggest that consequences of this property of light should be observed in the photoelectric effect, in which light causes the emission of electrons from a metal. At that time very little was known about this effect. It had been observed that light shining on a metal caused the electrostatic potential of the metal to increase (as a result of electron emission), up to a certain limit. The potential for a given metal went to a higher limit when light of shorter wavelength was used, but no quantitative measurements had been made.

Einstein's suggestion was that an electron escaped from a metal as a result of the energy it received in absorbing a single quantum of light (now called a "photon"). An electron that absorbed a photon of energy $h\nu$ would escape the metal and would retain a maximum kinetic energy $T_{\max}$ given by

$$T_{\max} = h\nu - W \quad , \tag{8}$$

where W, the work function of the metal, is the minimum energy that an electron must acquire in order to escape from the metal. (Some electrons must acquire an energy greater than W to escape. Such electrons have kinetic energy of less than $T_{\max}$ after they escape.) One can verify relation (8) by applying a retarding potential V that is just large enough to prevent all the electrons from reaching the collector. One should then find that the retarding potential is linearly dependent on ν:

$$V = (h/e)\nu - W/e \quad .$$

In this case W is the work function of either the emitter or the collector, whichever is greater. (See sec. 11.4 for clarification of this point.)

In terms of the wavelength, we may write the energy of a photon as

$$E = hc/\lambda = \frac{12{,}400 \text{ eV}-\text{Å}}{\lambda} \quad ,$$

writing hc in the convenient units of eV$-$Å. Thus if λ equals 1 Å, the photon has an energy of 12,400 eV (to three significant figures). (1eV = 1 electron-Volt = electron charge $\times$ 1 Volt $\approx 1.6 \times 10^{-19}$ J.)

The whole idea behind Eq. (8) is totally at variance with the prevailing concept of light as a wave. According to classical theory, electrons are accelerated by the oscillating electric field in the light wave. If the intensity of the light were increased, the amplitude of the electric field oscillations would increase so that an electron, following the field, would have greater acceleration and would thus acquire more energy. But according to Einstein, the energy acquired by a single electron depends *only* on the *frequency* of the light, *not* on the *intensity*. Of course the *total* energy absorbed by *all* the electrons increases as the intensity increases, because more photons are absorbed.

Philipp Lenard had shown experimentally in 1902 that the energy of the ejected electron is independent of the light intensity, but the experimental verification of a linear relationship between V and v, with the correct slope, presented a challenge, which was met by R. A. Millikan. By setting up what he described as a "machine shop *in vacuo*," Millikan produced a clean alkali-metal surface, which he then irradiated with monochromatic light of various frequencies. For each frequency he observed the retarding potential needed to stop the photoelectric current. (The photoelectrons were actually caught on a quadrant electrometer, also in the vacuum; the potential of the electrometer with respect to the alkali metal could be varied until the electrometer ceased to acquire a charge.)

Millikan found that the plot of V versus v was indeed a straight line. By assuming the slope to be h/e, and using the value of e which he had found in the oil-drop experiment, Millikan found h to be 6.56×10^{-27} erg-sec, in excellent agreement with the value found by Planck in an entirely different way, and only 1 percent below the currently accepted value of 6.626×10^{-27}.

Both Einstein and Millikan received the Nobel Prize for their work, Einstein in 1921 for the theory, Millikan in 1923 for this experiment and the oil-drop experiment. Like many other Nobel Prize-winning experiments, the photoelectric effect experiment is now performed by students in many undergraduate laboratories.[6]

3.3 *LINE SPECTRA AND THE BOHR ATOM*

When a gas is heated, it emits light; but, unlike black-body radiation, this light does not have a continuous spectrum of frequencies. Instead, certain discrete frequencies are emitted, which appear as bright lines when the light is analyzed by prism or a grating. Each element has its own characteristic "line spectrum" consisting of a unique set of frequencies. These spectra have long been used to identify elements (and even to discover a new element, helium, in the sun). Lines appear in the ultraviolet and infrared, as well as in the visible frequency range, and they form "series" in which the lines become more closely spaced with decreasing wavelength, approaching a "series limit" at which the line spacing converges to zero. (See Fig. 4.)

[6] For a good description of the experiment as performed in undergraduate laboratories today, see A. Melissinos, "Experiments in Modern Physics." Academic Press, New York, 1966.

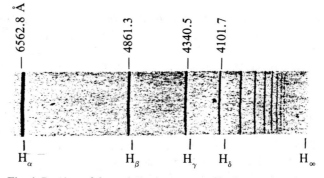

Fig. 4. *Portion of the emission spectrum of hydrogen gas, show-ing the wavelengths (in Angstroms) of the Balmer series lines. H_∞ shows the theoretical position of the series limit. (From Chapter 1 of G. Herzberg, "Atomic Spectra and Atomic Struc-ture," translated by J. W. T. Spinks, Dover Publications, Inc., N.Y., 1945, used with the permission of the publisher.)*

In the nineteenth century, many efforts were made to work out formulas that would give relationships among the wavelengths of a given series. But almost all of these attempts failed, even for hydrogen, the simplest element, because people thought that the frequencies should be related like the harmonics of a classical oscillating system. However, in 1885, J. J. Balmer, a geometry instructor who was not handicapped by preconceived notions, produced the formula

$$\lambda_n = 3645.6[n^2/(n^2 - 4)] \times 10^{-8} \quad \text{cm} \quad (n = 3, 4, 5,\ldots)$$

which gave the correct wavelengths for one series of atomic hydrogen (and almost as important, did *not* give wavelengths of lines which are *not* observed).

In 1890, Rydberg saw how the Balmer formula could be generalized to predict more lines. He rewrote it as

$$k_n = \frac{1}{\lambda_n} = 27430\left(1 - \frac{4}{n^2}\right) \quad \text{cm}^{-1}$$

or

$$k_n = R\left(\frac{1}{2^2} - \frac{1}{n^2}\right) \quad (n = 3, 4, 5, \ldots)$$

where k_n is called the *wave number* and R is Rydberg's constant,[7] originally given by Rydberg as 109,720 cm^{-1}. When the formula is written in this form, one sees the possibility that there may be other series for which the formula would be

$$k_{n', n} = R\left(\frac{1}{n'^2} - \frac{1}{n^2}\right) \quad (n = n' + 1, n' + 2, \ldots) \tag{9}$$

In 1908 Lyman discovered the hydrogen series for $n' = 1$ (in the ultraviolet), and Paschen discovered the $n' = 3$ series (in the infrared).

[7] Fortunately there is no equation relating the Rydberg constant R to the gas constant R!

In order to explain these formulas, and to seek a way of predicting other relationships between spectral lines, it was necessary to have a model to describe how the atom is constructed. One of the first reasonable models was J. J. Thomson's "raisin cake" model, in which electrons are imbedded in a large positively charged sphere. Using this model, Thomson could calculate modes of vibration of the electrons, and he found some agreement between the frequencies of these modes and the observed frequencies of the spectral lines. He could also show that as more electrons are added, the electrons split into groups, and from this he could show a connection with the periodic table of the elements. But unfortunately for this model, there began to appear evidence which indicated that the positive charge of the atom resided in a nucleus of much smaller volume than that of the whole atom. Thomson's successes had been mere coincidence.

We shall discuss the evidence for the nuclear atom in more detail later (Section 13.1). For the present, it suffices to say that Lord Rutherford settled the issue rather conclusively, by analyzing the behavior of alpha particles as they pass through matter. It was found that these particles were occasionally deflected through angles of 90° or more in passing through thin foils of metal. It was not possible, statistically, that these large deflections could be a result of several smaller deflections in succession, for in that case an extremely large number of smaller deflections (in the 10–90° range) would also be observed, and a sufficiently large number was not observed. A large-angle deflection must therefore have been the result of a single encounter, relatively rare, in which the atom exerts a huge force on the alpha particle.

Such a large force could simply be the result of the inverse-square law of electrostatics, provided that the positive charge of the atom, and most of its mass, resided in a small nucleus, of radius less than 10^{-12} cm. The force exerted by the positive charge follows the inverse square law *outside* the charge distribution, but the force becomes smaller as the particle penetrates inside the distribution, the force having its maximum value at the surface of the distribution. If the positive charge distribution were as large as the whole atom ($r \cong 10^{-8}$ cm), the maximum force would be much too small to account for the large-angle deflections. We shall see in Section 13.1 how the detailed scattering formula, derived by Rutherford on the assumption of a *point* nucleus, was completely verified by Geiger and Marsden in 1913. (The assumption that the nucleus is concentrated at a point agreed with the observations, because none of the alpha particles was energetic enough to penetrate the nucleus; the closest approach was to about 10^{-12} cm from the center.)

The Bohr Atom. If the atom has such a structure, with the positive charge in a small nucleus and negative electrons spread out over a much larger volume, the obvious question is, "Why don't the electrons fall into the nucleus?" The obvious answer is that the electrons move in orbits just as the planets in the solar system do, because the electrostatic force, like the gravitational force, follows the inverse-square law. But this answer is insufficient, for an electron in such an orbit is accelerated, and so, according to classical electrodynamics, it should radiate electromagnetic waves, thereby losing energy and spiraling into the nucleus.

In 1913 Niels Bohr produced a set of hypotheses which circumvented the above objection and which enabled him to derive the Rydberg formula for the spectrum of hydrogen.[8] Bohr made three basic assumptions.

1. The electrons move in circular orbits about the center of mass of the atom.

2. The only allowed orbits are such that the angular momentum of the atom about its center of mass is an integral multiple of $h/2\pi$.

3. Radiation occurs only when an electron "jumps" from one of the allowed orbits to one of lower energy. The difference in energy ΔE is then radiated as a photon of frequency $v = \Delta E/h$, in accordance with the Einstein condition.

In addition, Bohr had previously pointed out that the hydrogen atom evidently contains only one electron, a fact not so obvious then as it is now. The evidence for this came mainly from experiments by J. J. Thomson on positive ions produced by cathode rays. The hydrogen ion was the only one which never appeared doubly ionized.

The derivation of the Rydberg formula proceeds from the above assumptions as follows. First one uses assumptions 1 and 2, combined with Newton's second law for the centripetal force, to find the radius of the nth allowed orbit. Let M be the mass of the nucleus (a proton in this case) and m the electron mass. Both nucleus and electron revolve around the center of mass of the atom, at respective distances R_n and r_n, with angular velocity ω_n (see Fig. 5). From the definition of center of mass,

$$MR_n = mr_n \qquad \text{or} \qquad R_n = \frac{mr_n}{M} \tag{10}$$

The angular momentum is

$$m\omega_n r_n{}^2 + M\omega_n R_n{}^2 = n\hbar \tag{11}$$

where $\hbar$ is defined as $h/2\pi$, and n is an integer (assumption 2). Setting the force on the electron, as given by Coulomb's law, equal to the mass times the centripetal acceleration, we have

$$m\omega_n{}^2 r_n = \frac{Ze^2}{4\pi\varepsilon_0(r_n + R_n)^2} \qquad \text{or} \qquad \frac{Ze^2}{(r_n + R_n)^2} \tag{12}$$

$$\text{mks units} \qquad\qquad\qquad \text{cgs units}$$

[8] N. Bohr, *Phil. Mag.* **26**, 1 (1913).

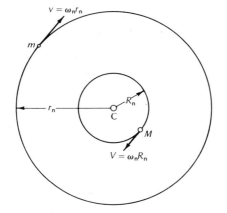

Fig. 5. *Motion of nucleus M and electron m about center of mass C. Motion of nucleus exaggerated.*

The numerator Ze^2 is the product of the charge Ze on the nucleus and the charge e on the electron. For hydrogen Z is simply 1, but it is written explicitly here so that the result may be applied to other one-electron systems, e.g., ionized helium.

Using Eq. (10) to eliminate R we may write Eqs. (11) and (12) as

$$m\omega_n r_n^2 \left(1 + \frac{m}{M}\right) = n\hbar \tag{13}$$

$$m\omega_n^2 r_n = \frac{Ze^2}{4\pi\varepsilon_0 r_n^2 \left(1 + \dfrac{m}{M}\right)^2} \quad \text{or} \quad \frac{Ze^2}{r_n^2 \left(1 + \dfrac{m}{M}\right)^2} \tag{14}$$

mks units   cgs units

Squaring Eq. (13) and dividing by Eq. (14) eliminates the unknown ω and gives

$$r_n = \frac{4\pi\varepsilon_0 n^2\hbar^2}{mZe^2} \quad \text{or} \quad r_n = \frac{n^2\hbar^2}{mZe^2} \tag{15}$$

mks units   cgs units

Substitution of the values of the constants into Eq. (15) yields

$$r_n = a_0 n^2/Z = (5.29177 n^2/Z) \times 10^{-9} \text{ cm} \tag{16}$$

where a_0, the radius of the normal state of the hydrogen atom (the "ground state"), is called the first Bohr radius. Notice that r_n depends only on m, not on M, which means that the orbit radii are the same for all isotopes of hydrogen.

The next step is to calculate the energy, which is the sum of the kinetic energy of nucleus and electron plus the (negative) electrostatic potential energy:

$$E_n = \frac{m\omega_n{}^2 r_n{}^2}{2} + \frac{M\omega_n{}^2 R_n{}^2}{2} - \frac{Ze^2}{4\pi\varepsilon_0(r_n + R_n)} \qquad \text{mks units}$$

or   (17)

$$E_n = \frac{m\omega_n{}^2 r_n{}^2}{2} + \frac{M\omega_n{}^2 R_n{}^2}{2} - \frac{Ze^2}{(r_n + R_n)} \qquad \text{cgs units}$$

If we again use Eq. (10) to eliminate R, this becomes

$$E_n = \left(\frac{m\omega_n{}^2 r_n{}^2}{2}\right)\left(1 + \frac{m}{M}\right) - \frac{Ze^2}{4\pi\varepsilon_0 r_n\left(1 + \dfrac{m}{M}\right)} \qquad \text{mks units}$$

or

$$E_n = \left(\frac{m\omega_n{}^2 r_n{}^2}{2}\right)\left(1 + \frac{m}{M}\right) - \frac{Ze^2}{r_n\left(1 + \dfrac{m}{M}\right)} \qquad \text{cgs units}$$

By comparing these expressions with Eq. (14), we see that, in either mks or cgs, the first term (kinetic energy) is one-half the magnitude of the second term (potential energy), a result which, of course, holds for all circular orbits. Therefore we may write

$$E_n = \frac{-Ze^2}{8\pi\varepsilon_0 r_n\left(1 + \dfrac{m}{M}\right)} \qquad \text{mks units}$$

or   (18)

$$E_n = \frac{-Ze^2}{2r_n\left(1 + \dfrac{m}{M}\right)} \qquad \text{cgs units}$$

Substituting into Eq. (18) the value of r_n from Eq. (15), we have the result

$$E_n = \frac{-m_r Z^2 e^4}{32\pi^2\varepsilon_0^2 n^2 h^2} \qquad \text{mks units}$$

or   (19)

$$E_n = \frac{-m_r Z^2 e^4}{2n^2 h^2} \qquad \text{cgs units}$$

where we have abbreviated the quantity $m/(1 + m/M)$ as m_r, which is called the reduced mass. As $M \to \infty$, $m_r \to m$, so that the energy in Eq. (19) is the same as the energy of an electron of mass m_r revolving around an infinitely massive nucleus. Since $M = 1836\,m$ for hydrogen, the reduced mass is $m_r =$

$1836m/1837$, and use of m instead of m_r in Eq. (19) would result in an error of only one part in 1837.

In the Bohr picture, therefore, the possible energies of a hydrogen atom may be represented by an energy-level diagram like that of Fig. 6, which shows the energy corresponding to the various values of n. Arrows on the diagram indicate transitions from one level to another, leading to emission of photons which produce the spectral lines. (The horizontal axis has no significance; each level is simply drawn as a horizontal line to make it easier to indicate the transitions which occur.)

The lowest level E_1 is the level which is normally occupied by the electron, and its energy is therefore equal in magnitude to the energy required to free the electron from the proton. This energy, in electron volts, is numerically equal to the ionization potential, in volts, of atomic hydrogen. The energy levels merge into a continuum for positive values of E; these are energy levels of the free electron, which are not restricted to discrete values.

The value of E_1 may be found by inserting values of the fundamental constants into Eq. (19). In doing this, it is useful to know that the combination $e^2/\hbar c$ in cgs, or $e^2/4\pi\varepsilon_0\hbar c$ in mks, is a dimensionless number, very close to

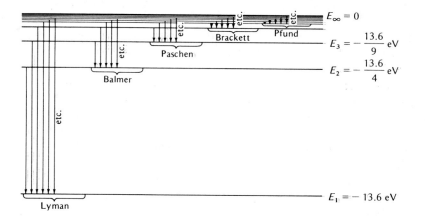

Fig. 6. Energy levels of the hydrogen atom, according to Bohr, showing transitions which give rise to the Lyman, Balmer, and other series of spectral lines. (Adapted from R. M. Eisberg, "Fundamentals of Modern Physics." Wiley, New York, 1961.)

1/137, known as the fine structure constant and denoted by the symbol α. In terms of α, Eq. (19) may be written

$$E_n = -\frac{m_r c^2}{2}\frac{\alpha^2 Z^2}{n^2}$$

$$= - \frac{510725.25 \text{ eV}}{2 \times (137.036)^2} \frac{Z^2}{n^2}$$

$$= - 13.5984 \ (Z^2/n^2) \text{ eV}.$$

It is now simple to use assumption 3 to find the frequencies of the radiation, and thereby to reproduce the Rydberg formula. In a transition from level n to level n', the frequency radiated is

$$\nu_{n',n} = \frac{(E_n - E_{n'})}{h} = \frac{m_r c^2 \alpha^2 Z^2}{2h} \left[\frac{1}{n'^2} - \frac{1}{n^2} \right] = \nu_0 \left[\frac{1}{n'^2} - \frac{1}{n^2} \right] \qquad (20)$$

where ν_0 is the frequency corresponding to $n' = 1$ and $n = \infty$. ($h\nu_0 =$ ionization energy.) The wave number $k_{n',n}$ is equal to $\nu_{n',n}/c$, so Eq. (20) agrees with Rydberg's formula (9) if we put Z equal to 1 and

$$R = m_r c^2 \alpha^2 / 2hc \qquad (21)$$

Bohr was able to use the known values of the constants in Eq. (21) to calculate the Rydberg constant, thereby converting R from an empirical constant to a derived quantity.[9] The Rydberg constant is now known so accurately from spectroscopic measurements that it can be used to improve the precision on the other constants.

Equation (20) describes the spectrum of any one-electron system—for example, ^{2}H (or deuterium, D), ^{3}H (or tritium, T), singly ionized helium, doubly ionized lithium. Spectra of these have been observed and are in agreement with theory; in fact, deuterium was discovered, in 1931, by observation of "satellite" lines in the hydrogen spectrum. The lines of deuterium of course differ from the lines of ordinary hydrogen only because of the difference in the reduced mass m_r. For deuterium, the reduced mass is $3672m/3673$, so each deuterium line is displaced from the corresponding hydrogen line by about 1 part in 3670. Such a difference is easily detected by spectroscopic methods, the Rydberg constant R_H for hydrogen is now known to a few parts in 10^8: $R_H = 109677.656 \pm 0.008 \text{ cm}^{-1}$.

3.4   THE CORRESPONDENCE PRINCIPLE

We are now in the curious position of having two completely different radiation laws, each of which seems to work in its own domain. How are we to decide upon the boundaries of these domains—to know when to use Bohr's law and when to use classical electrodynamics to compute the properties of

[9] Rydberg's original value, quoted previously, did not quite agree with that of Bohr, but later spectroscopic results agree extremely well with the Bohr value of R.

the radiation from an accelerated charge? For example, suppose that in interstellar space, where the distance between atoms may approach a centimeter, there is an electron in orbit around a proton. If the orbit radius is close to a centimeter, should we expect the frequency of the radiation to be that given by classical theory or by Bohr's theory?

Bohr's correspondence principle tells us the answer. The correspondence principle states that the result of any quantum theory must approach the classical result as the quantum number approaches infinity. Just as classical mechanics is an approximation valid when $v \ll c$, classical theories of mechanics and electrodynamics are also approximations which are valid in the limit $n \to \infty$. So in the above example, Bohr's theory still gives the correct result, but the frequency predicted by classical theory is extremely close to the correct frequency.

Classical theory predicts that the frequency radiated by a charged particle in a circular orbit should equal the frequency of revolution of the particle in the orbit. So, for any orbit which is allowed by Bohr's theory, we may find the radiation frequency expected *classically*, simply by solving Eqs. (13) and (14) for ω_n; the radiation frequency v_n then equals $\omega_n/2\pi$. The solution for $Z = 1$ is

$$\omega_n = \frac{m_r e^4}{16\pi^2 \varepsilon_0^2 n^3 \hbar^3} = \frac{m_r c^2 \alpha^2}{n^3 \hbar}$$

so that

$$v_n = \frac{\omega_n}{2\pi} = \frac{m_r c^2 \alpha^2}{n^3 h} \tag{22}$$

(Do not be disturbed by the presence of Planck's constant and the quantum number n in a classical equation. They are there only because we are referring to an orbit in which the angular momentum is *given* as $n\hbar$. Such an orbit is not *required* classically, but it is certainly *permitted*.)

To compare Eq. (22) with the frequency given by Bohr, consider a transition from the nth orbit to the $(n - 1)$th orbit. From Eq. (20), with $Z = 1$

$$v_{n-1,\,n} = \frac{m_r c^2 \alpha^2}{2h} \left[\frac{1}{(n-1)^2} - \frac{1}{n^2} \right]$$

$$= \frac{m_r c^2 \alpha^2}{2h} \left(\frac{2n-1}{n^4 - 2n^3 + n^2} \right)$$

As n increases, the expression in parentheses approaches $2n/n^4$, so that

$$v_{n-1,\,n} \to \frac{m_r c^2 \alpha^2}{n^3 h}$$

which is just the classical result.

In general, the frequency predicted by Bohr in a transition from the nth orbit to the $(n-1)$th orbit lies between the frequencies of revolution in the two orbits. This follows from the inequality

$$\frac{m_r c^2 \alpha^2}{n^3 h} < \frac{m_r c^2 \alpha^2}{2h}\left(\frac{2n-1}{n^4 - 2n^3 + n^2}\right) < \frac{m_r c^2 \alpha^2}{(n-1)^3 h}$$

which may easily be verified by transforming it to

$$n^3 > \frac{2(n^4 - 2n^3 + n^2)}{2n-1} > (n-1)^3$$

Classically, the electron spirals in from the nth orbit to the $(n-1)$th orbit, emitting radiation whose frequency continuously increases from $m_r c^2 \alpha^2 / n^3 h$ to $m, c^2 \alpha^2 / (n-1)^3 h$, while in the Bohr theory the orbit changes suddenly from n to $(n-1)$ and a photon of the intermediate frequency

$$\frac{m_r c^2 \alpha^2}{2h}\left(\frac{2n-1}{n^4 - 2n^3 + n^2}\right)$$

is emitted. The ratio of the two classical frequencies is $(n-1)^3 / n^3 \approx 1 - 3/n$, while the Bohr frequency differs from the classical frequency for the nth orbit by the factor

$$\frac{2n^4 - 4n^3 + 2n^2}{2n^4 - n^3} \approx \frac{2n-4}{2n-1} \approx 1 - \frac{3}{2n}$$

placing the Bohr frequency squarely between the two classical frequencies.

For the example originally quoted, an orbit of about 1 cm radius, the value of n may be found from Eq. (16):

$$n = \left(\frac{r}{5.29177 \times 10^{-9}\text{ cm}}\right)^{1/2}$$

$$= \left(\frac{10^9}{5.29177}\right)^{1/2} = 13{,}747$$

to the nearest integer. (As we might have guessed, an orbit radius of *exactly* 1 cm is not allowed. For $n = 13{,}747$, r_n is actually 1.00004 cm, while for $n = 13{,}746$, $r_n = 0.99989$ cm.) The two classical frequencies therefore differ by about 3 parts in 13,747, while the Bohr frequency differs from each by about 3 parts in 27,494, or about 0.01 percent.

The correspondence principle is useful in extending the quantum theory into new areas. For example, the Bohr theory says nothing about the length of time one electron spends in an orbit before making a transition to a lower orbit, but we can estimate this time by using the classical theory of radiation and the correspondence principle. In the limit of large quantum numbers, the rate at which energy is radiated by an accelerated charge should be correctly given by the classical expression

$$\frac{2e^2}{3c^3}(\ddot{r})^2 \qquad \text{or} \qquad \frac{e^2}{6\pi\varepsilon_0 c^3}(\ddot{r})^2 \qquad\qquad (23)$$

$$\text{cgs} \qquad\qquad\qquad \text{mks}$$

where $\ddot{r}$ is the acceleration. It is reasonable to assume, on the basis of the correspondence principle, that the *average* time which an electron spends in an orbit is the time which would be required, at the classical rate determined by the acceleration in the orbit, to radiate an amount of energy equal to the energy of the photon which must be emitted in the transition to the lower orbit. It must be emphasized, however, that the actual time of emission of a photon from any particular atom is random. It is a general feature of all quantum theories that they cannot make definite predictions of the time of any single quantum transition.

Elliptical Orbits. If you have read this discussion of the correspondence principle with a critical eye, you may have noticed that so far we have carefully avoided the question of transitions in which n changes by more than one. These transitions certainly occur and are observed in atomic spectra. What does the correspondence principle have to say about that? According to the correspondence principle, the frequencies emitted in such transitions should, in the limit $n \to \infty$, approach some frequency which is given by classical theory. Let us see what that limit is:

$$
\begin{aligned}
\nu_{n-\delta n,\, n} &= \frac{m_r c^2 \alpha^2}{2h}\left[\frac{1}{(n-\delta n)^2} - \frac{1}{n^2}\right] \\
&= \frac{m_r c^2 \alpha^2}{2h}\left[\frac{2n\,\delta n - (\delta n)^2}{n^4 - 2n^3\,\delta n + n^2(\delta n)^2}\right] \\
&\underset{n\to\infty}{\longrightarrow} \frac{m_r c^2 \alpha^2\,\delta n}{n^3 h}
\end{aligned}
$$

So in the limit $n \to \infty$, the frequencies emitted in transitions from n to $n - \delta n$ are integral multiples, or harmonics, of the fundamental frequency emitted in the transition n to $n - 1$. But classically a particle in a circular orbit emits *only* the fundamental frequency; the oscillation along any axis is purely sinusoidal. Thus Bohr's assumption that only circular orbits are allowed is in conflict with the correspondence principle, as long as one allows transitions from a circular orbit to another orbit such that $\delta n > 1$. But we can resolve the conflict by modifying the Bohr theory to permit *elliptical* orbits as well as circular ones. It is obvious that such a modification should be required by the correspondence principle, because elliptical orbits are certainly possible for large quantum numbers; they are observed on a macroscopic scale. Permitting elliptical orbits removes the difficulty with transitions for which $\delta n > 1$; the higher harmonics are to be expected then, because these harmonics are present in the motion of the particle—the oscillation along any axis is *not* purely sinusoidal when the particle moves on an elliptical orbit. We should then expect that transitions for which $\delta n > 1$ must start from an elliptical orbit; in the language of quantum theory, we say that there is a *selection rule* which requires that $\delta n = 1$ for transitions from a circular orbit.

Bohr did not have to consider elliptical orbits, because without them he could derive the Rydberg formula and the value of R, and that was certainly an impressive achievement. It worked out that way because the quantization of angular momentum, which Bohr introduced and which is a feature of all subsequent theories, yields the same energies for elliptical orbits as for circular orbits, as long as small relativistic effects are neglected. But when the spectrum of hydrogenlike atoms is examined closely, it is seen that each line is actually a set, or "multiplet," consisting of two or more lines of very nearly equal wavelength. In 1916, Arnold Sommerfeld[10] explained this fine structure as a relativistic effect resulting from the existence of elliptical orbits.

The effect of elliptical orbits on the allowed energy levels and frequencies may be understood qualitatively as follows: For each value of the quantum number n, there is not only a circular orbit with angular momentum $n\hbar$, but also a set of elliptical orbits whose major axis equals the diameter of the circular orbit, but whose angular momentum is smaller than that of the circular orbit. The angular momentum of each elliptical orbit, like that of each circular orbit, must be an integral multiple of $\hbar$. We may write the angular momentum of each orbit as $k\hbar$, where $0 < k \leqslant n$. The case $k = n$ gives the circular orbit; the case $k = 0$ is excluded, in order to give agreement with the observed fine structure, and also on the grounds that an electron in such an orbit would strike the nucleus.[11] Each ellipse is thus characterized by two quantum numbers: n, which determines the major axis of the ellipse, and k, which determines the electron's angular momentum.

In the absence of relativistic corrections, all orbits with the same major axis have the same energy; thus the requirement that orbits with the same n value have the same major axis makes the theory *almost* agree with the original Bohr theory, as it must. The relativistic correction comes about because an electron in a more eccentric orbit reaches higher speeds, because of its closer approach to the nucleus, so that the expression $\frac{1}{2}mv^2$ becomes less accurate as an approximation to the kinetic energy. The result is a slight difference in energy between orbits of the same n and different k. These differences in energy were calculated by Sommerfeld, and were found to explain perfectly the observed fine structure of many spectral lines of hydrogen. In order to obtain perfect agreement with the *number of lines* in a multiplet, Sommerfeld postulated a "selection rule" that $\delta k = \pm 1$ in any transition;

[10] A. Sommerfeld, *Ann. Phys. (Leipzig)* **51**, 1 (1916).
[11] We know now that the nucleus is quite transparent to electrons, and that the electron does penetrate the nucleus, but the argument for excluding $k = 0$ seemed reasonable at the time.

this rule can be seen to be a generalization of the rule that $\delta n = 1$ in a transition which begins with a circular orbit. For example, an electron in an orbit with $n = 3$, $k = 3$ can make a transition *only* to the orbit with $n = 2$, $k = 2$; but an electron in the *elliptical* orbit with $n = 3$, $k = 2$ could make transitions to $n = 2$, $k = 1$ or to $n = 1$, $k = 1$.

Unfortunately Sommerfeld's theory, like the original Bohr theory, could not be extended successfully to multi-electron atoms. When a more comprehensive theory of quantum mechanics was developed, it was found that the explanation of fine structure was completely different from that of Sommerfeld.[12] Sommerfeld's success had been nothing more than an amazing coincidence! However, the concept of an elliptical orbit remains a useful one which can help us to visualize certain features of atoms and to understand some effects which we shall discuss later.

EXAMPLE PROBLEM 2.   Derive the relation $b/a = k/n$, where b is the semiminor axis and a the semimajor axis of an elliptical orbit (Fig. 6), by making the following two assumptions:

1. The angular momentum is $k\hbar$.

2. The energy is the same as the energy of a circular orbit of quantum number n.

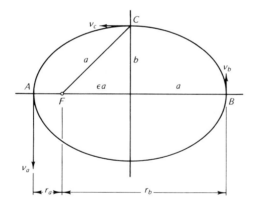

Fig. 7. *Elliptical orbit of electron, with nucleus at focus F, semimajor axis a, semiminor axis b, and eccentricity ε.*

Solution.   For simplicity, let us assume that the nucleus is fixed; it is, of course, located at one focus of the ellipse, point *F*. First let us prove that

[12] See Section 7.3.

the energy depends only on a, the semimajor axis, and *not* on the eccentricity ϵ (defined by Fig. 7). The two quantities which remain constant during the motion are the energy E and the angular momentum L. The energy is (in cgs units)

$$E = \frac{mv^2}{2} - \frac{Ze^2}{r}$$

At point A or point B the velocity is perpendicular to the radius vector from F, so the angular momentum may be written

$$L = mv_a r_a$$
$$= mv_b r_b$$

Therefore we may write the energy as

$$E = \frac{mv_a^2}{2} - \frac{Ze^2}{r_a}$$
$$= \frac{mv_b^2}{2} - \frac{Ze^2}{r_b}$$

or

$$E = \frac{L^2}{2mr_a^2} - \frac{Ze^2}{r_a}$$
$$= \frac{L^2}{2mr_b^2} - \frac{Ze^2}{r_b}$$

obtaining identical quadratic equations for r_a and r_b:

$$2mEr_{a,b}^2 + 2mZe^2 r_{a,b} - L^2 = 0$$

Since $r_b > r_a$, the solutions are

$$r_a = \frac{-Ze^2}{2E} - \left(\frac{Z^2 e^4}{4E^2} + \frac{L^2}{2mE} \right)^{1/2}$$

$$r_b = \frac{-Ze^2}{2E} + \left(\frac{Z^2 e^4}{4E^2} + \frac{L^2}{2mE} \right)^{1/2}$$

The major axis is then $2a = r_a + r_b = -Ze^2/E$, so that $E = -Ze^2/2a$, and E does depend only on a and not on the eccentricity.

In the special case of a circular orbit, a is the radius, which from Eq. (15) equals $n^2 a_0$. The elliptical orbits with the same energy as the nth circular orbit (assumption 2) have the same value of a, so, since $a_0 = h^2/mZe^2$,

$$a = \frac{n^2 \hbar^2}{mZe^2}$$

To introduce the semiminor axis b, we notice that when the electron is at point C, the angular momentum may be written

$$L = mv_c b$$

while the energy is

$$E = \frac{mv_c^2}{2} - \frac{Ze^2}{a}$$

$$= \frac{L^2}{2mb^2} - \frac{Ze^2}{a}$$

But it was shown above that the energy is $-Ze^2/2a$, so that

$$\frac{L^2}{2mb^2} = \frac{Ze^2}{2a}$$

Inserting the value of a from above, and letting $L = k\hbar$, we find

$$b = \frac{kn\hbar^2}{mZe^2}$$

Therefore

$$\frac{b}{a} = \frac{\dfrac{kn\hbar^2}{mZe^2}}{\dfrac{n^2\hbar^2}{mZe^2}} = \frac{k}{n}$$

Knowing this result, we can easily draw sets of elliptical orbits. The set of orbits for $n = 4$ is shown in Fig. 8, for a nucleus located at point F.

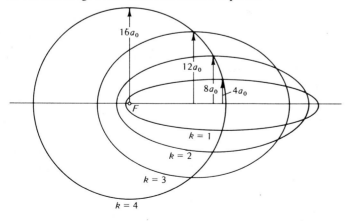

Fig. 8. *Set of allowed orbits for* $n = 4$.

3.5 THE FRANCK–HERTZ EXPERIMENT

We have seen that the existence of discrete *frequencies* in atomic radiation is proof of the existence of discrete *energy levels* in atoms, provided that Einstein's equation $E = h\nu$ is accepted. But in 1914, James Franck and Gustav Hertz obtained *direct* evidence of the discreteness of atomic energy levels, in an experiment which displayed the effect of collisions between atoms and electrons. Even though radiation was not involved in the transfer of energy, the results showed that the electrons lost energy in *discrete* amounts in inelastic collisions with atoms. Thus it was clear that the process was governed by the existence of discrete energy levels in atoms. This is another of those classic experiments which are now performed in most undergraduate laboratories; it won the 1925 Nobel Prize for Franck and Hertz.

The basic idea of Franck–Hertz experiment may be understood by reference to Fig. 9a. Electrons in a vapor-filled tube are accelerated by the

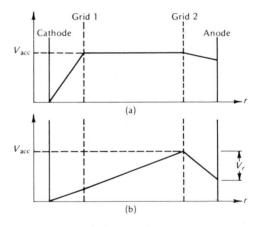

Fig. 9. *Different arrangements of the potential in a Franck–Hertz tube.*

grid potential V_{acc}. After passing Grid 2, the electrons are retarded by a reversed voltage V_r between grid and anode. A meter in the anode circuit measures the current produced by electrons which reach the anode. If there were no vapor present, an increase in grid voltage V_{acc} should always result in an increase in current, as it does in any vacuum tube. The same thing happens when the vapor is present, as long as V_{acc} is smaller than the energy needed to excite a vapor atom from its ground state to its first excited state. But as V_{acc} is increased, eventually the accelerating potential gives electrons a kinetic energy equal to this excitation energy. At this point, electrons are able to lose energy in *inelastic collisions* which excite atoms of the vapor. After such a collision an electron cannot overcome the retarding potential V_r, so many

electrons do not reach the anode, and the current falls. As V_{acc} is increased
still further, the current increases again, as illustrated in Fig. 9, but it falls
again when V_{acc} is large enough so that an electron can make *two* inelastic
collisions with atoms of the vapor. Figure 10 shows a whole series of drops in
current, equally spaced at intervals of about 5 V in V_{acc}, indicating that the

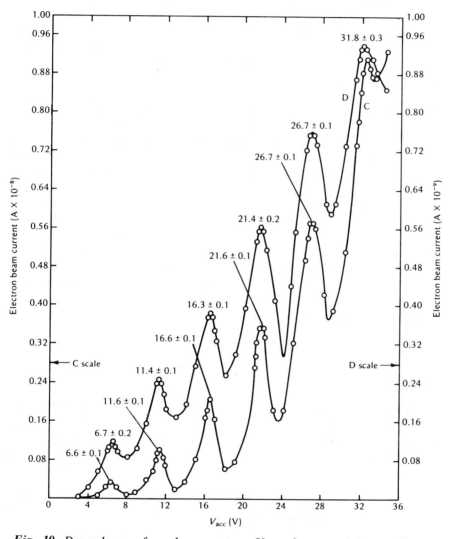

Fig. 10. *Dependence of anode current on V_{acc}, for potentials as in Fig. 8b.*
(Data were obtained by a student; both figures are from A. Melissinos, "Experi-
ments in Modern Physics," Chapter 1. Academic Press, New York, 1966). The
tube contained mercury vapor. Curve C (left-hand scale) was obtained with fila-
ment at 2.5 V. Curve D (right-hand scale) was obtained with filament at 1.85 V.

first excited state of a vapor atom must be about 5 eV above the ground state; this result is in reasonable agreement with other observations: the vapor atoms were mercury, for which there is spectroscopic evidence that the first excited state lies at 4.86 eV.

Two other features of Fig. 10 are of some interest. First, you will notice that the first drop in current does not occur at about 5 V, but at 6.6 or 6.7 V instead. The extra 1.7 V must indicate the work function of the cathode (or more precisely, the contact potential between cathode and anode—Section 11.4c); the kinetic energy acquired by the electrons is reduced by the amount of energy needed to free an electron from the cathode. Second, you see that all of the drops in current can be attributed to excitation of the first excited state of mercury. Apparently when the electrons reached the energy needed to excite the first excited state, they collided so quickly with a mercury atom that they did not have time to acquire enough energy to excite any higher energy levels. This happens when the electrons are accelerated slowly, by grid potentials like those of Fig. 9b. But with an arrangement of grid potentials like that shown in Fig. 9a, the electrons gain energy faster, and one can see additional current drops which result from excitation of higher energy levels.

To make the connection between these observations and the line spectra doubly certain, Hertz in 1924 observed the spectrum of light emitted by mercury vapor in such a tube. He saw that the appearance of certain spectral lines depended on the value of V_{acc}; a given line did not appear until the voltage V_{acc} was large enough to excite the state which was believed to give rise to that line. When the value of V_{acc} was sufficiently large, the atoms thus excited emitted the line in falling back to the ground state (or in some cases, to a lower excited state).

PROBLEMS

1. The earth receives energy from the sun at a rate of 0.135 W/cm^2. On the assumption that the sun radiates as a black body, calculate its surface temperature according to the Stefan–Boltzmann law (Example Problem 1). Useful information: the angular diameter of the sun as seen from the earth is $0.53°$.

2. If one plots the intensity of the sun's radiation as a function of *wavelength*—that is, one plots $f(\lambda)$ such that $f(\lambda) \, d\lambda$ is the total intensity of the radiation whose wavelength is between λ and $\lambda + d\lambda$—the maximum intensity is at $\lambda_m = 5000$ Å. Is this result consistent with Eq. (7) and the answer to Problem 1? (*Hint*: The value of λ_m is *not* equal to c/v_m.)

3. Verify Eq. (7) by numerical analysis of Planck's formula [Eq. (5)].

4. (a) Show from Eq. (6) that the constant σ is equal to 5.67×10^{-8} W-m^{-2} = K^{-4}

   (b) Suppose that the Joule heat I^2R developed by a current I in an aluminum wire is removed only by black-body radiation. If the wire has a radius of 1.0 mm, find the current it carries when it is in equilibrium at $T = 900$ K (just below the melting point). Resistivity of aluminum is 2.7×10^{-8} ohm-m.

5. Electrons are ejected from a certain metal only when the wavelength of incident light is less than or equal to 4000Å. What wavelength would be required to eject electrons with a kinetic energy of 2 eV from this metal?

6. The shortest wavelength of visible radiation is about 4000 Å. What is the maximum kinetic energy with which an electron can escape from

   (a) Cs (work function $W = 1.9$ eV)

   (b) Na (work function $W = 2.3$ eV)

   (c) Al ($W = 4.1$ eV)

   when it is irradiated with light of this wavelength?

7. An atom of positronium (Ps) is composed of a positron and an electron, bound in the same way as the proton and electron in the H atom. (The positron has the same charge as the proton, but has a mass equal to the electron mass.)

   (a) Determine the radii of the Bohr orbits for Ps, and make a scale drawing comparing the ground state orbits in Ps with those in the H atom.

   (b) Determine the energy levels of Ps and compare with those of the H atom.

8. The negative muon has a charge equal to the electron's charge and a mass of about 207 electron masses. If a muon replaces the electron in an H atom, muonic hydrogen is formed. Compute the reduced mass, the energy levels, and the orbit radii for muonic ^{1}H and muonic ^{2}H.

9. The H$_\alpha$ line of the Balmer series is emitted in the transition from $n = 3$ to $n = 2$. Compute the wavelength of this line for ^{1}H and for ^{2}H.

10. Calculate the limit of the Brackett series (Fig.6) in eV and in Angstroms. What is the longest wavelength in this series?

11. Use the correspondence principle and the classical rate of radiation [Eq. (23)] to estimate the average lifetime of the $n = 2$ state for the H atom.

12. If one shines a continuous spectrum of light through a gas of hydrogen

atoms, discrete frequencies will be absorbed, corresponding to the energies needed to excite the atoms to their various excited states. At room temperature one sees only the Lyman series in this absorption spectrum. Explain why the other series are absent, and determine the temperature at which 1 percent of the atoms would be able to absorb the frequencies of the Balmer series. (*Hint*: Use the Boltzmann factor; see Section 1.1.)

13. Calculate the maximum energy which an electron can lose in an elastic collision with a mercury atom.

14. Mercury has a second excited state at 6.7 eV. If you were to use the grid potentials of Fig. 9a, so that you could excite this state, at what additional voltages (in addition to those shown in Fig. 10) should you see a drop in current? What wavelengths of radiation would you then expect the tube to emit?

15. Eq. (15) shows that the radius of the orbit in a hydrogen-like atom or ion depends only on the electron mass, not on the nuclear mass M or the reduced mass m_r. Show that the "radius" that one would find from Eq. (15), if one replaced m by m_r in that equation, is equal to the distance $r + R$ from the nucleus to the electron.

4

Waves and Particles

The quantum hypothesis has implications which are even more far-reaching than those discussed in the previous chapter. Einstein's hypothesis that radiation has a corpuscular nature is merely the beginning of a subtle chain of reasoning concerning particles and waves in nature. The conclusion, amply supported by experiment, is that there is a wave–particle "duality" in all of nature—that is, that matter has wave properties, just as radiation has particle properties.

We shall begin our study of wave–particle duality by looking at additional evidence for this duality in radiation —specifically in X-rays. Then we shall see that the same sort of duality is observed in the "particles" of nature —electrons, protons, neutrons, etc.—and we shall see how these two natures can be combined in one entity without any logical contradiction.

4.1 X-RAYS

The particle nature of radiant energy becomes most apparent in experiments with X-rays. Because of the high frequency of these rays, the radiation is more "lumpy" than visible light; a given quantity of X-ray energy is composed of fewer photons, each photon having more energy than a photon of visible light.

Wilhelm K. Röntgen discovered the X-rays in 1895, while working with cathode rays. He noticed a fluorescent glow in a mineral located near an operating cathode-ray tube in a dark room, and he proceeded very systematically to investigate the cause. (Other persons working with cathode-ray equipment had previously seen effects which must have been caused by X-rays, but they did not investigate them as thoroughly as Röntgen did; many simply were careful not to leave their photographic plates lying near the cathode-ray tube, because they became fogged.) Röntgen, after a six-week series of experiments, published an article listing many fundamental properties of the rays. He found:

1. The rays are produced when the cathode rays strike a solid surface, such as the anode of the cathode-ray tube.
2. The rays are not deflected by magnetic fields (and hence they must be uncharged).
3. Passage of the rays through a gas increases the conductivity of the gas.
4. Lighter elements are more transparent to the rays than are heavier elements.
5. Photographic emulsions are sensitive to the rays.
6. The rays are not refracted to an appreciable degree in passing through matter.

In studying the transparency of various materials to the rays, Röntgen took the first X-ray photographs: of bones of the hand, of objects enclosed in a box, and of inhomogeneities in a piece of metal. The applications of this work were immediate; in January, 1896, the month following Röntgen's publication of his observations, a broken arm was set with the aid of X-rays.[1]

X-Ray Diffraction. A step forward in understanding X-rays was taken when H. Haga and C. Wind observed (in 1899) that the rays were diffracted when they passed through a narrow slit. Haga and Wind concluded from their observations that the rays were like light waves, but of a much shorter

---

[1] This has often been cited as an example of the value of basic research. If Röntgen had initially set out to discover ways to improve surgical techniques, it seems unlikely that he would have studied the cathode rays which led to his discovery of X-rays. After all, what possible use could anyone have for a strange glow inside an evacuated tube?

wavelength, of the order of 1 Å or 10^{-8} cm. However, the easily observed particle nature of the rays tended to confuse the issue. It was observed that at a distance of several meters from the point of original impact of an electron beam on a plate, another electron could suddenly appear with the same energy as the original electrons in the beam. This is, of course, simply the photoelectric effect, with X-rays instead of light. The lack of any attenuation in energy from dispersion as the X-ray traversed the intervening distance was evidence of the corpuscular nature of X-rays. Since these observations were made before Einstein's photoelectric equation had been verified for light waves, they were puzzling to many who felt that the X-rays had to be either waves or particles, but not both.

The wave nature of X-rays was finally established beyond doubt when Max von Laue suggested that the atoms in a crystal should make a good diffraction grating for the rays, and von Laue, Friedrich, and Knipping obtained a diffraction pattern from X-rays passing through a crystal of ZnS. However, a three-dimensional crystal is considerably more complicated than the usual diffraction grating, so the pattern which was observed was difficult to interpret. Although von Laue gave a detailed analysis of the pattern, a clearer and more general explanation was provided by W. L. Bragg. It then became apparent that X-rays could be used to study the structure of crystals.

To understand Bragg's explanation, one may visualize the atoms of a crystal as arrayed in various planes which "reflect" X-rays (see Fig. 1). That

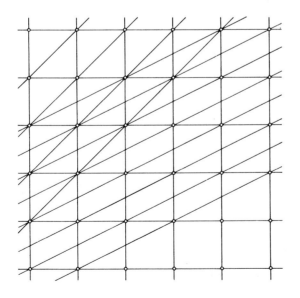

Fig. 1. *Edgewise view of some planes formed by a cubic array of atoms.*

is, the effect of a single plane of atoms on an X-ray is to scatter it in such a way that the usual law of reflection is obeyed: angle of incidence = angle of reflection. The effect of planes of atoms parallel to the first plane is to produce interference of the scattered X-ray waves, so that only rays reflected at certain angles interfere constructively and are observed. The condition for constructive interference is that rays reflected from successive planes travel path lengths which differ by an integral number of wavelengths. Therefore an X-ray of wavelength λ, striking a set of planes whose spacing is d (Fig. 2)

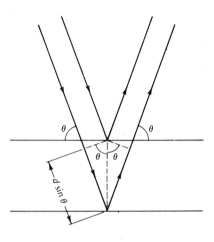

Fig. 2. Bragg diffraction, somewhat oversimplified. The path of the ray "reflected" from the lower plane is longer than the path of the ray "reflected" from the upper plane, by the length $2d \sin \theta$, which must equal $n\lambda$ for constructive interference.

will be reflected, with the angle of incidence equal to the angle of reflection, if the angle θ between the X-ray direction and the plane satisfies the "Bragg condition":

$$2d \sin \theta = n\lambda$$

where n is an integer.

The above discussion provides a good way to *remember* the Bragg law, but it is not a good derivation, for it is based on the *assumption* that the angles of incidence and reflection are equal when the wave is "reflected" from a single plane of atoms. Actually, the whole phenomenon is one of diffraction rather than reflection, so there is no reason to make this assumption. A more satisfying derivation of the Bragg law is given in Appendix B.

The Bragg condition seems simple enough, but the situation can become complicated when we attempt to use it to determine the structure of a crystal containing many planes in various directions and with various spacings. However, X-rays are now used routinely to determine crystal structures and even to analyze the structure of large molecules. Further discussion of these matters would take us too far from the main topic of this chapter; we shall return to problems involving crystal structure in Chapter 12, when we discuss electronic properties of solids.

X-Ray Production. Just as we can use X-rays to study crystals, we can use a crystal to study X-rays. The crystal structure of a simple crystal can be determined by observing the general features of the X-ray diffraction pattern; the grating spacing d for a given set of planes can then be found if the density of the crystal, the atomic mass of the atoms in the crystal, and Avogadro's number are known. Knowing d we can use the crystal to analyze the frequency spectrum of the X-rays produced by a cathode-ray tube and to study the effects of changes in the accelerating voltage and in the composition of the anode of the tube.

Figure 3 shows the results of such an analysis of X-rays produced when cathode rays strike a target. Two features of the spectrum are of interest: the sharp cutoff of the continuous spectrum at the short wavelength end, and the presence of the sharp peaks labeled K_α and K_β. Other peaks appear at longer wavelengths. The wavelength of the cutoff of the continuous spectrum

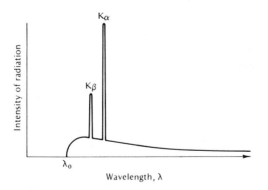

Fig. 3. *Spectrum of X-rays produced when a cathode-ray beam strikes a target made of a single element.*

depends on the accelerating potential of the cathode-ray tube and is independent of the target material, while the position of the peaks depends on the target material and is independent of the accelerating potential. (Of course, a peak does not appear if its wavelength is shorter than the cutoff wavelength λ_0 for a given accelerating potential.) Let us examine the reasons for these two effects.

The continuous spectrum, called "bremsstrahlung" ("braking radiation"), is produced by the acceleration of the electron as it stops in the target. According to classical electrodynamics, the high-frequency components[2] of the radiation are produced when the electron undergoes a large acceleration as a result of passing very close to a nucleus. But the classical analysis breaks down for very high-frequency radiation, because of the high energy of the photons involved and the fact that the photon energy cannot exceed the original kinetic energy of the electron.[3] Therefore the maximum photon energy from a cathode-ray tube whose accelerating voltage is V may be written

$$E_{max} = h v_0 = \frac{hc}{\lambda_0} = eV$$

so that

$$\lambda_0 = \frac{hc}{eV}$$

Since $hc \approx 12,400$ eV-Å, a tube operating at 12,400 volts produces X-rays whose shortest wavelength is 1 Å. This dependence of λ_0 on accelerating voltage was verified in the early experiments, and it provided further confirmation of Einstein's theory of the photoelectric effect, for bremsstrahlung production is simply the *inverse* photoelectric effect.

The sharp peaks in the X-ray spectrum were the subject of a systematic study by H. G. J. Moseley in 1913.[4] The Bohr theory had just been published, and Moseley correctly deduced that the peaks were produced by the same process which produces the optical spectra of the elements: Emission of a photon as an electron makes a transition from a high energy level to a lower energy level in an atom. In this case, the transition is made possible when an electron from the cathode-ray beam strikes the atom and knocks an electron out of a lower energy level in the atom.

Moseley's work resulted in the first completely convincing determination of the atomic *numbers* of the elements. It was already known that the periodic

---

[2] One could, in principle, make a Fourier analysis (Section 4.3) of the velocity of the electron as it stops in the target; the frequencies found would be present in the same proportion in the emitted radiation, according to classical theory.
[3] The photon energy could be *slightly* greater than this if the final state of the electron is a bound state, of negative total energy.
[4] H. G. J. Moseley, *Phil. Mag.* **26**, 1024 (1913); **27**, 703 (1914).

table of the elements could not be properly constructed if the elements were listed simply in the order of increasing atomic *mass*. For example, the alkali metals follow the "inert" gases in the table, so potassium, obviously an alkali metal, should follow argon; but the atomic mass of argon (39.944) is greater than the atomic mass of potassium (39.100). Moseley found, however, a perfect correlation between the frequency of the K_α X-ray line and the order of the elements as determined by their chemical properties. This regularity in the X-ray spectra of the elements occurs because the inner electron shells involved in the transitions do not vary very much from one atom to the next; the only change which affects the energy levels of these electrons is the change in the charge of the nucleus.

In his first paper, Moseley gave the wavelengths produced by ten different target elements; his results are shown in Table 1. Moseley discovered that the

Table 1

Wavelengths of the K_α and K_β Radiation, as Reported by Moseley

Element	λ_α (Å)	λ_β (Å)	$\sqrt{(\nu_\alpha/\tfrac{3}{4}\nu_0)}$ [a]	Z
Ca	3.357	3.085	19.00	20
Ti	2.766	2.528	20.99	22
V	2.521	2.302	21.96	23
Cr	2.295	2.088	22.98	24
Mn	2.117	1.923	23.99	25
Fe	1.945	1.765	24.99	26
Co	1.796	1.635	26.00	27
Ni	1.664	1.504	27.04	28
Cu	1.548	1.403	28.01	29
Zn	1.446	—	29.01	30

[a] See text for explanation of this quantity.

atomic number Z indicated by an element's logical position in the periodic table (when this was known) increased in direct proportion to the square root of the frequency of the element's K_α line, or

$$Z = A\sqrt{\nu_\alpha} + s \tag{1}$$

where A and s are constants. The form of this equation and the value of the constant A could be explained perfectly by analogy to the Bohr theory for the hydrogen atom, as follows:

It is assumed that the states of the inner electrons in a multi-electron atom are similar to the electronic states in the hydrogen atom, the energy of a state being determined primarily by the "principal quantum number" n. The difference between the multi-electron atom and the single-electron atom is in the mutual repulsion of the electrons. This repulsion can be accounted for in an empirical way by saying that each electron feels an attractive charge of $(Z - s)e$, rather than Ze, in an atom of atomic number Z. The number s is a "shielding number," which presumably is different for each electron in the atom, but which should be approximately the same for all electrons with the same value of n. The energy emitted in a transition from one value of n to another can be related to an average s, determined empirically by Moseley. Moseley's determination of s was possible because s for a given transition does not change noticeably as outer electrons are added to form heavier atoms.[5]

To see how s and Z are related to the X-ray wavelengths, we recall that the frequency ν emitted in a transition in the *hydrogen atom* is given by

$$h\nu = h\nu_0 \left[\frac{1}{n'^2} - \frac{1}{n^2} \right] \tag{2}$$

where $h\nu_0$ is the ionization energy of the hydrogen atom (Eq. 20, sec. 3.3). We know that for a *one-electron ion,* of nuclear charge Ze, the transition energy is proportional to Z^2, so the frequency is

$$\nu = \nu_0 Z^2 \left[\frac{1}{n'^2} - \frac{1}{n^2} \right]$$

For a neutral *atom,* the effective nuclear charge is $(Z - s)e$, so we expect the frequency to be

$$\nu = \nu_0 (Z - s)^2 \left[\frac{1}{n'^2} - \frac{1}{n^2} \right] \tag{3}$$

We may solve Eq. (3) for Z to obtain

$$Z = \left[\nu_0 \left(\frac{1}{n'^2} - \frac{1}{n^2} \right) \right]^{-1/2} \sqrt{\nu} + s$$

an equation of the same form as Eq. (1).

---

[5] This is to be expected from Gauss's law, as long as the added electrons remain outside the "orbit" of the electron making the transition. We shall see in Chapter 7 that this condition is not strictly satisfied, but Table 1 shows that the effect on s is small enough to be neglected. For a further discussion, see A. M. Lesk, *Am. J. Phys.* **48**, 492 (1980).

Moseley assumed that the K_α line results from a transition from one of the $n = 2$ states (which form what is called the L shell) to an $n = 1$ state (in the K shell), so he set n' equal to 1 and n equal to 2, obtaining $z = (v_\alpha/\frac{3}{4}v_0)^{1/2} + s$. The quantity $(v_\alpha/\frac{3}{4}v_0)^{1/2}$ should thus increase by exactly one unit as we go from one element to the next in the periodic table, and Table 1 confirms this idea. We see that $s = 1$ for electrons making such transitions.

In a second paper, Moseley presented results of measurements of spectra of heavier atoms. In these results, another series of lines, the L series, showed the same sort of regularity that the K series had shown. The quantity $(v_L/\frac{5}{36}v_0)^{1/2}$ was found to increase by one unit as one goes from one element to the next in the periodic table. This indicated that these lines result from transitions from the M shell ($n = 3$) to the L shell, for $(1/2^2) - (1/3^2) = \frac{5}{36}$. The screening number s is about 7.4 in this case, reflecting the fact that the L and M shells are farther from the nucleus than the K shell.

By now it should be possible for you to deduce for yourself the origin of the K_β lines. The analysis of Moseley's data for these lines is left as an exercise.

Later work with better resolution has shown that many of the X-ray "lines" observed by Moseley are actually multiple lines; there are really three energy levels, L_I, L_{II}, and L_{III}, in the L shell, and five levels in the M shell. The differences between these levels may only be explained by a complete theory which takes into account all quantum numbers, including electron spin. Nevertheless, Moseley's work was extremely important as confirmation of essential features of the Bohr theory and in clarifying the periodic table of the elements. His results established the proper positions of cobalt and nickel, whose atomic masses are almost equal, and they showed that elements were missing at $Z = 43$, 61, and 75.

The Compton Effect.   The Compton effect, first observed with X-rays, provides the most striking evidence of the corpuscular nature of radiation. The main features of the effect were reported[6] by J. A. Gray in 1920, as follows:

> "It has been proved, when an ordinary beam of X- or γ-rays[7] is scattered, that the rays scattered in any definite direction are of a quite different type to that of the primary rays. The effect is such that the scattered rays become less penetrating than the primary rays, the greater the angle of scattering. The distribution of the scattered radiation for all types of X- and γ-radiation is similar."

The appearance of less penetrating radiation (radiation of longer wavelength) is not in itself surprising; photoelectric absorption of X-rays can leave vacancies in inner electron shells of atoms, and X-rays of longer wavelength

---

[6] J. A. Gray, *J. Franklin Inst.* **190**, 633-655 (1920).
[7] Gamma rays are electromagnetic radiation, just like X-rays. The two kinds of rays differ only in their modes of production.

(fluorescent radiation) are emitted as the outer electrons make transitions to the inner shells. In that case, however, the wavelength of the secondary radiation would be characteristic of the absorber, and would be independent of the angle of "scattering."

Gray was not able to produce an explanation of the phenomenon, because his attention was fixed on the wave nature of the rays. In 1922, A. H. Compton gave the correct explanation and verified it by making quantitative measurements.[8] Compton supposed that the X-ray photon *really* behaves like a particle, so that it can "bounce" off an electron and transfer energy and momentum to it. The final *wavelength* of a photon rebounding in a given direction should then be determined simply by the conservation of momentum and energy. To apply these laws, Compton assumed that the momentum of a photon of frequency v should be hv/c, because the momentum density carried by an electromagnetic wave is equal to the energy density divided by c. Compton then derived the formula relating λ', the wavelength of the scattered radiation, to the wavelength λ of the incident radiation and the angle θ through which the radiation is scattered.

The derivation is straightforward. A photon of energy hv, moving along the x axis, strikes a free electron of mass m which is at rest (see Fig. 4). The electron recoils with momentum p at an angle ϕ with the x axis, while a

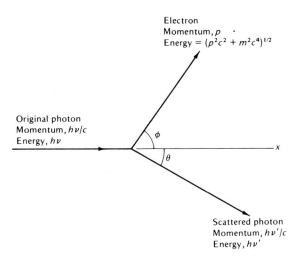

Fig. 4. *Angles of recoil of photon and electron in the Compton effect.*

[8] A. H. Compton, *Bull. Nat. Res. Council (U.S.)* **20**, 18 (1922); *Phys. Rev.* **21**, 715 (1923); *Phys. Rev.* **22**, 409 (1923).

photon of frequency v' departs at an angle θ with the x axis. Applying the conservation laws, we have

$$\text{Conservation of } p_x: \qquad \frac{hv}{c} - \frac{hv'}{c} \cos \theta = p \cos \phi \qquad (4)$$

$$\text{Conservation of } p_y: \qquad \frac{hv'}{c} \sin \theta = p \sin \phi \qquad (5)$$

$$\text{Conservation of energy:} \qquad hv - hv' + mc^2 = (m^2 c^4 + p^2 c^2)^{1/2} \qquad (6)$$

Squaring Eqs. (4) and (5) and adding corresponding sides eliminates the angle ϕ. Squaring Eq. (6) removes the $\frac{1}{2}$ power, and we have the two equations:

$$\frac{h^2 v^2}{c^2} + \frac{h^2 v'^2}{c^2} - \frac{2h^2 vv' \cos \theta}{c^2} = p^2$$

$$h^2 (v - v')^2 + 2h(v - v')mc^2 = p^2 c^2$$

It is now easy to eliminate p. After combining terms, we have

$$(v - v')mc^2 = hvv'(1 - \cos \theta) \qquad (7)$$

We may write Eq. (7) in terms of wavelengths by dividing by $mcvv'$, with the final result that

$$\boxed{\lambda' - \lambda = \frac{h}{mc} (1 - \cos \theta)} \qquad (8)$$

We see that the *change* in wavelength is independent of the original X-ray wavelength; it depends only on θ, and it has a maximum value of $2h/mc$ when $\theta = 180°$—that is, when the X-ray is scattered directly backwards. The quantity h/mc is called the *Compton wavelength* of the electron; it is the wavelength of a photon whose energy equals the rest energy of the electron:

$$\lambda = \frac{c}{v} = \frac{hc}{hv} = \frac{hc}{mc^2} = \frac{h}{mc}$$

Since $hc = 1.24 \times 10^4$ eV-Å, and $mc^2 = 5.11 \times 10^5$ eV (useful numbers to remember),

$$\frac{h}{mc} = \frac{1.24 \times 10^4}{5.11 \times 10^5} = 0.024 \quad \text{Å}$$

The shortness of this wavelength explains why this effect was never observed with light, although it should occur; a change of 0.024 Å in a wavelength of thousands of angstroms is not noticeable. However, the effect is enormous in many cases which are now commonly observed. For example, atomic nuclei often emit photons whose energy is equal to or greater than the electron rest energy mc^2. Suppose that a photon whose energy equals mc^2 is scattered backwards. Then $\lambda' - \lambda = 2h/mc$; since $\lambda = h/mc$, $\lambda' = 3h/mc$, and the wavelength has *tripled*, so that the final energy is only $mc^2/3$.

> You may have wondered about the assumption in the previous derivation —that the electron involved in the collision is originally free and at rest. The X-ray energies with which we are usually concerned are quite large compared to the energy required to free an electron from a solid or from an outer atomic shell, so the assumption that such electrons are free has little effect on the result. However, the X-ray can also interact with tightly bound electrons in inner shells of atoms; in such a case, the electron may remain in the shell, and the X-ray, effectively, is scattered from the atom as a whole. The scattered radiation is then virtually unchanged in frequency, because the mass appearing in the denominator of Eq. (8) should be the mass of the *atom* rather than the mass of the electron. This results in the appearance of an "unmodified line" at each angle in addition to the line of longer wavelength.
>
> The assumption that the electron is at rest is not strictly correct either. The motion of the electron does have a noticeable effect, which may simply be considered to be a Doppler shift in the wavelength of the scattered photon. But the spread in wavelengths introduced by the motion of the electron is, for X-rays, considerably smaller than the total wavelength change $\lambda' - \lambda$. Because of this fact and because, for any given speed, as many electrons move in one direction as in the opposite direction, the *average* wavelength of the photons scattered at a given angle is still correctly given by Compton's formula. The *distribution* of wavelengths produced by the motion of the electron may be studied in order to gain information about the motions of electrons in solids.

In summary, X-rays have given us convincing evidence that radiant energy is emitted, absorbed, and scattered in the manner of corpuscles of energy $h\nu$ and momentum $h\nu/c$. We have also seen convincing evidence that the radiation which displays this corpuscular behavior is diffracted in the manner of waves. It appears that we must accept the fact that radiation is *both* corpuscular and wavelike in nature.

4.2 MATTER WAVES

In 1923, inspired by the knowledge that light has a corpuscular nature as well as a wave nature, Louis de Broglie suggested that matter might have wave properties. This suggestion was a brilliant stroke, which marked the beginning of the modern quantum theory. By attributing a wave nature to matter, de Broglie was immediately able to explain, at least qualitatively, the existence of discrete *energy levels* in atoms, from the fact that only discrete *frequencies* are permitted for waves subject to various boundary conditions. Most of the quantum mechanics in this book will be a study of the implications of the wave equation developed by Erwin Schrödinger on the basis of de Broglie's hypothesis.

The best test of de Broglie's hypothesis would be to see if a beam of particles (for example, electrons) can be diffracted in the same way as a light beam. But before making this test, we should try to determine what wavelength we expect to find in the beam. A *quantitative* prediction, which is verified by experiment, should make us very confident that the hypothesis is correct.

A simple analogy with light might enable one to guess at the wavelength of a matter wave. A photon whose momentum is p has a wavelength of h/p, so we might assume that the same is true for a "particle" of matter. In fact, this gives us the relation suggested by de Broglie:

$$\lambda = \frac{h}{p} \tag{9}$$

the de Broglie wavelength of a "particle." At the moment this seems to be an arbitrary assumption, so you may well wonder whether a different guess might be just as reasonable. For example, since a photon of *energy E* has a wavelength of hc/E, you might suggest that hc/E is the wavelength of the wave associated with a particle of energy E. Clearly this gives a result in disagreement with Eq. (9), for $p = E/c$ only for particles of zero rest mass. However, de Broglie suggested that Eq. (9) is correct, because it satisfies the requirements of the theory of relativity.

The argument of de Broglie begins with the assumption that the rest energy of a particle is equal to the energy of some internal vibration of frequency v, so that

$$hv = mc^2$$

The "displacement" associated with this vibration may be written as

$$\psi = A \sin 2\pi vt \tag{10}$$

Of course we cannot say just what it is that is being "displaced," but we write Eq. (10) so that we may determine how this internal vibration appears in a frame of reference in which the particle is moving with velocity v. In that frame, the time coordinate is different; the Lorentz transformation [Eq. (17a.), Chapter 2] tells us that the time in the rest frame is

$$t = \frac{t' - x'v/c^2}{(1 - \beta^2)^{1/2}}$$

in terms of the coordinates x' and t' of the moving frame. The displacement may therefore be written in terms of x' and t' as

$$\psi = A \sin \left[2\pi v \left(\frac{t' - x'v/c^2}{(1 - \beta^2)^{1/2}} \right) \right] \tag{11}$$

But we know that the standard form of the displacement in a traveling sine wave of period T is

$$\psi = A \sin 2\pi \left(\frac{t'}{T} - \frac{x'}{\lambda} \right) \tag{12}$$

so we can set the coefficient of x' in Eq. (11) equal to the coefficient of x' in Eq. (12), to obtain

$$\frac{2\pi v v}{c^2 (1 - \beta^2)^{1/2}} = \frac{2\pi}{\lambda}$$

or

$$\lambda = \frac{c^2 (1 - \beta^2)^{1/2}}{v v}$$

But we originally assumed that $h v = m c^2$, or $v = m c^2 / h$. Therefore

$$\lambda = \frac{h(1 - \beta^2)^{1/2}}{m v}$$

$$= \frac{h}{p}$$

because $p = m v / (1 - \beta^2)^{1/2}$.

We have not proved that the de Broglie relation is correct, but we have shown that, if there is a wave associated with the motion of a particle, the relation between the wavelength of that wave and the momentum of the particle should be given by Eq. (9), in order to satisfy the requirements of relativity. But why propose such a wave in the first place?

One advantage of the wave idea is that it is a step toward finding a more fundamental reason for the apparently arbitrary rules governing electron orbits in atoms. For example, let us try to derive the Bohr condition from a requirement that an electron forms a standing wave in the atom. If there is a standing wave, the maxima and minima of the wave must occur at fixed points on the "orbit," so that the circumference of the orbit must be an integer multiple of the electron's wavelength (see Fig. 5). Thus

$$2\pi r = n\lambda = \frac{nh}{p}$$

so that the angular momentum is $pr = nh/2\pi$, just the value prescribed by Bohr.

After reasoning like this had suggested the possible existence of matter waves, direct evidence of the wave nature of electrons was produced remarkably soon, as a result of investigations begun in 1920 by C. J. Davisson. Davisson was scattering electrons from metal targets, not to look for wave properties of electrons, but rather as a follow-up to Rutherford's alpha-particle scattering experiments, to see if electron scattering would yield further information about atomic structure.

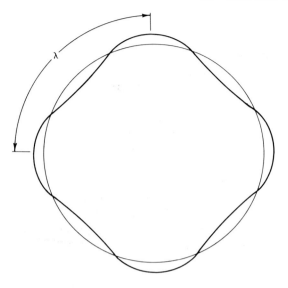

Fig. 5. *Standing wave produced by an electron in a Bohr orbit. The difference between the heavy line and the circle indicates the wave amplitude at each point. In this case, n = 4, so the circumference of the circle is 4λ.*

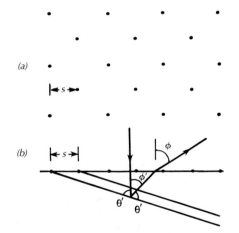

Fig. 6. *Electron diffraction by a nickel lattice whose top surface is a [111] plane. (a) Top view. (b) Side view. Spacing between rows of atoms is s = 2.16 Å. Electron wave enters vertically and is Bragg reflected by crystal planes making an angle θ' with the top surface. The reflected wave is then refracted, to emerge at an angle of φ to the vertical.*

Davisson's experiment took a dramatic turn when his nickel target was apparently "ruined" by a break in his vacuum tube.[9] When the target was reannealed, the reflected electron beam showed a number of strong peaks as a function of angle, suggestive of Bragg diffraction.

Davisson and L. H. Germer reported the results and their wave interpretation in 1927.[10] As the electron beam energy was varied, the beam was scattered more intensely into just the angles that one would expect if electrons were waves with the wavelength given by Eq. (9). For example, 54-eV electrons gave a scattering peak at 50°. The wave theory explains these numbers as follows:

In the annealing of the nickel target, large single crystals were formed. A crystal whose top surface is a [111] plane would appear as in Fig. 6a; the [111] plane, which cuts diagonally through the face-centered-cubic lattice,[11] has atoms on the corners of equilateral triangles, forming rows that are spaced a distance $s = 2.16$ Å apart in nickel. The electron beam penetrates into the crystal from above, and is then reflected from a set of planes that make an angle θ' with the top [111] plane. The beam is scattered in accordance with the Bragg equation (1) $2d \sin \theta' = n\lambda'$, where λ' is the wavelength of the electron inside the crystal.

From Fig. 6b, we see that $d = s \cos \theta'$, so

$$2s \sin \theta' \cos \theta' = n\lambda'$$

or
$$s \sin 2\theta' = n\lambda' \tag{13}$$

When the scattered beam emerges from the crystal, its wavelength changes (because the potential energy changes), so it is refracted to emerge at an angle ϕ such that $\lambda \sin \phi' = \lambda' \sin \phi$ (Snell's Law). Using the fact that $\phi' = \pi - 2\theta'$, we can substitute into (13) to obtain the relation between the observed angle ϕ and the wavelength λ in vacuum:

$$s \sin \phi = n\lambda \tag{14}$$

Eq. (14) is satisfied by scattering from any set of planes within the crystal; the orientation of the planes determines the angle ϕ, and hence the wavelength λ for which there will be a peak. Since the planes are discrete, a peak that appears for one value of λ should disappear (rather than simply shifting position) when λ is changed, and this was observed by Davisson and Germer. Eq. (14) would also apply to a situation in which scattering took place from the top plane of atoms only, but in this experiment that is not the case. If it were, there would be a peak regardless of the electron energy; the peak would simply appear at different values of ϕ as the electron energy changed.

Fig. 6b applies only to a crystal whose top surface is a [111] plane, and for

---

[9] R.K. Gehrenbeck, *Phys. Today* **31**, No. 1, 34 (Jan., 1978).

[10] C.J. Davisson and L.H. Germer, *Phys. Rev.* **30**, 705 (1927).

[11] A perspective view of a face-centered cubic lattice is shown in Fig. 16c, Chapter 12. The cube edge is $a = 3.524$ Å, and the face diagonal is $a\sqrt{2}$, making s equal to $a(\sqrt{2}/2)(\sqrt{3}/2)$, or 2.158 Å.

electrons scattered from planes that are perpendicular to the plane of the paper. Davisson and Germer also observed other peaks, at different azimuthal angles, which could be interpreted as resulting from other crystal orientations and other planes. They found that most of the angles at which these peaks appeared could also be derived from the wavelengths given by Eq. (9). A few of the angles apparently resulted from surface effects; electron scattering has become an important method for studying surfaces.

"Electron diffraction tubes" are now commercially available, so that students can easily observe diffraction patterns for themselves. In these tubes, the crystal which diffracts the electrons is mounted inside a cathode-ray tube, and the electron beam passes through the crystal, producing a diffraction pattern on the face of the tube.

There no longer is any reasonable doubt that the de Broglie relation correctly describes a fundamental property of all matter. Diffraction of protons, neutrons, and even entire atoms has been observed.[12] In each case the diffraction pattern has been in agreement with the wavelength predicted by the de Broglie relation. Solid-state physicists have put these facts to good use by using neutron diffraction to study the structure of crystals. Because the neutron has an intrinsic magnetic moment, the structure of many magnetic materials can best be studied by neutron diffraction.[13]

Connection between the Wave and the Particle. The analysis of an experiment with matter waves is identical to that of experiments with any other kind of wave. The observed intensity at any point is proportional to the square of the wave amplitude at that point. But it is important to remember that the intensity of the wave is detected only by the transfer of energy in *discrete* quantities to the detector. For example, photons may be detected by the photoelectric effect, either in a photocell or in your eye; electrons can produce a current or can deposit a charge on an electroscope, but a fraction of an electron is never observed. Thus in all cases the intensity is proportional to the *number* of *particles* detected.

When the intensity of the wave is small enough, one can count the particles individually. One is then struck by the fact that the particles are not detected at a uniform rate, or in any regular pattern at all. The particles come in at random, but in such a way that the average rate is what one would expect from the wave intensity, after one allows for normal statistical fluctuations. So it seems that the *wave* determines the *probability* of detecting a *particle*. (See Appendix A.) We shall return to this point after we develop our mathematical apparatus.

[12] Diffraction of He atoms was reported in 1931 by Estermann, Frisch, and Stern, Z. *Phys. (Leipzig)* **73**, 348.
[13] X-ray diffraction would reveal the overall lattice structure of such materials, but it would be insensitive to the orientation of the magnetic moment at each lattice point.

4.3   WAVE PACKETS

We must now attempt to reconcile the wave and the particle aspects of matter. When we analyze a diffraction experiment, we assume an *infinite* train of waves to be striking the grating, or crystal, as in Fig. 6. On that assumption, we can find a set of wave fronts, or surfaces such that the wave at each point on the surface is in phase with the wave at each other point. These wave fronts are planes which determine the direction for maximum intensity in the diffraction pattern. But if the incident wave is not infinite, or if it does not repeat itself *exactly* from one crest to the next, the wave scattered in a given direction from one lattice point does not have a fixed amplitude and phase relationship with the wave scattered from another lattice point. The resulting diffraction pattern then cannot be analyzed in terms of a single, precise wavelength for the wave. Thus when we speak of a wave of definite wavelength, we are always referring to a wave of infinite extent in space.

But how can an infinite wave represent a particle, which by definition occupies only one point in space at a given time? Obviously, it cannot. Nature simply does not contain objects which fit that definition of a particle. If we persist in imagining the "particles" of nature to be something like baseballs, only smaller, we are conjuring up images which exist only in our minds.

Nevertheless, there are certain features of the behavior of subatomic particles, such as the electron, which agree with this mental picture. In free space, the trajectory of an electron may be calculated from the forces acting on it, just like the trajectory of a baseball. The speed of an electron can be determined by observing it as it passes two different points, and measuring the time required for it to pass from one point to the other.[14] Clearly, the electron does move from one place to another; it cannot be represented correctly by an infinite wave of a single frequency.

Therefore, if we must consider the electron to be associated in some way with a wave, that wave must be localized, like the wave of Fig. 7, and hence it does not have a single, precise wavelength. However, a localized wave, or "wave packet," can be built up by adding together infinite waves which interefere destructively everywhere except in one small region. The procedure for doing this was worked out long ago. Although it is purely a mathematical

---

[14] One method of "observing" the electron as it passes a point would be to detect the charge induced by the electron as it passes a conductor; the sudden appearance of the induced charge produces a pulse which can be observed on an oscilloscope. A sufficiently large pulse can be produced if one has a group of electrons traveling together; such a group can be obtained from an electron accelerator. This method of measuring electron velocity has been used in a demonstration that electrons obey the relativistic relation between energy and speed. (See the Physical Science Study Committee film "The Ultimate Speed" produced by William Bertozzi.)

problem, the superposition of waves is so fundamental in quantum theory that it is worthwhile at this point to go into the details (in a nonrigorous way).

Fourier Series. We begin with the Fourier series, which enables one to express a *periodic* function as a sum of sine and cosine functions. Suppose

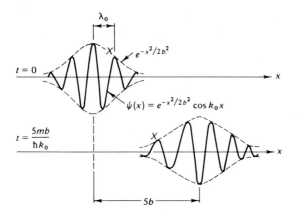

Fig. 7. *A " wave packet " representing a particle whose position at t = 0 lies approximately between x = + b and x = −b, and whose momentum is approximately h/λ₀. At a later time, when the center of the packet has moved to x = +5b, the packet is broader, because the various component waves move with different speeds. Notice that the crest marked X moves only half as far as the wave packet as a whole; the phase velocity is one-half of the group velocity. (See Problems 17 and 20 at the end of the chapter.)*

we are given $\psi(x)$, a function of x, which is periodic, with period $2a$. Then, by the definition of a periodic function,

$$\psi(x + 2a) = \psi(x) \qquad \text{for } \textit{all values of } x \qquad (15)$$

Let us *assume* that $\psi(x)$ can be written as the sum of an infinite series of sine and cosine functions. We choose the sine and cosine functions to have period $2a$, $2a/2$, $2a/3$, ..., so that they all have period $2a$ as well [by definition (15)], and therefore the series itself has period $2a$, as it must. The problem then is only to find the amplitude of each required sine or cosine function. We write

$$\psi(x) = A_0 + \sum_{n=1}^{\infty} \left[A_n \cos \frac{n\pi x}{a} + B_n \sin \frac{n\pi x}{a} \right] \tag{16}$$

and we attempt to find the A_n and the B_n; if we can do this, our assumption will have been justified.

The subsequent developments will be simplified if we rewrite Eq. (16). We know that

$$\cos \frac{n\pi x}{a} = \frac{e^{in\pi x/a} + e^{-in\pi x/a}}{2}$$

and

$$\sin \frac{n\pi x}{a} = \frac{e^{in\pi x/a} - e^{-in\pi x/a}}{2i}$$

Therefore we can write Eq. (16) as

$$\psi(x) = A_0 + \sum_{n=1}^{\infty} \left\{ A_n \left[\frac{e^{in\pi x/a} + e^{-in\pi x/a}}{2} \right] + B_n \left[\frac{e^{in\pi x/a} - e^{-in\pi x/a}}{2i} \right] \right\}$$

$$= A_0 + \sum_{n=1}^{\infty} \left[\left(\frac{A_n}{2} + \frac{B_n}{2i} \right) e^{in\pi x/a} + \left(\frac{A_n}{2} - \frac{B_n}{2i} \right) e^{-in\pi x/a} \right]$$

$$= \sum_{n=-\infty}^{\infty} C_n e^{in\pi x/a} \tag{17}$$

where

$$\left. \begin{aligned} C_n &= \frac{A_n}{2} + \frac{B_n}{2i} \\ C_{-n} &= \frac{A_n}{2} - \frac{B_n}{2i} \end{aligned} \right\} \quad (n > 0) \tag{18}$$

$$C_0 = A_0$$

so that, if we find all of the coefficients C_n, we can, if we choose, use those coefficients to find the A_n and the B_n. Notice that the C_n are defined for negative n as well as positive n, so the series in Eq. (17) runs from $-\infty$ to $+\infty$, instead of from 0 to ∞.

Now it is a simple matter to find a specific coefficient C_m. We multiply both sides of Eq. (17) by $e^{-im\pi x/a}$, and we integrate both sides on x from $-a$ to $+a$:

$$\int_{-a}^{a} \psi(x) e^{-im\pi x/a} \, dx = \int_{-a}^{a} \left[\sum_{n=-\infty}^{\infty} e^{-im\pi x/a} C_n e^{in\pi x/a} \right] dx \tag{19}$$

We assume that it is legitimate to integrate the series term by term, obtaining a series of integrals of the form $\int_{-a}^{a} e^{-im\pi x/a} C_n e^{in\pi x/a} \, dx$. Remember that

m is the *specific* integer we chose, while n varies, taking on all integer values. But if $n \neq m$, the integral is zero. If $n = m$, C_n becomes C_m, and the integral becomes simply

$$\int_{-a}^{a} C_m \, dx = 2aC_m$$

Equation (19) then becomes

$$C_m = \frac{1}{2a} \int_{-a}^{a} \psi(x) e^{-im\pi x/a} \, dx \tag{20}$$

and if the right-hand side is integrable [as it will be with any $\psi(x)$ in which we might be interested], we have found the coefficient we set out to find.

EXAMPLE PROBLEM 1.   Compute the coefficients in the Fourier series for the function

$$\begin{array}{ll} \psi(x) = 1 & (-b < x < +b) \\ \psi(x) = 0 & (b < |x| < a) \\ \psi(x + 2a) = \psi(x) & (\text{for all } x) \end{array}$$

Solution: We simply substitute the above $\psi(x)$ into Eq. (20), to obtain

$$C_m = \frac{1}{2a} \int_{-b}^{b} e^{-im\pi x/a} \, dx$$

$$= \frac{1}{2a} \left[\frac{e^{-im\pi x/a}}{-im\pi/a} \right]_{-b}^{+b}$$

$$= \frac{1}{2im\pi} (e^{im\pi b/a} - e^{-im\pi b/a})$$

$$= \frac{1}{m\pi} \sin \frac{m\pi b}{a} \qquad (m \neq 0)$$

and

$$C_0 = \frac{1}{2a} \int_{-b}^{b} dx = \frac{b}{a}$$

We may write the resulting Fourier series in real form by solving Eqs. (18) for A_n and B_n. We find that

$$\left. \begin{array}{l} A_n = C_n + C_{-n} = \dfrac{2}{n\pi} \sin \dfrac{n\pi b}{a} \\[2mm] B_n = i(C_n - C_{-n}) = 0 \end{array} \right\} \quad (n \neq 0)$$

$$A_0 = C_0 = \frac{b}{a}$$

Therefore

$$\psi(x) = \frac{b}{a} + \frac{2}{\pi}\sin\frac{\pi b}{a}\cos\frac{\pi x}{a} + \frac{2}{2\pi}\sin\frac{2\pi b}{a}\cos\frac{2\pi x}{a} + \cdots$$

$$= \frac{b}{a} + \sum_{n=1}^{\infty}\frac{2}{n\pi}\sin\frac{n\pi b}{a}\cos\frac{n\pi x}{a}$$

To be more specific, let $b/a = \frac{1}{4}$. Then

$$\psi(x) = \frac{1}{4} + \frac{\sqrt{2}}{\pi}\cos\frac{\pi x}{a} + \frac{2}{2\pi}\cos\frac{2\pi x}{a} + \frac{\sqrt{2}}{3\pi}\cos\frac{3\pi x}{a}$$

$$+ 0 \cdot \cos\frac{4\pi x}{a} - \frac{\sqrt{2}}{5\pi}\cos\frac{5\pi x}{a} - \frac{2}{6\pi}\cos\frac{6\pi x}{a} - \frac{\sqrt{2}}{7\pi}\cos\frac{7\pi x}{a} + \cdots$$

A graph of the exact $\psi(x)$ is shown in Fig. 8, together with the curve obtained from the first six nonzero terms of the Fourier series, as shown above.

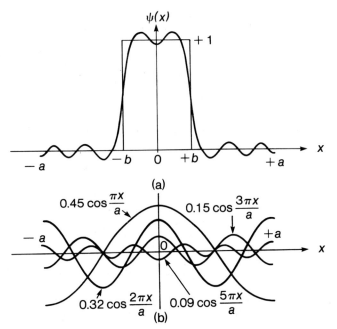

Fig. 8. *The function $\psi(x)$ of Example Problem 1, for $b/a = 1/4$.* (a) *The sum of the first six nonzero terms in the Fourier series for* $\psi(x)$. (b) *A few of the individual terms in this series.*

The Fourier Integral. The coefficients found by the above procedure enable one to analyze the behavior of any *periodic* wave, of any shape. For example, for a diffraction experiment, one could Fourier–analyze the incident wave, find the amplitudes at any point produced by the diffraction of each of the component sine waves, and add the amplitudes with "weights" determined by Eqs. (18) and (20). The resulting diffraction pattern will always be in agreement with experiment.

But in the most general case, a wave is not periodic. For example, the wave packet of Fig. 7 does not fit the definition (15) for any value of a. We could say that the period $2a$ is infinite in this case, and by letting $a \to \infty$ we may be able to adapt the previous results to cover this situation. Of course, if a goes to infinity while the wave packet remains of finite length, the function $\psi(x)$ is zero over an infinite range and nonzero over a finite range, and you might envision some difficulty in arranging the interference of waves to produce this result. But we saw in our example problem that we could construct a Fourier series for a function which is nonzero for only a small fraction of its period, and there was nothing to prevent us from making that fraction as small as we wished. The smaller the fraction b/a, the more terms we must use to make the series fit the function well, but by using enough terms we can always make the Fourier series equal zero, to any degree of accuracy desired, over any specified range.

We begin our treatment of the nonperiodic function by defining a quantity $k_n \equiv n\pi/a$. The difference between successive values of k_n is $\delta k \equiv \pi/a$. We can now write Eq. (20) as

$$C_n = \frac{\delta k}{2\pi} \int_{-a}^{a} \psi(x') e^{-ik_n x'} \, dx' \tag{21}$$

where x' has been substituted for x as the "dummy variable" in the integral, to avoid confusion when we insert this expression into Eq. (17). We may now write Eq. (17) as

$$\psi(x) = \sum_{n=-\infty}^{\infty} C_n e^{ik_n x}$$

which becomes, when we substitute Eq. (21) for C_n,

$$\psi(x) = \frac{1}{2\pi} \sum_{n=-\infty}^{\infty} \left[e^{ik_n x} \, \delta k \int_{-a}^{a} \psi(x') e^{-ik_n x'} \, dx' \right]$$

As $a \to \infty$, the difference $\delta k = \pi/a \to 0$ and the sum becomes an integral. The result is the Fourier integral theorem:

$$\psi(x) = \frac{1}{2\pi} \int_{-\infty}^{\infty} e^{ikx} \, dk \int_{-\infty}^{\infty} \psi(x') e^{-ikx'} \, dx' \tag{22}$$

We may define

$$\phi(k) = \frac{1}{(2\pi)^{1/2}} \int_{-\infty}^{\infty} \psi(x) e^{-ikx} \, dx \tag{23}$$

so that

$$\psi(x) = \frac{1}{(2\pi)^{1/2}} \int_{-\infty}^{\infty} \phi(k)e^{ikx}\, dk \tag{24}$$

and we have a certain symmetry between $\psi(x)$ and $\phi(k)$. The function $\phi(k)$ is a continuous amplitude function which has taken the place of the discrete amplitudes C_n; $\phi(k)$ is called the *Fourier transform* of $\psi(x)$. The only difference between this result and our previous one for a periodic function is that $\psi(x)$ now must be built up from waves with a *continuous range* of wavelengths instead of the discrete set $2a, 2a/2, 2a/3, \ldots$, and therefore we need the *continuous amplitude function* $\phi(k)$ to determine the amplitudes of these component waves.

Group Velocity.   Bear in mind that we have not yet described a *wave* because we have no time dependence in our equations. In general, a wave traveling in the positive x direction with speed v is described by a function of $x - vt$ (see Fig. 9). This suggests that we simply substitute $x - vt$ for x in Eqs. (23) and (24). But this must be done with care because v is not the same for all of the waves that are superposed, via Eq. (24), to form ψ. By the de Broglie theory, these component waves, having different wavelengths, must correspond to different momenta, and hence *must* have *different* velocities. In other words, v is a function of k. The wave packet therefore does not have a unique velocity, and it even changes shape as it moves, because of the changes in the phase relations among the different component waves, as they move relative to one another (see Fig. 7).

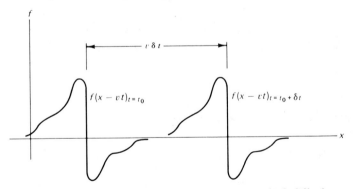

Fig. 9. *Plot of a function $f(x - vt)$ at two times which differ by an amount δt. If we increase x by an amount $v\,\delta t$ while increasing t by an amount δt, we leave the quantity $x - vt$ unchanged; hence the value of the function is unchanged. Therefore the curve showing the function at time $t_0 + \delta t$ is identical in shape to the curve showing the function at time t_0, but the former is displaced a distance $v\,\delta t$ from the latter. Since the disturbance represented by the function travels a distance $v\,\delta t$ in time δt, its speed is v.*

We therefore introduce a new symbol, v_p, for the velocity of each of the individual waves comprising $\psi(x, t)$. This velocity is called the *phase velocity*. The function ψ is not a function of $x - v_p t$ because there is no unique value of v_p for all the waves of which ψ is composed. We now rewrite Eq. (24) in terms of v_p as

$$\psi(x, t) = \frac{1}{(2\pi)^{1/2}} \int_{-\infty}^{\infty} \phi(k) e^{ik(x - v_p t)} dk$$

where v_p is, as we have indicated above, a function of k. We may then define $\omega \equiv k v_p$, so that we have

$$\psi(x, t) = \frac{1}{(2\pi)^{1/2}} \int_{-\infty}^{\infty} \phi(k) e^{i(kx - \omega t)} dk \tag{25}$$

When the component waves have very nearly the same values of k, the wave packet retains its shape well enough that one can observe the motion of the packet as a whole. Even under these conditions, however, the speed of the overall packet, called the *group velocity*, is not equal to the phase velocity. An example will illustrate the difference between v_p, and the group velocity v_g, and will suggest how we may find v_g in the general case.

Let $\psi(x, t)$ be the sum of two waves which differ very slightly in k value:

$$\psi(x, t) = \sin(kx - \omega t) + \sin[(k + \delta k)x - (\omega + \delta\omega)t]$$

Straightforward trigonometry shows that the above is equivalent to

$$\psi(x, t) = 2 \cos\left[\left(\frac{\delta k}{2}\right)x - \left(\frac{\delta\omega}{2}\right)t\right] \sin\left[\left(k + \frac{\delta k}{2}\right)x - \left(\omega + \frac{\delta\omega}{2}\right)t\right]$$

If $\delta k \ll k$ and $\delta\omega \ll \omega$, we have

$$\psi(x, t) \approx 2 \cos\left[\left(\frac{\delta k}{2}\right)x - \left(\frac{\delta\omega}{2}\right)t\right] \sin(kx - \omega t).$$

Since $\delta k \ll k$, the cosine factor varies much more slowly than the sine factor, and $\psi(x, t)$ is essentially a sine wave with slowly varying amplitude (see Fig. 10). The individual crests of this sine wave move with the phase velocity v_p, which is determined by the factor $\sin (kx = \omega t)$, so that $v_p = \omega/k$, in accordance with the definition of ω leading to Eq. (25). But the envelope of the wave is given by the cosine factor, and it is clear from inspection of this factor that it is a function of $x - v_g t$, provided that $v_g = \delta\omega/\delta k$. If v_p is different for waves of different k, then

$$\frac{\omega + \delta\omega}{k + \delta k} \neq \frac{\omega}{k}, \qquad \text{so that} \qquad \frac{\delta\omega}{\delta k} \neq \frac{\omega}{k}$$

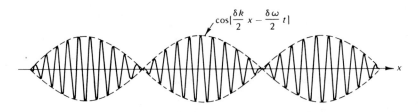

Fig. 10. *Sum of two sine waves, of equal amplitude, differing in frequency by* $\delta\omega$. *The sum oscillates between* $+2\cos[(\delta k/2)x - (\delta\omega/2)t]$ *and* $-2\cos[(\delta k/2)x - (\delta\omega/2)t]$, *shown by the broken lines. As time passes this envelope moves with a speed of* $\delta\omega/\delta k$, *while the crests within the envelope move with speed* ω/k.

The envelope then moves with a speed different from the speed of the individual crests. This sort of behavior can be observed in water waves; throw a rock into a lake and see for yourself.

The above discussion suggests that we define the group velocity of a wave packet to be

$$v_g = \left(\frac{d\omega}{dk}\right)_{k=k_0}$$

where k_0 is the average k value for waves in the packet. When the wave packet consists mostly of waves whose k values lie near k_0, the envelope of the packet moves with velocity v_g. The group velocity as defined above has a particularly satisfying application to the de Broglie wave. It is clear from comparison of the expression $\sin(kx - \omega t)$ with the standard form Eq. (12) that

$$\omega = 2\pi\nu \quad \text{and} \quad k = \frac{2\pi}{\lambda}$$

But for a de Broglie wave, $h\nu = E$ and $h/\lambda = p$. Therefore

$$\omega = \frac{2\pi E}{h} \quad \text{and} \quad k = \frac{2\pi p}{h}$$

so that

$$v_g = \left(\frac{d\omega}{dk}\right)_{k=k_0} = \left(\frac{dE}{dp}\right)_{p=p_0}$$

Thus we can express the group velocity of the de Broglie wave in terms of the momentum as follows:

$$E^2 = p^2c^2 + m^2c^4$$

$$2E\,dE = 2pc^2\,dp$$

$$v_g = \left(\frac{dE}{dp}\right)_{p=p_0} = \frac{p_0\,c^2}{E_0}$$

which is just the velocity of a particle of momentum p_0 and energy E_0, according to relativity theory. Therefore the group velocity of a wave packet is just the velocity which one observes for the associated particle.

Other interesting features of the behavior of a wave packet appear when we use Eq. (23) to calculate the Fourier transform of a specific function. Suppose that the wave packet of an electron is given, at $t = 0$, by the formula

$$\psi(x) = e^{-x^2/4\sigma^2}e^{ik_0x} \tag{26}$$

whose real part is the cosine function with Gaussian amplitude plotted in Fig. 7 (with $\sigma = b/\sqrt{2}$). The fact that $\psi(x)$ is complex does not disqualify it as an allowed function to describe an electron's motion because $\psi(x)$ itself cannot be observed directly. We recall from the discussion at the end of Section 4.2 that the intensity of a wave at any point is proportional to the square of the wave amplitude at that point. When the wave is complex, as in this case, we simply require that the intensity, the *observable* quantity, be the absolute square, $|\psi(x)|^2$, a quantity that is always real. In this case,

$$|\psi(x)|^2 \equiv \psi^*(x)\psi(x) = e^{-x^2/4\sigma^2}e^{-ik_0x}\,e^{-x^2/4\sigma^2}e^{+ik_0x} = e^{-x^2/2\sigma^2} \tag{27}$$

If σ were infinite, then $\psi(x)$ would be simply e^{ik_0x}. It would have a unique wavelength equal to $2\pi/k_0$, and the Fourier transform of $\psi(x)$ would be zero everywhere except at $k = k_0$. But the presence of the Gaussian amplitude factor introduces other wavelengths, or k values, whose amplitude distribution is given by the Fourier transform (Eq. 23):

$$\phi(k) = \frac{1}{\sqrt{2\pi}}\int_{-\infty}^{\infty} e^{-x^2/4\sigma^2}e^{i(k_0 - k)x}dx$$

We can evaluate this integral most easily by "completing the square" in the exponent, obtaining the form

$$\phi(k) = \frac{1}{\sqrt{2\pi}}\int_{-\infty}^{\infty} e^{-[(x/2\sigma) - i(k_0 - k)\sigma]^2}e^{-(k_0 - k)^2\sigma^2}dx$$

The second exponential, being independent of x, can be removed from the integral. Then we have

$$\phi(k) = \frac{2\sigma}{\sqrt{2\pi}}e^{-(k_0 - k)^2\sigma^2}\int_{-\infty}^{\infty} e^{-u^2}du$$

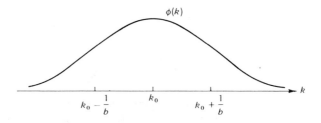

Fig. 11. *The function* $\phi(k)$*, the Fourier transform of the function* $\psi(x)$ *shown in Fig. 7, plotted near* $k = k_0$.

where $u = (x/2\sigma) - i(k_0 = k)\sigma$, and $du = dx/2\sigma$. The value of the last integral is $\sqrt{\pi}$, so the final result is

$$\phi(k) = \sigma\sqrt{2}\ e^{-(k_0 - k)^2\sigma^2}$$

Thus $\phi(k)$ is also a Gaussian function, peaked at $k = k_0$. A graph of $\phi(k)$ vs. k is shown in Fig. 11, where $b = \sigma\sqrt{2}$ is the same parameter used in Fig. 7. Notice that, if b increased, $\phi(k)$ would become narrower, while $\psi(x)$ would spread out. If the position of the particle is to be more precisely defined, b must be smaller, but that will make $\phi(k)$ broader, which means that the momentum of the particle will be less precisely defined. (Remember that $p = h/\lambda = \hbar k$.)

To be more quantitative, we note that the wave intensity [Eq. (27)] is of the same form as the standard Gaussian (normal) probability distribution given in Appendix A, Eq. (4), where σ is the standard deviation. This is appropriate because we recall from Section 4.2 that the intensity $|\psi(x)|^2$ does determine the probability of finding the particle. More precisely, $|\psi(x)|^2$ is a *probability density,* which can be integrated between x_1 and x_2 to find the probability of finding the particle between those values of x (just as, in Appendix A, the normal probability distribution is integrated to find the probability that the variable lies between two limits). A measurement of the position of the particle described by this wave will usually (about two-thirds of the time) yield a result within one standard deviation of the mean value of x, which in this case is $x = 0$. It is thus customary to say that the uncertainty in the position of the particle is $\delta x = \sigma$.

Now suppose that one measures the wavelength of the wave instead of the position of the particle. The wave is a superposition of various component waves whose *amplitude* distribution is $\phi(k)$. The *intensity* distribution of these component waves is, following the previous arguments,

$$|\phi(k)|^2 = \frac{\sigma^2}{2}\ e^{-2(k_0 - k)^2\sigma^2}$$

and the probability that the measurement will yield a value of k between k_1 and k_2 is found by integrating this intensity between $k = k_1$ and $k = k_2$. In the form of the normal probability distribution, the intensity would be written

$$|\phi(k)|^2 = \frac{\sigma^2}{2}\, e^{-(k_0-k)^2/2(\delta k)^2}$$

where δk is the standard deviation in k. We see by comparing the two expressions that $\delta k = 1/2\sigma$.

We find that the product of the uncertainties δx and δk is $\frac{1}{2}$, independent of σ. Since $p = \hbar k$, $\delta p = \hbar(\delta k)$, and we have

$$\boxed{(\delta p)\cdot(\delta x) = \hbar(\delta k)\cdot(\delta x) = \hbar/2} \tag{28}$$

We have here an important relation between the position uncertainty and the momentum uncertainty, whose significance goes beyond what you might infer from the nature of this special case. The generalization of Eq. (28) by Heisenberg, and the experimental implications of this generalization, will be discussed in Section 4.4. For now, we summarize the features of the relationship between $\psi(x)$ and $\phi(k)$ as follows:

1. In order to have wavelike properties, an electron must be spread out over some nonzero volume; that is, it must simultaneously "occupy" a range of positions, rather than only one definite point.
2. In order to have particlelike properties, i.e., in order to be localized (while retaining its wavelike properties), an electron must simultaneously possess a range of values of momentum, rather than one definite momentum.
3. The range of positions is inversely proportional to the range of momentum values.

Nothing in this discussion prevents us from seeing a particle that is temporarily localized at one specific point; however, the momentum of such a particle would be completely uncertain. We shall now see how to describe such a situation mathematically.

The Delta Function.   As mentioned above, the Fourier transform of an infinitely long wave of the form $e^{ik_0 x}$ is zero everywhere except at $k = k_0$. Conversely, a "wave" could be zero everywhere except at a single point in space, and its Fourier transform would extend over an infinite range of values of k. These are simply special cases of Eq. (26), in the limits $\sigma \to \infty$ and $\sigma \to 0$, respectively.

It is helpful to have a special symbol for a wave that is localized at one point. Such a wave is described by the *Dirac delta function*, $\delta(x)$, whose properties are

$$\delta(x) = 0 \quad \text{if } x \neq 0 \; ; \quad \delta(x) = \infty \quad \text{if } x = 0$$

$$\int_{x_1}^{x_2} \delta(x)dx = 1 \quad \text{and} \quad \int_{x_1}^{x_2} f(x)\delta(x)\,dx = f(0) \; , \quad \text{if } x_1 < 0 < x_2$$

Thus the delta function is the limit of the Gaussian function as the standard deviation σ goes to zero (Appendix A, Eq. (4), with $m = 0$) because the Gaussian has a factor $1/\sigma$ that causes the function to go to infinity at $x = 0$ as $\sigma \to 0$, while the *area* under the Gaussian curve remains the same for all values of σ.

To describe a function that is localized at, say, $x = x_0$, one simply writes a delta function in such a way that its *argument* is zero at $x = x_0$; that is, the function is $\delta(x - x_0)$. Similarly, if $\phi(k) = \delta(k - k_0)$, then $\phi(k)$ is zero everywhere except at $k = k_0$, and the Fourier transform of this $\phi(k)$ is a pure, infinitely long sinusoidal wave with $k = k_0$.

The symbol δ, as used here to describe a *function*, is not to be confused with the symbol used in preceding subsections to describe an *uncertainty*; $\delta(x)$ is a *function* of x, while δx is the *uncertainty* in x.

4.4 THE UNCERTAINTY PRINCIPLE

In 1927, Heisenberg proposed that Eq. (28) sets an ultimate limit on the accuracy of a set of physical measurements. The Heisenberg uncertainty principle states that, in any simultaneous measurement of the position and the momentum of a particle,

$$(\delta x)(\delta p_x) \geqslant \hbar/2 \tag{29}$$

where δx is the uncertainty[15] in the measurement of a particular coordinate x, and δp_x is the uncertainty in the measurement of the corresponding momentum.

Since 1927, there have been many attempts to devise methods by which this principle might be violated, but none has succeeded. Let us examine a simple sort of measurement, to see how these uncertainties actually come about. Imagine that we are able to use a microscope to measure the position of an electron, and that we can detect a single photon scattered from the electron (see Fig. 12). After striking the electron, the photon passes through a lens and produces a spot on a photographic plate. We wish to deduce the position of the electron by observing the position of the spot on the plate. We can certainly do this, as far as geometrical optics is concerned, as long as the plane of the incident photons is in focus at the photographic plate. But error is introduced by diffraction of light as it passes through the lens aperture. Light of wavelength λ may be diffracted through an angle as large as $\delta\theta \approx \lambda/D$, where D is the lens diameter.[16] Therefore, if the electron is at a distance l from the lens, its x coordinate is uncertain by approximately $l\lambda/D$. However, we can, in principle, determine the position of the electron to any desired degree of accuracy, simply by increasing D, or by decreasing λ or l.

---

[15] To be more precise, we may define the "uncertainty" statistically. We imagine the same experiment to be repeated many times, with identical initial conditions; the uncertainty is then the root-mean-square deviation of the individual results from the mean. In other words, it is the standard deviation.

[16] See F. Jenkins and H. White, "Fundamentals of Optics," 3rd ed. 302, McGraw-Hill, New York, 1957.

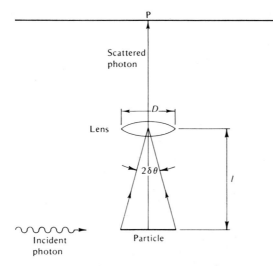

Fig. 12. *Use of a microscope to detect the position of a particle. Light from any part of the extended area (marked "particle") may reach point P as a result of diffraction.*

Having measured the electron's position, we could repeat the measurement at a later time, and then divide the displacement by the time in order to find the electron's velocity, and hence its momentum. It would seem that we could in this way attain any desired degree of precision in the momentum as well as in the position of the electron. But what we set out to do was to determine the momentum and the position *as they were at the instant we began to observe*, whereas the momentum as found above is the momentum which the electron has *after* the first photon strikes it. In order to answer our original question, we must determine how much momentum was transferred from the first photon to the electron, and subtract that from the measured electron momentum. We can do this if we know the final momentum of this *photon*, whose original momentum we may assume to be known. But the *x* component of this final momentum is uncertain because we do not know the direction in which the photon went immediately after it was scattered; it would have been "focused" at point P as long as it passed through any part of the lens. Therefore, if the photon has frequency v, its final component of momentum could lie anywhere between $(hv \sin \phi)/c$ and $(-hv \sin \phi)/c$, where $2\phi \approx D/l$ is the angle subtended by the lens diameter at the electron. Our determination of the momentum of the electron is thus uncertain by the amount $\delta p_x = (hv \sin \phi)/c$. If ϕ is small, $\sin \phi \approx \phi \approx D/2l$, so that

$$\delta p_x \approx \frac{hvD}{2lc}$$

In order to increase our precision in the measurement of p_x, we may make D smaller, but we have seen that this decrease in D decreases the precision in the measurement of x. We are left with a *fundamental* uncertainty in position and a *fundamental* uncertainty in momentum, whose product is, roughly,

$$(\delta x)(\delta p_x) \approx \left(\frac{l\lambda}{D}\right)\left(\frac{h\nu D}{2\,lc}\right) = \frac{h}{2}$$

a quantity which is independent of the size of the lens, the frequency of the light, or the distance l.

There is no restriction on a simultaneous measurement of x and p_y or p_z; the uncertainty relation applies only to "conjugate" coordinates: x and p_x, y and p_y, z and p_z. One can show that it also applies to other pairs of conjugate variables, such as angular position and angular momentum with respect to a given axis.

We can imagine many other experiments by which one attempts to measure both the position and the conjugate momentum of a "particle"; in each case the accuracy of the position measurement is inversely proportional to the accuracy of the momentum measurement. Even though the experiments which illustrate these fundamental error limits are "thought experiments"— experiments which we imagine but do not actually perform—we are confident of the results, because these results are deduced from effects which have been observed experimentally. In the example we have just discussed, the result follows directly from well-known laws of optics, combined with the equally well-verified de Broglie relation.

The de Broglie relation, connecting the particle with a wave, suggests another way to view the uncertainty principle. Consider a wave of the form

$$\psi(x) = \sin \frac{2\pi x}{\lambda} \qquad\qquad\qquad 0 < x < N\lambda$$

$$\psi(x) = 0 \qquad\qquad\qquad\qquad x < 0 \quad \text{or} \quad x > N\lambda$$

If this wave extended over all space with the same form that it has in the region between $x = 0$ and $x = N\lambda$, it would have a definite wavelength equal to λ. But what happens if we try to measure the wavelength of this limited wave packet, which is only N "wavelengths" long? Figure 13 shows that only N slits can be effective in making this measurement. Thus it is just as if the grating only *had N* slits, and we know from elementary physics[17] that such a grating has a resolving power given by $\lambda/\delta\lambda = N$, where $\delta\lambda$ is the uncertainty in the measurement of the wavelength λ. If the particle described by this wave packet has momentum $p_x = h/\lambda$, then p_x has an uncertainty of magnitude $h\delta\lambda/\lambda^2$ (because the differential of $1/\lambda$ is $-d\lambda/\lambda^2$). This result can also be obtained from the Fourier transform of such a wave (Problem 16). On the other hand, the center of this wave

---

[17] See, for example, D. Halliday and R. Resnick, "Physics," 3rd ed., 1054. John Wiley, New York, 1978.

packet is at $x = N\lambda/2$, so we can say that the particle described by the packet is at $x = N\lambda/2 \pm \delta x$, where δx can be estimated[18] to be $N\lambda/2$. Thus the uncertainty product is, with the substitution $N = \lambda/\delta\lambda$,

$$(\delta p_x)(\delta x) \simeq (h\delta\lambda/\lambda^2)(N\lambda/2) = (h\delta\lambda/\lambda^2)(\lambda^2/2\delta\lambda) = h/2$$

a result consistent with (29).

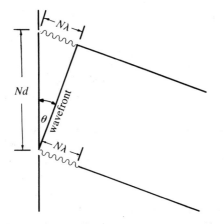

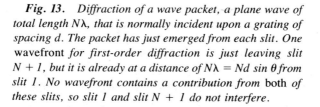

Fig. 13. *Diffraction of a wave packet, a plane wave of total length Nλ, that is normally incident upon a grating of spacing d. The packet has just emerged from each slit. One wavefront for first-order diffraction is just leaving slit N + 1, but it is already at a distance of Nλ = Nd sin θ from slit 1. No wavefront contains a contribution from both of these slits, so slit 1 and slit N + 1 do not interfere.*

We shall see other examples of the working of the uncertainty principle in subsequent chapters. For example, in Section 5.7 we see that a harmonic oscillator cannot have zero energy, because the oscillating mass would then have $\delta x = 0$ and $p_x = 0$ simultaneously. In Chapter 9, we find that a molecule, for the same reason, cannot have zero *internal* energy. Problem 11 of that chapter shows how to deduce the values of the minimum internal energy (the so-called *zero-point* energy) of hydrogen and deuterium molecules by comparing their dissociation energies. So the principle is well established in the "real world."

Complementarity. When we discover an ultimate limitation to our powers of observation, the very existence of that which we wanted to observe is called into question. We encountered this situation before, in Chapter 2. There we argued that the concept of absolute time is useless, because absolute

---

[18] This estimate is obviously somewhat larger than the root-mean-square deviation, but it is sufficient for an order-of-magnitude calculation.

time is inaccessible to direct observation, and its existence is not required in the development of a consistent theory of motion.

The situation with position and momentum is somewhat different, for both of these can be observed, to any desired degree of accuracy. But here again there is something one can *not* observe, and that is the exact *trajectory* of a particle. In order to know an exact trajectory, one must know the position of a particle at *all* times; to know that, one must also know the momentum,[19] in violation of the uncertainty principle.

It is easy to recognize that quantization prevents us from determining an exact trajectory. We always disturb the trajectory by our attempt to observe it, and, because of quantization, we cannot make the disturbance arbitrarily small. However, we might imagine that a particle still *has* a definite trajectory, even though we are not able to determine what that trajectory is. This view leads us into a serious dilemma, as the following "thought experiment" shows.

Consider the diffraction of electrons by two parallel slits (Fig. 14). Again, as in our previous thought experiments, we are in no doubt about the result, for we have observed electron diffraction in many other ways. Just as in experiments with interference of light, a classical interference pattern should form on the screen. This pattern can be built up even though the electrons are observed individually.[20] For example, if one electron per second strikes the screen, and each electron is detected in one of a series of detectors A, B, C, D, etc., one could, in time, determine that interference had occurred, simply by counting the number of electrons detected by the various detectors.

The presence of an interference pattern makes it clear that somehow *both* slits are involved, and since the electrons are detected one at a time, one electron cannot interfere with another. *Each* electron is in some way influenced by both slits. We can further demonstrate this fact by closing one slit; the pattern now changes, and some of the detectors count *more* electrons per unit time than they did before. This would be incomprehensible if the electrons were classical particles which must go through one slit or the other.

We cannot avoid the conclusion that a single electron, as a wave, goes through both slits; yet, as a particle, it can be detected only as an indivisible entity—we never see a fractional electronic charge. It might seem that the wave nature of the electron is incompatible with the particle nature, at least in this case. But what happens if we try to *observe* the particle nature, by detecting the electron's passage through one slit or the other? We can do this by illuminating the slits. As an electron passes through a slit, a photon may

---

[19] We can only *observe* the position at a discrete set of points. The position at in-between points must be *deduced* from the momentum. Therefore to know the complete trajectory requires knowledge of both momentum and position at certain points.

[20] An interference pattern can also be built up from observations of single *photons*. See S. Parker, "Double-Slit Experiment with Single Photons," *Am. J. Physics.* **40**, 1003 (1972).

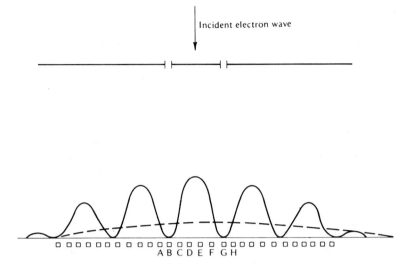

Fig. 14. *Intensity distribution that might be seen in diffraction of electrons by narrow slits. Solid line is intensity distribution seen when both slits are open; dashed line shows the pattern when only the right-hand slit is open. Boxes indicate electron detectors which could detect the patterns by counting individual electrons. Detectors B and G, near nodes, count more electrons when one slit is closed than when both are open.*

be scattered from it, and we can deduce where the photon came from, as in the microscope "experiment."

Imagine that we succeed in doing this; we will then find that we have destroyed the interference pattern! Each photon which scatters from an electron causes the electron to recoil, and the amount of recoil is sufficiently uncertain to prevent our determining where the electron would have gone if the photon had not struck it. The end result is a fairly uniform distribution of counts among our detectors A, B, C, D, E, ... If we use photons of longer wavelength (and hence smaller momentum) in order to reduce the recoil, we find a difficulty in locating the electrons as they pass through the slits. As soon as the photons have such a long wavelength that the interference pattern is preserved, the diffraction of the photons themselves becomes so great that we can no longer use them to determine which slit each electron passes through.

The uncertainty principle ensures that the above argument is quantitatively correct; it thus guarantees that we do not observe a contradiction between the wave and particle properties. When we observe the wave nature of electrons, as in this two-slit experiment, we cannot at the same time observe their particle nature. The same arguments would apply to a two-slit experiment with light waves. Bohr described this situation by saying that the wave and particle aspects of nature are *complementary*. Each aspect exists only as a *potentiality*; an experiment is required to convert the potentiality into something which we can observe. The act of observing the one aspect then suppresses the other aspect.

It is therefore useless to attempt to visualize an electron trajectory in a situation of this sort. We cannot "explain" (picture) what is going on by drawing on our ordinary experience, because these situations lie totally outside that experience. We must simply observe what happens and then draw up a set of consistent rules to describe the observed behavior. These rules, unlike the rules of classical physics, do not allow one to predict *exactly* what will happen to a given electron in a given experimental situation. When the electrons are detected, the number of electrons seen by each detector is proportional to the intensity (or amplitude squared) of the electron wave at that detector, but this number is built up by a series of unpredictable interactions of individual electrons with each detector. These interactions are unpredictable in the sense that the wave intensity only determines the *probability* that an individual electron will be detected by a detector. However, once an electron *is* seen by one detector, the probability of its being detected by another detector is zero. Thus the occurrence of one of these unpredictable interactions must change the electron wave; this is also what happens when we destroy the interference pattern by our detection of the electrons as they go through the slits.

Hidden Variables. Although the view outlined above is the orthodox view of physicists at present, it is philosophically unsatisfactory in some respects, and many distinguished physicists, beginning with Einstein, have voiced objections to it. All are agreed that there is a wave nature to matter, and that the de Broglie wave model accurately gives the probability of various experimental results. The quarrel is over the description of the process of measurement, and the unpredictable, mysterious nature of the "collapse" of a wave when a particle is observed by a detector.

The most recent discussions have centered on the possible existence of so-called "hidden variables." To understand the basis of the discussion, consider the flipping of a coin. The result is unpredictable—there is equal probability for the coin to come up heads or tails—because we do not know the *exact* initial state of the coin. Presumably if we knew the precise position, velocity, angular momentum, etc., of the coin when it was flipped, we could predict whether the coin would land heads or tails. But suppose we do the analogous thing with electrons; we send an electron beam through a narrow slit, and we detect whether each electron is diffracted to the left or to the right (as in Fig. 14). Again the result is unpredictable for a given electron; it goes to the left and to the right with equal probability. But in this case we cannot find an "exact" initial state which will remove this uncertainty. If we send two electrons through the slit, even though each of them is described by an *identical wave*, one may be detected by a detector on the right while the other is detected by a detector on the left.

Some investigators have considered the possibility that the waves representing these two electrons *do* differ in some way *before* the electrons are detected by different detectors—that there is some hidden variable which has one value for the electron which is diffracted toward the left, and another value for the electron which is diffracted toward the right.[21] Then, because the value of this variable is randomly distributed among electrons with identical waves, the usual laws of probability determine the results of experiments with large numbers of electrons. The introduction of the hidden variable removes the mystery from the collapse of the wavefunction during the measurement process; it becomes a change which is predetermined by the value of the hidden variable.

In Chapter 8, we shall take up this matter again, and we shall see that there are experiments in the quantum domain for which standard quantum theory and "hidden variable" theory predict different results. But our present task is to develop a consistent dynamical theory *without* hidden variables—a theory in which one cannot in principle make the exact predictions which one could make in classical dynamics. This task turns out to be less difficult than it sounds. We must simply alter our idea of a complete specification of a dynamical state, and say that we will settle for a knowledge of the wavefunction. In the next chapter we shall see how to compute some of these wavefunctions.

PROBLEMS

1. The negative muon is a particle with a charge equal to the electron's charge and a mass of about 207 electron masses. Suppose that a negative muon is captured by an aluminum atom ($Z = 13$) and it falls down through its various Bohr orbits. What would be the energy of the X-ray emitted when the muon goes from the $n = 2$ orbit to the $n = 1$ orbit? Why should you *not* use a screening factor like the s of Eq. (1) in computing the energy of the muon in each of these orbits?

2. Deduce from Moseley's data [Table 1 and Eq. (1)] the wavelength of the K_α line of aluminum ($Z = 13$).

3. What transition produces the K_β lines observed by Moseley? For those lines, find the expression analogous to the expression $(v_\alpha/\frac{3}{4}v_0)^{1/2}$ which Moseley computed for the K_α lines, and evaluate this expression for

[21] See B. d'Espagnat, "The Quantum Theory and Reality," *Scientific American*, **241**, No. 5, 158-170 (Nov., 1979).

several of the K_β wavelengths reported by Moseley. Does the value of this expression vary as it should from one element to the next?

4. Sulfur emits X-rays of 5.07 and 5.37 Å when bombarded by electrons. An "absorption edge" for X-rays in sulfur is at 5.02 Å. This means that X-rays of slightly shorter wavelength than 5.02 Å are absorbed much more strongly than X-rays of slightly longer wavelength. The reason for this is that X-rays of wavelength shorter than 5.02 Å have enough energy to eject electrons from the K shell of sulfur, while longer wavelength X-rays cannot be absorbed in this way. On the basis of this absorption edge and the X-ray emission data given above, find the approximate energy levels of the K and L shells of sulfur, and use your result to predict the wavelength at which another absorption edge (the L absorption edge) should appear. You may check your answer by referring to the "Handbook of Chemistry and Physics."

5. Compute the percentage change in wavelength and the change in the energy of a photon when (a) yellow light ($\lambda = 5000$ Å), and (b) a 100-MeV X-ray, is Compton scattered through a 90° angle.

6. A photon whose initial energy is 50,000 eV strikes an electron whose kinetic energy is 8×10^5 eV. The photon and electron were traveling in exactly opposite directions before the "collision." Use the conservation laws directly to compute the final photon wavelength if the photon is scattered through 180°. Verify your answer by calculating the same thing in another way: Use the Doppler shift formula (Chapter 2, Problem 14) to find the original photon's frequency in the frame of reference in which the electron is at rest; then use the Compton formula to find the frequency of the scattered photon in this frame; finally, use the Doppler formula again to find the frequency of the scattered photon in the original reference frame.

7. It has been suggested that high energy photons might be found in cosmic radiation, as a result of the "inverse Compton effect": A photon of visible light (energy ≈ 2 eV) *gains* energy by scattering from a high energy *proton*. Compute the maximum final photon energy which can result from such an interaction when the proton's momentum is (a) 10^9 eV/c (b) 10^{11} eV/c. *Hint:* The *second* method of Problem 6 is easier to apply in this case.

8. For the Compton scattering event of Problem 5(b), find the angle at which the electron recoils.

9. If a photon of energy E_0 is Compton scattered by an electron at rest, show that the final photon energy is $E_f = E_0[1 + (E_0/mc^2)(1 - \cos\theta)]^{-1}$.

10. A photon whose energy is $mc^2/2$ is Compton scattered at $90°$ from an electron (mass m) at rest. By direct application of energy and momentum conservation, find the final photon energy, the angle at which the electron recoils, and the kinetic energy of the recoiling electron. Verify that the change in photon wavelength agrees with that given by Eq. (8).

11. Show that the wavelength of a particle of rest energy E_0 and kinetic energy E_k is given by

$$\lambda = hc(E_k^2 + 2E_0E_k)^{-1/2}$$

12. Compute the energy of a photon whose wavelength is the same as that of a 50-eV electron.

13. Compute the speed and the de Broglie wavelength of
    (a) an electron whose kinetic energy is 5 eV,
    (b) an electron whose kinetic energy is 5 MeV,
    (c) a proton whose kinetic energy is 5 MeV, and
    (d) a baseball (mass $\approx$ 150 gm) whose kinetic energy is 5 MeV.

14. Show that the scattering peak at $50°$, observed by Davisson and Germer for 54-eV electrons incident on nickel, is consistent with Eq. (14).

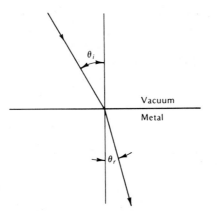

15. If one does not attempt to localize the particles, the wave picture gives the same results as the particle picture. For example, an electron beam is "refracted" when it encounters a sudden change in potential energy. A beam of electrons of kinetic energy E crosses a boundary where the potential energy suddenly decreases by an amount V (as might happen when the electron enters a metal), so that the kinetic energy becomes $E + V$. The electrons, as particles, receive an impulse perpendicular to

the surface, as a result of the sudden change of potential energy, so the component of velocity perpendicular to the surface increases while the other component is unchanged, and the " ray " is bent toward the normal. On the other hand, if the electron is a de Broglie wave, the wavelength decreases when the momentum increases, so the ray is again bent toward the normal. Show that either picture gives the same result, by deriving the relationship between θ_r and θ_i in both ways.

16. (a) Find the Fourier transform $\phi(k)$ of the function

$$\psi(x) = \sin (2\pi x/a) \qquad\qquad 0 < x < na$$
$$\psi(x) = 0 \qquad\qquad\qquad\quad x < 0, \quad x > na$$

(b) Show that $|\phi(k)|^2$ is peaked at $k = \pm (2\pi/a)$, and that the width of each peak is proportional to $1/n$. Notice the connection between this result and the discussion of Section 4.4 concerning the resolving power of a diffraction grating.

17. (a) Use the relativistic energy-momentum relation to show that, for a de Broglie wave, $v_g v_p = c^2$. Thus, since $v_g < c$, it follows that $v_p > c$.

(b) The phase velocity v_p is not directly observable, because one cannot measure the frequency of the de Broglie wave. The frequency is given by the relation $E = \hbar\omega$, where E is the total energy, but the total energy has an arbitrary zero level, because only changes in total energy are observed in any experiment. Therefore one could define ω in terms of the total energy *minus* the rest energy, so that

$$\hbar\omega = E - mc^2 \approx p^2/2m$$

for a particle of low speed. Show that this definition leads to $v_p = p/2m$ and $v_g = p/m$.

18. Suppose in a two-slit experiment on diffraction of light, we place photon indicators by the slits to tell us through which slit each photon passes on its way to the screen. If these indicators (free electrons, for example) are to be effective, their recoil must tell us the y coordinate of the photon at least to accuracy $a/2$, where a is the distance between slits. Show that when this happens, the uncertainty principle ensures that the interference pattern is destroyed. (The angle θ between interference maxima and minima is equal to $\lambda/2a$, if the photon wavelength is λ).

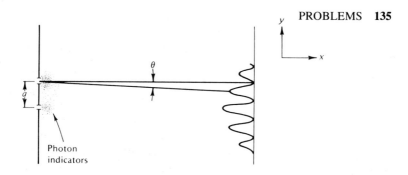

19. Compute the Fourier transform of the function

$$\psi(x) = 1 \qquad (-b < x < b)$$
$$\psi(x) = 0 \qquad (|x| > b)$$

Show that the series found in Example Problem 1 may be obtained from your Fourier transform $\phi(k)$, by making each coefficient equal to

$$\frac{1}{(2\pi)^{1/2}} [\phi(k) + \phi(-k)] \, \delta k$$

with $k = n\pi/a$ and $\delta k = \pi/a$.

20. Consider a wave packet whose form at $t = 0$ is

$$\psi(x,0) = (e^{-x^2/2b^2} e^{ik_0 x}) b$$

a function whose real part is shown in Fig. 7. Assume that this packet has $v_p = \hbar k_0/2m$ and $v_g = \hbar k_0/m$, as given by the result of Problem 17b.

(a) Show that at time t this packet is described by the function

$$\psi(x, t) = \exp\left[-(x - v_g t)^2 (1 - iat)/2b^2(1 + a^2 t^2)\right]$$
$$\exp\left[ik_0(x - v_p t)\right]$$

where $a = \hbar/mb^2$. (*Hint:* Find the Fourier transform of $\psi(x, 0)$ and thus show that

$$\psi(x, 0) = \frac{1}{(2\pi)^{1/2}} \int_{-\infty}^{\infty} \exp\left[-(k - k_0)^2 b^2/2\right] \exp\left[+ikx\right] dk$$

Then obtain $\psi(x, t)$ by replacing x by $x - vt$, where $v = \hbar k/2m$ inside the integral. Why is this procedure valid, in physical terms?)

(b) Notice that $b = \lambda_0$ for the wave packet shown in Fig. 7 ($\lambda_0 = 2\pi/k_0$). Then use the result of part (a) to show that Fig. 7 is drawn correctly. In particular, show that the standard deviation of the Gaussian envelope has become approximately $1.28b$ at $t = 5mb/\hbar k_0$, and show that the phase of the wavefunction—the coefficient of i in the exponent—changes more rapidly with x as x increases, so that the leading part of the wavefunction oscillates with a shorter wavelength than the trailing part. The physical interpretation of this is that the high-momentum part of the packet travels faster.

21. Electrons are boiled off a thin wire filament and are attracted to a flat plate in which there is a narrow slit parallel to the wire. (See figure. The wire is perpendicular to plane of figure.) In this way a nearly flat beam of electrons is formed. The beam diverges slightly because of the thickness of the wire and the width of the slit, but it might seem that one could make a beam with as little divergence as one chooses, simply by making the distance d sufficiently large. Explain how the uncertainty principle places an ultimate limitation on this possibility, and calculate (a) the "geometrical" divergence θ and (b) the divergence θ required by the uncertainty relation, when the wire diameter and slit width are each 10^{-5} cm, $d = 1$ cm, and there is a 5-V accelerating potential between the wire and the slit.

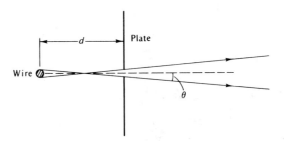

22. Use the uncertainty principle to deduce the size of the smallest possible circular electron "orbit" in hydrogen. Take the radius of the orbit to be δx, and the electron's momentum to be δp_x. (Notice that although p has a definite value in a circular orbit, p_x varies from $+p$ to $-p$ as the electron moves on the orbit, so $\delta p_x = p$.)

23. Suppose that you are dropping marbles from a height of 1 m onto a horizontal target. How large must the target be, in order that your accuracy will not be limited by the uncertainty principle? (Assume each marble to have a mass of 5 gm.)

CHAPTER 5

Schrödinger Equation I: One Dimension

The de Broglie relation, $\lambda = h/p$, is the starting point for a wave theory of matter. As we have seen in Section 4.2, this relation can explain electron diffraction results, and it can be used to ''derive'' the Bohr condition that the angular momentum of a hydrogen atom is $\hbar$ times an integer. But it is not obvious how such a simple relation can be applied to situations in which the potential energy varies with position. In such situations, the *kinetic* energy must depend on the position, as must the momentum; thus, applying the de Broglie relation would mean finding a wavefunction that has the appropriate ''wavelength'' at each point in space.

Rather than wrestle with the conceptual (or semantic) difficulty of defining a wavelength at a single point, it is better to have a differential equation that will deal with such matters by redefining the relationship between the momentum and the wave in terms of the shape of the wavefunction. The solution to this differential equation will be a wavefunction that is appropriate for the potential that is inserted into the equation. Schrödinger developed such a wave equation shortly after the de Broglie hypothesis appeared. His equation is not only consistent with that hypothesis, it also correctly describes phenomena whose occurrence was not previously suspected.

137

For example, consider the one-dimensional motion of a particle whose potential energy V is constant in a limited range, say $|x| < a$, and infinite outside that range (like the "square well" of Fig. 3, but with $V_0 = \infty$). We know that the particle's wavefunction must be zero where V is infinite because this represents an impenetrable barrier. Where the wavefunction is not zero, the de Broglie condition tells us that the wavelength must be constant because the potential energy is constant. Therefore the wavefunction must have one of a set of wavelengths that make the wave equal to zero at $x = \pm a$ (as in Fig. 6). The Schrödinger equation leads to the same result, with solutions whose wavelengths are equal to $4a/n$, where n is an integer. But we shall see in Section 5.4 that the Schrödinger equation gives a *completely new* result when the barrier is *not* impenetrable—when V is finite for $|x| \geq a$. Where the total energy E is less than V, so that the kinetic energy $E - V$ is negative, the wavefunction, as determined by the Schrödinger equation, is not zero. This means that there is a nonzero probability of finding the particle in a region where it could never be found according to classical theory. This is not a flaw in the Schrödinger equation; it is a feature that is necessary to explain observed phenomena, as we shall see in Sections 5.5 and 15.1.

Within a few years of Schrödinger's publication of his equation, it was used in the computation of the energy levels of a great variety of atomic systems. The results have been in excellent agreement with experiment; there is no doubt that the equation correctly describes nonrelativistic systems on the atomic scale.

5.1 CONSTRUCTION OF THE SCHRÖDINGER EQUATION

To construct[1] the Schrödinger equation, we may begin by considering a particle of definite momentum $p = h/\lambda = \hbar k$ (where $\hbar = h/2\pi$ by definition). The wavefunction for such a particle is a wave of infinite extent, with constant wavelength $\lambda = 2\pi/k$, whose equation may be written as

$$\psi = A \cos(kx - \omega t)$$

Although this a very special case, it is reasonable to construct a wave equation on the basis of this case, because we know, from the Fourier integral theorem, that *any* wave with physical meaning may be built up from a superposition of such infinite waves. Our only problem is to ensure that this superposition results in a wave which is also a solution to our wave equation. To accomplish this, we simply make the wave equation *linear*. Then if a number of infinite waves satisfy our equation, the superposition of these waves also does, for the sum of any number of solutions of a linear differential equation is also a solution of the equation.

---

[1] Notice that the word is "construct," not "derive." There are many possible choices for a wave equation consistent with de Broglie's hypothesis. The "proof" of Schrödinger's equation is that it gives the right results, when put to experimental tests.

The cornerstone in the construction of the wave equation is the idea that the *total energy* of the particle—kinetic energy plus potential energy—is equal to Planck's constant times the frequency of the wave, or to $\hbar\omega$. Therefore we write

$$E = \hbar\omega$$

$$= \frac{p^2}{2m} + V$$

$$= \frac{\hbar^2 k^2}{2m} + V \tag{1}$$

where V is the potential energy, and we use the nonrelativistic expression for kinetic energy (for simplicity in dealing with atomic problems, where the electron speeds are almost always much less than c; see Chapter 4, Problem 17).

We now use our assumed wavefunction ψ to see what sort of differential equation could lead to Eq. (1). First we take derivatives of ψ with respect to x. Since k is constant,

$$\frac{\partial \psi}{\partial x} = -Ak \sin(kx - \omega t)$$

$$\frac{\partial^2 \psi}{\partial x^2} = -Ak^2 \cos(kx - \omega t)$$

$$= -k^2 \psi \tag{2}$$

But, from Eq. (1),

$$k^2 = \frac{2m(E - V)}{\hbar^2} \tag{3}$$

Therefore

$$\frac{\partial^2 \psi}{\partial x^2} = -\frac{2m}{\hbar^2}(E - V)\psi \tag{4}$$

Equation (4) is the time-independent Schrödinger equation in one dimension.

Because we assumed k to be constant, and because E is also constant, our construction of this equation is only valid when V is constant. If the equation were only valid for such cases, it would be no more than a reformulation of de Broglie's momentum-wavelength relation. The great step forward was to postulate that this equation should also be valid when V is *not* constant. Like many great theoretical advances, this step can only be "proven" by experimental tests. These tests have taken a multitude of forms, and no evidence has been found to contradict any predictions of the Schrödinger equation in the nonrelativisitic domain.

With this equation we can already find the discrete energy levels in a few simple systems, as we shall see in a moment. Although a one-dimensional problem is not very realistic, we shall work some examples in one dimension in order to see, in the simplest possible way, the manner in which discrete energy levels emerge as a consequence of the wave equation.

The wave equation must also be able to describe the development of a system in time. To obtain a time-dependent equation, we proceed as we did with the x coordinate. We take derivatives of the same ψ with respect to t:

$$\psi = A \cos(kx - \omega t)$$

$$\frac{\partial \psi}{\partial t} = \omega A \sin(kx - \omega t) \tag{5}$$

$$\frac{\partial^2 \psi}{\partial t^2} = -\omega^2 A \cos(kx - \omega t)$$

$$= -\omega^2 \psi \tag{6}$$

We wish to combine the space and time derivatives into one equation. Remembering that $\omega = E/\hbar$, we may rewrite Eq. (4) as

$$E\psi = -\frac{\hbar^2}{2m} \frac{\partial^2 \psi}{\partial x^2} + V\psi$$

or

$$\hbar\omega\psi = -\frac{\hbar^2}{2m} \frac{\partial^2 \psi}{\partial x^2} + V\psi \tag{7}$$

We might be tempted to incorporate Eq. (6) into Eq. (7), but this would require squaring both sides of Eq. (7) to obtain

$$\hbar^2 \omega^2 \psi^2 = \left[-\frac{\hbar^2}{2m} \frac{\partial^2 \psi}{\partial x^2} + V\psi \right]^2$$

or, from Eq. (6)

$$-\hbar^2 \psi \frac{\partial^2 \psi}{\partial t^2} = \left[-\frac{\hbar^2}{2m} \frac{\partial^2 \psi}{\partial x^2} + V\psi \right]^2$$

which fails to meet our need for a *linear* wave equation. Therefore we turn to Eq. (5), which contains the first power of ω. The difficulty with Eq. (5) is that it contains a new function, the sine, rather than the cosine with which we started. It would be helpful if $\partial\psi/\partial t$, which gives us the necessary *first* power of ω, could yield a result which contains the original function ψ. But we can arrange that; we let ψ equal $Ae^{i(kx-\omega t)}$ instead of $A \cos(kx - \omega t)$. Then

$$\frac{\partial \psi}{\partial t} = -i\omega Ae^{i(kx-\omega t)}$$

$$= -i\omega\psi$$

or

$$\frac{\partial \psi}{\partial t} = - \frac{iE}{\hbar} \psi \tag{8}$$

Now we may substitute Eq. (8) into Eq. (7) to obtain

$$i\hbar \frac{\partial \psi}{\partial t} = - \frac{\hbar^2}{2m} \frac{\partial^2 \psi}{\partial x^2} + V\psi \tag{9}$$

Equation (9) is the time–dependent Schrödinger equation in one dimension.

You may be disturbed by the imaginary coefficient in this equation, or by the fact that the solution $\psi = Ae^{i(kx - \omega t)}$ has an imaginary part. But this will cause no difficulty, for the wave equation and the wavefunction are only the means to an end, not an end in themselves. The wavefunction itself is not observable; it is simply used to compute observable quantities, which always turn out to be real. For example, as mentioned in Chapter 4, the number of particles counted by a detector is proportional to the *intensity* of the wave, which for a real wave is proportional to the square of the wave amplitude. For the complex ψ, we write the intensity as $\psi^*\psi$, where ψ^*, the complex conjugate of ψ, is found by simply replacing i by $-i$ everywhere in the expression for ψ. Then

$$\text{Intensity} = |\psi|^2 \equiv \psi^*\psi = A^* e^{-i(kx - \omega t)} \cdot Ae^{i(kx - \omega t)} = A^*A = |A|^2$$

When the amplitude A is real, the intensity is still the square of the amplitude; when A is complex, the intensity is the square of the absolute value of A, or A^*A, which is still a real number. (Allowing A to be complex provides an easy way to handle phase factors between interfering waves.)

5.2 BOUNDARY CONDITIONS

Another step forward in the Schrödinger theory is the introduction of ''boundary conditions'' that govern the behavior of the wave at a ''boundary'' where the potential energy is discontinuous.

We can deduce the boundary conditions from the reasonable requirement that ψ be finite. [We have allowed for an infinite ψ in discussing the delta function (Section 4.3), but such a function is zero except at one point, so it is not relevant to the boundary-conditions problem.] We also assume that in any physical problem E and V will be finite. Eq. (4) then tells us that $\partial \psi^2/\partial x^2$ must be finite. Consequently ψ and $\delta\psi/\delta x$ must be continuous everywhere, even if V is discontinuous.

To see the significance of these boundary conditions, consider an electron wave traveling from left to right across a potential "step," as depicted in Fig. 1. The potential energy is

$$V = \begin{cases} 0 & (x < 0) \\ -V_0 & (x > 0) \end{cases}$$

If the electrons are incident from the left with kinetic energy E, their kinetic energy suddenly increases to $E + V_0$ at the step. This means that the electron

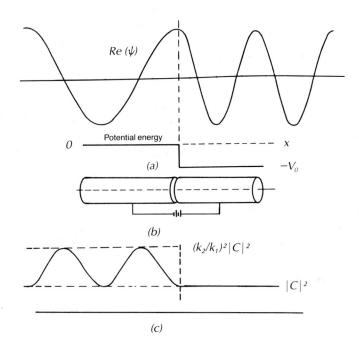

Fig. 1. (a) Behavior of the real part of the wavefunction, $Re(\psi)$, at a potential step. When the potential energy suddenly decreases, the kinetic energy increases and the wavelength decreases, while ψ and its slope are continuous. (b) Physical setup. An electron traveling along the axis of the cylinder sees a constant potential until it passes from one cylinder to the other, where there is a sudden drop in potential energy. (c) Behavior of $|\psi|^2$ at the same step, showing how the particle density changes. As explained in Section 5.3, $|\psi|^2$ is constant for $x > 0$, so it must have zero slope at the step. The wavelength of $|\psi|^2$ is one-half the wavelength of $Re(\psi)$ plotted in (a).

momentum also increases, so that the wavelength of the electron wave decreases, as indicated in the figure.

Something of this sort happens when electrons are incident on the surface of a metal. An electron is attracted to a metal, and the drop in potential near the metallic surface is rather sharp. (See Chapter 4, Problem 14; see also Section 11.4d.) Classically, the potential step only acts to increase the speed of the electrons as they cross it. But the electron *wave* must satisfy the boundary conditions, which it cannot do unless there is also a reflected wave traveling away from the boundary toward the left. Thus some electrons are reflected and cannot enter the metal, even though, at the step, there is an impulse which tends to pull electrons *into* the metal. The effect occurs for the same reason that a light wave is reflected when there is a discontinuity in the index of refraction.

The amplitude of the reflected wave, and hence the probability that an incident electron will be reflected, is found by solving the Schrödinger equation for all values of x. In doing this, we consider the region of negative x and the region of positive x separately, because within each region V is constant. Then we use the boundary conditions to match up the solutions at $x = 0$, so that we can find the relative amplitudes of the waves in the two regions.

You may easily verify that

$$\psi_a = Ae^{+i(k_1 x - \omega t)}$$
$$= Ae^{+ik_1(x - \omega t/k_1)}$$

and

$$\psi_b = Be^{-i(k_1 x + \omega t)}$$
$$= Be^{-ik_1(x + \omega t/k_1)}$$

with $k_1 = (2mE/\hbar^2)^{1/2}$ and $\omega = E/\hbar$, are both solutions of Eq. (9) for $V = 0$.[2] These are therefore the solutions for $x < 0$. Solution ψ_b, being a function of $x + vt$ (with $v = \omega/k_1$), describes a wave traveling toward the left, and solution ψ_a, a function of $x - vt$, describes a wave traveling toward the right. If electrons come in from the left and are reflected at $x = 0$, we expect to find waves traveling in *both* directions for $x < 0$, so we write for the solution in this region

$$\psi_1 = Ae^{+i(k_1 x - \omega t)} + Be^{-i(k_1 x + \omega t)} \qquad (x < 0)$$

For $x > 0$, there is only a wave traveling toward the right, so we write

$$\psi_2 = Ce^{+i(k_2 x - \omega t)} \qquad (x > 0)$$

---

[2] Remember that we began with the function ψ_a when we constructed the equation in the first place; ψ_b is also a solution because the term ikx in the exponent may have either sign—in both cases $\partial^2 \psi/\partial x^2 = -k^2 \psi$.

where $k_2 = (2m(E + V_0)/\hbar^2)^{1/2}$, because the kinetic energy is $E + V_0$ in this region. In both regions there is the same ω, equal to $E/\hbar$, because the *total* energy E is the same in both regions. We see here a similarity to classical wave behavior, in which the frequency remains unchanged, although the wavelength changes, as the wave crosses a boundary.

We now relate the coefficients A, B, and C through the boundary conditions. Since ψ is continuous, ψ_1 and ψ_2 must be equal where they meet, at $x = 0$, so

$$A + B = C \tag{10}$$

Also, $\partial\psi_1/\partial x]_{x=0} = \partial\psi_2/\partial x]_{x=0}$, so that

$$ik_1 A - ik_1 B = ik_2 C \tag{11}$$

(The factors $e^{-i\omega t}$, common to all terms, have been divided out.)

Solving for B in terms of A, we obtain

$$\frac{B}{A} = \frac{k_1 - k_2}{k_1 + k_2} \tag{12}$$

Since k_1 and k_2 are real, A, B, and C can all be real. In that case, when $t = 0$, the real part of ψ, $Re(\psi)$, is a maximum at the step (as shown in Fig. 1a), because

$$Re(\psi_1)_{t=0} = A \cos k_1 x + B \cos k_1 x$$

and

$$Re(\psi_2)_{t=0} = C \cos k_2 x$$

both of which are maximum when $x = 0$. For a different time t, $Re(\psi)$ need not be maximum at $x = 0$, and Fig. 1a could have a different appearance. But Fig. 1c, which shows the observable quantity $|\psi|^2$, would not change, as we shall see.

The probability R that an incident electron will be reflected at the potential step must be equal to the ratio of the leftward intensity to the rightward intensity for $x < 0$, or

$$R = \frac{|B|^2}{|A|^2} = \frac{(k_1 - k_2)^2}{(k_1 + k_2)^2} \tag{13}$$

Notice that when $k_1 = k_2$, there is no step, and $R = 0$.

We can check this result by solving for C in terms of A, in order to find the *transmission* coefficient—the probability that an electron is transmitted across the step. The sum of the reflection and transmission probabilities must equal unity. But we must be careful, for C^2/A^2 is the ratio of the *particle densities* in the transmitted and incident waves, and this ratio does *not* equal the transmission probability. The transmission probability must equal the ratio

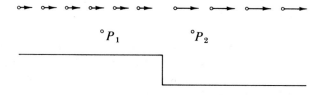

Fig. 2. *Classical behavior of a stream of particles crossing a potential step. As they cross the step, their velocity increases, and the distance between particles increases in proportion to the speed.*

of the wave *intensities* I_1 and I_2 on the two sides of the barrier, because the intensity is the number of particles per second *passing* a given point; in a steady state, with no reflection, the number per second passing point P_1 (Fig. 2) must equal the number per second passing point P_2. But the *density* of *particles* at each point depends on the *velocity* as well as the *intensity*. In general, it is easy to see that the intensity must equal the *product* of the particle density and the velocity, so we may write the transmission coefficient T as

$$T = \frac{I_2}{I_1} = \frac{C^2 v_2}{A^2 v_1} \tag{13}$$

where v_1 and v_2 are the particle speeds for $x < 0$ and $x > 0$, respectively. Since $v = \hbar k/m$, we may write T as

$$T = \frac{k_2}{k_1} \frac{C^2}{A^2}$$

Solving Eqs. (10) and (11) for C^2/A^2 yields

$$\frac{C^2}{A^2} = \frac{4k_1^2}{(k_2 + k_1)^2}$$

so that

$$T = \frac{4k_1 k_2}{(k_2 + k_1)^2} \tag{14}$$

If our theory is correct, the probability of reflection plus the probability of transmission must equal 1. We find from Eqs. (12) and (14) that

$$R + T = \frac{(k_1 - k_2)^2}{(k_2 + k_1)^2} + \frac{4k_1 k_2}{(k_1 + k_2)^2} \equiv 1$$

a reassuring result.

5.3 PROBABILITY CURRENT

In the preceding discussion we have made use of the idea, first brought out in Chapter 4, that the absolute square of the wavefunction is proportional to the density of the particles associated with the wave. But wave behavior is observed even when there is only one particle present at a given time. We cannot very well speak of a particle density when there is only one particle, but we can consider $|\psi|^2$ in this case to be a *probability density*, as suggested by Born in 1926. That is,

$$\int_{x_1}^{x_2} \psi^*\psi \, dx = \text{probability of finding the particle between } x_1 \text{ and } x_2$$

Then, if we were to measure the position of a particle n times, each time arranging the initial conditions so that the particle had the same ψ before each measurement, the number of times that the particle was found between x_1 and x_2 would be $n \int_{x_1}^{x_2} \psi^*\psi \, dx$, and $n\psi^*\psi$ would be a particle density. We have seen in the potential step problem that this view gives consistent results, as we verified that the sum of the transmission and reflection probabilities is 1.

If $|\psi|^2$ is to be a probability density for a single particle, we must require that

$$\int_{-\infty}^{\infty} |\psi(x)|^2 \, dx = 1 \tag{15}$$

because there is unit probability of finding the particle somewhere. A wave-function which satisfies Eq. (15) is called a *normalized* function. In most cases, given an unnormalized wavefunction ψ_u, it is a simple matter to normalize it by multiplying it by a normalizing factor N. The normalized function is

$$\psi_N = N\psi_u$$

so that substitution of $\psi = \psi_N$ into Eq. (15) yields

$$|N|^2 \int_{-\infty}^{\infty} |\psi_u|^2 \, dx = 1$$

and

$$|N|^2 = \left[\int_{-\infty}^{\infty} |\psi_u|^2 \, dx \right]^{-1}$$

It was mentioned in the discussion of boundary conditions (Section 5.2) that it was "reasonable" to require the wavefunction to be finite. Now this requirement can be stated more precisely by considering the normalization of the wavefunction. In order that $|\psi|^2$ be the probability density, we must require that $\int |\psi|^2 dx$ be finite over any range of values of x. This would permit $|\psi|^2$ to be a delta function, because the integral of the delta function is finite even though $\delta(0)$ is infinite. (See Section 4.3.)

Some wavefunctions cannot be normalized. For example, if $\psi_u = e^{i(kx-\omega t)}$, then $\int_{-\infty}^{\infty} |\psi_u|^2 \, dx$ goes to infinity; thus N goes to zero and nothing remains to work with. However, we can use a wavefunction of this sort if we assume that ψ_u equals zero outside some large but finite region, so that N is small but not zero. This is clearly a reasonable thing to do, for no wave in an experiment ever extends to infinity. The normalizing factor N, depending as it does on the size of our assumed region, is somewhat arbitrary, but as long as we use the same volume for all wavefunctions in a given problem, we will obtain consistent results. In our treatment of the potential step problem, we simply let N equal 1, and there was no difficulty, because we ultimately used only the ratios of the wavefunctions.

Let us examine these ideas in a more general way, without reference to a specific potential. We write the Schrödinger equation

$$i\hbar \frac{\partial \psi}{\partial t} = -\frac{\hbar^2}{2m} \frac{\partial^2 \psi}{\partial x^2} + V\psi \tag{9}$$

and we take the complex conjugate of both sides, to obtain

$$-i\hbar \frac{\partial \psi^*}{\partial t} = -\frac{\hbar^2}{2m} \frac{\partial^2 \psi^*}{\partial x^2} + V\psi^* \tag{16}$$

We multiply Eq. (9) by ψ^* and Eq. (16) by ψ, and we subtract the latter from the former, to obtain

$$i\hbar \left(\psi^* \frac{\partial \psi}{\partial t} + \psi \frac{\partial \psi^*}{\partial t} \right) = -\frac{\hbar^2}{2m} \left(\psi^* \frac{\partial^2 \psi}{\partial x^2} - \psi \frac{\partial^2 \psi^*}{\partial x^2} \right)$$

or

$$\frac{\partial}{\partial t} (\psi^* \psi) = \frac{i\hbar}{2m} \frac{\partial}{\partial x} \left(\psi^* \frac{\partial \psi}{\partial x} - \psi \frac{\partial \psi^*}{\partial x} \right)$$

Let us define the quantity

$$S_x = -\frac{i\hbar}{2m} \left(\psi^* \frac{\partial \psi}{\partial x} - \psi \frac{\partial \psi^*}{\partial x} \right)$$

so that the above equation becomes

$$\frac{\partial}{\partial t} (\psi^* \psi) = -\frac{\partial}{\partial x} S_x \tag{17}$$

This can be generalized to three dimensions as

$$\frac{\partial}{\partial t} (\psi^* \psi) = -\operatorname{div} \mathbf{S} \tag{17a}$$

an equation analogous to the equation of continuity of flow of fluid or charge, in hydrodynamics or electrodynamics:

$$\frac{\partial \rho}{\partial t} = - \operatorname{div} \mathbf{J}$$

where ρ is the charge density or fluid density, and $\mathbf{J}$ is the current density. It expresses the fact that a net flow of current out of a region results in a decrease in the charge or fluid in that region. The quantum-mechanical equations (17) and (17a) express the same thing, but the "fluid" in this case is *probability,* and S is the *probability current,* which as a three-dimensional vector becomes

$$\mathbf{S} = - \frac{i\hbar}{2m} (\psi^* \nabla \psi - \psi \nabla \psi^*)$$

To understand how $\mathbf{S}$ describes a probability current, let us consider the special case of a plane wave:

$$\psi = N e^{i(kx - \omega t)}$$

Differentiation yields $\dfrac{\partial \psi}{\partial x} = ik\psi$, so $\dfrac{\partial \psi^*}{\partial x} = -ik\psi^*$, and

$$S_x = - \frac{i\hbar}{2m} (ik\psi^* \psi + ik\,\psi^* \psi)$$

$$= \frac{\hbar k}{m} (\psi^* \psi)$$

We know that $\hbar k$ equals the momentum, or mv_x, where v_x is the particle velocity, so

$$S_x = v_x \psi^* \psi$$

Thus the probability *current* is equal to the product of the *velocity* and the *probability density,* as one should expect.

If we integrate both sides of Eq. (17) between the limits $x = x_1$ and $x = x_2$, we obtain

$$\frac{\partial}{\partial t} \int_{x_1}^{x_2} \psi^* \psi \, dx = - \int_{x_1}^{x_2} \frac{\partial S_x}{\partial x} \, dx = S_x(x_1) - S_x(x_2)$$

The integral $\int_{x_1}^{x_2} \psi^* \psi \, dx$ is the total amount of "fluid"—in this case the total probability—between $x = x_1$ and $x = x_2$. The fluid flows into the region between x_1 and x_2 at a rate equal to $S_x(x_1)$, and it flows out of this region at a rate equal to $S_x(x_2)$. The difference between these rates is the rate at which total probability builds up in the region.

We can apply these concepts to the square potential step of Fig. 1. For $x > 0$, the wavefunction is $\psi_2 = C e^{i(k_2 x - \omega t)}$, and

$$|\psi|^2 = |\psi_2|^2 = |C|^2, \text{ a constant.}$$

The probability density is constant in both time and space, and the probability current is

$$S = v_2|C|^2 = (\hbar k_2/m)\, |C|^2 = (\hbar k_2/m)\, |\psi_2|^2$$

For $x < 0$,

$$|\psi|^2 = |\psi_1|^2 = |A|^2 + |B|^2 + AB^*e^{2ik_1x} + A^*Be^{-2ik_1x},$$

A and B being complex constants, in general. But $|\psi|^2$ must be *real*; we can show this explicitly by writing

$$A = \mathcal{C}e^{i\phi} \qquad \text{and} \qquad B = -\mathcal{B}e^{i\phi}$$

where $\mathcal{C}$, $\mathcal{B}$, and ϕ are real, positive constants. The minus sign is needed because the ratio B/A is negative in this case; see Eq. (12). Then $AB^* = A^*B = -\mathcal{C}\mathcal{B}$, and

$$|\psi_1|^2 = \mathcal{C}^2 + \mathcal{B}^2 - \mathcal{C}\mathcal{B}(e^{2ik_1x} + e^{-2ik_1x})$$
$$= \mathcal{C}^2 + \mathcal{B}^2 - 2\mathcal{C}\mathcal{B}\cos 2k_1x$$

This is true for all time t, because t appears in ψ only in the factor $e^{-i\omega t}$, and $|e^{-i\omega t}|^2 = 1$. As a function of x, $|\psi|^2$ oscillates between the values

$$|\psi|^2_{\max} = \mathcal{C}^2 + \mathcal{B}^2 + 2\mathcal{C}\mathcal{B} = (\mathcal{C} + \mathcal{B})^2$$

and

$$|\psi|^2_{\min} = \mathcal{C}^2 + \mathcal{B}^2 - 2\mathcal{C}\mathcal{B} = (\mathcal{C} - \mathcal{B})^2$$

producing the standing-wave pattern shown in Fig. 1c. The wavelength, determined by the factor $\cos 2k_1x$, is π/k_1, just half the wavelength of ψ itself. At $x = 0$, $|\psi_1|^2 = |\psi|^2_{\min}$. (For a *positive* step, as long as $E > V$, k_2 would be *smaller* than k_1, B/A would be *positive*, and $|\psi|^2$ would equal $|\psi|^2_{\max}$ at $x = 0$.)

The function $|\psi|^2$ and its first derivative are continuous because ψ and ψ^* have those properties. We can satisfy both conditions only by setting $|\psi_2|$ equal to $|\psi|_{\min}$. (See Fig. 1c.)

The probability current must be the same for all x, because there are no sources or sinks. For $x < 0$, S is the sum of two currents, a rightward current of $v_1\mathcal{C}^2$ and a leftward current of $v_1\mathcal{B}^2$. The net current is

$$S = v_1(\mathcal{C}^2 - \mathcal{B}^2) = (\hbar k_1/m)(\mathcal{C} + \mathcal{B})(\mathcal{C} - \mathcal{B}) \tag{18}$$
$$= (\hbar k_1/m)\, |\psi|_{\max}\, |\psi|_{\min}$$

Equating this to the current for $x > 0$, we have

$$(\hbar k_1/m)\, |\psi|_{\max}\, |\psi|_{\min} = (\hbar k_2/m)\, |\psi_2|^2 = (\hbar k_2/m)\, |\psi|^2_{\min}$$

Therefore

$$\frac{|\psi|_{\text{max}}}{|\psi|_{\text{min}}} = \frac{k_2}{k_1}$$

as indicated in Fig. 1c. For further discussion, see the article by James E. Draper in *American Journal of Physics*, vol. 47, page 525 (1979).

5.4 ONE-DIMENSIONAL SQUARE WELL POTENTIAL

We are now in a position to see in detail how quantized energy levels are predicted by the Schrödinger equation. Let us solve a somewhat artificial example whose simplicity allows us to concentrate on the fundamentals. Again limiting ourselves to one dimension, we suppose the potential to be given by (see Fig. 3)

$$V = +V_0 \qquad (|x| > a)$$
$$V = 0 \qquad (|x| < a)$$

and we consider values of the particle's energy E such that $0 < E < V_0$.

Classically, E can have any value in this range, and the particle is confined to the region between $x = -a$ and $x = +a$, because its kinetic energy would be negative outside that range. The particle bounces back and forth from one side of the well to the other, always moving with the same speed as it remains in the "box" defined by the potential.

But in quantum theory we must determine the behavior of the particle by solving the Schrödinger equation. We wish to find the wavefunction for a state of energy E. Since the wave has angular frequency $\omega = E/\hbar$, the wavefunction contains the factor $e^{-i\omega t}$ or $e^{-iEt/\hbar}$, so that we may write the wavefunction as $\psi(x, t) = u(x)e^{-iEt/\hbar}$, the spatial dependence being contained entirely in the factor $u(x)$. Substitution of this form of ψ into Schrödinger's equation (9) yields

$$\frac{d^2u}{dx^2} = -\frac{2m}{\hbar^2}(E - V)u$$

This equation is just the time-independent Schrödinger equation again, except that now it involves only the space-dependent part of ψ, $u(x)$. To determine $u(x)$ for all values of x, we substitute for V the values given above. For $|x| > a$,

$$\frac{d^2u}{dx^2} = -\frac{2m}{\hbar^2}(E - V_0)u$$

$$= \frac{2m}{\hbar^2}(V_0 - E)u \tag{19}$$

while for $|x| < a$,

$$\frac{d^2u}{dx^2} = -\frac{2mE}{\hbar^2} u \tag{20}$$

Before going through the mathematics, let us make some general observations. The solution must go to zero as $|x| \to \infty$, for otherwise $\int_{-\infty}^{\infty} \psi^*\psi \, dx$ would be infinite, and the probability of finding the particle within the well would be zero. The solution must also obey the continuity requirements: $u(x)$ and du/dx must be continuous at $x = \pm a$. The result of these conditions is to make certain values of E impossible. Let us demonstrate this graphically.

In Fig. 3a, we begin at the left with an attempt to construct the graph of $u(x)$ for a very small value of E. For $x < -a$, d^2u/dx^2 is positive (assuming that u is positive), and u curves upward as x increases. When we reach $x = -a$, d^2u/dx^2 becomes negative, but is very small in magnitude. Thus the slope of $u(x)$ decreases very slowly, becoming negative just before we reach $x = +a$. At this point, d^2u/dx^2 becomes positive again, causing the slope, and eventually u itself, to increase without limit as $x \to \infty$. This u, the inevitable result of our choice of E, is not an acceptable wavefunction, and we must conclude that the particle simply cannot possess that value of the energy.

In Fig. 3b we see the graph for the smallest acceptable value of E. Here d^2u/dx^2 in the region $-a < x < +a$ is just great enough so that the slope at $x = +a$ is the negative of the slope at $x = -a$. This ensures that $u(x)$ will asymptotically approach zero as $x \to \infty$. Obviously, only one precise value of E can produce this result. For a somewhat larger E, we have the situation shown in Fig. 3c, where u goes to $-\infty$ as $x \to \infty$. But if the well is deep enough, we can have a second acceptable value of E, for which du/dx changes so rapidly in the well that it is positive again at $x = +a$, and u then approaches zero from the negative side as $x \to \infty$.

Inspection of the acceptable wavefunction reveals behavior which is decidedly peculiar by classical standards. The wavefunction extends into the "forbidden" region $|x| > a$, where the *kinetic* energy $(E - V)$ is negative. This means that there is some probability of observing the particle *outside* the well. But it is clearly impossible to *observe* a particle with negative kinetic energy; quantum theory may seem strange, but it cannot provide for an observation which is not definable experimentally. How can we extricate ourselves from this paradox?

The resolution of the paradox, as you might guess, depends on the uncertainty principle. The process of measuring the particle's position necessarily gives it enough energy so that a subsequent measurement of its kinetic energy yields a positive result. Of course, the energy which is given to the particle depends on the precision of the position measurement. But a certain precision is necessary in order to locate the particle in the limited region just outside the well where the wavefunction is still appreciable. When one attains that precision, the particle's momentum, and hence its kinetic energy, is sufficiently *im*precise to permit a positive kinetic energy after the measurement. In other words, the particle acquires enough energy, in the process of

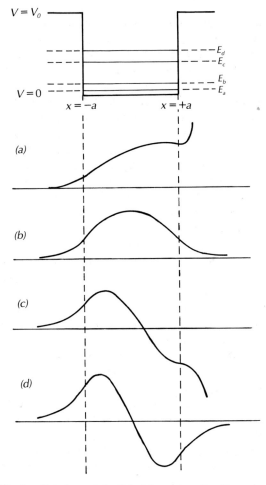

Fig. 3. *Solutions of the Schrödinger equation for various energies in a square well of depth V_o and width 2a. Graphs (a) and (c) show unacceptable solutions which go to infinity; the corresponding energies are not allowed. Graph (b) shows the solution for the lowest allowed energy E_b, and (d) shows the solution for the next-lowest allowed energy E_d. Notice that (d) shows a greater penetration than (b) into the forbidden region, because $E_d > E_b$.*

measurement of its position, so that its kinetic energy is positive *when it is observed*. As the particle energy E becomes smaller, the wavefunction drops off to zero more rapidly outside the well, so that a greater precision is needed to locate the particle outside the well, and this allows the particle to acquire more energy during the position measurement. Quantitative details are left as an exercise for the student.

Calculation of Energy Levels. Let us define

$$\alpha \equiv \left[2m(V_0 - E)/\hbar^2\right]^{1/2}$$

so that Eq. (19) becomes

$$\frac{d^2u}{dx^2} = \alpha^2 u$$

whose solution is a linear combination of $e^{+\alpha x}$ and $e^{-\alpha x}$. For a bound state, α is a real number because $V_0 > E$. We can choose the positive value of α, in which case $e^{+\alpha x} \to \infty$ as $x \to +\infty$, and $e^{-\alpha x} \to \infty$ as $x \to -\infty$. This requires that we use the solutions

$$u = Ae^{+\alpha x} \qquad (x < -a) \tag{21}$$

$$u = Be^{-\alpha x} \qquad (x > +a) \tag{22}$$

where A and B are constants whose values are determined by boundary conditions and normalization of the wavefunction.

Similarly, Eq. (20) may be written

$$\frac{d^2u}{dx^2} = -k^2 u$$

where $k \equiv (2mE/\hbar^2)^{1/2}$, again a real number, because E is positive. The solution to Eq. (20) is therefore

$$u = C \cos kx + D \sin kx \qquad (|x| < a) \tag{23}$$

For the special case $V_0 \to \infty$, these solutions become very simple. In that case, α also becomes infinite; therefore, from Eqs. (21) and (22), $u = 0$ for $|x| > a$. Therefore, since u must be continuous, the solution for the region between $-a$ and $+a$ (Eq. 23) must go to zero at $x = a$. Setting x equal to $+a$ we have

$$u = C \cos ka + D \sin ka = 0$$

and setting x equal to $-a$ we have [since $\sin(-ka) = -\sin ka$]

$$u = C \cos ka - D \sin ka = 0$$

These two equations can be added to give

$$2C \cos ka = 0$$

and they can be subtracted to give

$$2D \sin ka = 0$$

We can satisfy *both* of these equations in either of two possible ways:

Either $\cos ka = 0$ and $D = 0$
or $\sin ka = 0$ and $C = 0$

If $D = 0$, then ka must be an odd multiple of $\pi/2$, and $u = C \cos kx$. If $C = 0$, then ka must be an even multiple of $\pi/2$, and $u = D \sin kx$. Combining both possibilities gives us the full set of possible values of ka:

$$ka = n\pi/2 \qquad\qquad n = \text{any positive integer}$$

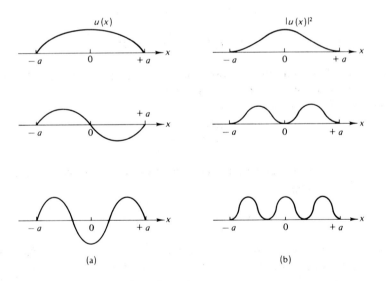

(a)   (b)

Fig. 4. (a) *Wavefunctions for the three lowest energy states in an infinitely deep, square potential well.* (b) *Probability densities for finding the particle at different positions inside the well, as given by the wavefunctions shown in (a).*

The corresponding functions are shown in Fig. 4, for the lowest three energy levels ($n \leqslant 3$). The energies are found from the definition

$$k^2 = 2mE/\hbar^2 \quad \text{which gives} \quad E = \hbar^2 k^2/2m$$

Knowing that the wave must go to zero at $x = \pm a$, one could have deduced the energy levels directly from $p = \hbar/\lambda$, because λ must equal $4a/n$, and $E = p^2/2m$.

The solution for the general case when V_0 is finite is somewhat tedious, but it provides a good example of the use of boundary conditions. The continuity of u gives

$$\text{at } x = -a: \quad Ae^{-\alpha a} = C \cos ka - D \sin ka \tag{24}$$

$$\text{at } x = +a: \quad Be^{-\alpha a} = C \cos ka + D \sin ka \tag{25}$$

Adding Eqs. (24) and (25) yields

$$(A + B)e^{-\alpha a} = 2C \cos ka \tag{26}$$

Subtracting Eq. (24) from Eq. (25) yields

$$(B - A)e^{-\alpha a} = 2D \sin ka \tag{27}$$

We obtain two more equations from the continuity of du/dx. Differentiation of Eqs. (21), (22), and (23) gives us

$$\text{at } x = -a: \quad \alpha Ae^{-\alpha a} = \quad Ck \sin ka + Dk \cos ka \tag{28}$$

$$\text{at } x = +a: \quad -\alpha Be^{-\alpha a} = -Ck \sin ka + Dk \cos ka \tag{29}$$

Adding Eqs. (28) and (29) yields

$$\alpha(A - B)e^{-\alpha a} = 2Dk \cos ka \tag{30}$$

Subtracting Eq. (29) from Eq. (28) yields

$$\alpha(A + B)e^{-\alpha a} = 2Ck \sin ka \tag{31}$$

Now we are in a position to eliminate the constants, using Eqs. (26), (27), (30), and (31). We can divide corresponding sides of Eq. (30) by Eq. (27), to obtain

$$-\alpha = k \cot ka \quad \text{(if } A \neq B, \text{ and } D \neq 0) \tag{32}$$

We can also divide Eq. (31) by Eq. (26), to obtain

$$\alpha = k \tan ka \quad \text{(if } A \neq -B, \text{ and } C \neq 0) \tag{33}$$

It is clearly impossible for *both* Eq. (32) and Eq. (33) to be valid, for in that case $k \cot ka$ would equal $-k \tan ka$, or $\tan^2 ka$ would equal -1, which is impossible, because k and a are real. But either Eq. (32) or Eq. (33) could be wrong only if it had been obtained by dividing by zero, which is possible if either C or D equals zero. Therefore we have two types of solution:

Type 1: Equation (32) is *wrong*, because $A = B$ and $D = 0$. Therefore

$$\alpha = k \tan ka \tag{33}$$

and the solution is

$$
\begin{aligned}
u &= Ae^{\alpha x} & (x < -a) \\
u &= Ae^{-\alpha x} & (x > +a) \\
u &= C \cos kx & (|x| < a)
\end{aligned}
$$

In this case $u(x)$ is an *even function*: $u(x) = u(-x)$ for all x.

Type 2: Equation (33) is *wrong*, because $A = -B$, and $C = 0$. Therefore

$$-\alpha = k \cot ka \tag{32}$$

and the solution is

$$
\begin{aligned}
u &= Ae^{\alpha x} & (x < -a) \\
u &= -Ae^{-\alpha x} & (x > +a) \\
u &= D \sin kx & (|x| < a)
\end{aligned}
$$

In this case $u(x)$ is an *odd function*: $u(x) = -u(-x)$ for all x.

Notice that Fig. 3b shows an even solution, while Fig. 3d shows an odd solution. Solutions of the two types alternate as E increases. In either case we must solve a transcendental equation, either Eq. (32) or Eq. (33), to find the energy levels. To solve for k or α, we need a second equation relating the two; that equation may be found from the definitions of k and α. Since $k = (2mE/\hbar^2)^{1/2}$ and $\alpha = [2m(V_0 - E)/\hbar^2]^{1/2}$,

$$k^2 + \alpha^2 = \frac{2mV_0}{\hbar^2} \equiv \beta^2 \tag{34}$$

At this point the problem is solved, as far as the physics is concerned. We could easily program a computer to solve Eq. (34), with either Eq. (32) or Eq. (33), to find k to any desired degree of accuracy for any given values of V_0 and a. But to gain more understanding of the way the energy levels are determined, and to find how many energy levels there are for a given well, it is best to use a graphical method. A standard approach is simply to use Eq. (34) and either Eq. (32) or Eq. (33) to draw curves of k versus α, and then to look for the points of intersection. But it is instructive to first transform these equations into a more convenient pair of equations to plot. We substitute Eq. (33) into Eq. (34) to obtain

$$k^2 + k^2 \tan^2 ka = \beta^2$$

or

$$k^2 \sec^2 ka = \beta^2$$

so that

$$\cos ka = \pm \frac{ka}{\beta a} \qquad \text{(and } \tan ka > 0) \qquad \text{for the even solution} \qquad (35)$$

By a similar sequence of steps, using Eq. (32) instead of Eq. (33), we obtain

$$\sin ka = \pm \frac{ka}{\beta a} \qquad \text{(and } \cot ka < 0) \qquad \text{for the odd solution} \qquad (36)$$

Now we can find the allowed values of ka simply by looking for the intersections of the graph of $\cos ka$ (or of $\sin ka$) and the graph of $\pm ka/\beta a$, both plotted versus ka. We may restrict ourselves to positive values of ka, for the negative values yield no additional energy levels. Then, for the *even solution*, the condition that $\tan ka > 0$ (which follows from Eq. (33) and the fact that α is positive) means that the values of ka lie in the *odd quadrants*—i.e., between 0 and $\pi/2$, or between π and $3\pi/2$, etc. Similarly, for the *odd solution*, the fact that $\cot ka < 0$ means that the values of ka lie in the *even quadrants*. Thus when we draw our graphs we must label only those points of intersection which satisfy these conditions. Fig. 5 shows the result for the even solution. Seven points of intersection are seen, but the even numbered points do not give acceptable values of k, because ka is in the wrong quadrant for this class of solutions (even solutions).

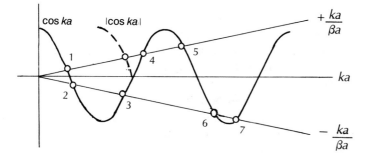

Fig. 5. *Graphs of cos ka and $\pm$ ka/βa, showing possible values of ka for the "even solution" of the square-well problem. Dotted curve shows |cos ka| in the third quadrant. (See text for discussion.)*

At this point a very simple trick makes the task of finding k much easier. Notice that point 3, the intersection of cos ka with the line $-ka/\beta a$, has the same value of k as point 3′, the intersection of $-\cos ka$, or $|\cos ka|$, with the graph of $+ka/\beta a$. Therefore we could obtain the values of ka for this solution *without drawing the graph of* $-ka/\beta a$, simply by graphing $|\cos ka|$ in the *odd quadrants*, and marking its intersections with the graph of $+ka/\beta a$. Similarly, we can also obtain the *odd* solutions by drawing a graph of $|\sin ka|$ in the *even* quadrants, again marking the intersections with the graph of $+ka/\beta a$ (see Fig. 6a). But now we really can save ourselves a lot of work, for we notice that all the quadrants are identical, as far as the sine or cosine curve goes, so we only have to draw *one* cosine curve in *one* quadrant, as long as we draw additional straight lines to represent the continuation of the curve $+ka/\beta a$. We then can read off all the roots, if we remember that a point on the lowest

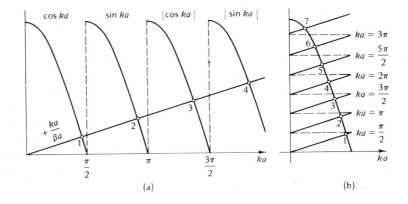

Fig. 6. *(a) Figure 5 redrawn to show odd solutions as well as even solutions. (b) Figure 6a remapped into the first quadrant.*

straight line segment means that ka is between 0 and $\pi/2$, a point on the second segment means that ka is between $\pi/2$ and π, etc. In our example, there are seven points, so there are seven possible energy levels for the particle in this particular well.

Notice that $\beta a \to \infty$ if the well is infinitely deep. The line $+ka/\beta a$ then has zero slope, there are infinitely many levels, and the values of ka for these allowed levels become $n\pi/2$, where n is any positive integer, in agreement with our previous result for this special case.

EXAMPLE PROBLEM 1. Find the energies of the two lowest states of an electron in a potential well of depth 10 eV and width $2a = 2 \times 10^{-7}$ cm. Also determine how many bound energy levels there are.

Solution. First we calculate βa. In doing this sort of calculation, it is convenient to work with units of electron-volts and angstroms. In these units, $mc^2 = 5.1 \times 10^5$ eV, and $hc = 1.24 \times 10^4$ eV-Å. Therefore

$$\beta a = \left(\frac{2mV_0 a^2}{\hbar^2}\right)^{1/2}$$

$$= \left[\frac{2mc^2 V_0 a^2}{(\hbar c)^2}\right]^{1/2}$$

$$= \left[\frac{2 \times 5.1 \times 10^5 \times 10 \times 10^2 \times (2\pi)^2}{(1.24 \times 10^4)^2}\right]^{1/2}$$

$$= 16.2$$

This value tells us directly how many energy levels there are. We know from Eq. (35) or Eq. (36) that $ka \leqslant \beta a$, and there is one level for $0 \leqslant ka \leqslant \pi/2$, another for $\pi/2 \leqslant ka \leqslant \pi$, another for $\pi \leqslant ka \leqslant 3\pi/2$, until the last level is for $n\pi/2 \leqslant ka \leqslant \beta a$, making $n + 1$ levels in all, where n is the largest integer such that $n\pi/2 \leqslant \beta a$. (See Fig. 6 again; there $n = 6$ and there are seven levels.) For the data in this problem, $2\beta a/\pi = 32.4/\pi = 10.3$, so that $n = 10$ and there are 11 bound levels.

We could now construct a graph like that of Fig. 6b and read the values of ka from it. Instead of doing that, let us observe that the lowest two values of ka lie very close to $\pi/2$ and π, respectively, because the straight line $ka/\beta a$ has such a small slope in this case. Near $ka = \pi/2$, we can approximate the cosine curve as follows:

$$\cos ka = \sin\left(\frac{\pi}{2} - ka\right) \approx \frac{\pi}{2} - ka$$

so that Eq. (35) becomes

$$\frac{\pi}{2} - (ka)_1 = \frac{(ka)_1}{\beta a}$$

and

$$(ka)_1 = \frac{\dfrac{\pi}{2}}{1 + \dfrac{1}{\beta a}} = 1.48$$

The second value of ka is still in the region where the cosine curve is linear, so

$$(ka)_2 = 2(ka)_1 = 2.96$$

The energy levels are found from the definition of k:

$$k = \left(\frac{2mE}{\hbar^2}\right)^{1/2}$$

Rather than plug in the values of a, m, and $\hbar$ all over again, we recall that

$$\beta = \left(\frac{2mV_0}{\hbar^2}\right)^{1/2}$$

so that

$$\frac{k}{\beta} = \left(\frac{E}{V_0}\right)^{1/2}$$

and

$$E = V_0\left(\frac{ka}{\beta a}\right)^2$$

Therefore

$$E_1 = 10\left(\frac{1.48}{16.2}\right)^2 \text{eV} = 0.083 \text{ eV}$$

and

$$E_2 = 10\left(\frac{2.96}{16.2}\right)^2 \text{eV} = 0.334 \text{ eV}$$

Notice that these energies are just slightly lower than the lowest energies in an infinitely deep well of the same width, for which the energy would be, in general,

$$E = \frac{\hbar^2 k^2}{2m} = \frac{\hbar^2 \left(\frac{n\pi}{2a}\right)^2}{2m} \tag{37}$$
$$= \frac{n^2 h^2}{32ma^2}$$

The two lowest levels would have

$$(ka)_1 = \pi/2 \; ; \qquad E_1 = 0.094 \text{ eV}$$
$$(ka)_2 = \pi \; ; \qquad E_2 = 0.376 \text{ eV}$$

Each of these energies is about 12% larger than the corresponding energy in the example problem. The difference is easily related to the wavefunctions. In an infinitely deep well, an integer number of half-wavelengths must fit between the boundaries, whereas in the well of finite depth, each wavelength must be somewhat longer than the corresponding wavelength in the infinitely deep well, to allow the wavefunction to expand into the forbidden region outside the well and go smoothly to zero instead of having a discontinuous slope at the edge of the well. (A discontinuous slope is only permitted when the potential energy is infinite, because it means that $\partial^2\psi/\partial x^2$ is infinite.)

Fig. 4b shows, for the infinitely deep well, the probability density $|u|^2$ as a function of x. You will notice that, in all states except the lowest one, the probability density goes to zero at certain "nodes" inside the well. We have come very far from the classical idea of a particle bouncing back and forth inside a box. If you wonder how the electron can go from side to side in the well without being found near a node,[3] it is helpful to think of a standing wave on a vibrating string. Energy flows back and forth along the string, but there are nodes at which one cannot *extract* this energy from the string. Similarly, in *interacting with* the electron in the box, one does not often find it near a node.

Parity of Solutions. It is not merely an accident that $u(x) = \pm u(-x)$ for all of the solutions of the square-well problem; it is a necessary consequence of the symmetry of the potential—the fact that $V(x) = V(-x)$. We can prove this by writing the Schrödinger equation with x replaced by $-x$ everywhere, and comparing this with the original equation. We have

$$\frac{d^2u(x)}{dx^2} - \frac{2m}{\hbar^2} V(x)u(x) = -\frac{2m}{\hbar^2} Eu(x) \qquad (38)$$

and

$$\frac{d^2u(-x)}{dx^2} - \frac{2m}{\hbar^2} V(-x)u(-x) = -\frac{2m}{\hbar^2} Eu(-x) \qquad (39)$$

If $V(-x) = V(x)$, Eq. (39) becomes

$$\frac{d^2u(-x)}{dx^2} - \frac{2m}{\hbar^2} V(x)u(-x) = -\frac{2m}{\hbar^2} Eu(-x) \qquad (40)$$

Equation (40) is identical to Eq. (38), except that the wavefunction is $u(-x)$ instead of $u(x)$. It is clear from our graphical discussion of the square well that there is only one wavefunction for a given energy E;[4] that is, $u(-x)$ can differ from $u(x)$ only by a multiplicative constant:

$$u(-x) = au(x) \qquad (41)$$

Since Eq. (41) must hold for *all* values of x, it must be valid if we replace x by $-x$. This yields

---

[3] We say "near" a node rather than "at" a node, because, classically or quantum mechanically, the probability of finding a particle *at* a precise point is always zero. There are an infinite number of possible points. The only meaningful number is the probability of finding the electron in a given *interval* or volume.

[4] In many problems, it is possible that there is more than one wavefunction for a given energy. In that case, the solutions are said to be "degenerate." It is then possible, even with a symmetrical potential, to have wavefunctions which are neither even nor odd. However, we can still combine the wavefunctions for a given energy in such a way as to produce new wavefunctions which are even or odd.

$$u(x) = au(-x)$$

Therefore

$$a^2 = 1$$
$$a = \pm 1$$

and $u(x)$ is either even or odd. If $u(x) = +u(-x)$, we say that the solution has *even parity*; if $u(x) = -u(-x)$, the solution has *odd parity*.

It is often very helpful in solving a problem to know that the solutions must be either even or odd. Suppose that you wished to find the wavefunctions for a particle in the well of Fig. 7a:

$$V = V_0, \qquad |x| < a$$
$$V = 0, \qquad a < |x| < b$$
$$V = +\infty, \qquad |x| > b$$

for energies $E < V_0$.

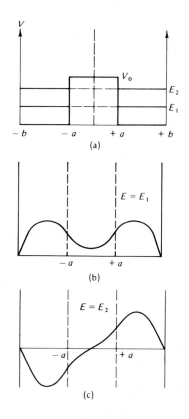

(a)

(b)

(c)

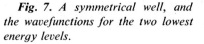

Fig. 7. *A symmetrical well, and the wavefunctions for the two lowest energy levels.*

From our experience with the square well, we know that the solution in the "forbidden" region $|x| < a$ is an exponential: either $e^{\alpha x}$, $e^{-\alpha x}$, or a linear combination of these. (As before, $\alpha = [2m(V_0 - E)/\hbar^2]^{1/2}$.) But because the potential is symmetrical, the wavefunction must be either

or

$$\left.\begin{array}{l} u(x) = A \cosh \alpha x \\[1em] u(x) = A \sinh \alpha x \end{array}\right\} \quad (|x| < a)$$

because these are the even and odd combinations of $e^{\alpha x}$ and $e^{-\alpha x}$, respectively.

In the regions $a < |x| < b$, we know that the solution is a linear combination of $\sin kx$ and $\cos kx$ [with $k = (2mE/\hbar^2)^{1/2}$], which may be written in one of the forms $B \sin(kx \pm \phi)$ or $B \cos(kx \pm \phi)$. To make the solution an *even* function of x, we write it as

$$u(x) = B \cos(kx + \phi) \quad (-b < x < -a)$$
$$u(x) = B \cos(kx - \phi) \quad (+a < x < +b)$$

Inspection of Fig. 8a, or substitution of $-x$ for x in the above equations, shows that $u(x)$ is indeed even. For the *odd* solution we may write

$$u(x) = B \sin(kx + \phi) \quad (-b < x < -a)$$
$$u(x) = B \sin(kx - \phi) \quad (a < x < b)$$

Inspection of Fig. 8b shows that this solution is odd.

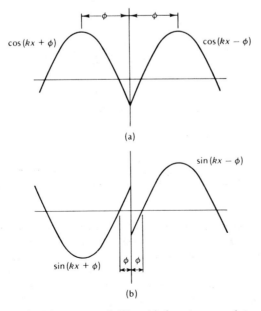

(a)

(b)

Fig. 8. (a) *Even and* (b) *odd functions used in solving the potential well of Fig. 7.*

The problem has now been considerably reduced in complexity. The two lowest-energy solutions are shown in Fig. 7b and 7c. The two boundary conditions at $x = a$ (continuity of u and of du/dx), together with the condition that $u = 0$ at $x = b$, are sufficient to determine the wavefunction; two equations may be used to eliminate two of the three constants, A, B, and ϕ, and the third equation then determines the allowed values of k. (The third of the three constants is determined by normalizing the wavefunction.) Details are left as an exercise for the student.

5.5   BARRIER PENETRATION

Our solution to the above problem seems to indicate that a particle can pass back and forth through a potential barrier, even though its energy is less than the potential energy at the top of the barrier. Let us investigate this possibility more thoroughly, by considering a stream of particles incident from the left upon the square potential barrier shown in Fig. 9. We wish to find, as a function of the barrier dimensions and the particle energy, the probability that a particle can penetrate the barrier and emerge to be observed on the other side.

Classically, when $E < V_0$, all of the particles would be reflected, and when $E > V_0$, all would be transmitted past the barrier. We have already seen

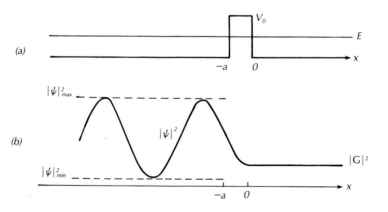

Fig. 9. (a) Square potential barrier of width a and height $V_0 = \hbar^2/2ma^2$. (b) Probability density, drawn to scale for particles incident from the left with energy $E = V_0/2$; $|\psi|^2$ oscillates for $x < -a$, where the reflected wave interferes with the incident wave, and it decays inside the barrier in such a way as to have zero slope at $x = 0$, where it becomes constant.

(Section 5.2) that the classical prediction is wrong, at least for particles of energy $E > V_0$, incident on an infinitely long barrier. What happens when E is less than V_0? In that case, just as in the square well, the wavefunction has an exponential form in the "forbidden" region inside the barrier. But there is also an oscillatory wave to the right of the barrier, which is required in order that the boundary conditions be satisfied. Thus we must conclude that some particles "tunnel" through the barrier and appear on the other side. This tunneling process is not allowed classically, because the particles have to traverse a region where their kinetic energy is negative, but the wave nature of matter forces us to admit this possibility. We shall find that tunneling is well verified experimentally.

There is an analogous situation in optics. Consider a beam of light which undergoes total internal reflection in a piece of glass. Even though no visible light emerges from the glass, there is an electromagnetic field in the region just outside the glass. The existence of this field may be demonstrated by bringing a second piece of glass near the face of the first piece, without allowing the two pieces to touch. A beam of light appears in the second glass (Fig. 10), as the standing electromagnetic field is converted back into a traveling

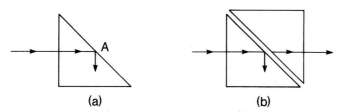

(a)   (b)

Fig. 10. (a) *Light wave is totally reflected inside a prism, at point A.* (b) *When second prism is brought near enough, the light wave continues in its original direction into the second prism. Even though there is no light wave in the space between the prisms, there is an electromagnetic field in that region, and the boundary conditions on this field require that a traveling light wave be produced in the second prism.*

wave. This effect was observed by Newton, who considered it evidence for the corpuscular theory of light![5] The effect has since been studied very carefully, and it has been observed that the intensity of the second light beam decreases exponentially with increasing distance between the prisms, as required by the wave theory.[6]

[5] I. Newton, "Opticks," 4th ed. (1730), Book III, Query 29 (pp. 194 and 205 in 1952 edition by Dover, New York).
[6] D. D. Coon, *Amer. J. Phys.* **34**, 240 (1966).

Fig. 9b should be compared with Fig. 1c. In both figures, $|\psi|^2$ is constant for $x > 0$ (where there is no leftward-traveling wave), so $|\psi|^2$ must have zero slope at $x = 0$. For $-a < x < 0$, the solution is of the form

$$u(x) = Ce^{\alpha x} + De^{-\alpha x}$$

where $\alpha^2 = 2m(V_0 - E)/\hbar^2$. Clearly, neither C nor D can be zero if $|u|^2$ is to have zero slope at $x = 0$. For $x < -a$, $|\psi|^2$ oscillates as in Fig. 1c, but it is not equal to $|\psi|^2_{max}$ at the boundary of this region.

To calculate the transmission probability, we must use the continuity requirement on u and du/dx at both $x = -a$ and $x = 0$, beginning by writing the solutions as

$$u = Ae^{ikx} + Be^{-ikx} \qquad (x < -a)$$

$$u = Ce^{\alpha x} + De^{-\alpha x} \qquad (-a < x < 0)$$

$$u = Ge^{ikx} \qquad (x > 0)$$

Multiplication of the above functions by the time factor $e^{-i\omega t}$ will yield wavefunctions of the form used in Section 5.2, for the region outside the barrier: $e^{i(kx - \omega t)}$ for a wave traveling toward the right, or $e^{-i(kx + \omega t)}$ for a wave traveling toward the left. Notice that we allow for waves traveling in both directions for $x < -a$, to represent the incident particles and the reflected particles, but that we only have a wave traveling toward the right for $x > 0$. Notice too that, as mentioned above, both C and D are nonzero; this would not be the case if the barrier were infinitely thick, like the walls of our square well, for then the condition that $u \to 0$ as $x \to \infty$ would take the place of the boundary conditions at $x = 0$, and would force the constant C to equal zero.

The boundary conditions at $x = 0$ and $x = -a$ yield four equations, which we can solve for the four unknown ratios B/A, C/A, D/A, and G/A:

Continuity at $x = -a$	Continuity at $x = 0$
u: $\quad Ae^{-ika} + Be^{ika} = Ce^{-\alpha a} + De^{\alpha a}$	$C + D = G$
$\dfrac{du}{dx}$: $\quad ik(Ae^{-ika} - Be^{ika}) =$ $\qquad = \alpha(Ce^{-\alpha a} - De^{\alpha a})$	$\alpha(C - D) = ikG$

After some straightforward (but tedious) algebra, we find that

$$\frac{B}{A} = \left[\frac{(k^2 + \alpha^2) \sinh \alpha a}{2ik\alpha \cosh \alpha a + (k^2 - \alpha^2) \sinh \alpha a} \right] e^{-2ika} \qquad (42)$$

The reflection coefficient is therefore

$$R = \frac{BB^*}{AA^*} \equiv \frac{|B|^2}{|A|^2} = \left[1 + \frac{4k^2\alpha^2}{(k^2 + \alpha^2)^2 \sinh^2 \alpha a} \right]^{-1}$$

Recalling the definitions $k^2 = 2mE/\hbar^2$ and $\alpha^2 = 2m(V_0 - E)/\hbar^2$, we may now write

$$R = \left[\, 1 + \frac{4E(V_0 - E)}{V_0^2 \sinh^2 \alpha a}\, \right]^{-1}$$

Similarly, we may solve for G/A and find the transmission coefficient. The result is

$$T = \frac{|G|^2}{|A|^2} = \left(\, 1 + \frac{V_0^2 \sinh^2 \alpha a}{4E(V_0 - E)}\, \right)^{-1}$$

You may verify that $R + T = 1$. (Notice that no velocity factor enters here, as it did in Section 5.2, because the potential, and hence the speed, is the same on both sides of the barrier.)

We may apply the above results to particles of energy $E > V_0$ as well. The equations are unchanged except that α becomes imaginary. Thus, since $\sinh x = -i\sin ix$,

$$R = \left(\, 1 + \frac{4E(E - V_0)}{V_0^2 \sin^2 i\alpha a}\, \right)^{-1} \quad \text{and} \quad T = \left(\, 1 + \frac{V_0^2 \sin^2 i\alpha a}{4E(E - V_0)}\, \right)^{-1}$$

where the argument of the sine function is the real quantity

$$i\alpha a \equiv \sqrt{2ma^2\,(E - V_0)/\hbar^2}$$

You can see that $T = 1$ when $i\alpha a = n\pi$; this shows up clearly in Fig. 11. An effect of this sort is well known in the transmission of light through a thin layer. The reason for the effect is illustrated in Fig. 12.

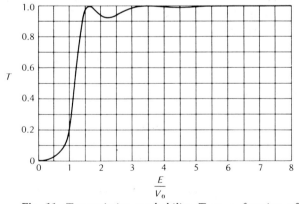

Fig. 11. *Transmission probability T as a function of energy E, for a rectangular barrier of width a and height V_0 equal to $8\hbar^2/ma^2$. (Adapted from L. I. Schiff, "Quantum Mechanics," 2nd ed. Copyright McGraw-Hill, New York, 1955. Used with permission of McGraw-Hill Book Company.)*

We can gain more insight into this phenomenon, and into the appearance of Figs. 9 and 12, by studying the probability current S, which is given by Eq. (18) as

$$S = (\hbar k/m)|\psi|_{max}|\psi|_{min}$$

for any region in which there are traveling waves; $|\psi|_{max}$ and $|\psi|_{min}$ are the maximum and minimum values of $|\psi|$ in that region. For $x > 0$, $|\psi|$ is constant and S reduces to

$$S = (\hbar k/m)|\psi|^2 = (\hbar k/m)|G|^2$$

Since S is the same everywhere (no sources or sinks), we can deduce that

$$|G|^2 = |\psi|_{max}|\psi|_{min}$$

In other words, |G| is the geometric mean of the maximum and minimum values of |ψ| in the region to the left of the barrier.

One can deduce the reflection coefficient directly from an accurately drawn graph of |ψ|². R is the ratio of the leftward intensity to the rightward intensity on the left side of the barrier, or

$$R = \frac{|B|^2}{|A|^2} = \frac{\mathcal{B}^2}{\mathcal{A}^2}$$

where, as in Section 5.3, we define $\mathcal{A}$ and $\mathcal{B}$ as real, positive quantities. The only difference here is that we need a phase angle between $\mathcal{A}$ and $\mathcal{B}$, because Eq. (12) no longer holds. So we define

$$A = \mathcal{A}e^{i\phi} \qquad \text{and} \qquad B = \mathcal{B}e^{i\phi}e^{i\theta}$$

Then (compare with Sec. 5.3)

$$\begin{aligned} |\psi|^2 &= |A|^2 + |B|^2 + AB^*e^{2ikx} + A^*Be^{-2ikx} \qquad\qquad x < -a\\ &= \mathcal{A}^2 + \mathcal{B}^2 + \mathcal{A}\mathcal{B}\{e^{2ikx}e^{-i\theta} + e^{-ikx}e^{i\theta}\}\\ &= \mathcal{A}^2 + \mathcal{B}^2 + 2\mathcal{A}\mathcal{B}\cos(2kx - \theta) \end{aligned}$$

As before, $|\psi|_{min} = \mathcal{A} - \mathcal{B}$, $|\psi|_{max} = \mathcal{A} + \mathcal{B}$, and $R = \dfrac{\mathcal{B}^2}{\mathcal{A}^2}$, so

$$|\psi|_{min}/|\psi|_{max} = \frac{\mathcal{A} - \mathcal{B}}{\mathcal{A} + \mathcal{B}} = \frac{1 - \sqrt{R}}{1 + \sqrt{R}}$$

Solving for R, we have

$$R = \left(\frac{1 - |\psi|min/|\psi|max}{1 + |\psi|min/|\psi|max}\right)^2$$

This relation holds for both Fig. 9b and Fig. 12c. In Fig. 9b, the ratio of $|\psi|_{min}$ to $|\psi|_{max}$ is about ¼, so $R \approx (\frac{3}{5})^2 = 0.36$. In Fig 12c, this ratio is one, so $R = 0$.

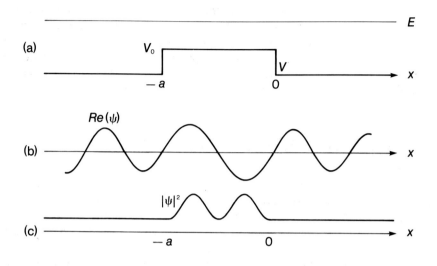

Fig. 12. *Perfect transmission past a potential barrier.*
(a) Total energy E and potential energy V. (b) Real part of
wavefunction, Re(ψ), shown at one particular time; at some
other time, the nodes could be at different values of x, but
the magnitude of Re(ψ) at x = 0 will always equal the
magnitude at x = −a, because the wavelength is 2a/n (n =
2 in this case). Notice that the wavelength and amplitude
outside the barrier are smaller than they are inside, because
the kinetic energy is greater outside. (c) Probability density
$|\psi|^2$, which must have zero slope at x = 0 if there is no
reflected wave (because in that case $|\psi|^2$ is constant for x
> 0). When $|\psi|^2$ has a wavelength of a/n (n being an
integer), it also has zero slope at x = −a, joining smoothly
to the same constant value that it has for x > 0. This means
that there is no reflected wave for x < −a. As in Fig. 1c,
the wavelength of $|\psi|^2$ is half the wavelength of Re(ψ).

The ability of the particle to penetrate the barrier even when $E < V_0$ can be considered to be another manifestation of the uncertainty principle. In order to know whether a particle is on one side of the barrier or the other, one must know its position with an uncertainty δx of less than the barrier thickness a. But a measurement of x to this precision introduces an uncertainty into the momentum, and hence into the kinetic energy, so that the particle may occasionally acquire sufficient energy to pass the barrier, even though, on the average, its energy is too small.

Quantum mechanical barrier penetration has been observed in many different situations—for example, in the oscillations of atoms in molecules, in the tunneling of electrons through a potential barrier between two super-conductors, or in the emission of alpha particles by atomic nuclei. The quantum mechanical theory not only explains these classically inexplicable occurrences, it also enables one to calculate quantitatively the rate at which these processes occur. (See the analysis of alpha decay in Chapter 15.)

5.6   EXPECTATION VALUES AND OPERATORS

Now that we have had some practice with wavefunctions, let us try to acquire a deeper understanding of the meaning of the wavefunction and the wave equation before we tackle our next (and last) one-dimensional problem.

Although we are not able to use, in general, the wavefunction to predict the exact result of a measurement of the x coordinate of a particle, we *can* predict the *average* result of a large number of measurements. Suppose that one measures the position of a particle in a number of systems in which the wavefunctions are identical. The arithmetic mean of all the values of x, called the *expectation value* of x, is, by definition,

$$\langle x \rangle \equiv \bar{x} = \int_{-\infty}^{\infty} x P(x)\, dx,$$

where $\int_{x_1}^{x_2} P(x)\, dx$ is the probability of finding a value of x between x_1 and x_2. But we know that $P(x)$ is given by $\psi^*(x)\psi(x)$, so we may write[7]

$$\boxed{\bar{x} = \int_{-\infty}^{\infty} \psi^*(x) x \psi(x)\, dx} \qquad (43)$$

(The x is written between the ψ^* and the ψ for consistency with later develop-ments.) It is not too hard to deduce that the same relation holds for any power of x:

$$\langle x^n \rangle \equiv \overline{x^n} = \int_{-\infty}^{\infty} \psi^*(x) x^n \psi(x)\, dx$$

---

[7] Of course, since ψ depends on t as well as x, the expectation value $\bar{x}$ depends on the time at which the measurements were made. But we do not need to exhibit the time dependence explicitly here, since none of our integrations involves t.

and therefore one may also write

$$\langle f(x) \rangle \equiv \overline{f(x)} = \int_{-\infty}^{\infty} \psi^*(x) f(x) \psi(x) \, dx$$

where $f(x)$ is any function of x which may be expanded in a power series.

One may also use the wavefunction to find the expectation value of the momentum. We recall that the Fourier transform of the wavefunction is the probability amplitude for the various values of k, just as the wavefunction itself is the probability amplitude for values of x. This means that $\int_{k_1}^{k_2} \phi^*(k)\phi(k) \, dk$ is the probability that the k-value lies between k_1 and k_2, or that a measurement of k will yield such a value.[8] Measuring k is equivalent to measuring p_x, since $p = \hbar k$. Therefore

$$\bar{k} = \int_{-\infty}^{\infty} \phi^*(k) k \phi(k) \, dk$$

by definition of the mean value, and

$$\bar{p}_x = \hbar \bar{k} = \int_{-\infty}^{\infty} \phi^*(k) \hbar k \phi(k) \, dk \tag{44}$$

But it is not necessary to find $\phi(k)$ in order to compute p_x; we can find p_x directly from the wavefunction. The formula for doing this is found by substituting the expression for $\phi(k)$ in terms of $\psi(x)$ (Chapter 4, Eq. (23)),

$$\phi(k) = \frac{1}{(2\pi)^{1/2}} \int_{-\infty}^{\infty} \psi(x) e^{-ikx} \, dx$$

and the complex conjugate equation[9]

$$\phi^*(k) = \frac{1}{(2\pi)^{1/2}} \int_{-\infty}^{\infty} \psi^*(x') e^{ikx'} \, dx'$$

into Eq. (44) to obtain

$$\bar{p}_x = \frac{\hbar}{2\pi} \int_{-\infty}^{\infty} dk \int_{-\infty}^{\infty} \psi^*(x') e^{ikx'} \, dx' \int_{-\infty}^{\infty} k \psi(x) e^{-ikx} \, dx$$

But $k e^{-ikx} = i(\partial e^{-ikx}/\partial x)$, so we may write

$$\bar{p}_x = \frac{\hbar i}{2\pi} \int_{-\infty}^{\infty} dk \int_{-\infty}^{\infty} \psi^*(x') e^{ikx'} \, dx' \int_{-\infty}^{\infty} \psi(x) \frac{\partial}{\partial x} (e^{-ikx}) \, dx$$

[8] Of course, $\phi(k)$ must be normalized ($\int_{-\infty}^{\infty} \phi^*(k)\phi(k) \, dk = 1$) so that the probability for an event is always ≤ 1. But it can be shown from the Fourier integral theorem that $\phi(k)$ is normalized if $\psi(x)$ is.

[9] The dummy variable is x', to avoid confusion in the subsequent equations.

and integration by parts on x yields (since $\psi(\infty) = 0$),

$$\bar{p}_x = -\frac{\hbar i}{2\pi} \int_{-\infty}^{\infty} dk \int_{-\infty}^{\infty} \psi^*(x')e^{ikx'} \, dx' \int_{-\infty}^{\infty} e^{-ikx} \frac{\partial \psi(x)}{\partial x} \, dx \qquad (45)$$

We may rearrange the terms in the above equation to obtain

$$\bar{p}_x = \frac{\hbar}{i} \int_{-\infty}^{\infty} \left[\frac{1}{2\pi} \int_{-\infty}^{\infty} e^{-ikx} \, dk \int_{-\infty}^{\infty} \psi^*(x')e^{ikx'} \, dx' \right] \frac{\partial \psi(x)}{\partial x} \, dx$$

(Be assured that this rearrangement is legitimate. It essentially consists of (a) placing factors which are independent of x inside the integral on x, and (b) integrating on k before integrating on x.)

The quantity in brackets can now be identified simply as $\psi^*(x)$, according to Eq. (22) of Chapter 4 (the Fourier integral theorem). The final result is therefore

$$\langle p_x \rangle \equiv \bar{p}_x = \int_{-\infty}^{\infty} \psi^*(x) \frac{\hbar}{i} \frac{\partial \psi(x)}{\partial x} \, dx \qquad (46)$$

Notice the similarity between Eq. (43) for x and Eq. (46) for p_x. Equation (46) is identical to Eq. (43) except for the replacement of the factor x by the *operator* $(\hbar/i)/(\partial/\partial x)$. The operator $(\hbar/i)(\partial/\partial x)$ is called the *momentum operator*; it yields the expectation value of p_x when used according to the prescription of Eq. (46). This result should not be too surprising. Suppose that $\psi(x)$ is $e^{ik_0 x}$, a wavefunction for a state of definite momentum $p_0 = \hbar k_0$. Application of the momentum operator yields

$$\frac{\hbar}{i} \frac{\partial}{\partial x} \psi(x) = \hbar k_0 \, e^{ik_0 x}$$
$$= p_0 \psi(x)$$

and there is the same connection between the *operator* $(\hbar/i)/(\partial/\partial x)$ and the variable p_x. In this case $\bar{p}_x = p_0$; only one value of p_x is possible as a result of a measurement.

We see therefore that the wavefunction is not merely a probability amplitude for the *position* of a particle; the rate at which the wavefunction changes with x gives the x component of the *momentum* of the particle. This is a logical extension of the connection between momentum and wavelength in a de Broglie wave.

Many other dynamical variables are represented by operators in quantum theory. In general, the average value of the operator o is given, by analogy to Eq. (46), as

$$\boxed{\langle o \rangle = \int_{-\infty}^{\infty} \psi^* \, o\psi \, dx}$$

For example, applying p_x twice *as an operator* gives the operator for p_x^2:

$$p_x^2 = p_x p_x = \frac{\hbar}{i}\frac{\partial}{\partial x}\left(\frac{\hbar}{i}\frac{\partial}{\partial x}\right) = -\hbar^2\frac{\partial^2}{\partial x^2}$$

In general,

$$p_x^n = \frac{\hbar^n}{i^n}\frac{\partial^n}{\partial x^n} \qquad \text{and} \qquad \overline{p_x^n} = \int_{-\infty}^{\infty}\psi^*(x)\frac{\hbar^n}{i^n}\frac{\partial^n}{\partial x^n}\psi(x)\,dx$$

We have already used the p^2 operator; it appears in the Shrödinger equation. The total energy of a particle is $(p^2/2m) + V$, and the corresponding energy operator[10] is $-(\hbar^2/2m)/(\partial^2/\partial x^2) + V$. Applying this operator to ψ yields the time-independent Schrödinger equation:

$$\left(-\frac{\hbar^2}{2m}\frac{\partial^2}{\partial x^2} + V\right)\psi = E\psi \tag{47}$$

Expectation Values and Uncertainty.   The quantities δx and δp that appear in the uncertainty relation (Chapter 4, Eq. (27)) are defined experimentally, as the root-mean-square (r.m.s.) deviation (from the mean) of the results of individual measurements on identical systems. Now we are in a position to calculate these deviations by using expectation values. Let us define Δx to be the deviation in an individual measurement; thus

$$\Delta x \equiv x - \langle x \rangle$$

Then the r.m.s. deviation is

$$\delta x \equiv \langle(\Delta x)^2\rangle^{1/2} = \langle(x - \langle x\rangle)^2\rangle^{1/2}$$

with $\langle x \rangle$ as given by Eq. (43), and $\langle(\Delta x)^2\rangle$ computed in the same way:

$$\langle(\Delta x)^2\rangle = \int_{-\infty}^{\infty}\psi^*(x)\,(\Delta x)^2\psi(x)\,dx \tag{48}$$

For example, if ψ is a normalized Gaussian wavefunction, then $|\psi|^2 = (1/\sigma\sqrt{2\pi})\,e^{-x^2/2\sigma^2}$, where σ is the standard deviation. (See Appendix A and Section 4.4) We can see by inspection that $\langle x \rangle$ equals zero for this wavefunction. Then $\Delta x = x$, and

$$\langle(\Delta x)^2\rangle = \frac{1}{\sigma\sqrt{2\pi}}\int_{-\infty}^{\infty}x^2\,e^{-x^2/2\sigma^2}\,dx$$

---

[10] This operator is often called the Hamiltonian, a term carried over from classical mechanics. Do not worry if you never heard this term before: familiarity with the jargon is not essential for understanding the physics.

The value of this integral is unchanged if we change the lower limit to zero and multiply by two. If we also change the variable to $u = x^2/2\sigma^2$, then $du = xdx/\sigma^2$, so $xdx = \sigma^2 du$, $x = u^{1/2}\sigma\sqrt{2}$, and we obtain

$$\langle(\Delta x)^2\rangle = \frac{2}{\sigma\sqrt{2\pi}}\,\sigma^3\,\sqrt{2}\int_0^\infty u^{\frac{1}{2}}\,e^{-u}\,du$$

We recall from Section 3.1 that the gamma function is defined as

$$\Gamma(n) = \int_0^\infty u^{n-1}\,e^{-u}\,du$$

so we have

$$\langle(\Delta x)^2\rangle = (2\sigma^2/\sqrt{\pi})\,\Gamma\left(\tfrac{3}{2}\right)$$

and since $\Gamma(\tfrac{3}{2}) = \tfrac{1}{2}\sqrt{\pi}$, the final result is

$$\delta x \equiv \langle(\Delta x)^2\rangle = \sigma$$

as we should expect from the definition of σ.

Using the same wavefunction, we can also calculate the momentum uncertainty, with the analogous definitions

$$\Delta p_x \equiv p_x - \langle p_x\rangle \quad \text{and} \quad \delta p_x \equiv \langle(\Delta p_x)^2\rangle^{1/2}$$

Symmetry again tells us that $\langle p_x\rangle = 0$, so we have, using the p_x operator

$$\langle(\Delta p_x)^2\rangle = \frac{1}{\sigma\sqrt{2\pi}}\int_{-\infty}^\infty e^{-x^2/4\sigma^2}\,(\hbar^2/i^2)\,\frac{\partial^2}{\partial x^2}\,(e^{-x^2/4\sigma^2})\,dx$$

$$= -\frac{\hbar^2}{2\sigma^3\sqrt{2\pi}}\int_0^\infty (x^2 - 2)\,e^{-x^2/2\sigma^2}\,dx$$

The same change of variable to $u = x^2/2\sigma^2$ reduces this integral to the sum of two gamma functions, with the final result that $\langle(\Delta p_x)^2\rangle = \hbar^2/4\sigma^2$

or

$$\delta p_x = \hbar/2\sigma$$

Therefore we find that the uncertainty product for this ψ is equal to $\hbar/2$, in agreement with the result obtained in Section 4.4 by use of the Fourier transform. But now we see precisely what is meant by δx and δp_x in terms of experimental results. Furthermore, we can now evaluate δx and δp_x for any other wavefunction. (See Problem 13.) If we do that, we see in general that the uncertainty product is greater than $\hbar/2$ when the wavefunction is not Gaussian.[11]

---

[11] For a mathematical proof of this inequality, see L. I. Schiff, "Quantum Mechanics," 54, McGraw-Hill, New York, 1949.

Eigenfunctions and Eigenvalues. The time-independent Schrödinger equation is an example of an *eigenvalue* equation, which may be defined as an equation of the form

$$\text{operator} \times \text{function} = \text{number} \times \text{function}$$

where the function is the same on both sides of the equation. The "number" is called the eigenvalue, and a function which satisfies such an equation for a given operator is called an *eigenfunction* of that operator. As another example, the equation

$$\frac{\hbar}{i}\frac{\partial}{\partial x}e^{ikx} = \hbar k e^{ikx}$$

is an eigenvalue equation, in which e^{ikx} is the eigenfunction and $\hbar k$ is the eigenvalue, of the operator $(\hbar/i)/(\partial/\partial x)$. Notice that $\sin kx$ is *not* an eigenfunction of this operator, for

$$\frac{\hbar}{i}\frac{\partial}{\partial x}\sin kx = \left(\frac{\hbar k}{i}\right)\cos kx$$

and the function $\sin kx$ is not reproduced after application of the operator.

Because we are able to produce operators corresponding to any power of p, we can produce operators for any dynamical variable that is a function of position and momentum simply by combining the variable x (or y or z, in three dimensions) with the appropriate $p_x, p_y,$ or p_z operator(s). For example, we shall, in Chapter 6, use the angular momentum operator. The operator for the z component of angular momentum is found directly from the classical expression for this quantity:

$$L_z = xp_y - yp_x$$
$$= \frac{x\hbar}{i}\frac{\partial}{\partial y} - \frac{y\hbar}{i}\frac{\partial}{\partial x}$$

We are now in a position to restate in more general form the ideas behind the wave equation. We state the following two *fundamental postulates*.

POSTULATE 1.   Any dynamical variable describing the motion of a particle can be represented by an operator.

POSTULATE 2.   The only possible result of a measurement of a dynamical variable is one of the eigenvalues of the corresponding operator. Immediately after measurement of a dynamical variable, the wavefunction must be an eigenfunction of this operator, and the corresponding eigenvalue must equal the measured value of the variable. [12]

---

[12] For example, if one measures the energy of a particle which is bound in a square well, the measurement must yield one of the levels which we found in Section 5.4, which are, by definition, the eigenvalues of the energy operator for the square well.

Being postulates, these statements cannot be "proved." But we can judge them in terms of their consistency with the theory and experiment. It is clear that they are consistent with what we know about the energy operator and the solutions of the Schrödinger equation; in fact, *we could start with these postulates or similar ones and from them "derive" the Schrödinger equation.* Postulates 1 and 2 are also in agreement with the discussion of expectation values in this section, at least for the case where the wavefunction ψ is an eigenfunction of the operator; in that case the eigenvalue and the expectation value are identical. (We shall, in Chapter 6, explore the situation in which the wavefunction for a system is *not* one of the eigenfunctions of the dynamical operator that we wish to study.)

To obtain further confirmation of the validity of these postulates, we must use them to make predictions which can be tested experimentally. So far, they have passed every test.

5.7 THE HARMONIC OSCILLATOR

As a further illustration of the principles of quantum mechanics, let us find the energy levels of the harmonic oscillator. Knowledge of the harmonic oscillator provides the basis for understanding much more complicated systems because any finite potential can be approximated by the harmonic oscillator potential for small oscillations of a particle about the minimum.

The energy levels of the harmonic oscillator are found by solving the time-independent Schrödinger equation, with the potential energy $V = \frac{1}{2}Kx^2$. The equation is

$$\left(-\frac{\hbar^2}{2m}\frac{d^2}{dx^2} + \tfrac{1}{2}Kx^2\right)u = Eu,$$

which, by means of the definitions

$$a^2 \equiv mK/\hbar^2 \quad \text{and} \quad b \equiv 2mE/\hbar^2$$

may more simply be written

$$\left(\frac{d^2}{dx^2} - a^2x^2\right)u = -bu \tag{49}$$

Because classical mechanics tells us that $(K/m)^{1/2}$ is the angular frequency ω, we may write a in terms of ω as $a^2 = m^2\omega^2/\hbar^2$, or $a = m\omega/\hbar$.

We may proceed, as we did with the square well, to sketch graphs of the possible wavefunctions and show that only a discrete set of energy levels is possible (see Fig. 13). The wavefunctions are somewhat similar in appearance to those for the square well. Inside the well, where $E > V$, the wavefunction

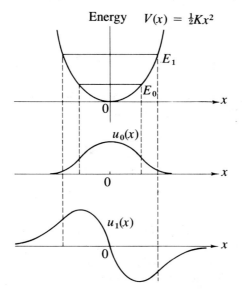

Fig. 13. *Harmonic oscillator potential and the wavefunctions for the two lowest allowed energy levels.*

curves toward the axis and tends to oscillate; and outside the well the wavefunction curves away from the axis like a decaying exponential. Again we see that an energy $E = 0$ is not allowed; the lowest possible level has a minimum energy, called the "zero-point energy," which is required by the uncertainty principle. (If E were zero, the only allowed position of the particle would be $x = 0$, and its momentum would be exactly zero.) Our rough sketch shows that the lowest energy wavefunction resembles a Gaussian, and we shall find that it *is* a Gaussian.

The difference between this case and the square well is that, as $E - V$ varies, the value of $(d^2u/dx^2)/u$ varies continuously, instead of having a single sudden change at the side of the well. Thus we might expect $u(x)$ to be more complicated than the wavefunction for the square well.

We could go ahead and solve Eq. (49) by a standard technique; we could assume that $u(x)$ can be represented by a power series in x with undetermined coefficients, and we could find relations among the coefficients by substituting this series for $u(x)$ in the differential equation. This procedure would enable us to find the eigenvalues, but it does not tell us much more about the physics. Let us instead find the energy levels by another technique, which further illustrates the role of operators in quantum mechanics. (The power series method is adequately illustrated in Appendix C in the solution of the hydrogen atom.)

We begin by "factoring" the operator in Eq. (49). We write the equation as[13]

$$\left[\left(\frac{d}{dx} - ax\right)\left(\frac{d}{dx} + ax\right) - a\right]u = -bu \tag{50}$$

Expanding the left-hand side of Eq. (50) verifies that it is equivalent to the left-hand side of Eq. (49). The expansion yields

$$\frac{d^2u}{dx^2} - ax\frac{du}{dx} + \frac{d}{dx}(axu) - a^2x^2u - au$$

or

$$\frac{d^2u}{dx^2} - ax\frac{du}{dx} + au + ax\frac{du}{dx} - a^2x^2u - au$$

All of the terms cancel except the first and fifth, leaving $(d^2u/dx^2) - a^2x^2u$, which is just the left-hand side of Eq. (49).

We may consider Eq. (50) to be a new eigenvalue equation with the eigenvalue $-b + a$, for Eq. (50) may be written as

$$\left(\frac{d}{dx} - ax\right)\left(\frac{d}{dx} + ax\right)u = (-b + a)u \tag{51}$$

Let us now apply the operator $(d/dx + ax)$ to both sides of Eq. (51). We obtain

$$\left(\frac{d}{dx} + ax\right)\left(\frac{d}{dx} - ax\right)\left(\frac{d}{dx} + ax\right)u = -(b - a)\left(\frac{d}{dx} + ax\right)u$$

which could be written

$$\left(\frac{d}{dx} + ax\right)\left(\frac{d}{dx} - ax\right)u' = -(b - a)u' \tag{52}$$

where

$$u' \equiv \left(\frac{d}{dx} + ax\right)u.$$

But Eq. (52) is equivalent to

$$\frac{d^2u'}{dx^2} - a^2x^2u' + ax\frac{du'}{dx} - \frac{d}{dx}(axu') = -(b - a)u'$$

or

$$\frac{d^2u'}{dx^2} - a^2x^2u' = -(b - 2a)u' \tag{53}$$

---

[13] There is no general rule for factoring operators, to take us from Eq. (49) to Eq. (50); one can only try various factors and see if they work, as we have done here.

After all that manipulation, we are back to the harmonic oscillator equation again! Equation (53) has exactly the same form as Eq. (49), but the eigenfunction is now u' instead of u, and the eigenvalue is $-(b - 2a)$ instead of $-b$. Thus if we define $b - 2a \equiv 2mE'/\hbar^2$, just as we defined $b = 2mE/\hbar^2$, the energy E' must be the energy of a harmonic oscillator whose wavefunction is u'. From the definitions of b and a we find that

$$\frac{2mE}{\hbar^2} - \frac{2m\omega}{\hbar} = \frac{2mE'}{\hbar^2}$$

so that

$$E' = E - \hbar\omega$$

Thus we have shown that if $u(x)$ is an eigenfunction of the energy operator for the harmonic oscillator, the function $u'(x) = (d/dx + ax)u$ is also an eigenfunction of that same operator, with an eigenvalue which is less than the eigenvalue for $u(x)$, by an amount $\hbar\omega$. We can therefore generate a whole series of eigenvalues, or energy levels, equally spaced in steps of $\hbar\omega$. It only remains to find one level, and we will know all the others.

We might have deduced from the correspondence principle that the spacing between levels is $\hbar\omega$, or $h\nu$, before we went through all of these manipulations. Classically, a charged particle oscillating with frequency ν would radiate electromagnetic waves of the same frequency; thus it must, from the correspondence principle, emit photons of energy $h\nu$. If a particle of energy E emits a photon of energy $h\nu$, the particle's final energy is $E - h\nu$, so $E - h\nu$ must be an allowed energy level (if E is an allowed level). Since no photons of lower energy are ever emitted, we conclude that there is no allowed level between energy E and energy $E - h\nu$. (We shall see later that if a level exists, there is always some probability of a transition of a particle to that level from a higher level, with the energy difference being emitted as a photon.) We might say that the equal spacing of the levels is a consequence of the fact that the frequency of a harmonic oscillator is independent of its energy. In the Bohr theory of the hydrogen atom we saw that the frequency of the emitted photon lies *between* the frequencies of revolution (oscillation) in the upper and lower states; in the harmonic oscillator the frequencies of oscillation of the upper and lower states are equal, so the correspondence principle requires the photon frequency to be exactly that frequency of oscillation.

We still have the task of finding the actual energy of each level and of obtaining an expression for the wavefunctions. We can do this if we observe that all of the allowed energies must be positive, and because they are equally spaced, there must be a *lowest* level. (The energies are all positive because $V \geqslant 0$ everywhere. If E were negative, the kinetic energy of the particle would be negative *everywhere*, and such a particle could not be observed.) Let the wavefunction of the lowest level be u_0, and let the energy of this level be E_0. What happens when we apply the operator $(d/dx + ax)$ to u_0? Normally, if

u is the wavefunction for energy E, then $(d/dx + ax)u$ is the wavefunction for energy $E - \hbar\omega$, as we have just seen. But $(d/dx + ax)u_0$ cannot be a wavefunction for energy $E_0 - \hbar\omega$, because E_0 is the lowest possible energy. The only other possibility is that this "wavefunction" is zero; that is,

$$\left(\frac{d}{dx} + ax\right)u_0 = 0 \tag{54}$$

Equation (54) is not nearly so difficult to solve as Eq. (49) would have been. You may easily separate the variables to obtain

$$\frac{du_0}{u_0} = -ax\,dx$$

or

$$\ln u_0 = \frac{-ax^2}{2} + \text{constant}$$

Thus

$$u_0 = Ae^{-ax^2/2}$$

where the constant A may be chosen to normalize the function. The complete wavefunction has the time dependence required by Eq. (8):

$$\psi_0(x, t) = u_0(x)e^{-iE_0 t/\hbar} = Ae^{-ax^2/2}e^{-iE_0 t/\hbar}$$

By inserting $u_0(x)$ into Eq. (49), we may find the value of E_0. We have

$$\frac{du_0}{dx} = -axAe^{-ax^2/2}$$

$$\frac{d^2u_0}{dx^2} = -aAe^{-ax^2/2} + a^2x^2Ae^{-ax^2/2}$$

so that Eq. (49) becomes

$$-aAe^{-ax^2/2} + a^2x^2Ae^{-ax^2/2} - a^2x^2Ae^{-ax^2/2} = -bAe^{-ax^2/2}$$

which reduces to

$$a = b$$

or

$$\frac{m\omega}{\hbar} = \frac{2mE_0}{\hbar^2}$$

with the final result that

$$E_0 = \frac{\hbar\omega}{2}$$

We have found the lowest energy level, and incidentally verified that Eq. (54) gave us the correct eigenfunction of Eq. (49). [If u_0 were not an eigenfunction of Eq. (49), we could not have used it to find the eigenvalue as we did.] If we wish to find the other eigenfunctions, we proceed as follows: We apply the operator $(d/dx - ax)$ to both sides of Eq. (52), and we find that this equation reduces to

$$\left(\frac{d^2}{dx^2} - a^2x^2\right)u'' = -bu''$$

where u'' is defined as $(d/dx - ax)u'$. Thus u'' is an eigenfunction of the original equation (49), with the *original* eigenvalue $-b$, so that the energy eigenvalue for u'' is $\hbar\omega$ *greater* than that for u'. In other words, given an eigenfunction and its eigenvalue, we can always generate the eigenfunction with the next *higher* eigenvalue, by applying the operator $(d/dx - ax)$. Thus, by starting with u_0, we can work our way up through the whole series of eigenfunctions.

In summary, the energy levels of the harmonic oscillator are

$$E_n = \frac{\hbar\omega}{2} + n\hbar\omega$$

and the corresponding eigenfunctions are (except for the normalizing factor)

$$\psi_n = \left(\frac{\partial}{\partial x} - ax\right)^n e^{-ax^2/2} e^{-iE_n t/\hbar}$$

the first three of which are (normalized)

$$\psi_0 = (a/\pi)^{1/4} e^{-ax^2/2} e^{-i\omega t/2}$$

$$\psi_1 = (4a^3/\pi)^{1/4} x e^{-ax^2/2} e^{-3i\omega t/2}$$

$$\psi_2 = (a/4\pi)^{1/4}(2ax^2 - 1) e^{-ax^2/2} e^{-5i\omega t/2}$$

where ω is the classical frequency of oscillation ($\omega = (K/m)^{1/2}$), and $a = m\omega/\hbar$.

The lowest level has the zero-point energy which is required by the uncertainty principle, and the spacing between levels is $\hbar\omega$, as required by the correspondence principle.

EXAMPLE PROBLEM 2. Find the zero-point energy of a harmonic oscillator on the assumption that it is the minimum energy required by the uncertainty principle.

Solution. The average potential energy is equal to the average kinetic energy or to one-half the total energy E, so $\frac{1}{2}K\overline{x^2} = (1/2m)\overline{p^2} = \frac{1}{2}E$. But the mean value of x is zero; therefore the mean value of x^2 is the mean-square deviation, and is equal to $(\delta x)^2$ by definition, since δx is the *root*-mean-square deviation. Similarly, $\overline{p^2} = (\delta p)^2$. Therefore $K(\delta x)^2 = (\delta p)^2/m = E$. But we know that $K = m\omega^2$, so $\delta p = m\omega(\delta x)$. If the uncertainty product $(\delta p)(\delta x)$ has its minimum value, we may write $\delta x = \hbar/2(\delta p)$, and combining this with the previous equation we have $\delta p = m\omega\hbar/2(\delta p)$, so that

$$(\delta p)^2 = \frac{m\omega\hbar}{2}$$

and the total energy is

$$E = \frac{(\delta p)^2}{m} = \frac{\hbar\omega}{2}$$

Notice that this procedure gives *exactly* the correct ground-state energy for the harmonic oscillator. This means that the uncertainty product for this system is as small as it can ever be, and the uncertainty principle is solely responsible for the energy of the ground state. In general, $(\delta x)(\delta p) \geq \hbar/2$, and the *equality* holds *only* for a Gaussian wavefunction (of the form discussed at the end of Section 4.3). But the ground-state wavefunction u_0 *is* of Gaussian form.

PROBLEMS

1. An electron of kinetic energy 4 eV suddenly reaches a boundary where its potential energy decreases by 5 eV, so that its kinetic energy increases to 9 eV. Find the probability that the electron will be reflected at this potential step.

2. Do the "exercise for the student" mentioned in Section 5.4. Let δx be the reciprocal of the decay constant α outside the well, and let δp be the additional momentum the particle would need in order to escape from the well. Show that $\delta x \cdot \delta p \approx \hbar$.

3. Verify that Fig. 3 is drawn correctly for the case $\beta a = \pi$, by checking, as well as you can, how the values of E_b/V_0 and E_d/V_0 agree with the wavelengths inside the well and the decay constants outside the well.

4. Calculate the minimum depth V_0, in electron-volts, required for a square well to contain two allowed energy levels, if the width of the well is $2a = 4.0 \times 10^{-13}$ cm and the particle has a mass of 2.0×10^9 eV/c^2.

5. Draw a graph of a wavefunction for an *unbound* state in the square-well

potential of Section 5.4 (Fig. 3). Is there any restriction on the possible values of E in this case?

6.   For the potential

$$V(x) = +\infty \qquad\qquad x < 0$$
$$V(x) = 0 \qquad\qquad 0 < x < a$$
$$V(x) = +V_0 \qquad\qquad x > a \;,$$

write a general expression for the wave function for an electron in a bound state, and solve for the three lowest energy levels when $V_0 = 10\,\mathrm{eV}$ and $a = 50\,\text{Å}$. How many bound levels are there? For what range of values of a are there no bound levels at all?

7.   For the square well with a barrier inside, discussed at the end of Section 5.4, when $b = 2a$, for solutions of energy $E < V_0$,

(a)   show that the lowest-energy solution (even parity) and the next-lowest-energy solution (odd parity) require

$$\alpha a \,\tanh(\alpha a) = - \,ka \,\cot(ka) \;,$$

and

$$\alpha a \,\coth(\alpha a) = - \,ka \,\cot (ka) \;,$$

respectively;

(b)   show that there is no solution for $ka < \pi/2$, and demonstrate graphically why this is so;

(c)   find the lowest numerical solution for ka, and the corresponding energy in terms of V_0, when

$$V_0 = \frac{3\hbar^2}{2ma^2}$$

(d)   show that, for the value of V_0 given in part (c), there is only one solution with $E < V_0$;

(e)   show that, when $\beta^2 a^2 \equiv 2mV_0 a^2/\hbar^2 \to \infty$, $ka \to \pi$ for the two lowest energy levels. Also show specifically that in this limit

$$ka \cot ka \approx - \alpha a$$

for both the even and the odd solution, so that the two energy levels merge into a single level at the energy of a square well of width a.

8. Find the probability that an electron will be found within 0.01 Å of the wall, in an infinitely deep potential well of width 2.00 Å, when the electron is in the ground state.

9. For the ground state of Example Problem 1, find the probability that the electron will be found outside the well.

10. The potential

$$V(x) = \frac{\hbar^2}{2m} \left[\frac{4 \sinh^2 x}{225} - \frac{2 \cosh x}{5} \right]$$

is a "double well," with two minima.

(a) Draw a graph of $V(x)$ and locate the minima.

(b) Show that

$$\psi(x) = (1 + 4 \cosh x) \exp \left(\frac{-2 \cosh x}{15} \right)$$

is a solution of the Schrödinger equation for this potential, find the corresponding energy level, and show this level on your graph of $V(x)$.

(c) Draw a graph of $\psi(x)$ and show that it has the proper behavior at the classical turning points and in the classically forbidden regions.

11. A particle of mass m is trapped in the potential

$$V(x) = 500 \, \hbar^2 x/ma^3 \qquad 0 < x < a$$
$$V(x) = \infty \qquad\qquad x > a \text{ or } x < 0$$

By numerical integration of the Schrödinger equation, find

(a) the lowest energy level E such that $E > V(a)$

(b) the greatest energy level such that $E < 0.8 \, V(a)$

Sketch the wave function ψ for each of these levels. How many allowed energy levels lie between these two? For the higher-energy wavefunction, observe the variation of amplitude of the various antinodes, and compare this with what you might expect from the correspondence principle. (The probability density, in the classical limit, should be inversely proportional to the velocity of the particle.) *Hint:* to integrate the Schrödinger equation, start at $x = 0$ with $\psi = 0$ and $d\psi/dx = 1.0$. You can achieve sufficient accuracy by dividing the range from zero to a into 1000 segments, finding the value of ψ at the end of each segment by adding $(a/1000)d\psi/dx$ to the value of ψ at the beginning of that segment. You can find the new value of

$d\psi/dx$ in the same way, using $d^2\psi/dx^2$ as determined by the Schrödinger equation and a trial value of E. After you have determined $\psi(a)$, repeat the procedure with a new value of E, until you find a value of E for which the condition $\psi(a) = 0$ is satisfied.

12. Find the transmission coefficient for

(a) a 2-eV electron incident on a barrier 3 eV high and 10^{-8} cm thick;

(b) a 2-MeV alpha particle (helium nucleus) incident on a barrier 4 MeV high and 10^{-12} cm thick; and

(c) a 5-gm marble, of kinetic energy 1 erg, incident on a barrier 1.000001 erg high and 0.001 mm thick. (Express T in the form 10^{-A} where A itself is a large power of 10.)

13. For the wavefunction

$$\psi = A(\cos \frac{\pi x}{a} + 1) \qquad |x| \leq a$$
$$\psi = 0 \qquad |x| > a \ ,$$

(a) Show that A must equal $1/\sqrt{3a}$ to normalize ψ.

(b) Evaluate δx and δp_x.

(c) Find the boundaries of the classically allowed region (where the kinetic energy is positive).

14. Extend the treatment of the barrier penetration problem (Section 5.5) to the case of a *negative* "barrier" ($V_0 < 0$); make a graph of T versus E, similar to Fig. 11, for $V_0 = -8\hbar^2/ma^2$.

15. Use the definition of the expectation value (Section 5.6) to find the expectation value of the potential energy $(Kx^2/2)$ for the lowest energy level of a harmonic oscillator, and compare with the total energy of this level. Also evaluate $\langle p_x^2 \rangle$ for this state, to find the expectation value of the kinetic energy.

16. Show that the expectation value of p is zero for a particle in a square well, and explain the result.

17. Use the prescription given in Section 5.7 to find the harmonic oscillator wavefunctions for energy $E_1 = 3\hbar\omega/2$ and for energy $E_2 = 5\hbar\omega/2$. Verify that both wavefunctions satisfy Eq. (49), with the correct eigenvalue for each.

18. Find the wavefunction for the ground state of a particle of mass m bound by the potential $V(x) = -b\,\delta(x)$ where b is a positive constant, and $\delta(x)$ is the delta function defined on page 123. Show that the ground-state energy is $E = -mb^2/2\hbar^2$. Hint: Evaluate the change in $d\psi/dx$ in terms of $d^2\psi/dx^2$, between $x = -\varepsilon$ and $x = +\varepsilon$.

CHAPTER 6

Schrödinger Equation II: Three Dimensions

We are now ready to study the three-dimensional world. In this chapter, we extend the Schrödinger equation to three dimensions, and we investigate the properties of the solutions for cases in which the potential energy is spherically symmetric. A most important property of any solution is its angular momentum, so we shall spend considerable time on the quantum mechanical description of angular momentum. We shall then use our knowledge to solve three-dimensional counterparts to some of the problems of Chapter 5, and to study some features of scattering of particles from a spherically symmetric potential.

187

6.1 EXTENSION OF THE SCHRÖDINGER EQUATION TO THREE DIMENSIONS

In three dimensions, the total energy of a particle is $(p_x^2 + p_y^2 + p_z^2)/2m + V(x, y, z)$. We have seen (Section 5.6) that the Schrödinger equation may be constructed by replacing p_x by the operator $(\hbar/i)(\partial/\partial x)$. There is no essential difference between the x coordinate and the y and z coordinates, so we do the same for p_y and p_z, replacing them by $(\hbar/i)(\partial/\partial y)$ and $(\hbar/i)(\partial/\partial z)$ respectively, to obtain the Schrödinger equation in three dimensions:

$$-\frac{\hbar^2}{2m}\left(\frac{\partial^2}{\partial x^2} + \frac{\partial^2}{\partial y^2} + \frac{\partial^2}{\partial z^2}\right)\psi + V(x, y, z)\psi = E\psi$$

$$= i\hbar \frac{\partial\psi}{\partial t} \qquad (1)$$

As before, the wavefunction ψ for a definite energy may be written as a product of a space-dependent part and a time-dependent part:

$$\psi(x, y, z, t) = u(x, y, z)e^{-iEt/\hbar}$$

Particle in a Box. To illustrate how the three components of momentum fit together, let us find the wavefunction for a particle which is confined to the inside of a rectangular box. This apparently artificial example has a fruitful application to the theory of electrons in metals, as we shall see in Section 11.4.

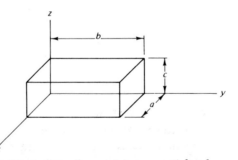

Fig. 1. *"Box" containing a particle whose potential energy is zero inside and infinite outside it.*

Let the walls of the box be the planes $x = 0$, $x = a$, $y = 0$, $y = b$, $z = 0$, $z = c$, and assume that the potential energy is zero inside the box and infinite outside the box (see Fig. 1). As in the one-dimensional square well (Section 5.4), the function u must go to zero when the potential step is infinite, so $u = 0$ at the walls of the box.

We shall solve the problem by the standard technique of *separation of variables*, which is applicable to a great variety of partial differential equations. We assume that $u(x,y,z)$ may be written as a product of three functions: $X(x)$, a function of x only, $Y(y)$, a function of y only, and $Z(z)$, a function of z only:

$$u(x, y, z) = X(x) \cdot Y(y) \cdot Z(z)$$

Equation (1) now becomes, for the region inside the box,

$$-\frac{\hbar^2}{2m}(X''YZ + XY''Z + XYZ'') = EXYZ \qquad (2)$$

where $X'' \equiv d^2X/dx^2$, $Y'' \equiv d^2Y/dy^2$, and $Z'' \equiv d^2Z/dz^2$. If we divide both sides of Eq. (2) by the product XYZ, we obtain

$$-\frac{\hbar^2}{2m}\left(\frac{X''}{X} + \frac{Y''}{Y} + \frac{Z''}{Z}\right) = E \qquad (3)$$

Suppose we evaluate the left side of Eq. (3) at two different values of x, leaving y and z fixed. When x changes, the only term which could possibly change is the first term, but since the sum of the three terms is the *constant E*, the first term must remain unchanged also. But this term is a function of x only; if it does not change when x is varied, it must never change; i.e., it is equal to a constant, and we may write

$$\frac{X''}{X} = -k_x^2 \qquad (4a)$$

Similarly, we may deduce that the second term equals another constant:

$$\frac{Y''}{Y} = -k_y^2 \qquad (4b)$$

and finally that

$$\frac{Z''}{Z} = -k_z^2 \qquad (4c)$$

Thus we have completed the separation of variables, obtaining three ordinary differential equations, one for each coordinate. The constants are written in this form because we know that they should be negative; a negative constant makes each equation equivalent to the equation of simple harmonic motion, whereas a positive constant would lead to an exponential solution which cannot satisfy the boundary condition that u be zero at each wall. Thus the solutions to Eqs. (4) are

$$X = A \sin(k_x \cdot x)$$
$$Y = B \sin(k_y \cdot y)$$
$$Z = C \sin(k_z \cdot z)$$

We may now evaluate k_x, k_y, and k_z by using the boundary conditions, which become, in terms of X, Y, and Z:

$$X(0) = 0, \quad Y(0) = 0, \quad Z(0) = 0, \quad X(a) = 0, \quad Y(b) = 0, \quad Z(c) = 0$$

We have already satisfied the first three conditions by choosing the sine rather than the cosine for our three functions. The last three conditions are satisfied if

$$k_x a = n_x \pi, \quad k_y b = n_y \pi, \quad k_z c = n_z \pi$$

where n_x, n_y, and n_z are integers. The complete solution to the problem is therefore

$$\psi(x, y, z, t) = \text{constant} \times \sin \frac{n_x \pi x}{a} \sin \frac{n_y \pi y}{b} \sin \frac{n_z \pi z}{c} e^{-iEt/\hbar}$$

where

$$E = \frac{\hbar^2}{2m} (k_x^2 + k_y^2 + k_z^2)$$

$$= \frac{\hbar^2}{2m} \left[\left(\frac{n_x \pi}{a}\right)^2 + \left(\frac{n_y \pi}{b}\right)^2 + \left(\frac{n_z \pi}{c}\right)^2 \right] \tag{5}$$

In other words, we have three independent wavenumbers, k_x, k_y, and k_z, which take the place of the single wavenumber k that we had in the one-dimensional case; these wavenumbers determine the momentum components:

$$p_x^2 = \hbar^2 k_x^2, \quad p_y^2 = \hbar^2 k_y^2, \quad p_z^2 = \hbar^2 k_z^2$$

6.2 *SPHERICALLY SYMMETRIC POTENTIALS AND ANGULAR MOMENTUM*

The example of the particle in a box was particularly simple because the potential energy was constant in a region whose boundaries were parallel to the coordinate axes. We cannot usually expect such simple situations in nature, although we can use the above result to make a fairly good approximation to the behavior of electrons in a metal (see Chapter 11).

A situation which is not quite so simple, but still manageable, arises when the potential energy is a function of only one of the three coordinates. A very important special case is that of a central force field, in which the potential is a function only of r (in spherical coordinates, r, θ, and ϕ). For example, this is true for any two particles which interact by the Coulomb force only, provided that we use the center-of-mass coordinate system, in which r is the distance between the two particles.

Thus the theory developed in this section is an essential part of the theory of the hydrogen atom, where the dominant force is the Coulomb force between electron and proton. However, we shall first develop the theory in a form that applies to *all* central forces, before taking up the hydrogen atom in detail in Chapter 7.

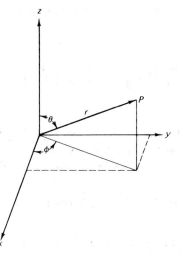

Fig. 2. *Spherical and rectangular coordinates of a point* **P.**

We begin by writing the Schrödinger equation in spherical coordinates, to take advantage of the fact that the potential energy is a function of r only, and to pave the way for the separation of variables. By referring to Fig. 2, we see that the spherical coordinates may be written in terms of cartesian coordinates as

$$r^2 = x^2 + y^2 + z^2$$

$$\cos \theta = \frac{z}{r} = \frac{z}{(x^2 + y^2 + z^2)^{1/2}} \tag{6}$$

$$\tan \phi = \frac{y}{x}$$

The space derivatives in Eq. (1) may be converted to derivatives with respect to r, θ, and ϕ by using the "chain rule":

$$\frac{\partial u}{\partial x} = \frac{\partial u}{\partial r}\frac{\partial r}{\partial x} + \frac{\partial u}{\partial \theta}\frac{\partial \theta}{\partial x} + \frac{\partial u}{\partial \phi}\frac{\partial \phi}{\partial x}$$

with corresponding equations for $\partial u/\partial y$ and $\partial u/\partial z$. The necessary partial derivatives can be shown, with the aid of Eqs. (6), to be

$$\frac{\partial r}{\partial x} = \sin \theta \cos \phi; \qquad \frac{\partial r}{\partial y} = \sin \theta \sin \phi; \qquad \frac{\partial r}{\partial z} = \cos \theta$$

$$\frac{\partial \theta}{\partial x} = \frac{1}{r}\cos \theta \cos \phi; \qquad \frac{\partial \theta}{\partial y} = \frac{1}{r}\cos \theta \sin \phi; \qquad \frac{\partial \theta}{\partial z} = -\frac{1}{r}\sin \theta$$

$$\frac{\partial \phi}{\partial x} = -\frac{1}{r}\frac{\sin \phi}{\sin \theta}; \qquad \frac{\partial \phi}{\partial y} = \frac{1}{r}\frac{\cos \phi}{\sin \theta}; \qquad \frac{\partial \phi}{\partial z} = 0$$

To find $\partial^2 u/\partial x^2$ one simply substitutes $\partial u/\partial x$ for u into the chain rule, obtaining

$$\frac{\partial^2 u}{\partial x^2} = \left[\frac{\partial}{\partial r}\left(\frac{\partial u}{\partial x}\right)\right] \cdot \frac{\partial r}{\partial x} + \left[\frac{\partial}{\partial \theta}\left(\frac{\partial u}{\partial x}\right)\right] \cdot \frac{\partial \theta}{\partial x} + \left[\frac{\partial}{\partial \phi}\left(\frac{\partial u}{\partial x}\right)\right]\frac{\partial \phi}{\partial x}$$

and one finds $\partial^2 u/\partial y^2$ and $\partial^2 u/\partial z^2$ in corresponding fashion. It is understood that the first partial derivatives are already expressed in terms of spherical coordinates and derivatives with respect to those coordinates. Thus

$$\frac{\partial u}{\partial x} = \sin\theta\cos\phi\,\frac{\partial u}{\partial r} + \frac{1}{r}\cos\theta\cos\phi\,\frac{\partial u}{\partial \theta} - \frac{\sin\phi}{r\sin\theta}\frac{\partial u}{\partial \phi}$$

Completing these operations for all three coordinates, one finally obtains the time-independent Schrödinger equation in spherical coordinates:

$$-\frac{1}{r}\frac{\partial^2}{\partial r^2}(ru) - \frac{1}{r^2\sin\theta}\frac{\partial}{\partial\theta}\left(\sin\theta\,\frac{\partial u}{\partial\theta}\right) - \frac{1}{r^2\sin^2\theta}\frac{\partial^2 u}{\partial\phi^2} = \frac{2m}{\hbar^2}[E - V(r)]u \quad (7)$$

We can now separate the variables by writing $u(r,\theta,\phi)$ as a product of two functions, one a function of r only, the other a function of θ and ϕ:

$$u(r, \theta, \phi) = R(r)Y(\theta, \phi)$$

Making this substitution in Eq. (7), and multiplying both sides by r^2/u, we obtain

$$-\frac{r}{R}\frac{d^2}{dr^2}(rR) - \frac{1}{Y}\left[\frac{1}{\sin\theta}\frac{\partial}{\partial\theta}\left(\sin\theta\,\frac{\partial Y}{\partial\theta}\right) + \frac{1}{\sin^2\theta}\frac{\partial^2 Y}{\partial\phi^2}\right] = \frac{2mr^2}{\hbar^2}[E - V(r)]$$

$$(8)$$

The dependence on the variables θ and ϕ is all contained in the second term in Eq. (8). Therefore, if θ or ϕ is varied, the other terms in the equation cannot change, and we conclude that the second term in the equation must be a constant, according to the same reasoning used in Section 6.1:

$$-\frac{1}{Y}\left[\frac{1}{\sin\theta}\frac{\partial}{\partial\theta}\left(\sin\theta\,\frac{\partial Y}{\partial\theta}\right) + \frac{1}{\sin^2\theta}\frac{\partial^2 Y}{\partial\phi^2}\right] = \alpha$$

or

$$\Omega Y = \alpha Y \quad (9)$$

where we have defined the *operator* Ω as

$$\Omega \equiv -\frac{1}{\sin\theta}\frac{\partial}{\partial\theta}\left(\sin\theta\frac{\partial}{\partial\theta}\right) - \frac{1}{\sin^2\theta}\frac{\partial^2}{\partial\phi^2}$$

The eigenvalue equation (9) must be obtained from the Schrödinger equation every time that the potential energy depends on r only; the eigenfunctions of (9) must describe the angular dependence of the wavefunction in all such cases. Before attempting to find these eigenfunctions, let us inquire into their physical meaning, and the meaning of the operator Ω and its eigenvalue(s) α.

We can obtain a clue to the meaning of the operator Ω by writing the kinetic energy in spherical coordinates as [1]

$$T = \frac{p_r^2}{2m} + \frac{L^2}{2mr^2} \tag{10}$$

where L^2 is the square of the total angular momentum of the particle. ($\mathbf{L} = \mathbf{r} \times \mathbf{p}$.) Equation (10) tells us that the time independent Schrödinger equation should be

$$\left(\frac{p_r^2}{2m} + \frac{L^2}{2mr^2}\right)u = (E - V)u$$

where p_r^2 and L^2 are *operators*. If we compare this equation with Eq. (7), we see that they agree, provided that

$$p_r^2 u = -\frac{\hbar^2}{r}\frac{\partial^2}{\partial r^2}(ru) \tag{11}$$

and

$$L^2 u = -\frac{\hbar^2}{\sin\theta}\frac{\partial}{\partial\theta}\left(\sin\theta\frac{\partial u}{\partial\theta}\right) - \frac{\hbar^2}{\sin^2\theta}\frac{\partial^2 u}{\partial\phi^2} \tag{12}$$

(Notice that the operator for p_r^2 is not simply $-\hbar^2\,\partial^2/\partial r^2$, as it would be if r were a Cartesian coordinate like x, y, or z.) Comparison of Eq. (12) with the definition of the operator Ω shows that $L^2 = \hbar^2\Omega$.

We may also deduce the above result directly from the definition of $\mathbf{L}$. Since $\mathbf{L} = \mathbf{r} \times \mathbf{p}$, the operators for the rectangular components of $\mathbf{L}$ should be

$$L_x = yp_z - zp_y = -i\hbar\left(y\frac{\partial}{\partial z} - z\frac{\partial}{\partial y}\right)$$

$$L_y = zp_x - xp_z = -i\hbar\left(z\frac{\partial}{\partial x} - x\frac{\partial}{\partial z}\right)$$

---

[1] It is easy to verify that Eq. (10) is the correct expression for T in terms of L. The velocity vector can be resolved into two components, v_r and $v_\perp$, which are respectively parallel and perpendicular to the radius vector r. Then $T = mv_r^2/2 + mv_\perp^2/2$, and $L^2 = (mv_\perp r)^2$, so that $v_\perp^2 = L^2/m^2r^2$. Substituting for $v_\perp$ into T yields $T = mv_r^2/2 + L^2/2mr^2$, which is equivalent to Eq. (10), with $p_r = mv_r$.

$$L_z = xp_y - yp_x = -i\hbar \left(x \frac{\partial}{\partial y} - y \frac{\partial}{\partial x} \right)$$

which become, in spherical coordinates,

$$L_x = i\hbar \left(\sin \phi \frac{\partial}{\partial \theta} + \cot \theta \cos \phi \frac{\partial}{\partial \phi} \right)$$

$$L_y = -i\hbar \left(\cos \phi \frac{\partial}{\partial \theta} - \cot \theta \sin \phi \frac{\partial}{\partial \phi} \right) \tag{13}$$

$$L_z = -i\hbar \frac{\partial}{\partial \phi}$$

as you may verify yourself by using the chain rule and the transformation equations given above. Equations (13) may now be used to compute $L^2 = L_x^2 + L_y^2 + L_z^2$; if you work it out, you will find, in agreement with our previous result, that $L^2 = h^2 \Omega$.

Having determined the meaning of the operator Ω, we can proceed to find the eigenvalues—the values of α which yield an acceptable function Y when we solve Eq. (9). These eigenvalues will tell us the possible results of a measurement of L^2. To find them, we complete the separation of variables as follows. We write

$$Y(\theta, \phi) = \Theta(\theta)\Phi(\phi)$$

Using the definition of Ω, we may write Eq. (9) as

$$\frac{\Phi}{\sin \theta} \frac{d}{d\theta} \left(\sin \theta \cdot \frac{d\Theta}{d\theta} \right) + \frac{\Theta}{\sin^2 \theta} \frac{d^2 \Phi}{d\phi^2} = -\alpha \Theta \Phi$$

or, after multiplication by $(\sin^2 \theta)/\Theta\Phi$,

$$\frac{1}{\Theta} \sin \theta \frac{d}{d\theta} \left(\sin \theta \frac{d\Theta}{d\theta} \right) + \frac{1}{\Phi} \frac{d^2 \Phi}{d\phi^2} = -\alpha \sin^2 \theta$$

By the same reasoning used in our previous separation of variables, we deduce that the only ϕ-dependent term in the equation must equal a constant:

$$\frac{d^2 \Phi}{d\phi^2} = \text{constant} \times \Phi \tag{15}$$

A very simple condition on the wavefunction tells us what this constant must be. We require that the wavefunction be *single valued*—that there be just one value of ψ for a given point in space at a given time. (Although it gives the right answer in this case, this requirement is not strictly necessary. See Problem 8.) As a consequence,

$$\Phi(\phi + 2\pi) = \Phi(\phi)$$

since the angle $\phi + 2\pi$ describes the same points in space as the angle ϕ. We can satisfy this condition if we write Eq. (15) as

$$\frac{d^2\Phi}{d\phi^2} = -m^2\Phi \tag{16}$$

where m is an integer,[2] so that the equation becomes identical to the standard equation of simple harmonic motion, whose solution[3] has period 2π and may be written as

$$\Phi = Ae^{im\phi}$$

where A is an arbitrary constant which may be chosen to normalize the wave-function.

Notice that this solution is an eigenfunction of the L_z operator, for

$$L_z\Phi = -i\hbar \frac{\partial}{\partial\phi}\Phi = -i\hbar\frac{\partial}{\partial\phi}Ae^{im\phi} = m\hbar\Phi$$

and the eigenvalue is $m\hbar$. Therefore, according to the eigenvalue postulate (Section 5.6), the only possible results of a measurement of a component of angular momentum are integral multiples of $\hbar$. It seems that after all that work we are right back where we started, with the Bohr condition! However, our next step, the investigation of the eigenvalues of L^2, will show that there is a significant difference between the results of the Bohr theory and those of the present theory.

Continuing our search for the eigenvalues α, we substitute from Eq. (16) into Eq. (14) to eliminate Φ, and obtain an equation in the single variable θ:

$$\frac{\sin\theta}{\Theta}\frac{d}{d\theta}\left(\sin\theta\frac{d\Theta}{d\theta}\right) - m^2 = -\alpha\sin^2\theta \tag{17}$$

Equation (17) appears formidable because of the sine functions scattered through it, but these may be eliminated by the substitution $\cos\theta = x$. With the function $y(x)$ replacing the function $\Theta(\theta)$, the equation becomes

$$\frac{d}{dx}\left[(1 - x^2)\frac{dy}{dx}\right] - \frac{m^2y}{1 - x^2} = -\alpha y$$

which may be solved by assuming y to be a power series in x. The details of

---

[2] This is the standard notation, which could possibly lead to confusion between the quantum number m and the mass m, but the meaning should always be clear from the context.
[3] Of course there are *two* independent solutions of a second-order equation, but the other solution, $e^{-im\phi}$, becomes the same as the first solution when we consider negative values of m as well as positive values.

the general solution need not concern us here[4]; the important feature is that the solution is finite for all values of θ—that is, for all values of x between -1 and $+1$—*only* when the constant α is of the form

$$\alpha = l(l+1)$$

where l is an integer or zero, and $l \geq |m|$.

Thus Eq. (9) becomes

$$\Omega Y = l(l+1)Y$$

but we found that the L^2 operator is equal to $\hbar^2 \Omega$, so that

$$\boxed{L^2 Y(\theta, \phi) = l(l+1)\hbar^2 Y(\theta, \phi)}$$

In other words, the eigenvalues of the L^2 operator are $l(l+1)\hbar^2$, so that these values: $0, 2\hbar^2, 6\hbar^2, 12\hbar^2, \ldots$ are the only possible results of a measurement of L^2. Contrast these values with the prediction of the Bohr theory, where one expects that L^2 could have, for example, the value $\hbar^2$, which is simply the square of an allowed value of a component of $\mathbf{L}$. We shall see in Section 6.3 that the value of $\hbar^2$ is excluded because the uncertainty principle prevents our knowing the exact *direction* of $\mathbf{L}$; if L_z is known to be exactly equal to $\hbar$, then there must be another nonzero component of $\mathbf{L}$, making L^2 greater than $\hbar^2$. We shall also see that the mysterious-looking eigenvalues $l(l+1)\hbar^2$ can be deduced from a straightforward analysis of how L^2 is determined experimentally.

Eigenfunctions of L^2.　The eigenfunctions of L^2 are the products of the functions Θ and Φ, solutions of Eq. (17) and Eq. (16), respectively, corresponding to possible combinations of the integers l and m. The functions Φ, as we have seen, depend only on m, but the functions Θ depend on l and on the *absolute value of m*, because Eq. (17) contains m^2 as well as l. Thus we write the eigenfunctions of L^2 as

$$Y_{l,m}(\theta, \phi) = P_l^{|m|}(\cos \theta)e^{im\phi} \qquad (l \geq |m|)$$

where the functions $P_l^{|m|}(\cos \theta)$, found by solving Eq. (17), are called *associated Legendre polynomials*. The functions $Y_{l,m}(\theta, \phi)$ are called *spherical harmonics*, and are well known in classical theory; the first few are:

---

[4] The method is very similar to that used for the solution of the radial part of the Schrödinger equation for the hydrogen atom, which is fully worked out in Appendix C. For details of the solution of Eq. (17), see, for example, R. Leighton, "Principles of Modern Physics," pp. 166–171. McGraw–Hill, New York, 1959.

$$Y_{2,2} = \left(\frac{15}{32\pi}\right)^{1/2} \sin^2 \theta \, e^{2i\phi}$$

$$Y_{1,1} = -\left(\frac{3}{8\pi}\right)^{1/2} \sin \theta \, e^{i\phi} \qquad Y_{2,1} = -\left(\frac{15}{8\pi}\right)^{1/2} \sin \theta \cos \theta \, e^{i\phi}$$

$$Y_{0,0} = \left(\frac{1}{4\pi}\right)^{1/2} \qquad Y_{1,0} = \left(\frac{3}{4\pi}\right)^{1/2} \cos \theta \qquad Y_{2,0} = \left(\frac{5}{16\pi}\right)^{1/2} (3 \cos^2 \theta - 1)$$

$$Y_{1,-1} = \left(\frac{3}{8\pi}\right)^{1/2} \sin \theta \, e^{-i\phi} \qquad Y_{2,-1} = \left(\frac{15}{8\pi}\right)^{1/2} \sin \theta \cos \theta \, e^{-i\phi}$$

$$Y_{2,-2} = \left(\frac{15}{32\pi}\right)^{1/2} \sin^2 \theta \, e^{-2i\phi}$$

(The minus signs in front of $Y_{1,1}$ and $Y_{2,1}$ are not essential, and they do not always appear in tables of spherical harmonics. They are included here in order to simplify the matrix representation of the angular momentum operators, to be developed in the problems of Chapter 7.)

Notice that the θ dependence of $Y_{l,-m}$ is the same as that of $Y_{l,m}$, because the function Θ depends on $|m|$ rather than on m itself. The numerical factors shown in the above functions are simply normalization factors derived from a requirement that the integral of $|Y_{l,m}(\theta. \phi)|^2$ over an entire sphere be equal to unity:

$$\int_0^{2\pi} \int_0^{\pi} |Y_{l,m}|^2 \sin \theta \, d\theta \, d\phi = 1 \qquad (18)$$

This requirement enables one to interpret $|Y_{l,m}|^2$ as a probability density for the particle's angular coordinates. In other words, if a particle has a wavefunction whose angular dependence is $Y_{l,m}$, the quantity

$$\int_{\phi_1}^{\phi_2} \int_{\theta_1}^{\theta_2} |Y_{l,m}|^2 \sin \theta \, d\theta \, d\phi$$

is equal to the probability of finding the coordinates θ and ϕ of the particle between the limits

$$\phi_1 < \phi < \phi_2 \qquad \text{and} \qquad \theta_1 < \theta < \theta_2$$

when one measures the position of the particle.

Because of the exponential form of the ϕ dependence, the absolute square $|Y_{l,m}|^2$ is a function of θ only. A polar plot of $|Y_{l,m}|^2$ versus θ can give us some insight into the connection between spherical harmonics and the angular momentum of a particle. Figure 3 shows polar plots of $|Y_{2,0}|^2$, $|Y_{2,1}|^2$, and $|Y_{2,2}|^2$. Each of these functions represents a particle with the same value of L^2, but the values of L_z are 0, $+\hbar$, and $+2\hbar$, respectively. As L_z increases, the

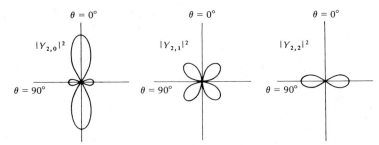

Fig. 3. *Polar plots of some probability densities* $|Y_{l,m}|^2$ *as functions of* θ.

probability density shifts from $\theta = 0$ (the z axis) toward $\theta = 90°$. For a given value of r, an orbit in the xy plane ($\theta = 90°$) must correspond to the greatest value of L_z, classically as well as quantum mechanically; and of course a particle with $\theta = 0$ is located on the z axis and must have $L_z = 0$.

Let us recapitulate the somewhat tedious operations of this section, in order to bring out the essential points.

1. A straightforward transformation into spherical coordinates yields a time-independent Schrödinger equation of the form

$$-r \frac{\partial^2}{\partial r^2}(ru) + \Omega u = \frac{2mr^2}{\hbar^2}[E - V(r)]u$$

   where Ω is an operator involving derivatives with respect to θ and ϕ.

2. By means of the definition $\mathbf{L} = \mathbf{r} \times \mathbf{p}$, and substitution of the operators for p_x, p_y, and p_z, expressions for the operators L_x, L_y, L_z, and L^2 can be found, and the operator L^2 is found to be equal to $\hbar^2\Omega$.

3. Separation of the variables in the Schrödinger equation, by means of the substitution $u(r, \theta, \phi) = R(r)Y(\theta, \phi)$, yields the eigenvalue equation $\Omega Y = \alpha Y$. To be acceptable as a factor in the wavefunction, the functions $Y(\theta, \phi)$ must have a ϕ dependence of the form $e^{im\phi}$, where m is a positive or negative integer, or zero; these functions are then eigenfunctions of the L_z operator, with eigenvalues $m\hbar$. Furthermore, the eigenvalue α must have the form $l(l + 1)$, where l is an integer greater than or equal to $|m|$. Thus each function Y is an *eigenfunction of both* L_z *and* L^2, and it is characterized by two indices, l and m:

$$L^2 Y_{l,m}(\theta, \phi) = l(l + 1)\hbar^2 Y_{l,m}(\theta, \phi)$$

   and

$$L_z Y_{l,m}(\theta, \phi) = m\hbar Y_{l,m}(\theta, \phi)$$

These functions $Y_{l,m}(\theta, \phi)$ are extremely important because they form the angular part of the wavefunction every time the potential energy is spherically symmetric. Because the L^2 operator involves only θ and ϕ, the complete wave-function $u(r, \theta, \phi) = R(r)Y_{l,m}(\theta, \phi)$ is also an eigenfunction of L^2 and L_z as well as an eigenfunction of energy. Thus a particle may always be in a state which has a definite value of L^2 as well as a definite value of energy.[5] This fact, of course, has its classical basis in the fact that angular momentum is conserved when a particle moves under the influence of a central force. It is left as an exercise for the reader to verify that each of the $Y_{l,m}$ listed in this section is indeed a solution of Eq. (14), with $\alpha = l(l + 1)$.

6.3   MEASUREMENT OF ANGULAR MOMENTUM

   a. Theory.   In the preceding section, we showed that the functions $Y_{l,m}(\theta,\phi)$ are eigenfunctions of the L_z operator. They are *not* eigenfunctions of L_x or L_y, a fact you can easily verify by applying either of these operators [Eq. (13)] to any of the $Y_{l,m}$. Except for the trivial case of $Y_{0,0}$, the operation fails to yield a constant multiple of the original $Y_{l,m}$.

But there cannot be an *essential* difference between the z axis and the x or y axis. We can measure a component of **L** with respect to any direction in space, so for any particular value of L^2, the eigen*values* of L_x and L_y must be the same as those of L_z, even though the eigen*functions* are different. For example, if $l = 1$, then L_z can be $+\hbar$, 0, or $-\hbar$, so the same must be true of L_x or L_y. The fact that the eigenfunctions of L_z are not also eigenfunctions of L_x or L_y means that one can *measure only one component* of **L** *at a time*; when we write the wavefunction as an eigenfunction of L_z, we are in effect *defining* the z axis as the axis along which we have chosen to measure a component of **L**.

The method of measurement of L_z will be discussed in Section 6.3b. Let us now turn to L^2, to inquire into the verification of those seemingly peculiar eigenvalues $l(l + 1)\hbar^2$. We can imagine the following thought experiment to determine the value of L^2:

We assemble in free space a large number of systems that all have the same value of L^2, but for each of which the direction of **L** is completely undetermined. The systems are otherwise identical. We measure L_z for each of these systems, verifying that L_z is always an integer times $\hbar$. We define the integer l as the maximum observed value of $L_z/\hbar$. Each of the possible values of L_z ($+ l\hbar$, $+ (l - 1)\hbar, \ldots - l\hbar$) must occur with equal frequency, because in the absence of an

---

[5] There is a possibility that eigenfunctions corresponding to different values of L^2 may accidentally have the same energy eigenvalue. In this case, one could form an energy eigenfunction from a superposition of eigenfunctions with different values of L^2, thereby creating an eigenfunction of energy which is *not* an eigenfunction of L^2. However, such situations do not actually occur in nature; the nearest thing to such a situation occurs in the hydrogen atom, for which eigenfunctions of different L^2 value can have *almost* the same energy—to one part in 10^4.

external force field, there is no energy difference between states with different values of L_z.

From these values, we can easily compute $\langle L_z^2 \rangle$. There are $2l + 1$ equally-likely values of L_z^2, so we sum these values and divide by $2l + 1$, obtaining

$$\langle L_z^2 \rangle = (2l + 1)^{-1} \hbar^2 \sum_{m = -l}^{l} m^2 = 2(2l + 1)^{-1} \hbar^2 \sum_{m = 1}^{l} m^2$$

The series sum is easily shown by mathematical induction to be equal to $l(l + 1)(2l + 1)/6$, so $L_z^2 = l(l + 1)\hbar^2/3$.

Since there is no preferred direction, we know that

$$\langle L_x^2 \rangle = \langle L_y^2 \rangle = \langle L_z^2 \rangle .$$

We also know that L^2 is the same for each system, so

$$L^2 = \langle L^2 \rangle = \langle L_x^2 \rangle + \langle L_y^2 \rangle + \langle L_z^2 \rangle = 3\langle L_z^2 \rangle = l(l + 1)\hbar^2.$$

Thus we have deduced the eigenvalues of L^2 from a statistical analysis of the observed values of L_z.

Having measured L^2 and L_z, we can visualize our knowledge of the angular momentum vector as follows: $\mathbf{L}$ lies somewhere on a cone whose axis is the z axis and whose apex angle is equal to $\cos^{-1}(L_z/|\mathbf{L}|) = \cos^{-1}(m/[l(l + 1)]^{1/2})$ (see Fig. 4). We assume that all azimuthal angles for $\mathbf{L}$ are equally likely, be-

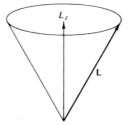

Fig. 4. *Visual represent-ation of the angular momentum vector* **L**.

cause we do not have any additional information about L_x or L_y; the eigenfunction $Y_{l,m}$ provides no additional information because $|Y_{l,m}|^2$ is independent of ϕ.

In many introductions to the subject, the theory of angular momentum is left at this point, because the eigenvalues have been determined and no further information can be obtained. Let us examine more carefully the reason *why* no further information can be obtained, and let us investigate what happens if we do try to measure *two* components of angular momentum. Study of these points will give us much more insight into the meaning of superposition in quantum theory, and will also show how the uncertainty principle affects all of these measurements.

In attempting to measure two components of angular momentum, we must first measure one component and then measure the second component. (In part *b* of this section we shall describe how this is done experimentally.) The measurement of the first component *defines our z axis*; so we must give a different label to the second axis. Let us call it the *x* axis, and let us therefore determine the eigenfunctions of the L_x operator. This would be a rather tedious chore if we attempted to compute these eigenfunctions directly from expression (13) for L_x, but fortunately there is a much simpler way to do it.

We can convert eigenfunctions of L_z into eigenfunctions of L_x simply by rotating the coordinate system. This is done most easily if we write the $Y_{l,m}$ in rectangular coordinates first; then, if we rotate the coordinate system by 90° about the *y* axis, *z* is replaced by *x* and *x* is replaced by $-z$, leaving *y* unchanged (see Fig. 5). Let us perform these operations for the

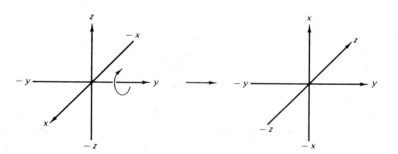

Fig. 5. *A 90° rotation about the y axis causes the x axis to replace the z axis. The z coordinate becomes the x coordinate, and the x coordinate becomes the −z coordinate.*

three functions with $l = 1$. In rectangular coordinates, we have

$$Y_{1,1} = -\left(\frac{3}{8\pi}\right)^{1/2}(\sin\theta\cos\phi + i\sin\theta\sin\phi) = -\left(\frac{3}{8\pi}\right)^{1/2}\frac{x+iy}{r}$$

$$Y_{1,0} = \left(\frac{3}{4\pi}\right)^{1/2}\cos\theta \qquad\qquad = \left(\frac{3}{4\pi}\right)^{1/2}\frac{z}{r} \qquad (19)$$

$$Y_{1,-1} = \left(\frac{3}{8\pi}\right)^{1/2}(\sin\theta\cos\phi - i\sin\theta\sin\phi) = \left(\frac{3}{8\pi}\right)^{1/2}\frac{x-iy}{r}$$

After the substitutions $x \rightarrow -z$, $y \rightarrow y$, and $z \rightarrow x$, these functions become the eigenfunctions of L_x which we denote as $Y_{1,mx}$:

$$Y_{1,1x} = \left(\frac{3}{8\pi}\right)^{1/2} \frac{+z - iy}{r}$$

$$Y_{1,0x} = \left(\frac{3}{4\pi}\right)^{1/2} \frac{x}{r} \tag{20}$$

$$Y_{1,-1x} = \left(\frac{3}{8\pi}\right)^{1/2} \frac{-z - iy}{r}$$

We may think of the operation which generated the $Y_{1,mx}$ as simply a *relabelling*. The functions $Y_{1,mx}$ have exactly the same spatial distribution as the functions $Y_{1,m}$; only the coordinates used to describe this distribution have changed. Thus these functions are still eigenfunctions of L^2, with $l = 1$; and their respective eigenvalues of L_x are the same as the L_z eigenvalues of the original functions. As an exercise, you may verify by direct application of the L_x operator (13) and the L^2 operator that these functions do indeed have the eigenvalues which we claim.

Now suppose that we have measured L^2 and L_z, finding that $l = 1$ and $m = 1$; that is, $L^2 = l(l + 1)\hbar^2 = 2\hbar^2$, and $L_z = m\hbar = +\hbar$. What will happen when we now attempt to measure L_x? Since $l = 1$, there are three possible results: $L_x = +\hbar$, 0, or $-\hbar$. But the wavefunction's angular part is $Y_{1,1}$ before the measurement, so the wavefunction is not an eigenfunction of L_x. Can we make any prediction at all about the values of L_x which are likely to result from the measurement under these circumstances? Certainly. Just as we can Fourier analyze a wave packet to determine the probabilities of obtaining various values of momentum in a measurement (even though the wave packet is not an eigenfunction of momentum), we can analyze the function $Y_{1,1}$ and determine it to be a *superposition* of *various eigenfunctions* of L_x. That is, we can write

$$Y_{1,1} = aY_{1,1x} + bY_{1,0x} + cY_{1,-1x} \tag{21}$$

and the coefficients a, b, and c should then determine the probabilities of obtaining the results $+\hbar$, 0, and $-\hbar$, respectively, when L_x is measured. As before, the square of the amplitude determines the probability; that is, $|a|^2$ equals the probability of obtaining the result $+\hbar$, just as $|\phi(k)|^2$ equals the probability of obtaining a value of p equal to $\hbar k$ in a momentum measurement.

Before performing a computation of a, b, and c for this example, let us try to deduce the probabilities from a knowledge of the expectation values. We know that $L_x^2 + L_y^2 = L^2 - L_z^2$, which equals $\hbar^2$ in this case. The x and y directions are equivalent, so $\langle L_x^2 \rangle = \langle L_y^2 \rangle = \hbar^2/2$. There are only two possible values for

L_x^2: either $\hbar^2$ or 0. Therefore, to have the proper expectation value, L_x^2 must equal zero half of the time, and $\hbar^2$ half of the time. When L_x^2 is $\hbar^2$, L_x is either $+\hbar$ or $-\hbar$, with equal probability, so each of these probabilities must be $\frac{1}{4}$. That means that the three probabilities are $\frac{1}{4}$, $\frac{1}{2}$, and $\frac{1}{4}$, for obtaining eigenvalues of $+\hbar$, 0, and $-\hbar$, respectively.

This result is confirmed when we compute a, b, and c. Substitution from Eqs. (19) and (20) into Eq. (21) yields

$$- \frac{x + iy}{r} = \frac{a(z - iy)}{r} + \frac{b\sqrt{2}x}{r} + \frac{c(-z-iy)}{r}$$

This equation must hold for all values of x, y, or z, so we may find a, b, and c by equating the coefficients of x, y, and z in turn, obtaining

$$b\sqrt{2} = 1, \quad a + c = 1, \quad \text{and} \quad a - c = 0$$

with the final result that

$$a = \tfrac{1}{2}, \quad b = \frac{1}{\sqrt{2}}, \quad \text{and} \quad c = \tfrac{1}{2}$$

or

$$|a|^2 = \tfrac{1}{4}, \quad |b|^2 = \tfrac{1}{2}, \quad \text{and} \quad |c|^2 = \tfrac{1}{4},$$

in agreement with the previous result.

This sort of procedure is of such general use in quantum theory that it is worthwhile to generalize it by stating a third fundamental postulate (to go with the two postulates stated in Section 5.6).

POSTULATE 3. *Any* acceptable wavefunction ψ can be expanded in a series of the eigenfunctions of the operator corresponding to *any* dynamical variable. The probability of finding a particular value of a dynamical variable as a result of a measurement on a system with normalized wavefunction ψ is equal to the sum of the absolute squares of the coefficients of the corresponding normalized eigenfunctions in the expansion of the function ψ.

In other words, any set of eigenfunctions, like the spherical harmonics, forms a "complete set" of functions, so that by combining members of the set in various ways one can form any other function which is reasonably well behaved—that is, any function which satisfies the requirements of continuity, etc., for an acceptable wavefunction. Of course, one could always prove *mathematically* that any *particular* set of functions forms a complete set; the postulate is a *physical* one, regarding the completeness of *any* set of functions which form the eigenfunctions of an operator for a *dynamical variable*.

The statement of the postulate regarding the *sum* of the squares of the coefficients recognizes the possibility that more than one independent eigenfunction may correspond to the same eigenvalue. Further discussion of the spherical harmonics will illustrate this point. If the angular dependence of the wavefunction is contained in a factor $f(\theta, \phi)$, one can always write $f(\theta, \phi)$ as a superposition of $Y_{l,m}$ with constant coefficients $c_{l,m}$:

$$f(\theta, \phi) = \sum_{l=0}^{\infty} \sum_{m=-l}^{m=l} c_{l,m} Y_{l,m}(\theta, \phi) \tag{22}$$

Each term in the series (22) has a particular pair of values for the variables L^2 and L_z, and the probability $|c_{l,m}|^2$, with a *given* value of l *and* m, is the probability of finding the *pair* of values $l(l+1)\hbar^2$ and $m\hbar$, respectively, when one measures L^2 and L_z. For a *given* l, there are $2l+1$ terms (with $m = l$, $m_2 = l - 1, \ldots, m = -l$), and the probability of finding the corresponding value for L^2, without regard to the value of L_z, is thus equal to the *sum* of the squares of the coefficients $c_{l,m}$ in those $2l+1$ terms, or $\sum_{m=-l}^{l} |c_{l,m}|^2$.

We have interpreted the squares of the coefficients as probabilities by analogy to the fact that the intensity of a *wave* is the square of its amplitude. This analogy is not optional; it is a strict mathematical necessity, if we are to be consistent with what we have done before, in particular in our treatment of expectation values (Section 5.6). For example, we write the expectation value of L^2 as

$$\langle L^2 \rangle = \int_0^{2\pi} \int_0^{\pi} f^*(\theta, \phi) L_{op}^2 f(\theta, \phi) \sin \theta \, d\theta \, d\phi$$

whenever $f(\theta, \phi)$ is the angular part of the wavefunction. Substitution of the expansion (22) into this integral yields

$$\langle L^2 \rangle = \sum_{l,m} \sum_{l',m'} \int_0^{2\pi} \int_0^{\pi} c_{l',m'}^* Y_{l',m'}^* L_{op}^2 c_{l,m} Y_{l,m} \sin \theta \, d\theta \, d\phi$$

and performing the indicated operation with the L^2 operator, we obtain

$$\langle L^2 \rangle = \sum_{l,m} \sum_{l',m'} \int_0^{2\pi} \int_0^{\pi} c_{l',m'}^* Y_{l',m'}^* \hbar^2 l(l+1) c_{l,m} Y_{l,m} \sin \theta \, d\theta \, d\phi$$

To evaluate the integral, one makes use of the fact that the $Y_{l,m}$, like the sine and cosine, are all orthogonal to one another. This means that

$$\int_0^{2\pi} \int_0^{\pi} Y_{l',m'}^* Y_{l,m} \sin \theta \, d\theta \, d\phi = 0$$

unless $l' = l$ *and* $m' = m$. Thus every term of the infinite series vanishes, except the terms for which $l' = l$ and $m' = m$. For the nonvanishing terms, we use the fact that the $Y_{l,m}$ are normalized, and we finally obtain

$$\langle L^2 \rangle = \sum_{l,m} |c_{l,m}|^2 l(l+1)\hbar^2$$

$$= \sum_{l} l(l+1)\hbar^2 \sum_{m=-l}^{+l} |c_{l,m}|^2$$

Thus it is *necessary* to identify the quantity $\sum_{m=-l}^{l} |c_{l,m}|^2$ as the probability of obtaining the result $L^2 = l(l+1)\hbar^2$, because the average value of L^2 is *by definition* equal to the sum of the possible results of the measurement, weighted by the *probability* of each result.

We can take advantage of the fact that the $Y_{l,m}$ are orthogonal and normalized, to find a general formula which gives the coefficients $c_{l,m}$ in the expansion of any *given* function $f(\theta, \phi)$. The procedure is simply to multiply both sides of Eq. (22) by a *particular* spherical harmonic $Y_{l',m'}^*$, and then to integrate both sides of the equation over all angles. (This is a perfect analogy to the procedure used in finding the coefficients in a Fourier series (Section 4.3); here we integrate over the surface of a sphere, which is analogous to integrating over a complete period for the sine and cosine series.) Because of the orthogonality of the $Y_{l,m}$, all terms except one drop out of the series, and we obtain the formula

$$c_{l',m'} = \int_0^{2\pi} \int_0^{\pi} f(\theta, \phi) Y_{l',m'}^*(\theta, \phi) \sin\theta \, d\theta \, d\phi \tag{23}$$

which enables us to find the coefficient for any *given* indices l' and m'.

Let us now explore further implications of our discussion of the measurement of L_z and L_x. We assumed that we had measured L_z, and we computed the probabilities of various results of a subsequent measurement of L_x. Now suppose that we go ahead and measure L_x. Can we then say that we know *both* L_z and L_x? No, we cannot, because after the measurement of L_x, the angular part of the wavefunction becomes one of the functions $Y_{l,mx}$ rather than one of the functions $Y_{l,m}$. It therefore consists of a *superposition* of different eigenfunctions of L_z, so that L_z is *no longer known* with certainty. Thus we conclude that the process of measuring L_x *must disturb* L_z; in part b of this section we shall discuss the *physical mechanism* which causes this disturbance.

In quantum theory one encounters many similar cases in which measurement of one variable disturbs another variable so that one cannot know the value of both variables simultaneously. There is a general rule for knowing whether or not a pair of variables can be measured simultaneously: If it is possible to measure one variable without disturbing the value of another variable, the operators O_a and O_b representing the two variables must *commute*: that is,

$$O_a O_b \Psi = O_b O_a \Psi$$

for *any* wavefunction Ψ, where we assume that the operator on the right is applied to Ψ first, and the other operator is then applied to the result.

To prove this we note first that, if the value of one variable is to be unchanged when a second variable is measured, it must be possible to expand any eigenfunction of the first variable in a series of eigenfunctions of *both* variables.[6] Since any wavefunction may be written as a series of eigenfunctions of the first variable, and each of the latter wavefunctions may be written as a series of eigenfunctions of both variables, we see that *any* wavefunction may be written as a series of eigenfunctions of both variables. Suppose that Ψ_{ij} is one such eigenfunction:

$$O_a \Psi_{ij} = a_i \Psi_{ij}$$

and

$$O_b \Psi_{ij} = b_j \Psi_{ij}$$

Then

$$O_a O_b \Psi_{ij} = O_a b_j \Psi_{ij} = a_i b_j \Psi_{ij}$$

and

$$O_b O_a \Psi_{ij} = O_b a_i \Psi_{ij} = a_i b_j \Psi_{ij}$$

so that

$$O_a O_b \Psi_{ij} = O_b O_a \Psi_{ij}$$

that is O_a and O_b commute when applied to any of the eigenfunctions Ψ_{ij}. But in that case, these operators must commute when applied to *any* wavefunction Ψ, for we may expand Ψ as

$$\Psi = \sum_i \sum_j c_{ij} \Psi_{ij}$$

so that

$$O_a O_b \Psi = \sum_i \sum_j c_{ij} a_i b_j \Psi_{ij} = O_b O_a \Psi$$

It is also possible to prove that the commutation of these operators is a *sufficient* as well as a *necessary* condition for the corresponding variables to be simultaneously measurable.

An example of a pair of commuting operators is the energy operator H for a spherically symmetric potential and the operator L^2. H may be written as (Section 6.2)

$$H = \frac{p_r^2}{2m} + \frac{L^2}{2mr^2} + V(r)$$

---

[6] Suppose that $O_a \Psi = A\Psi$, where A is a constant, so that Ψ is an eigenfunction of variable a. Since *any* function can be expanded in eigenfunctions of variable b, we may write $\Psi = \sum c_n \Psi_{nb}$ where the Ψ_{nb} are eigenfunctions of variable b. After we make a measurement of variable b, the wavefunction of the system must be one of the Ψ_{nb}. But if we can make the measurement without disturbing the value of variable a, then this wavefunction (one of the Ψ_{nb}) must be an eigenfunction of variable a as well as variable b. Thus for this type of measurement, *all* of the Ψ_{nb} are eigenfunctions of a as well as b.

L^2 clearly commutes with this operator, because L^2, being a function of θ and ϕ only, must commute with p_r^2, which is a function of r only. Therefore both the energy and the total angular momentum may be known simultaneously for a particle in a spherically symmetric potential (as we have already assumed). On the other hand, the operators L_x and L_z clearly do not commute. In fact, you may verify from Eqs. (13) that

$$L_x L_z = L_z L_x - i\hbar L_y$$

b. Experiment.   Direct evidence of quantization of angular momentum was obtained in 1922 by Stern and Gerlach.[7] They made use of the fact that a circulating electron in an atom forms a tiny loop of electric current, upon which a magnetic field can exert a force. The influence of a magnetic field upon a current loop is most simply described in terms of the magnetic dipole moment of the loop, defined as a vector $\mathbf{\mu}$ perpendicular to the plane of the loop, whose magnitude is equal (in mks) to IA, where I is the current and A is the area of the loop. (A magnetic dipole so defined behaves in a magnetic field $\mathbf{B}$ just as an electric dipole behaves in an electric field; that is, there is a torque on the dipole equal to $\mathbf{\mu} \times \mathbf{B}$, which tends to align $\mathbf{\mu}$ parallel to $\mathbf{B}$, and the potential energy of this dipole is equal to $-\mathbf{\mu} \cdot \mathbf{B}$. For further details see any text on electricity and magnetism.[8]) From this definition one can easily show that an orbiting electron has a moment $\mathbf{\mu}$ which is proportional to its orbital[9] angular momentum $\mathbf{L}$, and in mks units $\mathbf{\mu} = -e\mathbf{L}/2m$. Therefore, if the values of a component of $\mathbf{L}$ are quantized, the same must be true for a component of $\mathbf{\mu}$, and the experiment of Stern and Gerlach, which showed that μ_z is quantized, is a verification of the quantization of L_z.

Stern and Gerlach saw the effect of the moment $\mathbf{\mu}$ by passing a beam of atoms through an *inhomogeneous* magnetic field (see Fig. 6). In such a field a magnetic dipole feels a net force whose z component is

$$F_z = \mu_z \frac{\partial B_z}{\partial z}$$

$$= -\left(\frac{eL_z}{2m}\right) \frac{\partial B_z}{\partial z}$$

(Fig. 6(b) indicates the origin of this net force.) Therefore each atom of the beam was deflected in passing through the magnet; the deflection of the beam was observed by catching the atoms on a photographic plate. The atoms were

---

[7] O. Stern and W. Gerlach, Z. Phys. **8**, 110; **9**, 349 (1922).

[8] For example, Chapter 10 of E. M. Purcell, "Electricity and Magnetism," Vol. 2 of the Berkeley Physics Course. McGraw-Hill, New York, 1965.

[9] In the actual experiment of Stern and Gerlach, the angular momentum resulted from internal spin of the electron rather than from its orbital motion; but the general idea is the same, because all kinds of angular momentum are quantized. We shall discuss spin in Section 7.3.

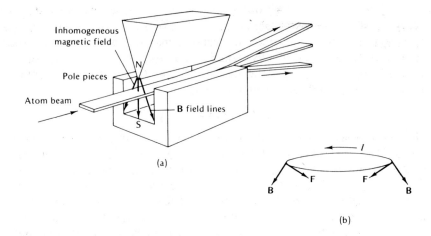

(a)

(b)

Fig. 6. (a) *Schematic view of Stern–Gerlach apparatus. Atomic beam travels along sharp pole piece, which produces a field stronger near the north pole (N). As a result, field lines converge and are not exactly vertical. (b) Edgewise view of a current loop in the* **B** *field of (a); current is into paper at right. Because the* **B** *field lines are not parallel, the forces* **F** *on opposite sides of the loop are not in opposite directions; each has a downward component, so there is a net downward force on the loop.*

found in distinct clusters, indicating that the angular momentum possessed only discrete values rather than the continuous range of values which one would expect classically.

Let us now try to understand, on the basis of the mechanism of the Stern–Gerlach experiment, why we cannot measure one component of **L** without disturbing the other components. Suppose we do *two consecutive* Stern–Gerlach experiments, one with the **B** field along the z axis, and a second one with the **B** field along the x axis. Let $L^2 = 2\hbar^2$ (that is, $l = 1$) throughout the experiments. As a result of passing through the B_z field, the beam splits into three beams, which can be identified with the values $L_z = +\hbar$, $L_z = 0$, and $L_z = -\hbar$, respectively. We can split off the beam in which $L_z = +\hbar$ and send it through the B_x field, to measure the value of L_x for each atom of this beam. We know from the preceding theoretical discussion that the beam should split into three components, with one-fourth of the atoms having $L_x = +\hbar$, one-half having $L_x = 0$, and one-fourth having $L_x = -\hbar$.

Consider the atoms for which $L_x = +\hbar$ (and for which we already found L_z to be $+\hbar$). Theory tells us that L_z is no longer equal to $+\hbar$ for all of these atoms. What happened to change L_z while we were measuring L_x? We can understand classically that the measurement of L_x must change L_z to some extent. The magnetic field **B** causes a torque equal to $\boldsymbol{\mu} \times \mathbf{B}$ on each atom, and by classical mechanics

$$\text{torque} = d\mathbf{L}/dt$$

Since $\boldsymbol{\mu} = -e\mathbf{L}/2m$,

$$\frac{d\mathbf{L}}{dt} = \boldsymbol{\mu} \times \mathbf{B} = -\left(\frac{e\mathbf{L}}{2m}\right) \times \mathbf{B}$$

In the measurement of L_x, **B** is in the x direction. If we assume that **L** initially has components $L_x = +\hbar$ and $L_z = +\hbar$, then

$$\frac{d\mathbf{L}}{dt} = -\left(\frac{e}{2m}\right)L_z B_x \mathbf{j} = -\left(\frac{e\hbar}{2m}\right)B_x \mathbf{j}$$

where **j** is a unit vector in the y direction. Thus **L** acquires a y component. In a small time δt, the change in **L** is

$$\delta\mathbf{L} = \frac{d\mathbf{L}}{dt}\delta t = -\left(\frac{e\hbar}{2m}\right)B_x(\delta t)\mathbf{j}$$

As time goes on, the vector **L** precesses about the x axis; the tip of the **L** vector describes a circle about the x axis. In time t, the angle through which the tip of **L** moves is $|\delta L/L_z| = (e/2m)B_x t$ (see Fig. 7). The angular velocity of the tip of the **L** vector is $\omega = eB_x/2m$, as this point moves on a circle centered on the x axis.

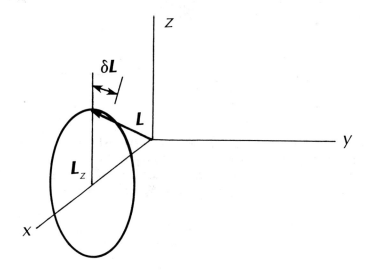

Fig. 7. *Precession of the tip of the* **L** *vector in time* δt.

This discussion, assuming as it does that L_x *and* L_z are known initially, is entirely based on classical theory. It shows that measurement of L_x does disturb L_z, even in classical theory, for as L precesses, L_z must change. But according to classical theory, we still retain *knowledge* of L_z after the measurement of L_x, because the *amount of precession, and hence the amount of change in L_z, may be precisely calculated* from the knowledge of the initial conditions. On the other hand, when we consider the limitations imposed by the uncertainty principle, we find that we cannot design a Stern–Gerlach experiment to measure L_x in such a way that we can *predict* the resulting change in L_z. Let us examine this statement closely.

In order to predict the change in L_z, we must know the field B_x which acts on each atom. But because of the presence of the field gradient $\partial B_x/\partial x$, and because we cannot know the x coordinate of each atom precisely, we do not know B_x for each atom. Thus each atom's angular momentum vector precesses by an *uncertain* amount, and we lose our knowledge of L_z.

It remains to be seen that the precession is sufficiently great that we are justified in saying that our knowledge of L_z is completely lost. Let us make a rough calculation of the uncertainty in the precession angle for the above case, in which L_z was originally $+\hbar$, and L^2 is $2\hbar^2$. As mentioned above, measurement of L_x splits the beam into three parts; the $L_x = 0$ part is undeflected, while the $L_x = +\hbar$ part is deflected, acquiring a momentum of $p_x = F_x t = \mu_x(\partial B_x/\partial x)t$, where t is the time required to traverse the magnetic field. If this component is to be distinguishable from the $L_x = 0$ component, p_x must be greater than twice the uncertainty in p_x demanded by the uncertainty principle. Thus, if the beam initially had a spread of $\pm \delta x$ in its x coordinate,

$$p_x > 2\delta p_x = \hbar/\delta x$$

so that

$$\mu_x \frac{\partial B_x}{\partial x} t\, \delta x > \hbar \qquad (24)$$

The product $(\partial B_x/\partial x)\delta x$ is the uncertainty in the value of B_x for an atom in the beam, the uncertainty resulting from the spread of the beam over an area in which the field is not uniform. There must be a corresponding uncertainty in the rate of precession of **L**. Since $\omega = eB_x/2m$,

$$\delta\omega = \frac{e\, \delta B_x}{2m}$$

$$= \frac{e}{2m} \frac{\partial B_x}{\partial x} \delta x$$

The angle of precession is then uncertain by an amount

$$\delta\theta = \frac{e}{2m} t \frac{\partial B_x}{\partial x} \delta x$$

But from Eq. (24), with $\mu_x = (e\hbar/2m)$, we have

$$\frac{e\hbar}{2m} \frac{\partial B_x}{\partial x} t \, \delta x > \hbar$$

making

$$\delta\theta > 1 \quad \text{rad}$$

The precise value of 1 rad should not be taken too seriously, because it resulted from a somewhat arbitrary assumption about the value of p_x required to separate the beams, and we have not justified our classical approach to the measurement process. But we do see that the basic uncertainty principle for position and momentum, combined with classical laws, predicts an uncertainty in the direction of $\mathbf{L}$ which justified our statement that the knowledge of L_z is destroyed by measurement of L_x.

6.4   AN EXAMPLE: THE THREE-DIMENSIONAL HARMONIC OSCILLATOR

The potential energy of a spherically symmetric three-dimensional harmonic oscillator may be written

$$V(x, y, z) = \tfrac{1}{2}Kr^2 = \tfrac{1}{2}m\omega^2(x^2 + y^2 + z^2)$$

The time-independent Schrödinger equation is therefore

$$\left[-\frac{\hbar^2}{2m}\left(\frac{\partial^2}{\partial x^2} + \frac{\partial^2}{\partial y^2} + \frac{\partial^2}{\partial z^2}\right) + \frac{1}{2}m\omega^2(x^2 + y^2 + z^2) \right]u = Eu$$

which may be written

$$(H_x + H_y + H_z)u = Eu \tag{25}$$

where

$$H_x = -\frac{\hbar^2}{2m}\frac{\partial^2}{\partial x^2} + \frac{1}{2}m\omega^2 x^2$$

$$H_y = -\frac{\hbar^2}{2m}\frac{\partial^2}{\partial y^2} + \frac{1}{2}m\omega^2 y^2$$

$$H_z = -\frac{\hbar^2}{2m}\frac{\partial^2}{\partial z^2} + \frac{1}{2}m\omega^2 z^2$$

H_x, H_y, and H_z are simply the energy operators (or Hamiltonian operators) for one-dimensional oscillation along the respective axes. We know, therefore, from the result of Section 5.7, that

$$H_x u_q(x) = (q + \tfrac{1}{2})\hbar\omega u_q(x)$$

where q is an integer and $u_q(x)$ is one of the eigenfunctions found in Section 5.7:

$$u_q(x) = \left(\frac{\partial}{\partial x} - ax\right)^q e^{-ax^2/2} \qquad (a = m\omega/\hbar) \tag{26}$$

and a similar equation holds for the variables y and z.

We can construct an eigenfunction of Eq. (25) simply by taking the product of the eigenfunctions of H_x, H_y, and H_z:

$$u_{qst}(x,y,z) = u_q(x)u_s(y)u_t(z) \tag{27}$$

where q, s, and t are integers, and the functions on the right are defined in Eq. (26). Application of the energy operator then yields

$$(H_x + H_y + H_z)u_{qst}(x, y, z) = (q + \tfrac{1}{2} + s + \tfrac{1}{2} + t + \tfrac{1}{2})\hbar\omega u_{qst}(x, y, z) \tag{28}$$

so that $u_{qst}(x,y,z)$ is indeed an eigenfunction of the energy operator, and the energy levels are

$$E_n = (n + \tfrac{3}{2})\hbar\omega \qquad (n = q + s + t)$$

where n is any positive integer, or zero.

There are three linearly independent eigenfunctions for $n = 1$: u_{100}, u_{010}, and u_{001}. We see here an example of *degeneracy*; independent eigenfunctions with the same eigenvalue are said to be degenerate. Level E_2 is sixfold degenerate, for there are six eigenfunctions for $n = 2$: u_{200}, u_{020}, u_{002}, u_{110}, u_{101}, and u_{011}. Level E_3 is tenfold degenerate, for there are ten ways to combine the integers q, r, and s so that they total three.

It is interesting to examine the angular momentum of these eigenfunctions. To do this, we examine the angular dependence of each function and try to express it as a combination of spherical harmonics; according to postulate 3, this should be possible. We start with $n = 0$; there is one eigenfunction with energy eigenvalue $E_0 = \tfrac{3}{2}\hbar\omega$:

$$u_{000} = u_0(x)u_0(y)u_0(z)$$
$$= e^{-ar^2/2}$$

This eigenfunction is spherically symmetrical, so its angular dependence, a constant, can be expressed as the single spherical harmonic $Y_{0,0}$, with total angular momentum of zero. (Clearly *any* spherically symmetrical wavefunction has total angular momentum zero, because the L^2 operator yields zero when it operates on a spherically symmetrical function.)

The three eigenfunctions for $n = 1$ may be written, according to Eqs. (26) and (27), as

$$u_{100} = u_1(x)u_0(y)u_0(z) = \left(\frac{\partial}{\partial x} - ax\right)e^{-ax^2/2}e^{-ay^2/2}e^{-az^2/2}$$

$$= -2axe^{-ar^2/2}$$

$$u_{010} = -2aye^{-ar^2/2}$$

$$u_{001} = -2aze^{-ar^2/2}$$

The angular dependence of these wavefunctions is all contained in the respective factors x, y, and z. But Eqs. (19) show that x may be written in terms of the sum of the two spherical harmonies $Y_{1,1}$ and $Y_{1,-1}$, because

$$Y_{1,1} + Y_{1,-1} = \left(\frac{3}{8\pi}\right)^{1/2}\frac{2x}{r}$$

so that

$$x = \frac{r}{2}\left(\frac{8\pi}{3}\right)^{1/2}(Y_{1,1} + Y_{1,-1})$$

and

$$u_{100} = Nre^{-ar^2/2}(Y_{1,1} + Y_{1,-1})$$

where N is a normalization constant. Therefore u_{100} is an eigenfunction of L^2, with $l = 1$, composed of equal parts of $m = +1$ and $m = -1$; a measurement of L_z for a particle whose wavefunction is u_{100} would yield either $L_z = +\hbar$ or $L_z = -\hbar$ with equal probability.

In a similar way, we find that

$$u_{010} = f(r)(Y_{1,1} - Y_{1,-1})$$

and

$$u_{001} = g(r)Y_{1,0}$$

where the functions $f(r)$ and $g(r)$ contain the appropriate normalization constants as well as the factors r and $e^{-ar^2/2}$. Thus the three linearly independent eigenfunctions of the energy operator for $n = 1$ are all eigenfunctions of L^2, with $l = 1$, but only one of them, u_{001}, is also an eigenfunction of L_z; the other two are mixed states, as far as L_z is concerned. (It should be clear that u_{100} is an eigenfunction of L_x and u_{010} is an eigenfunction of L_y.)

It is often more convenient to work with functions which are eigenfunctions of L_z. We can construct linear combinations of u_{100}, u_{010}, and u_{001} which are linearly independent and are eigenfunctions of L_z, as follows:

$$\begin{array}{ll} u_a = u_{100} + iu_{010} & (n = 1, \quad l = 1, \quad m = +1) \\ u_b = u_{001} & (n = 1, \quad l = 1, \quad m = 0) \\ u_c = u_{100} - iu_{010} & (n = 1, \quad l = 1, \quad m = -1) \end{array}$$

The functions u_a, u_b, and u_c are simultaneous eigenfunctions of energy, of L^2, and of L_z, with eigenvalues indicated by the quantum numbers given above. Of course, they all have the same n and l quantum numbers because these numbers are characteristic of the three original functions u_{100}, u_{010}, and u_{001}. The factor i comes in with u_{010} because $Y_{1,\pm1}$ contains the factor $x \pm iy$.

It becomes a little more interesting if we go through the same procedure for $n = 2$. The six linearly independent eigenfunctions for $n = 2$ contain quadratic terms such as x^2, y^2, z^2, xy, xz, yz, as well as terms which are spherically symmetrical. In order to construct six independent functions as superpositions of spherical harmonics, we need six spherical harmonics. We cannot use any with $l > 2$, for these contain cubic or higher order terms. Neither can we use $l = 1$, for these contain only linear terms, which are absent from the $n = 2$ eigenfunctions. This leaves us with the five functions for $l = 2$ plus the function for $l = 0$. Thus when $n = 2$, L^2 may be either $2(2 + 1)\hbar^2 = 6\hbar^2$, or zero, and L_z may be $+2\hbar$, $\hbar$, 0, $-\hbar$, or $-2\hbar$. Just as we did for $n = 1$, one may construct new energy eigenfunctions which are eigenfunctions of L^2 and L_z, by using appropriate linear combinations of the six functions u_{200}, u_{020}, u_{002}, u_{110}, u_{101}, and u_{011}. Similarly, for $n = 3$, the ten eigenfunctions are linear combinations of the seven spherical harmonics for $l = 3$ and the three for $l = 1$; for $n = 4$ there are fifteen eigenfunctions, which are linear combinations of the nine spherical harmonics for $l = 4$, the five for $l = 2$, and the one for $l = 0$, etc.

6.5 THE RADIAL EQUATION

We have not yet dealt with the radial part of Eq. (8). Let us eliminate the angular part of that equation by substituting the eigenvalue $l(l + 1)$ for the operator Ω. We obtain

$$-\frac{r}{R}\frac{d^2}{dr^2}(rR) + l(l + 1) = \frac{2mr^2}{\hbar^2}[E - V(r)] \tag{29}$$

Consider first the case $l = 0$. We may then rewrite Eq. (29) as

$$\frac{d^2}{dr^2}(rR) = -\frac{2m}{\hbar^2}[E - V(r)](rR) \tag{30}$$

Equation (30) is identical to the one-dimensional Schrödinger equation, Chapter 5, Eq. (4), except that the variable is r rather than x and the eigenfunction is $rR(r)$ rather than $u(x)$.

We may use Eq. (30) to find the wavefunction for a free particle with $l = 0$. We let $V(r) = 0$ and we obtain

$$\frac{d^2}{dr^2}(rR) = -\frac{2mE}{\hbar^2}(rR)$$

whose solution is

$$rR = e^{\pm ikr} \qquad \left(\text{with } k = \left(\frac{2mE}{\hbar^2}\right)^{1/2}\right)$$

so that the wavefunction, neglecting the normalization factor, is

$$u(r, \theta, \phi) = R(r)Y_{0,0}(\theta, \phi)$$

$$\propto e^{\pm ikr}/r \tag{31}$$

since $Y_{0,0}(\theta, \phi)$ is a constant.

If we multiply $u(r, \theta, \phi)$ by the time dependent factor $e^{-i\omega t}$, we see that the solution e^{+ikr}/r represents a wave traveling outward from the origin (in the positive r direction), and the solution e^{-ikr}/r represents a wave traveling inward toward the origin. To obtain a solution that is acceptable in a region that includes the origin, we must combine these waves to form a standing wave (see Sec. 5.2) of the form $(\sin kr)/r$.

Equation (30) may also be used to find the spherically symmetric $(l = 0)$ solutions for any spherically symmetric potential energy. For example, if we let $V(r)$ be the harmonic oscillator potential[10]:

$$V(r) = \frac{m\omega^2 r^2}{2}$$

we find that corresponding to each solution of the one-dimensional harmonic oscillator equation (Section 5.7) there is a spherically symmetric solution for the three-dimensional harmonic oscillator. There is only one difference; the lowest eigenvalue is $\frac{3}{2}\hbar\omega$ in the three-dimensional case, rather than $\frac{1}{2}\hbar\omega$. How can this be, when the equations are identical? Suppose we put $E = \frac{1}{2}\hbar\omega$ in Eq. (30), to see what happens. The product $rR(r)$ should then be the same *function* as the eigenfunction u_0 found for the one-dimensional case; that is, since

$$u_0(x) = e^{-ax^2/2}$$

then

$$rR(r) = e^{-ar^2/2}$$

However, this function is not acceptable for $rR(r)$, because there is another *boundary condition*, that $rR(r)$ must vanish at the origin. This condition is necessary to ensure that the expectation values of all dynamical variables remain finite, when they are computed by the rules of Section 5.6. Therefore we rule out the eigenvalue $\frac{1}{2}\hbar\omega$ for the three dimensional case, in agreement with the result of Section 6.4. (Notice also that the above-mentioned functions of the form $e^{\pm ikr}/r$ can only be used in regions of space that exclude the origin.)

---

[10] In this paragraph, unlike the previous paragraph, ω is the classical oscillator frequency, and is *not* equal to the frequency of the time-dependent part of the wavefunction.

It is easy to see that Eq. (30) gives us the same answer that we found in Section 6.4 for the eigenvalue $\frac{3}{2}\hbar\omega$ also. In this case the product $rR(r)$ should be the same function as the eigenfunction u_1 for the one-dimensional case, so, since

$$u_1(x) = xe^{-ax^2/2}$$

$$rR(r) = re^{-ar^2/2}$$

Therefore $R(r) = e^{-ar^2/2}$, which is the eigenfunction u_{000} we found in Section 6.4 to correspond to an energy of $\frac{3}{2}\hbar\omega$. (Remember that $u(r, \theta, \phi) = R(r)$, because the wavefunction is spherically symmetric.)

The Centrifugal Potential. If we rewrite Eq. (29) with $l \neq 0$, we can obtain an equation of the same form as Eq. (30):

$$\frac{d^2}{dr^2}(rR) = -\frac{2m}{\hbar^2}\left[E - V(r) - \frac{l(l+1)\hbar^2}{2mr^2}\right](rR) \tag{32}$$

Equation (32) is also of the same form as the one-dimensional Schrödinger equation, if we consider the potential energy to be the *sum*

$$V(r) + \frac{l(l+1)\hbar^2}{2mr^2}$$

The additional term is called the centrifugal potential; it is analogous to the fictitious potential energy which appears when one uses a rotating coordinate system in classical mechanics. To see that this is reasonable, notice that this energy is equivalent to $L^2/2mr^2$. As we saw in Section 6.2, the kinetic energy in spherical coordinates is

$$E - V(r) = \frac{p_r^2}{2m} + \frac{L^2}{2mr^2}$$

so that

$$\frac{p_r^2}{2m} = E - V(r) - \frac{L^2}{2mr^2} \tag{33}$$

while in one dimension,

$$\frac{p_x^2}{2m} = E - V(x) \tag{34}$$

Therefore, to write a radial equation in the form of the one-dimensional equation, it makes sense to replace the right-hand side of Eq. (34) by the right-hand side of Eq. (33), replacing the p_x operator by the p_r operator.

6.6 THE THREE-DIMENSIONAL SQUARE WELL

As an example of the application of the radial equation, let us consider the three-dimensional counterpart to the "square well" treated in Section 5.4. The potential is

$$V(r) = +V_0 \qquad (r > a)$$
$$V(r) = 0 \qquad (r < a)$$

When $l = 0$, the radial equation (30) can be written, for $r < a$, as

$$\frac{d^2}{dr^2}(rR) = -\frac{2mE}{\hbar^2}(rR) = -k^2(rR)$$

where we have again used the definition $k^2 = 2mE/\hbar^2$. The solution is

$$rR(r) = A \cos kr + B \sin kr \qquad (35)$$

or

$$u(r, \theta, \phi) = R(r)$$

$$= \frac{A \cos kr}{r} + \frac{B \sin kr}{r}$$

We write u in this form rather than the exponential form, because u must be finite at $r = 0$, and we can ensure this by making $A = 0$. Therefore

$$u(r, \theta, \phi) = \frac{B \sin kr}{r} \qquad (r < a)$$

Outside the well, the radial equation is

$$\frac{d^2}{dr^2}(rR) = -\frac{2m(E - V_0)}{\hbar^2}(rR)$$

For a bound state, $(E - V_0)$ is negative, so the solution must be a decaying exponential, as in Section 5.4:

$$rR = Ce^{-\alpha r} \qquad (r > a) \qquad (36)$$

with α defined as $(2m(V_0 - E)/\hbar^2)^{1/2}$, as before. Now we can proceed as in Section 5.4, matching the wavefunction and its derivative at $r = a$. Because the functions involved are the same, the result must be exactly as in Section 5.4 *for the odd parity solution*, because rR, the function analogous to $u(x)$, is the odd function $\sin kr$; the boundary condition that $rR = 0$ at $r = 0$ eliminates the even solution.[11] Therefore the energy levels are given by the solutions of Eq. (36), Chapter 5:

$$\sin ka = \pm \frac{ka}{\beta a} \qquad (\beta^2 = k^2 + \alpha^2) \qquad (37)$$

---

[11] See the discussion at the bottom of page 215.

Because we are restricted to the odd solutions, it is possible that no bound state exists at all, for a shallow well (or a narrow well). (Remember that the lowest energy solution for the one-dimensional case was an even parity state.) The lowest-energy odd solution has

$$\frac{\pi}{2} < ka < \pi$$

and since $ka \le \beta a$ [to satisfy Eq. (37)], we see that if $\beta a < \pi/2$, that is, if $2mV_0 a^2/\hbar^2 < (\pi/2)^2$, there is no bound state. There is one bound state if $\pi/2 < \beta a < 3\pi/2$, two if $3\pi/2 < \beta a < 5\pi/2$, and so on.

If $l \ne 0$, the presence of the centrifugal potential makes the radial equation a bit more complicated:

$$\frac{d^2}{dr^2}(rR) = -\frac{2m}{\hbar^2}\left[E - \frac{l(l+1)\hbar^2}{2mr^2}\right](rR)$$

or

$$\frac{d^2}{dr^2}(rR) = -k^2(rR) + \frac{l(l+1)}{r^2}(rR) \qquad (r < a) \qquad (38)$$

and

$$\frac{d^2}{dr^2}(rR) = -\frac{2m}{\hbar^2}\left[E - V_0 - \frac{l(l+1)\hbar^2}{2mr^2}\right](rR)$$

or

$$\frac{d^2}{dr^2}(rR) = -(i\alpha)^2(rR) + \frac{l(l+1)}{r^2}(rR) \qquad (r > a) \qquad (39)$$

where k and α are defined as before.[12]

With the addition of the centrifugal potential, the effective well is as shown in Fig. 8. The well is effectively narrower, because the particle must stay away from the origin if it is to have some angular momentum. Therefore we might expect fewer bound levels for successively higher values of l, in a given well.

The functions $R(r)$ which are solutions of Eq. (38) or Eq. (39) are not hard to work out for a given l; they belong to a class of functions called spherical Bessel functions $j_l(kr)$ and spherical Neumann functions $n_l(kr)$.[13]

---

[12] Notice that Eq. (39) is identical to Eq. (38), except that k is replaced by $i\alpha$.

[13] Obviously, differential equations of the form of Eqs. (38) and (39) were studied and solved long before quantum theory was invented. Professor Bessel died in 1846. The general formula for $j_l(x)$ is $j_l(x) = (\pi/2x)^{1/2}J_{(l+\frac{1}{2})}(x)$, where $J_{(l+\frac{1}{2})}(x)$ is the ordinary Bessel function, defined by the infinite series

$$J_n(x) = \frac{x^n}{2^n\Gamma(n+1)}\left\{1 - \frac{x^2}{2(2n+2)} + \frac{x^4}{2\cdot 4(2n+2)(2n+4)}\right.$$
$$\left. - \frac{x^6}{2\cdot 4\cdot 6(2n+2)(2n+4)(2n+6)} + \cdots\right\}$$

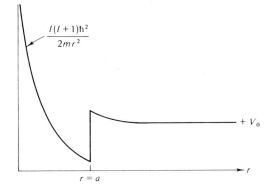

Fig. 8. *The effective potential well which appears in the radial equation, when one adds the
centrifugal potential to a square-well potential.*

(Of course, for Eq. (39) the argument of the functions is $i\alpha r$ rather than kr.) For
each value of l there are two linearly independent solutions—a spherical
Bessel function and a spherical Neumann function. The first three of each are

$$j_0(kr) = \frac{\sin kr}{kr}$$

$$j_1(kr) = \frac{\sin kr}{(kr)^2} - \frac{\cos kr}{kr}$$

$$j_2(kr) = \left[\frac{3}{(kr)^3} - \frac{1}{kr}\right] \sin kr - \frac{3 \cos kr}{(kr)^2}$$

$$n_0(kr) = -\frac{\cos kr}{kr}$$

$$n_1(kr) = -\frac{\cos kr}{(kr)^2} - \frac{\sin kr}{kr}$$

$$n_2(kr) = \left[-\frac{3}{(kr)^3} + \frac{1}{kr}\right] \cos kr - \frac{3 \sin kr}{(kr)^2}$$

We have already seen that $j_0(kr)$ and $n_0(kr)$, as given above, are solutions
of the radial equation (38) when $l = 0$, and you may easily verify that the other j
and n functions given here are also solutions of Eq. (38) with the appropriate
values of l.

The functions $j_l(kr)$ are important aside from their application to the special case of a square-well potential. If a particle is completely free and has a definite angular momentum, its radial equation is Eq. (38) for all r, and the radial part of its wavefunction must therefore be $j_l(kr)$ for all r. (Consider $a \to \infty$). Thus, in the expansion of a plane wave in a series of eigenfunctions of L^2, the coefficients are multiples of the $j_l(kr)$ (see Section 6.7, Eq. (49)).

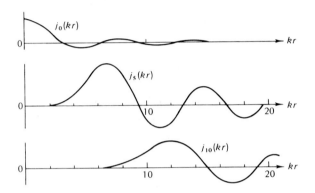

Fig. 9. *Graphs of $j_0(kr)$, $j_5(kr)$, and $j_{10}(kr)$. Vertical scale is 4 units per inch for j_0, and 0.4 units per inch for j_5 and j_{10}. (Plotted from values tabulated in Tables of Spherical Bessel Functions, National Bureau of Standards, Columbia Univ. Press, New York, 1947.)*

Figure 9 shows a few of the $j_l(kr)$. Notice how the function is "pushed away" from the origin as l increases. In general, for $l \neq 0$, $j_l(kr)$ is maximum at kr just greater than l, and is quite small for $kr < l$. This is the behavior that one should expect; a particle with momentum $p = \hbar k$ and angular momentum $|L| = [l(l+1)\hbar^2]^{1/2} \approx l\hbar$ should not be found closer to the origin than $r \approx l/k$, because

$$|\mathbf{L}| = |\mathbf{r} \times \mathbf{p}| \leq |\mathbf{r}||\mathbf{p}| = r\hbar k$$

so

$$l\hbar \leq r\hbar k \qquad \text{or} \qquad r \geq l/k$$

To find the energy levels for the square well, one proceeds in much the same way as before. The solution for a given l for $r < a$ must be $j_l(kr)$, because the functions $n_l(kr)$ go to infinity at $r = 0$. For $r > a$, we require a solution which goes to zero as $r \to \infty$, and we can obtain such a solution from a linear combination of $j_l(i\alpha r)$ and $n_l(i\alpha r)$. Guided by the knowledge that

$$\cos(i\alpha r) + i \sin(i\alpha r) = e^{i(i\alpha r)} = e^{-\alpha r}$$

we construct the combination

$$h_l^{(1)}(i\alpha r) = j_l(i\alpha r) + in_l(i\alpha r)$$

which is called a spherical Hankel function of the first kind.[14] In particular,

$$h_1^{(1)}(i\alpha r) = i\left(\frac{1}{\alpha r} + \frac{1}{(\alpha r)^2}\right)e^{-\alpha r}$$

and

$$h_2^{(1)}(i\alpha r) = \left(\frac{1}{\alpha r} + \frac{3}{(\alpha r)^2} + \frac{3}{(\alpha r)^3}\right)e^{-\alpha r}$$

These functions clearly have the right behavior as $r \to \infty$, so we use them for the region $r > a$. Thus, for example, the complete solution for $l = 1$ is

$$R(r) = Aj_1(kr) \qquad (r < a)$$

$$R(r) = Bh_1^{(1)}(i\alpha r) \qquad (r > a)$$

where A and B are arbitrary constants to be determined by the boundary conditions and normalization. We may eliminate A and B by the requirement that the ratio $(dR/dr)/R$ be continuous at $r = a$. Therefore

$$\frac{\dfrac{d}{dr}[j_1(kr)]_{r=a}}{j_1(ka)} = \frac{\dfrac{d}{dr}[h_1^{(1)}(i\alpha r)]_{r=a}}{h_1^{(1)}(i\alpha a)}$$

or

$$\frac{\dfrac{d}{dr}\left[\dfrac{\sin kr}{(kr)^2} - \dfrac{\cos kr}{kr}\right]_{r=a}}{\dfrac{\sin ka}{(ka)^2} - \dfrac{\cos ka}{ka}} = \frac{\dfrac{d}{dr}\left[ie^{-\alpha r}\left(\dfrac{1}{\alpha r} + \dfrac{1}{(\alpha r)^2}\right)\right]_{r=a}}{ie^{-\alpha a}\left(\dfrac{1}{\alpha a} + \dfrac{1}{(\alpha a)^2}\right)}$$

or

$$\frac{-\dfrac{2\sin ka}{k^2 a^3} + \dfrac{2\cos ka}{ka^2} + \dfrac{\sin ka}{a}}{\dfrac{\sin ka}{(ka)^2} - \dfrac{\cos ka}{ka}} = \frac{-\dfrac{1}{a} - \dfrac{2}{\alpha a^2} - \dfrac{2}{\alpha^2 a^3}}{\dfrac{1}{\alpha a} + \dfrac{1}{(\alpha a)^2}}$$

---

[14] There is a spherical Hankel function of the second kind, $h_l^{(2)}$, which is also a solution of Eq. (39) (with argument $i\alpha r$) or Eq. (38) (with argument kr):

$$h^{(2)}(x) = j_l(x) - in_l(x)$$

This function is linearly independent of $h^{(1)}$, but we do not use it because $h_l^{(2)}(i\alpha r)$ contains the factor $e^{+\alpha r}$, so that any wavefunction that contains $h_l^{(2)}(i\alpha r)$ would blow up as $r \to \infty$. You may wish to verify that $j_l(i\alpha r)$, which is a superposition of $h_l^{(1)}(i\alpha r)$ and $h_l^{(2)}(i\alpha r)$, goes to infinity as $r \to \infty$. (See the asymptotic expression for $j_l(kr)$ given in Section 6.7.)

After multiplication by $-a$, and performing the division on each side, we have

$$-2 + \left[\left(\frac{1}{ka}\right)^2 - \frac{\cot ka}{ka}\right]^{-1} = -2 - \left[\frac{1}{\alpha a} + \left(\frac{1}{\alpha a}\right)^2\right]^{-1}$$

or

$$\frac{\cot ka}{ka} - \left(\frac{1}{ka}\right)^2 = \frac{1}{\alpha a} + \left(\frac{1}{\alpha a}\right)^2 \tag{40}$$

Equation (40), together with the condition that

$$(ka)^2 + (\alpha a)^2 = \frac{2mV_0 a^2}{\hbar^2} \tag{41}$$

may be solved graphically to find the possible values of k, and hence the energy levels, for $l = 1$.

EXAMPLE PROBLEM 1.   Use Eq. (40) to show that there is no bound state for $l = 1$ when

$$V_0 a^2 < \frac{\pi^2 \hbar^2}{2m}$$

Solution.   For $ka < \pi$, $\cot ka < 1/ka$. Therefore the left side of Eq. (40) is negative for these values of ka. But the right side of Eq. (40) is always positive,[15] so there is no solution for $ka < \pi$. But Eq. (41) tells us that

$$\frac{2mV_0 a^2}{\hbar^2} \geq (ka)^2$$

so that, when $ka \geq \pi$—that is, when there *is* a solution—we have

$$\frac{2mV_0 a^2}{\hbar^2} \geq \pi^2$$

or

$$V_0 a^2 \geq \frac{\pi^2 \hbar^2}{2m}$$

Q.E.D.

We can have a solution with $ka = \pi$. In that case, $\cot ka \to \infty$, $\alpha a = 0$, and

$$\frac{2mV_0 a^2}{\hbar^2} = \pi^2$$

---

[15] Remember that α was defined to be a real, positive quantity; if it were not, our solution which goes as $e^{-\alpha r}$ would not behave properly at infinity.

A second bound state appears when

$$\frac{2mV_0 a^2}{\hbar^2} = (2\pi)^2$$

Notice that the first bound level for $l = 1$ appears at a value of $V_0 a^2$ which is four times the value required in order that there be a bound level for $l = 0$.

6.7 SCATTERING OF PARTICLES FROM A SPHERICALLY SYMMETRIC POTENTIAL

To conclude our general discussion of three-dimensional problems, we consider the three-dimensional counterpart of the problem of transmission of particles past a potential barrier (Section 5.5). The problem, of course, has a completely different character in three dimensions, for the particles which strike the barrier (or well) can go off in any direction, instead of simply being transmitted or reflected.

Suppose that we send a "beam" of particles—a plane wave of the form[16] $\Psi_{inc} = Ae^{ikz}$—along the z axis toward a "scattering center," and suppose that the potential energy is nonzero over a limited region, $r < a$, surrounding the scattering center (see Fig. 10). The density of particles in the beam is $|\Psi_{inc}|^2 = |A|^2$, and the intensity of the beam—the number of particles crossing a unit area in a unit of time—is, as usual, the product of particle density and particle velocity, or $v|A|^2$.

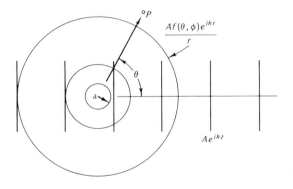

Fig. 10. *Scattering of a plane wave from a scattering center, resulting in a spherical scattered wave. The interaction which produces the scattered wave occurs only in the region $r < a$.*

[16] Throughout this section we will omit the time-dependent factor $e^{-i\omega t}$ in writing the wavefunctions.

A certain fraction of the particles will be scattered, forming a wave which emanates from the scattering center, while the remainder continue on as the plane wave. At large r, the scattering potential and the centrifugal potential both go to zero, so we know from Eq. (31) that the radial part of the scattered wave is e^{+ikr}/r. (We use e^{+ikr} rather than e^{-ikr}, because we want a wave which is traveling *outward*, toward positive r.) Therefore we write the scattered wave as $\Psi_{sc} = Af(\theta, \phi)e^{ikr}/r$, where the factor A expresses the fact that the scattered wave amplitude is proportional to the incident wave amplitude. The particle density in the scattered wave is $|A|^2|f(\theta, \phi)|^2/r^2$, and the intensity of the scattered wave is $v|A|^2|f(\theta,\phi)|^2/r^2$, in the region of large r.

Scattering Cross Section.   Given a scattering potential, we shall see that it is straightforward, although perhaps tedious, to compute $f(\theta, \phi)$ and hence to determine the scattered wavefunction. In practice, one has the more difficult task of deducing the scattering potential after measuring $|f(\theta, \phi)|^2$. To measure $|f(\theta, \phi)|^2$, one places a detector at a point P (Fig. 10) where it can detect particles in the scattered wave but not those in the incident wave.[17] The number N_d of particles seen by the detector per unit time is equal to the product of the intensity of the scattered wave and the area dA of the detector, so

$$N_d = \left[\frac{v|A|^2|f(\theta, \phi)|^2}{r^2}\right] dA$$

But dA/r^2 is just the solid angle $d\Omega$ subtended by the detector at the scattering center, and $v|A|^2$ is the intensity I_{inc} of the incident beam, so

or

$$N_d = |f(\theta, \phi)|^2 I_{inc}\, d\Omega \qquad (d\Omega = \sin\theta\, d\theta\, d\phi)$$

$$|f(\theta, \phi)|^2 = \frac{d\sigma}{d\Omega}$$

where the differential $d\sigma$ is defined as

$$d\sigma = \frac{N_d}{I_{inc}} = \frac{\text{Number of particles/sec striking detector}}{\text{Number of particles/sec-cm}^2 \text{ in incident beam}} \qquad (42)$$

The ratio $d\sigma/d\Omega$, which has the dimensions of an area, is called the differential cross section for scattering from the scattering center. Its integral over a sphere is the *total* scattering cross section σ:

$$\sigma = \int d\sigma = \int \frac{d\sigma}{d\Omega}\, d\Omega = \int_0^{2\pi} \int_0^{\pi} |f(\theta, \phi)|^2 \sin\theta\, d\theta\, d\phi$$

[17] The incident beam cannot be a *pure* plane wave, because it has a finite width, but we assume that the width is much greater than the wavelength, so that we do not have to worry about diffraction effects. The detector is placed in the region where the incident wave is zero.

Thus by counting particles, one can determine $|f(\theta, \phi)|^2$ experimentally for all values of θ and ϕ. The problem then is to fit the experimental results to a theoretical determination of $f(\theta, \phi)$ on the basis of an assumed potential function. We shall see one way to go about this in this section.

There is a physical significance to the fact that σ has the dimensions of an area. If the scattering were classical—for example, that of a uniform stream of bullets bouncing off a piece of armor plate—the integral of $d\sigma$ as defined by Eq. (42) would be just the cross sectional area of the plate; that is, the total number of bullets scattered per second would be the product of the area of the plate and the number per unit area per second striking the plate (assuming the plate's area to be smaller than the cross sectional area of the stream of bullets). In quantum theory, however, the cross section, although still an area, is not directly given by the area covered by the scattering potential, because diffraction effects complicate the situation.

Partial Waves. So far the discussion has been quite general. But this is not the place for a full exposition of scattering theory[18]; we simply wish to show one treatment as an application of the concepts of angular momentum and the solutions of the Schrödinger equation which have been developed in this chapter.

Because we assume the scattering potential to be spherically symmetric, we can simplify the analysis in two ways. First, there is symmetry with respect to rotation about the z axis, so the ϕ dependence disappears, and $f(\theta, \phi)$ becomes $f(\theta)$. Second, angular momentum is conserved, so that we can decompose the incoming wave into components (partial waves) with different l-values, and can then calculate the scattering of each component separately.

We begin by writing the complete wavefunction in the region of large r:

$$\Psi \to A\left\{ e^{ikz} + f(\theta)\frac{e^{ikr}}{r} \right\} \qquad (r \to \infty) \tag{43}$$

This is to be compared with the actual solution of the Schrödinger equation, which in general is a linear combination of spherical harmonics, each multiplied by a radial factor $R_l(r)$. Since there is no ϕ dependence, we have

$$\Psi = \sum b_l R_l(r) P_l(\cos \theta) \tag{44}$$

$P_l(\cos \theta)$ is identical[19] to $P_l^0(\cos \theta)$ and is called the Legendre polynomial of order l.

We determine the scattering amplitude $f(\theta)$ by matching expression (43) to the *limit* of expression (44) as r goes to infinity. In order to find this limit,

---

[18] For a more complete discussion of scattering theory, see any standard text on quantum mechanics, for example, L. I. Schiff, "Quantum Mechanics," 3rd Edition. McGraw-Hill, New York, 1968.
[19] Except for a normalizing factor. See Eq. (48) and Section 6.2.

we make use of the fact that the scattering potential $V(r)$ is zero at large r ($r > a$), so that R_l is a linear combination of solutions of Eq. (38), namely

$$R_l(r) = \cos \delta_l j_l(kr) - \sin \delta_l n_l(kr) \qquad (r > a) \qquad (45)$$

The sum of the squares of the coefficients in this combination must be 1, so that $R_l(r)$ will be normalized (as j_l and n_l are). Hence the two constant coefficients are not independent and may be written as we have done, expressing the fact that for $r > a$ the solution for a given l value is completely specified by the single parameter δ_l, called the *phase shift* of the lth partial wave. The name "phase shift" is appropriate for reasons which will become clear in a moment. We may now find the limit of $R_l(r)$ as $r \to \infty$, because the limits of the functions $j_l(kr)$ and $n_l(kr)$ are known to be given by

$$\left.\begin{array}{l} j_l(kr) \to \dfrac{\sin\left(kr - \dfrac{l\pi}{2}\right)}{kr} \\[2em] n_l(kr) \to -\dfrac{\cos\left(kr - \dfrac{l\pi}{2}\right)}{kr} \end{array}\right\} \qquad (r \to \infty)$$

so that

$$R_l(r) \to \cos \delta_l \frac{\sin\left(kr - \dfrac{l\pi}{2}\right)}{kr} + \sin \delta_l \frac{\cos\left(kr - \dfrac{l\pi}{2}\right)}{kr}$$

which reduces to

$$R_l(r) \to \frac{\sin\left(kr - \dfrac{l\pi}{2} + \delta_l\right)}{kr} \qquad (r \to \infty) \qquad (46)$$

The quantity δ_l thus appears as a phase factor in the asymptotic limit of $R_l(r)$. This phase factor is a convenient measure of the effect of the scattering potential on the wavefunction; if the scattering potential is zero, then $\delta_l = 0$, because if $V(r) = 0$ for all r, the same solution $R_l(r)$ must hold all the way in to $r = 0$, and thus $R_l(r)$ cannot contain the function $n_l(kr)$, which goes to infinity at $r = 0$. The effect of a potential on the lth partial wave at large distances is a shift in phase by the amount δ_l. We shall see that there is a simple expression for the scattering cross section σ in terms of the various phase shifts δ_l.

Now we can equate expression (43) to the limit of expression (44) as $r \to \infty$, obtaining[20]

$$e^{ikz} + f(\theta)\frac{e^{ikr}}{r} = \sum b_l P_l(\cos\theta) \frac{\sin\left(kr - \frac{l\pi}{2} + \delta_l\right)}{kr} \tag{47}$$

In order to solve this equation for $f(\theta)$, we must find the coefficients b_l, and to do that we expand the left side in a series of Legendre polynomials, in order to match the two sides term by term. We know that this expansion is possible; the angular dependence of any wavefunction may be expressed as a series of spherical harmonics, which reduce to the Legendre polynomials when $m = 0$ (and m is zero in this case because the function $e^{ikz} = e^{ikr\cos\theta}$ has no ϕ dependence). Thus we write

$$e^{ikz} = \sum_{l=0}^{\infty} a_l(r)P_l(\cos\theta)$$

The coefficients $a_l(r)$ may be evaluated in the same way as the coefficients in a Fourier series, by using the fact that

$$\int_{\cos\theta=-1}^{\cos\theta=+1} P_l(\cos\theta)P_{l'}(\cos\theta)\,d(\cos\theta) = \begin{cases} 0 & (l \neq l') \\ \dfrac{2}{2l+1} & (l = l') \end{cases} \tag{48}$$

The result is

$$e^{ikz} = \sum_{l=0}^{\infty}(2l+1)i^l j_l(kr)P_l(\cos\theta) \tag{49}$$

Substitution of this expression for e^{ikz} into Eq. (47) yields

$$\sum_{l=0}^{\infty}(2l+1)i^l \frac{\sin\left(kr - \frac{l\pi}{2}\right)}{kr}P_l(\cos\theta) + f(\theta)\frac{e^{ikr}}{r}$$

$$= \sum_{l=0}^{\infty} b_l \frac{\sin\left(kr - \frac{l\pi}{2} + \delta_l\right)}{kr}P_l(\cos\theta)$$

where we have replaced the function $j_l(kr)$ by its limit as $r \to \infty$. If we rewrite the sine functions in exponential form [using the identity $\sin x \equiv (e^{ix} - e^{-ix})/2i$], the last equation becomes

$$e^{ikr}\left\{2ikf(\theta) + \sum_{l=0}^{\infty}(2l+1)i^l e^{-il\pi/2}P_l(\cos\theta)\right\} - e^{-ikr}\sum_{l=0}^{\infty}(2l+1)i^l e^{il\pi/2}P_l(\cos\theta)$$

$$= e^{ikr}\sum_{l=0}^{\infty} b_l e^{-il\pi/2}e^{i\delta_l}P_l(\cos\theta) - e^{-ikr}\sum_{l=0}^{\infty} b_l e^{il\pi/2}e^{-i\delta_l}P_l(\cos\theta)$$

[20] The constant A of Eq. (43) is no longer needed, because it can be "absorbed" into the constant coefficients b_l.

The constants b_l are found by equating the coefficients of e^{-ikr} on the two sides, obtaining

$$b_l = (2l + 1)i^l e^{i\delta_l}$$

We may then find $f(\theta)$ by equating the coefficients of e^{ikr}:

$$f(\theta) = (2ik)^{-1} \sum (2l - 1)(e^{2i\delta_l} - 1)P_l(\cos \theta)$$

or

$$f(\theta) = \frac{1}{k} \sum (2l + 1)e^{i\delta_l} \sin \delta_l P_l(\cos \theta)$$

Therefore

$$\frac{d\sigma}{d\Omega} = |f(\theta)|^2$$

$$= \left(\frac{1}{k^2}\right) |\sum (2l + 1)e^{i\delta_l} \sin \delta_l P_l(\cos \theta)|^2 \tag{50}$$

and

$$\sigma = \int \frac{d\sigma}{d\Omega} d\Omega$$

$$= \frac{2\pi}{k^2} \sum \left[\int (2l + 1) \sin \delta_l P_l(\cos \theta) \sin \theta \, d\theta\right]^2$$

the cross products in the summation having vanished because of the orthogonality of the $P_l(\cos \theta)$ [Eq. (48)]. Equation (48) may now be used to evaluate the integrals, with the final result that

$$\sigma = \frac{4\pi}{k^2} \sum_{l=0}^{\infty} (2l + 1) \sin^2 \delta_l \tag{51}$$

Thus we have reduced the problem of finding σ to one of finding the phase shifts produced by the potential in the various angular momentum components of the incident plane wave. This is a remarkable simplification, but it appears that we are still faced with a formidable problem, because Eq. (51) contains an infinite series. To analyze a scattering experiment, we must assume a scattering potential $V(r)$, use it to calculate all of the δ_l, use Eq. (51) to find σ, and then compare this with the observed σ. We can repeat the process using several different assumed potentials, to see which one gives the best fit to the data. But if we have to compute an infinite number of phase shifts in order to use Eq. (51), there is not much point in commencing the calculation.

There are many situations, fortunately, in which only one or two of the phase shifts are nonzero. To understand this fact, we may visualize the decomposition of the incident plane wave into partial waves as follows (Fig. 11):

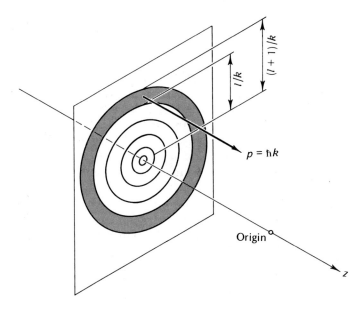

Fig. 11. *Illustration of the decomposition of a beam of particles into separate annular beams, each with a definite angular momentum about the origin. Particles passing through the shaded area have angular momentum between $l\hbar$ and $(l+1)\hbar$.*

The plane wave is a stream of particles of momentum $p = \hbar k$, moving parallel to the z axis. If the particles were localized, the angular momentum of a given particle about the origin would be $|L| = pd = \hbar kd$, where d is the distance of that particle from the z axis. Thus in this semiclassical view, particles with angular momentum between $l\hbar$ and $(l+1)\hbar$ pass through an annulus whose radius lies between l/k and $(l+1)/k$. [Notice that the number of particles passing through such an annulus is proportional to its area, which is proportional to $(2l+1)$; since these particles contribute to the total cross section in proportion to their number, we see why there is a $(2l+1)$ factor in each term of the series in Eq. (51).] If l/k is greater than the radius a at which the scattering potential becomes zero, these particles cannot be influenced by the potential, and the corresponding phase shift δ_l must be zero.

So if $ka \ll 1$, even nonlocalized particles with $l = 1$ are but slightly influenced by the potential, and σ is given to good accuracy by the first term in Eq. (52):

$$\sigma = \frac{4\pi}{k^2} \sin^2 \delta_0 \qquad (ka \ll 1) \qquad (52)$$

Of course, if one is trying to find out something about the shape of the scattering potential, one does *not* use incident particle energies so low that $ka \ll 1$. Obviously one cannot measure the shape of something by doing an experiment whose result is determined by only one parameter; a physician cannot determine the condition of a patient simply by weighing him.[21] To put it another way, a particle whose de Broglie wavelength is λ cannot respond to details of dimensions much less than λ in the scattering potential. So it is often necessary to analyze experiments in which several phase shifts are involved, in order to explore details of various potentials.

Study of Fig. 12 may give you a better idea of the connection between the scattering potential and the phase shifts. Because the radial equation is mathematically identical to the one-dimensional Schrödinger equation, we can sketch the function $rR_l(r)$, which is mathematically analogous to $u(x)$, guided by the same principles we used in Section 5.4 in discussing Fig. 3 of Chapter 5. In the region where the effective potential—the scattering potential plus the centrifugal potential—is greater than the particle's total energy E, the function $rR_l(r)$ curves away from the axis. There is a point of inflection at $r = r_c$, where $E = l(l+1)\hbar^2/2mr^2 + V(r)$; for $r > r_c$, the total energy is greater than the effective potential, and the function curves toward the axis, eventually becoming sinusoidal as $r \to \infty$.

The region $r < r_c$ is the classically forbidden region, and the wavefunction does not amount to much in this region. Therefore if the scattering potential cuts off at a value of r much less than $r = r_c$, it can have very little influence on the wavefunction, and the phase shift must be zero for the particular l value in question.

The present argument gives the same criterion as the semiclassical view illustrated in Fig. 11, for determining which partial waves contribute to the cross section. We may find r_c by setting V_{eff} equal to E, obtaining

$$\frac{l(l+1)\hbar^2}{2mr_c^2} = \frac{\hbar^2 k^2}{2m}$$

or

$$r_c = \frac{[l(l+1)]^{1/2}}{k} \approx \frac{l}{k}$$

[21] Sometimes the physician can tell that you are "out of shape," but he must at least know your height as well as your weight.

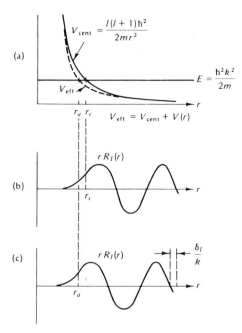

Fig. 12. *Connection between scattering potential and phase shift. (a) Total energy E, centrifugal potential V_{cent}, and effective potential V_{eff} (dotted line); $V_{eff} = V_{cent} + V(r)$, where $V(r)$ is the scattering potential, which is negative in this case. (b) The product $rR_l(r)$ versus r, for the case when $V(r) = 0$. In this case $R_l(r) = j_l(kr)$. (c) The product $rR_l(r)$ versus r for the effective potential shown in (a). At larger r, this function has the same form as the function plotted in (b), but it is shifted toward smaller r, because the whole curve has been "pulled in" toward the origin by the attractive potential, the point of inflection having moved from r_c, where $E = V_{cent}$, in to r_a, where $E = V_{eff}$. (Shift shown is δ_l/k rather than δ_l itself, because the functions are plotted versus r rather than kr.) The phase shift is positive, by definition, in this case; remember that the asymptotic form of $R_l(kr)$ was written $(\sin(kr - (l\pi/2) + \delta_l))/kr$. If $V(r)$ were positive, the curve would be pushed away from the origin, and the phase shift δ_l would be negative.*

In order for a given partial wave to contribute, r_c for that wave must be not much larger than the radius a at which $V(r)$ cuts off. That is, $l/k \lesssim a$ or $l \lesssim ka$ for a wave of a given l value to be able to contribute appreciably to the cross section. (Also see the discussion associated with Fig. 9.)

EXAMPLE PROBLEM 2. Find σ and $d\sigma/d\Omega$ for the scattering of particles from a "perfectly rigid" sphere (infinitely repulsive potential) of radius a, for the case $ka \ll 1$.

Solution. The potential energy is

$$V(r) = +\infty \qquad (r < a)$$
$$V(r) = 0 \qquad (r > a)$$

Therefore the wavefunction must vanish at $r = a$, just as it did in the one-dimensional square well with rigid walls. For $r \geq a$, the radial function must be a combination of $j_0(kr)$ and $n_0(kr)$:

$$R(r) = \frac{\sin(kr + \delta_0)}{kr}$$

because none of the partial waves for $l > 0$ should be expected to contribute to the scattering in this case. We find the phase shift δ_0 by using the boundary condition that $R(r) = 0$ at $r = a$:

$$\frac{\sin(ka + \delta_0)}{ka} = 0$$

so that $ka + \delta_0 = 0$ and $\delta_0 = -ka$. Equation (52) gives the cross section:

$$\sigma = \frac{4\pi}{k^2} \sin^2(-ka) \approx \frac{4\pi}{k^2}(-ka)^2 = 4\pi a^2$$

Because $l = 0$, the scattered wave is isotropic. Therefore

$$\frac{d\sigma}{d\Omega} = \frac{\sigma}{4\pi} = a^2$$

Notice that σ is just four times the geometrical cross section which the sphere presents to the beam.

It is instructive to see how small ka must be so we can safely neglect the $l = 1$ wave. For this wave, the radial function is [from Eq. (45)]

$$R_1(r) = \cos \delta_1 j_1(kr) - \sin \delta_1 n_1(kr) \qquad (r \geq a)$$

Setting $R(a)$ equal to zero, we have

$$\tan \delta_1 = \frac{j_1(ka)}{n_1(ka)}$$

$$= \frac{\sin(ka) - (ka)\cos(ka)}{-\cos(ka) - (ka)\sin(ka)}$$

Using the series expansions of sine and cosine, we obtain

$$\tan \delta_1 = \frac{\left\{(ka) - \dfrac{(ka)^3}{6} + \cdots\right\} - ka\left\{1 - \dfrac{(ka)^2}{2} + \cdots\right\}}{-\left\{1 - \dfrac{(ka)^2}{2} + \cdots\right\} - ka\left\{(ka) - \dfrac{(ka)^3}{6} + \cdots\right\}}$$

which becomes

$$\tan \delta_1 = -\frac{(ka)^3}{3}$$

when we retain only the lowest-order terms in ka in numerator and denominator. Using Eq. (51) to find σ, we have

$$\sigma = \frac{4\pi}{k^2}(\sin^2 \delta_0 + 3\sin^2 \delta_1) \approx 4\pi a^2\left\{1 + \frac{(ka)^4}{3}\right\}$$

and the error resulting from neglect of the $l = 1$ term is seen to be about $(ka)^4/3 \times 100$ percent.

The effect of the $l = 1$ term shows up at *smaller k* values in the *differential* cross section, which is given by Eq. (50) as

$$\frac{d\sigma}{d\Omega} = \frac{1}{k^2} |e^{i\delta_0} \sin \delta_0 + 3e^{i\delta_1} \sin \delta_1 \cos \theta|^2$$

$$= \frac{1}{k^2} \{\sin^2 \delta_0 + 3e^{i(\delta_1 - \delta_0)} \sin \delta_1 \sin \delta_0 \cos \theta$$

$$+ 3e^{-i(\delta_1 - \delta_0)} \sin \delta_1 \sin \delta_0 \cos \theta + 9\sin^2 \delta_1 \cos^2 \theta\}$$

$$= \frac{1}{k^2} \{\sin^2 \delta_0 + 6\sin \delta_1 \sin \delta_0 \cos(\delta_1 - \delta_0)\cos \theta + 9\sin^2 \delta_1 \cos^2 \theta\}$$

With the substitutions $\sin \delta_0 = -ka$, $\cos \delta_0 = 1$, $\sin \delta_1 = \tan \delta_1 = -(ka)^3/3$, and $\cos \delta_1 = 1$, we have

$$\frac{d\sigma}{d\Omega} = \frac{1}{k^2}\left\{(ka)^2 + 6ka\frac{(ka)^3}{3}\cos \theta + 9\frac{(ka)^6}{9}\cos^2 \theta\right\}$$

$$\approx a^2\{1 + 2(ka)^2 \cos \theta\}$$

Thus at $\theta = 0$ or $180°$, the percent error in $d\sigma/d\Omega$ caused by neglecting the $l = 1$ term is $2(ka)^2 \times 100$ percent—considerably larger than the error in σ. If we begin to bombard a scattering center with low energy particles, and then increase the energy, we see a *difference* between the *forward* and *backward differential* cross sections much sooner than we see the effect of the $l = 1$ contribution to the total cross section.

PROBLEMS

1. Evaluate the three lowest energy levels for an electron in a "box" whose dimensions are $3.0 \text{ Å} \times 3.0 \text{ Å} \times 4.0 \text{ Å}$. Give the degeneracy of each level.

2. Verify by direct substitution that $Y_{1,1}(\theta,\phi)$ is a solution of Eq. (9), with the expected eigenvalue ($\alpha = l(l + 1) = 2$).

3. For a particle in a state for which $l = 1$ and $m = 1$, compute $\langle L_x^2 \rangle$ directly from the three-dimensional definition of expectation value:

$$\langle L_x^2 \rangle = \int_0^{2\pi} \int_0^\pi Y_{1,1}^* L_x^2 Y_{1,1} \sin\theta \, d\theta \, d\phi \ ,$$

using expression (13) for the L_x operator. Is the result consistent with the values of L^2 and L_z?

4. Using Eqs. (13), verify that $L_x L_y - L_y L_x = i\hbar L_z$.

5. Suppose that we have a beam of particles in which $L^2 = 2\hbar^2$ ($l = 1$) and $L_z = +\hbar$. What will be the result of measuring the angular momentum $L_{z'}$, about an axis z' which makes an angle of α with the z axis? Write expressions in terms of α for the probability of finding $L_{z'}$ to be equal to $+\hbar$, 0, and $-\hbar$, respectively. (*Hint*: Write the three $l = 1$ eigenfunctions of $L_{z'}$ in rectangular coordinates x, y, and z by rotating the coordinate system as we did in finding the eigenfunctions of L_x in Section 6.3. Then write the appropriate spherical harmonic as a linear combination of these eigenfunctions.)

6. Extend the discussion of Section 6.3a to the case $l = 2$: By a rotation of axes, find an expression for $Y_{2,2x}$, the eigenfunction of L_x with eigenvalue $+2\hbar$. Then express $Y_{2,2x}$ as a linear combination of the functions $Y_{2,m}$. Finally, from the coefficients in this expression, deduce the probabilities of the various possible results of a measurement of L_z on a system for which $L_x = +2\hbar$ and $L^2 = 6\hbar^2$.

7. Use the probabilities found in Problem 6 to compute the value of $\langle L_z^2 \rangle$ for that system. Explain why $\langle L_z^2 \rangle$ should be equal to $\hbar^2$.

8. (a) Equation (17) requires that l be an integer because m is an integer. If m were a half-integer, Eq. (17) would yield half-integer values of l as well. Thus there exist spherical harmonics such as

$$Y_{1/2,1/2} = \text{constant} \times (\sin \theta \, e^{i\phi})^{1/2} = \Theta(\theta) \, \Phi(\phi)$$

Verify by direct substitution that $\Theta(\theta)$ for this spherical harmonic is a solution to Eq. (17) with $m = \frac{1}{2}$, and that the eigenvalue l equals $l(l + 1) = \frac{1}{2}(\frac{1}{2} + 1) = \frac{3}{4}$.

(b) We ruled out solutions such as $Y_{1/2,1/2}$ because they are not single-valued. However, the wavefunction itself is not observable—the observable is $|\psi|^2$—so it is unduly restrictive to require ψ itself to be single valued. But we can rule out solutions of noninteger m in another way, illustrated by the following example: The functions $Y_{1/2,1/2}$ and $Y_{1/2,-1/2}$ are the only two solutions for $l = \frac{1}{2}$. Therefore it should be possible to find two eigenfunctions of L_x that are linear combinations of these two solutions, just as we found three eigenfunctions of L_x for $l = 1$ (Eqs. 20). Find one of these functions, $Y_{1/2,1/2x}$, in the same way that we found $Y_{1,1x}$ in Eqs. (20), namely, by writing $Y_{1/2,1/2}$ in cartesian coordinates and making the substitutions $x \rightarrow -z$ and $z \rightarrow x$. Show that $Y_{1/2,1/2x}$ cannot be written as a linear combination of $Y_{1/2,1/2}$ and $Y_{1/2,-1/2}$. Thus the functions with $l = \frac{1}{2}$ do not behave properly (they violate Postulate 3) and they must be discarded.

9. One can generate all of the spherical harmonics from the formula

$$Y_{l,m} = \sqrt{\frac{(2l + 1)(l + |m|)!}{4\pi(l - |m|)!}} \, \frac{1}{2^l \, l! \, (\sin \theta)^{|m|}} \left(\frac{d}{d \cos \theta}\right)^{l - |m|} (\sin \theta)^{2l} \, e^{im\phi}$$

Use this formula to generate $Y_{3,2}$, and verify that it is a solution of Eq. (17) with the correct eigenvalue.

10. Show that the sum of the probability density functions

$$\sum_{m = -2}^{+2} |Y_{2,m}(\theta,\phi)|^2$$

is spherically symmetric. (The general result, that

$$\sum_{m = -l}^{l} |Y_{l,m}|^2$$

is spherically symmetric for any value of l, was proven by A. Unsöld, *Annalen der Phys.* **82**, 355 (1927).)

11. Consider an *anisotropic* three-dimensional harmonic oscillator, with a different frequency of oscillation along each axis, so that the potential energy is

$$V(x, y, z) = \frac{m}{2}(\omega_x^2 x^2 + \omega_y^2 y^2 + \omega_z^2 z^2)$$

Write an expression for the energy levels, in terms of ω_x, ω_y, and ω_z. Find the four lowest energy levels, for the case $\omega_y = \omega_x = 2\omega_z/3$, and determine the degeneracy of each level. Do you expect the energy eigenfunctions to be eigenfunctions of L^2? Explain.

12. Construct the six linear combinations of u_{200}, u_{020}, u_{002}, u_{110}, u_{101}, and u_{011} which are eigenfunctions of L^2 and L_z. (See Section 6.4.)

13. Extend the analysis of Sec. 6.4 to the case $n = 2$. Write the function u_{200} as a linear combination of eigenfunctions of L^2 and L_z, and find the coefficients, which are functions of r in this case. Use these coefficients to find the probability of obtaining each possible combination of l and m in a measurement on a system whose wavefunction is u_{200}. (Hint: You must do the appropriate integral on r to eliminate the dependence of the result on the r coordinate.) Use your results to compute $\langle L^2 \rangle$ and $\langle L_z^2 \rangle$ for this system.

14. Verify that $j_1(kr)$ and $j_2(kr)$ as given in Section 6.6 are solutions of Eq. (38).

15. Find the lowest energy level for an electron trapped in a spherically symmetric well for which $V = 0$ when $1 \text{ Å} < r < 4 \text{ Å}$, and V is infinite everywhere else. (*Hint*: The ground-state wavefunction must be a superposition of $j_0(kr)$ and $n_0(kr)$, and it must be zero where V is infinite.)

16. Find the probability that the electron of the previous problem will be found within a spherical shell of inside diameter 3.9 Å and outside diameter 4.0 Å.

17. One can generate all of the spherical Bessel functions by means of the formula

$$j_l(kr) = \left(-\frac{r}{k}\right)^l \left(\frac{1}{r}\frac{d}{dr}\right)^l j_0(kr)$$

Show that this is true by using the formula to find $j_1(kr)$ and $j_2(kr)$. Also find $j_3(kr)$ and show that it is a solution of Eq. (37) for $l = 3$.

18. Use the first three partial waves ($l = 0$, $l = 1$, and $l = 2$) to compute σ and $d\sigma/d\Omega$ for scattering from a perfectly rigid sphere when $ka = \frac{1}{4}$. Compare with the result that you would obtain by using only one ($l = 0$) or two ($l = 0$, $l = 1$) partial waves.

19. A particle of mass m is scattered from a spherically symmetric potential $V(r)$, (see figure) where

$$V(r) = \begin{cases} -V_0 & (0 < r < a) \\ \dfrac{-V_0}{4} & (a < r < 2a) \\ 0 & (r > 2a) \end{cases}$$

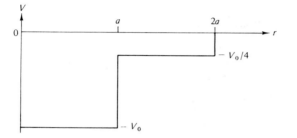

Assume that the incident particle has very low energy, so that $ka \ll 1$, where k is the wavenumber for $r > 2a$. Find σ and $d\sigma/d\Omega$ for the case where $(2mV_0/\hbar^2)^{1/2} = \pi/2a$. (*Hint*: Since $E \ll V_0$, the wavenumber for $r < a$ is $k_1 \approx \pi/2a$, and the wavenumber for $a < r < 2a$ is $k_2 \approx \pi/4a$. The radial functions are therefore of the form

$$R_l(r) = \begin{cases} \dfrac{A \sin k_1 r}{k_1 r} & (r < a) \\ \dfrac{B \sin(k_2 r + \delta')}{k_2 r} & (a < r < 2a) \\ \dfrac{C \sin(kr + \delta)}{kr} & (r > 2a) \end{cases}$$

and one can eliminate the arbitrary constants A, B, and C, and find δ, by using the boundary conditions on the wavefunction at $r = a$ and $r = 2a$.)

20. If $\delta_0 = 180°$ and all the other phase shifts are quite small, the cross section is very small for incident particles of just the right energy, and σ becomes much larger for particles of slightly different energy, because $\sin^2 \delta_0$, which provides the sole contribution to σ, increases whether δ_0 increases or decreases. A sharp minimum in σ for the scattering of electrons from rare-gas atoms (called the Ramsauer-Townsend effect) was explained in this way by Niels Bohr. Show that the effect can occur with an attractive potential but not with a repulsive potential.

7

The Hydrogen Atom

One of the greatest and earliest triumphs of the Schrödinger equation was the solution of the problem of the hydrogen atom, which was achieved without the assumption of arbitrary rules like those of the old quantum theory. Although it was still impossible to obtain exact solutions for atoms containing more than one electron, the new quantum theory permitted approximations which could be applied, in principle, to any problem and to any desired degree of accuracy. The ultimate result was almost unbelievable accuracy in theoretical calculations of hydrogen energy levels, taking into account such factors as the spin of the electron, a quantity which was unheard of in the older theory.

Experimenters have responded with equally accurate experiments to test these calculations. This accuracy has not been sought simply to demonstrate the prowess of physics and physicists; proof of the existence of each small contribution to the energy in the amount predicted by the theory is an indication of the correctness of our fundamental ideas concerning the nature of matter.

In 1930, Dirac showed that electron spin emerges as a *natural* consequence of a relativistic wave equation for the electron. We close this chapter with a brief account of the Dirac equation. We cannot treat the Dirac equation in the detail with which we have treated the Schrödinger equation, but we can see that two fundamental properties of matter—internal spin of particles and the existence of antimatter—can be "derived" directly from this equation.

7.1 WAVEFUNCTIONS FOR MORE THAN ONE PARTICLE

If the proton were of infinite mass, it would be a fixed center of force for the electron in the hydrogen atom, and we could solve the problem by the methods of Chapter 6. The wavefunction would be a function of the co-ordinates of a single particle, the electron.

However, because the proton also moves, we must incorporate this fact into our wavefunction. According to our postulates (Section 5.6), each dynamical variable for each particle must be represented by an operator whose eigenvalues are the allowed values of the variable. It is a logical extension of these postulates to state that there must be a wavefunction for a *system* which is capable of generating all of the dynamical variables of the system when it is operated upon by the appropriate operators. This wavefunction must therefore be a function of the coordinates of *each particle* in the system, so that for a hydrogen atom the wavefunction may be written as

$$\psi(x_p, y_p, z_p, x_e, y_e, z_e, t)$$

where x_p is the x coordinate of the proton, x_e is the x coordinate of the electron, etc.

It is now clear that the wavefunction is simply a mathematical construction; there is no physical " wave " in the sense of a simple displacement that exists at each point of space and time, for the wavefunction is a function of *six* space coordinates in this case, rather than three. Our single-particle problems of Chapters 5 and 6 enabled us to make useful analogies with conventional waves, but we must now go into a more abstract realm of theory. This does not mean that the problems are necessarily more difficult to solve. It simply means that we must beware of *visualizations* of the solutions; the limited experience that we have gained through our senses is not sufficient to permit this.

As before, we find the wavefunction by solving the Schrödinger equation. The Schrödinger equation, in turn, is found by writing the operator for the total energy of the system, according to the rules developed in Chapter 5. The total energy of a proton–electron system is

$$\frac{p_p^2}{2m_p} + \frac{p_e^2}{2m_e} + V(r)$$

which is written in operator form as

$$-\frac{\hbar^2}{2m_p} \nabla_p^2 - \frac{\hbar^2}{2m_e} \nabla_e^2 + V(r)$$

where r is the distance between the proton and the electron, m_p is the proton mass, m_e is the electron mass, and $V(r)$ is the Coulomb potential, equal to $-e^2/r$ in cgs units, or to $-e^2/4\pi\epsilon_0 r$ in mks (SI) units.

We have here introduced the standard symbol ∇^2 to represent $\partial^2/\partial x^2 + \partial^2/\partial y^2 + \partial^2/\partial z^2$ or the equivalent expression in spherical coordinates; the operators ∇^2_p and ∇^2_e operate on the proton and electron coordinates, respectively, in the wavefunction.

The Schrödinger equation for the hydrogen atom is therefore

$$\left(-\frac{\hbar^2}{2m_p}\nabla^2_p - \frac{\hbar^2}{2m_e}\nabla^2_e + V(r)\right)\psi = i\hbar\frac{\partial\psi}{\partial t}$$

$$= E_T\psi \qquad (1)$$

but when the equation is written in this form, the energy eigenvalue E_T includes the kinetic energy of translation of the center of mass of the whole atom, a quantity in which we are not interested at the moment. We want to know the *internal* energy states—states of the *relative* motion of proton and electron—in order to compare the theory with the results of measurements on the hydrogen spectrum. Fortunately, the potential energy is a function only of the relative coordinates of proton and electron, so it is easy to rewrite Eq. (1) in terms of the coordinates X, Y, and Z of the center of mass and the coordinates x, y, and z of the electron relative to the proton. These coordinates are related to the coordinates of the individual particles as follows:

$$X = \frac{m_e x_e + m_p x_p}{m_e + m_p}, \qquad x = x_e - x_p$$

$$Y = \frac{m_e y_e + m_p y_p}{m_e + m_p}, \qquad y = y_e - y_p \qquad (2)$$

$$Z = \frac{m_e z_e + m_p z_p}{m_e + m_p}, \qquad z = z_e - z_p$$

It is not difficult to use Eqs. (2) to write the kinetic energy in terms of the relative velocity and the velocity of the center of mass; the result is

Total kinetic energy = $\frac{1}{2}(m_e + m_p)(\dot{X}^2 + \dot{Y}^2 + \dot{Z}^2) + \frac{1}{2}m_r(\dot{x}^2 + \dot{y}^2 + \dot{z}^2)$,

where $m_r = m_e m_p/(m_e + m_p)$ is the reduced mass, which we previously encountered in discussing the Bohr model of hydrogen (Chapter 3).

In terms of momentum variables defined as $P_x = (m_e + m_p)\dot{X}$, $p_x = m_r\dot{x}$, ..., the kinetic energy may be written as

$$\frac{P_x^2 + P_y^2 + P_z^2}{2(m_e + m_p)} + \frac{p_x^2 + p_y^2 + p_z^2}{2m_r}$$

so that, according to the rules for writing the momentum operator, the Schrödinger equation becomes

$$\left[-\frac{\hbar^2}{2(m_e + m_p)}\nabla^2_{\text{c.m.}} - \frac{\hbar^2}{2m_r}\nabla^2 + V(r)\right]\psi = i\hbar\frac{\partial\psi}{\partial t} \tag{3}$$

where the operators $\nabla^2_{\text{c.m.}}$ and ∇^2 operate on center of mass and relative coordinates, respectively.

This equation could also have been obtained by direct transformation of the partial derivatives in Eq. (1), making use of Eqs. (2) to find the derivatives with respect to the new variables.

The wavefunction ψ is now a function of X, Y, Z, x, y, z, and t. If we separate the variables by writing

$$\psi(X, Y, Z, x, y, z, t) = u(x, y, z)U(X, Y, Z)e^{-i(E + E')t/\hbar}$$

where E is the energy of the relative motion and E' the energy of translation of the center of mass, Eq. (3) may be separated into the two equations

$$-\frac{\hbar^2}{2(m_e + m_p)}\nabla^2_{\text{c.m.}}U = E'U$$

$$\left[-\frac{\hbar^2}{2m_r}\nabla^2 + V(r)\right]u = Eu \tag{4}$$

But Eq. (4) is simply the Schrödinger equation for a single particle of mass m_r moving in a *fixed* potential energy field; the fact that both proton and electron are moving has been completely accounted for by the use of m_r instead of m_e (or m_p) for the mass of the particle. (You will recall that the same thing happened in the analysis of the Bohr atom.) The eigenvalues E are the energy levels of the hydrogen atom in the frame of reference in which the center of mass of the atom is at rest.

7.2   ENERGY LEVELS OF THE HYDROGEN ATOM

Because Eq. (4) is identical to the one-particle equation treated in Chapter 6, we already know the solution of the angular part of the equation. To find the energy levels, we need only to solve the radial equation, which according to Eq. (32), Chapter 6, must be

$$\frac{d^2}{dr^2}(rR) = -\frac{2m_r}{\hbar^2}\left[E - V(r) - \frac{l(l + 1)\hbar^2}{2m_r r^2}\right](rR) \tag{5}$$

where the term $-[l(l + 1)\hbar^2/2m_r r^2]$ is the centrifugal potential whose introduction is, as we have previously seen, the result of eliminating the angular dependence in the equation.

Equation (5) is identical to the equation for one-dimensional motion of a particle in the potential field $V(r) + l(l + 1)\hbar^2/2m_r r^2$. This "effective poten-

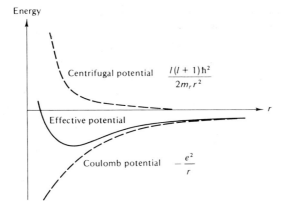

Fig. 1. *The effective potential (solid line), the sum of the centrifugal and Coulomb potentials for the hydrogen atom.*

tial," the sum of the Coulomb potential and the centrifugal potential, is sketched in Fig. 1. An electron of negative total energy is trapped in the "potential well" formed by the effective potential. Classically, the particle would describe an elliptical orbit under these conditions; it would oscillate between the two values of r at which its total energy would be equal to the potential energy. In quantum theory, we may find a wavefunction for this well just as we did for the one-dimensional wells considered in Chapter 5.

The solution of Eq. (5) by the method of infinite series is carried out in Appendix C, which treats the general case of a one–electron ion with nuclear charge Ze. The solution is a polynomial containing $n - l$ terms, where l is the angular momentum quantum number and n is a new quantum number, the radial quantum number. The statement that $R(r)$ contains $n - l$ terms may be considered to be the definition of n.

It is a curious feature of the solutions that the energy depends only on n, not on l. For example, the energy is the same for the $l = 1$ solution with a two-term radial solution and for the $l = 2$ solution with a one-term radial solution; in both cases $n = 3$. The energy levels which result are identical to the levels predicted by Bohr for the hydrogen atom or any one-electron ion:

$$E_n = \frac{-m_r c^2 \alpha^2 Z^2}{2n^2}$$

$$= -13.60 \frac{Z^2}{n^2} \text{ eV} \qquad (6)$$

although the number n now has a completely different interpretation from that given by Bohr.

Figure 2a shows the effective potentials for $l = 0$, 1, and 2, and the energy levels for $n = 1$, 2, 3, and 4. Because the energy depends only on n and not on l, the same levels which are allowed for any given l are also allowed for *all lower* values of l. The only effect of l on the levels appears through the condition that $n \geqslant l + 1$, so that lower energy levels are possible for smaller l, because smaller n values are then possible. Thus, for example, E_3 is an energy eigenvalue for all three effective wells shown in Fig. 2a, but E_2 is an eigenvalue only for $l = 1$ and $l = 0$. The fact that all of these different wells have the same set of energy levels is a remarkable property peculiar to the Coulomb potential.

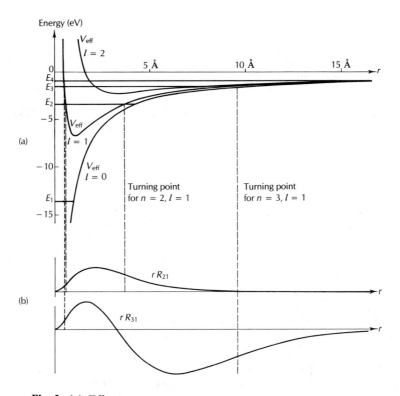

Fig. 2. (a) *Effective potential and energy levels of hydrogen for $l = 0$, 1, and 2. Only the lowest four levels are shown; the number of levels is infinite, because the effective well becomes wider, without limit, as the energy approaches zero.* (b) *Eigenfunctions of the radial equation. Because the radial equation has the same form as the one-dimensional Schrödinger equation, the main features of these eigenfunctions may be deduced directly from Fig. 2(a). Note the points of inflection at the turning points, and the curvature on either side of these points.*

Because Eq. (5) is identical to the one-dimensional Schrödinger equation, we can use a graphical analysis, as in Chapter 5, to gain more understanding of the form of the eigenfunctions. Figure 2b shows graphs of the eigenfunctions $rR_{nl}(r)$ for the states ($n = 2$, $l = 1$) and ($n = 3$, $l = 1$), whose energy eigenvalues are shown in Fig. 2a as E_2 and E_3, respectively. Notice that, as in the one-dimensional examples of Chapter 5, the eigenfunction curves away from the axis in the region where $E < V_{eff}$ (the classically forbidden region) and it curves toward the axis, tending to oscillate, where $E > V_{eff}$; the classical turning point, where $E = V_{eff}$, is a point of inflection for the eigenfunction. You may also notice that the function rR_{31} curves more rapidly than rR_{21} in the allowed region, because the curvature is proportional to the kinetic energy. As usual, each eigenfunction contains one more node than the eigenfunction immediately lower in energy.

The functions $rR(r)$ are of interest because the square $|rR(r)|^2$ is the probability density $P(r)$ for finding the electron at a radius between r and $r + dr$. We may show this simply by integrating the probability density $|u_{n, l, m}(r, \theta, \phi)|^2$ over the surface of a sphere of radius r, so that the angular dependence is eliminated. This yields

$$P(r) = \int_\theta \int_\phi |R_{nl} Y_{l, m}(\theta, \phi)|^2 r^2 \sin \theta \, d\theta \, d\phi$$

or, because the $Y_{l, m}$ are normalized,

$$P(r) = r^2 |R_{nl}|^2$$

Thus Fig. 2 shows where the electron is likely to be, as far as the r coordinate is concerned. It indicates how the average radius of the "orbit" increases as n increases; it is evident from the figure that this average radius must be close to the value given by the original Bohr theory.

The complete normalized wavefunctions $u_{n, l, m}(r, \theta, \phi)$ for the lowest energy states of the hydrogen atom are given in Table 1. These wavefunctions also apply to any one-electron ion, if one uses the appropriate values for the atomic number Z and the nuclear mass M. The probability densities associated with some of these wavefunctions are shown in Fig. 3.

Spectroscopic Symbols.   For historical reasons associated with observation of the various series of lines in atomic spectra, the l value of each state is designated by a letter, as follows:

Letter:	s	p	d	f	g	h	i
l:	0	1	2	3	4	5	6

Table 1

Wavefunctions for Hydrogen Atoms and Hydrogenlike Ions

$$u_{100} = \frac{1}{(\pi)^{1/2}} \left(\frac{Z}{a'_0}\right)^{3/2} e^{-\rho}$$

$$u_{200} = \frac{1}{(32\pi)^{1/2}} \left(\frac{Z}{a'_0}\right)^{3/2} (2-\rho) e^{-\rho/2}$$

$$u_{210} = \frac{1}{(32\pi)^{1/2}} \left(\frac{Z}{a'_0}\right)^{3/2} \rho \cos\theta \, e^{-\rho/2}$$

$$u_{21\pm1} = \frac{1}{(64\pi)^{1/2}} \left(\frac{Z}{a'_0}\right)^{3/2} \rho \sin\theta \, e^{-\rho/2 \pm i\phi}$$

$$u_{300} = \frac{1}{81(3\pi)^{1/2}} \left(\frac{Z}{a'_0}\right)^{3/2} (27 - 18\rho + 2\rho^2) e^{-\rho/3}$$

$$u_{310} = \frac{2^{1/2}}{81(\pi)^{1/2}} \left(\frac{Z}{a'_0}\right)^{3/2} (6-\rho)\rho \cos\theta \, e^{-\rho/3}$$

$$u_{31\pm1} = \frac{1}{81(\pi)^{1/2}} \left(\frac{Z}{a'_0}\right)^{3/2} (6-\rho)\rho \sin\theta \, e^{-\rho/3 \pm i\phi}$$

$$u_{320} = \frac{1}{81(6\pi)^{1/2}} \left(\frac{Z}{a'_0}\right)^{3/2} \rho^2(3\cos^2\theta - 1) e^{-\rho/3}$$

$$u_{32\pm1} = \frac{1}{81(\pi)^{1/2}} \left(\frac{Z}{a'_0}\right)^{3/2} \rho^2 \sin\theta \cos\theta \, e^{-\rho/3 \pm i\phi}$$

$$u_{32\pm2} = \frac{1}{162(\pi)^{1/2}} \left(\frac{Z}{a'_0}\right)^{3/2} \rho^2 \sin^2\theta \, e^{-\rho/3 \pm 2i\phi}$$

$$\rho = \frac{Zr}{a'_0} \qquad\qquad a'_0 = a_0\left(\frac{m_e + M}{M}\right)$$

where a_0 = first Bohr radius $\qquad M$ = mass of nucleus
$\quad$ = 5.29×10^{-9} cm

The letters go in alphabetical order for $l > 3$. Each state is then identified by the number for n followed by the letter for l: for example, 3d for $n = 3$, $l = 2$.[1]

---

[1] There is no "code" to distinguish states of different m value. The m value indicates only the orientation of the state in space, and since the z direction is arbitrary, transitions to states of different m value are not distinguishable, unless one applies a magnetic field or otherwise establishes a unique direction. Historically, therefore, there was no need to label these states.

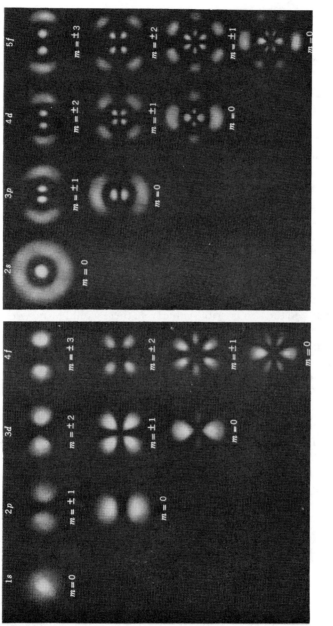

Fig. 3. *Photographic representation of the electron probability density distribution u^*u for several energy eigenstates. These may be regarded as sectional views of the distributions in a plane containing the polar axis which is vertical and in the plane of the paper. The scale varies from figure to figure. [From "Principles of Modern Physics" by R. B. Leighton. Copyright McGraw-Hill, New York, 1959. Used by permission of McGraw-Hill Book Company.]*

7.3   FINE STRUCTURE IN THE HYDROGEN SPECTRUM

It is not strictly true that the energy levels of the hydrogen atom depend only on the quantum number n, as implied by Eq. (6). We saw in discussing the Bohr theory (page 86) that certain levels must be split into two or more sublevels, because of the splitting of lines in the hydrogen spectrum. We also saw that Sommerfeld introduced a second quantum number to explain this splitting. The effect was supposed to be relativistic, arising because the electron travels at a much greater speed as it approaches the nucleus when the orbit is eccentric. This idea successfully explained the fine structure of the hydrogen atom, but inconsistencies arose when it was applied to alkali metals, whose spectra, as we shall see in Chapter 9, should be quite similar to the hydrogen spectrum.

In 1925, Goudsmit and Uhlenbeck[2] proposed an alternative explanation, which was supported by the results of the Stern–Gerlach experiment.[3] The explanation was that the electron possesses an intrinsic angular momentum, or "spin," which is independent of its state of motion. In the Stern–Gerlach experiment, a beam of neutral silver atoms split into *two* components after passing through an inhomogeneous magnetic field. If the magnetic moment of the silver atom resulted from *orbital* angular momentum of its electrons, the beam would have split into an *odd* number of components, because L_z has $2l + 1$ possible values (from $+l\hbar$ to $-l\hbar$ in steps of $\hbar$). But two components are possible if there are only two possible values for the z component of this *new kind* of angular momentum—*spin* angular momentum. As long as the *difference* between the two possible values is $\hbar$, there would be no contradiction with our previous ideas concerning angular momentum in general. The two possible values for this z component must therefore be $+\hbar/2$ and $-\hbar/2$. (Clearly, for each possible positive value, there must be a negative value of equal magnitude, because space is isotropic, so there is no other way to allow two values which differ by $\hbar$.)

The intrinsic magnetic moment of the electron, which results from its spin, leads to a splitting of the energy levels of hydrogen into sublevels, *provided that there is a magnetic field in the atom*, because in each of the states determined by the Schrödinger equation the electron would then have two possible values for its magnetic potential energy, according to the value of the z component of its magnetic moment. But why should there be a magnetic field in the hydrogen atom? The only other particle present is the proton. One might expect the proton, like the electron, to have internal spin and magnetic moment, and it does, but it turns out that the proton's intrinsic magnetic moment produces a field which is too small to account for the observed fine structure in the hydrogen spectrum.[4] However, as far as the electron is concerned, the proton's *electric charge* leads to the existence of a magnetic field of sufficient strength, because *in the electron's rest frame* the *proton* moves in an orbit around the electron. To put it another way, a body moving through an *electric* field also "sees" a magnetic field, according to the theory of relativity.

[2] S. Goudsmit and G. E. Uhlenbeck, *Nature*, **117**, 264 (1926).
[3] See Section 6.3.
[4] Instead, it produces "hyperfine" structure, as we shall see at the end of this section.

Energy of the Spin–Orbit Interaction.   Let us try to estimate the magnitude of the splitting of the hydrogen levels, on the basis of the above idea. First we must know the magnetic moment of the electron, which may be deduced from the magnitude of the splitting of the atomic beam in the Stern–Gerlach experiment. This moment is very close to

$$\boldsymbol{\mu} = \frac{-e}{m_e} \mathbf{S} \text{ (mks)} \qquad \text{or} \qquad \boldsymbol{\mu} = \frac{-e}{m_e c} \mathbf{S} \text{ (cgs)}$$

where $\mathbf{S}$ is the spin angular momentum of the electron, and e is the magnitude of the electronic charge. Since $S_z = \pm\hbar/2$, we have[5]

$$\mu_z = \pm \frac{e\hbar}{2m_e} = \pm 1 \text{ "Bohr magneton"}$$

which is often denoted by the symbol μ_B and simply called the magnetic moment of the electron. This is, strictly speaking, not *the* magnetic moment, but merely a *component* of the magnetic moment, but it is the only quantity that one ever observes when one measures the magnetic moment.

The magnetic energy of the moment $\boldsymbol{\mu}$ in a magnetic field $\mathbf{B}$ is simply $-\boldsymbol{\mu} \cdot \mathbf{B}$, and $\mathbf{B}$ may be related to the electron's *orbital* angular momentum in the following way: If the velocity of the electron is $\mathbf{v}$, the velocity of the proton *relative to the electron*[6] is $-\mathbf{v}$. The magnetic field produced by the *proton's charge e* is therefore

$$\mathbf{B} = \frac{\mu_0 e(-\mathbf{v}) \times \mathbf{r}}{4\pi r^3} \equiv e \frac{(-\mathbf{v}) \times \mathbf{r}}{4\pi r^3 \varepsilon_0 c^2} \text{ where } \mu_0 \equiv 1/\varepsilon_0 c^2 = 4\pi \times 10^{-7} \frac{\text{henry}}{\text{m}}$$

This may be written in terms of the *electron's* angular momentum $\mathbf{L} = \mathbf{r} \times m\mathbf{v}$ as

$$\mathbf{B} = \frac{e\mathbf{L}}{4\pi r^3 m \varepsilon_0 c^2}$$

The interaction energy is then[7]

$$W = -\boldsymbol{\mu} \cdot \mathbf{B}$$

$$= \frac{e^2}{4\pi \varepsilon_0 m^2 c^2 r^3} (\mathbf{S} \cdot \mathbf{L})$$

in the electron's rest frame. If we remember that the component of $\mathbf{S}$ in *any* direction, such as the direction of $\mathbf{L}$, must be $\hbar/2$ in magnitude, and we let $|\mathbf{L}|$ be approximately $\hbar$, we can compute the value of W; it is of the order of 10^{-4} eV, which is the correct order of magnitude to explain the observed fine structure.

[5] For the remainder of this section we shall use mks units.
[6] We neglect the small motion of the proton in the center-of-mass frame.
[7] For cgs units, simply remove the $4\pi\varepsilon_0$ from the denominator.

The above expression is not the correct one to use for W, however, because the energy splitting seen in the lab is not the same as that seen in the electron's rest frame. The reason for this is rather complicated, involving a phenomenon known as the Thomas precession,[8] which is related to the fact that the time scale is not the same in the laboratory as in the rest frame of the electron. The proper relativistic treatment shows that the energy shift seen in the laboratory is

$$W = \frac{e^2}{8\pi\varepsilon_0\, m^2 c^2 r^3} (\mathbf{S} \cdot \mathbf{L}) \tag{7}$$

just one-half of the value found in the electron's rest frame.

Combining Spin and Orbital Angular Momentum. If we wish to make an accurate calculation of the value of W, we cannot simply guess at a value for $\mathbf{S} \cdot \mathbf{L}$, as we did above; we must find out what its value is in the states which are allowed by quantum theory.

The postulates of quantum mechanics tell us that the spin–orbit energy, like any other dynamical variable, must be represented by an operator, which can be included in the operator for the total energy as it appears in the Schrödinger equation. The energy is then found by finding the eigenvalues of the Schrödinger equation in this form, with the variables $\mathbf{S}$ and $\mathbf{L}$ represented as operators:

$$-\frac{\hbar^2}{2m_r}\nabla^2 u - \frac{e^2}{4\pi\varepsilon_0\, r} u + \frac{e^2}{8\pi\varepsilon_0\, m^2 c^2 r^3}(\mathbf{S} \cdot \mathbf{L})_{op} u = Eu$$

Two difficulties now arise. The first is that we do not know what operator to use for $\mathbf{S} \cdot \mathbf{L}$; the second is that even knowing this, we cannot solve the equation exactly. The latter difficulty happens to be a minor one, because the spin–orbit energy, as we saw above, is only about 10^{-4} eV, or about one part in 10^4 of the total energy. We shall see (in Chapter 8) that effects of this size may be calculated by the methods of perturbation theory. We shall find that as a first approximation one can use the eigenfunctions of the unperturbed Schrödinger equation (the equation without the $\mathbf{S} \cdot \mathbf{L}$ term) to calculate the energy shift caused by perturbation. The perturbing energy is then simply the expectation value of the $\mathbf{S} \cdot \mathbf{L}$ term in each of the states of the unperturbed system; that is,

$$W = \iiint u^* \frac{e^2}{8\pi\varepsilon_0\, m^2 c^2 r^3}(\mathbf{S} \cdot \mathbf{L})_{op}\, u \; dx\, dy\, dz \tag{8}$$

where u is an eigenfunction of the unperturbed equation (4). We should therefor be able to express this energy in terms of the quantum numbers n, l, and m for these eigenfunctions, if we can find out what the $\mathbf{S} \cdot \mathbf{L}$ operator does.

In order to determine what the $\mathbf{S} \cdot \mathbf{L}$ operator should be, or at least to

---

[8] See R. Eisberg, "Fundamentals of Modern Physics," pp. 341–346. Wiley, New York, 1961.

determine the result of the operation $(S \cdot L)_{op} u$, we begin with the observation that the vector **L** is no longer a constant of the motion, when there is an interaction which depends on $S \cdot L$. Because there are no *external* torques on the system, the *total* angular momentum is constant during the motion, but this total is the sum of **L** and **S**. The vector **L** itself is constant *only* when the force between electron and nucleus is a *central* force; but in this case the **B** field caused by the electron's orbital motion *exerts a torque* on the *spin* magnetic moment; when this magnetic moment changes direction, **L** must also change direction, in order that **L** + **S** remain constant. The vector **S**, of course, remains constant in *magnitude*, because this is a fundamental property of the electron, but the *direction* of **S** changes.

In classical theory, we simply define the total angular momentum vector **J** = **L** + **S**, and there is no difficulty in performing the vector addition of **L** and **S**. But when the components of **L** and **S** are quantized, we must consider what it means to say **J** = **L** + **S**, particularly when **S** is a nonclassical quantity which cannot be defined in terms of coordinates and momenta (by an equation such as **L** = **r** × **p**). We cannot proceed by using the classical definition of intrinsic angular momentum for a body as the integral of **r** × **p** over the volume of the body; the wavefunction contains only information about the position and momentum of the *whole* electron, not about the coordinates of various parts of it, because electrons, as far as we know, are indivisible; one has no possible way of observing a *part* of an electron.

We seem to have reached an impasse; how can the **S** operator operate on the wavefunction at all? It cannot, unless we *assume* that the wavefunction is a function of *one* additional coordinate, the spin coordinate, which has only two possible values, because the spin has only two possible orientations. We may describe this situation by saying that there are *two wavefunctions* with a given *space–time* dependence, which differ only in the value of the *spin* coordinate. One of these wavefunctions may be written as $e^{-i\omega t}u(x, y, z)(+)_s$, and the other as $e^{-i\omega t}u(x, y, z)(-)_s$, where $(+)_s$ and $(-)_s$ indicate the two possible "eigenfunctions" of the spin operator. The spin operator S_z operates on the "functions" $(+)_s$ and $(-)_s$, respectively, to yield the two possible eigenvalues $\pm\hbar/2$ as follows:

$$S_z(+)_s = +\frac{\hbar}{2}(+)_s, \qquad S_z(-)_s = -\frac{\hbar}{2}(-)_s \qquad (9)$$

Let us now further assume that the vector **J** can be defined as in classical theory, even though our conception of it is rather hazy. In that case, we can rewrite the product $S \cdot L$ in terms of the quantities J^2, S^2, and L^2, by means of the equation

$$J^2 \equiv J \cdot J = (L + S) \cdot (L + S)$$
$$= L^2 + S^2 + S \cdot L + L \cdot S$$

But the **L** operator must commute with the **S** operator, because L_{op} and S_{op}

operate on different coordinates. Thus $\mathbf{S} \cdot \mathbf{L} = \mathbf{L} \cdot \mathbf{S}$, and we can solve for $\mathbf{S} \cdot \mathbf{L}$, obtaining

$$\mathbf{S} \cdot \mathbf{L} = \tfrac{1}{2}(J^2 - L^2 - S^2) \tag{10}$$

and the problem of evaluating $(\mathbf{S} \cdot \mathbf{L})_{op} u$ becomes the simpler one of evaluating $\tfrac{1}{2}(J^2 - L^2 - S^2)_{op} u$.

We already know how to evaluate $L_{op}^2 u$; it is simply equal to $l(l + 1)\hbar^2 u$, if u is one of the eigenfunctions of Eq. (4). By analogy with the result for L^2, we could simply assume that $S^2 u = s(s + 1)\hbar^2 u$, with $s = \tfrac{1}{2}$, because the maximum value of S_z should equal $s\hbar$, just as the maximum value of L_z is $l\hbar$. We could likewise assume that $J^2 u = j(j + 1)\hbar^2 u$. But what value do we assume for j? Perhaps we can start with a more fundamental assumption, which will enable us to *derive* rather than assume the above eigenvalue equations, and which may help us to determine the connection between the values of j, l, and s.

The most fundamental assumption that we can make is that any type of angular momentum—$\mathbf{L}$, $\mathbf{S}$, or $\mathbf{J}$—must satisfy the uncertainty principle; that is, one cannot simultaneously determine two components of angular momentum to arbitrary precision. That this assumption is fundamentally necessary is apparent from the analysis of the double Stern–Gerlach experiment (Section 6.3). When this assumption is translated into mathematics, it is equivalent to a requirement that the operators for the different components of $\mathbf{L}$ (or of $\mathbf{S}$ or $\mathbf{J}$) do not commute; we saw in Section 6.3 that if two variables are to be simultaneously observable, it is necessary and sufficient that the corresponding operators commute.

It is convenient, at this point, to introduce the symbol $[L_x, L_y]$ to represent $L_x L_y - L_y L_x$. It is easy to show (Problem 4, Chapter 6) that the L operators obey the relations

$$[L_x, L_y] = i\hbar L_z$$
$$[L_y, L_z] = i\hbar L_x$$
$$[L_z, L_x] = i\hbar L_y$$

Since the uncertainty is directly related to the commutation relation, and the uncertainty should be the same in measuring all kinds of angular momentum, we may safely assume that similar commutation relations apply to $\mathbf{J}$ and to $\mathbf{S}$; that is, $[J_x, J_y] = i\hbar J_z$, $[S_x, S_y] = i\hbar S_z$, and so on. It turns out that this assumption permits us to derive everything that we need to know about all angular momentum operators.

In particular, we may derive the eigenvalues of J^2 and J_z, as follows. Let the function u_{ab} be a simultaneous eigenfunction of J^2 and J_z, with eigenvalues a^2 and b, respectively; that is,

$$J^2 u_{ab} = a^2 u_{ab} \tag{11}$$

and

$$J_z u_{ab} = b u_{ab} \tag{12}$$

To evaluate these eigenvalues, we use a trick similar to the one used in finding the harmonic oscillator wavefunctions. We construct the operator $J_+ = J_x + iJ_y$, and we demonstrate that the function $J_+ u_{ab}$ is also an eigenfunction of J_z, as follows:

$$J_z J_+ u_{ab} = (J_z J_x + iJ_z J_y)u_{ab}$$

From the commutation relations,

$$J_z J_x = J_x J_z + i\hbar J_y \quad \text{and} \quad J_z J_y = J_y J_z - i\hbar J_x,$$

so

$$J_z J_+ u_{ab} = (J_x J_z + i\hbar J_y + iJ_y J_z + \hbar J_x)u_{ab}$$

On the right-hand side of this equation, J_z operates directly on u_{ab}, so we may rewrite the equation, as

$$J_z J_+ u_{ab} = (J_x b + i\hbar J_y + iJ_y b + \hbar J_x)u_{ab}$$

which reduces to

$$J_z J_+ u_{ab} = (b + \hbar)(J_x + iJ_y)u_{ab}$$

so from the definition of J_+ we have

$$J_z J_+ u_{ab} = (b + \hbar)J_+ u_{ab}$$

Thus we have proven that the function $J_+ u_{ab}$ is an eigenfunction of J_z, with eigenvalue $b + \hbar$.

In a similar fashion, we can show that $J_- u_{ab}$, where $J_- = J_x - iJ_y$, is an eigenfunction of J_z, with eigenvalue $b - \hbar$. Therefore we can generate from u_{ab} a whole set of eigenfunctions of J_z, whose eigenvalues are b, $(b \pm \hbar)$, $(b \pm 2\hbar)$, ... Each of these functions is also an eigenfunction of J^2; all have the *same* eigenvalue, namely a^2, for the operator J^2, as you may easily verify. [Simply apply the operator $J^2 = J_x^2 + J_y^2 + J_z^2$ to the function $J_+ u_{ab}$, and work it out, using the commutation relations and Eqs. (11) and (12).]

But the set of eigenvalues of J_z cannot be infinite, because we know that the eigenvalues of J_z^2 should always be less[9] than the eigenvalue of J^2, and the latter eigenvalue is a^2 for all eigenfunctions in the set. Thus there must be a largest eigenvalue of J_z, which we denote by B, and $B^2 < a^2$. We write the eigenfunction corresponding to this eigenvalue as u_{aB}; that is,

$$J_z u_{aB} = B u_{aB}$$

It is now clear that

$$J_+ u_{aB} = 0 \tag{13}$$

for otherwise $J_+ u_{aB}$ would be an eigenfunction of J_z corresponding to eigenvalue $B + \hbar$, in contradiction to the assumption that B is the largest eigenvalue. We may use Eq. (13) to find the relation between a and B as follows: We apply the operator $J_x - iJ_y$ to Eq. (13), obtaining

$$(J_x - iJ_y)(J_x + iJ_y)u_{aB} = 0$$

---

[9] Obviously, J_z^2 cannot be greater than J^2. And if $J_z^2 = J^2$, then $J_x^2 = J_y^2 = 0$, and all components of J would be known, in contradiction to the uncertainty principle.

or
$$[J_x^2 + J_y^2 + i(J_x J_y - J_y J_x)]u_{aB} = 0$$

which becomes, from the $J_x J_y$ commutation rule,
$$(J_x^2 + J_y^2 - \hbar J_z)u_{aB} = 0$$

or, since $J_z u_{aB} = Bu_{aB}$,
$$(J_x^2 + J_y^2)u_{aB} = \hbar B u_{aB} \tag{14}$$

But $J_x^2 + J_y^2 = J^2 - J_z^2$, and $(J^2 - J_z^2)u_{aB} = (a^2 - B^2)u_{aB}$. Therefore, from Eq. (14),
$$a^2 - B^2 = \hbar B$$

or
$$a^2 = B^2 + B\hbar \tag{15}$$

We can go through the same sort of argument to show that if the *smallest* eigenvalue of J_z is B', then
$$a^2 = B'^2 - B'\hbar \tag{16}$$

Therefore, from Eqs. (15) and (16),
$$B'^2 - B'\hbar = B^2 + B\hbar \tag{17}$$

so that[10]
$$B' = -B$$

We might have realized that this was so anyway, simply by symmetry, or the isotropy of space; for each positive eigenvalue of J_z there must be a negative eigenvalue of equal magnitude. But any two eigenvalues of J_z differ by an integer times $\hbar$; this is clear[11] from the sequence of eigenvalues which we generated. Therefore
$$B - B' = B + B = \text{integer} \cdot \hbar$$

and we may write $B = j\hbar$, where j is either an integer or an integer $+\frac{1}{2}$. Equation (15) then tells us that
$$a^2 = (j^2 + j)\hbar^2$$
$$= j(j + 1)\hbar^2$$

Thus, recalling the definition of a^2 (Eq. 11), we have shown *directly from the commutation relations* that the eigenvalues of J^2 are $j(j + 1)\hbar^2$, and that the states with a given eigenvalue of J^2 form a set for which the eigenvalue of J_z is one of a sequence
$$+j\hbar, +(j - 1)\hbar, \ldots, -j\hbar$$

These eigenvalues may be written as $m_j \hbar$, in analogy to our notation for orbital angular momentum; j is then the maximum value of m_j:
$$m_j = -j, -j + 1, -j + 2, \ldots, +j$$

---

[10] Another solution of Eq. (17) would be $B' = B + \hbar$; this is out because we know $B' < B$, by definition.

[11] This is clearly true for any eigenvalues in the sequence, and because the maximum eigenvalue is unique, no other sequence of eigenvalues exists.

Notice that this derivation did not depend at all on the form of the eigenfunctions; we never had to know what u_{ab} is. We could have used this procedure, rather than the differential equation, for obtaining the eigenvalues of L^2 in Chapter 6.[12] However, in that case, we wanted to know more than just the possible results of measuring L^2 or L_z; for example, it was of interest to know the probability distribution as a function of the angle, for particles of a given angular momentum, so that we could visualize the structure of the hydrogen atom.

We still have to determine the connection between the value of j and the values of the quantum numbers l and s. We can do this by considering the z *components* of the vectors **J**, **L**, and **S**, because we can always determine this component precisely, and if $\mathbf{J} = \mathbf{L} + \mathbf{S}$, then $J_z = L_z + S_z$. Therefore $m_j = m_l + m_s = m_l \pm \frac{1}{2}$. The maximum value of m_l is l, so

$$m_j \leqslant l + \tfrac{1}{2}$$

The quantum number j is the maximum value of m_j, so there must be a *set* of states for which $j = l + \frac{1}{2}$: one state with $m_j = l + \frac{1}{2}$, one with $m_j = l - \frac{1}{2}$, one with $m_j = l - \frac{3}{2}$, and so on until we reach the state with $m_j = -l - \frac{1}{2}$.

For this set of states (with $j = l + \frac{1}{2}$), we may say that the spin vector **S** is parallel to **L**, at least to the extent permitted by the uncertainty principle, because the magnitude of **J** has its greatest possible value. Is any other value possible for j? We certainly should have a value of j which corresponds to **L** and **S** being *anti*parallel; in this case, the component of **S** *in the direction of* **L** would be $-\hbar/2$ instead of $+\hbar/2$, and the j value would therefore be $l - \frac{1}{2}$ instead of $l + \frac{1}{2}$. Corresponding to this j value there should, as usual, be a *set* of states with values of m_j ranging from $+j$ to $-j$ in unit steps. Therefore there are *two sets* of states for a given l and a spin of $\frac{1}{2}$, which may be described by the quantum numbers

Set 1: $j = l + \frac{1}{2}$; $m_j = l + \frac{1}{2}, l - \frac{1}{2}, l - \frac{3}{2}, \ldots, -(l + \frac{1}{2})$

Set 2: $j = l - \frac{1}{2}$; $m_j = l - \frac{1}{2}, l - \frac{3}{2}, \ldots, -(l - \frac{1}{2})$

Thus there are *two independent states* for *each* value of m_j from $m_j = l - \frac{1}{2}$ down to $m_j = -(l - \frac{1}{2})$. We could have arrived at this conclusion in another way, by considering the possible ways of combining m_l and m_s to form a resultant m_j. For example, there are two different ways to combine m_l and m_s to form $m_j = l - \frac{1}{2}$; one can either let m_l equal l and let m_s equal $-\frac{1}{2}$, or let m_l equal $l - 1$ and let m_s equal $+\frac{1}{2}$. The wavefunctions of these two states could be written in terms of spherical harmonics and spin functions as $Y_{l,l}(-)_s$ and $Y_{l,l-1}(+)_s$, respectively. You may easily verify that there are also two (and only two) ways of forming states with each of the other m_j values except the largest and the smallest values; the latter two can each be formed in only one way. Therefore the sets listed above contain as many

---

[12] The statistical method used in Section 6.3, to find the eigenvalues of $\mathbf{L}^2$, can also be used to show that the eigenvalues of J^2 are $j(j + 1)\hbar^2$, if J_z has a maximum value of $j\hbar$.

independent states as are possible; there are no other possible values of j, because any other state must be a linear combination of the states already listed.

Neither of the functions $Y_{l,l}(-)_s$ and $Y_{l,l-1}(+)_s$ is an eigenfunction of the J^2 operator, although they both are eigenfunctions of J_z with eigenvalue $m_j = l - \frac{1}{2}$. Each of these functions is a *superposition* of *two* eigenfunctions of J^2, with the two eigenvalues $j = l \pm \frac{1}{2}$. One can, however, construct two independent linear combinations of $Y_{l,l}(-)_s$ and $Y_{l,l-1}(+)_s$, one of which is an eigenfunction of J^2 with eigenvalue $j = l + \frac{1}{2}$, and the other of which is an eigenfunction of J^2 with eigenvalue $j = l - \frac{1}{2}$.[13] (See problem 11.) In general, we cannot assign values of all three quantum numbers m_l, m_s, and m_j to a specific state; the quantities L_z, S_z, and J^2 are not simultaneously observable, because the operator of J^2 does not commute with the L_z and S_z operators.

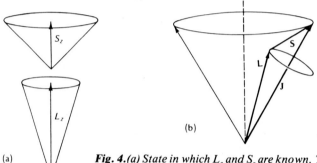

Fig. 4. *(a) State in which L_z and S_z are known. There is no definite value for J^2, because the direction of L relative to S is not known. (b) State in which J^2 is known. Although J_z is also known, L_z and S_z are now not known, because the triangle formed by L, S, and J is free to rotate about J. (In the case shown here, L and S are "parallel".)*

Figure 4 illustrates this point. Fig. 4a shows the situation in which L_z and S_z are both known. Each vector lies on a cone whose axis is the z axis, but the position of each on its cone is completely undetermined. Therefore the direction of S relative to L is unknown, $S \cdot L$ is unknown, and so J^2 is unknown. (Remember that $J^2 = L^2 + S^2 + 2S \cdot L$.) Figure 4b shows the converse situation, in which J^2 is known. In this case $S \cdot L$ must be known, so that S lies somewhere on a cone whose axis is *the direction of* L rather than the z axis. The triangle formed by L, S, and J is completely determined in size and shape, but this triangle is free to rotate about the J vector, so that L_z and S_z are undetermined, although J^2 and J_z are known.

It is easy to generalize the discussion to the case where one must add any

---

[13] If this seems strange, the analogy with vectors may be helpful. If i and j are unit vectors along the x and y axes, respectively, and i' and j' are unit vectors along the x' and y' (rotated) axes, then each of the vectors i and j is a linear combination of i' and j', and conversely, each of the vectors i' and j' is a linear combination of i and j. The two eigenfunctions of J^2 are analogous to one pair of unit vectors, and the two eigenfunctions of L_z and S_z are analogous to the other pair of unit vectors. The use of the term "orthogonal" to describe a set of eigenfunctions of a given variable has its origin in this analogy to vectors.

two angular momentum vectors J_1 and J_2. (For example, one might be interested in the total angular momentum of a two-electron atom.) If

$$J = J_1 + J_2$$

where $J_1^2 = j_1(j_1 + 1)\hbar^2$ and $J_2^2 = j_2(j_2 + 1)\hbar^2$, then

$$J^2 = j(j + 1)\hbar^2$$

where j is one of the sequence $j_1 + j_2, j_1 + j_2 - 1, \ldots, |j_1 - j_2|$.

This may be easily demonstrated by reference to m_j, as we did before in considering $L + S$. The largest possible m_j value must be $j_1 + j_2$, and this must therefore be the largest possible j value. For $m_j = j_1 + j_2 - 1$, there are two independent states, which may be constructed by combining the state $(m_{j1} = j_1; m_{j2} = j_2 - 1)$ and the state $(m_{j1} = j_1 - 1; m_{j2} = j_2)$ in two different ways. One of these combinations must be the $j = j_1 + j_2$ state, and the other combination must be the $j = j_1 + j_2 - 1$ state. For $m_j = j_1 + j_2 - 2$, it is easy to see that there may be *three* independent possibilities. We can continue increasing the number of possibilities until we reach $m_j = |j_1 - j_2|$, at which point we have the maximum number of independent states. Therefore this is the smallest value of j needed to account for all the states. These limits on j are the quantum mechanical analog of the triangle inequality $|J_1| + |J_2| \geqslant |J_1 + J_2| \geqslant ||J_1| - |J_2||$, derived from the fact that J_1, J_2, and $J_1 + J_2$ form the sides of a triangle.

To illustrate these points, suppose that $j_1 = 2$ and $j_2 = 1$. Then we have the possibilities shown in Table 2.

Table 2

Possible Ways to Combine Quantum Numbers m_{j1} and m_{j2}

m_{j1}	m_{j2}	Total m_j	Total j
+2	+1	+3	3
+2	0	+2	3 or 2
+1	+1	+2	3 or 2
+2	−1	+1	3, 2, or 1
+1	0	+1	3, 2, or 1
0	+1	+1	3, 2, or 1
+1	−1	0	3, 2, or 1
0	0	0	3, 2, or 1
−1	+1	0	3, 2, or 1
0	−1	−1	3, 2, or 1
−1	0	−1	3, 2, or 1
−2	+1	−1	3, 2, or 1
−1	−1	−2	3 or 2
−2	0	−2	3 or 2
−2	−1	−3	3

There are three ways to combine m_{j1} and m_{j2} to yield $m_j = +1$, and there are also three ways to obtain $m_j = 0$ or $m_j = -1$. Thus there are three j values corresponding to these three m_j values. A state with a single j value can be constructed in each case, by making the appropriate linear combination of the three states with the same m_j value.

Evaluation of the Spin–Orbit Energy.   We can now evaluate the spin–orbit energy as given by Eq. (8):

$$W = \iiint u^* \frac{e^2}{8\pi\varepsilon_0 \, m^2 c^2 r^3} (\mathbf{S} \cdot \mathbf{L})_{\text{op}} \, u \, dx \, dy \, dz$$

We have seen that $\mathbf{S} \cdot \mathbf{L} = [J^2 - L^2 - S^2]/2$. Therefore

$$W = \iiint u^* \frac{e^2}{8\pi\varepsilon_0 \, m^2 c^2 r^3} \left(\frac{J^2 - L^2 - S^2}{2} \right)_{\text{op}} u \, dx \, dy \, dz$$

This expression can be calculated directly if u is an eigenfunction of J^2, L^2, and S^2. Furthermore, if the state of the system is an eigenstate of these operators (rather than the operators L_z and S_z, for example), the system will remain in that state, because J^2 is constant. Because of the spin–orbit interaction, $\mathbf{L}$ and $\mathbf{S}$ change direction, precessing around $\mathbf{J}$, while J^2 remains constant. Therefore the functions u which are used in Eq. (8) are eigenfunctions of J^2 which are linear combinations of the functions $Y_{l,\,m}(-)_s$ and $Y_{l,\,m-1}(+)_s$. These linear combinations, when multiplied by the appropriate radial functions, are also eigenfunctions of the "unperturbed" Schrödinger equation, so that they may legitimately be used in Eq. (8). Thus we may write

$$(J^2 - L^2 - S^2)_{\text{op}} u = \hbar^2 [j(j + 1) - l(l + 1) - s(s + 1)]u$$

and W becomes (after the substitution $s = \tfrac{1}{2}$)

$$W = \frac{Ze^2\hbar^2}{16\pi\varepsilon_0 \, m^2 c^2} [j(j + 1) - l(l + 1) - \tfrac{3}{4}]$$

$$\times \iiint \left(\frac{u^* u}{r^3} \right) r^2 \sin\theta \, dr \, d\theta \, d\phi \tag{18}$$

It is now a simple enough matter to evaluate the integral and find W for any *particular* state, simply by plugging into the integral the appropriate radial function, which may be found by reference to Table 1 or Appendix C. (The *angular* part of u, of course, integrates to 1, because of the normalization of the Y_{lm}.) Finding a *general formula* for the result is a bit more difficult; it would not be worthwhile to develop the equations needed for this, so we simply quote the result[14]:

$$\iiint \left(u^* \frac{u}{r^3}\right) r^2 \sin\theta \, dr \, d\theta \, d\phi = \frac{Z^3}{[a_0'^3 n^3 l(l + \frac{1}{2})(l + 1)]} \quad (l \neq 0) \quad (19)$$

where a_0' is defined as in Table 1. When $l = 0$, the integral is infinite, but the factor $[j(j + 1) - l(l + 1) - \frac{3}{4}]$ is zero in this case, so the product is indeterminate. Thus the results below are valid only when $l \neq 0$.

It is convenient to write W, the fine structure energy, in terms of the fine structure constant $\alpha = e^2/4\pi\epsilon_0\hbar c \approx 1/137$. In terms of α and the energy E_n of the unperturbed state $(E_n = -m_r c^2 Z^2 \alpha^2/2n^2)$, we obtain from Eqs. (18) and (19)

$$W = \frac{Z^2 |E_n| \alpha^2 [j(j + 1) - l(l + 1) - \frac{3}{4}]}{2nl(l + \frac{1}{2})(l + 1)}$$

We may be more explicit by using the fact that j must equal either $l + \frac{1}{2}$ or $l - \frac{1}{2}$. If $j = l + \frac{1}{2}$,

$$W = \frac{Z^2 |E_n| \alpha^2}{n} \frac{[(l + \frac{1}{2})(l + \frac{3}{2}) - l(l + 1) - \frac{3}{4}]}{2l(l + \frac{1}{2})(l + 1)}$$

or

$$\boxed{W = \frac{Z^2 |E_n| \alpha^2}{n(2l + 1)(l + 1)} \quad (j = l + \frac{1}{2}, \quad l \neq 0)} \quad (20)$$

If $j = l - \frac{1}{2}$,

$$W = \frac{Z^2 |E_n| \alpha^2}{n} \frac{[(l - \frac{1}{2})(l + \frac{1}{2}) - l(l + 1) - \frac{3}{4}]}{2l(l + \frac{1}{2})(l + 1)}$$

or

$$\boxed{W = -\frac{Z^2 |E_n| \alpha^2}{nl(2l + 1)} \quad (j = l - \frac{1}{2}, \quad l \neq 0)} \quad (21)$$

When W is written in this form we see how small the effect is; the magnitude of W is smaller than that of E_n by a factor of order α^2, or about 1/19,000. (Remember that E_n in turn is smaller than the electron's rest energy by a factor of α^2, so we now have the first three terms in an expansion of the energy in a power series in α^2—except that we have not quite finished the calculation of the fine structure energy.)

[14] The general case is solved by using a "generating function" for $R_{nl}(r)$. These functions are discussed in L. Pauling and E. B. Wilson, "Introduction to Quantum Mechanics," Appendix VII. McGraw-Hill, New York, 1935.

EXAMPLE PROBLEM 1. Verify that Eq. (19) correctly gives the value of the integral when u is the function $u_{210}(r, \theta, \phi)$ given in Table 1.

Solution. With $n = 2$ and $l = 1$, Eq. (19) gives a value of $Z^3/24a_0'^3$ for the integral. By direct substitution from Table 1, the integral is (with $\rho = Zr/a_0'$)

$$\frac{1}{32\pi} \left(\frac{Z}{a_0'}\right)^3 \int_0^{2\pi} \int_0^{\pi} \int_0^{\infty} \frac{\rho^2 \cos^2 \theta e^{-\rho} r^2 \sin \theta \, dr \, d\theta \, d\phi}{r^3}$$

$$= -\frac{2\pi}{32\pi} \left(\frac{Z}{a_0'}\right)^3 \int_0^{\pi} \int_0^{\infty} \rho e^{-\rho} \, d\rho \cos^2 \theta \, d(\cos \theta)$$

$$= \frac{1}{16} \left(\frac{Z}{a_0'}\right)^3 \frac{\cos^3 \theta}{3}\Big]_{\pi}^{0} \int_0^{\infty} \rho e^{-\rho} \, d\rho$$

$$= \frac{1}{24} \left(\frac{Z}{a_0'}\right)^3$$

Q.E.D.

Relativistic Correction. Although Sommerfeld's original explanation of fine-structure splitting as a relativistic effect was not correct, you must not conclude that there is no relativistic *contribution* to this splitting. The kinetic energy is not exactly $p^2/2m$ as one assumes in constructing the Schrödinger equation, so that equation does not give exactly correct energy levels. But for hydrogen atoms and hydrogenlike ions, we know that the Schrödinger result is *almost* correct, so it is logical simply to seek a first-order correction to the result.

We may "correct" the expression $p^2/2m$ by expanding the relativistic expression for the kinetic energy in a power series in p^2. The kinetic energy T is found by subtracting the rest energy from the total energy E [Chapter 2 Eq. (39)], so

$$T = E - mc^2 = (p^2c^2 + m^2c^4)^{1/2} - mc^2$$

$$= mc^2 \left(1 + \frac{p^2}{m^2c^2}\right)^{1/2} - mc^2$$

$$= mc^2 \left(1 + \frac{p^2}{2m^2c^2} - \frac{p^4}{8m^4c^4} + \cdots\right) - mc^2$$

$$= \frac{p^2}{2m} - \frac{p^4}{8m^3c^2} + \cdots \tag{22}$$

The ratio of successive terms in the series is of order p^2/m^2c^2, which may be written

$$\frac{\dfrac{p^2}{2m}}{\dfrac{mc^2}{2}} \approx \frac{\text{kinetic energy}}{\frac{1}{2}(\text{rest energy})}$$

which for a typical state is of order $10 \, \text{eV}/10^5 \, \text{eV} = 10^{-4}$, about the same order of magnitude as α^2. Thus the series converges very rapidy, and we may cut it off at two terms. The "corrected" Schrödinger equation is therefore[15]

$$\left[\frac{p^2}{2m} - \frac{p^4}{8m^3c^2} + V(r) \right] u = Eu \tag{23}$$

where p^2 and p^4 are, of course, operators determined by the usual rule

$$\mathbf{p} \to -i\hbar \mathbf{\nabla} \quad \text{or} \quad p^2 \to -\hbar^2 \nabla^2.$$

We can easily solve Eq. (23) to the same degree of accuracy attained in the truncated series (22). We do this by evaluating the eigenvalues of the p^4 operator in the following way. We rewrite Eq. (23) in the form

$$\frac{p^2}{2m} u = \left[E - V(r) + \frac{p^4}{8m^3c^2} \right] u \tag{24}$$

and we apply the operator $p^2/2m$ to each side, obtaining

$$\frac{p^4}{4m^2} u = \left\{ \frac{p^2}{2m} [E - V(r)] + \frac{p^6}{16m^4c^2} \right\} u \tag{25}$$

Using Eq. (24), we may substitute for the $p^2/2m$ operator in Eq. (25) to obtain

$$\frac{p^4}{4m^2} u = \left\{ \left[E - V(r) + \frac{p^4}{8m^3c^2} \right] [E - V(r)] + \frac{p^6}{16m^4c^2} \right\} u$$

$$= \left\{ [E - V(r)]^2 + \frac{p^4 \, [E - V(r)]}{8m^3c^2} + \frac{p^6}{16m^4c^2} \right\} u$$

The second and third terms on the right-hand side are of order p^6; since we cut off the series (22) at p^4, we are justified in neglecting terms of order p^6 here, so that

$$\frac{p^4}{4m^2} u \approx [E - V(r)]^2 u \tag{26}$$

With the use of Eq. (26), Eq. (23) becomes, for a reduced mass m_r,

$$\frac{\hbar^2}{2m_r} \nabla^2 u + [E - V(r)]u + \frac{1}{2m_r c^2} [E - V(r)]^2 u = 0 \tag{27}$$

This equation leads to the radial equation

$$-\frac{\hbar^2}{2m_r} \frac{d^2R}{dr^2} = \left\{ [E - V(r)] + \frac{1}{2m_r c^2} [E - V(r)]^2 - \frac{\hbar^2 l(l+1)}{2m_r r^2} \right\} R \tag{28}$$

---

[15] Here we neglect the spin–orbit interaction term; we shall add that energy as another correction at the end of the calculation.

which, with the substitution of the Coulomb potential for $V(r)$, becomes an equation of the form

$$\frac{d^2R}{dr^2} + \left[A + \frac{B}{r} + \frac{C}{r^2}\right]R(r) = 0$$

where the constants A, B, and C may be deduced from Eq. (28). As it happens, this equation may be solved by the power series method, just as the uncorrected radial equation was solved. The result[16] may be written in powers of the fine structure constant α as

$$E = -\frac{Z^2\alpha^2 m_r c^2}{2n^2}\left[1 + \frac{\alpha^2 Z^2}{n^2}\left(\frac{n}{l+\frac{1}{2}} - \frac{3}{4}\right) + \cdots\right] \tag{29}$$

The first term in this expansion is simply the energy E_n found in solving the uncorrected equation. The second term is the relativistic correction, which in terms of E_n is

$$\Delta E_r = \frac{|E_n|\alpha^2 Z^2}{n^2}\left[\frac{3}{4} - \frac{n}{l+\frac{1}{2}}\right] \tag{30}$$

Notice that this correction is of the same order of magnitude as the spin–orbit correction [Eq. (20) or (21)]. If we add the two corrections, we obtain the total shift of each level of hydrogen relative to the uncorrected level. For $j = l + \frac{1}{2}$,

$$\Delta E = W + \Delta E_r = \frac{Z^2\alpha^2|E_n|}{n^2}\left[\frac{n}{(2l+1)(l+1)} + \frac{3}{4} - \frac{n}{l+\frac{1}{2}}\right]$$

$$= \frac{Z^2\alpha^2|E_n|}{n^2}\left[\frac{3}{4} - \frac{n}{l+1}\right]$$

while for $j = l - \frac{1}{2}$,

$$\Delta E = \frac{Z^2\alpha^2|E_n|}{n^2}\left[-\frac{n}{l(2l+1)} + \frac{3}{4} - \frac{n}{l+\frac{1}{2}}\right]$$

$$= \frac{Z^2\alpha^2|E_n|}{n^2}\left[\frac{3}{4} - \frac{n}{l}\right]$$

If written in terms of j, both cases may be expressed by the same formula:

$$\Delta E = \frac{Z^2\alpha^2|E_n|}{n^2}\left[\frac{3}{4} - \frac{n}{j+\frac{1}{2}}\right] \tag{31}$$

[16] For further details of the solution, see E. U. Condon and G. H. Shortley, "The Theory of Atomic Spectra," pp. 118–120. Cambridge Univ. Press, London and New York, 1935.

As derived here, from the corrected Schrödinger equation, Eq. (31) does not apply when $l = 0$, because Eqs. (20) and (21) do not apply. However, a completely relativistic treatment using the Dirac equation (Section 7.4) shows that Eq. (31) is nevertheless correct[17] for all values of l. Thus a given j value is associated with the same energy, independently of the value of l. The energy levels which result are shown in Fig. 5. Contrast Fig. 5 with Fig. 6, Chapter 3, which shows only the dependence of the energy on n. Now *for a given n there are two energy levels for each non-zero value of l.* However, because the energy depends on j rather than l, levels of different l value coincide, so that the total number of levels is equal to the number of possible values[18] of l, which is simply n.

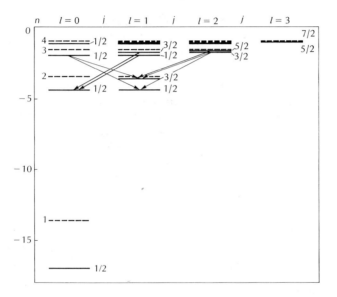

Fig. 5. *Energy levels of the hydrogen atom, showing fine structure. Uncorrected levels are shown by dashed lines, corrected levels by solid lines. The energy scale is for the uncorrected levels; the displacement of the corrected levels from the uncorrected levels is exaggerated by a factor of $\alpha^{-2} \approx 19,000$.*

This also happens to be the number of possible values of k in the old Sommerfeld theory, so that the number of levels is the same as that originally given by Sommerfeld. The coincidence in energy between states of different l and the same j does not occur in atoms other than hydrogen, so for those atoms the number of levels is greater than that predicted by Sommerfeld. Sommerfeld's theory in any case could not predict the correct number of *states* involved, because it did not allow for the existence of electron spin.

[17] Except for the "hyperfine" energy splitting and the "Lamb shift" described below. See ref. 16.
[18] Each additional l value introduces *one* additional j value.

Fig. 5 also shows the seven distinct transitions that produce the first "line" of the Balmer series. Because the $2s_{1/2}$ and $2p_{1/2}$ levels have the same j value (indicated by the subscript), they have the same energy, and the same is true for $3s_{1/2}$ and $3p_{1/2}$; thus only five distinct energies are emitted in these transitions.

Other transitions involving the $n = 3$ and $n = 2$ levels do not occur. We shall find that there are *selection rules* which forbid such transitions. For a one-electron atom, the selection rules permit only transitions such that

$$\Delta l = \pm 1$$
$$\Delta j = \pm 1 \quad \text{or} \quad 0$$
$$\Delta m_j = \pm 1 \quad \text{or} \quad 0$$

Thus, for example, $3p_{3/2} \to 2p_{3/2}$ is forbidden because $\Delta l = 0$, and $3d_{5/2} \to 2p_{1/2}$ is forbidden because $\Delta j = -2$. These selection rules, which are related to the fact that a photon has angular momentum of $\hbar$, will be discussed more fully in Chapter 10.

Experimentally, the early measurements showed this line (the so-called H_α line) to consist of two lines, primarily produced by the transitions $3d_{5/2} \to 2p_{3/2}$ and $3d_{3/2} \to 2p_{1/2}$. (The other three lines are less intense, and each is relatively close to one or the other of the two brighter lines, so they were not resolved.) We can calculate the splitting of these lines from Eq. (31), by computing the shift of each level involved in the transitions. The result is that the *wavelengths* differ by

$$\Delta\lambda \approx 0.142 \text{ A} \tag{32}$$

The wavelength of each line is approximately

$$\lambda \approx \frac{hc}{E_3 - E_2} \approx \frac{12,400}{\left(\dfrac{13.6}{4}\right) - \left(\dfrac{13.6}{9}\right)} = 6560 \quad \text{Å}$$

so the ratio $\Delta\lambda/\lambda$ is approximately 1/46,000.

Although this value seems extremely small, it can be determined experimentally to an accuracy of better than 1 percent.[19] The elementary theory of the diffraction grating shows[20] that the resolving power $\lambda/\Delta\lambda$ is equal to N, the total number of lines in the grating. As early as 1882, H. A. Rowland developed a ruling engine which gave 14,438 lines per inch and could produce gratings nearly 6 in. wide,[19] so that N was about 85,000. Thus he could easily resolve the first Balmer lines. Having resolved the lines, one can measure the relative positions of the components to an accuracy much better than the width of each line.

[19] C. Candler, "Atomic Spectra and the Vector Model," 2nd ed., pp. 30–31. Van Nostrand, Princeton, New Jersey, 1964.
[20] See, for example, D. Halliday & R. Resnick, "Physics," Third Edition, p. 1054, John Wiley, N. Y., 1978.

Hyperfine Structure. The *proton* also has a spin and a magnetic moment, as mentioned at the beginning of this section, so there is an additional extremely tiny energy shift caused by the interaction between this magnetic moment and the magnetic fields resulting from the orbital and spin moments of the electron. We cannot neglect this shift, any more than we could neglect the motion of the nucleus in determining the energy levels. The energy levels which determine the spectral lines are levels of the entire atom; the *entire atom* radiates, and the motions of *both* the electron and the nucleus change in an atomic transition.

The magnitude of this "hyperfine" energy splitting may be determined by an extension of the methods developed earlier in this section. In general, the *total* angular momentum of the *atom* is defined as $\mathbf{F} = \mathbf{J} + \mathbf{I}$, where $\mathbf{I}$ is the internal angular momentum of the *nucleus*, resulting from the intrinsic spin of each particle in the nucleus (neutrons as well as protons) *and* from the orbital motion of protons *within* the nucleus. Of course, in the nucleus of normal hydrogen there is just the intrinsic spin of the proton, which is the same as that of an electron, but in heavier nuclei the motion of the protons and neutrons within the nucleus contributes to the angular momentum of the nucleus, and each neutron has an intrinsic magnetic moment which contributes to the total magnetic moment of the nucleus. (See Chapter 14 for details.)

The total angular momentum $\mathbf{F}$ must obey the same rules as any other angular momentum vector. The possible values of $|\mathbf{F}|^2$ are $F(F + 1)\hbar^2$, where the quantum number F takes on integrally spaced values between $J + I$ and $|J - I|$. Here the letter I refers to the *quantum number* for the nuclear angular momentum $\mathbf{I}$; I may be an integer or a half-integer, and as usual $|\mathbf{I}|^2 = I(I + 1)\hbar^2$. The magnetic interaction energy resulting from the nuclear magnetic moment is proportional to $\mathbf{I} \cdot \mathbf{J}$, so it has as many possible values as there are values of F—namely, either $2I + 1$, or $2J + 1$, whichever is smaller. (You may verify this for yourself.)

Each spectral line is therefore split by this interaction and exhibits hyperfine structure (unless either I or J is equal to 0). By simply counting the number of lines in the hyperfine structure of the spectrum of an atom, one can usually determine the value of I for the nucleus of that atom. We shall discuss the implications of this determination in Chapter 14.

The magnetic moment of the proton is $2.79 \, e\hbar/2M_p$, where M_p is the proton mass. In analogy to the Bohr magneton ($\mu_B = e\hbar/2m_e$), the quantity $e\hbar/2M_p$ is called one nuclear magneton (μ_N). The fact that μ_N is equal to $\mu_B/1830$ explains why the nuclear moment produces such a small effect. The effect is too small to observe optically in the hydrogen spectrum, but the energy splitting has been detected at radio frequencies. The $1s_{1/2}$ level of hydrogen is split into two levels by the hyperfine splitting, and a transition between one of these levels and the other leads to emission of a photon of wavelength 21 cm. This "21-cm radiation" is quite prominent in radiation from atomic hydrogen throughout the galaxy,[21] and it has provided a means for determining the structure of the galaxy. Notice that the energy of such a photon is less than 10^{-5} eV, while the fine structure splitting is of order 10^{-4} eV; the energy is in perfect agreement with the theory.

You may notice that the transition between the two substates of the $1s_{1/2}$ level is in violation of the selection rule $\Delta l = \pm 1$. However, the *total* angular momentum of the atom *does* change by $\hbar$; the transition is "forbidden" by the selection rule only because it is a "magnetic dipole" transition, which involves a change in *magnetic* dipole moment of the atom, rather than a change in its *electric* dipole moment, as in the allowed transitions. But a change in magnetic dipole moment is *possible*; it is simply far less *probable* than a change in electric dipole moment. In other words, it is much easier for an electron to change orbits than it is for it to "flip" its spin. Thus the selection rule does not say that such a transition *never* occurs, but it indicates that you are not likely to observe one. It is only the enormous number of isolated hydrogen atoms in the galaxy (of the order of one per cubic centimeter) which makes the observation possible. Again, you are referred to Chapter 10 for further discussion of selection rules.

We have now carried our analysis of the hydrogen energy levels as far as is practical here, from the 10-eV energy range down to the 10^{-5}-eV range. The agreement between theory and experiment over this enormous range indicates that our basic quantum theory is valid and that we understand the electromagnetic interaction very well on the atomic level. But we should not feel that the last word has been said. A complete relativistic quantum treatment, including the quantization of the energy levels of the electromagnetic field (quantum electrodynamics) predicts further shifts in the levels. For example, the $2s_{1/2}$ and $2p_{1/2}$ levels, which have the same energy according to the treatment given here, are shown to differ in energy by 4.3743×10^{-6} eV, corresponding to a photon frequency of 1.0578×10^{9} $\sec^{-1}$. This splitting is called the Lamb shift. The photons are in the radio frequency (microwave) range, and the theory has been verified by the resonance experiments of Lamb and Retherford.[22]

To conclude this section, let us note that the success of quantum theory and electromagnetic theory on the atomic level by no means guarantees their success on the subatomic level. Nevertheless, there is no doubt that on the scale of the hydrogen atom, our theories are essentially correct.

7.4   SPIN AND RELATIVITY

So far we have worked exclusively with the Schrödinger equation, which is based on the nonrelativistic expression $p^2/2m$ for the kinetic energy. This is

---

[21] This radiation was first observed by H. I. Ewen and E. M. Purcell, *Phys. Rev.* **83**, 881 (1951).
[22] W. E. Lamb and R. C. Retherford, *Phys. Rev.* **72**, 241 (1947).

reasonable, because the kinetic energy of an electron in an atom is always much smaller than the electron's rest energy. However, if we are looking for small effects, we can no longer ignore the possibility of relativistic corrections to the results of the Schrödinger equation. We have just developed one such correction, in the calculation of the fine-structure energy.

Rather than attempt to correct a theory which is an approximation to begin with, it would obviously be more satisfactory to have a completely relativistic theory which is based on the energy-momentum relation $E^2 = p^2c^2 + m^2c^4$. In 1926, Schrödinger had proposed a relativistic form of his equation, by writing the equation for a free particle as $E^2\psi = (p^2c^2 + m^2c^4)\psi$, and making the usual operator substitutions $E \rightarrow i\hbar(\partial/\partial t)$ and $\mathbf{p} \rightarrow -i\hbar\nabla$, so that he obtained

$$-\hbar^2\frac{\partial^2\psi}{\partial t^2} = -\hbar^2c^2\nabla^2\psi + m^2c^4\psi$$

One can incorporate a potential energy term into this equation in order to obtain correct results with bound systems. However, one must be careful whenever the potential energy involves a *vector*, because relativistic equations must be based on *four*-vectors rather than *three*-vectors.[23] For example, the spin–orbit energy is proportional to $\mathbf{S} \cdot \mathbf{L}$, and the spin operators form a *three-dimensional* vector. Any attempt to incorporate such a term into the above equation would yield an equation with the wrong transformation properties; it would not be transformed correctly by a Lorentz transformation, so that it would not be valid in all inertial frames of reference. (One could deduce a logical fourth component for the $\mathbf{L}$ operator, but it is difficult to see how this could be done for the spin operators.) Thus the Schrödinger relativistic equation can only be applied to particles without spin.

In 1930, P. A. M. Dirac published a different relativistic wave equation which solves the problem of spin. In the Dirac equation there is no special term which provides for a spin-dependent energy; rather, the equation *automatically* includes the existence of spin, as an inevitable consequence of the fundamental equation itself. The full implications of the Dirac equation are beyond the scope of this text, but there are a number of interesting facets, such as the automatic inclusion of spin, and the existence of antimatter, which should be discussed here.

Dirac's equation differs from the Schrödinger relativistic equation in having an operator for $E = (p^2c^2 + m^2c^4)^{1/2}$, rather than an operator for E^2. There is no postulate of quantum theory which tells us how to make an operator out of this square root, so Dirac was free to devise his own rule and calculate the consequences. He assumed that the operator must be *linear* in the momentum components, so that

[23] See Section 2.4.

$$H_{op} = (p^2c^2 + m^2c^4)^{1/2} = \boldsymbol{\alpha} \cdot \mathbf{p}c + \beta mc^2 \tag{33}$$

where $\boldsymbol{\alpha}$ and β may be operators, and the factors c and mc^2 are inserted to make both $\boldsymbol{\alpha}$ and β dimensionless.

It is important to realize that $\boldsymbol{\alpha}$ and β *must not depend on the time or on the position of the particle*, because Eq. (33) represents a *free* particle, whose behavior must be invariant with respect to translation or rotation of the coordinate system. Therefore $\boldsymbol{\alpha}$ and β *must commute with* $\mathbf{p}$. (As usual, $\mathbf{p}$ is the differential operator $-i\hbar\nabla$.) Further information about $\boldsymbol{\alpha}$ and β is obtained by squaring both sides of Eq. (33). Equating the coefficients of p_x^2, p_y^2, p_z^2, and m^2c^4, respectively, we find that

$$\alpha_x^2 = \alpha_y^2 = \alpha_z^2 = \beta^2 = 1 \tag{34a}$$

We also find, from the coefficients of terms such as $p_x p_y$, that

$$\begin{aligned}
\alpha_x \alpha_y + \alpha_y \alpha_x &= 0, & \alpha_x \alpha_z + \alpha_z \alpha_x &= 0 \\
\alpha_y \alpha_z + \alpha_z \alpha_y &= 0, & \alpha_x \beta + \beta \alpha_x &= 0 \\
\alpha_y \beta + \beta \alpha_y &= 0, & \alpha_z \beta + \beta \alpha_z &= 0
\end{aligned} \tag{34b}$$

We can summarize Eqs. (34) by saying that the four entities α_x, α_y, α_z, and β each have unit magnitude and they *anticommute* in pairs. Thus they must be operators[24] of some sort, which operate on the wavefunction. Since these operators *cannot* depend on the *space–time coordinates* of the particle, they must operate on *some other variable* in the wavefunction, which describes some *intrinsic* property of the particle, such as spin.

It is not difficult, and we leave it as an exercise for the reader, to show from information in Section 7.3 that the angular momentum operators, and in particular the *spin operators, also anticommute*; that is,

$$S_x S_y + S_y S_x = 0$$
$$S_y S_z + S_z S_y = 0$$
$$S_z S_x + S_x S_z = 0$$

Furthermore, we saw in Section 7.3 that the commutation relations can be used to derive all the important properties of angular momentum. Thus we begin to suspect that the operators α_x, α_y, α_z, and β, which arise naturally in the Dirac equation, *automatically* account for the spin of the electron.

The fact that there are four of these operators, rather than three, also resolves the dilemma about incorporating the spin into a four-vector. But we must not jump to the conclusion that the operators α_x, α_y, and α_z can be

---

[24] They obviously cannot be *numbers*, for then $\alpha_x \alpha_y + \alpha_y \alpha_x \neq 0$, unless $\alpha_x = \alpha_y = 0$.

identified with the spin operators S_x, S_y, and S_z. There is a very good reason why they *cannot* be so identified: the operator β *anticommutes* with *all* of the α operators, but there is no operator which anticommutes with all of the $\mathbf{S}$ operators, as long as the $\mathbf{S}$ operators are defined as operating only on the spin function. We show below that there can be *no more than three anticommuting operators* which operate only on the spin part of the wavefunction.[25] Therefore the existence of the fourth operator requires that the wavefunction involve an additional internal variable, and that the α operators operate on a function of this variable.

The additional variable happens to be the *algebraic sign* of the *energy* of the particle. A given function $u(x, y, z)$ of the space coordinates corresponds either to a positive energy eigenvalue or to a negative energy eigenvalue of the same magnitude. This comes about because the energy operator $(p^2c^2 + m^2c^4)^{1/2}$ is a square root which may be either positive or negative. Since the two eigenvalues of the same magnitude but opposite sign correspond to *absolutely identical* space functions $u(x, y, z)$ it is clear that it is left to the operators α and β, *operating on another factor in the wavefunction*, to determine whether a given wavefunction has a positive or a negative energy eigenvalue.

We shall discuss the physical meaning of a negative energy eigenvalue in a moment; for the present, let us conclude our analysis of the operators in the Dirac equation. The fact that there are two energy eigenvalues for each function of the space and spin coordinates means that there are *four* complete wavefunctions for *each space function*—two for each orientation of the spin. We may express this fact by saying that each wavefunction is a product of a space part $u(x,y, z)$, a spin part which is a combination of the two spin functions $(+)_s$ and $(-)_s$, and a part which is a combination of two other functions which we may denote as $(+)_E$ and $(-)_E$, corresponding to the two possible signs for the energy of the particle. The operators α_x, α_y, α_z, and β operate on the functions $(+)_s$, $(-)_s$, $(+)_E$, and $(-)_E$ to produce linear

---

[25] Each operator, when applied to a spin function $(+)_x$ or $(-)_x$, can yield only a linear combination of these two functions, because these functions form a complete set of functions for the spin variable. Thus, for example,

$$S_x(+)_s = a(+)_s + b(-)_s$$

and

$$S_x(-)_s = c(+)_s + d(-)_s$$

where a, b, c, and d are constants. Specification of these four constants suffices to define S_x completely (see problem 7). Thus, so far as operations on the spin functions are concerned, *any operator O may be written as a linear combination of the three independent operators* S_x, S_y, and S_z, plus the unit operator 1, because these four operators, being linearly independent, may be combined to yield any combination of the four constants which completely specify the operator O. An operator which is expressible as a linear combination of the operators S_x, S_y, and S_z clearly cannot anticommute with all of these operators. Thus if one has a *fourth* anticommuting operator, one or more of the four operators *must involve some additional variable* other than spin; that is, one must require more than the four constants a, b, c, and d in order to define one or more of these operators completely.

combinations of the same four functions. Further development of the algebra of these operators is left for the problems.

Let us now demonstrate more precisely the connection between the $\boldsymbol{\alpha}$ operators and the spin of the electron. First we show that the Dirac energy operator H, unlike the Schrödinger energy operator, does *not* commute with L_z, so that there are no functions which are simultaneously eigenfunctions of L_z and of the energy (see Section 6.3a). Thus one cannot simultaneously measure L_z and energy, as one can do in the nonrelativistic case. The algebra is as follows: We start with the definitions

$$L_z = -i\hbar\left(x\frac{\partial}{\partial y} - y\frac{\partial}{\partial x}\right)$$

$$H = c(\alpha_x p_x + \alpha_y p_y + \alpha_z p_z + \beta mc)$$

$$= -ci\hbar\left(\alpha_x\frac{\partial}{\partial x} + \alpha_y\frac{\partial}{\partial y} + \alpha_z\frac{\partial}{\partial z} + \frac{i\beta mc}{\hbar}\right)$$

Then, remembering that $\boldsymbol{\alpha}$ and β are independent of x, y, and z, and noting that L_z is independent of z, we see that L_z commutes with the last two terms of H, and the commutator may be written

$$L_z H - HL_z = -c\hbar^2\left[\left(x\frac{\partial}{\partial y} - y\frac{\partial}{\partial x}\right)\left(\alpha_x\frac{\partial}{\partial x} + \alpha_y\frac{\partial}{\partial y}\right)\right.$$
$$\left. - \left(\alpha_x\frac{\partial}{\partial x} + \alpha_y\frac{\partial}{\partial y}\right)\left(x\frac{\partial}{\partial y} - y\frac{\partial}{\partial x}\right)\right]$$

which becomes, after the operations are carried out,

$$L_z H - HL_z = -c\hbar^2\left[\left(\alpha_x x\frac{\partial^2}{\partial x\,\partial y} + \alpha_y x\frac{\partial^2}{\partial y^2} - \alpha_x y\frac{\partial^2}{\partial x^2} - \alpha_y y\frac{\partial^2}{\partial x\,\partial y}\right)\right.$$
$$- \left(\alpha_x\frac{\partial}{\partial y} + \alpha_x x\frac{\partial^2}{\partial x\,\partial y} + \alpha_y x\frac{\partial^2}{\partial y^2}\right.$$
$$\left.\left. - \alpha_x y\frac{\partial^2}{\partial x^2} - \alpha_y\frac{\partial}{\partial x} - \alpha_y y\frac{\partial^2}{\partial x\,\partial y}\right)\right]$$

or

$$L_z H - HL_z = -c\hbar^2\left(\alpha_y\frac{\partial}{\partial x} - \alpha_x\frac{\partial}{\partial y}\right) \neq 0$$

What does this mean? As mentioned above, it means that one cannot measure L_z and energy simultaneously. If a particle is in a state of definite energy, its value of L_z must be uncertain. This implies that L_z *must change with time*; L_z is not a constant of the motion.[26] But the only way for a *free* particle—a particle subject to *no* external forces—to change its value of L_z is by *transferring* some of its *orbital* angular momentum to *internal* angular momentum, that is, to *spin*. In such a transfer, L_z and S_z change, but $J_z = L_z + S_z$ must remain constant. With a slight further development of the algebra of the α and β operators, it is not difficult to show that J_z *does* commute with the Dirac energy operator (see Problem 15).

Thus the important property of spin, which was introduced as an ad hoc hypothesis to explain the splitting of spectral lines, emerges as a natural consequence of the Dirac equation.

Antimatter.   The existence of negative energy states is another equally important feature of relativistic quantum theory. *Classical* relativity theory also permits the energy to be negative, because one can always take the negative square root in the equation $E = (p^2 c^2 + m^2 c^4)^{1/2}$. But one ignores this possibility in *classical* problems, because the positive energies range from $+mc^2$ to $+\infty$, and the negative energies range from $-mc^2$ to $-\infty$, leaving a gap of impossible energy values between $-mc^2$ and $+mc^2$. Classically, the energy must change in a continuous manner, so that there is no way for a particle to cross this gap and go from a positive energy to a negative energy.

But quantum theory permits "quantum jumps," or discrete changes in energy of any magnitude, so if a negative energy state exists, there is always the possibility of a transition from a positive energy state to a negative energy state, with the emission of the energy difference of $2mc^2$ or more as radiation. Yet we know that ordinary matter never makes such transitions (well, hardly ever); if it did, then it would have long ago disappeared into the infinite "sea" of negative energy states.

What is it that prevents a particle from falling into this sea? One could simply *postulate* that there is a *selection rule* which prevents it, but that would be an inelegant solution which would have no connection with the rest of physics. A far more fruitful alternative was proposed by Dirac. He said that the *exclusion principle* keeps particles from falling into this sea; he assumed that the negative energy states are ordinarily *occupied*, so that positive-energy particles have no empty states to fall into. At first sight this seems like a preposterous assumption which has no advantage over a simple postulate of a selection rule; it means that all space, even what appears to be a perfect vacuum, is filled with an infinitely dense collection of particles which occupy the infinite variety of negative energy states. But the assumption begins to make sense when you consider how one could *observe its consequences*.

---

[26] It can be shown that *any* dynamical variable F is a constant of the motion only if the F *operator* commutes with H. See, for example, L. I. Schiff, "Quantum Mechanics," Ch. 6. McGraw-Hill, New York, Third edition, 1968.

The fact that a vacuum *appears* to be *empty* is not a contradiction to the assumption of Dirac. How does one ordinarily detect the presence of a particle? One can only do it if the particle is free to interact with other particles or with radiation—that is, *if the particle is free to change its state of motion*. A particle which is "locked" in the infinite sea of negative energy states is *not free* to change its state of motion, because there are no empty states to which it can change. Thus you can walk right through this infinite sea of particles with no hindrance, because the particles cannot be affected by your presence. (Remember that a particle is not a solid object like a macroscopic ball; the exclusion principle prevents two particles from occupying the same *state*, but they are *not* prevented from occupying the same *place*. An *infinite number* of particles can occupy the same *place*, as long as they all have different energies or momenta.)

But there is one way in which the presence of this infinite set of particles may be detected. If one gives one of them enough energy so that it can *transfer to an unoccupied state*, then an effect can be seen. For a particle of mass m, the lowest unoccupied (positive energy) state has energy $+mc^2$, and the highest state in the negative energy sea has energy $-mc^2$, so one must supply an energy of $2mc^2$ to permit a particle in the negative energy sea to transfer to an unoccupied state. But if two positive-energy particles *collide* with a *kinetic energy of $2mc^2$* or more, and if Dirac's assumption has any meaning, it will then be possible for a particle in the negative energy sea to *acquire this energy* and to emerge from the sea. It is also possible for a particle to be excited from a negative energy state via the photoelectric effect, if it interacts with a gamma ray of energy $2mc^2$ or more. (As in other cases of the photoelectric effect, a third body must be present in order that momentum and energy be conserved; a gamma ray can collide with a nucleus, so that the recoiling nucleus absorbs the momentum of the gamma ray, while a negative-energy electron absorbs its energy.)

When a particle—an electron, for example—emerges from this negative energy sea, what should we expect to observe? Obviously we observe the electron, where no electron was observable before; the electron appears to be created. But we also should observe the *unoccupied negative energy state* which the electron vacates when it acquires its positive energy; the presence of this unoccupied state allows other negative energy electrons to *change state*, so that the negative energy sea *suddenly becomes capable of interacting with the rest of the world*.

One can describe this interaction between the sea and the world very simply by considering what happens to the unoccupied state, or "hole" in the sea, Filling the hole would produce an inert sea again, which behaves like a perfect vacuum; by definition, such a system has zero mass and zero charge. Since the electron which fills the hole has *negative* charge and *negative* energy (which is equivalent to negative mass), the hole must behave *like a particle of positive energy and positive mass*—a particle whose properties, except for sign, must be exactly the same as those of the electron. The existence of such a particle, called an antielectron by Dirac,[27] was first suggested in 1931, and it was seen in cosmic rays a year later[28]; the particle is now commonly observed in the decay of artificially produced radioisotopes, and it is called the *positron.* (See Section 13.3) Dirac also pointed out that gamma rays of sufficiently high energy could create antiprotons, but these were not observed until 1955. It is now clear that every particle of nonzero spin has an antiparticle. When a particle encounters its antiparticle, it becomes possible for the particle to drop into the unoccupied state, so that both particle and antiparticle are *annihilated,* and the energy of $2mc^2$ or more is emitted as radiation.

Notice that there is perfect symmetry between the particles and the antiparticles in the Dirac theory. It is conceivable that there are regions of the universe where antimatter—mostly positrons, antiprotons, and antineutrons—exists in abundance, and what we call matter is rare, because it is quickly annihilated by antimatter. Inhabitants of such a region might well have developed the "anti-Dirac" equation, and they would doubtless then describe the electron as a "hole" and the positron as a particle. (See Chapters 15 and 16 for further discussion of the symmetry between matter and antimatter.)

Although we do not have the opportunity at this level to go into the further development of the consequences of the Dirac equation, we hope that this brief introduction has shown you the connection between this equation and the important phenomena of spin and antimatter.

PROBLEMS

1. By direct substitution into the Schrödinger equation for the hydrogen atom, verify that several of the wavefunctions given in Table 1 are solutions of that equation, with the energy eigenvalue which is appropriate for the *n* quantum number of each function.

2. By reference to Appendix C, construct the (unnormalized) functions u_{400}, u_{410}, and u_{411}.

3. Compute the average value (expectation value—Section 5.6) of *r* for

[27] It seems obvious now that the existence of positrons is implied by the Dirac equation, but it was not *immediately* obvious. Dirac at first said that the negative energy states could be *almost* all filled, and that the empty states, having positive charge and positive mass might be protons. Oppenheimer then showed that electrons and protons would annihilate one another too rapidly if that were the case, and he suggested that the negative energy states are *all* filled. Weyl then proved that an empty state, if it existed, would behave like a particle with the *same mass* as an electron, and Dirac then suggested that such particles could be produced by high energy gamma rays [*Proc. Roy. Soc.* **A133**, 60, (1931)].

[28] C. D. Anderson, *Science* **76**, 238 (1932).

hydrogen atoms whose wavefunctions are u_{100}, u_{210}, and u_{320}, respectively. (See Table 1.) Compare your result with the Bohr formula for the radius of the orbit; consider both the magnitude of the result and its Z dependence. (Notice that these three wavefunctions each have the greatest possible l value for the given n, so that they correspond to circular orbits in the old picture.)

4. On the same scale as Fig. 2, sketch graphs of the functions rR_{41} and rR_{42}, *without calculating* these functions. (The drawing need not be too accurate; simply make sure that the behavior is correct at the turning points, that there are the proper number of nodes, and that the curvature varies in the qualitatively correct way as $E - V$ changes.) What can you say in general about the change in the r dependence of the wavefunction as l varies, with n fixed? Can you relate this to the Bohr–Sommerfeld picture of elliptic orbits?

5. A proton has a radius of about 10^{-13} cm. Find the probability than an electron whose wavefunction is u_{200} (Table 1) will be found inside the proton. Compute the same probability for the wavefunction u_{210}, and comment on the difference in terms of classical orbits.

6. Prove that, as stated on page 253, the functions $J_+ u_{ab}$ and $J_- u_{ab}$ are eigenfunctions of J^2, with eigenvalue a^2, provided that u_{ab} is such an eigenfunction.

7. Matrix algebra provides a convenient way of representing the spin operators S_x, S_y, and S_z. One considers the spin functions $(+)_s$ and $(-)_s$ to be "unit vectors" in a hypothetical two-dimensional "space." A wavefunction for a particle whose S_z eigenvalue is $+\hbar/2$ is represented then as a vector whose direction is the $(+)_s$ "axis," and a wavefunction for a particle whose S_z eigenvalue is $-\hbar/2$ is represented as a vector along the $(-)_s$ axis. Any other wavefunction may be represented as a vector with two components in this "spin space." The components of such a vector are often written columnwise; that is, $u(x, y, z) \cdot \begin{pmatrix} \alpha \\ \beta \end{pmatrix}$ is a way of writing the wavefunction $u(x, y, z)[\alpha(+)_s + \beta(-)_s]$, whose spin part is $\alpha(+)_s + \beta(-)_s$, where α and β are constants. Such a function is, of course, not an eigenfunction of S_z; a measurement of S_z will yield the result $S_z = +\hbar/2$ with a probability of $|\alpha|^2$, and the result $-\hbar/2$ with probability $|\beta|^2$, in accordance with Postulate 3 in Chapter 6.

The operators S_x, S_y, and S_z operate on the spin functions $(+)_s$ and $(-)_s$ to produce linear combinations of the same spin functions; therefore we can say that these operators operate on the *vector* $\begin{pmatrix} \alpha \\ \beta \end{pmatrix}$ to produce a new vector $\begin{pmatrix} \alpha' \\ \beta' \end{pmatrix}$, where the components α' and β' are each a linear combination of α and β:

$$\begin{pmatrix} \alpha' \\ \beta' \end{pmatrix} = \begin{pmatrix} a\alpha + b\beta \\ c\alpha + d\beta \end{pmatrix}$$

Specification of the constants a, b, c, and d suffices to determine the operator completely; thus an operator in a two-dimensional space is often represented as the *matrix* of these coefficients: $\begin{pmatrix} a & b \\ c & d \end{pmatrix}$, to imply that the expression

$$\text{operator} \cdot \begin{pmatrix} \alpha \\ \beta \end{pmatrix} = \begin{pmatrix} \alpha' \\ \beta' \end{pmatrix}$$

is equivalent to the expression

$$\begin{pmatrix} a & b \\ c & d \end{pmatrix}\begin{pmatrix} \alpha \\ \beta \end{pmatrix} = \begin{pmatrix} \alpha' \\ \beta' \end{pmatrix} = \begin{pmatrix} a\alpha + b\beta \\ c\alpha + d\beta \end{pmatrix}$$

Now to the problem. We wish to use the information given in the text to determine the matrix form of the operators S_x, S_y, and S_z. We shall do the easiest one, S_z, to show how it is done. We know that $S_z(+)_s = (+\hbar/2)(+)_s$, and $S_z(-)_s = (-\hbar/2)(-)_s$, or in vector form

$$S_z\begin{pmatrix} 1 \\ 0 \end{pmatrix} = +\frac{\hbar}{2}\begin{pmatrix} 1 \\ 0 \end{pmatrix} \quad \text{and} \quad S_z\begin{pmatrix} 0 \\ 1 \end{pmatrix} = -\frac{\hbar}{2}\begin{pmatrix} 0 \\ 1 \end{pmatrix}$$

But by definition

$$\begin{pmatrix} \alpha \\ \beta \end{pmatrix} = \alpha\begin{pmatrix} 1 \\ 0 \end{pmatrix} + \beta\begin{pmatrix} 0 \\ 1 \end{pmatrix}$$

so we may write

$$S_z\begin{pmatrix} \alpha \\ \beta \end{pmatrix} = \frac{\hbar}{2}\left[\alpha\begin{pmatrix} 1 \\ 0 \end{pmatrix} - \beta\begin{pmatrix} 0 \\ 1 \end{pmatrix}\right]$$

$$= \frac{\hbar}{2}\begin{pmatrix} \alpha \\ -\beta \end{pmatrix}$$

and it is clear that $a = +\hbar/2$, $b = 0$, $c = 0$, and $d = -\hbar/2$, or

$$S_z = \frac{\hbar}{2}\begin{pmatrix} 1 & 0 \\ 0 & -1 \end{pmatrix}$$

Your problem is to use the commutation relations $S_x S_y - S_y S_x = i\hbar S_z$, ..., and the fact that

$$S_x^2 + S_y^2 + S_z^2 = S^2 = s(s + 1)\hbar^2 = \tfrac{3}{4}\hbar^2$$

to show that

$$S_x = \frac{\hbar}{2}\begin{pmatrix} 0 & b \\ c & 0 \end{pmatrix} \quad \text{and} \quad S_y = \frac{\hbar}{2}\begin{pmatrix} 0 & -ib \\ ic & 0 \end{pmatrix}$$

with $bc = 1$. For a hint, consider the results of the operations $(S_x + iS_y)(+)_s$, $(S_x - iS_y)(+)_s$, $(S_x + iS_y)(-)_s$, and $(S_x - iS_y)(-)_s$, according to the arguments on pages 252–254.

The precise representation of S_x and S_y is arbitrary beyond this point, but the customary choice is to let $b = c = 1$, so that

$$S_x = \frac{\hbar}{2}\begin{pmatrix} 0 & 1 \\ 1 & 0 \end{pmatrix} \quad \text{and} \quad S_y = \frac{\hbar}{2}\begin{pmatrix} 0 & -i \\ i & 0 \end{pmatrix}$$

The operators S_x, S_y, and S_z are often written in the form

$$S_x = \frac{\hbar}{2}\sigma_x, \qquad S_y = \frac{\hbar}{2}\sigma_y, \qquad \text{and} \qquad S_z = \frac{\hbar}{2}\sigma_z$$

where the symbols σ_x, σ_y, and σ_z represent dimensionless operators which may be written in matrix form as

$$\sigma_x = \begin{pmatrix} 0 & 1 \\ 1 & 0 \end{pmatrix}, \qquad \sigma_y = \begin{pmatrix} 0 & -i \\ i & 0 \end{pmatrix}, \qquad \sigma_z = \begin{pmatrix} 1 & 0 \\ 0 & -1 \end{pmatrix}$$

These matrices are called the *Pauli spin matrices*; the operators σ are the Pauli spin operators.

8. Find the linear combinations of $(+)_s$ and $(-)_s$ that are eigenfunctions of the operators S_x and S_y, and show that the eigenvalues are the same as the eigenvalues of S_z.

9. The operator $S_{z'}$ is the component of **S** along the z' axis. The z' axis lies in the xz plane, at a $45°$ angle to the z axis.

   (a) Write $S_{z'}$ as a linear combination of the operators S_x and S_z (components of **S**) given in problem 7.

   (b) Write the eigenfunctions of $S_{z'}$ in terms of the eigenfunctions of S_z.

   (c) For an electron whose spin component along z' is $+\hbar/2$, find the probability that a measurement of the spin component along z will yield the result $+\hbar/2$.

10. The operators L_x, L_y, and L_z may be represented by matrices similar to those for S_x, S_y, and S_z (problem 7). For $l = 1$, there are three eigenfunctions of L_z, so the matrix for each component of **L** must be a 3×3 matrix, and the three eigenfunctions are "unit vectors." Then, in analogy to problem 7, one can show that

$$L_x = \frac{\hbar}{\sqrt{2}} \begin{pmatrix} 0 & 1 & 0 \\ 1 & 0 & 1 \\ 0 & 1 & 0 \end{pmatrix} \quad ;$$

$$L_y = \frac{i\hbar}{\sqrt{2}} \begin{pmatrix} 0 & -1 & 0 \\ 1 & 0 & -1 \\ 0 & 1 & 0 \end{pmatrix} \quad ;$$

$$L_z = \hbar \begin{pmatrix} 1 & 0 & 0 \\ 0 & 0 & 0 \\ 0 & 0 & -1 \end{pmatrix} \quad ,$$

which means, for example, that the operation $L_x Y_{1,1}$ may be represented as

$$\frac{\hbar}{\sqrt{2}} \begin{pmatrix} 0 & 1 & 0 \\ 1 & 0 & 1 \\ 0 & 1 & 0 \end{pmatrix} \begin{pmatrix} 1 \\ 0 \\ 0 \end{pmatrix} \quad \text{with the result} \quad \frac{\hbar}{\sqrt{2}} \begin{pmatrix} 0 \\ 1 \\ 0 \end{pmatrix} ,$$

so that $L_x Y_{1,1} = (\hbar/\sqrt{2}) \, Y_{1,0}$. By carrying out the matrix operations, write the results of the operations $L_x Y_{1,0}$, $L_x Y_{1,-1}$, $L_y Y_{1,1}$, $L_y Y_{1,0}$ and $L_y Y_{1-1}$. Verify these results by direct application of the differential operators [Eq. (13), Chapter 6].

11. (a) Using the results of problems 7 and 10, find two linear combinations of $Y_{1,1}(-)_s$ and $Y_{1,0}(+)_s$ that are eigenfunctions of $\mathbf{L \cdot S}$ (and hence are eigenfunctions of J^2; see Section 7.3).

    (b) Find the value of J^2 for each of these combinations.

12. A point charge q at a distance x from the surface of a dielectric is attracted because it induces a charge of opposite sign on the surface of the dielectric. It is shown in electromagnetism texts[29] that the attractive force is equal to $q^2(1 - \epsilon)/(2x)^2 (1 + \epsilon)$, where ϵ is the dielectric constant (cgs units). However, if the point charge is an electron, it cannot enter the dielectric because the low-lying electron states inside the dielectric are filled. We can say that there is a high potential barrier at the surface of the dielectric.

    (a) Write the Schrödinger equation for an electron near the surface of the dielectric. Assume that the surface of the dielectric extends to infinity in the yz plane, so the problem is effectively one dimensional.

    (b) Solve for the energy levels of the system, by applying the boundary condition that the wavefunction vanishes at the dielectric surface. (Notice that the equation is very similar to the radial equation for the hydrogen atom.)

13. Use Eq. (31) to calculate the shift ΔE in each of the levels involved in the emission of the "H_α line." From this result determine the wavelength of

[29] For example, G. P. Harnwell, "Principles of Electricity and Electromagnetism" McGraw-Hill, New York, 1949.

each of the five lines in the fine structure of H_α, and verify that Eq. (32) gives the correct $\Delta\lambda$ for the splitting of the lines specified here.

14.   Calculate classically the maximum interaction energy between a Bohr magneton and a nuclear magneton when they are separated by a distance of one Bohr radius. Compare this energy with the energy of the hyperfine splitting of the ground state of hydrogen. (The field $\mathbf{B}$ at a point P, produced by a dipole moment $\boldsymbol{\mu}$ may be written (mks units)

$$\mathbf{B} = -\frac{\mu}{4\pi\varepsilon_0 c^2}\, \mathbf{V}\, \frac{\cos\theta}{r^2}$$

where the vector operator $\mathbf{V}$ is given in spherical coordinates as

$$\mathbf{V} = \hat{\mathbf{r}}\frac{\partial}{\partial r} + \frac{\hat{\boldsymbol{\theta}}}{r}\frac{\partial}{\partial \theta}$$

$\hat{\mathbf{r}}$ and $\hat{\boldsymbol{\theta}}$ being unit vectors in the direction of increasing r and increasing θ, respectively (see figure).

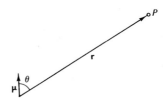

15.   Each of the Dirac operators α and β may be written as a product of two operators; one of these is an operator σ_x, σ_y, or σ_z which operates only on the spin functions $(+)_s$ and $(-)_s$, and the other is an operator ρ_x, ρ_y, or ρ_z which operates only on the energy functions $(+)_E$ and $(-)_E$. The $\boldsymbol{\sigma}$ operators are the Pauli spin operators, defined by $\mathbf{S} = (\hbar/2)\boldsymbol{\sigma}$ (Problem 7). The $\boldsymbol{\rho}$ operators operate on the energy functions in exactly the same way that the $\boldsymbol{\sigma}$ operators operate on the spin functions; thus the components of $\boldsymbol{\rho}$ must satisfy the same commutation relations as the components of $\boldsymbol{\sigma}$.

   (a)   Show from the results of Problem 7 that the components of $\boldsymbol{\sigma}$ anticommute; for example, $\sigma_x\sigma_y + \sigma_y\sigma_x = 0$.

   (b)   Show from (a) that Eqs. (34) are satisfied if

   $$\alpha_x = \rho_x\sigma_x, \qquad \alpha_y = \rho_x\sigma_y, \qquad \alpha_z = \rho_x\sigma_z, \qquad \text{and} \qquad \beta = \rho_z.$$

   (c)   Using the results of (a) and (b), and the commutation rules $[S_x, S_y] = i\hbar S_z$, $[S_y, S_z] = i\hbar S_x$, $[S_z, S_x] = i\hbar S_y$, show that

   $$[S_z, H] = c\hbar^2\left(\alpha_y\frac{\partial}{\partial x} - \alpha_x\frac{\partial}{\partial y}\right),$$

   and that therefore $[J_z, H] = 0$, when H is the Dirac Hamiltonian.

Atomic Structure

The triumphs of the new quantum theory, or "quantum mechanics" as it is now called, described in the previous two chapters, leave no doubt that this theory correctly describes atomic phenomena. Unfortunately, there are very few systems for which the Schrödinger equation can be solved exactly; even the hydrogen atom requires an approximation treatment if one is to account for the effects of electron spin (Section 7.3).

Therefore we begin our study of atomic structure by introducing approximation methods that permit us to obtain, to a high degree of accuracy, the energy levels of a three-body system (the helium atom), and we shall then see, qualitatively, how the structure of heavier atoms is determined.

In the course of this study, we shall see that there is a fundamental difference between the classical and the quantum view of systems containing two or more identical particles. The quantum-mechanical view of identical particles is required not only for atomic physics, but also for the explanation of a wide variety of phenomena in solid-state and molecular physics.

8.1 TIME-INDEPENDENT PERTURBATION THEORY

Suppose that the potential energy function of a system is such that we cannot solve the Schrödinger equation exactly, but that we can solve it exactly for a potential $V(x)$ which differs slightly[1] from the actual potential. We can write the potential energy of the system as

$$V(x) + v(x)$$

where $v(x)$ is a small[1] "perturbation" added to the "unperturbed" potential $V(x)$.

The Hamiltonian operator of the unperturbed system is[2]

$$H_0 = -\frac{\hbar^2}{2m}\frac{\partial^2}{\partial x^2} + V(x)$$

so that the Schrödinger equation of that system is

$$H_0\psi_l = E_l\psi_l$$

with a known set of eigenvalues E_l and eigenfunctions ψ_l.

The Schrödinger equation of the "perturbed" system contains the additional term $v(x)$ in the Hamiltonian:

$$[H_0 + v(x)]\psi_n' = E_n'\psi_n' \tag{1}$$

and as a result we cannot find the perturbed eigenfunctions ψ_n' and eigenvalues E_n' directly. However, we know that any acceptable wavefunction may be expanded in a series of eigenfunctions of the unperturbed Hamiltonian (Postulate 3, Section 6.3); therefore, we may write each function ψ_n' as a series of the functions ψ_l with constant coefficients:

$$\psi_n' = \sum_l a_{nl}\psi_l \tag{2}$$

so that Eq. (1) becomes

$$[H_0 + v(x)]\sum_l a_{nl}\psi_l = E_n'\sum_l a_{nl}\psi_l \tag{3}$$

Equation (2) may be thought of as a generalized Fourier series, but in this case we are not expanding a *known* function—the eigenfunctions ψ_n' as well as the coefficients a_{nl} being unknown—so we cannot operate on Eq. (2) directly to find the coefficients a_{nl}. However, the technique used in finding Fourier coefficients does suggest a possibility (see Section 4.3). We multiply

---

[1] The meaning of the words "slightly" and "small" in this context will become clear as we go along.
[2] For simplicity, we shall work in only one dimension here.

each side of Eq. (3) by ψ_m^*—the complex conjugate of a particular unperturbed eigenfunction ψ_m—and we integrate both sides over all values of x, to obtain

$$\int \psi_m^*[H_0 + v(x)](\sum_l a_{nl}\psi_l) \, dx = \int E_n' \psi_m^*(\sum_l a_{nl}\psi_l) \, dx \qquad (4)$$

Assuming that it is legitimate to integrate each side of Eq. (4) term by term, and using the fact that $H_0\psi_l = E_l\psi_l$, we obtain

$$\sum_l a_{nl} \int \psi_m^* v(x)\psi_l \, dx = \sum_l (E_n' - E_l)a_{nl} \int \psi_m^*\psi_l \, dx \qquad (5)$$

and because the functions ψ are normalized and orthogonal, each term in the right-hand series is zero, except the one for which $l = m$. We finally have

$$\sum_l a_{nl} v_{ml} = (E_n' - E_m)a_{nm} \qquad (6)$$

where the symbol v_{ml} has been introduced as an abbreviation for

$$\int \psi_m^* v(x)\psi_l \, dx,$$

a quantity often called the "*matrix element*" of the perturbing potential $v(x)$, between the states m and l.

Equation (6) *is exact*; we have not yet introduced any approximations. Equation (6) expresses one particular unknown coefficient a_{nm} of the expansion (2) in terms of the other coefficients in that expansion, the known functions $v(x)$, ψ_m^*, and ψ_l, and the unknown energy difference $E_n' - E_m$ between a perturbed level E_n' and an unperturbed level E_m. As there are still too many unknown quantities here, we must proceed further by a method of successive approximations.

The first quantity that we want to find is the energy, so for a first approximation we guess at the coefficients a_{nl}. The simplest guess is that each wavefunction of the perturbed system is almost identical to one of the wavefunctions of the unperturbed system, so that $a_{nl} = 1$ if $n = l$, and zero otherwise. The series in Eq. (6) collapses to a single term, and if we set m equal to n the equation becomes

$$E_n' - E_n = v_{nn}$$

$$= \int \psi_n^* v(x)\psi_n \, dx \qquad (7)$$

This approximation gives a reasonable value for the perturbed levels if the levels of the unperturbed system and those of the perturbed system correspond closely (Fig. 1), so that the difference between E_n' and E_n is much smaller than

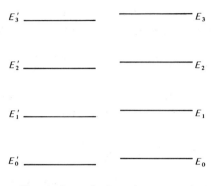

Fig. 1. *Perturbed and unperturbed levels compared, for a typical "small" perturbation.*

the difference between successive levels of the unperturbed system. It is reasonable in that case to assume that each wavefunction of the perturbed system resembles most closely the unperturbed wavefunction corresponding to the closest energy level of that system.

Equation (7) is a result which one might have guessed intuitively; we have already used it to calculate the fine-structure splitting in the energy levels of hydrogen. Now we are in a position to determine the accuracy of the result that we obtain in this way. We can do this by obtaining a second approximation, which we do by finding the changes in the wavefunctions that result from the change in the potential, and using these more accurate wavefunctions instead of the unperturbed wavefunctions in calculating the matrix elements v_{nn} (which are, of course, just the expectation values of $v(x)$ in the various states). In other words, we calculate the coefficients a_{nl}.

Of course, we still cannot solve Eq. (6) for these coefficients without making an approximation, so we continue to assume that each coefficient except a_{nn} is quite small, and that $a_{nn} \sim 1$. We can now solve Eq. (6) for *each particular* coefficient a_{nm} in turn, by assuming that each *other* coefficient, except a_{nn}, may be neglected. With this assumption, Eq. (6) becomes

$$v_{mn} + a_{nm} v_{mm} = (E'_n - E_m) a_{nm}$$

or

$$a_{nm} = \frac{v_{mn}}{(E'_n - E_m - v_{mm})}$$

From Eq. (7), we see that $v_{mm} = E'_m - E_m$, so

$$a_{nm} = \frac{v_{mn}}{(E'_n - E'_m)} \qquad (n \neq m) \tag{8}$$

The coefficients found from Eq. (8) can be inserted into Eq. (2), and the resulting wavefunctions can be used (instead of the unperturbed wavefunctions) in Eq. (7) to obtain the second-order approximation E_n'' to the energy. The result is

$$E_n'' = E_n' + \sum_{s \neq n} \frac{|v_{sn}|^2}{E_n - E_s} \tag{9}$$

This second-order perturbation result is important in cases where the matrix elements v_{nn} vanish, so that there is no first-order perturbation in the energy levels. For example, application of a uniform electric field $\mathscr{E}$, in the z direction, to an atom other than hydrogen[3], produces a first-order shift of

$$E_n' - E_n = \int |\psi_{nlm}|^2 V \, d\tau = \int |\psi_{nlm}|^2 \mathscr{E} z \, d\tau$$

in the level whose unperturbed wavefunction is ψ_{nlm}. With $z = r \cos \theta$ and $d\tau = r^2 \sin \theta \, d\theta \, d\phi \, dr = -r^2 \, d\phi \, dr \, d(\cos \theta)$, the θ part of the integral becomes one on $\cos \theta$, with limits -1 and $+1$. Since the wavefunction is either an even or an odd function of $\cos \theta$, the square of the wavefunction is an even function, and the integrand, with the $r \cos \theta$ factor, is an odd function. Thus the integral vanishes, and there is no *first*-order shift. However, there is a nonzero *second*-order shift in the energy levels, which is given by Eq. (9). The term $|v_{sn}|^2$ contains the square of the field $\mathscr{E}$, so the shift in energy (called the *Stark effect*) is proportional to $\mathscr{E}^2$, and this is what is observed experimentally.

EXAMPLE PROBLEM 1. Use first-order perturbation theory to find the lowest energy level of the "stepped" potential well shown in Fig. 2, where

$$V(x) = \begin{cases} +\infty & (|x| > d) \\ 0 & \left(\dfrac{d}{3} < |x| < d \right) \\ +\delta & \left(|x| < \dfrac{d}{3} \right) \end{cases}$$

assuming that δ is small compared to the energy of the lowest level. Also find the wavefunction for this level, to first order.

---

[3] In hydrogen, the wavefunctions for different l (but the same n) are degenerate, so the perturbation treatment must be different. See the discussion of perturbation of degenerate eigenfunctions at the end of this section.

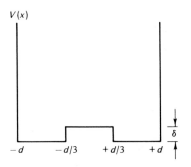

V(x)

−d   −d/3   +d/3   +d

δ

Fig. 2. *"Stepped" potential well.*

Solution. We consider the step for $|x| < d/3$ to be a small perturbation on a square-well potential. The unperturbed wavefunctions are simply those of the infinitely deep square well (Section 5.4):

$$\psi_n = \begin{cases} N_n \cos \dfrac{n\pi x}{2d}, & n \text{ odd} \\[2mm] N_n \sin \dfrac{n\pi x}{2d}, & n \text{ even} \end{cases} \quad (|x| \leqslant d) \\[2mm] \quad\; 0 \hspace{5.8cm} (|x| \geqslant d)$$

where N_n is a normalizing constant which is found from the equation (Section 5.3) $\int_{-\infty}^{\infty} \psi_n^* \psi_n \, dx = 1$, which leads to the result that $N_n^2 = 1/d$ for all n. The unperturbed energy levels are $E_n = \hbar^2 k^2 / 2m$, with $k = n\pi/2d$, so that

$$E_n = \frac{n^2 h^2}{32 m d^2}$$

(Compare with Chapter 5, Eq. (37).)

The perturbed energy levels are now found from Eq. (7). For the state of lowest energy, the perturbation causes a change in energy of

$$E_1' - E_1 = \int_{-\infty}^{\infty} \psi_1^* v(x) \psi_1 \, dx$$

which becomes

$$E_1' - E_1 = \delta \int_{-d/3}^{d/3} \psi_1^* \psi_1 \, dx$$

because the perturbation $v(x)$ is equal to δ in the region $-d/3 < x < +d/3$ and is zero elsewhere. This result, not surprisingly, is simply the product of

δ and the probability of finding the particle in the region where $V(x) = \delta$. Completing the calculation, we have

$$E_1' - E_1 = \frac{\delta}{d} \int_{-d/3}^{+d/3} \cos^2 \frac{\pi x}{2d} \, dx$$

$$= \frac{\delta}{d} \left[\frac{d}{\pi} \sin \frac{\pi}{3} + \frac{d}{3} \right]$$

$$= 0.61\delta$$

In other words, the probability of finding the particle between $x = +d/3$ and $x = -d/3$ is 0.61, a reasonable result, because this region occupies one-third of the well, and the probability density $\psi^* \psi$ is larger than average throughout this region.

To find the perturbed wavefunction ψ_1', we write

$$\psi_1' = \psi_1 + a_{12} \psi_2 + a_{13} \psi_3 + \cdots$$

and we apply Eq. (8) to find a_{12}, a_{13}, etc., in turn:

$$a_{12} = \frac{\int_{-\infty}^{+\infty} \psi_2^* v(x) \psi_1 \, dx}{E_1' - E_2'}$$

$$= \frac{N_1 N_2 \delta \int_{-d/3}^{+d/3} \sin \frac{\pi x}{d} \cos \frac{\pi x}{2d} \, dx}{E_1' - E_2'}$$

which equals zero, because the integrand is an odd function. We should have expected a_{12} to be zero, because the sum of the even function ψ_1 and the odd function $a_{12} \psi_2$ would yield a wavefunction with no definite parity, in violation of the rule (Section 5.4) that the wavefunction in a symmetrical potential must have a definite parity. Let us therefore find a_{13}:

$$a_{13} = \frac{N_1 N_3 \delta \int_{-d/3}^{+d/3} \cos \frac{3\pi x}{2d} \cos \frac{\pi x}{2d} \, dx}{E_1' - E_3'}$$

$$\approx \frac{\delta \left[\dfrac{\sin(\pi/3)}{\pi/d} + \dfrac{\sin(2\pi/3)}{2\pi/d} \right]}{d \left[\dfrac{h^2}{32md^2} (1 - 9) \right]}$$

$$= -\frac{4\delta md^2}{h^2} \left[\frac{\sqrt{3}}{2\pi} - \frac{\sqrt{3}}{4\pi} \right]$$

$$= -\frac{4\delta md^2}{h^2} \left(\frac{\sqrt{3}}{4\pi} \right)$$

Calculation of the remaining coefficients a_{15}, a_{17}, ..., is straightforward. Notice that they become progressively smaller, because of the increasing energy difference in the denominator.

The perturbed wavefunction ψ'_1 is compared with the unperturbed function ψ_1 in Fig. 3. Because a_{13} is negative, ψ'_1 is smaller than ψ_1 in the middle of

Fig. 3. *Wavefunction for lowest energy in the well of Fig. 2. Dotted line shows unperturbed wavefunction, for infinite square well.*

the well, and larger than ψ_1 at the sides of the well. This is what we should expect, because it means that ψ'_1 is flatter (that is, it has a smaller second derivative) than ψ_1 near the center of the well, where the perturbed *kinetic* energy $E'_1 - \delta$ is smaller than the unperturbed kinetic energy E_1. Similarly, ψ'_1 has a greater second derivative than ψ_1 near the sides of the well, where the perturbed kinetic energy E'_1 is greater than the unperturbed kinetic energy E_1.

Perturbation of Degenerate Eigenfunctions. Up to now, we have considered only the very special case in which there is just one eigenfunction for each eigenvalue, and the eigenvalues are well separated. In that case, it is quite reasonable to assume that a single wavefunction makes an overwhelming contribution to the perturbed wavefunction when the perturbed energy levels are very close to the unperturbed levels.

But in most cases of practical interest, the unperturbed levels are degenerate, so that each perturbed energy level may be close to an unperturbed level for which there are two or more independent wavefunctions. In such a case, we can only say as a first approximation that each perturbed eigenfunction must be some linear combination of the unperturbed eigenfunctions corresponding to the closest unperturbed level. Instead of setting all but one of the coefficients a_{nm} equal to zero, we must allow as many nonzero coefficients as there are independent eigenfunctions for the corresponding unperturbed level. We then must use Eq. (6) to set up as many *simultaneous* equations as are necessary to solve for the various coefficients and the corresponding perturbed energy levels.

We shall not go into the details of solving such simultaneous equations, because this procedure is often not necessary. For a given eigenvalue, one has some choice as to the set of degenerate unperturbed eigenfunctions to start with, because any linear combination of the eigenfunctions for a given eigenvalue is also an eigenfunction with the same eigenvalue. If we start with the right set of unperturbed eigenfunctions, we do not have to use more than one eigenfunction for a first approximation to each perturbed eigenfunction. It might seem to be terribly difficult to decide in advance which set of unperturbed functions to use, but in practice we can decide very easily, by considering the symmetry of the perturbing potential. We can then return to the assumption that all but one of the coefficients a_{nm} are zero, and we can apply Eq. (7) directly to find the shift in energy for each eigenfunction.

An example should clarify this discussion. We have already dealt with a degenerate case (in Section 7.3) when we considered the fine-structure splitting of the levels of the hydrogen atom. In that case, a typical pair of orthogonal degenerate eigenfunctions could be written as the product of a radial function and the functions $Y_{l, l}(\theta, \phi)(-)_s$ and $Y_{l, l-1}(\theta, \phi)(+)_s$, respectively. In the absence of the spin–orbit interaction, both of these eigenfunctions correspond to the same energy level, because the energy depends only on the radial quantum number n. But the spin–orbit energy is proportional to $\mathbf{L} \cdot \mathbf{S}$, so that the correct perturbed wavefunctions must be eigenfunctions of the $\mathbf{L} \cdot \mathbf{S}$ operator. Neither of the above functions meets this requirement, but there are two orthogonal linear combinations which do. Each of these combinations is an acceptable first approximation to one of the perturbed wavefunctions, and when substituted into Eq. (7), each yields a good approximation to the energy shift resulting from the spin–orbit interaction. But the problem was even simpler than it appears to be here, because we did not even have to find the actual linear combinations which were substituted into Eq. (7); we were able to evaluate the integral simply from knowledge of the eigenvalues of these functions when operated on by the $\mathbf{L} \cdot \mathbf{S}$ operator.

It happens in the spin–orbit example that the perturbation removes the degeneracy because the eigenvalues of $\mathbf{L} \cdot \mathbf{S}$ are different for the two eigenfunctions. We shall encounter other examples of perturbations which split degenerate energy levels, when we discuss the Zeeman effect (Section 9.2) and when we discuss energy levels in a periodic potential (Section 12.3). The general procedure followed is always the same; from the infinite number of sets of orthogonal degenerate eigenfunctions of the unperturbed potential, we select that set which has the symmetry demanded by the perturbation (in the spin–orbit case, the set which are eigenfunctions of $\mathbf{L} \cdot \mathbf{S}$), and we use each of those eigenfunctions in Eq. (7) in turn to find the various energy shifts caused by the perturbation.

8.2 IDENTICAL PARTICLES

Indistinguishability. When we study atoms heavier than hydrogen we will find that they differ from hydrogen in a fundamental way—not just because they contain more particles, but because they contain a number of *identical* particles. The treatment of identical particles involves further principles which are just as fundamental, just as deceptively simple, and just as difficult to appreciate, as any of the basic laws of quantum mechanics which we have already encountered.

The basic principle, which one may take as obvious and yet be unaware of its significance, is this: Two states which differ only in the interchange of identical particles are really one and the same state; there can be no way to distinguish between such states.

For example, if two electrons come close enough together so that their wavefunctions overlap, and then they fly apart again, there is no way after the interaction to determine which electron came from the left and which one came from the right initially. It is as if two drops of water coalesced and the resulting drop later split again into two drops; it is meaningless to ask which of the two final drops was originally on the left, because the drops lost their identity when they mixed. (The only difference between the two situations is that "drops" of electric charge—electrons—always come in the same size, and cannot be separated further into smaller "drops.")

Symmetry of Wavefunctions. How do we express this principle mathematically? We saw in Section 7.1 that the wavefunction of a system must be a function of all the coordinates of all the particles in the system. Thus the spatial part of the wavefunction for a system of N particles is a function of $3N$ coordinates, $u_T(x_1, y_1, z_1, \ldots, x_N, y_N, z_N)$, which we abbreviate as $u_T(1, 2, \ldots, N)$. If in this expression we interchange the respective coordinates x_1, y_1, z_1 and x_2, y_2, z_2, then u_T has the same functional dependence on the "1" coordinates that it formerly had on the "2" coordinates, and we have mathematically interchanged the positions of particle 1 and particle 2. But according to our basic principle, if the particles are identical, the state resulting from this interchange must be identical to the original state, or

$$u_T(2, 1, \ldots, N) = A u_T(1, 2, \ldots, N) \tag{10}$$

where A is a constant. Interchanging particles 1 and 2 a second time must have the same effect on the wavefunction, so we can also write

$$u_T(1, 2, \ldots, N) = A u_T(2, 1, \ldots, N) \tag{11}$$

Combining Eqs. (10) and (11), we see that $A^2 = 1$, so that $A = +1$ or -1.

If $A = +1$, we say that the wavefunction is symmetric with respect to interchange of identical particles; if $A = -1$, we say that the function is antisymmetric. Figure 4 illustrates the two situations. It is rather difficult

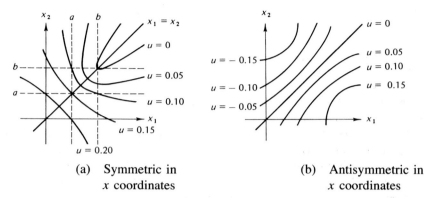

(a) Symmetric in
x coordinates

(b) Antisymmetric in
x coordinates

Fig. 4. *"Contour map" showing values of possible symmetric (a) and anti-symmetric (b) wavefunctions u as functions of the x coordinates x_1 and x_2 of two identical particles. Contours connect points at which u has a constant value, as indicated on each line. Notice the line of symmetry, $x_1 = x_2$, where u must be zero in the antisymmetric case.*

to graph a function of six variables, so the figure shows a one-dimensional case, where u is a function of two variables only: the coordinate x_1 of the first particle and the coordinate x_2 of the second particle. The line $x_1 = x_2$ is a line of symmetry; interchanging x_1 and x_2 is equivalent to reflecting the figure across this line.

In interpreting the physical meaning of Fig. 4, we must always bear in mind the indistinguishability of the particles. We affix labels 1 and 2 to our coordinates because we have no other way to write down a mathematical function of two coordinates, or to plot it; but the physics of a situation does not depend on our mode of communication—in other words, any physical result must be independent of our artificial labels 1 and 2. For example, consider the probability of finding one of the two particles between the positions $x = a$ and $x = b$. Physically, it is meaningless to specify which particle we find. The function u^*u is a probability density for *both* particles, its value at a point being the probability density for finding a particle at *each* of the *two* coordinates specified by the location of the point. Thus to find the probability that *either one* (or both) of the particles is located between $x = a$ and $x = b$, we integrate u^*u over the entire region for which *either* $a < x_1 < b$ or $a < x_2 < b$; this is the region enclosed by the dashed horizontal and vertical lines in Fig. 4a.

One feature of Fig. 4b deserves special comment. Notice that $u = 0$ when $x_1 = x_2$. It is easy to prove that this must be true if u is antisymmetric. Let $x_1 = x_2 = c$. Then, from Eq. (9),

$$u(x_2, x_1) = -u(x_1, x_2)$$

or

$$u(c, c) = -u(c, c)$$

which can only be true if $u(c, c) = 0$.

Separation of Variables.   Of course, there is still the fundamental requirement that u be a solution of the Schrödinger equation. The problem of solving the Schrödinger equation for a wavefunction of several particles is a formidable one, as we might expect. For a start on the problem, we make the simplifying assumption that we can neglect the interactions between the particles; we can later introduce these interactions as perturbations. In other words, we assume that each particle is moving in a given external potential which is independent of the positions of the other particles.

The Schrödinger equation for the case of two particles may then be written

$$\left[-\frac{\hbar^2}{2m} (\nabla_1^2 + \nabla_2^2) + V(1) + V(2) \right] u(1, 2) = E_T u(1, 2) \qquad (12)$$

where ∇_1^2 operates on the coordinates of particle 1, and $V(1)$, a function of x_1, y_1, and z_1, is the potential energy of particle 1, etc.[4] We find it simplist to approach the solution as if the particles were distinguishable; later we introduce the requirement of indistinguishability. Our first step is to separate the variables by writing $u(1, 2) = u_a(1)u_b(2)$, where u_a and u_b may be different functions. Substitution into Eq. (12) and regrouping terms yields

$$\left[-\frac{\hbar^2}{2m} \nabla_1^2 + V(1) \right] u_a(1)u_b(2)$$

$$+ \left[-\frac{\hbar^2}{2m} \nabla_2^2 + V(2) \right] u_a(1)u_b(2) = E_T u_a(1)u_b(2) \qquad (13)$$

As we did before in separating variables, we now divide all terms by the wavefunction $u_a(1)u_b(2)$, to obtain

$$\frac{1}{u_a(1)} \left[-\frac{\hbar^2}{2m} \nabla_1^2 + V(1) \right] u_a(1) + \frac{1}{u_b(2)} \left[-\frac{\hbar^2}{2m} \nabla_2^2 + V(2) \right] u_b(2) = E_T$$

The first term is the only term which depends on the coordinates of particle 1, and the second term is the only term which depends on the coordinates of

---

[4] This is reminiscent of the treatment of the hydrogen atom (Section 7.1), but here, instead of the interaction between the particles, we have the *external* potentials $V(1)$ and $V(2)$.

particle 2, so by the reasoning of Section 6.2, each term must equal a constant. Call the respective constants E_a and E_b. Then we have, after further rearrangement of terms,

$$\left[-\frac{\hbar^2}{2m} \nabla_1^2 + V(1) \right] u_a(1) = E_a u_a(1) \tag{14}$$

$$\left[-\frac{\hbar^2}{2m} \nabla_2^2 + V(2) \right] u_b(2) = E_b u_b(2) \tag{15}$$

$$E_a + E_b = E_T \tag{16}$$

Except for the labels, Eqs. (14) and (15) are really the same equation: the one-particle Schrödinger equation. Therefore u_a and u_b belong to the same set of eigenfunctions, when particles 1 and 2 are subject to the same potential V. In such a case, Eqs. (13) through (16) tell us that the wavefunction for the system can be constructed from two independent wavefunctions for a single particle, and the resulting energy eigenvalue is the sum of the energy eigenvalues corresponding to the respective single-particle wavefunctions. For example, one can, as a first approximation, assume that *each* electron in a two-electron atom (helium) separately occupies one of the states of an electron in a helium *ion*, whose hydrogenlike wavefunctions are already known (Section 7.2). In this approximation, the total energy E_T of each state of the helium *atom* is the sum of two energy eigenvalues E_a and E_b for the helium *ion*. It remains to be seen how much error is introduced by neglecting the interaction between the two electrons in the atom, or how we can correct the theory to account for this interaction.

Before we proceed further with study of a real atom, we must take account of the indistinguishability of the electrons. This requires that we make the wavefunction of the system either symmetric or antisymmetric; the function $u_a(1)u_b(2)$ is in general neither symmetric nor antisymmetric, so it cannot be an acceptable wavefunction. However, we can easily use this function to construct an acceptable wavefunction; we simply add $u_a(1)u_b(2)$ to the function $u_a(2)u_b(1)$ which is obtained by interchanging the particles, and we obtain the *symmetric wavefunction*:

$$u_S(1, 2) = \frac{1}{\sqrt{2}} \left[u_a(1)u_b(2) + u_a(2)u_b(1) \right]$$

We can also take the difference of these two functions to obtain the *antisymmetric wavefunction*:

$$u_A(1, 2) = \frac{1}{\sqrt{2}} \left[u_a(1)u_b(2) - u_a(2)u_b(1) \right]$$

The factor $1/\sqrt{2}$ is required to preserve the normalization of the wavefunctions. With each of these wavefunctions, we have one particle in the single-particle state whose wavefunction is u_a, and one particle in the state whose wavefunction is u_b, but we have no specification of which particle is in which state. You may easily verify that

$$u_S(1, 2) = u_S(2, 1)$$

and

$$u_A(1, 2) = -u_A(2, 1)$$

in agreement with Eq. (9).

Pauli Exclusion Principle. Since both the antisymmetric and the symmetric functions are acceptable solutions of the Schrödinger equation, and both may be constructed from the same single-particle wavefunctions (with one important exception, as we shall see directly), you might be surprised to learn that it makes a considerable difference whether we use one or the other. A great variety of phenomena depend on the fact that *all particles with half-integer spin* (electrons, protons, neutrons, muons, neutrinos, and an assortment of "strange" particles) *must have antisymmetric wave functions*. Particles with integer spin (photons, pions, and others) must have symmetric wavefunctions.

Notice that the antisymmetric wavefunction $u_A(1, 2)$ vanishes when both particles are in the same state:

$$u_a(1)u_a(2) - u_a(2)u_a(1) \equiv 0$$

Thus the antisymmetry requirement is equivalent to a requirement that no two electrons can occupy the same state; in other words, no two electrons can have the same complete set of quantum numbers. In this form, the requirement was first enunciated by W. Pauli in 1925, to explain features of atomic spectra and the periodic table of the elements, and it is therefore known as the Pauli exclusion principle.

The antisymmetry requirement applies to the exchange of *all* coordinates, including spin. Thus the spatial part of the wavefunction is symmetric when the spin part is antisymmetric, and vice versa. For example, there are four different completely antisymmetric wavefunctions for two electrons in states *a* and *b*:

$$\frac{1}{\sqrt{2}} [u_a(1)u_b(2) + u_a(2)u_b(1)] \frac{1}{\sqrt{2}} [(+)_{s1}(-)_{s2} - (-)_{s1}(+)_{s2}]$$

$$\frac{1}{\sqrt{2}} [u_a(1)u_b(2) - u_a(2)u_b(1)][(+)_{s1}(+)_{s2}]$$

$$\frac{1}{\sqrt{2}} [u_a(1)u_b(2) - u_a(2)u_b(1)] \frac{1}{\sqrt{2}} [(+)_{s1}(-)_{s2} + (+)_{s2}(-)_{s1}] \qquad (17)$$

$$\frac{1}{\sqrt{2}} [u_a(1)u_b(2) - u_a(2)u_b(1)][(-)_{s1}(-)_{s2}]$$

The first of these is symmetric with respect to exchange of the space coordinates, but it is antisymmetric with respect to exchange of the spin coordinates, so it is antisymmetric with respect to exchange of *all* coordinates of the two particles. The other three are antisymmetric in the space coordinates, but symmetric in the spin coordinates. The first combination, with the antisymmetric spin function, is called the singlet state; the other three combinations, with the symmetric spin functions, are called the triplet.

The latter three are the states in which the two spins are "parallel," so that the quantum number for the total spin angular momentum is $S = 1$, and there are three possible values for the component S_z: $S_z = +\hbar$, 0, and $-\hbar$, respectively, for the three states in the order in which they are listed above. The singlet state is the state in which the spins are "antiparallel"; the total spin is zero, and of course S_z is also zero. (You may verify all these statements by applying the spin operators to the functions. See Problem 4.)

This division of states into a triplet with $S = 1$ and a singlet with $S = 0$ is characteristic of the states of any two particles of spin $\frac{1}{2}$, whether or not the particles are identical; the quantization of angular momentum requires that the states be classified in this way. This fact has a direct nonclassical consequence[5] which is well verified experimentally: If two particles of spin $\frac{1}{2}$ come together at random, their spins are "parallel" three-quarters of the time, and their spins are "antiparallel" one-quarter of the time. For example, in a random collection of hydrogen molecules, three-quarters of the molecules are "ortho-hydrogen," with the resultant spin of the two protons being $S = 1$, and the other one-quarter are "para-hydrogen," with $S = 0$.

Exchange Energy. The requirement that the total wavefunction, including spin dependence, be antisymmetric manifests itself as a new kind of "force" which is unknown in classical physics. This "force" is not a force in the classical sense at all. It is called a force because there is a correlation between the positions of the electrons in a system; the electrons move about as if there were another force between them, in addition to the Coulomb force. Of course, we do not actually follow the motions of the electrons, but from measurements of energy levels, we deduce that this correlation is present.

The effect arises in the following way: If the space part of the wavefunction is symmetric, the particles tend to be closer together, on the average, than when the space part of the wavefunction is antisymmetric. Study of Fig. 4 helps one to understand how this comes about; notice that in the antisymmetric case, there is zero probability density for the two particles to be found with the same coordinate, while this is not true for the symmetric case. One

---

[5] That is, it is a logical consequence if one assumes that each quantum state is, *a priori,* equally likely, in the absence of any energy considerations. We make this assumption because it is the simplest possible assumption, and also because it works; it leads to consequences which agree with experiment (see Chapter 10).

can show in general that the expectation value of $(r_2 - r_1)^2$, the square of the distance between the particles, is greater for an antisymmetric wavefunction than for a symmetric one. This correlation between the positions of the particles causes the energy of two electrons in the antisymmetric state to be lower than that of two electrons in the symmetric state *constructed from the same two single-particle wavefunctions*, for the simple reason that two electrons have a lower energy when they are farther apart. The energy difference between the two types of state, which is simply a result of the Coulomb potential, is called the *exchange energy*.[6]

Notice that the exchange energy appears to be spin dependent, because the symmetric space function can only occur with the antisymmetric spin function ($S = 0$), and the antisymmetric space function can only occur with the symmetric ($S = 1$) spin functions. Thus, in the absence of other spin-dependent effects, the $S = 1$ states of two electrons have lower energy than the $S = 0$ state. This effect is much larger than the effect of the magnetic dipole interaction between the two electrons.

Local Realism. Consider two protons in the singlet state. If we measure the spin component of one of these protons along an arbitrary axis, we obtain one of the two possible results $+\hbar/2$ or $-\hbar/2$. If we measure the other proton, the result must agree with that of the first measurement; that is, the total spin being zero in a singlet state, if proton 1 has spin ''up,'' proton 2 must have spin ''down,'' and vice versa. This is true regardless of the direction of the measurement axis; the two protons are in *one state*, whose characteristic is that the spins are opposite, although each individual spin is unknown (or undetermined) until a measurement is made.

This fact raises questions similar to those raised in Section 4.4. Does the direction of each individual spin have any reality, independent of the measurement? And if it does not, how is it that the measurement on one proton *creates* a reality for the second proton? Quantum mechanics does not address such questions; it simply says what result you will get if you make the measurement.

But it is possible to adopt (provisionally) the view that was favored by Einstein—that quantum mechanics is just a partial description of reality, and that values of unmeasured variables do exist. It is even possible to test this point of view by an experiment!

The experiment consists of preparing a large batch of pairs of protons in the singlet state, and then measuring the spin of each proton along one of the three

---

[6] We sometimes refer to an "exchange *force*," simply because we always associate energy with some kind of force. But the force is actually just the Coulomb force. The only new, quantum mechanical "force" is the "force of law" which prevents the electrons from having totally symmetric wavefunctions.

axes A, B or C.[7] For each singlet pair, we may represent the result of the two measurements by a symbol of the type A^+B^-, where the letter refers to the axis of measurement of the spin of each proton of that pair, and the superscript $+$ or $-$ shows the spin direction that was found for that proton.

If we obtain the result $A^+ B^-$ from a measurement on a given pair, then we can say with certainty, concerning *hypothetical* alternative measurements, that

1. *if* we had measured the first proton along B instead of along A, the result would have been B^+, OR
2. *if* we had measured the second proton along A instead of along B, the result would have been A^-.

So it seems that we have found a way to measure two spin components of a single particle simultaneously. But we have not actually done this because only one direct measurement on a given proton has been made, and a second measurement on the *same* proton will always agree with the first measurement on *that* proton, regardless of the result of the measurement on the second proton. For example, after the result A^+B^-, a second measurement on proton 1 along axis A will always yield A^+, but a subsequent measurement on proton 1 along axis B will yield either B^+ or B^-, with a probability determined by the angle between the A and B axes, just as if proton 2 had not been measured. The first measurement on proton 1 "decouples" it from proton 2.

Before the first measurement was made, however, the protons *were* coupled, and statements 1 and 2 about hypothetical initial measurements must be valid. This opens the door to an interpretation which says that the first proton possesses the values A^+ and B^+ even before the measurement. Quantum mechanics does not allow this interpretation; according to quantum mechanics, the measurements interfere with each other, even though they are carried out on separate protons and the mechanism of the interference cannot be described further. The protons are part of a single quantum mechanical state, and any measurement on this state affects all of its parts.

It is easy to see why Einstein objected to the quantum-mechanical interference picture. Consider a thought experiment in which the two protons have separated to a distance of one light year before each spin is measured, but the protons remained undisturbed in the singlet state while they were separating. If one says that measuring one of these protons now interferes with the other proton, one seems to be invoking action at a distance, which is rejected by the theory of relativity.

But the ultimate test is the experiment, to which we now return. On the *assumption* that each proton does possess a definite value of each component along each of the axes A, B, and C, we can show that

---

[7] The experiment is more fully described by B. d'Espagnat, in "The Quantum Theory and Reality," *Scientific American*, **241**, No. 5, 158—170 (Nov., 1979).

$$n(A^+B^+) \leq n(B^+C^+) + n(A^+C^+)$$

where $n(A^+B^+)$ is the number of *pairs* for which the experimental result is A^+B^+, etc. This inequality is called the *Bell inequality,* after John S. Bell, who derived it in 1964. To prove it, we introduce the definition $N(A^+B^-)$ as the number of *single protons* that possess the spin components A^+ and B^- simultaneously, and we further define $N(A^+B^-C^+)$ as the number of protons, each of which possesses the stated values—A^+, B^-, and C^+—for its spin components along the three axes A, B, and C. Since each proton must be either C^+ or C^-, it is clear that

$$N(A^+B^-) = N(A^+B^-C^+) + N(A^+B^-C^-) \tag{18}$$

provided, of course, that these definitions are allowable at all.

We can also say that $N(A^+B^-C^+) \leq N(B^-C^+)$ because the former is a subset of the latter. Similarly, $N(A^+B^-C^-) \leq N(A^+C^-)$, for the same reason. Thus Eq. (18) can be replaced by the inequality

$$N(A^+B^-) \leq N(B^-C^+) + N(A^+C^-) \tag{19}$$

Now when we find the measured values to be A^+ and B^+, respectively, for the two protons of a pair, we know that the first proton was of type A^+B^-, belonging to the population $N(A^+B^-)$; therefore, the number of results $n(A^+B^+)$ must be proportional to the number of protons $N(A^+B^-)$. Similarly, we can say that $n(B^+C^+)$ is proportional to $N(B^-C^+)$, and $n(A^+C^+)$ is proportional to $N(A^+C^-)$, with the same constant of proportionality in each case. Thus, in inequality (19) we can replace $N(A^+B^-)$ by $n(A^+B^+)$, replace $N(B^-C^+)$ by $n(B^+C^+)$, and replace $N(A^+C^-)$ by $n(A^+C^+)$, to obtain the Bell inequality.

The significance of the Bell inequality is that, for certain orientations of the axes A, B, and C, quantum mechanics predicts results that violate this inequality. (Problem 5 gives one such set of orientations.) Thus, if quantum mechanics in its present form is correct, the assumption leading to the Bell inequality must be wrong, and each proton does *not* possess a definite value of each of the spin components. Experiments are now in progress to test the Bell inequality by measuring spin components of protons in singlet pairs, using axes such that the Bell inequality will be violated if quantum mechanics is correct. The results so far uphold quantum mechanics.

8.3 THE HELIUM ATOM

The helium atom, with only two electrons, offers us the best chance to test the principles we have just discussed. If we neglect the interaction between the two electrons, the wavefunction for a He atom is an antisymmetric combination like one of the expressions (17), u_a and u_b each being one of the wavefunctions for the helium *ion* (Chapter 7, Table 1, with $Z = 2$).

The lowest energy level is found by letting the quantum number n equal 1 for both electrons, so that u_a and u_b are both the same eigenfunction u_{100}; the spin function must then be the antisymmetric combination. This state is designated $1s^2$, to indicate that two electrons are in the 1s state; according to Eq. (16), its energy should be $E_T = E_a + E_b = 2E_{100} = -108.8$ eV. (E_{100} is given by Eq. (6) in Chapter 7, with $n = 1$ and $Z = 2$.) E_T should be equal in magnitude to the energy required to separate *both* electrons from a helium nucleus, but the measured value of this energy is only 79.0 eV.

It is easy to see why the actual energy level should be higher than our first estimate; the electron–electron repulsion must cause the He^+ ion to attract the second electron less strongly than the He nucleus attracts the first electron. (Remember the screening number s used in Moseley's work—Section 4.1.) We can use perturbation theory to calculate the approximate magnitude of this effect. The perturbing potential v is simply the electrostatic potential energy of two electrons at positions r_1 and r_2:

$$v = \frac{e^2}{4\pi\varepsilon_0 |r_1 - r_2|}$$

The energy shift caused by this perturbation is then, to first order,

$$E' - E = \iint |u_{100}(1)|^2 |u_{100}(2)|^2 \frac{e^2}{4\pi\varepsilon_0 |r_1 - r_2|} d\tau_1 d\tau_2$$

where $u_{100}(1)$ and $u_{100}(2)$ are simply the *hydrogen* atom wavefunction u_{100} (Chapter 7, Table 1) written in terms of the coordinates r_1 and r_2 of particle 1 and particle 2, respectively; and $d\tau_1$ and $d\tau_2$ are volume elements of the respective coordinates. With the substitutions

$$\rho_1 = \frac{Zr_1}{a_0'}, \qquad \rho_2 = \frac{Zr_2}{a_0'}, \qquad \rho_{12} = \frac{Z|r_1 - r_2|}{a_0'}, \qquad d\rho_1 = (Z/a_0')\, dr_1, \ldots,$$

the integral becomes

$$I = \frac{Ze^2}{\pi^2 a_0'} \iiiint\!\!\iint \frac{e^{-2\rho_1} e^{-2\rho_2}}{4\pi\varepsilon_0 \rho_{12}} \rho_1^2 \sin\theta_1\, d\rho_1\, d\theta_1\, d\phi_1\, \rho_2^2 \sin\theta_2\, d\rho_2\, d\theta_2\, d\phi_2$$

which (except for the factor $Ze^2/\pi^2 a_0'$), is identical to the integral for the mutual interaction energy of two spherically symmetric charge distributions whose charge densities are $e^{-2\rho_1}$ and $e^{-2\rho_2}$, respectively. The evaluation of this energy is a standard exercise in electrostatic theory.[8] (If you are not clever enough to evaluate it analytically, you can always program such an integral for a computer.) The result is

$$I = (\tfrac{5}{4}Z)\frac{m_r c^2 \alpha^2}{2} = (\tfrac{5}{2}) \times 13.60 \text{ eV} = 34.0 \text{ eV} \qquad (20)$$

The total energy of the ground state of the helium atom is found by adding this "perturbation" energy to the "unperturbed" energy—the energy that the atom would have if there were no interaction between the electrons. The unperturbed energy E is just twice the energy of the helium *ion*, or

$$E = -2Z^2(m_r c^2 \alpha^2/2)$$

so the "perturbed" energy is

$$E' = E + I = \left(-2Z^2 + \frac{5Z}{4}\right)\left(\frac{m_r c^2 \alpha^2}{2}\right) \qquad (21)$$

When $Z = 2$, $E' = -74.8$ eV. This should be the energy required to separate both electrons from a helium nucleus. According to Table 1, the experimental value for the sum of the first and second ionization potentials of helium is 78.98 volts, so the calculation is inaccurate by about 6%. That is a remarkably good result, when you consider that a perturbation of 34 eV on a total energy of 79 eV can hardly be called "small."

Table 1

Ionization Potentials of the Elements, in Volts[a]

		Stage of ionization					
Z	Element	I	II	III	IV	V	VI
1	Hydrogen	13.595					
2	Helium	24.580	54.400				
3	Lithium	5.390	75.619	122.42			
4	Beryllium	9.320	18.206	153.85	217.657		

[a] From "Handbook of Physics" E. U. Condon and H. Odishaw, eds. Copyright McGraw-Hill, New York, 1958. Used by permission of McGraw-Hill Book Company.

[8] It can be evaluated by integrating over the *first* distribution to obtain the total potential field of *all* elements in that distribution first. This potential field may then be written as a function of ρ_2, and the energy of the *second* distribution in the field of the first may be calculated as an integral on ρ_2. The trick is in summing energy elements in the most efficient order; for further details see L. Pauling and E. B. Wilson, "Introduction to Quantum Mechanics," Appendix V. McGraw-Hill, New York, 1935.

Table 1 *(Cont.)*

Z	Element	Stage of ionization					
		I	II	III	IV	V	VI
5	Boron	8.296	25.149	37.920	259.298	340.127	
6	Carbon	11.264	24.376	47.864	64.476	391.986	489.84
7	Nitrogen	14.54	29.605	47.426	77.450	97.863	551.92
8	Oxygen	13.614	35.146	54.934	77.394	113.873	138.08
9	Fluorine	17.42	34.98	62.646	87.23	114.214	157.12
10	Neon	21.559	41.07	64	97.16	126.4	157.91
11	Sodium	5.138	47.29	71.65	98.88	138.60	172.36
12	Magnesium	7.644	15.03	80.12	109.29	141.23	186.86
13	Aluminum	5.984	18.823	28.44	119.96	153.77	190.42
14	Silicon	8.149	16.34	33.46	45.13	166.73	205.11
15	Phosphorus	10.55	19.65	30.156	51.354	65.01	220.41
16	Sulfur	10.357	23.4	35.0	47.29	72.5	88.03
17	Chlorine	13.01	23.80	39.90	53.5	67.80	96.7
18	Argon	15.755	27.62	40.90	59.79	75.0	91.3
19	Potassium	4.339	31.81	46	60.90	. . . . . .	99.7
20	Calcium	6.111	11.87	51.21	67	84.39	
21	Scandium	6.56	12.89	24.75	73.9	92	111.1
22	Titanium	6.83	13.63	28.14	43.24	99.8	120
23	Vanadium	6.74	14.2	29.7	48	65.2	128.9
24	Chromium	6.763	16.49	30.95	49.6	73	90.6
25	Manganese	7.432	15.64	33.69	. . . . . .	76	
26	Iron	7.90	16.18	30.64			
27	Cobalt	7.86	17.05	33.49			
28	Nickel	7.633	18.15	36.16			
29	Copper	7.724	20.29	36.83			
30	Zinc	9.391	17.96	39.70			
31	Gallium	6.00	20.51	30.70	64.2		
32	Germanium	7.88	15.93	34.21	45.7	93.4	
33	Arsenic	9.81	20.2	28.3	50.1	62.6	127.5

Table 1 (Cont.)

Z	Element	Stage of ionization					
		I	II	III	IV	V	VI
34	Selenium	9.75	21.5	32.0	42.9	73.1	81.7
35	Bromine	11.84	21.6	35.9			
36	Krypton	13.99	24.56	36.9			
37	Rubidium	4.176	27.5	40			
38	Strontium	5.692	11.027	......	57		
39	Yttrium	6.5	12.4	20.5	......	77	
40	Zirconium	6.95	14.03	24.8	33.97	......	99

Z	Element	Stage of ionization		Z	Element	Stage of ionization	
		I	II			I	II
41	Niobium	6.77	14	57	Lanthanum	5.61	11.43
42	Molybdenum	7.06		58	Cerium	6.54	
43	Technetium			59	Praseodymium	(5.8)	
44	Ruthenium	(7.5)	(16)	60	Neodymium	(6.3)	
45	Rhodium	7.46	(18)	61			
46	Palladium	8.33	19.42	62	Samarium	(5.6)	(11.4)
47	Silver	7.574	21.48	63	Europium	5.64	11.2
48	Cadmium	8.991	16.90	64	Gadolinium	6.7	
49	Indium	5.785	18.86	65	Terbium	6.7	
50	Tin	7.332	14.628	66	Dysprosium	(6.8)	
51	Antimony	8.639	16.5	67			
52	Tellurium	8.96	(19)	68			
53	Iodine	10.44	19.0	69			
54	Xenon	12.13	21.2	70	Ytterbium	6.22	(12)
				71	Casseopium		
55	Cesium	3.893	25.1	72	Hafnium		(14.8)
56	Barium	5.210	10.001	73	Tantalum		
				74	Tungsten	7.94	

Table 1 *(Cont.)*

Z	Element	Stage of ionization		Z	Element	Stage of ionization	
		I	II			I	II
75	Rhenium	(8)	(13)	84	Polonium	(8.2)	(19)
76	Osmium	(8.7)	(15)	85			
77	Iridium	(9.2)	(16)	86	Radon	10.75	(20)
78	Platinum	(8.9)	18.5	87			
79	Gold	9.23	20.0	88	Radium	5.27	10.1
80	Mercury	10.44	18.8	89	Actinium		
81	Thallium	6.12	20.3	90	Thorium		
82	Lead	7.42	15.0	91	Protactinium		
83	Bismuth	(8.8)	(17)	92	Uranium	(4)	

Just as the results for the hydrogen atom can be applied to any one-electron ion, this result for the helium atom can be applied to any two-electron ion, if the proper value of Z is used. It will yield the energy required to separate the *last two* electrons; that is, it will give the sum of the second and third ionization potentials of lithium, or the sum of the third and fourth ionization potentials of beryllium, etc. One should expect the calculation to become more accurate as the value of Z increases and the perturbing potential makes a relatively smaller contribution to the total energy. For example, for beryllium ($Z = 4$) the perturbation energy is, from Eq. (20), $I = (5Z/4) \times 13.60 \text{ eV} = 68.0 \text{ eV}$. The unperturbed energy is twice the energy of triply ionized Be, or $2 \times 4^2 \times (-13.60 \text{ eV}) = -435.2 \text{ eV}$. The total energy of -367.2 eV is only about 1% different from the value found from table 1 by adding together the third and fourth ionization potentials.

The Variation Method. If we can do that well with heavier atoms by using the *hydrogen-atom* wavefunctions, it should not surprise us to learn that the ground-state energy levels can be calculated much more accurately (to within two parts in 10^5 for helium) simply by making a better approximation to the wavefunctions. The *variation method* provides a systematic way to alter the wave function in the proper direction.

The basis of the method is the fact that the allowed wavefunctions of the system must form a complete, orthogonal set of functions, so that any well-behaved function can be expanded in a series of functions belonging to this set. (Postulate 3, Section 6.3) Therefore if we use an *arbitrary* normalized function ψ' to find the expectation value of the energy, the result will be a weighted sum of all the energy eigenvalues of the system. In symbols, if

$$\psi' = \Sigma \, a_n \, \psi_n \quad \text{and} \quad H \, \psi_n = E_n \psi_n$$

where the ψ_n are the normalized eigenfunctions, the E_n are the energy eigenvalues, and H is the Hamiltonian operator for the complete system, including perturbations, then the expectation value of the energy will be

$$E' = \int \psi'^* H \, \psi' \, d\tau = \Sigma \, |a_n|^2 \, E_n$$

Since $|a_n|^2$ is always positive or zero, and $\Sigma |a_n|^2 = 1$, the weighted sum E' must always be greater than or equal to the smallest energy E_1, the actual ground-state energy of the system. When $E' = E_1$, the "arbitrary" function ψ_1 is the ground-state eigenfunction.

Therefore, if one uses a series of trial wavefunctions for ψ', obtaining various values for E', the smallest value of E' must be closest to the correct ground-state energy. A simple way to make use of this result is to write a ψ' containing one or more parameters that may be varied until a minimum in E' is found.

For example, to compute the ground-state energy of the helium atom, one could use the same product of hydrogen-atom wavefunctions that was used above, with one small change: Z in the exponents is changed to a variable parameter Z' instead of being fixed at 2. Using this trial wavefunction, one may then break up E' into the sum of three parts—the kinetic energy E_k of each electron, the potential energy V_p of the interaction between each electron and the nucleus, and the potential energy V_e of the interaction between the two electrons. That is,

$$E' = \int \psi'^* H \, \psi' \, d\tau = E_k + V_p + V_e$$

where

$$E_k = \iint \psi'^* \left[-\frac{\hbar^2}{2m} (\nabla_1^2 + \nabla_2^2) \right] \psi' \, d\tau_1 \, d\tau_2$$

$$V_p = -\iint \psi'^* \left(\frac{Ze^2}{4\pi\epsilon_0 r_1} + \frac{Ze^2}{4\pi\epsilon_1 r_2} \right) \psi' \, d\tau_1 \, d\tau_2$$

and V_e is the integral I evaluated previously (Eq. (20), but with Z replaced by Z'. (Notice that the factor Z is *not* replaced by Z' in the *potential* appearing in V_p, because Ze is the fixed charge on the nucleus.)

Each operator ∇^2 must yield the same result that it does for the hydrogen atom, except for a factor of Z'^2 resulting from the Z' factors in the exponents in the wave function. Thus E_k is the sum of two kinetic energies of $Z'^2 \times 13.60 \text{ eV}$ or

$$E_k = 2 Z'^2 \times 13.60 \text{ eV}$$

because the electron in the hydrogen atom has a kinetic energy of 13.60 eV.

Finally, the value of V_p must equal Z' times twice the potential energy of the electron in a hydrogenlike ion of nuclear charge Ze, because the factor of Z' in the exponents pulls each electron closer to the nucleus by a factor of Z': Thus

$$V_p = 2\,ZZ' \times (-27.2\text{ eV})$$

Adding together E_k, V_p, and V_e gives us

$$E' = (2\,Z'^2 - 4\,ZZ' + 5Z'/4) \times 13.60\text{ eV} \qquad (22)$$

For helium, we set the actual charge Z equal to 2, so

$$E' = \left(2\,Z'^2 - \frac{27}{4}\,Z'\right) \times 13.60\text{ eV} \qquad (23)$$

If we also were to set Z' equal to 2 in Eq. (22), we would obtain the same result that we previously found from first-order perturbation theory. But by differentiating E' with respect to Z', we can find the value of Z' that yields the *smallest* E', and thus improve on the previous result. We find that E' is a minimum when $Z' = \frac{27}{16}$, and the minimum value of E' is -77.46 eV. This is less than 2% above the experimentally measured value of -78.98. For a summary of further improvements in the trial function ψ', see L. Pauling and E. B. Wilson, Jr., ''Introduction to Quantum Mechanics,'' McGraw-Hill, New York (1935).

Higher Energy Levels.   Consider the 1s2s and 1s2p states (one electron in the 1s state and the other in the 2s or 2p state). If one neglects the electron–electron interaction, these states are separated only by the minute difference in spin–orbit energy. Inclusion of the electron–electron interaction raises the energy of both of these states, and it also greatly increases the splitting between them, because the average distance between a 1s and a 2s electron is greater than the average distance between a 1s and a 2p electron, making the average Coulomb potential energy smaller for the 1s2s state. (Refer to Chapter 7, Fig. 2; it is clear that the 2s wavefunction must extend to larger r than the 2p wavefunction, while the 1s wavefunction is confined to very small values of r. The θ and ϕ dependence of the 2p wavefunction does not affect the average distance, because the 1s wavefunction is spherically symmetric.)

There is a further splitting of the 1s2s and the 1s2p states into two levels each, because either state may be constructed with either a symmetric or an antisymmetric combination of the spatial part of the two single-particle wavefunctions. You will recall from Section 8.2 that the symmetric space combination, which requires that the total spin quantum number be $S = 0$, has a higher energy than the antisymmetric space combination requiring $S = 1$, because the two electrons are closer together, on the average, in the former case. Finally, the spin-orbit interaction splits the 1s2p, $S = 1$, level into three closely spaced levels, one for each possible orientation of **S** relative to **L**. (The 1s2s level with $S = 1$ is not split further in this manner, because **L** $= 0$ in this level.) This splitting, like the spin-orbit splitting in the hydrogen atom, is only of the order of 10^{-4} eV. Thus, with these values of n, we have a total of six distinct energy levels—two 1s2s levels and four 1s2p levels. (See Fig. 5.)

The fact that the 1s2s and 1s2p levels of the helium atom are split in this way was one of the pieces of evidence which led Pauli to formulate the exclusion principle. Although the hypothesis of electron spin had not yet appeared when Pauli enunciated his principle, it was known that a fourth quantum number, with two possible values, was needed to explain atomic spectra. Pauli reasoned that the helium ground state is unsplit because the two electrons, having identical values for the first three quantum numbers,[9] are required to have unequal values for the fourth quantum number (and there is only one set of unequal values); on the other hand, each excited state (1s2s or 1s2p) is split because the two electrons have different sets of values for the first three quantum numbers, and thus may have either equal or unequal values for the fourth quantum number.

8.4 THE PERIODIC TABLE OF THE ELEMENTS

In principle, there is no reason why we could not continue the perturbation calculation of the helium atom, calculating the wavefunctions as expansions of the hydrogen atom wavefunctions according to the technique of Section 8.1, and thus determining the energy levels to arbitrary precision. And if we can do this for a two-electron atom, we can do it for larger atoms, or even molecules, so that all of chemistry becomes an exercise in quantum mechanical perturbation theory.

Obviously this "exercise" has not been carried out, but there are theoretical chemists who do perturbation calculations to determine properties of atoms and molecules. These calculations often involve "trial wavefunctions" which are quite different from the hydrogen atom functions; but the actual wavefunctions of these more complicated systems still bear some resemblance to the hydrogen atom wavefunctions, and must be characterized by the same set of quantum numbers. Therefore, by knowing the properties of the hydrogen atom wavefunctions, and by invoking the Pauli exclusion principle, we can explain some of the features of even the heaviest atoms. For example, one can begin to explain the major features of the periodic table of the

---

[9] Before the development of the Schrödinger equation, the first three quantum numbers were the Bohr quantum number n, the Sommerfeld quantum number k, and a "magnetic" quantum number m needed to explain the Zeeman effect.

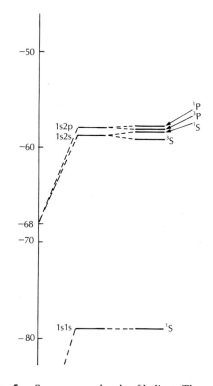

Fig. 5.   *Some energy levels of helium. The mark at −68 eV shows where the 1s2s and 1s2p levels would lie if there were no interaction between the two electrons. (The 1s1s level would be at −108.8 eV.) The left-hand levels show the energies that would result if there were no exchange energy. The right-hand levels are experimentally determined. Spectroscopic symbols on the right indicate the spin state via the superscript (3 for triplet state, 1 for singlet state), and the total orbital angular momentum by the capital letters S and P. The level marked ^3P is actually <u>three</u> levels because of the spin-orbit splitting, which is too small to see on this scale.*

elements (Fig. 6) simply by counting the number of possible hydrogen atom wavefunctions with given values of n and l, and then using the exclusion principle.

Although the periodic table was first conceived (by Mendeleev, in 1869) as a periodicity in the chemical properties of the elements with respect to their atomic masses, the periodicity is much more striking if one plots the ionization potentials as functions of Z, as in Fig. 7 (see also Table 1). We see that

s^1	s^2		$3d$	$4d$	$5d$	$6d$
1 H	4 Be					
3 Li	12 Mg					
11 Na	20 Ca					
19 K	38 Sr					
37 Rb	56 Ba					
55 Cs	88 Ra					
87 Fr						

d-block:

d^1	d^2	d^3	d^4	d^5	d^6	d^7	d^8	d^9	d^{10}
21 Sc	22 Ti	23 V	24 Cr $4s3d^5$	25 Mn	26 Fe	27 Co	28 Ni	29 Cu $4s^1 3d^{10}$	30 Zn
39 Y	40 Zr	41 Nb $5s^1 4d^4$	42 Mo $5s^1 4d^5$	43 Tc	44 Ru $5s^1 4d^7$	45 Rh $5s^1 4d^8$	46 Pd $5s^0 4d^{10}$	47 Ag $5s^1 4d^{10}$	48 Cd
57 La R.E.*	72 Hf	73 Ta	74 W	75 Re	76 Os	77 Ir	78 Pt $6s^1 5d^9$	79 Au $6s^1 5d^{10}$	80 Hg
89 Ac	90 Th H.E.†								

p-block:

	p^1	p^2	p^3	p^4	p^5	p^6
						2 He $1s^2$
$2p$	5 B	6 C	7 N	8 O	9 F	10 Ne
$3p$	13 Al	14 Si	15 P	16 S	17 Cl	18 Ar
$4p$	31 Ga	32 Ge	33 As	34 Se	35 Br	36 Kr
$5p$	49 In	50 Sn	51 Sb	52 Te	53 I	54 Xe
$6p$	81 Tl	82 Pb	83 Bi	84 Po	85 At	86 Rn

*Rare earths $4f$

f^1	f^2	f^3	f^4	f^5	f^6	f^7	f^8	f^9	f^{10}	f^{11}	f^{12}	f^{13}	f^{14}
58 Ce $5d^1$	59 Pr $5d^0$	60 Nd $5d^0$	61 Pm $5d^0$?	62 Sm $5d^0$	63 Eu $5d^0$	64 Gd $5d^1 4f^7$	65 Tb $5d^1 4f^8$	66 Dy $5d^0$?	67 Ho $5d^0$?	68 Er $5d^0$?	69 Tm $5d^0$?	70 Yb $5d^0$	71 Lu $5d^1 4f^{14}$

†Heaviest elements $5f$

91 Pa $6d^1$?	92 U $6d^1$	93 Np $6d^1$?	94 Pu $6d^1$?	95 Am $6d^1$?	96 Cm $6d^1$?	97 Bk $6d^1$?	98 Cf $6d^1$?	99 Es	100 Fm	101 Md	102 No

Examples:
22 Ti: $1s^2$ $2s^2$ $2p^6$ $3s^2$ $3p^6$ $3d^2$ $4s^2$
42 Mo: " " " " " $4p^6$ $4d^5$ $5s$
64 Gd: " " " " " " $3d^{10}$ $4d^{10}$ $4f^7$ $5s^2$ $5p^6$ $5d$ $6s^2$
74 W: " " " " " " " " $4f^{14}$ $5d^4$
94 Pu: " " " " " " " " $5d^{10}$ $5f^5$ $6p^6$ $7s^2$ $6d$

Fig. 6. The electron configurations of the elements. Deviations from a regular order of filling of the shells are indicated in the boxes of the elements concerned. [From "Principles of Modern Physics" by R. B. Leighton. Copyright 1959 McGraw-Hill, New

the first and second ionization potentials follow the same pattern, but that the pattern of the latter is displaced one unit toward larger Z. We also see that the first ionization potential tends to increase with increasing Z, except for sharp drops at $Z = 3$, 11, 19, 37, and 55, corresponding to the alkali metals lithium, sodium, potassium, cesium, and rubidium, respectively. The elements with the maximum first ionization potentials are the noble gases helium, neon, argon, krypton, and xenon.

This behavior of the ionization potentials is easy to correlate with the number of possible hydrogenlike states, as indicated in Table 2. Because there are two spin states, the number of states shown in the table is in each case equal to twice the number of possible values of m_l.

If the energy depends primarily on the principal quantum number n, as it does in hydrogen, then the ionization potential must drop at $Z = 3$ and at $Z = 11$, which are the first Z-values at which $n = 2$ and $n = 3$, respectively, become occupied; an electron in such a state, having higher energy than electrons in atoms of lower Z, must have a lower ionization potential. For the same reason, the *second* ionization potential drops at $Z = 4$ and at $Z = 12$; for example, singly ionized beryllium contains two 1s electrons and one 2s electron, so it has a smaller ionization potential than singly ionized lithium, with two 1s electrons.

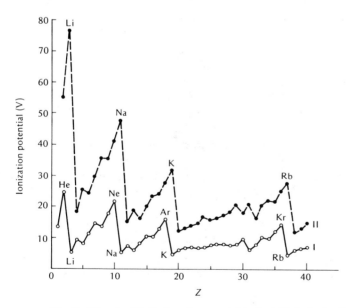

Fig. 7. First (solid line) and second (dashed line) ionization potentials of the elements as a function of Z, up to $Z = 40$.

Table 2

Electron States

n	l	m_l	Label	Number of states
1	0	0	1s	2
2	0	0	2s	2
2	1	$\pm 1, 0$	2p	6
3	0	0	3s	2
3	1	$\pm 1, 0$	3p	6
3	2	$\pm 2, \pm 1, 0$	3d	10
4	0	0	4s	2
4	1	$\pm 1, 0$	4p	6
4	2	$\pm 2, \pm 1, 0$	4d	10
4	3	$\pm 3, \pm 2, \pm 1, 0$	4f	14

As Z increases from 3 to 10, the L shell, consisting of two 2s and six 2p electrons, is filled. The ionization potential gradually increases as the increased Coulomb force pulls the shell in more tightly; it is also observed that the atoms become progressively smaller as Z increases in this region.

The noble gas neon, at $Z = 10$, has a "closed shell"; it is spherically symmetric and has a total angular momentum of zero, because the six occupied p states include all possible orientations of the spin and orbital angular momentum vectors. For this reason neon (as well as the other noble gases) has no external electric or magnetic fields, and it is very inactive chemically, although it has been possible to form chemical compounds containing neon as well as other noble gases.[10]

At $Z = 11$, a 3s electron is added to the neon shell to form sodium, whose chemical properties, being determined by the odd 3s electron, are very similar to those of lithium with an odd 2s electron. As Z increases further, the 3s and 3p states fill, so that argon, with a filled 3p shell, is another noble gas.

After argon, it might seem logical that the next element, potassium, would contain a 3d electron, because the 3d state is the next state above the 3p state in hydrogen. But spectroscopic observations[11] indicate that the potassium atom contains a 4s electron outside the argon shell, and this indication is

---

[10] See H. Selig, J. G. Malm, and H. H. Claassen, *Sci. Amer.* **210**, 66 (1964).
[11] See Section 9.1.

confirmed by the chemical behavior of potassium; its electron configuration is $1s^2 2s^2 2p^6 3s^2 3p^6 4s$ in the ground state. Thus at $Z = 19$, it is not even a good first approximation to neglect interactions between the electrons; the interactions among the 19 electrons are so large that they affect the *order* of the energy levels.

The effect of electron–electron interactions is, in general, quite complicated, but we can at least estimate the collective effect of the *complete* inner shells on the energy levels of an outer electron. In 1922, Bohr pointed out that this effect should cause the energy to depend strongly on l as well as on n because electrons of lower l travel on more eccentric orbits and hence penetrate the inner shells more than electrons with higher l. Bohr's analysis can be translated into the quantum-mechanical theory: Wavefunctions for states of lower l extend closer to the origin than those for states of higher l. As a consequence, the high-l electron is more shielded from the nucleus and is thus less tightly bound than an electron of lower l-value; that is, the high-l electron has a higher energy.[12] This results in such a large energy difference between a d state and a p state that the filling of *every* p shell creates a noble-gas atom which is very difficult to raise to an excited state; and in each case the d state has such a high energy that the next electron to be added prefers to occupy an s state, even though n is one unit larger. These facts, of course, account for the sharp drop in ionization potential at $Z = 19$, because the 4s electron is in a much higher energy level than the 3p electrons of the argon shell.

At $Z = 20$, a second 4s electron is added; beyond this point, the 3d shell begins to fill. But there is a complication, because the energy difference between the 3d and the 4s states is quite small, and the 3d state does not always lie above the 4s state in these elements. The closeness of the 3d and the 4s states is illustrated by comparison of the calcium *atom* and the scandium *ion* Sc^+, each of which contain 20 electrons; outside the argon shell, the calcium atom has two 4s electrons, but the Sc^+ ion has one 3d and one 4s electron. For another example, the titanium atom has the configuration $3d^2 4s^2$, whereas singly ionized vanadium, with the same number of electrons, has the configuration $3d^4$. The difference between the atom and the ion in these situations is basically that Z is larger for the ion, while the number of electrons is not; as Z increases, the effect of the nucleus becomes larger, relative to the electron–electron interaction, so that the wavefunctions and the energy levels become more like those of the hydrogen atom. In hydrogen, of course, the 3d level is much lower than the 4s. Thus as Z increases the 3d level drops, relative to the 4s, and the electrons tend to go into the 3d states rather than the 4s states.

When we reach zinc ($Z = 30$), the 3d states are filled. But filling the 3d states does not have the same effect as filling the 3p states; zinc behaves not

---

[12] We have already seen this effect in the energy levels 1s2s and 1s2p of helium (Fig. 5).

at all like a noble gas! The difference, of course, is in those two 4s electrons. In order to ionize zinc, you do not have to remove one of the electrons from a "closed shell" of 3d electrons; you remove a 4s electron instead, because, as mentioned above, the 3d level in the ion lies below the 4s level. (In this region of the table the 3d level is also lower in the neutral atom, as well as in the ion; the electron configuration of copper is $3d^{10}4s$, the tenth 3d state being filled instead of the second 4s state.) In zinc the energy of the 4s state is considerably higher that that of the 3d state, so that the ionization potential of zinc, 9.3 V, does not approach that of a noble gas. In general, the atoms from $Z = 21$ to $Z = 30$ are ionized by the removal of the 4s electrons rather than the 3d electrons, and their chemical properties are determined by the former rather than the latter, so all of these elements have similar properties and are listed in a separate group (the "transition elements") in the periodic table.

Beyond $Z = 30$, the 4p states fill until we reach the noble gas krypton at $Z = 36$. This point marks the end of the first "long period", containing 18 electrons. The cycle is then repeated; the 5s states fill, then the 4d states, and finally the 5p states, to complete another long period and bring us to the noble gas xenon, $Z = 54$.

The only remaining complication occurs when we begin the next period. After filling the two 6s states and one 5d state, we find that we must go back and fill a group of $n = 4$ states—the fourteen 4f states ($l = 3$), which were previously empty because the l dependence of the energy places them even above the 6s states in energy. The fourteen elements thus formed comprise the "lanthanides," or "rare earths," and they all are extremely similar in chemical properties. All are trivalent, because all contain the odd 5d electron and two 6s electrons which are relatively easy to remove. (As in the case of the 3d–4s states in the transition elements, the 4f state lies below the 5d or 6s state in an *ion*, even though it is higher in the *atom*, so that the ground state of a triply ionized lanthanide contains no 5d or 6s electrons.) Notice that the same eccentricity of the "orbit" of a 5d or 6s electron, which causes it to penetrate the inner shell and have lower energy than the 4f electron, also causes it to extend further from the nucleus and thus be perturbed more by neighboring atoms, so that these electrons are the ones which determine the chemical properties of the element.

After the 4f states are filled, the remainder of the 5d shell is filled, followed by the 6p shell, until another rare gas, radon, is reached at $Z = 86$. A seventh period then begins in the same way as the sixth; the two 7s states are filled, followed by a 6d state, and then the 5f states begin to fill, producing the actinide elements which are analogous to the lanthanides. The first actinide, actinium, has $Z = 89$; all are radioactive, and those past uranium ($Z = 92$) have not been found in nature for reasons to be discussed in Chapter 13. However, all have been produced artificially, including lawrencium ($Z = 103$), the last of the actinides. As Z increases, these elements become progressively more difficult to produce, because each is built up from previous elements in the series, and all have very unstable nuclei. Thus the heaviest elements of the seventh period have not yet been produced.

PROBLEMS

1. Use first-order perturbation theory to find the lowest energy level for a particle in a "V-bottom" well given by

$$V(x) = \frac{2|x|\delta}{a} \qquad (|x| \leqslant a/2)$$

$$= \infty \qquad (|x| \geqslant a/2)$$

Also find the coefficients a_{13} and a_{15} of the first-order expansion of the eigenfunction $\psi'_1(x)$ for the lowest level:

$$\psi'_1(x) = \psi_1(x) + a_{13}\psi_3(x) + a_{15}\psi_5(x) + \cdots$$

where $\psi_1(x)$, $\psi_3(x)$, and $\psi_5(x)$ are wavefunctions for the square well. (Why are $\psi_2(x)$ and $\psi_4(x)$ absent from the expansion?)

2. A one-dimensional harmonic oscillator of charge e is perturbed by the application of an electric field $\mathscr{E}$ in the positive x direction, so that the potential energy becomes

$$V(x) = \tfrac{1}{2} m\omega^2 x^2 - e\mathscr{E}x$$

(a) Show that, according to first-order perturbation theory, the shift in the energy levels, as a result of this perturbation, is *zero*. (See Section 5.7 for harmonic-oscillator wavefunctions.)

(b) Show that second-order perturbation theory gives a shift of $-e^2\mathscr{E}^2/2m\omega^2$ in the energy of the lowest level.

(c) Show that the Schrödinger equation may be solved exactly for this system, and find the lowest energy level. (*Hint*: The potential may be written in the form $V(x) = \tfrac{1}{2}m\omega^2(x - x_0)^2 + V_0$, where V_0 and x_0 are constants. This leads to an equation of the same form as that for the simple harmonic oscillator.)

3. Do a numerical integration to find the lowest-energy wavefunction for the potential of Problem 1. Compare the energy with the first-order perturbation result. (*Hint*: The method is the same as that of Problem 11, Chapter 5, but the boundary conditions are different. Start with $\psi = 1.0$ and $d\psi/dx = 0$ at $x = 0$, and find the lowest energy level such that $\psi = 0$ at $x = a/2$. Choose a value of $\delta \ll E_1$.)

4. Verify, by direct application of the six spin operators (three for each elec-

tron) to the two-particle spin functions of expression (17), that those functions have the eigenvalues given in the text—namely, that the square of the total spin is $S(S + 1)\, \hbar^2 = 2\hbar^2$ for the triplet states and zero for the singlet state. (*Hint:* The operator $\mathbf{S}^2$ may be written as

$$(\mathbf{S}_1 + \mathbf{S}_2) \cdot (\mathbf{S}_1 + \mathbf{S}_2) = \mathbf{S}_1{}^2 + \mathbf{S}_2{}^2 + 2S_{1x}S_{2x} + 2S_{1y}S_{2y} + 2S_{1z}S_{2z}$$

where $\mathbf{S}_1$ operates only on the spin function of particle 1, etc.)

5. Show that quantum mechanics predicts a violation of the Bell inequality if axis A is the $+x$ axis, axis B is the $+y$ axis, and axis C has its positive direction at a 45° angle to axis A and to axis B. (Use the spin functions from part (b), Problem 9, Chapter 7, to find the quantum mechanical probabilities for measurements along axis C, for particles in eigenstates along B or A.)

6. Compute the percentage errors in the energies of the ground states of Li^+, Be^{2+}, B^{3+}, and C^{4+}, as given by Eq. (21), by comparing with the measured values given in Table 1. Does the error become smaller as Z increases, as stated in the text?

7. Show that the minimum value for E' in Eq. (22) is found when $Z' = Z - \frac{5}{16}$. Use this result to compute the ground-state energy for Li^+, Be^{2+}, B^{3+}, and C^{4+}. Compare the result with the answer to Problem 6 and with Table 1.

8. Determine the *seventh* ionization potential of nitrogen from the theory of the *one*-electron ion. Then use this result and the value of Z' from Problem 7 to compute the *sixth* ionization potential [using Eq. (22)] and compare with Table 1.

9. The H^- ion consists of one proton bound to two electrons. Experiment shows that 0.80 eV is required to separate one of these electrons. What, then, is the ground-state energy of this ion? Compare this energy with the energy predicted by Eq. (22). (See Problem 7 for the value of Z'.) Does Eq. (22) predict that an electron will be bound to a neutral H atom to form the H^- ion?

10. From *American Institute of Physics Handbook*, Third Edition, p. 7–14, (McGraw-Hill, N.Y., 1971), find the energies and wavelengths of several allowed transitions in helium, and show them on an energy-level diagram similar to Fig. 5.

11. As elements of higher and higher Z are produced, at what value of $Z > 100$ might you expect to find another noble gas? Discuss any assumptions you must make in order to arrive at an answer.

Atomic and Molecular Spectra

Much of the experimental information referred to in previous chapters has been obtained by spectroscopic observations, which were very thoroughly made long before the
underlying theory was known. The unraveling of the complex spectrum of a multielectron atom or molecule is a
formidable task, even though the theory is now presumably
completely known. Our aim here, therefore, is simply to
illustrate general principles by applying them to some typical cases, showing how the features of the spectra are related to electronic transitions or to changes in the state of
motion of atoms within a molecule.

313

9.1 ATOMIC SPECTROSCOPY

Alkali-Metal Spectra. Alkali-metal spectra, the simplest of all, resemble the hydrogen atom spectrum because there is only one electron outside a noble-gas core, so that one normally observes only the transitions of this single electron, with the core remaining undisturbed. But there are important differences between alkali-metal spectra and the spectrum of the hydrogen atom. In hydrogen, the dependence of the energy on n alone is broken only by the minute magnetic (spin–orbit) and relativistic effects discussed in Section 7.3; but in the alkali metals, the energy depends strongly on l because of the Coulomb interaction between the outer electron and the inner electron shells (discussed previously in Section 8.4), which tends to lower the energy of the low-l electrons which penetrate these shells.

You might think that this great increase in the number of distinct energy levels would make the observed spectrum enormously complicated, but many of the transitions which are energetically possible are not actually observed; a "selection rule" permits only transitions in which the quantum number l changes by one unit:[1]

$$\Delta l = \pm 1$$

The result of this rule is to allow only the transitions depicted in Fig. 1 for a typical alkali metal, sodium.

The overall spectrum of sodium, like that of hydrogen, thus consists of lines belonging to a number of different series, although there is no neat formula which exactly expresses the wavelengths of all the lines in each series. Each series consists of the transitions *to* a particular level; long before the process was understood, four series were identified and given names:

the sharp series, produced by transitions from s states to the 3p state,

the principal series, produced by transitions from p states to the 3s state,

the diffuse series, produced by transitions from d states to the 3p state, and

the fundamental series, produced by transitions from f states to the 3d state.

It was the existence of these names which led to the letters s, p, d, f for the various angular momentum states.[2] You will notice that all of these transitions obey the selection rule $\Delta l = \pm 1$.

---

[1] In certain cases a forbidden transition may occur. See Chapter 10 for a theoretical discussion of selection rules.

[2] The names sharp, diffuse, and principal arose from the appearance of the lines. For example, in lithium the lines of the principal series are very intense. The name "fundamental" arose from the agreement of the wavelengths in this series with an empirical equation which has a superficial resemblance to the Rydberg formula for hydrogen.

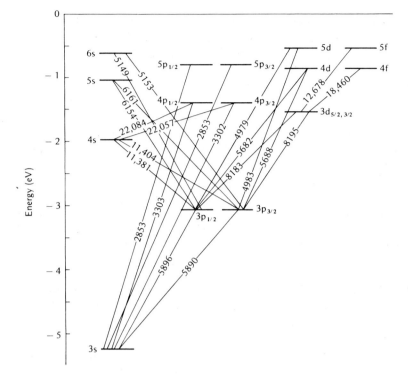

Fig. 1 *Energy levels of the sodium atom, labeled according to the hydrogenlike state of the odd electron, and showing the wavelengths of the major spectral lines (allowed transitions). Notice that the 4s level lies considerably below the 3d level. The spin-orbit splitting ($3p_{1/2} - 3p_{3/2}$) is too small to see as a displacement on the diagram, but it can be deduced from the wavelengths indicated.*

Just as in the hydrogen and helium spectra, each line has a ''fine structure'' which results from the splitting of each level, except the s levels, by the spin-orbit interaction. Each line of the *sharp* and *principal* series is a *doublet,* because the p states are split into two levels, $p_{1/2}$ and $p_{3/2}$, while the s states are unsplit. For example, the transitions $3p_{1/2}-3s$ and $3p_{3/2}-3s$ lead to radiation of 5896 and 5890 Å, respectively, the 6-Å difference occurring because the two upper levels differ in energy by about 0.002 eV. (These two lines, the most striking feature of the sodium spectrum, are called the D lines.) For the transitions of the *diffuse* and *fundamental* series, *both* the initial level and the final level are split into two, so that there are *four* conceivable transitions. However, *one* of these transitions is forbidden by another selection rule: $\Delta j = \pm 1$ or 0. For example, in the diffuse series, the transitions $d_{5/2}$–$p_{3/2}$, $d_{3/2}$–$p_{3/2}$, and $d_{3/2}$–$p_{1/2}$ occur, but the transition $d_{5/2}$–$p_{1/2}$ is forbidden and is not observed. Thus the lines of this series are *triplets*, as are the lines of the fundamental series.

Now you can see how it would be possible to determine the ground-state configuration of an element such as potassium, which we discussed in Section 8.4. For any given value of l, the energy levels approach zero as n increases, so that the wavelengths in each series must converge to a limit, or minimum wavelength λ_m, which is determined by the energy E_f of the final state involved in the transitions of that series: $\lambda_m = hc/E_f$. Study of the potassium spectrum shows that the series with the shortest wavelength limit is the "principal" series, and E_f in this case turns out to be the ionization potential of potassium. Therefore the s state, which is the final state for all of the transitions in the principal series, must be the state of lowest energy in the atom.[3]

Alkaline Earths. The elements in the second column of the periodic table, called alkaline earths, also have spectra which are similar to one another, but these spectra are much more complicated than those of the alkali metals. The greater complexity follows from the fact that these atoms each have two "active" electrons instead of one: They have two electrons which are outside a closed shell and can be easily excited. Thus there are many more possible energy levels, because there are levels at which both electrons are excited, as well as levels of excitation of only one electron. Figure 2 shows a typical energy level diagram for an element of this group, calcium.

Each level in Fig. 2 is labeled as we labeled the states of the helium atom; that is, according to the single-particle hydrogenlike state occupied by each electron. But as we saw in discussing the helium atom in Section 8.3, the energy is not uniquely determined by these spatial single-particle quantum numbers. For example, in analogy to the 1s2s and 1s2p levels of He, the 4s5s level splits into two levels, and the 4s4p level splits into four levels, arranged according to the value of the resultant spin S and the value of $S \cdot L$.

---

[3] It is not difficult to distinguish the principal series in the potassium atom, in order to establish these facts. We have seen that the principal series is made up of doublets, and the only other doublet series is the sharp series. But in the sharp series the initial state, an s state, is unsplit, and all of the transitions go to the *same* final state, so that the doublets of this series have a *constant* separation in energy, determined by the splitting of the final state. On the other hand, the splitting of the doublets of the principal series is determined by the splitting of the *initial* state, a p state, and this splitting *decreases in magnitude with increasing* n (as the electron "orbit" moves farther away from the nucleus) so that the doublets of the principal series have a *steadily decreasing* separation in energy as their wavelength decreases.

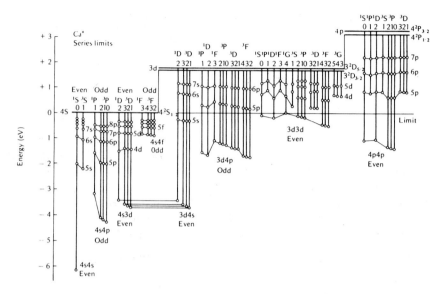

Fig. 2 *Complete energy level diagram of calcium. States on a vertical line all have the same value of j, shown at the top of the line. [From "Introduction to Atomic Spectra" by H. E. White. Copyright McGraw-Hill, New York, 1934. Used by permission of McGraw-Hill Book Company.]*

In addition to the "spin–spin" and "spin–orbit" interactions which cause these splittings, there is an "orbit–orbit" effect which splits the levels according to the relative orientation of the *orbital* angular momentum of the two electrons. This effect has not been previously discussed, because it is not present when one of the single-particle states is an s state, but it must be considered for states in which both electrons possess orbital angular momentum.

In order to gain a coherent picture of all of these effects, we must consider the general problem of how to combine the angular momenta of two electrons to form a resultant angular momentum vector. The resultant angular momentum may be written as the sum of the individual spin and orbital angular momentum vectors:

$$\mathbf{J} = \mathbf{L}_1 + \mathbf{S}_1 + \mathbf{L}_2 + \mathbf{S}_2$$

We saw in Section 7.3 how to combine *two* angular momentum vectors to form a resultant, but what do we do with *four*?

We can always add pairs of vectors by the rules of Section 7.3, and eventually combine all the vectors into one resultant. There are two logical choices for the manner of choosing the pairs:

1. We could write a resultant *orbital* angular momentum:

$$\mathbf{L} = \mathbf{L}_1 + \mathbf{L}_2$$

and a resultant *spin* angular momentum:

$$\mathbf{S} = \mathbf{S}_1 + \mathbf{S}_2$$

and then combine these into a resultant total angular momentum:

$$\mathbf{J} = \mathbf{L} + \mathbf{S}$$

This is commonly done for the light elements, in which it is found that the energy of interaction between $\mathbf{S}_1$ and $\mathbf{S}_2$, the exchange energy, is greater than the spin–orbit interaction, so that even after the spin–orbit splitting, states with a given value of the quantum number S are widely separated from states with a different value of S. This method of combining angular momenta is called the LS coupling scheme.

2. We could write a resultant angular momentum for particle 1:

$$\mathbf{J}_1 = \mathbf{L}_1 + \mathbf{S}_1$$

and a resultant angular momentum for particle 2:

$$\mathbf{J}_2 = \mathbf{L}_2 + \mathbf{S}_2$$

and then combine these into a resultant total angular momentum:

$$\mathbf{J} = \mathbf{J}_1 + \mathbf{J}_2$$

This scheme, called *jj* coupling, is useful for the heavier elements, in which it happens that the spin–orbit interaction is much stronger than the interactions between the individual spins or the interaction between the individual orbital angular momenta.

The LS, or Russell–Saunders coupling scheme, as it is often called, is somewhat simpler in concept; the "ideal" LS coupling case is illustrated in Fig. 3. Let us examine each stage of the splitting in turn, in order to understand what distinguishes LS coupling from other conceivable schemes:

1. The largest splitting is between the $S = 0$ states and the $S = 1$ states, which results from the exchange energy which we have previously discussed (Section 8.2).

2. The next largest effect splits states of a given S value according to the value of L. (The *quantum number L* is defined by the equation $|\mathbf{L}|^2 = L(L + 1)\hbar^2$.) This is another Coulomb effect, because the average distance between the two electrons depends on the relative orientation of their orbits, that is, on $\mathbf{L}_1 \cdot \mathbf{L}_2$. (Of course, $|\mathbf{L}|^2 = |\mathbf{L}_1|^2 + |\mathbf{L}_2|^2 + 2\mathbf{L}_1 \cdot \mathbf{L}_2$, so that the L quantum number, for a given pair of values of the individual quantum numbers l_1 and l_2, does depend on $\mathbf{L}_1 \cdot \mathbf{L}_2$.)

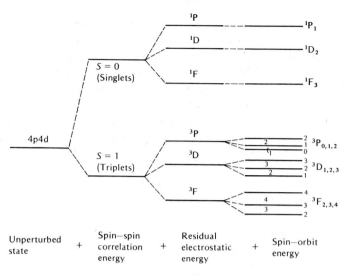

Fig. 3. Schematic diagram illustrating the fine-structure split-
ting of a level corresponding to a 4p and a 4d valence electron. The
Landé interval rule is illustrated in the spacings of the triplet levels.
[From "Principles of Modern Physics" by R. B. Leighton. Copy-
right McGraw-Hill, New York, 1959. Use by permission of
McGraw-Hill Book Company.]

3. Finally, there is a small spin–orbit splitting, magnetic in origin, between
    states of different J value, because the energy depends on the orientation
    of the *resultant* $\mathbf{L}$ relative to the *resultant* $\mathbf{S}$, that is, on $\mathbf{L} \cdot \mathbf{S}$. (Again,
    you will recall that $|\mathbf{J}|^2 = |\mathbf{L}|^2 + |\mathbf{S}|^2 + 2\mathbf{L} \cdot \mathbf{S}$.)

Thus a given energy level is distinguished by *seven* quantum numbers:
n_1, n_2, l_1, l_2, S, L, and J. The values of the last three quantum numbers are
indicated in the following (rather indirect) way, by a symbol of the form
$^{2S+1}L_J$. In words, the initial superscript gives the value of $2S + 1$, which is
the multiplicity of the state; the value of L is indicated by a *capital letter*,
according to the spectroscopic "code" used previously for the values of l; and
the value of J is written directly as a subscript.[4] Thus, for example,

$^3P_2$ denotes a state with $S = 1$, $L = 1$, and $J = 2$;

$^1F_3$ denotes a state with $S = 0$, $L = 3$, and $J = 3$; and

$^3F_1$ denotes an *impossibility*: $S = 1$, $L = 3$, and $J = 1$.

---

[4] The terminology is unfortunate, but we are stuck with it. An "S state" is one with $L = 0$; the S here
has nothing to do with the S quantum number.

(Remember the rule for the possible values of the resultant of two angular momenta.) The complete designation of a state might then, for example, be 3d4p $^3P_1$, where the symbols 3d and 4p define the quantum numbers n and l for the single-particle state of each electron, and the symbol $^3P_1$ indicates the way in which these single-particle states are combined to form resultant angular momentum vectors **S**, **L**, and **J**.

A noteworthy feature of Fig. 3 illustrates a rule, the *Landé interval rule*, which has been quite helpful in analyzing complicated spectra. The rule applies to the energy of levels with the same values of all quantum numbers except J, and it states:

The energy difference between levels is proportional to the larger of the two J values involved.

Thus,

$$\frac{\text{Energy difference between } ^3F_4 \text{ and } ^3F_3}{\text{Energy difference between } ^3F_3 \text{ and } ^3F_2} = \frac{4}{3}$$

You may easily prove this rule for ideal LS coupling. The energy is proportional to $\mathbf{L} \cdot \mathbf{S} = (|\mathbf{J}|^2 - |\mathbf{L}|^2 - |\mathbf{S}|^2)/2$, and you can show that the difference between the value of $\mathbf{L} \cdot \mathbf{S}$ for the quantum number J and the value of $\mathbf{L} \cdot \mathbf{S}$ for the quantum number $J + 1$ is proportional to $J + 1$.

Unfortunately, the *ideal LS* coupling scheme is not applicable to a very large number of cases. For example, Fig. 2 shows that in the calcium spectrum there is not a clear separation between the $S = 0$ states and the $S = 1$ states for the higher levels such as 3d4p; some of the $S = 0$ states actually lie below some of the $S = 1$ states, because the exchange energy is not the dominant contribution to the splitting. But the LS scheme is still quite important, and even where it breaks down, the levels are often labeled (sometimes erroneously) as if the ideal LS coupling scheme were valid.

You may ask what difference it makes how you label the states, and the answer lies in the selection rules governing the transitions. In the ideal case, there is a rule that

$$\Delta L = \pm 1 \text{ or } 0, \quad \text{and} \quad \Delta S = 0$$

in a radiative transition. In the departure from the ideal, these rules become less binding, until finally in the heavier elements the extreme of jj coupling is reached, and it is impossible even to identify a state by its L or S value, the well-defined quantum numbers being j_1 and j_2, which refer to the resultant angular momentum of each electron. In the intermediate cases, although one can label states by the value of L or the value of S, the spin–orbit interaction is sufficiently large that states with different values of L and S become "mixed," and the selection rule given above cannot hold rigorously. (Remember the discussion of degenerate perturbation theory, in which it was pointed out that the perturbed states may be *linear combinations* of the unperturbed states corresponding to a given set of quantum numbers.) The transition from *LS* to *jj* coupling, as one goes to heavier elements, is illustrated in Fig. 4.

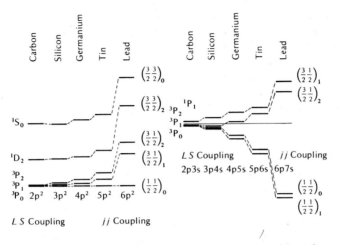

Fig. 4. *LS to jj coupling, as shown by the normal and first excited states in the carbon group of elements. [From "Introduction to Atomic Spectra" by H. E. White. Copyright McGraw-Hill, New York, 1934. Used by permission of the McGraw-Hill Book Company.]*

Positive Energy Levels. Figure 2 illustrates a feature of the two-electron system which does not appear in a one-electron system. Some of the energy levels are shown at *positive* values of energy—that is, at an energy sufficient to ionize the atom. Yet these are levels of the *neutral* atom, with both electrons still present. The two electrons *share* the excitation energy, so that neither one individually has enough energy to escape from the atom.[5] An atom in such an excited state may radiate a photon in the normal way as one of the electrons "jumps" to a lower level, or it may undergo a process known as "auto-ionization,"[6] in which one electron, in jumping to a lower level, transfers enough energy to the other electron to enable the second electron to escape from the atom. Any excess energy appears as kinetic energy of the ejected electron. *No photon is created* in this process; the energy is transferred directly from one electron to the other.

---

[5] The spectral lines emitted in deexcitation of these levels had been known for some time, and were called "anomalous," when Bohr and Wentzel made the suggestion that they arose from positive energy levels [*Phys. Z.* **24**, 106 (1923)].

[6] Also known as the Auger effect. A similar effect, known as internal conversion, can occur in the deexcitation of nuclear states (see Chapter 14).

Notice that the states 3d4s, 3d5s, 3d6s, . . . form a series whose limit is a calcium *ion* with the odd electron in the 3d state. Similarly, the series 4p4p, 4p5p, 4p6p, . . . form a series whose limit is a calcium ion in a 4p state. Thus by studying the spectrum of *neutral* calcium and finding the limits of these series, one can deduce the energy levels of the calcium *ion*. This has been done, and the result is quite satisfactory, so that one can have confidence that the identification of the various levels is correct. (The series limit for the 3d4p, 3d5p, . . . sequence was calculated by Russell and Saunders, for the ^{3}P states, to be at an energy of $+1.74$ eV, where zero energy, of course, is the ground state of the calcium ion. The first excited state of the ion is observed to be at $+1.70$ eV. The agreement is within the error of the calculation.)

Effect of the Exclusion Principle in Two-Electron Systems.   In the example of Fig. 3, each of the quantum numbers L, S, and J takes on all values within its allowed range: S is 0 or 1, L ranges from 1 to 3, and for each pair of values of L and S, J ranges from $L - S$ to $L + S$. But in some cases, the two electrons have equal values of n and equal values of l. In such cases, the exclusion principle acts to limit the number of allowed combinations of the two states to form sets of the L, S, and J quantum numbers. For example, in Fig. 2 the ^{3}S state is missing from the 4s4s level, although it does appear in the 4s5s level.

It is easy to see why the 4s4s ^{3}S state would be forbidden. In the 4s4s level the n, l, and m_l quantum numbers have the same values for both electrons, so the ^{3}S combination, for which the m_s quantum numbers would also be equal, is a clear violation of the exclusion principle. Of course, the 4s5s ^{3}S state is permitted, because the n quantum numbers are different for the two electrons.

Figure 2 also shows that, of the ten possibilities which appear for the 4p5p states, only five are permitted for the 4p4p states; the combinations 4p4p^{1}P, 4p4p ^{3}S, and 4p4p ^{3}D are missing. The working of the exclusion principle in this case becomes a bit complicated, but it may be summarized as follows: When $l_1 = l_2$, the combinations for which the quantum number L is *even* must be *symmetric* in the interchange of two electrons, whereas the combinations for which the quantum number L is *odd* must be *antisymmetric* in the interchange of two electrons.[7] In the former case, the wavefunction is symmetric except for its spin part, so the spin part must be antisymmetric; in the latter case, the spin part must be symmetric. Thus $S = 0$ when $L = 0, 2, 4, \ldots$, and $S = 1$ when $L = 1, 3, 5, \ldots$; the ^{1}P, ^{3}S, and ^{3}D states are not

[7] We state this without proof. The result is analogous to the result for the spin functions, for which the combination with $S = 1$ is symmetric, and the combination with $S = 0$ is antisymmetric. Notice that the largest possible S value comes from a symmetric combination, and the largest possible L value also comes from a symmetric combination; the latter is always an even number, when $l_1 = l_2$.

allowed. You may verify from Fig. 2 that the "missing" 3d3d states also follow this rule; the states 3d3d ^{1}P, 3d3d ^{1}F, 3d3d ^{3}S, 3d3d ^{3}D, and 3d3d ^{3}G are not allowed.

The connection between the symmetry of the wavefunction and the value of L, which we stated without proof, is not immediately obvious, to say the least. Therefore it may be instructive to examine in more detail the effect of the exclusion principle on the 4p4p states. First, we can easily see that the ^{3}D$_3$ state must be excluded, because in this state m_s must equal $+\frac{1}{2}$ and m_l must equal $+1$ for both electrons, in order that m_J equal 3; in that case each of the two electrons would have the same set of the four quantum numbers, in violation of the exclusion principle.[8] Furthermore, if ^{3}D$_3$ is excluded, then ^{3}D$_2$ and ^{3}D$_1$ must also be excluded; if the pair of values $L = 2$, $S = 1$ is allowed at all, it must be allowed to combine in all possible ways, to yield $J = 3$ as well as $J = 2$ or $J = 1$, so that the exclusion of $J = 3$ requires the exclusion of $J = 2$ or 1. In other words, the exclusion principle restricts the possible choices of the *magnitudes* of the L and S vectors, but it cannot restrict the relative *orientation* of the vectors, so if a certain L and S pair is not permitted in one relative orientation—one J value—then it is not permitted at all.

To see why the ^{1}P and ^{3}S states are excluded, it is best to work out which states are *permitted*, by considering all allowed pairs of values of m_l for the two electrons. When the electrons are in p states, m_l can be $+1, 0$, or -1 for each electron, and there are four spin states of the two electrons, so there are $3 \times 3 \times 4 = 36$ possible combinations of L and S. But if the values of n are equal, as in the 4p4p states, the wavefunction is symmetric in the r and θ coordinates, and it must therefore be antisymmetric in the ϕ and spin coordinates. Therefore the states in which the m_l values of the two electrons are equal must combine only with the antisymmetric spin state $S = 0$. With the symmetric $S = 1$ state, the m_l values must be unequal, *and they must combine in an antisymmetric way*. Thus of the 36 possible combinations, only the following pairs of m_l values remain with the spin states indicated:

For $S = 1$		For $S = 0$	
m_l values in antisymmetric combination	$+1, 0$ $+1, -1$ $-1, 0$	m_l values in symmetric combination	$+1, +1$ $0, 0$ $-1, -1$ $+1, 0$ $+1, -1$ $-1, 0$

[8] Of course, with $J = 3$ there must be a state with $m_J = 3$, that is, with $J_z = 3\hbar$.

This makes a total of 15 allowed combinations (three $S = 1$ spin states with three m_l combinations yield nine states, plus the six m_l combinations with $S = 0$.) Let us consider the possible values of L for each group of states.

When $S = 0$, the quantum number m_L (the sum of the two m_l values) takes on the values $+2$, 0, -2, $+1$, 0, and -1 in that order, as one reads down the list. Associated with the group of five values $+2$, $+1$, 0, -1, -2, there must be a value of L equal to 2. There is also a second state with $m_L = 0$, and this state standing alone can only have $L = 0$. Thus the states $^1D_2$ and $^1S_0$ are permitted.

When $S = 1$, there are only three possible m_l values: $+1$, 0, or -1; these must be associated with $L = 1$. When $S = 1$ and $L = 1$, the possible values of J are 0, 1, or 2; these, with their various m_J values, account for the nine states with $S = 1$:

$^3P_2$, a set of five states, with $m_J = \pm 2$, ± 1, or 0;

$^3P_1$, a set of three states, with $m_J = \pm 1$ or 0; and

$^3P_0$, a single state, with $m_J = 0$.

Thus the levels labeled $^1D_2$, $^1S_0$, $^3P_2$, $^3P_1$, and $^3P_0$ are observed, and the others are forbidden by the exclusion principle.

Other Atomic Systems. The same general rules described here for the two-electron system may also be applied to systems containing three or more "active" electrons. There are, of course, more possible *states* as the number of electrons increases, but no additional *principles* are involved. For example, the three-electron system may be analyzed by considering combinations of two-electron states and one-electron states. The four spin states of two electrons ($S = 1$ or 0) may be combined with the two spin states of one electron in eight independent ways, to form a resultant spin whose quantum number is $S = \frac{1}{2}$ or $S = \frac{3}{2}$. Of these eight states, four have $S = \frac{3}{2}$, forming a "quartet," all of which are symmetric in the interchange of *any* two electrons.[9] The other four states, with $S = \frac{1}{2}$, cannot be antisymmetric; we know from the exlusion principle that there is no antisymmetric way to put three identical particles into two states. The $S = \frac{1}{2}$ states can be written as two doublets, each of which is antisymmetric in the interchange of two *specific* electrons. The difference between the $S = \frac{3}{2}$ and the $S = \frac{1}{2}$ states shows up in atomic spectra, where evidence of "quartets" and "doublets" is seen in the spectral series for three-electron systems.

The resultant spin vector **S** combines with the resultant orbital angular momentum vector **L** to form **J**, with the quantum numbers L, S, and J obeying the rule $|L - S| < J < L + S$ as before. The number of possible J-values is

[9] The isotropy of space requires that if one state with $S = \frac{3}{2}$ is symmetric, all four states must be symmetric, because the symmetry of a state must be independent of its orientation in space—that is, of its value of m_z.

equal to $2S + 1$ or to $2L + 1$, whichever is smaller. Thus if $S = \frac{3}{2}$ and $L = 3$, one has the quartet $^4F_{9/2}$, $^4F_{7/2}$, $^4F_{5/2}$, and $^4F_{3/2}$.

Figure 5 shows a typical group of spectral lines arising from transitions ^4F–^4F and ^4G–^4F between the $3d^24p$ and $3d^24s$ states in neutral scandium. The figure illustrates several points. First, a selection rule permits only transitions in which

$$\Delta J = \pm 1 \quad \text{or} \quad 0$$

Thus there are only ten allowed lines in the ^4F–^4F group, and only nine lines in the ^4G–^4F group. Second, the spin–orbit interaction splits the states so that states with higher J-value have higher energy, just as in the two-electron case. Third, among the initial states the ^4F states have higher energy than the ^4G states; this splitting, which results from the same orbit–orbit correlation discussed for two-electron systems, is much greater than the spin–orbit splitting. (This is clear from the fact that all of the ^4F–^4F lines are much shorter in wavelength than the ^4G–^4F lines.) Finally, the entire set of lines covers a very narrow range of wavelengths: 5670 to 5740, about a 1 percent variation, for ^4F–^4F, and 5080 to 5090, about a 0.2 percent variation, for ^4G–^4F, so that we are seeing very small effects, and yet the lines are resolved without any difficulty.

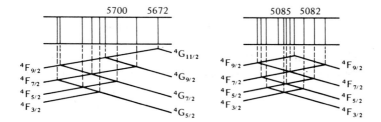

Fig. 5. Spectral lines resulting from the transitions $3d^24p$ ^4G–$3d^24s$ ^4F and $3d^24p$ ^4F–$3d^24s$ ^4F in the spectrum of the scandium atom. Transitions are associated with the corresponding lines by means of the intersecting lines below the spectrum; for example, the transition $^4G_{11/2}$–$^4F_{9/2}$ produces the line at 5672 Å.

As the number of active electrons increases, the complexity of the spectra increases to some extent, but there is a limit. Because of the exclusion principle, the number of ways of arranging the electrons in any given shell increases only until the shell is half filled, and then it decreases again. For example, there are only six single-electron states in the 4p shell, and as we have seen, we can fill two of these states in 15 different ways.[10] There are also exactly 15 ways to fill *four* of these states, because the two *unoccupied* states combine in exactly the same way as two electrons in occupied states. The reason for this is quite simple: If we were to fill the two unoccupied states, the result would be a closed shell, with zero resultant angular momentum. Therefore the angular momentum of the two added electrons must *always* be equal in magnitude to that of the original four electrons, and for each four-electron state there is a corresponding two-electron state. Thus there must be just as many four-electron states as there are two-electron states, namely 15. The argument can, of course, be generalized to any shell.

9.2   THE ZEEMAN EFFECT REVISITED

In the absence of a magnetic field, the energy levels referred to in the previous section are still degenerate, because even after all the spin–spin, orbit–orbit, and spin–orbit splittings, the energy still depends only on the *magnitude* of **J**, and not on its *orientation*. But when an external magnetic field is applied, the degeneracy is broken by the interaction energy $-\mathbf{\mu} \cdot \mathbf{B}$ between the magnetic field and the resultant magnetic moment of the atom. This interaction energy has as many possible values as there are possible components of the magnetic moment $\mathbf{\mu}$, and thus it has a different value for each value of J_z. This means that a level whose angular momentum quantum number is J should split into $2J + 1$ levels in the presence of a magnetic field.

We have already seen in Section 1.2 that a magnetic splitting of atomic spectral lines was quite large enough to be observed as early as 1896. As we saw, the effect has a classical explanation in terms of the change in frequency of an orbiting electron when the external magnetic field changes the centripetal force on the electron. This classical explanation accounts for the splitting of a spectral line into *three* lines in the presence of a magnetic field; it was in agreement with Zeeman's original observations, and it yielded a moderately good value for the charge-to-mass ratio of the electron. However, it was observed very early that there were some lines which split into more than three components in an external magnetic field: these lines were said to exhibit the "anomalous" Zeeman effect, because the splitting was inexplicable by classical theory. The word anomalous has turned out to be somewhat of a misnomer, because the anomalous effect is not at all uncommon, and with the correct quantum theory it can be explained just as easily as the "normal" Zeeman effect.

---

[10] We enumerated these 15 states on page 323, but we could arrive at the number of states by combinatorial analysis also. The "first" electron can occupy any of six different states, and the "second" can occupy any of the five remaining states; but since the order of the two electrons is immaterial, we have $6 \times 5/2$ ways in which the two electrons can go into the six states.

The quantum mechanical explanation of the Zeeman effect is based upon transitions between split energy levels, as illustrated in Fig. 6. In the case shown, with four initial states and two final states, there are $4 \times 2 = 8$ conceivable transitions, so that one might expect eight spectral lines instead of three as in the normal Zeeman pattern. However, there is a selection rule

$$\Delta m_J = \pm 1 \text{ or } 0$$

but

$$m_J = 0 \quad \text{to } m_J = 0$$

forbidden if

$$\Delta J = 0$$

which reduces the number of possible *transitions* to six in this case. The number of actual *spectral lines* observed could be fewer than six; if, for example, the energy splitting between the successive upper ($^2P_{3/2}$) levels is equal to the splitting between the two $^2S_{1/2}$ levels, then different transitions yield the same photon energy, and one observes *three spectral lines*. (Transition 1 involves the same energy difference as transition 4, etc.)

Thus whether one observes the "anomalous" or "normal" Zeeman effect

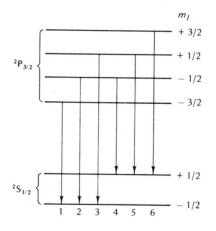

Fig. 6. Splitting of atomic states in an external magnetic field. Value of m_J is shown beside each energy level. The selection rule permits six possible transitions between upper and lower states, as shown; under certain conditions, different transitions may involve the same energy difference, so that as few as three distinct spectral lines may result. (See text).

in any given case usually depends on the splitting of the initial state relative to the splitting of the final state; when the splittings are equal, the "normal" effect results. Our next task, therefore, is to compute the magnitude of the splitting for any given state.

To do this, we need to evaluate $\boldsymbol{\mu} \cdot \mathbf{B}$ for each of the atomic states permitted in the presence of $\mathbf{B}$, the external magnetic field. The moment $\boldsymbol{\mu}$ is the vector sum of the moments

$$\boldsymbol{\mu}_{\text{spin}} = \frac{-e\mathbf{S}}{m}$$

and

$$\boldsymbol{\mu}_{\text{orbital}} = \frac{-e\mathbf{L}}{2m}$$

which result from the electrons' spin and orbital motion, respectively.[11] Therefore

$$-\boldsymbol{\mu} = \frac{e\mathbf{S}}{m} + \frac{e\mathbf{L}}{2m}$$

$$= \frac{e}{2m}(2\mathbf{S} + \mathbf{L})$$

$$= \frac{e}{2m}(\mathbf{S} + \mathbf{J}) \qquad (1)$$

Equation (1) indicates that $\boldsymbol{\mu}$ *is not parallel* to $\mathbf{J}$ (unless $S = 0$). We shall see in a moment that this fact is directly responsible for the existence of the anomalous Zeeman effect. (It is not surprising that classical theory could not explain this effect, which is a direct result of the nonclassical phenomenon of electron spin.)

To proceed further, we must consider the relative magnitudes of the external field $\mathbf{B}$ and the *internal* magnetic field which couples $\mathbf{L}$ and $\mathbf{S}$ within the atom. The internal field causes $\mathbf{L}$ and $\mathbf{S}$ to precess around the direction of $\mathbf{J}$, while the external field causes the vector $\boldsymbol{\mu}$ (or $\mathbf{J} + \mathbf{S}$) to precess around the direction of $\mathbf{B}$ (see Fig. 7). If the external field is small in comparison to the internal field (that is, if it is of the order of 1000G (0.1 T) or less), then the precession of $\boldsymbol{\mu}$ around $\mathbf{B}$ is slow, relative to the precession of $\mathbf{L}$ and $\mathbf{S}$ around $\mathbf{J}$. Of course, $\boldsymbol{\mu}$ precesses around $\mathbf{J}$ as $\mathbf{L}$ and $\mathbf{S}$ precess, and in a weak external field $\boldsymbol{\mu}$ makes *many* revolutions around $\mathbf{J}$ in the course of a *single* revolution around $\mathbf{B}$.

Because the precession of $\boldsymbol{\mu}$ around $\mathbf{J}$ is so fast, we can evaluate the average value of $\boldsymbol{\mu} \cdot \mathbf{B}$ by *averaging* the vector $\boldsymbol{\mu}$ over one full revolution around the vector $\mathbf{J}$. During one revolution, the component of $\boldsymbol{\mu}$ perpendicular to the $\mathbf{J}$ direction has an average value of zero, and the component of $\boldsymbol{\mu}$ parallel to $\mathbf{J}$

---

[11] μ_{spin} is given in Section 7.3; μ_{orbital} is given in Section 6.3.

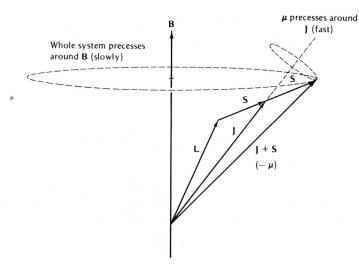

Fig. 7. *Relationships among the vectors* **L**, **S**, **J**, **B**, *and* **μ**, *and the precessions in a weak magnetic field* **B**.

has the constant magnitude $\boldsymbol{\mu} \cdot \mathbf{J}/|\mathbf{J}|$. The vector $\boldsymbol{\mu}_{\text{av}}$ is therefore a vector of this magnitude, whose direction is the direction of **J**, and we may write

$$\boldsymbol{\mu}_{\text{av}} = \frac{\boldsymbol{\mu} \cdot \mathbf{J}}{|\mathbf{J}|^2}\,\mathbf{J}$$

an expression which you may easily verify to have the correct magnitude and direction. The average value of the interaction energy is then

$$\langle \text{Energy} \rangle = -\boldsymbol{\mu}_{\text{av}} \cdot \mathbf{B} = -\frac{(\boldsymbol{\mu} \cdot \mathbf{J})(\mathbf{J} \cdot \mathbf{B})}{|\mathbf{J}|^2} \tag{2}$$

We may now substitute expression (1) for μ into the right-hand side of Eq. (2), to obtain

$$\langle \text{Energy} \rangle = \frac{+e}{2m}\frac{[(\mathbf{J} + \mathbf{S}) \cdot \mathbf{J}][\mathbf{J} \cdot \mathbf{B}]}{|\mathbf{J}|^2}$$

or, if **B** is in the z direction,

$$\langle \text{Energy} \rangle = \frac{+eBJ_z}{2m}\frac{(|\mathbf{J}|^2 + \mathbf{J} \cdot \mathbf{S})}{|\mathbf{J}|^2}$$

We may evaluate $\mathbf{J} \cdot \mathbf{S}$ from the equation

$$|\mathbf{L}|^2 = (\mathbf{J} - \mathbf{S}) \cdot (\mathbf{J} - \mathbf{S}) = |\mathbf{J}|^2 + |\mathbf{S}|^2 - 2\mathbf{J} \cdot \mathbf{S}$$

or

$$\mathbf{J} \cdot \mathbf{S} = \frac{(|\mathbf{J}|^2 + |\mathbf{S}|^2 - |\mathbf{L}|^2)}{2}$$

so that we finally obtain

$$\langle \text{Energy} \rangle = \frac{+ eBJ_z}{2m} \left[1 + \frac{|\mathbf{J}|^2 + |\mathbf{S}|^2 - |\mathbf{L}|^2}{2|\mathbf{J}|^2} \right] \tag{3}$$

The allowed states must therefore be eigenstates of the operators $|\mathbf{J}|^2$, $|\mathbf{S}|^2$, $|\mathbf{L}|^2$, and J_z; and the energy levels are given by substituting the eigenvalues of these operators into Eq. (3), with the result that

$$\text{Energy} = \frac{+ eB\hbar}{2m} \, m_J g \tag{4}$$

where the factor g, a pure number called the Landé g factor, is given by

$$g = \left[1 + \frac{J(J + 1) + S(S + 1) - L(L + 1)}{2J(J + 1)} \right]$$

The g factor depends on the relative weights of $\mathbf{L}$ and $\mathbf{S}$ in determining the vector $\mathbf{J}$. If $L = 0$, then $J = S$ and the g factor is 2; if $S = 0$, then $J = L$ and the g factor is 1. In both of these extreme cases, $\boldsymbol{\mu}$ is parallel to $\mathbf{J}$; the difference between the g factors is caused by the difference between the gyromagnetic ratios for spin and orbital angular momentum (e/m for spin and $e/2m$ for orbital angular momentum). When neither L nor S is zero, g can have a fractional value, and it can even have a value smaller than 1. (For example, if $L = 2$, $S = 1$, and $J = 1$, we have $g = 1 + (2 + 2 - 6)/4 = \frac{1}{2}$. This apparently strange result occurs because the vectors $\mathbf{L}$ and $\mathbf{S}$ are in opposite directions in this case, and the spin angular momentum offsets the orbital angular momentum to a smaller degree than the spin magnetic moment offsets the orbital magnetic moment, so that the ratio of magnetic moment to angular momentum is smaller than it would be if the spin were zero.)

The linear dependence of the energy on m_J [Eq. (4)] shows that the B field splits each level into a number of equally spaced levels, with a different energy for each m_J value. The spacing between levels is proportional to g, so we should expect the normal Zeeman effect whenever g is the same for both states involved in a transition. This would, of course, happen if the electron had no spin, so we see why the classical theory should predict only three lines. Because there are many levels in which either L or S is zero, there are many examples of the normal Zeeman effect, but of course there are many more examples of the anomalous effect. The important fact is that quantum mechanics correctly predicts the splitting of every single line of any atomic spectrum in an applied magnetic field.

The linear dependence of the energy on B, m_J, and g is found to hold as long as our assumption that **B** is much weaker than the internal field remains valid. If **B** is too large, the precession of μ around the z axis becomes too rapid for our averaging procedure to be valid; our expression for μ_{av} is then incorrect, and the equations following Eq. (2) are invalid. In that case we are dealing with the *Paschen-Back effect*, in which the split levels are no longer equally spaced. This effect must be analyzed by considering the precession of **L** and **S** *independently* around the direction of **B**; the vector **J** does not have a constant magnitude in this case. In such a situation, the atomic transitions identify states with a given value of L_z and S_z, rather than states with a given value of J; one is "measuring" the values of the quantum numbers m_l and m_s, rather than the quantum number J. (See Section 7.3 for a discussion of the alternate ways of characterizing a quantum state.) We shall not pursue these complications further; the interested reader is referred to H. E. White's classic text, "Introduction to Atomic Spectra" (McGraw-Hill, New York, 1934).

9.3   MOLECULAR STRUCTURE

Before we can understand molecular spectra, we must know something about the structure of molecules, and to begin with, we can ask why molecules exist at all. The answer to this question will also tell us something about the structure of solids, as we shall see in Chapter 12.

Ionic Binding.   We have seen (Chapter 8, Fig. 7) that the valence electron in an alkali-metal atom is relatively loosely bound; the ionization energy typically is only about 5 eV. A halogen *ion* holds its *extra* electron almost as tightly as an alkali-metal *atom* holds its *valence* electron. For example, the binding energy of the eighteenth electron in the Cl^- ion is 3.82 eV. The extra electron is bound because of the polarization of Cl when an electron approaches; the incomplete M shell ($n = 3$ states) is rearranged by the repulsion of this electron, so that the neutral atom becomes a dipole, whose positive end is closer to the electron and can hold it.

Now consider a system consisting of a potassium atom and a chlorine atom, with the atoms widely separated. If an electron is gently removed from the potassium atom, the system's energy increases by 4.34 eV, the ionization energy of potassium. If this electron is now added to the chlorine atom, the system's energy decreases by the above-mentioned 3.82 eV, making the net gain in energy only 0.52 eV. But now we have a pair of ions that attract each other like point charges, as long as they are far enough apart to remain spherically symmetric. If we let these ions "fall" toward each other, their potential energy at a separation of r Angstroms will be $=14.4$ eV/r. When r is less than 27.7 Å, this potential energy is less than -0.52 eV, and the total energy of the system is less than that of the original two atoms. At these values of r, an input of energy is required to return to the original state of two separate neutral atoms.

If we follow such an ion pair as the ions continue to fall toward each other, we find that the potential energy ceases to be given by the point-charge expression when the ions are close enough to polarize each other. The complete potential energy curve is shown in Fig. 8. At $r \leqslant 2.8$ Å, the curve turns sharply upward. This happens because of the exclusion principle; when the 3p electrons of the K^+ ion reach the space occupied by the 3p electrons of the Cl^- ion, it is impossible for the electrons to penetrate one another's space unless some of them are excited to higher energy states. A similar curve is followed by other ionic pairs, producing molecules at equilibrium separations of a few Angstroms in most cases.

In the ground state, the two ions of such a molecule oscillate about the equilibrium separation, with the "zero-point energy" that is always possessed by an oscillator (Section 5.7). The ions can also be excited to a spectrum of vibrational states of higher energy, forming the vibrational spectrum to be discussed in Section 9.4.

Covalent Binding.   An electron can bind one atom to another without the complete electron transfer that occurs in ionic binding. We know that one nucleus can be bound to two electrons (as in He, Li^+, etc.), and a refinement of the calculations of Section 8.3 shows that even a single proton can be bound to two electrons, forming an H^- ion with a binding energy of 0.8 eV. In a similar fashion, one can also show that a single *electron* can be bound to two *protons*, forming an H_2^+ molecule-ion. The electron's probability density is concentrated between the two protons, attracting both of them. The total energy, as a function of the proton–proton separation, has a minimum at a value of 1.06 Å—very close to twice the first Bohr radius.

Similarly, in the neutral hydrogen molecule, the two electrons are concentrated between the two protons. The Schrödinger equation for the whole system may be written

$$\left[\frac{1}{M_p} (\nabla_1^2 + \nabla_2^2) + \frac{1}{m_e} (\nabla_3^2 + \nabla_4^2) \right] \psi = -\frac{2}{\hbar^2}(E - V)\psi \tag{5}$$

where ψ is a function of twelve coordinates (three for each particle), ∇_1^2 operates on the coordinates of the first proton, ∇_2^2 on those of the second proton, ∇_3^2 on those of the first electron, and ∇_4^2 on those of the second electron.

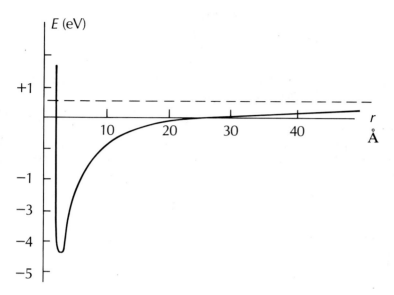

Fig. 8. *Potential energy of K⁺ and Cl⁻ as a function of internuclear separation r. Minimum of −4.42 eV is at r = 2.79 Å. At large r, E is equal to 0.52 − 14.4/r.*

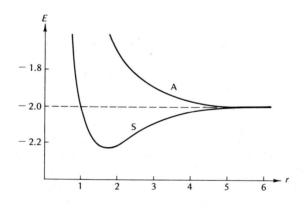

Fig. 9. *Calculated potential energy of two hydrogen atoms, in units of the H-atom ionization energy, as a function of the distance between the protons, in units of the Bohr radius. Curve A is for the antisymmetric spatial wavefunction. [From "Introduction to Quantum Mechanics" by L. Pauling and E. B. Wilson. Copyright McGraw-Hill, New York, 1935. Used by permission of McGraw-Hill Book Company.]*

The potential energy V may be written

$$V = \frac{e^2}{r_{12}} - \frac{e^2}{r_{13}} - \frac{e^2}{r_{14}} - \frac{e^2}{r_{23}} - \frac{e^2}{r_{24}} + \frac{e^2}{r_{34}}$$

where r_{12} is the distance between the two protons, r_{13} the distance between the "first" proton and the "first" electron, etc.

It is possible to rewrite the operations on the protons to put them in terms of the relative coordinates and the coordinates of the center of mass. When that is done, one can separate the equation as was done in Chapter 7 for the relative motion of proton and electron, and one finds one equation for the relative motion, involving the *reduced mass* m_r of the two protons (or other nuclei):

$$m_r = M_1 M_2/(M_1 + M_2)$$

where in the case of hydrogen,

$$M_1 = M_2 = M_p$$

As we saw in Section 8.2, where identical particles are involved, the distinction between the first and second proton (or first and second electron) is simply a mathematical convenience, with no observable consequence. To make the mathematics conform to reality, we always have to write ψ so that it is either symmetric or antisymmetric in the interchange of any two identical particles. We have already seen (also in Section 8.2) that the electrons are closer together when they are in the symmetric space state, which requires that the total spin S is equal to 0. Thus the total electron spin in the H_2 molecule is always $S = 0$ because it is energetically favorable to have both electrons between the two protons. Figure 9 gives the result of a simple perturbation calculation, showing how the energy varies with proton–proton separation, for both $S = 0$ and $S = 1$ electronic spin states. The $S = 0$ state yields a negative energy well that binds the two atoms together, but when the electrons are in the $S = 1$ state, the two atoms always repel each other. In the bound state, the equilibrium separation of the protons is 0.74 Å, considerably smaller than the separation in the H_2^+ ion (which should be no surprise). As in the case of ionic binding, the well permits a spectrum of higher-energy vibrational states with rotational energy as well (see Section 9.4).

Polyatomic Molecules.   As we add more atoms and more electrons to the system, Eq. (5) acquires more terms and becomes more and more difficult to handle; however, some approximations have been made that still permit useful conclusions to be drawn. For example, it can be shown, from an analysis of the equation for three hydrogen atoms, that a hydrogen molecule will repel a hydrogen atom, so that H_3 is not stable.[12] Intuitively, one can see this as a result of the fact that the third electron cannot be localized, in its lowest energy state, in the region of the first two (because of the exclusion principle). Thus the question raised by Avogadro's hypothesis (Section 1.1) has now been answered.

As we learn in elementary chemistry, other polyatomic molecules are stable whenever the electrons are able to be transferred (ionic binding) or shared (covalent binding) without the exclusion principle's requiring excitation to higher energy electronic states. For example, since oxygen is missing two electrons in its L shell, it has room in this shell for a share of an electron from two different H atoms, to form H_2O. Exact analysis of the wavefunctions in such systems is far beyond our capabilities, so it becomes a matter of deciding which approximation is most appropriate for understanding the important features of each molecule.

9.4 MOLECULAR SPECTRA

Vibrational States of a Diatomic Molecule. The wells shown in Figs. 8 and 9 are similar to the effective potential energy V_{eff} for the hydrogen atom (Fig. 2, Chapter 7). The wells have roughly the same depths, and the same $1/r$ dependence at large r. But the energy levels are quite different, because here we are dealing with nuclei rather than electrons. Because of the much greater mass involved, the uncertainty in position is much smaller and there are many more allowed states near the bottom of the well.

The bottom of the well can be approximated by a parabola. For KCl, the best-fitting parabola is given by

$$V = -4.42 + 2.66(r - 2.79)^2 \tag{6}$$

where r is in Angstroms and V is in eV. Recalling that the harmonic oscillator potential is $\frac{1}{2}m\omega^2 r^2$ (Section 6.4), we see that, for solutions near the bottom of the well, the energies should be approximately the solutions to the harmonic oscillator equation, as long as we set $\frac{1}{2}m_r\omega^2$ equal to the quadratic coefficient of 2.66 eV/Å^2 in Eq. (6). In this case, the reduced mass is

$$m_r = \frac{M_1 M_2}{M_1 + M_2} = \frac{39 \times 35}{39 + 35}\, \text{u}$$

where u is the atomic mass unit (Appendix D), and the atoms are ^{39}K and ^{35}Cl.

---

[12] See L. Pauling & E. B. Wilson, "Introduction to Quantum Mechanics," McGraw-Hill, New York, 368–373 (1935).

Solving for ω, and multiplying by $\hbar$, we find that $\hbar\omega$ equals 0.0347 eV. Thus the energy levels should be given by[13]

$$E_v = -4.42 + \hbar\omega(v + \tfrac{1}{2}) = -4.42 + 0.0347(v + \tfrac{1}{2})\,\text{eV} \qquad (7)$$

where v is the vibrational quantum number (a positive integer or zero), and the -4.42 eV is included because that is the level of the bottom of the well.

This analysis is supported by the observation in the KCl spectrum of an intense line whose wavelength is 35,700 Å, which is the wavelength of a photon whose energy is 0.0347 eV. The line is intense because, in a harmonic oscillator, transitions only occur for which the change in the quantum number is one unit; when the energy levels are equally spaced, all such transitions yield the same energy.

However, there is evidence that larger changes in v can occur. There are fainter lines at frequencies that are integral multiples of the fundamental frequency. These frequencies must result from transitions in which v changes by more than one unit, and they must result from the fact that the potential is not exactly that of a simple harmonic oscillator. Classically, this deviation would introduce higher harmonics—multiples of the fundamental frequency—into the motion and thus into the radiated spectrum. (See the discussion of elliptical orbits in the Bohr atom, Section 3.4.) A further effect of the actual potential is that the energy levels are more closely spaced as the energy increases and the well deviates more from the simple parabolic shape.

A better approximation to the real potential is given by the *Morse potential*

$$V(r) = D[1 - e^{-a(r - r_e)}]^2 \qquad (8)$$

where r_e is the equilibrium distance between the two nuclei and D is the dissociation energy. (Notice that $V(r)$ goes to D as $r \to \infty$, and $V(r) = 0$ when $r = r_e$.) For small values of $r - r_e$, $V(r)$ reduces to the harmonic oscillator potential:

$$V(r) \simeq D\{1 - [1 - a(r - r_e)]\}^2 = Da^2(r - r_e)^2$$

Thus for KCl, Da^2 must equal 2.66 eV/Å^2, to agree with Eq. (6). We know that $D = 4.42$ for KCl, so a^2 must equal 0.602 Å^{-2}.

It is possible to solve the Schrödinger equation exactly for this potential.[14] The resulting energy levels, when there is no rotational energy, are given by

---

[13] See Section 5.7. The zero-point energy here is that of a one-dimensional oscillator ($\hbar\omega/2$) rather than a three-dimensional one, because the potential is independent of θ and ϕ, leaving only one degree of freedom for *vibration* (as opposed to rotation).

[14] L. Pauling and E. B. Wilson (Reference 12), 271–274.

$$E_v \simeq \hbar a \sqrt{\frac{2D}{m_r}} \, (v + \frac{1}{2}) - \frac{\hbar^2 a^2}{2m_r} \, (v + \frac{1}{2})^2 \qquad (9)$$

$$= \hbar \omega (v + \frac{1}{2}) - \frac{\hbar^2 \omega^2}{4D} \, (v + \frac{1}{2})^2$$

where again v is the vibrational quantum number, and we have omitted additional terms which are negligible when $E_v << D$. Substituting numerical values for KCl, we find that

$$E_v = 0.0347(v + \frac{1}{2}) - 0.0000068(v + \frac{1}{2})^2 \text{ eV} \qquad \text{(KCl)}$$

For KCl and most other diatomic molecules, the calculated levels are quite close to the observed ones. At $v = 5$, the second term is only 1% of the first term (for KCl), and the neglected terms are much smaller still.

Rotational Spectra of Diatomic Molecules. The vibrational spectrum is associated with only one of the degrees of freedom of the relative motion of the two nuclei in a diatomic molecule. Taking the x axis to be the line joining the two nuclei, we have two other degrees of freedom associated with rotation about the y axis and the z axis. The energy resulting from these rotations is equal to $L^2/2m_r r^2$, where L is the angular momentum. As in the hydrogen atom, this energy enters the radial Schrödinger equation as a result of solving the angular part of the equation, and it can be thought of as a "centrifugal" contribution to the potential energy (even though it is actually a kinetic energy). As usual, the allowed values of L^2 are quantized, and may be written $L^2 = K(K + 1)\hbar^2$, where we now use K to denote an integer, to distinguish this angular momentum from others. The rotational energy is then

$$E_r = K(K + 1)\hbar^2/2m_r r^2 \ . \qquad (10)$$

As we did for the analysis of the hydrogen atom, we may attempt to solve the radial equation with the effective potential $V_{eff} = E + E_r$. But for a diatomic molecule, one can find the approximate effect on the energy levels with less effort. Because of the symmetry of the potential about r_e (for the low-lying states), the probability density $|\psi|^2$ is also symmetric about r_e, and it is strongly peaked at $r = r_e$. Therefore it is a good approximation to substitute r_e, a *constant*, for the *variable r* in Eq. (10), obtaining

$$E_r = K(K + 1)\hbar^2/2m_r r_e^2 \qquad (11)$$

If the molecule has a permanent electric dipole moment (which it will have unless it is homonuclear), it can radiate or absorb energy by making transitions in which K changes by one unit (just as l changes by one unit in an atomic transition, Section 8.5):

$$\Delta K = \pm 1$$

In a transition from K to $K - 1$, the emitted energy is

$$\Delta E = \hbar^2/2m_r r_e^2 \left\{ K(K + 1) - (K - 1)K \right\} = \hbar^2 K/m_r r_e^2$$

so that the emitted frequency is

$$\nu_K = \Delta E/h = \hbar K/2\pi m_r r_e^2 \tag{12}$$

These frequencies ν_K form a rotational spectrum of lines that are equally spaced in frequency, the difference between one frequency and the next being

$$\nu_{K+1} - \nu_K = \hbar/2\pi m r_e^2 \tag{13}$$

For KCl,

$$E_1 = \hbar\nu_1 = 2.91 \times 10^{-5} \text{ eV}$$

which is less than 0.1% of the vibrational energy of the lowest state. This places the "pure rotational" spectrum in the far infrared.

Vibrational-Rotational Spectra.   The above discussion has treated vibration and rotation as independent motions, but of course both can occur simultaneously, and each is influenced somewhat by the other.

Figure 10 shows a so-called vibrational *band* for HCl, and Table 1 lists the frequencies at which photons are absorbed to form this "band."[15] The band is produced by excitation of HCl molecules from the $v = 0$ vibrational state to the $v = 1$ state, accompanied by a change in the rotational state obeying the selection rule $\Delta K = \pm 1$. Each frequency is labeled by the value of K in the initial $v = 0$ state. The lines in the right-hand group ($\nu > \nu_0$) result from transitions in which K increases by one unit; those in the left hand group result from transitions in which K decreases by one unit. There is a gap between the two groups—no line at $\nu = \nu_0$—because such a line would require $\Delta K = 0$. (This "missing" line is at the frequency corresponding to the purely vibrational energy difference between $v = 0$ and $v = 1$.)

[15] Data are from N. J. Colthup, L. H. Daly, and S. E. Wiberley, "Introduction to Infrared and Raman Spectroscopy" Academic Press, New York (1964).

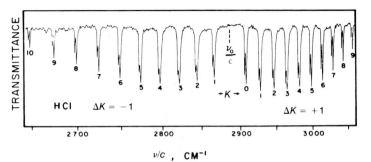

Fig. 10. *The HCl vibrational-rotational band. Each absorption line is labeled by the value of K in the initial state (v = 0). Lines of higher frequency have* $\Delta K = + 1$; *those of lower frequency (for the same K) have* $\Delta K = -1$. *(Adapted from Reference 15.)*

The differences between various pairs of frequencies in this spectrum can be used to determine the rotational energy levels in HCl, for both $v = 0$ and $v = 1$. The line labeled with any given integer, say K', in the right-hand series gives the energy of the transition from K' (with $v = 0$) to $K' + 1$ (with $v = 1$). The line whose label is $K' + 2$ in the left-hand series gives the energy of the transition from $K' + 2$ (with $v = 0$) to $K' + 1$ (with $v = 1$). The *difference* between these two energies is therefore equal to the difference in energy between the levels $K = K'$ and $K = K' + 2$, for $v = 0$, that is, to $E_r(K = K' + 2) - E_r(K = K')$. Evaluating E_r from Eq. (11), we have

$$\text{Difference} = [(K' + 2)(K' + 3) - K'(K' + 1)]\hbar^2/2m_r r_e^2$$
$$= (2K' + 3)\hbar^2/m_r r_e^2 \tag{14}$$

Table 1 shows that, for $K' = 1$, the frequency difference is $(2925.90 - 2821.56)c$, making the energy difference $104.34\,hc$. According to Eq. (14) with $K' = 1$, we must have, therefore,

$$5\hbar^2/m_r r_e^2 = 104.34\,hc$$

or

$$\hbar^2/m_r r_e^2 = 20.87\,hc \tag{15}$$

This result is in agreement with the wavelengths observed in the pure rotational spectrum of HCl, and it also agrees with Table 1 for other values of K'.

Figure 11 shows schematically the levels and transitions involved. In that figure each energy level, except for the lowest two in each group, has been deduced from the frequency differences in Table 1. This procedure, as described above, does not give the difference between a $K = 1$ and a $K = 0$ level, so those differences were taken to be $\frac{1}{3}$ of the difference between $K = 2$ and $K = 0$ in each case (because 1×2 is one-third of 2×3).

Table 1

Vibrational-Rotational Band of $H^{35}Cl$*

K of initial state	v/c ($\Delta K = +1$) (cm^{-1})	v/c ($\Delta K = -1$) (cm^{-1})
0	2906.24	
1	2925.90	2865.10
2	2944.90	2843.62
3	2963.29	2821.56
4	2981.00	2798.94
5	2998.04	2775.76
6	3014.41	2752.04
7	3030.09	2727.78
8	3045.06	2703.01
9	3059.32	2677.73

* From Reference 15. Values shown are for HCl containing only the isotope ^{35}Cl; frequencies for H^{37}Cl are slightly shifted, resulting in the double peaks seen in Figure 10 for the normal isotopic mixture of HCl.

If the average nuclear separation were the same constant r_e in each state, then the energy of each level would be exactly proportional to $K(K+1)$, and the level spacing in the $v = 1$ states would be the same as in the $v = 0$ states. But the average value of r is slightly greater in the $v = 1$ states, causing the spacing of those levels to be smaller than the spacing of the $v = 0$ levels. The average nuclear separation also increases as K increases because of the increasing centrifugal potential, which causes the minimum in V_{eff} to shift to larger r.

Electronic-Vibrational-Rotational Spectra.　All of the states discussed so far in this section have been ones in which the electrons are in their ground states, and the excitation energies are considerably smaller than typical electronic excitation energies. The largest excitation energy shown in Fig. 11 is 3452 hc, or about 0.43eV, and most diatomic molecules have vibrational and rotational energies much smaller than this (because m_r is greater). An electronic excitation adds an energy of order 1 eV to each of these levels, producing another set of levels that are 1 eV or more above the ones discussed above. (This second set of levels has similar, but not exactly the same, structure as the lower set, because the excitation of an electron changes the internuclear potential.)

In general, the electronic, molecular, and vibrational states can all change when a molecule absorbs or emits energy. When the energy is sufficient to cause

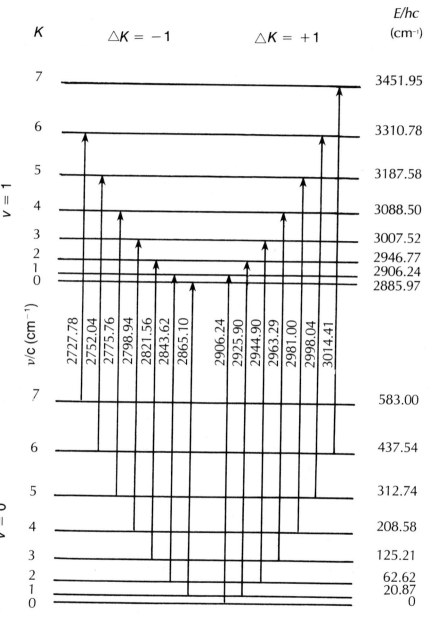

Fig. 11. *Vibrational and rotational levels of $H^{35}Cl$, deduced from Table 1. Vibrational energy difference not drawn to scale.*

a change in an electronic state, the frequency emitted or absorbed belongs to a set of lines centered on the frequency that would be emitted in a pure electronic transition. Because of the participation of the electron process, the rule $\Delta v = 1$ is no longer enforced, and Δv can have any value. Thus there is a band consisting of a series of equally-spaced lines, each one associated with a specific value of Δv. Each of these lines is in turn broadened into a sub-band by the rotational energy change.

PROBLEMS

1. Deduce, from the wavelengths given in Fig. 1, the energy difference between the $3p_{1/2}$ and $3p_{3/2}$ states in sodium. Calculate this difference by using two different pairs of transitions, to see that your answers agree with each other.

2. List the spectroscopic symbols for the states of two 4f electrons, indicating which states are permitted by the exclusion principle.

3. How many states are permitted by the exclusion principle for three equivalent p electrons? (Equivalent electrons are electrons having the same value for the quantum number n.)

4. Prove the Lande interval rule by the method outlined in Section 9.1.

5. When light shines through potassium vapor, absorption "lines" are observed (wavelengths at which light is absorbed very strongly). At moderate temperatures ($T \simeq 1000$ K), the eight longest-wavelength lines are at 7699, 7665, 4047, 4044, 3448, 3447, 3218, and 3217 Angstroms. Assuming that all of these lines result from excitation of atoms from their ground state, name the series that these lines belong to. At higher temperatures, another series appears, whose six longest wavelengths are 12434, 12523, 6911, 6939, 5782, and 5802 Angstroms. Use these wavelengths and the ionization energy of potassium (4.34 eV) to construct an energy-level diagram for potassium. Show that the lowest-energy state is an S state.

6. Determine the Zeeman splitting of the levels $^4F_{5/2}$ and $^4G_{7/2}$ in a magnetic field of 1000 G (0.1 Wb/m^2), and use your answer to predict the effect of such a magnetic field on the 5700-Å line of Sc (Fig. 5).

7. For ^{7}Li^{1}H (isotopes ^{7}Li and ^{1}H), $r_e = 1.596$ Å. Find the wavelengths of the absorption lines in the far infrared (pure rotational spectrum) for this molecule.

8. For ^{23}Na^{35}Cl, $r_e = 2.51$ Å, the dissociation energy is 3.58 eV, and ω [in Eq. (9)] is 7.16×10^{13} s^{-1}.

(a) use Eq. 9 to find the value of a that should appear in the Morse potential for this molecule.

(b) For $v = 10$, find the vibrational energy given by Eq. 7 and that given by Eq. 9, and compute the % difference.

(c) Find the rotational energy for $K = 2$.

9. From Fig. 11, compute the first six wavelengths in the pure rotational spectrum of $H^{35}Cl$.

10. To determine how much a diatomic molecule stretches as it rotates, one can find the effective potential V_{eff} for a given K (by adding E_r to the Morse potential) and then solve for the value of r that minimizes V_{eff}. Use this procedure to calculate the change in the value of r in KCl when the quantum number K increases from 0 to 10.

11. From Fig. 11, determine the average distance between the proton and the Cl nucleus in the ground state of $H^{35}Cl$, and find the change in this distance as v changes from 0 to 1. Also find the approximate change as K increases from 0 to 7, with v equal to 0.

12. The potential energy for H_2, HD, and D_2 ($D \equiv {}^2H$) is the same function $V(r)$ (see Fig. 9), but the dissociation energies differ, because the zero-point energy depends on m_r. Given that the dissociation energies are 4.477, 4.513, and 4.556 eV, for H_2, HD, and D_2, respectively, use the *ratios* of the zero-point energies (deduced from m_r and Eq. 9) to find each zero-point energy in eV. (Notice that these energies do not quite agree with the theoretical approximation graphed in Fig. 9.)

13. The Raman effect is a change in the frequency of light when it is scattered by a molecule. The energy lost by the light is given to the molecule to excite it to a higher energy state. This effect can be used to deduce rotational levels for molecules that do not radiate a rotational spectrum (because they have no electric dipole moment); the effect also is seen when vibration is involved. The selection rule for Raman scattering is $\Delta K = 0$, ± 2, in contrast to the rule $\Delta K = \pm 1$ for the rotational spectrum in emission or absorption. Given this selection rule, find the frequencies that should appear in the scattering of light of original frequency 1.00×10^{13} Hz from molecules of $^{23}Na^{35}Cl$ (see Problem 7).

Atomic Radiation

In previous chapters, we discussed the emission of radiation in atomic transitions without considering why these transitions should occur at all. We *assumed* that an atom may change from a higher energy state to a lower energy state, while emitting the energy difference in the form of radiation; and this assumption is certainly in accord with the experimental facts. But the Schrödinger equation contains no term which *requires* that such transitions occur; as far as this equation is concerned, an atom might remain in an excited state forever. Thus we need a theory which demands that transitions take place, and which tells us how to calculate their probability of occurrence.

In Section 3.4, we saw how to estimate a transition probability by making use of the classical expression for the rate of radiation of energy from an accelerated charge. Fundamentally we should do the same thing in developing a quantum theory of radiation. We must take the classical equations for the electric and magnetic fields—Maxwell's equations—and quantize them, so that these equations, which are known to be correct when large numbers of photons are present, may be extended into the domain in which very few photons are present.

345

Unfortunately, the process of quantizing Maxwell's equations is too sophisticated for this book, so we shall settle for an approximate treatment which gives good results. By means of a *time-dependent* perturbation theory, developed along the same lines as the time-independent theory of Section 8.1, we can calculate the effect of *external fields* in *inducing* transitions from one state to another. We can then use a thermodynamic argument to relate the induced transition rate to the probability that a *spontaneous* transition will occur in the absence of a field. This approach does not answer the fundamental question of why an atom makes a spontaneous transition, but it does enable us to calculate the transition rate with some degree of accuracy and assurance.

We shall find that the transition probabilities often turn out to be zero in such calculations. This fact leads us to the formulation of selection rules to determine which transitions are likely to occur. In many cases a transition probability is not absolutely zero, but it is calculated to be zero only because of the approximation made in the calculation. Such transitions are said to be "forbidden." We shall study the nature of these approximations, and thereby estimate the probabilities of various forbidden transitions relative to the probabilities of allowed transitions.

Finally, we shall see how these rules apply to the operation of the maser and the laser, and we shall study some of the properties of these remarkable devices.

10.1   TIME-DEPENDENT PERTURBATION THEORY; TRANSITION RATES

Time dependent perturbation theory is developed in a way which is quite similar to the development of time independent perturbation theory (Section 8.1). We assume that the potential energy contains a small time dependent perturbing term $v(x, t)$, and that without the addition of this term the Schrödinger equation may be solved exactly. Since the potential is time dependent, we begin with the *time-dependent* Schrödinger equation (instead of the time independent equation as in Section 8.1), and we write

$$[H_0 + v(x, t)]\psi'_n = i\hbar \frac{\partial \psi'_n}{\partial t} \tag{1}$$

where, as before, H_0 is the H operator (the Hamiltonian, or energy operator) for the unperturbed system, so that

$$H_0 \psi_l = i\hbar \frac{\partial \psi_l}{\partial t} \tag{2}$$

As in Section 8.1, we can now write each function ψ_n' as a linear combination of the functions ψ_l; but we must now allow the coefficients to be *time dependent*[1]:

$$\psi_n' = \sum_l a_{nl}(t)\psi_l \tag{3}$$

Thus Eq. (1) becomes, with the insertion of Eq. (3),

$$[H_0 + v(x, t)] \sum_l a_{nl}(t)\psi_l = i\hbar \frac{\partial}{\partial t} \sum_l a_{nl}\psi_l \tag{4}$$

[Compare with Eqs. (2) and (3) in Section 8.1.]

If we carry out the indicated differentiation term-by-term on the right-hand side, we obtain

$$H_0\left(\sum_l a_{nl}(t)\psi_l\right) + v(x, t)\left(\sum_l a_{nl}(t)\psi_l\right) = i\hbar\left[\sum_l\left(\dot{a}_{nl}\psi_l + a_{nl}\frac{\partial\psi_l}{\partial t}\right)\right] \tag{5}$$

The first series on the left may be written as $\sum_l a_{nl} H_0 \psi_l$, because the operator H_0 does not operate on the time variable. However, from Eq. (2) we see that this series is equal to $i\hbar \sum_l a_{nl}(\partial\psi_l/\partial t)$, so we may eliminate these two series from Eq. (5) to obtain

$$v(x, t) \sum_l a_{nl}\psi_l = i\hbar \sum_l \dot{a}_{nl}\psi_l \tag{6}$$

We are now in a position to find the rate of change of a_{nl}, by solving for $\dot{a}_{nl}$; this will eventually tell us the transition probability which we seek. We proceed as in finding the coefficients a_{ml} in Section 8.1; we multiply each side of Eq. (6) by ψ_m^*, a *particular* wavefunction, and we then integrate over all space, to obtain

$$\int \sum_l a_{nl} \psi_m^* v(x, t)\psi_l \, d\tau = i\hbar \int \sum_l \psi_m^* \dot{a}_{nl}\psi_l \, d\tau$$

which we integrate term-by-term as we did with similar expressions in Section 8.1. Because $\dot{a}_{nl}$ has no space dependence, and the wavefunctions ψ_l are orthogonal to one another, the only nonzero term on the right-hand side is the one for which $l = m$. (Remember that l is the running index in the summation, whereas m is a fixed number.) Inserting the time dependence of ψ_m,

$$\psi_m = u_m(x)e^{-iE_m t/\hbar}$$

we find that

$$\dot{a}_{nm} = -\frac{i}{\hbar} \sum_l a_{nl} e^{-i(E_l - E_m)t/\hbar} v_{ml} \tag{7}$$

where, as in Section 8.1, we use the abbreviation $v_{ml} = \int u_m^* v u_l \, d\tau$.

---

[1] At *any given time* t_1, ψ_n' is a linear combination of the ψ_l, with coefficients a_{nl}, which may be determined by time *in*dependent perturbation theory. But since $v(x, t)$ changes with time, the expansion coefficients will, in general, be different at some other time t_2.

Equation (7) is still exact; except for the time dependence, it is very similar to Eq. (6) in Section 8.1; and like that equation, it contains too many unknown quantities to be useful as it stands. So we make the *approximation* that $v(x, t)$ is a *small perturbation* of the original system. This means that the eigenfunctions of the perturbed system differ very slightly from the unperturbed eigenfunctions, so that, as in Section 8.1, one of the coefficients, a_{nn}, is approximately equal to 1 at all times, and the other coefficients are very small.[2] In that case, we can neglect all terms on the right-hand side of Eq. (7) except the one for which $l = n$, and we can thus find the time dependence of the small coefficient a_{nm}:

$$\dot{a}_{nm} \approx -\frac{i}{\hbar} e^{-i(E_n - E_m)t/\hbar} v_{mn} \tag{8}$$

If we know $v(x, t)$ for all x and t, we can now integrate Eq. (8) to find $a_{nm}(t)$, and the behavior of the system is determined, to the extent that our approximation is valid and each coefficient $a_{nm} \ll 1$ for $m \neq n$.

At this point, we should pause to point out the differences as well as the similarities between time-dependent and time-independent perturbation theory. The *methods* used are quite similar, but the *goals* are different. In Section 8.1, we were trying to calculate a shift in an energy level, but here we are not concerned with such shifts; rather, we are concerned with the *transitions* which the perturbation causes to occur between the unperturbed levels. It is possible to adopt this point of view only because the perturbation is not applied for an infinite period of time, so that after the perturbation is turned off, one can make a measurement which places the system in one of the *unperturbed* levels. We can say that the following sequence of events may occur:

1. The system is initially in an unperturbed state whose wavefunction is ψ_n; its energy is E_n, one of the energy eigenvalues of the unperturbed Schrödinger equation.

2. The perturbation is "turned on" at time $t = 0$, so that the system is, for $t > 0$, described by the *perturbed* Schrödinger equation, with a different set of eigenfunctions and energy levels. The wavefunction becomes ψ'_n, which may be expanded in terms of the unperturbed eigenfunctions by means of the coefficients $a_{nm}(t)$. Since the perturbation is small, the new wavefunction ψ'_n never differs greatly from the old wavefunction ψ_n during the time that the perturbation is on; that is, $a_{nn}(t) \simeq 1$, and the other coefficients a_{nm} are very small.

3. The perturbation is turned off at time $t = t'$, and the energy of the system is determined again. At this point, the system is again described by the

---

[2] Obviously, when the coefficients are time dependent, there is some question as to how long they can *remain* small. We shall discuss this implied limitation on the *time of application of the perturbation* in a moment.

unperturbed Schrödinger equation, so that the wavefunction must again be one of the unperturbed wavefunctions; it may be the original wavefunction ψ_n, or it may be a different wavefunction ψ_m. In the latter case, we say that the perturbation has *induced* a *transition*. According to Postulate 3 (Section 6.3), the probability that the wavefunction will be ψ_m is given by $|a_{nm}(t')|^2$—the absolute square of the coefficient of ψ_m in the expansion of the wavefunction ψ_m', which is the wavefunction of the system at the time of the measurement. Thus $|a_{nm}(t')|^2$ is the probability of a transition from an initial state with wavefunction ψ_n to a final state with wavefunction ψ_m.

Now we would like to integrate Eq. (8) in order to evaluate $a_{nm}(t')$ and thus determine the transition probability as a function of the time t'. To do this, we need to know the time-dependent potential $v(x, t)$ so that we can find the matrix element v_{nm}. We recall that we set out to solve the problem of an atom in an external electromagnetic field, so let us compute the potential energy of an atomic electron in such a field; this will enable us to derive many properties of atomic radiation.

Dipole Radiation.   Let us consider radiation whose wavelength is much greater than the diameter of the atoms involved. This is a reasonable thing to do, because we know that the shortest wavelength emitted by a hydrogen atom is more than 1000 times the diameter of the hydrogen atom. For such radiation, the field does not vary appreciably over the dimensions of the atom, and we can as a first approximation assume that the field is constant over the atom. That is, the field varies only in time, not space. (This is known as the dipole approximation, which gives results valid for electric dipole radiation; later we shall see what happens when this approximation is abandoned.) For simplicity, we shall consider monochromatic (single frequency) radiation, which is polarized along the x axis; it will be easy to generalize the results to unpolarized radiation with a continuous frequency spectrum. We therefore write the x component of the electric field[3] as

$$E_x = E_{0_x} \cos \omega t$$

where E_{0_x} is a constant. It is more convenient to rewrite this field in the complex notation:

$$E_x = \tfrac{1}{2}E_{0_x}(e^{i\omega t} + e^{-i\omega t})$$

By definition, the potential energy of a charge q in this field is $v(x, t) = -q \int E_x \, dx$; and since E_{0_x} is a constant, we can immediately evaluate this integral to obtain

$$v(x, t) = -\tfrac{1}{2}E_{0_x} qx(e^{i\omega t} + e^{-i\omega t})$$

---

[3] Here we neglect the interaction between the electron and the *magnetic* field. This is reasonable, because the speed of the electron is of order $0.01c$, and at such a speed the magnetic interaction energy is of order 0.01 times the electric interaction energy.

Inserting this expression into Eq. (8), we obtain (with q equal to the electron charge of $-e$)

$$\dot{a}_{nm}(t) = -\frac{ieE_{0x}}{2\hbar} \left[e^{i(E_m - E_n + \hbar\omega)t/\hbar} + e^{i(E_m - E_n - \hbar\omega)t/\hbar} \right] x_{mn} \qquad (9)$$

where we have used the abbreviation $x_{mn} = \int u_m^* x u_n \, d\tau$, the integral being over all space as before.

To find the value of $a_{nm}(t)$ at time t' (when the external field is "turned off"), we integrate Eq. (9) from $t = 0$ to $t = t'$. Using the initial condition $a_{nm}(0) = 0$ $(n \neq m)$, we have

$$a_{nm}(t') = \tfrac{1}{2} x_{mn} e E_{0x} \left[\frac{1 - e^{i(E_m - E_n + \hbar\omega)t'/\hbar}}{E_m - E_n + \hbar\omega} + \frac{1 - e^{i(E_m - E_n - \hbar\omega)t'/\hbar}}{E_m - E_n - \hbar\omega} \right] \qquad (10)$$

Because of the expressions $E_m - E_n + \hbar\omega$ and $E_m - E_n - \hbar\omega$ in the denominators, a graph of the coefficient $a_{nm}(t')$ would be peaked at $E_n - E_m = \pm\hbar\omega$. Thus we find a large transition probability when the energy difference between the two states is equal to the energy of a quantum of the applied field—that is, just when the Bohr condition is satisfied.

But we see here that E_m may be *greater* or *less* than E_n by the amount $\hbar\omega$. If E_m is greater than E_n, we have the well-known situation in which the atom has absorbed energy from the field; absorption of a photon has excited the atom from a lower energy state to a higher energy state. But if E_m is *smaller* than E_n, the atom has emitted energy as a result of the perturbation; a photon striking the atom has *stimulated* the emission of a *second* photon. This *stimulated emission*, which *deexcites* the atom and causes the external field to *gain* energy, is the phenomenon responsible for the operation of the maser and the laser.

To evaluate the probability $|a_{nm}(t')|^2 = a_{nm}^*(t')a_{nm}(t')$, let us consider the case in which $E_m - E_n \simeq \hbar\omega$, so that the first term of Eq. (10) is much smaller than the second term, and the first term may therefore be neglected. Direct multiplication then yields [after use of the formula $\cos\theta = 1 - 2\sin^2(\theta/2)$]

$$|a_{nm}(t')|^2 = e^2 E_{0x}^2 |x_{mn}|^2 \frac{\sin^2 \dfrac{(E_m - E_n - \hbar\omega)t'}{2\hbar}}{(E_m - E_n - \hbar\omega)^2} \qquad (11)$$

Two features of this expression require some explanation. One is the sinusoidal time dependence plotted in Fig. 1, which seems to indicate that the transition probability oscillates as time passes. However, because of the presence of the expression $(E_m - E_n - \hbar\omega)$ in the denominator, this oscillation is not quite what it seems, as we shall see.

The second feature, which is closely connected with the first, is the nonzero probability of transitions for which $E_m - E_n = \hbar\omega$—that is, transitions for which the change in energy of the atom is *not* equal to the energy of a photon for a field of angular frequency ω. Thus, either the law of conservation of energy is being violated, or else there is a possibility that the photons in the radiation field may have energies not equal to $\hbar\omega$.

Fortunately for the law of conservation of energy, we can easily see that the latter is the case. Our analysis requires that the perturbation—the radiation field—be turned on for a *finite time* t', and in that case the wave cannot possibly be monochromatic. A Fourier analysis[4] of a sine wave which is suddenly turned on at $t = 0$ and turned off again at $t = t'$ would show a sinusoidal distribution of frequencies which is consistent with expression (11). (See Problem 2.) Thus there is no conflict with either the law of conservation of energy or the condition that a photon's energy is $\hbar\omega$. Furthermore, if we let $t' \to \infty$ in Eq. (11), we see that the wave becomes truly monochromatic; the function $\sin^2[\frac{1}{2}(\omega_{nm} - \omega)t']/\hbar^2(\omega_{nm} - \omega)^2$ approaches zero for all frequencies except the frequency $\omega = \omega_{nm} = (E_m - E_n)/\hbar$. (Refer again to Fig. 1.)

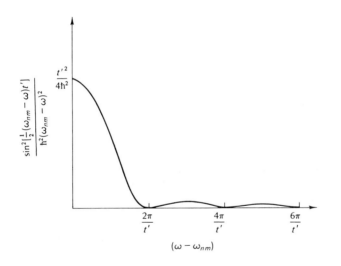

Fig. 1 *The function* $\sin^2[\frac{1}{2}(\omega_{nm} - \omega)t']/\hbar^2(\omega_{nm} - \omega)^2$, *which appears in Eq. (11), plotted versus angular frequency* ω. *Notice that the overwhelming contribution to the function comes from the region* $0 < |\omega - \omega_{nm}| < 2\pi/t'$, *and the width of this region goes to zero as* $t' \to \infty$. *Because the height of the curve is proportional to* t'^2, *and the width is proportional to* $1/t'$, *the area under the curve is proportional to* t'. *This area represents the total transition probability in a radiation field containing a uniform distribution of frequencies in the neighborhood of* $\omega = \omega_{nm}$.

[4] See Section 4.3.

Let us now consider the more general case, in which the radiation field cannot be represented by the function $E_{0_x} \cos \omega t$ at *any* time, but rather contains a *superposition* of such fields, with the various values of ω forming a *continuous spectrum*. In such a case, Eq. (11) can still be used to describe a component of that spectrum with any given frequency ω, but since the frequencies form a continuous spectrum, we cannot speak of the amplitude of any single frequency. (There are now an *infinite* number of possible frequencies, so the amplitude of any *single* frequency must be zero.) Therefore we rewrite Eq. (11) in terms of an energy density function $\rho(\omega)$, such that $\int_{\omega_1}^{\omega_2} \rho(\omega) \, d\omega$ is the energy density of radiation between the frequencies ω_1 and ω_2. It is easy to do this, because Eq. (11) is already expressed in terms of an energy density. According to classical electromagnetic theory, the quantity $\frac{1}{2}\varepsilon_0 E_{0_x}^2$ is the average energy density in an electric field of the form originally assumed.[5] Thus we set $\frac{1}{2}\varepsilon_0 E_{0_x}^2$ equal to $\rho(\omega) \, d\omega$, and we replace $E_{0_x}^2$ in Eq. (11) by $2\rho(\omega) \, d\omega/\varepsilon_0$. Then we find the total transition probability T_{nm}, resulting from the entire spectrum of radiation, by integrating the resulting expression over all frequencies:

$$T_{nm} = |a_{nm}(t')|^2 = 2e^2 |x_{mn}|^2 \int_0^\infty \frac{\sin^2[\frac{1}{2}(\omega - \omega_{nm})t']}{\varepsilon_0 \hbar^2 (\omega - \omega_{nm})^2} \rho(\omega) \, d\omega$$

The integrand here is the product of the energy density distribution $\rho(\omega)$ and the peaked function $\sin^2[\frac{1}{2}(\omega - \omega_{nm})t']/\hbar^2(\omega - \omega_{nm})^2$, which is characteristic of the atom and of the duration of the perturbation. Let us suppose that the distribution $\rho(\omega)$ varies much more slowly than the sharply peaked function by which it is multiplied. Then $\rho(\omega)$ is nearly constant over the small range of values of ω for which the integrand is nonzero, so that we can replace $\rho(\omega)$ by its value at $\omega = \omega_{nm}$, and remove it from the integral, with no loss of accuracy. With the further substitution $z = \frac{1}{2}(\omega - \omega_{nm})t'$, the expression becomes

$$|a_{nm}(t')|^2 = \frac{e^2 \rho(\omega_{nm}) |x_{mn}|^2 t'}{\hbar^2 \varepsilon_0} \int_{-\infty}^\infty \frac{\sin^2 z}{z^2} \, dz$$

or, since

$$\int_{-\infty}^\infty \frac{\sin^2 z}{z^2} \, dz = \pi$$

the transition probability is

$$T_{nm} = |a_{nm}(t')|^2 = \frac{\pi e^2 \rho(\omega_{nm}) |x_{mn}|^2 t'}{\hbar^2 \varepsilon_0} \tag{12}$$

[5] The energy density in a constant electric field of magnitude E_{0x} is equal to $\frac{1}{2}\epsilon_0 E_{0x}^2$ (mks). In a field which varies as $\cos \omega t$, the time average energy density is only one-half as great, but there is an equal amount of energy in the magnetic field which always accompanies an oscillating electric field.

We must remember that Eq. (12) holds for radiation which is polarized along the x axis. In the general case, when radiation is incident upon the atom from all directions and with random polarization, T_{nm} must include equal contributions from x_{mn}, y_{mn}, and z_{mn}, so that

$$T_{nm} = \frac{\pi e^2 \rho(\omega_{nm})t'}{3\hbar^2 \varepsilon_0} \left[|x_{mn}|^2 + |y_{mn}|^2 + |z_{mn}|^2 \right] \qquad (13)$$

where the factor of 3 has been introduced into the denominator because each polarization direction is assumed to contribute $\frac{1}{3}$ of the intensity.

Equation (13) and its derivation contain some features which you might have expected, and some other features which are not so obvious, as follows:

1. The induced transition probability is proportional to the energy density of the incident radiation. This is certainly reasonable.

2. T_{nm} is also proportional to the sum of the absolute squares of the matrix elements x_{mn}, y_{mn}, and z_{mn}. This is also reasonable, and it has an analogy in classical theory. The expression $ex_{mn} = e \int u_m^* x u_n \, d\tau$ is the x component of the so-called *dipole moment between states n and m*; the y and z components of the dipole moment are expressed in an analogous manner. The designation "dipole moment" comes from the classical expression for the dipole moment of a charge distribution of charge density ρ; the x component of the classical expression is $\int x\rho \, d\tau$.[6] We see the connection between the two expressions by writing the charge density produced by a single electron whose wavefunction is u_n; the charge density equals the electron charge e times the probability density $u_n^* u_n$, so that the dipole moment has x component $e \int x u_n^* u_n \, d\tau$. This expression differs from the one given above only in the fact that a single wavefunction u_n is involved, rather than the wavefunctions u_n and u_m. But under the influence of the perturbation, the electron in a sense *occupies both states* n and m simultaneously, so it is reasonable to find that an analogous expression for the dipole moment becomes $e \int u_m^* x u_n \, d\tau$. Because of this connection with the classical expression for dipole moment, the results obtained here are said to describe *electric dipole radiation*, which classically results from a changing dipole moment. (We shall see in Section 9.3 that the matrix elements x_{mn}, y_{mn}, and z_{mn} determine the electric dipole *selection rules* for absorption and emission of radiation by atoms; a zero matrix element indicates that the corresponding transition is "forbidden." In that section we shall also have something to say about the other possible types of radiation.)

---

[6] We use ρ here as a standard symbol for density; it is not to be confused with the *energy* density $\rho(\omega)$ of the radiation field.

3. T_{nm} is directly proportional to the time t' during which the perturbation is applied. Thus we can refer to T_{nm}/t' as a *constant transition probability per unit time*. This result requires some thought. Obviously, we cannot expect T_{nm} to increase indefinitely in this manner. The first-order perturbation theory used here is valid only for times such that $T_{nm} \ll 1$. But there are many real situations in which this condition is satisfied. For example, if the energy density of the radiation is small, we can suppose that the atom is exposed to only one photon at a time. Then the radiation impinging on the atom does not form a continuous wave, but is turned on when a photon reaches the atom and is turned off when the photon has passed. After the photon has passed, the atom is again unperturbed, and it must be found in one of its unperturbed states. If the time between turning on and off is short, the probability that a transition will occur is small enough so that first-order perturbation theory gives a valid result. (Of course, the transition probability also depends on the energy density of the radiation field; but this energy density can be quite small, because the energy of a photon can be spread out over many atoms, so that the probability of an interaction with any one atom is very small.)

Uncertainty Relation for Energy and Time. Let us return to the second feature of Eq. (11)—the fact that a field of the form $E_{0_x} \cos \omega t$, applied for a finite time t', can induce transitions such that $E_n - E_m \neq \hbar \omega$. We saw that, as t' is increased, the maximum difference between $\hbar \omega$ and the transition energy $(E_m - E_n)$ becomes smaller; Fig. 1 shows that

$$\hbar \omega - (E_m - E_n) \lesssim \frac{2\pi}{t'} \hbar$$

Now suppose that we attempt to measure the energy difference $E_m - E_n$ for a given atom. We might do this by applying a field of angular frequency ω and observing whether or not the atom absorbs energy from the field; we could vary ω and eventually observe an absorption of energy. The value of ω at which energy was absorbed would have to be considered equal to the quantity $(E_m - E_n)/\hbar$. But we see from Fig. 1 that we could in this way obtain a value of $E_m - E_n$ which is in error by as much as

$$\Delta E = \frac{2\pi}{t'} \hbar \simeq \frac{\hbar}{t'}$$

This example shows a special case of the so-called *uncertainty relation for energy and time*, which states that

$$(\Delta E)(\Delta t) \geqslant \hbar/2$$

where ΔE is the uncertainty in the energy of a system, and Δt is the time interval over which the energy is measured. This relation, which is parallel to the relation $(\Delta p_x)(\Delta x) \geq \hbar/2$ for momentum and position, has its origin in the same basic fact—that a wave packet of finite length must contain a superposition of different frequencies (wavelengths). In the case we are considering here, the wave packet is that of a *photon* of the radiation field which is inducing a transition, but the same mathematical laws must govern this wave packet as well as an electron wave packet. The principle is a general one which applies to any measurement of energy, just as the position–momentum uncertainty relation is valid for any method of measurement.

Notice that the energy–time uncertainty relation, like the position-momentum uncertainty relation, is in the form of an *inequality*. Equality is achieved only for a Gaussian wave packet in either case.

10.2 SPONTANEOUS TRANSITIONS

As we said at the beginning of this chapter, there is no mechanism in the Schrödinger equation to provide for a finite lifetime for an excited state. In Section 10.1, we found that an atom in an excited state could be induced to go to the ground state, if it is bathed in radiation of the resonant frequency. But we still have no mechanism for a transition to the ground state in the absence of an external field, although we know that such transitions must occur. As mentioned above, such a mechanism could be developed by quantization of the radiation field, but we are not about to do that here.

Instead, we shall rely on a trick, developed in 1917 by Einstein, which enables one to calculate the rate of spontaneous transitions from knowledge of the rate of induced transitions. The relation between the two rates is obtained from elementary thermodynamical considerations.

Consider a collection of atoms which can exchange energy only by means of radiation. The atoms are in thermal equilibrium, inside a cavity whose walls are kept at a constant temperature. Because the system is in thermal equilibrium, each atom must be emitting and absorbing radiation at the same rate, if one averages over a sufficiently long time (such as 1 sec). Let P_{nm} be the probability that a given atom will go from the nth state to the mth state in a short time dt. This probability must be proportional to the probability p_n that the atom is in the nth state to begin with, multiplied by the transition probability T_{nm} [which for dipole radiation is given by Eq. (13)]:

$$P_{nm} = T_{nm}\, p_n$$

More generally, P_{nm} may be written as

$$P_{nm} = A_{nm}\, \rho(\omega_{nm}) p_n \, dt \tag{14}$$

which expresses the fact that T_{nm} must be proportional to the radiation density $\rho(\omega_{nm})$, the time interval dt [which was denoted by t' in Eq. (13)], and other factors, incorporated into the coefficient A_{nm}, which depend on matrix elements.

Let us consider the case in which $E_n < E_m$, so that the transition from n to m involves absorption of a photon. We must recognize the possibility that the radiation field may induce a transition *downward* from state m to state n, *if* an atom is initially in state m. The probability of such a downward transition may be written as

$$P_{mn} = A_{mn}\,\rho(\omega_{mn})p_m\,dt$$

but because of the symmetry of the equations leading to Eq. (13), we know that $A_{mn} = A_{nm}$ and $\omega_{mn} = \omega_{nm}$, so that

$$P_{mn} = A_{nm}\,\rho(\omega_{nm})p_m\,dt \tag{15}$$

Notice that P_{nm} does *not* equal P_{mn}, because $p_n > p_m$; the nth state, being lower in energy, is more heavily populated, according to the Boltzmann factor (Section 1.1).

But we know, because the system is in equilibrium, that the *total* number of transitions from n to m must equal the *total* number of transitions from m to n. Since the *induced* transition probabilities are *un*equal, there must be additional transitions from state m to state n which are *spontaneous*. The spontaneous transition probability by definition does not depend on the energy density of the externally applied field; this probability may be written $B_{nm}p_m\,dt$, where B_{nm} is the quantity which we wish to calculate.

The total transition probability from m to n is therefore equal to $P_{mn} + B_{nm}p_m\,dt$, and this total should equal the transition probability P_{nm}. Thus, from Eqs. (14) and (15), we have

$$\rho(\omega_{nm})p_n A_{nm} = \rho(\omega_{nm})p_m A_{nm} + p_m B_{nm}$$

or

$$B_{nm} = \rho(\omega_{nm})A_{nm}\left[\frac{p_n}{p_m} - 1\right]$$

But the population of a state of energy E is proportional to the Boltzmann factor $e^{-E/kT}$, so the ratio p_n/p_m may be written

$$\frac{p_n}{p_m} = e^{(E_m - E_n)/kT}$$

$$= e^{\hbar\omega_{nm}/kT}$$

Thus

$$B_{nm} = \rho(\omega_{nm})A_{nm}(e^{\hbar\omega_{nm}/kT} - 1) \tag{16}$$

The factor $\rho(\omega_{nm})$ in Eq. (16) is illusory, because B_{nm} by definition does not depend on the energy density of the radiation field. We can eliminate this factor by recalling the assumption made at the beginning of this section that the atoms are in thermal equilibrium inside a cavity. In such a case we have an expression for the energy density inside the cavity; it was derived as a function of ν in Section 3.1 (Eq. 5). Writing it in terms of ω yields

$$\rho(\omega) = \frac{\hbar \omega^3}{\pi^2 c^3 (e^{\hbar \omega / kT} - 1)}$$

Substitution of this expression into Eq. (16) yields an expression for B_{nm} involving only A_{nm} and known constants:

$$B_{nm} = \frac{\hbar \omega_{nm}^3}{\pi^2 c^3} A_{nm} \tag{17}$$

Returning to Eq. (13) to evaluate A_{nm}, we finally find that the spontaneous transition probability in time dt is equal to

$$\frac{1}{3} \frac{e^2 \omega_{nm}^3}{\pi \varepsilon_0 c^3 \hbar} (|x_{mn}|^2 + |y_{mn}|^2 + |z_{mn}|^2) \, dt = \lambda \, dt \tag{18}$$

where λ, the decay constant, is the probability of decay per unit time (for times dt such that $\lambda \, dt \ll 1$).

Exponential Decay Law. In studying spontaneous transitions, we must remember that the emitted photon exhibits the same wave–particle duality that appears in other atomic phenomena. In making the transition, the atom oscillates back and forth between the initial state and the final state, and thus generates an electromagnetic wave. But it is characteristic in the analysis of such quantum processes that we cannot isolate the behavior of the system under study from the behavior of the *measuring instrument* used to observe that system. Thus the transition is *observed* as a *discontinuous* process, because the emitted *wave* interacts with a measuring instrument as a *photon*.

The term "measuring instrument" is to be interpreted in the broadest possible sense. The matter surrounding each atom acts as the measuring instrument, as it absorbs the photon emitted by the atom. Now consider the problem of determining the time at which each atom makes its transition. This determination is based upon the interaction of a photon with the measuring instrument. We can say that the instrument tests the atom at the end of each time interval dt; it either absorbs a photon or it does not, during that time interval. That is, the test shows one of only two possible results: Either the transition has occurred, or it has not. There is no way for the measuring instrument to find the atom in the in-between state of oscillation which produces the wave, because a measurement must always find the atom to be in an eigenstate. If the transition has *not* occurred at the time of the measurement, then the atom is *indistinguishable from other excited atoms*, regardless of how long it had been in the excited state before the mesurement was made.

Thus, although the transition probability as given by Eq. (18) is directly proportional to the time interval dt, the total number of transitions does *not* build up linearly with time. At the end of any given time interval, there are fewer atoms left in the excited state than there were at the beginning of that interval; each of these atoms has exactly the same transition probability in the subsequent time interval that it had in the first time interval, but since there are fewer atoms, there are fewer transitions per unit time as time goes by. We can express this quantitatively as follows: Let state m be the first excited state and state n be the ground state, and let N be the number of atoms in the excited state at any given time. Then in the next short time interval dt, N will change by an amount dN given by

$$dN = -N\lambda\, dt$$

where λ is the decay constant defined in Eq. (18). We can find N for any time t by integrating this equation as follows:

$$dN/N = -\lambda\, dt$$

or
$$\ln N = -\lambda t + \text{constant} \tag{19}$$
$$N = N_0\, e^{-\lambda t}$$

where N_0 must be the number of excited atoms at time $t = 0$. The form of this expression is independent of the choice of the zero of time; no matter how long the atoms have been in the excited state, one can arbitrarily set $t = 0$, and the number left at any subsequent time is related to the number at the (arbitrary) $t = 0$ by Eq. (19) (see Fig. 2).

In a time interval $t_{\frac{1}{2}}$ such that $e^{-\lambda t_{\frac{1}{2}}} = \frac{1}{2}$, the number of excited atoms is reduced by a factor of 2. Therefore the *half-life* of the excited state is given by $\lambda t_{\frac{1}{2}} = \ln 2$, or $t_{\frac{1}{2}} = 0.7/\lambda$. We often refer also to the *mean lifetime* of the state, which is the average, or mean, of the times which a large collection of atoms would spend in the excited state. By following the same mathematical pro-

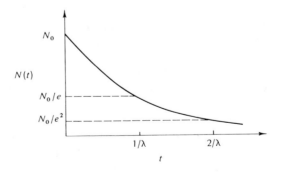

Fig. 2. *Number N of excited atoms remaining as a function of time, if N_0 are present at time $t = 0$. In each time interval of $1/\lambda$, the number remaining is divided by e.*

cedure used in Section 1.1 to find the mean free path of an atom, one can show that the mean lifetime τ is related to λ by

$$\tau = \frac{1}{\lambda}$$

You must realize, of course, that the decay process follows the laws of probability. It cannot be predicted exactly how many of a given collection of atoms will decay in any given time interval. For example, when there are only two excited atoms, the binomial distribution shows (Appendix A) that the probability is $\frac{1}{4}$ that both will decay in the next half-life, the probability is $\frac{1}{2}$ that one of them will decay, and the probability is $\frac{1}{4}$ that neither of them will decay.

Because an atom lives for a limited time in an excited state, the uncertainty relation $\Delta E \, \Delta t \geqslant \hbar/2$ imposes a fundamental limitation on the accuracy with which the energy of such a state can be defined. A short-lived state is said to be broader than a longer-lived state, because the atom can absorb a broader spectrum of frequencies in being excited to such a state. Thus each energy level in each atom is defined only to an accuracy which is permitted by the uncertainty relation for the lifetime of that state.[7] Observation of the " width " of a state can actually be used to obtain a rough estimate of the state's lifetime, in cases where the state is too short lived for its lifetime to be measured directly (see Chapters 15 and 16).

10.3   SELECTION RULES

Let us now turn our attention to the " dipole " matrix elements x_{mn}, y_{mn}, z_{mn}. If we can evaluate these, it is a simple matter to use Eq. (18) to determine the spontaneous transition probability, or to use Eq. (13) to find the induced transition probability. The evaluation is particularly easy for many pairs of states, because it happens that the matrix elements x_{mn}, y_{mn}, and z_{mn} are all equal to zero. In such cases the transition is said to be *forbidden* for dipole radiation. We have already seen that certain transitions were forbidden by selection rules developed empirically by spectroscopists (Section 7.3 and Chapter 9); now we are able to derive some of these rules.

The simplest example is the selection rule for the magnetic quantum number m:[8]

$$\Delta m = \pm 1 \text{ or } 0$$

[7] Of course, a definite amount of energy is absorbed and the same amount is re-emitted in each case (energy is conserved), but this amount can vary by ΔE from one event to another, or from one atom to another, because of the width of the energy level. The frequency corresponding to energy ΔE is called the *natural linewidth*.

[8] At this point, there is a slight risk of confusion between the quantum number m and the subscript m used as an index to identify the state. But we must run some such risk, because any letter that we could use as an index has already been assigned some other meaning.

which was necessary to explain the number of lines observed in the Zeeman effect (Section 9.2). This rule may be derived by considering only the wavefunction's ϕ dependence $e^{im\phi}$. If the wavefunction of the initial state is proportional to $e^{im\phi}$, and that of the final state is proportional to $e^{im'\phi}$, then the matrix element z_{mn} is proportional to

$$\int_0^{2\pi} e^{-im'\phi} z e^{im\phi} \, d\phi \tag{20}$$

But z is equal to $r \cos \theta$, so it may be removed from the integral, and the integral then vanishes at both limits, as long as $m' \neq m$. Thus transitions for which $m' \neq m$ cannot be induced by radiation which is polarized in the z direction, nor can such transitions occur spontaneously with the emission of light polarized in that direction; there is a *selection rule*

$$\Delta m = 0 \textit{ for light polarized in the z direction.}$$

For example, in the normal Zeeman effect (Section 1.2 and 9.2), the unshifted line results from a transition in which $\Delta m = 0$, so the light must be polarized in the z direction (the direction of $\boldsymbol{B}$). This line is not *emitted* along the z direction, because light waves are transverse.

If we replace z by x or y in expression (20), we find a different rule. With $x = r \sin \theta \cos \phi$, the integral may be written

$$\int_0^{2\pi} e^{-im'\phi} \cos \phi \, e^{im\phi} \, d\phi$$

We can evaluate this integral most easily by means of the substitution

$$\cos \phi = \tfrac{1}{2} (e^{i\phi} + e^{-i\phi})$$

which yields

$$\tfrac{1}{2} \left[\int_0^{2\pi} e^{i(m-m'+1)\phi} \, d\phi + \int_0^{2\pi} e^{i(m-m'-1)\phi} \, d\phi \right]$$

The first integral is zero unless $m - m' = -1$; the second integral is zero unless $m - m' = +1$. Thus, for radiation polarized in the x direction, the selection rule is

$$\Delta m = \pm 1$$

and it is easy to show that radiation polarized in the y direction obeys the same rule. Thus we have, considering all possible polarizations, the rule

$$\Delta m = \pm 1 \quad \text{or} \quad 0$$

for electric dipole radiation.

Again, the selection rule is manifest in the polarizations observed in the normal Zeeman effect. The shifted lines emitted in the y direction are polarized in the x direction; those emitted in the x direction are polarized in the y direction; and those emitted in the z direction are circularly polarized (see problem 6). (In the anomalous effect, *some* of the shifted lines come from $\Delta m = 0$ transitions and are polarized in the z direction.)

The selection rule on l, $\Delta l = \pm 1$, can be derived in much the same manner as was the rule for m. If $P_l^{|m|}(\cos\theta)$ is the θ dependence of the wavefunction for the initial state, and $P_{l'}^{|m|}(\cos\theta)$ is the θ dependence of the wavefunction for the final state, then for each matrix element the integral on θ vanishes unless $l' - l = \pm 1$. The general derivation of this result requires the use of two formulas involving the associated Legendre functions $P_l^{|m|}(\cos\theta)$; we state these without proof[9]:

$$\cos\theta \, P_l^{|m|} = \frac{(l - |m| + 1)P_{l+1}^{|m|} + (l + |m|)P_{l-1}^{|m|}}{2l + 1}$$

$$\sin\theta \, P_l^{|m|} = \frac{P_{l+1}^{|m+1|} - P_{l-1}^{|m+1|}}{2l + 1}$$

Given these formulas, derivation of the selection rule is straightforward, and is left as an exercise.

Parity.   The above selection rules were derived by using the specific wavefunctions of the hydrogen atom. But there is a more general rule which does not depend on any specific wavefunction; the rule is that *the parity of the wavefunction must change in an electric dipole transition.*[10] The reason for the rule is quite simple. If the initial state wavefunction and the final state wavefunction have the same parity, then the product of the two functions must have even parity (which is not changed by taking the complex conjugate of the final state wavefunction). Multiplying this function by x, y, or z yields a product which has odd parity, and the integral of a function of odd parity over all space is zero. Thus the matrix element always vanishes if the initial state and the final state have the same parity.

The rule that parity must change is, of course, satisfied in any one-electron transition for which $\Delta l = \pm 1$, because the parity of the associated Legendre function is even if l is even, and it is odd if l is odd. But there are other situations in which the parity selection rule imposes *additional restrictions* on the transitions. For example, in a multi-electron atom, when two electrons are involved in a transition, the sum of the l-values must change from odd to even or from even to odd. We may understand this rule by reference to the

[9] These formulas may be derived from a generating function for the Legendre polynomials. See, for example, L. Schiff, "Quantum Mechanics" p. 72 and p. 258. McGraw-Hill, New York, 1949.
[10] See Section 5.6 for a discussion of parity of wavefunctions.

Ca spectrum (Chapter 9, Fig. 2). There can be a transition from a 4p3d state (odd) to 3d4s or 4s4s (both even), but not to 4s4p (odd). Such a transition is forbidden even though it may satisfy the selection rule $\Delta L = \pm 1$ for the resultant angular momentum; for example, the transition 4p3d ^{3}D → 4s3p ^{3}P is forbidden. (A state such as 4p3d ^{3}D is often designated as ^{3}D°, the superscript o indicating odd parity.)

EXAMPLE PROBLEM 1 (*Selection Rules for the Simple Harmonic Oscillator*). Derive a selection rule for the simple one-dimensional harmonic oscillator by computing the matrix elements x_{mn} for the appropriate wavefunctions.

Solution. We require the matrix element $x_{mn} = \int u_m^* x u_n \, dx$. But it is not necessary to carry out a detailed computation, because we only need to know when x_{mn} is equal to zero. Therefore we make use of the orthogonality of the harmonic oscillator wavefunctions, by introducing the stepping operators $(d/dx) - ax$ and $(d/dx) + ax$, which convert one wavefunction into another (see Section 5.7). We can express the variable x in terms of these operators as

$$x = \frac{1}{2a}\left[\left(\frac{d}{dx} + ax\right) - \left(\frac{d}{dx} - ax\right)\right]$$

Thus we may write

$$x_{mn} = \frac{1}{2a}\left[\int u_m^*\left(\frac{d}{dx} + ax\right)u_n \, dx - \int u_m^*\left(\frac{d}{dx} - ax\right)u_n \, dx\right]$$

It was shown in Section 5.7 that

$$\left(\frac{d}{dx} - ax\right)u_n = \alpha u_{n+1}$$

and

$$\left(\frac{d}{dx} + ax\right)u_n = \beta u_{n-1}$$

where α and β are normalizing constants, and the subscripts indicate the value of the n quantum number. So we have

$$x_{mn} = \frac{1}{2a}\left[\alpha \int u_m^* u_{n+1} \, dx + \beta \int u_m^* u_{n-1} \, dx\right]$$

The orthogonality of the wavefunctions then requires that x_{mn} be zero unless $m = n + 1$ or $m = n - 1$. Thus the selection rule is $\triangle n = \pm 1$. Notice that this rule agrees with the comments in Section 5.7 about the correspondence principle; transitions for which Δn is greater than 1 would not involve photons whose frequency is equal to the frequency of oscillation.

Occurrence of Forbidden Transitions.   The selection rules which we have just examined are not absolute laws. It is found that some of the " forbidden " transitions actually do occur, although they occur far less often than the allowed transitions. How is this possible? We must not forget that these rules were derived on the basis of the dipole *approximation*, in which it is assumed that (1) the electric field is uniform over the entire atom, and (2) the magnetic energy has a negligible effect. These assumptions were made because a uniform electric field can have an effect which is much greater than the effect of the associated magnetic field or the effect of nonuniformity in the electric field. But if the effect of the uniform field on a given state is nil, then obviously other effects can no longer be neglected.

(a) *Electric Quadrupole and Higher Multipole Radiation.*   Let us first consider the effect of nonuniformity of the electric field. To do this, we must write the field as $E_{0_x} \cos(kx - \omega t)$, where as usual the wavelength λ is equal to $2\pi/k$. We can display the difference between this expression and the previous one ($E_{0_x} \cos \omega t$) most easily by writing it in exponential form as follows:

$$\cos(kx - \omega t) = Re[e^{i(kx - \omega t)}]$$

$$= Re[e^{ikx}e^{-i\omega t}]$$

$$= Re[\{1 + ikx + (ikx)^2/2! + (ikx)^3/3! + \cdots\}e^{-i\omega t}]$$

where Re [ ] denotes the real part of [ ]. The dipole approximation is now seen to be equivalent to replacing e^{ikx} by 1—that is, neglecting ikx and higher order terms in the series expansion of e^{ikx}. This is reasonable when the wavelength is long, for then $kx \ll 1$ over the dimensions of the atom. But if this procedure yields a null result, then it is logical to make the approximation $e^{ikx} \simeq 1 + ikx$; retaining the second term may lead to a nonzero result for the transition probability. To test this possibility, we must determine the potential energy when the electric field is proportional to x.

As before, we find the energy from the integral of the electric field. Integrating along the x direction, we find the potential energy to be proportional to x^2. Therefore the matrix element needed to determine T_{nm} [Eq. (12)] is of the form $\int u_m^* x^2 u_n \, d\tau$, or $(x^2)_{mn}$ rather than x_{mn}. When we consider all possible directions of polarization for the radiation, and when we integrate the field along all possible directions, we find that matrix elements of the form $(y^2)_{mn}$, $(x^2)_{mn}$, $(xy)_{mn}$, $(yz)_{mn}$, and $(zx)_{mn}$ also can contribute to T_{mn}. Just as the elements x_{mn}, y_{mn}, and z_{mn} are in the form of components of the electric *dipole* moment *vector*, the six elements listed above—$(x^2)_{mn}$, $(y^2)_{mn}$, $\ldots$—are in the form of the components of a second rank *tensor*, called the electric *quadrupole* moment between states n and m. When these matrix elements are nonzero, one can have electric quadrupole radiation.[11] The selection rules for electric quadrupole radiation can be worked out in the same manner as the rules for electric dipole radiation. The rules on m and l, for hydrogen-type wavefunctions, are

$$\Delta m = \pm 2, \quad \pm 1, \quad \text{or} \quad 0,$$

and

$$\Delta l = \pm 2 \quad \text{or} \quad 0,$$

for electric quadrupole radiation.

To understand the effect of these various selection rules, consider the 3d level of hydrogen, which can decay by dipole radiation to the 2p level ($\Delta l = -1$), or by quadrupole radiation to the 2s or 1s levels ($\Delta l = -2$). The probability of each transition is determined by the respective matrix elements. But the transition 3d $\rightarrow$ 2s, which is permitted only for quadrupole radiation, has a much smaller probability than 3d $\rightarrow$ 2p, because a typical quadrupole matrix element contains the fact kx, which is of order 10^{-2} for radiation considered here, and which is absent from the dipole matrix elements. Consequently we expect the probability of the 3d $\rightarrow$ 2s transition to be about $(10^{-2})^2 = 10^{-4}$ times the probability of the 3d $\rightarrow$ 2p transition (because the transition probability is proportional to the square of the matrix element). Thus the intensity of the spectral line for the 3d $\rightarrow$ 2p transition should be about 10,000 times the intensity of the lines for the 3d $\rightarrow$ 2s or 3d $\rightarrow$ 1s transitions; of 10,000 atoms in the 3d state, only one or two, on the average, will make the forbidden transition to either 2s or 1s.

If a transition is forbidden for both electric dipole and electric quadrupole radiation, there is still a possibility of radiation via higher order multipoles, whose matrix elements involve higher order terms in the expansion of e^{ikx}. For example, the term $(ikx)^2/2$ leads to electric octopole radiation, and the term $(ikx)^3/3!$ to electric hexadecapole radiation. (Of course an ideal octopole, classically, can be constructed from two equal and opposite quadrupoles; a hexadecapole from two equal and opposite octopoles, and so on.) Obviously, the intensity falls off very rapidly as one considers higher and higher poles; each order of multipole radiation is less intense, by a factor of order 10^{-4}, than the multipole of the next lower order.

(b) *Magnetic Dipole Radiation.* The magnetic interaction energy $-\mathbf{\mu} \cdot \mathbf{B}$ can also lead to transitions, when $\mathbf{B}$ is the oscillating magnetic field associated with electromagnetic radiation. If the electric field has amplitude E_0 the magnetic field has amplitude $B_0 = E_0/c$ in mks units. (In cgs units, $B_0 = E_0$.) If the $\mathbf{B}$ field is in the y direction, and we neglect spin, the electron energy in

---

[11] Just as an ideal dipole consists of two equal and opposite charges, an ideal quadrupole consists of two equal and opposite dipoles. Classically, the radiation from two equal dipoles oscillating 180° out of phase *almost* cancels; the result is quadrupole radiation.

such a field is proportional to $(e/2m)L_y(E_0/c)$, so that the matrix element is proportional to

$$\frac{eE_0}{2mc} \int u_m^* L_y u_n \, d\tau \tag{21}$$

where L_y is, of course, the operator for the orbital angular momentum of the electron. We would like to see what selection rules follow from this integral, and we would also like to compare the magnitude of the above expression with the magnitude of the corresponding expression

$$\tfrac{1}{2}eE_0 \int u_m^* x u_n \, d\tau \tag{22}$$

which determines the electric dipole transition probability.

We can determine the selection rules for m and l by rewriting L_y in terms of the "stepping operators" $L_x + iL_y$ and $L_x - iL_y$, as follows:

$$L_y = \frac{1}{2i} [(L_x + iL_y) - (L_x - iL_y)]$$

We recall from Section 7.3 that $L_x + iL_y$ is an operator which raises the value of the m quantum number by one unit while leaving the quantum number l unchanged. Similarly, the operator $L_x - iL_y$ lowers the value of m while leaving l unchanged. Therefore, by reasoning exactly analogous to that used in Example Problem 1, based upon the orthogonality of the wavefunctions, we find that

$$\Delta m = \pm 1 \qquad \text{and} \qquad \Delta l = 0$$

for magnetic dipole radiation.

To compare the magnitudes of electric dipole and magnetic dipole matrix elements, we take the ratio of expression (21) to expression (22), which is

$$\frac{\dfrac{1}{mc} \int u_m^* L_y u_n \, d\tau}{\int u_m^* x u_n \, d\tau}$$

The integral in the numerator, for an allowed transition, may be estimated to be equal to a typical value for the orbital angular momentum, or about $\hbar$. The integral in the denominator may be estimated to have the value of a typical atomic radius, or about $\tfrac{1}{2}$Å. Thus the ratio of the matrix elements is of order

$$\frac{\dfrac{\hbar}{mc}}{\tfrac{1}{2}\ \text{Å}} \simeq \frac{\hbar c}{mc^2 \times \tfrac{1}{2}\ \text{Å}}$$

$$\simeq \frac{2 \times 10^3\ \ \text{eV}}{2 \times 10^5\ \ \text{eV}}$$

$$\simeq 10^{-2}$$

So we see that matrix elements for magnetic dipole radiation, like those for electric quadrupole radiation, are of order 10^{-2} times those for electric dipole radiation, so that electric dipole radiation, when it is allowed, always makes the major contribution to the transition probability. But we see from the selection rules that there are transitions which are permitted for magnetic dipole radiation and forbidden for electric dipole, so that magnetic dipole radiation does have an important effect in some situations.

(c) Transitions Totally Forbidden for All Kinds of Radiation. In addition to the selection rules listed above for the various types of radiation, there is a rule that *transitions from $l = 0$ states to $l = 0$ states are always forbidden.* This rule could be proved mathematically, but it is also easy to understand from the nature of radiation. Electromagnetic radiation always carries angular momentum; each photon carries an angular momentum of $\hbar$. (The fact that radiation carries angular momentum as well as linear momentum is a direct consequence of Maxwell's equations, and was understood before the development of quantum theory; the amount of angular momentum predicted by classical theory comes out to $\hbar$ per photon.) It is possible for a photon to carry angular momentum away from an atom even when $\Delta l = 0$, as long as l is not zero, because the angular momentum vector of the electron can change direction. But a transition from $l = 0$ to $l = 0$ cannot provide a photon with the required angular momentum.

However, remember that we have neglected spin in this discussion. It is possible for a photon to obtain its angular momentum from a flip of the spin of the electron, in which case a transition from $l = 0$ to $l = 0$ could occur. The most famous example of this situation is the transition which leads to the 21-cm radiation from hydrogen, mentioned at the end of Chapter 7. These spin-flip transitions are, of course, very rare. The usual mechanism for de-excitation in such cases is via some nonradiative process, such as a collision with another atom.

10.4 AMPLIFICATION BY STIMULATED EMISSION OF RADIATION—THE MASER AND THE LASER

The ability of a photon to reproduce itself, by inducing an atomic transition to a lower energy state, was thought at one time to be of academic interest only.[12] However, since 1954 this ability has been exploited in an enormous variety of applications, including a frequency standard or "atomic clock," a low-noise amplifier for radio astronomy, and ultra high intensity, highly coherent light sources.

Devices based on stimulated emission of radiation are known as either masers or lasers. The nature of these devices is well described by their names; "MASER" is an acronym for "Microwave Amplification by Stimulated Emission of Radiation," and "LASER" is a similar acronym standing for "Light Amplification by Stimulated Emission of Radiation."[13]

Although masers and lasers function in quite different ways, for a variety of purposes, there are three fundamental requirements which are common to all such devices:

1. The existence, in the material filling the device, of *a pair of energy levels* such that transitions between these levels lead to emission of radiation of a desired frequency.

2. Some sort of *resonance* in the device, so that photons of the proper frequency spend sufficient time in the material to induce many transitions.

3. *Population inversion* of the two levels—a situation in which the *higher* of the two levels is *more heavily populated* than the *lower* level, so that, for radiation of the proper frequency, *stimulated emission* occurs more often than *absorption*, and the radiation is therefore amplified rather than

---

[12] In fact, four large U.S. industrial laboratories, which had sponsored a modest amount of research in this area, dropped the research around 1950 because it was felt to have little practical use. C. H. Townes, commenting on this fact in *Science* **159**, 699 (1968), said: "Consider the problem of a research planner setting out 20 years ago to develop any one of these technological improvements—a more sensitive amplifier, a more accurate clock, new drilling techniques, a new surgical instrument for the eye, more accurate measurement of distance, three-dimensional photography, etc. Would he have had the wit or courage to initiate for any of these purposes an extensive basic study of the interaction between microwaves and molecules? The answer is clearly No. . . . He would have tried other improvements of known techniques and very likely have achieved moderate success, but no breakthrough by orders of magnitude. It was the . . . atmosphere of basic research which seems clearly to have been needed for the real payoff."

[13] The laser actually functions more often as an oscillator rather than an amplifier, but one could hardly call it a "LOSER."

absorbed (as it would be if the population followed the Boltzmann distribution).

For a maser, the wavelength of the radiation is of the order of centimeters, so the energy levels involved are separated by an amount on the order of 10^{-4} eV. For a laser, the energy levels are separated by the 2 or 3 eV required for radiation of optical wavelengths. Thus the methods used to achieve resonance and population inversion differ widely for each device, and will be discussed individually.

The Ammonia Maser. The first device to operate on the principle of stimulated emission of radiation was the ammonia maser, developed by Gordon, Zeiger, and Townes in 1954.[14] It operates at the frequency of a so-called "inversion transition" of the ammonia (NH_3) molecule, at 2.387×10^{10} Hz, or a wavelength of 1.256 cm. This transition may be understood by reference to the structure of the NH_3 molecule (Fig. 3). The three hydrogen atoms form a triangle, and the nitrogen atom may lie on one side or the other of this plane. The nitrogen atom oscillates back and forth from one side of this plane to the other. But this oscillation is strictly a quantum-mechanical phenomenon because a position *in* the plane is classically forbidden for the nitrogen atom. Thus the potential energy of the system, when plotted against the displacement of the N atom from the plane, forms a potential well with a barrier inside, somewhat resembling Fig. 7 of Chapter 5. The inversion transition is a transition between two states with the respective symmetry shown in (b) and (c) of that figure. The energy difference between these two states is quite small because the wavefunction is very small in the forbidden region. You can see that, if there were no penetration into the forbidden region, the sym-

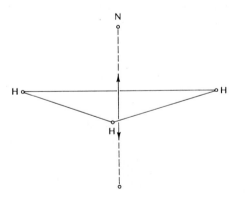

Fig. 3. The ammonia molecule.

[14] J. Gordon, H. Zeiger, and C. H. Townes, *Phys. Rev.* **95**, 282 (1954); **99**, 1264 (1955).

metric and antisymmetric wavefunctions, (b) and (c), respectively, would correspond to exactly the same energy.

The operation of the ammonia maser may be understood from Fig. 4. Population inversion is achieved by means of the electric field in the "focuser"; this is possible because molecules in the two energy states have slightly different charge distributions and thus are deflected differently by the electric field. The field is shaped so that molecules in the higher energy state enter the microwave cavity, while the other molecules are deflected away from the cavity. The cavity in turn is adjusted so that it resonates at the desired frequency of 2.387×10^{10} Hz. Thus a continuous cycle can be maintained—molecules enter, are stimulated by radiation in the cavity, and emit photons which in turn stimulate other molecules; as photons of the resonant frequency are emitted from the cavity, they are replaced by photons from entering molecules.

The principal application of this type of maser is as a frequency standard, or clock. The natural linewidth for the inversion transition is very small—about 4 kHz—and the radiation emerging from the cavity has an even narrower range of frequencies, because the cavity resonance can have a width as small as 10^{-2} Hz.[15] Thus the emitted frequency can be defined to one part in 10^{12}. This frequency does fluctuate as time passes, but over long periods of time a stability of one part in 10^{10} has been achieved. Such stability makes the ammonia maser valuable as a frequency standard or clock; a clock with this stability would gain or lose less than one second per century.

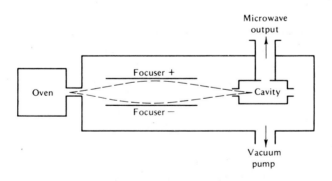

Fig. 4. *Schematic diagram of the ammonia maser.*

[15] See, for example, J. S. Thorp, "Masers and Lasers: Physics and Design," St. Martin's Press, New York, 1967. Spontaneous emission in the cavity occurs over the entire frequency range of 4 kHz, but photons which are not of the resonant frequency are quickly absorbed, while photons of the resonant frequency remain in the cavity for a long time, inducing the emission of additional photons identical to themselves. Thus the output is overwhelmingly in the narrow range selected by the cavity.

As a clock, the ammonia maser is superb, but as an amplifier it leaves much to be desired, for the following reasons:

1. The energy levels are fixed, so that one cannot "tune" the amplifier to any frequency other than $(2.387 \pm 0.0000004) \times 10^{10}$ Hz.

2. The bandwidth is limited to 4 kHz by the natural width of the line, so any signal with an appreciable amount of frequency modulation would be distorted.

3. Because a limited number of NH_3 molecules can be focused into the cavity, the power output is limited to about 10^{-9} W.

The Solid-State Maser.   If one uses a paramagnetic crystal instead of ammonia as the working substance, one can make a maser which has none of the three poor amplifier characteristics mentioned above. Such a maser can be tuned, because the energy levels can be members of a Zeeman multiplet whose separation changes when one changes the strength of an applied magnetic field. It has wide bandwidth, because nonuniform internal fields in the solid cause the energy levels to vary from one atom to another, so that different atoms in the same crystal can amplify different frequencies at the same time. Furthermore, there are many atoms available to amplify different frequencies as well as to provide a higher power output; the number of working atoms in the solid is enormous in comparison to the number of molecules in the ammonia beam. Thus it becomes quite practical to use such a maser to amplify weak microwave signals such as the signals from artificial satellites or interstellar gases.

However, another scheme for obtaining population inversion had to be devised in order to make such a maser work. In 1956, N. Bloembergen proposed a way to make use of *three* energy levels to maintain an inverted population across *two* of these levels.[16] Consider three levels of a Zeeman multiplet, and suppose that they are unequally spaced (Fig. 5). (In the simple theory of the Zeeman effect, the levels are equally spaced, but internal *electric* fields in a solid can cause additional energy shifts which destroy the equal spacing.)

If the splitting of the levels is much less than kT, the Boltzmann factor $e^{-E/kT}$ may be approximated as $1 - E/kT$, so that the equilibrium population of the levels varies linearly with E, as indicated in Fig. 5. Now suppose that the crystal is bathed in radiation of frequency $v_p = v_{13} = (E_3 - E_1)/h$, the so-called "pumping frequency." This radiation will induce rapid transitions between levels E_1 and E_3, in both directions. If the rate of these transitions greatly exceeds the spontaneous decay rate from E_3 to E_1, the populations of E_3 and E_1 will become equal; the levels are then said to be *saturated*. If the

[16] N. Bloembergen, *Phys. Rev.* **104**, 324 (1956).

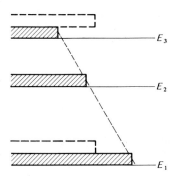

Fig. 5. *Population inversion. Solid bars indicate equilibrium populations of levels E_1, E_2, and E_3. If radiation causes rapid transitions between E_1 and E_3, the populations become as shown by the dashed bars; atoms transfer from E_3 to E_1 to equalize those populations while the population of E_2 is unchanged. Thus E_3 acquires a greater population than E_2.*

rate of these transitions is also much greater than the rate at which atoms can decay from level E_3 to level E_2, then the population of E_2 will be unaffected by the process. Since level E_2 originally had an equilibrium population smaller than the average population of E_3 and E_1, the equalization of populations in E_3 and E_1 causes E_3 to have a greater population than E_2. This inversion

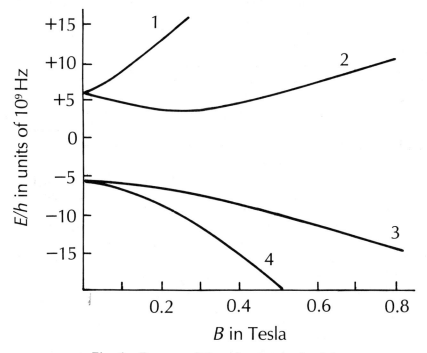

Fig. 6. *Zeeman splitting of energy levels of the Cr^{3+} ion in ruby, for magnetic field perpendicular to the crystalline c axis. Splitting between $\pm \frac{3}{2}$ levels and $\pm \frac{1}{2}$ levels at zero field is 1.14×10^{10} Hz.*

thus provides the possibility of maser action at the frequency $v = (E_3 - E_2)/h$; incoming radiation of this frequency can be amplified as it stimulates the transition from E_3 to E_2.

Bloembergen suggested that paramagnetic crystals might provide sets of levels which are suitable for this process, and it was not long before the possibility was realized in the ruby maser. Ruby is a crystal of Al_2O_3 in which some of the aluminum atoms have been replaced by chromium. The chromium appears in the lattice as a Cr^{3+} ion, with three $3d$ electrons outside a closed shell. In the ground state, the total orbital angular momentum of these electrons is zero, and their total spin angular momentum state is $S = 3/2$. Thus there are four substates, with $m_J = \pm 3/2$ and $\pm 1/2$, which would produce a set of levels like that of Fig. 6, Ch. 9, *if the ion were free*. But in the lattice, the states split in a different way; there are two energy levels even in zero magnetic field, because of the internal electric field of the crystal. The $\pm 3/2$ states lie 4.7×10^{-5} eV below the $\pm 1/2$ states when $\mathbf{B} = 0$, because of the difference in electric charge distribution between these sets of states.

When $\mathbf{B}$ is applied at right angles to the internal $\mathbf{E}$ field, the states split into four levels as shown in Fig. 6. Because both $\mathbf{B}$ and $\mathbf{E}$ are responsible for the energy of each state, there is no unique axis along which m_J can be defined, and each eigenstate of energy must be a superposition of states with different values of m_J.

To use these levels in a maser, one adjusts $\mathbf{B}$ to ''tune'' the energy difference between states 3 and 4 to the frequency that one wishes to amplify. Then one bathes the crystal in radiation at the frequency of the transition between states 2 and 4. When this radiation is sufficiently intense to equalize the populations in 2 and 4, the population in 4 becomes smaller than that in 3, and the ruby can amplify the desired frequency. This has been done in radio telescopes, to amplify the 21-cm radiation from interstellar hydrogen. (See Section 7.3 and Prob. 10).

The bandwidth of the ruby maser is of order 10^7 Hz—wide enough for a variety of applications, including the reception of FM signals from an artificial satellite. The wide bandwidth comes about because each Cr^{3+} ion is the source of an internal magnetic field which influences the energy levels of the neighboring Cr^{3+} ions. Because the ions are at random positions in the lattice, each one feels a different field, has a different set of energy levels, and thus responds to a different frequency. With a typical concentration of 0.05 percent Cr, the average potential energy between one ion and the next is of order 10^{-7} eV, and a given energy level of a given ion may differ from the average by a few times 10^{-8} eV. Such an energy difference corresponds to a frequency of the order of 10^7 Hz.

It was mentioned that the maser is a particularly good amplifier because of its low "noise". Noise in a conventional amplifier results from thermal agitation of electrons in the circuit elements. In the maser, spontaneous transitions between energy levels are equivalent to this agitation. But for microwave frequencies, the probability of a spontaneous transition is quite low, relative to the probability of an induced transition, even when the input signals are the very small ones received in radio astronomy. One can verify this assertion by use of Eq. (17). The reduction in noise is illustrated vividly in Fig. 7.

The Laser. Although the basic principles of a three-level laser are the same as those of a three-level maser, there are important differences in the applica-

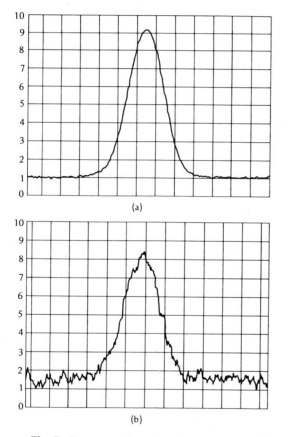

(a)

(b)

Fig. 7. Response of radio telescope as earth's rotation brought the telescope into line with the source Cassiopeia A. Both curves were obtained with the same system, except that (a) was obtained with a maser preamplifier, and (b) was obtained with a conventional preamplifier having the same gain. [From M. E. Bair et al., IRE Trans. Antennas Propagat. AP-9, 43 (1961).]

tion of those principles. Many of these differences arise because of the factor ω^3 in formula (18); this factor means that spontaneous transitions occur about 10^{18} times faster at optical frequencies than at microwave frequencies.

Because of this fact, a laser cannot serve as a low-noise amplifier for optical radiation. Instead, it serves simply as a high-intensity source of such radiation. (This is not to say that the laser is any less important for this reason; the coherence and extremely high intensity of laser light make the laser quite valuable.)

The high spontaneous transition rate also affects the pumping of a laser. Consider three levels $E_1 < E_2 < E_3$, whose energy separations correspond to optical frequencies. In general, an attempt to create an inverted population between E_2 and E_3 is doomed to failure; pumping at the frequency $v_{13} = (E_3 - E_1)/h$, as is done for a maser, will equalize *all three* populations, because of the rapid rate of spontaneous transitions between E_3 and E_2, and between E_2 and E_1.

But optical pumping is possible if one of the transitions is forbidden. For example, suppose that the quantum number l is equal to 0 for E_1, 1 for E_3, and 2 for E_2. Then the transition $E_2 \rightarrow E_1$, with $\Delta l = 2$, is forbidden, although $E_1 \rightarrow E_3$ and $E_3 \rightarrow E_2$ are allowed. Now application of high-intensity radiation of frequency v_{13} will cause level E_2 to acquire a large population, because atoms will go rapidly from E_1 to E_3 to E_2, and then be delayed at E_2. Thus there will be an inverted population between E_2 and E_1, with the possibility of laser action at frequency $v_{12} = (E_2 - E_1)/h$.

As in the maser, a resonance condition is needed to complete the action. This is achieved by making the active material into a cylindrical shape and placing parallel plane mirrors at each end. Then, after an inverted population is prepared by pumping, the action is initiated by spontaneous transitions between E_2 and E_1. (These transitions, although they are forbidden, do occur often enough to start the laser fairly promptly after inversion is achieved.) A photon emitted in a spontaneous transition may travel in any direction, but one which does not travel along the axis of the cylinder leaves the crystal very quickly and has little chance to stimulate further emissions (see Fig. 8). On the other hand, a photon traveling along the axis is reflected back and forth many times, stimulating the emission of other photons. These other photons, being identical to the photon which stimulated them, also travel along the axis, and in turn stimulate the emission of still more identical photons.

The output of the laser comes through one of the mirrors, which partially transmits the light. Because the vast majority of the photons are traveling along the axis of the cylinder, the beam is extremely well collimated and of a high intensity which does not diminish rapidly as it travels. The divergence of the beam is determined primarily by the wave nature of light, which dictates that the beam must diverge by an angle $\theta \gtrsim \lambda/D$, where D is the initial diameter

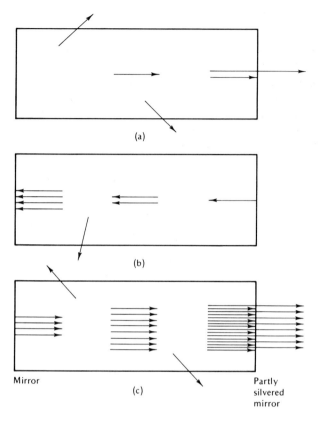

Fig. 8. *Buildup of coherent radiation in a laser. (a) Spontaneous emission of photons results in some which escape, with one photon traveling along axis of laser. Before reaching the end mirror, this photon has stimulated emission of second photon. One photon escapes, the other is reflected. (b) Reflected photon travels back along axis, stimulating emission of more photons. (c) Photons are reflected again, and more emissions are stimulated before some photons escape through right end. Intensity of escaping photons builds up until the rate of stimulated emissions equals the pumping rate. If pumping is maintained, laser can operate continuously.*

of the beam. Thus a laser beam of wavelength $\lambda = 5000$ Å, coming from a 1-mm-diameter laser, diverges by an angle $\theta = 5 \times 10^{-4}$ rad, or about $0.03°$. Such a beam is only a few feet in diameter after traveling one mile. Thus its intensity diminishes quite slowly with distance, and the beam is useful for long-range communications. It is possible to reduce the divergence of the beam still further by increasing D; this can be done with lenses or mirrors which diverge the beam and then recollimate it at a larger diameter. For example, a laser beam was directed at the moon through a 48-in. telescope acting in "reverse," so that it emitted a 48-in. diameter beam; such a beam ideally spreads by an angle of only 10^{-6} rad. It was estimated to be four miles in diameter when it reached the moon (May 9, 1962).

There are other ways to obtain a highly collimated light beam, although not one of such high intensity. But the laser beam is unique in its high coherence, a result of the fact that an entire beam is triggered by a single photon. The photons which are stimulated by this one photon all belong to a single wave, rather than forming separate waves with independent phases as in a conventional light source. Of course, as the laser operates, it emits different wave trains which result from different spontaneous emissions; but each *single wave train*, coming from a single initial photon, may be as much as several kilometers long. This property of high coherence simplifies many experiments in diffraction and interference of light, and it has made possible the production of three-dimensional pictures, or holograms. For further discussion of the multitude of developments in this field, we refer you to the vast literature which has developed, including entire journals devoted to the subject.

In summary, we have seen that very simple perturbation theory, when applied to the time-dependent Schrödinger equation, makes important predictions about selection rules and stimulated emission of radiation. Originally a laboratory curiosity, the latter has led to the development of industries and techniques that are now indispensable in modern technology.

PROBLEMS

1. Consider a particle of mass m in its ground state in an infinitely deep one-dimensional square potential well. A perturbing potential equal to a constant δ is suddenly turned on at time $t = 0$, and is turned off at $t = t'$. Use Eq. (8) to find the probability that the particle will be found in the first excited state at time $t = t'$.

2. A hydrogen atom is in the 1s state when an external electric field is applied along the z direction. The field is zero for time $t < 0$, and then becomes equal to $E_0 e^{-t/\tau}$. Find the first-order probability that the atom will be in the 2p state (u_{210}) for time $t \gg \tau$. (Neglect the spontaneous transition probability 2p $\rightarrow$ 1s.)

3. Find the Fourier transform $\phi(\omega)$ of a wave packet of the form

$$f(t) = E \cos \omega t \qquad (0 < t < t')$$
$$f(t) = 0 \qquad \text{otherwise}$$

(Use the equations of Section 4.3, with k and x replaced by ω and t, respectively.) Then compute $|\phi(\omega)|^2$, and thereby show that the distribution of frequencies in the wave packet is in agreement with Eq. (11), as stated in the text.

4. Compute the mean lifetime of the 2p state of the hydrogen atom, using expression (18), Eq. (19), and the appropriate wavefunctions from Table 1, Chapter 7. Compare with the result of Problem 10, Chapter 3.

5. Perform the exercise mentioned in Section 10.3 to show that $\Delta l = \pm 1$ for electric dipole radiation.

6. Use the wavefunctions from Table 1, Chapter 7 to calculate electric dipole and electric quadrupole matrix elements for transitions 3d → 2p, 3d → 2s, and 3p → 1s in the hydrogen atom. From your results verify that the nonzero electric quadrupole elements are of order 10^{-2} times the nonzero electric dipole matrix elements, as stated in the text.

7. Circularly polarized light can be considered to be a superposition of two components, linearly polarized at right angles to one another, whose amplitudes are equal and whose phases differ by $\pi/2$.
   (a) Repeat the steps leading to Eq. (10), and show that, in the electric dipole approximation, probabilities of transitions induced by circularly polarized light are determined by matrix elements of the form $(x \pm iy)_{mn}$.
   (b) Show that the matrix element $(x + iy)_{mn}$ is nonzero only when $\Delta m = +1$ (the magnetic quantum number increases by 1), and that $(x - iy)_{mn}$ is nonzero only when $\Delta m = -1$. Explain the connection between these facts and the polarizations observed in the normal Zeeman effect (Sections 1.2 and 9.2).

8. Compute the approximate uncertainty in the energy of the 2p state of the hydrogen atom, given that the mean lifetime of this state is of the order of 10^{-8} sec. Compare this uncertainty with the magnitude of the fine-structure splitting (Section 7.3).

9. For the system described in Problem 12, Chapter 7, determine the selection rule(s) and compute the mean lifetime of the first excited state.

10. In the normal Zeeman effect, how many lines are emitted at a 45° angle to the magnetic field direction? Describe the polarization of each of these lines (See problem 7.)

11. By direct measurement of Fig. 6, find
    (a) the value of B at which a transition between levels 3 and 4 involves a wavelength of 21 cm.

(b) the pumping frequency needed to equalize the populations in levels 2 and 4 at this value of B.

(c) the relative populations in levels 3 and 4 under these conditions, when the temperature is 2 K.

12. If the Hamiltonian changes suddenly from H_0 to H_1, and the eigenfunctions of both H_0 and H_1 are known, one can use the so-called *sudden approximation* to find the probabilities that the system is in the various possible final states (eigenstates of H_1). Suppose that

$$H_0 u_n = E_n u_n \quad \text{and} \quad H_1 v_m = E_m v_m$$

and let
$$\psi = u_n e^{-iE_n t/\hbar} \quad \text{for } t < 0$$

Then the final state must be a superposition of the eigenfunctions v_m multiplied by the appropriate time dependence, or

$$\psi = \Sigma b_m v_m e^{-iE_m t/\hbar} \qquad t > 0$$

where $|b_m|^2$ gives the probability that the system will be found in state v_m when it is observed. The time-dependent Schrödinger equation [Eq. (9), Ch. 5] shows that ψ must be a continuous function of time, so at $t = 0$,

$$\psi = u_n = \Sigma b_m v_m$$

We can find a given coefficient b_m by multiplying both sides of the above equation by $v_m{}^*$ and integrating over all space, obtaining

$$b_m = \int\!\int\!\int v_m{}^* u_n d\tau$$

Apply this equation to the beta-decay process, in which an atom of ³H suddenly changes to singly-ionized ³He by emitting a high-energy electron from its nucleus. If the original ³H electron was in the 1s state, find the probability that this electron now is in the 1s state in the ³H ion.

11

Quantum Statistics

The applications of quantum theory are not limited to extremely small systems such as isolated atoms or molecules. You already know that the quantum theory is *valid* for systems of large size as well as for small systems; but you might not realize that, for very large scale behavior of many systems, liquids or solids, the quantum theory makes predictions which sometimes differ in startling ways from those of classical theory.

At first glance, the task of constructing a quantum theory for a solid or liquid might seem hopeless. If we cannot even solve for the energy levels of a two-electron atom without making approximations, what can we do with a system containing 10^{23} particles? But we are encouraged by the example of classical theory; although the three-body problem of classical mechanics remains unsolved in general, classical theory has been quite successful in the use of statistical methods to predict the *average* behavior of systems containing many particles.

Classical statistical theory describes certain properties of gases quite successfully. We have seen (Chapter 1) that this theory involves the Boltzmann distribution function $n(\varepsilon)$, which is such that $\int_{\varepsilon_1}^{\varepsilon_2} n(\varepsilon)\, d\varepsilon$ is the average number of particles with energy between ε_1 and ε_2. But the Boltzmann distribution fails when it is applied to liquids or solids. As we shall see, it cannot be used to predict such a simple thing as the specific heat of a solid.

The reason for this failure lies in the indistinguishability of elementary particles. In a gas, although the atoms are identical, they are usually far enough apart so that they can be distinguished by their positions, and one could, in principle, "follow" a single atom as it wends its way through a gas. But the electrons in a solid or a liquid, even those belonging to different atoms, are often so close together that their wavefunctions overlap. The uncertainty principle thus prevents our defining their trajectories well enough to distinguish one from another. Or to put it another way, the motion is described by means of a single wavefunction for the *system*; this wavefunction can give probabilities for finding an electron in a certain region, but it cannot say *which* electron is found.

Except for the problem of indistinguishability, the classical derivation of the Boltzmann distribution does not run afoul of any principle of quantum theory. Let us therefore attempt to construct a distribution function which will be similar to the Boltzmann function but which will take proper account of the indistinguishability of particles. To see how we might do this, let us first analyze an example system containing only four particles.

11.1 THE THREE KINDS OF STATISTICS: AN EXAMPLE

Let us consider a system of four identical particles, and suppose that each particle can possess energy only in integral multiples of a quantity E; that is, the energy levels for each particle are $\varepsilon_1 = 0$, $\varepsilon_2 = E$, $\varepsilon_3 = 2E, \ldots$ (The assumption that the possible energies are discrete is made only to simplify the calculation, and has no quantum mechanical significance, because E can be made as small as we wish. The quantum theory enters only in the treatment of the fact that the particles are identical.)

For generality, we make the further assumption that there is more than one state for each energy level; that is, we say that some energy levels are *degenerate*. Again, the terminology sounds quantum mechanical, but it can be applied to classical theory as well. We wish to allow for the fact that certain energy levels are easier to populate than others; this is true classically as well as quantum mechanically. (For example, consider the Maxwell velocity distribution, $f(v) \propto v^2 e^{-mv^2/2kT}$. The probability that an individual molecule has a speed between v and $v + dv$ is proportional not only to the Boltzmann factor

$e^{-mv^2/2kT}$ but also the factor v^2, which is analogous to the degeneracy of a quantum-mechanical energy level.)

We shall use the symbol g_s to denote the degeneracy of the sth level. In our example we let $g_1 = 1$, $g_2 = 2$, $g_3 = 2$, $g_4 = 3$, $g_5 = 3$, and $g_6 = 4$. The energy levels and their degeneracies are shown graphically in Fig. 1; the states are depicted as boxes into which the particles are placed.

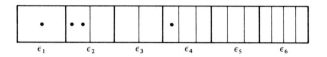

ϵ_1 $\qquad$ ϵ_2 $\qquad$ ϵ_3 $\qquad$ ϵ_4 $\qquad$ ϵ_5 $\qquad$ ϵ_6

Fig. 1. *Graphic representation of states of a particle as "boxes." Heavy lines denote boundaries between different energy levels; lighter lines show boundaries between different states having the same energy. The dots indicate the particles; in the case illustrated, one particle has energy ε_1, two have energy ε_2, both occupying the same state, and a fourth particle has energy ε_4.*

The physical and chemical properties of the system are related to the average distribution of the particles among the various energy levels. This distribution, of course, depends on the temperature of the system, so let us assign a temperature by setting the total energy of the system equal to $5E$.

Before proceeding further we must pause to make sure that our definitions are clear. We define a *distribution* to be the set of numbers n_s of particles in each *energy level* ε_s, and we define an *arrangement* as a definite way of putting the particles into each of the *states*. (An arrangement is sometimes called a microscopic distribution.) Thus there can be *several* arrangements which all correspond to the *same* distribution. For example, with reference to Fig. 1, if one of the two particles in the second box were moved over to the third box (so that it still had energy ε_2, but occupied a different state), we would have a different *arrangement* but the same *distribution*: $n_1 = 1$, $n_2 = 2$, $n_3 = 0$, $n_4 = 1$, $n_5 = 0$, and $n_6 = 0$.

Now in order to find an average distribution, we must take into account all possible distributions, and combine them with the appropriate weight for each one. We determine this weighting by assuming that each *arrangement* is equally probable, because we have no reason to believe that one arrangement should be more probable than any other arrangement. If each arrangement is equally probable, the probability that a given *distribution* is present is proportional to the *number of different arrangements* which correspond to that distribution. By counting the number of particles in a given level for each distribution, and then multiplying each number by the probability of occurrence of that distribution, we can obtain the contribution of each individual distribution to the average distribution. It is this average distribution which is governed by the Boltzmann factor in the case of a gas.

At this point the indistinguishability of the particles enters the problem. In counting up the number of arrangements, we count only those which are distinct from one another. If the particles are indistinguishable, two arrangements which differ only in the exchange of two particles would be counted as only *one* arrangement. Furthermore, some arrangements are not permitted at all, if the particles obey the Pauli exclusion principle. As we have seen, electrons do obey this principle, as do all other particles of half-integer spin, but particles with integer spin (such as photons or helium-4 nuclei) do not. Thus there are two kinds of quantum statistics, depending on the kind of particles involved; particles which obey the exclusion principle are said to be governed by *Fermi–Dirac statistics*, and other particles are governed by *Bose–Einstein statistics*.

Table 1

Possible Distributions of Four Particles

	Distribution						Arrangements			Probability		
	n_1	n_2	n_3	n_4	n_5	n_6	Boltz.[a]	FD[b]	BE[c]	Boltz.	FD	BE
(a)	3	0	0	0	0	1	16	0	4	16/464	0	4/39
(b)	2	1	0	0	1	0	72	0	6	72/464	0	6/39
(c)	2	0	1	1	0	0	72	0	6	72/464	0	6/39
(d)	1	2	0	1	0	0	144	3	9	144/464	3/5	9/39
(e)	0	3	1	0	0	0	64	0	8	64/464	0	8/39
(f)	1	1	2	0	0	0	96	2	6	96/464	2/5	6/39
							464	5	39			

[a] Boltzmann.
[b] Fermi–Dirac.
[c] Bose–Einstein.

Now we are ready to solve our example. Table 1 shows all of the possible distributions which obey the given conditions, together with the number of distinct arrangements for each distribution, according to each of the three types of statistics, including Boltzmann. The probability of each distribution, found by assuming each *arrangement* to be equally likely, is also shown in the table.

To understand this table, let us see how the first line was worked out. Energy level ε_6 has energy $5E$, which is the total energy of the four particles; therefore placing one particle in that level forces the other particles to go into the zero energy level. The degeneracy of level ε_6 is $g_6 = 4$, so there are four possible arrangements for Bose–Einstein (BE) statistics, corresponding to the four different states which the particle in level ε_6 may occupy. But for Boltzmann statistics, the particles are distinguishable, so a single BE arrangement becomes four Boltzmann arrangements, one for each possible particle in level ε_6; thus there are $4 \times 4 = 16$ arrangements for Boltzmann statistics. Finally, for Fermi–Dirac (FD) statistics, there are *no* arrangements, because the exclusion principle prevents three particles from occupying the one state in the zeroth level.

Line (d) illustrates the counting of arrangements for Fermi–Dirac statistics. There is only one way to place the two particles into the two states with energy ε_2; one particle must go into each state, because both particles cannot occupy the same state. There are three states available for the particle in level ε_4, so this accounts for the presence of three possible arrangements. In contrast, there are $3 \times 3 = 9$ arrangements for this distribution in Bose–Einstein statistics, because there are three ways to place the two particles into level ε_2; one in each state, two in the first state, or two in the second state.

The "probability" column in Table 1 is found simply by dividing the number of arrangements for a given distribution by the total number of arrangements. These probabilities may then be used to find the average distribution; we simply multiply the probability of each individual distribution by each n_s in that distribution, and we add the results for a given s, to find the n_s for the average distribution. You can easily verify that the average distributions are those shown in Table 2.

The total of all the values of n_s is, of course, 4 for each distribution. It may seem a bit strange that n_2 is greater than n_1 in each distribution, while the values of n_s otherwise decrease with increasing s. The reason, of course, lies in the degeneracy; g_2 is 2, whereas g_1 is only 1. It is useful to define the quantity n_s/g_s, called the *occupation index*, which is the average number of particles

Table 2

Average Distributions Derived from Table 1

	n_1	n_2	n_3	n_4	n_5	n_6
Boltzmann:	576/464	648/464	328/464	216/464	72/464	16/464
Bose–Einstein:	51/39	54/39	26/39	15/39	6/39	4/39
Fermi–Dirac:	1	8/5	4/5	3/5	0	0

per state in the sth level. The occupation index does decrease with increasing s, or increasing energy, and we shall soon see that this index is independent of g_s for all kinds of statistics. It is the occupation index that is given by the expression $e^{-\varepsilon_s/kT}$ in Boltzmann statistics. In our example, the occupation indices are as shown in Table 3.

Table 3

Average Occupation Indices Derived from Table 1

	n_1/g_1	n_2/g_2	n_3/g_3	n_4/g_4	n_5/g_5	n_6/g_6
Boltzmann	576/464	324/464	164/464	72/464	24/464	4/464
Bose–Einstein	51/39	27/39	13/39	5/39	2/39	1/39
Fermi–Dirac	1	4/5	2/5	1/5	0	0

The difference between the Boltzmann and the Bose–Einstein distributions happens to be slight in this case, but the Fermi–Dirac distribution is significantly different from either of the others.

This example should have been helpful in clarifying the manner of application of the basic principles, so that you will be prepared for the more general derivation which follows. Obviously, the distribution of four particles should not be precisely described by the statistics of large numbers, but after we have developed the form of the three general distribution laws, you may wish to return to these example distributions to test their similarity to the general laws.

11.2 DERIVATION OF THE GENERAL FORM FOR EACH DISTRIBUTION FUNCTION

We derive the general form for each of the three types of distribution by counting arrangements of N particles with total energy E, much as we did in the four-particle example. Obviously, when there are 10^{23} particles we cannot actually enumerate the arrangements, so we must be more clever; we must find a general formula for the number of arrangements in a given distribution.

Even then, there are an enormous number of distributions to consider in computing the average distribution, but we have the advantage that it is sufficient to find the *most probable* distribution rather than the average distribution. The most probable distribution, like any other *specific* distribution of 10^{23} particles, has a very small probability of occurrence. But all of the distributions which actually occur differ from the most probable one by a negligible amount. We may understand this fact by considering the standard deviation in the number of particles to be expected in any given energy level or set of levels. When there are of the order of 10^{23} particles, the number of particles in the states in any measurably large energy range could easily be of the order of 10^{16}; the standard deviation in 10^{16} is 10^8 (see Appendix A), a large number of ordinary standards, but only one part in 10^8. In other words, most of the possible distributions differ from the most probable distribution by less than one part in 10^8, for the number of particles in any particular energy range, and we are justified in computing the most probable distribution rather than the average distribution.

We begin the computation by finding P_s, the number of ways in which the n_s particles in the sth energy level may be put into the g_s states in that level. Then we find the *total number of arrangements* for the *whole set* of given numbers $n_1, n_2, \ldots, n_s, \ldots$ by taking the product of all the numbers $P_1, P_2, \ldots, P_s, \ldots$. This product equals the statistical *weight* $W(n_1, n_2, \ldots, n_s, \ldots)$ for the distribution $n_1, \ldots, n_s \ldots$. That is,

$$W(n_1, \ldots, n_s, \ldots) = \prod_{s=1}^{\infty} P_s$$

As mentioned above, the total number W is proportional to the probability of finding the distribution to be the set $n_1, n_2, \ldots, n_s, \ldots$ at any given time.

After finding an expression for $W(n_1, n_2, \ldots, n_s, \ldots)$, we shall find the set of numbers which maximizes W, subject to the two conditions

(1) that the total number N of particles is fixed:

$$\sum_{s=1}^{\infty} n_s = N \tag{1}$$

(2) that the total energy E is fixed:

$$\sum_{s=1}^{\infty} n_s \varepsilon_s = E \tag{2}$$

Computation of Statistical Weights.   The computation of P_s and W is shown below for each of the three types of statistics.

Boltzmann.   In this case only, we assume that the particles are distinguishable, so we are concerned with *which* particles we choose to put in each state. Let us consider level 1 first. From the N particles available, we can choose n_1 particles for this *level* in $N!/n_1!(N - n_1)!$ different ways.[1] These

---

[1] There are N choices for the first particle, $N - 1$ for the second, etc., so there are $N(N - 1)(N - 2) \ldots (N - n_1 + 1)$ ways to choose the n_1 particles in a *given order*. But we have the same n_1 particles regardless of the order of choosing them, so we must divide the above number by the number of possible orders of choice, which is $n_1!$. Thus the number of ways of choosing n_1 objects from N objects is $N!/n_1!(N - n_1)!$.

particles may then be distributed among the g_1 *states* of level 1 in $g_1{}^{n_1}$ different ways.[2] Thus there are a total of $P_1 = N!g_1{}^{n_1}/n_1!(N - n_1)!$ ways to put n_1 particles into level 1.

Having placed n_1 particles into level 1, we have $N - n_1$ remaining particles; we may choose n_2 of these for level 2 in $(N - n_1)!/n_2!(N - n_1 - n_2)!$ different ways. These particles may be distributed among the g_2 states of level 2 in $g_2{}^{n_2}$ ways, so there are a total of

$$P_2 = \frac{(N - n_1)!\, g_2^{n_2}}{n_2!(N - n_1 - n_2)!}$$

ways to put n_2 particles into level 2, *after* one has put n_1 particles into level 1.

By now it should be clear that, after one puts n_1 particles into level 1, n_2 into level 2, n_3 into level 3, and so on, up to n_{s-1} into level $s - 1$, there are in general

$$P_s = \frac{(N - n_1 - n_2 - \cdots - n_{s-1})!\, g_s^{n_s}}{n_s!(N - n_1 - n_2 - \cdots - n_s)!}$$

ways to put n_s particles into level s. The total number of ways to put N particles into the levels to produce a given distribution $n_1, n_2, \ldots, n_s \ldots$, must be the product $P_1 P_2 \cdots P_s \cdots$. It is easy to see that the intermediate factors $(N - n_1)!$, $(N - n_1 - n_2)!$, etc. cancel in numerator and denominator, so the final result is

$$W(n_1, \ldots, n_s, \ldots) = \prod_{s=1}^{\infty} P_s$$

$$= N! \prod_{s=1}^{\infty} \frac{g_s^{n_s}}{n_s!} \tag{3}$$

Bose–Einstein. In the quantum statistics, there is no factor analogous to the factor $N!/n_1!(N - n_1)!$ which appears in the Boltzmann statistics, because that factor involves a choice of particles for a given level; when the particles are indistinguishable, this choice is irrelevant. Now we need only know the factor analogous to $g_1{}^{n_1}$—the factor which counts the number of ways in which the n_1 particles can go into the g_1 states of level 1. Again, since the particles are indistinguishable, we do not count the choices for a given particle; instead, we consider the whole set of particles at once. We draw a picture similar to Fig. 1, but we show only the states in level s (Fig. 2). There are n_s particles in the g_s states. We obtain a different arrangement each time we change the number of particles in one or more states, by moving particles

---

[2] There are g_1 places for each of the n_1 particles, making $g_1{}^{n_1}$ possibilities in all.

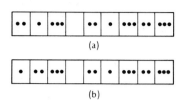

Fig. 2. (a) One arrangement of n_s particles in the g_s states of level s ($g_s = 9$). (b) A second arrangement of the same n_s particles; this arrangement is obtained from the arrangement of (a) simply by interchanging the second particle and the first partition.

from one state to another, but we obtain no new arrangement if we simply exchange two particles. We can count the arrangements by a simple trick, as follows: There are $g_s - 1$ partitions, or dividing lines, between the states, so that the total number of particles plus partitions is $n_s + g_s - 1$. We can obtain new arrangements by permuting particles and partitions; for example, the arrangement of Fig. 2b was obtained from Fig. 2a by interchanging the second particle and the first partition (between states 1 and 2). There are $(n_s + g_s - 1)!$ possible permutations of particles and partitions. However, many of these do *not* produce new arrangements; permutation of particles among themselves, or of partitions among themselves, changes nothing physically. The particles may be permuted in $n_s!$ ways, and the partitions in $(g_s - 1)!$ ways. Thus there are $n_s!(g_s - 1)!$ permutations of *each* arrangement, and the total number of arrangements must be the quotient

$$P_s = \frac{(n_s + g_s - 1)!}{n_s!(g_s - 1)!}$$

$$= \frac{\text{number of permutations}}{\text{number of permutations per arrangement}}$$

Therefore

$$W(n_1, \ldots, n_s, \ldots) = \prod_{s=1}^{\infty} P_s$$

$$= \prod_{s=1}^{\infty} \frac{(n_s + g_s - 1)!}{n_s!(g_s - 1)!} \tag{4}$$

Fermi–Dirac.   Here again we need only know how many arrangements there are of the n_s particles in the g_s states of level s. But we count the arrangements in a different way, because of the condition that there is no more than one particle in each state. Because of this condition, we can simply divide the states into two groups—the n_s occupied states and the $g_s - n_s$ unoccupied states. We can choose n_s objects from a total of g_s objects in $g_s!/n_s!(g_s - n_s)!$

different ways (see the footnotes under Boltzmann statistics), so this must be the number of ways of choosing n_s *states* to be occupied when there are g_s different states. (Remember that the *states* are distinguishable, even though the *particles* are not.) Therefore

$$W(n_1, \ldots, n_s, \ldots) = \prod_{s=1}^{\infty} P_s$$

$$= \prod_{s=1}^{\infty} \frac{g_s!}{n_s!(g_s - n_s)!} \tag{5}$$

Computation of n_s. We now have the three expressions for $W(n_1, \ldots, n_s, \ldots)$, the statistical weight of a given distribution $n_1, \ldots, n_s, \ldots$. In each case we can find the most probable distribution by finding the set of numbers n_s which maximizes W. We do this by maximizing the logarithm of W rather than W itself; this is equally effective and it simplifies the procedure by changing each product to a sum.

To find the maximum, we simply set the variation in $\ln W$ equal to zero as the numbers n_s are varied, just as one finds the maximum of a function of a single variable by setting the derivative equal to zero. Thus we write that the variation in $\ln W$ is

$$\delta(\ln W) = 0 \tag{6}$$

where $\delta(\ln W)$ is obtained by varying the numbers n_s by amounts δn_s and then finding the difference between the newly calculated value of W and the previous value of W. But we must remember that the variations δn_s in the numbers n_s are not arbitrary, because N and E are fixed [conditions (1) and (2)]. Therefore

$$\delta N = \sum_{s=1}^{\infty} \delta n_s = 0 \tag{7}$$

$$\delta E = \sum_{s=1}^{\infty} \varepsilon_s \, \delta n_s = 0 \tag{8}$$

Taking zero from zero leaves zero, so we may write, using Eqs. (6), (7), and (8):

$$\delta(\ln W) - \alpha \sum_{s=1}^{\infty} \delta n_s - \beta \sum_{s=1}^{\infty} \varepsilon_s \, \delta n_s = 0 \tag{9}$$

where α and β can be any numbers we choose.[3] We shall see the reason for

---

[3] This technique for maximizing a function of several variables while simultaneously satisfying certain conditions on these variables is quite widely used. The undetermined constants α and β are called Lagrange multipliers.

this step in a moment. Let us now introduce the expressions for W for the three types of statistics, and compute the occupation numbers n_s.

Boltzmann.

$$\ln W = \ln N! + \sum_{s=1}^{\infty} (n_s \ln g_s - \ln n_s !)$$

We wish to consider situations in which n_s is quite large, so we simplify this expression by using the Stirling approximation for $n!$, which for large n may be written[4]

$$\ln n! \to n(\ln n - 1) \quad \text{as} \quad n \to \infty$$

We may then write

$$\ln W = \ln N! + \sum_{s=1}^{\infty} (n_s \ln g_s - n_s \ln n_s + n_s)$$

Now, using the fact that

$$\delta(\ln W) = \sum_{s=1}^{\infty} \frac{\partial(\ln W)}{\partial n_s} \delta n_s$$

we obtain

$$\delta(\ln W) = \sum_{s=1}^{\infty} (\ln g_s - \ln n_s - 1 + 1) \delta n_s \tag{10}$$

Substitution of Eq. (10) into Eq. (9) yields

$$\boxed{\sum_{s=1}^{\infty} (\ln g_s - \ln n_s - \alpha - \beta \varepsilon_s) \delta n_s = 0} \tag{11}$$

Bose–Einstein.

$$\ln W = \sum_{s=1}^{\infty} [\ln(n_s + g_s - 1)! - \ln n_s! - \ln(g_s - 1)!]$$

Again using the Stirling approximation, we find that

$$\delta \ln W = \sum_{s=1}^{\infty} [\ln(n_s + g_s) - \ln n_s] \delta n_s$$

---

[4] There are more accurate forms of the Stirling approximation; see, for example, I. Sokolnikoff and R. Redheffer, "Mathematics of Physics and Modern Engineering." McGraw-Hill, New York, 1958. The expression given here is accurate to 1 percent at $n = 100$, and all expressions approach the same limit as n goes to infinity.

and substitution into Eq. (9) yields

$$\sum_{s=1}^{\infty} [\ln(n_s + g_s) - \ln n_s - \alpha - \beta\varepsilon_s] \, \delta n_s = 0 \tag{12}$$

Fermi–Dirac.

$$\ln W = \sum_{s=1}^{\infty} [\ln g_s! - \ln n_s! - \ln(g_s - n_s)!]$$

so that the use of the Stirling approximation gives us

$$\delta \ln W = \sum_{s=1}^{\infty} [-\ln n_s + \ln(g_s - n_s)] \, \delta n_s$$

and substitution into Eq. (9) yields

$$\sum_{s=1}^{\infty} [-\ln n_s + \ln(g_s - n_s) - \alpha - \beta\varepsilon_s] \, \delta n_s = 0 \tag{13}$$

The problem now is to eliminate the quantities δn_s from the equations. We can do this *if* the δn_s are arbitrary and can be varied *independently*, because in that case the coefficient of each δn_s must be zero in Eqs. (11), (12), and (13). Actually, all but two of the δn_s may be chosen arbitrarily, and the remaining two are then determined by conditions (1) and (2). But now the constants α and β come into play; we can *choose* these two constants so that the coefficients of the two final δn_s *are* zero. Then, since the other δn_s are all arbitrary, the coefficients of these are also zero, and we have, for *all* values of s,

Boltzmann

$$\ln \frac{n_s}{g_s} = -\alpha - \beta\varepsilon_s \qquad \boxed{\frac{n_s}{g_s} = e^{-\alpha - \beta\varepsilon_s}}$$

Bose–Einstein

$$\ln \frac{n_s}{n_s + g_s} = -\alpha - \beta\varepsilon_s \qquad \boxed{\frac{n_s}{g_s} = \frac{1}{e^{\alpha + \beta\varepsilon_s} - 1}}$$

Fermi–Dirac

$$\ln \frac{n_s}{g_s - n_s} = -\alpha - \beta\varepsilon_s \qquad \boxed{\frac{n_s}{g_s} = \frac{1}{e^{\alpha + \beta\varepsilon_s} + 1}}$$

In spite of the great differences in the assumptions used to derive them, the three distributions have a similar appearance. The Bose–Einstein and Fermi–Dirac occupation indices differ from the Boltzmann only in the presence of the -1 or the $+1$, respectively, in the denominator. But that little " 1 " can have enormous consequences; in order to see these consequences, we first must determine the significance of the constants α and β.

We already should know that $\beta = 1/kT$, at least in the case of Boltzmann statistics, because we have shown by other methods (Chapter 1) that the Boltzmann distribution is proportional to $e^{-\varepsilon_s/kT}$. To show that β has the same meaning in the other distributions, we consider a mixture of two kinds of particles, one obeying Boltzmann statistics, the other obeying one of the other kinds of statistics. The energy levels for the two kinds of particles may be labeled ε_s and ε_t', respectively, with occupation numbers n_s and n_t', respectively. The number of arrangements for a given distribution is $P(n_s, n_t') = P(n_s) \cdot P(n_t')$—the product of the number of arrangements of the two kinds of particle separately—and the overall distribution function for the mixture is found by maximizing $\ln P(n_s, n_t')$, in analogy to our previous procedure. The numbers n_s and n_t' are now subject to *three* conditions rather than two; we must have

$$\sum_{s=1}^{\infty} n_s = N_1$$

$$\sum_{t=1}^{\infty} n_t' = N_2$$

$$\sum_{s=1}^{\infty} n_s \varepsilon_s + \sum_{t=1}^{\infty} n_t' \varepsilon_t' = E$$

Therefore, we introduce *three* constants, α_1, α_2, and β, into the variation equation analogous to Eq. (6):

$$\delta[\ln W(n_1, \ldots, n_s, \ldots) + \ln W(n_1', \ldots, n_t', \ldots)]$$
$$- \alpha_1 \, \delta N_1 - \alpha_2 \, \delta N_2 - \beta \, \delta E = 0$$

or

$$\delta[\ln W(n_1, \ldots, n_s, \ldots) + \ln W(n_1, \ldots, n_t', \ldots)] - \alpha_1 \sum_{s=1}^{\infty} \delta n_s$$
$$- \alpha_2 \sum_{t=1}^{\infty} \delta n_t' - \beta \sum_{s=1}^{\infty} (\delta n_s)\varepsilon_s - \beta \sum_{t=1}^{\infty} \varepsilon_t' \, \delta n_t' = 0$$

This equation can be separated into two equations, each involving only one kind of particle, and we can then proceed to derive the same distribution functions as before. But notice that we have the *same* β for both kinds of particle. Thus if β equals $1/kT$ for Boltzmann statistics, it must equal $1/kT$ for the other kinds of statistics as well. Because only the total energy, and not the energy of each kind of particle, is absolutely fixed, we need only one β, or one temperature, to describe the system. The two kinds of particle are in thermal equilibrium, each behaving as if the other were not present.

The parameter α does not have such a simple interpretation, but it is clearly related to the number of particles in the system, as it was introduced in connection with the condition $\sum n_s = N$. The explicit evaluation of α in terms of the parameters of the system (g_s, ε_s, N, and T) depends on the type of statistics. For the Boltzmann distribution, it is easy to write an explicit expression for α, as follows:

$$N = \sum_{s=1}^{\infty} n_s$$

$$= \sum_{s=1}^{\infty} g_s e^{-\alpha} e^{-\varepsilon_s/kT}$$

therefore

$$e^{-\alpha} = \frac{N}{\displaystyle\sum_{s=1}^{\infty} g_s e^{-\varepsilon_s/kT}}$$

The quantity $\sum g_s e^{-\varepsilon_s/kT}$ is called the partition function of the system, and denoted by the symbol Z. Thus we may write the occupation indices for the Boltzmann distribution as

$$\frac{n_s}{g_s} = \frac{N}{Z} e^{-\varepsilon_s/kT}$$

Further inquiry into the meaning of α is best done in the course of studying applications to specific problems.

EXAMPLE PROBLEM 1. Consider a system containing only three equally-spaced energy levels, ε_1, ε_2, and ε_3, with degeneracies $g_1 = 1$, $g_2 = 2$, and $g_3 = 2$, at a temperature such that kT equals the difference between the first and second (or second and third) levels. If there are 100,000 distinguishable particles in the system, what is the most probable set of the occupation numbers n_1, n_2, and n_3?

Solution. We use Boltzmann statistics, because the particles are distinguishable. We let $\varepsilon_1 = 0$. (This choice is arbitrary.) Then $\varepsilon_2 = kT$ and $\varepsilon_3 = 2kT$. If we knew n_1, we could simply use the Boltzmann factor and the known degeneracies to find n_2 and n_3. But we are given N, so we use N and the partition function to find n_1, as follows:

$$Z = \sum g_s e^{-\varepsilon_s/kT} = 1 \cdot e^0 + 2 \cdot e^{-1} + 2 \cdot e^{-2}$$
$$= 1.000000 + 0.735759 + 0.270671$$
$$= 2.00643$$

Therefore

$$n_1 = g_1 \frac{N}{Z} e^{-\varepsilon_1/kT} = \frac{100,000e^0}{2.00643} = 49,840$$

$$n_2 = g_2 \frac{N}{Z} e^{-\varepsilon_2/kT} = \frac{200,000e^{-1}}{2.00643} = 36,670$$

$$n_3 = g_3 \frac{N}{Z} e^{-\varepsilon_3/kT} = \frac{200,000e^{-2}}{2.00643} = 13,490$$

$$n_1 + n_2 + n_3 = 100,000$$

Let us verify that the total number of arrangements, $W(n_1, n_2, n_3)$, is maximum for this set of the n_s relative to other possible sets. To do this, we wish to vary n_1, n_2, and n_3 in such a way that the total energy and the total N do not change. We can do this simply by making $\delta n_1 = \delta n_3 = -\delta n_2/2$. So let us set δn_1 first equal to $+1$ and then to -1, in order to generate two new sets of numbers—n_1', n_2', n_3', and n_1'', n_2'', n_3''—such that

$$n_1' = 49,841; \qquad n_2' = 36,668; \qquad n_3' = 13,491$$

and

$$n_1'' = 49,839; \qquad n_2'' = 36,672; \qquad n_3'' = 13,489$$

For the respective sets of numbers the values of W are given by

$$W(n_1, n_2, n_3) = \frac{N!}{n_1! n_2! n_3!} g_1^{n_1} g_2^{n_2} g_3^{n_3} = \frac{100,000!}{49840! \, 36670! \, 13490!} 2^{36670} 2^{13490}$$

$$W(n_1', n_2', n_3') = \frac{100,000!}{49841! \, 36668! \, 13491!} 2^{36668} 2^{13491}$$

$$W(n_1'', n_2'', n_3'') = \frac{100,000!}{49839! \, 36672! \, 13489!} 2^{36672} 2^{13489}$$

Obviously, we are not about to compute each of these W, but we can find their ratios without too much difficulty. You can easily see that

$$\frac{W(n_1, n_2, n_3)}{W(n_1', n_2', n_3')} = \frac{49841 \times 13491 \times 2}{36670 \times 36669} \approx \frac{134481}{134465} \approx 1.0001$$

and

$$\frac{W(n_1, n_2, n_3)}{W(n_1'', n_2'', n_3'')} = \frac{36672 \times 36671}{49840 \times 13490 \times 2} \approx \frac{134480}{134468} \approx 1.0001$$

Thus either increasing or decreasing n_2 leads to the same result, that the number of arrangements becomes smaller[5]; it seems that the distribution we found must indeed be the most probable one, given the assumption that each arrangement is equally likely.

11.3 APPLICATIONS OF BOSE–EINSTEIN STATISTICS

a. The Planck Black-Body Law Rederived.
There is a striking similarity between the Bose–Einstein distribution

$$n_s = \frac{g_s}{e^\alpha e^{\varepsilon_s/kT} - 1}$$

and the Planck expression for the average energy of the oscillators in the walls of a cavity (Section 3.1):

$$\bar\varepsilon = \frac{hv}{e^{hv/kT} - 1} \tag{14}$$

This is somewhat surprising, because you will recall that the Planck law was derived by application of *Boltzmann* statistics to the energy levels of oscillators in the walls of the cavity. But *Bose–Einstein* statistics can just as well be applied to the energy levels of the *radiation field* in the cavity, because *photons* obey Bose–Einstein statistics.

You will recall that the radiation field in the cavity consists of standing waves of various frequencies. A standing wave of frequency v can have energy nhv, where n is an integer; one can say that such a mode of oscillation "contains" n photons. Thus we can think of the mode as a *state* which is "occupied" by n photons, and we can apply statistics to the photons as if they were particles that are free to occupy various states. If a photon of frequency v is absorbed by a cavity wall, and a photon of frequency v' is emitted by the wall, the result is as if a single particle "dropped" from an energy level of hv to an energy level of hv', with the difference in energy, $h(v - v')$, being given to the wall of the cavity.

To describe the distribution of photons in the various energy states, or modes, we use Bose–Einstein statistics, as follows. First we note that $\alpha = 0$, because the number of photons is not fixed.[6] If N is not fixed, then $\sum \delta n_s \neq 0$,

---

[5] Notice that the three-level system is particularly well suited for this illustration, because only one n is independent when N and E are fixed. Thus we need not worry about the choice of possible variations among the n; the variation in any one of them, for example, in n_2, automatically determines the variations in the other two.

[6] If the cavity is to be at a fixed temperature, it is only necessary that *energy* be conserved. E may be conserved while N changes; for example, one photon of frequency v may be replaced by two photons of frequency $\frac{1}{2}v$.

and to make the term $\alpha \sum \delta n_s$ equal to zero [to satisfy Eq. (9)], we must set α equal to zero. The number of photons of energy ε_s is therefore, according to the Bose–Einstein distribution function,

$$n_s = \frac{g_s}{e^{\varepsilon_s/kT} - 1}$$

To go from here to Eq. (14) is quite simple; n_s/g_s is the number of photons per mode of oscillation, and the energy of each photon is $\varepsilon_s = hv$, so the average energy per mode is hvn_s/g_s, which from the above equation is just $hv/(e^{hv/kT} - 1)$, as required.

If we wish to consider the energy to be a continuous variable, we may write an equation for the number of photons per unit energy interval as

$$n(\varepsilon) = \frac{g(\varepsilon)}{e^{\varepsilon/kT} - 1} \tag{15}$$

where $\int_{E_1}^{E_2} n(\varepsilon)\, d\varepsilon$ is the number of photons with energy between E_1 and E_2, and $\int_{E_1}^{E_2} g(\varepsilon)\, d\varepsilon$ is the number of states with energy between E_1 and E_2.

Let us complete the black-body analysis by computing the density of states $g(\varepsilon)$ for the photons. The resulting black-body spectrum is, of course, the same as that derived in Section 3.1, but the point of view is different. The derivation is repeated here in this slightly modified form because this point of view is also applicable to subsequent topics in this chapter.

We begin the computation by observing that each photon has x, y, and z components of momentum, corresponding to wavelengths in the x, y, and z directions, and each wavelength must satisfy the boundary conditions on $\mathbf{E}$ (Section 3.1). Therefore, if the cavity is a cube of side a, each wavelength is equal to $2a$ divided by an integer, and the momentum components of a given photon may be written

$$p_x = \frac{h}{\lambda_x} = \frac{hl_x}{2a}$$

$$p_y = \frac{h}{\lambda_y} = \frac{hl_y}{2a} \tag{16}$$

$$p_z = \frac{h}{\lambda_z} = \frac{hl_z}{2a}$$

where l_x, l_y, and l_z are positive integers. The photon energy is therefore given by

$$\varepsilon^2 = c^2 p^2$$

$$= c^2(p_x^2 + p_y^2 + p_z^2)$$

$$= \frac{c^2 h^2}{4a^2}(l_x^2 + l_y^2 + l_z^2) \tag{17}$$

Comparison of Eq. (17) with Eq. (4) of Chapter 3 shows that this is the same result obtained there (with $\varepsilon = h\nu$).

Following the method of Section 3.1, we see from Eq. (17) that all sets of integers for which the photon energy is ε or less obey the relation

$$l_x^2 + l_y^2 + l_z^2 \leqslant \frac{4a^2\varepsilon^2}{h^2c^2} \tag{18}$$

If we consider the numbers l_x, l_y, and l_z to be coordinates of a point in three dimensions, the points corresponding to the numbers which satisfy Eq. (18) must lie within a sphere of radius $2a\varepsilon/hc$. The number of sets of positive integers within a sphere of radius R is just $\frac{1}{8}$ of the volume of the sphere, or $\frac{1}{8}4\pi R^3/3$, and there are two modes of oscillation (two polarization directions) for each set of integers, so the number G of photon states of energy ε or less is

$$G = 2 \cdot \frac{1}{8} \cdot \frac{4\pi}{3} \left(\frac{2a\varepsilon}{hc}\right)^3$$

The density of states must then be

$$g(\varepsilon) \equiv \frac{dG}{d\varepsilon}$$

$$= \frac{8\pi V\varepsilon^2}{h^3c^3} \qquad (V = a^3) \tag{19}$$

Therefore, from Eq. (15),

$$n(\varepsilon) = \frac{8\pi V\varepsilon^2}{h^3c^3(e^{\varepsilon/kT} - 1)}$$

The energy density dU in the cavity for radiation of frequency between ν and $\nu + d\nu$ is equal to the product of the number of photons per unit volume, $n(\varepsilon)\, d\varepsilon/V$, and the energy $\varepsilon = h\nu$ of each photon. (Remember that a *mode* has energy $nh\nu$, but a *photon* has energy $h\nu$.) Thus

$$dU = \frac{8\pi\varepsilon^3 \, d\varepsilon}{h^3c^3(e^{\varepsilon/kT} - 1)}$$

$$= \frac{8\pi h\nu^3 \, d\nu}{c^3(e^{h\nu/kT} - 1)}$$

and we see that the black-body spectrum can be derived either by applying Boltzmann statistics to the *oscillators*, as we did in Chapter 3, or by applying Bose–Einstein statistics to the photons of the radiation field. In our next example we shall see that we can do the same thing with the vibrations of a solid; the vibrations of the lattice are analogous to the electromagnetic waves in a cavity. The energy of a specific mode of the lattice vibrations is quantized, the energy levels again differing by hv (simply because these are the energy levels of a simple harmonic oscillator).

So we can say that a mode of frequency v, with energy nhv, is a state containing n particles called *phonons*, and we can treat the phonons just as we treated the photons in the cavity. A pho*n*on is exactly analogous to a pho*t*on; a phonon carries the energy and momentum of a lattice vibration—a sound wave—through a solid, just as a photon carries the energy and momentum of a light wave. The concept of a phonon is helpful in understanding our second application of Bose–Einstein statistics, which is the specific heat of a solid.

b. The Specific Heat of a Solid. Einstein was the first to point out, in 1907,[7] that the Planck quantization of the energy of oscillators in a cavity should be applicable to the atoms in any solid. This quantization should have an observable effect on the specific heat. It was already known that the classical specific heat law (law of Dulong and Petit) was not obeyed by all solids. According to this law, solids have a constant (temperature-independent) specific heat of $3R$ per mole, where R is the gas constant. This is to be expected classically, because one mole of a solid, with N_A atoms, has $3N_A$ modes of vibration of the lattice (because each atom is free to oscillate in three independent directions), and if each mode has an average energy of kT, the total energy is $3N_A kT$, making the specific heat $3N_A k$, or $3R$.[8] But, as Einstein pointed out, the specific heat is less than this for many solids at room temperature, and the specific heat becomes smaller as the temperature is reduced. Einstein saw the similarity of this behavior to the behavior of black-body radiation: as T becomes smaller, the spacing of energy levels becomes larger relative to kT, so that quantization is more effective in reducing the average energy per mode of oscillation below kT. As the average energy is reduced, the specific heat goes down. Thus the failure of the classical specific heat law as T is *reduced* is analogous to the failure of the Rayleigh–Jeans law as the frequency is *increased*; both laws fail when hv/kT becomes appreciable.

Einstein did not intend to develop a complete theory of solids. He merely wanted to show the similarity of this quantum effect to the effect seen in black-body radiation. So he simply assumed that the $3N_A$ modes of oscillation of the solid all have the same frequency v, and he replaced the classical average energy kT by the average energy given by the Planck formula (14):

$$\bar{\varepsilon} = \frac{hv}{e^{hv/kT} - 1}$$

[7] A. Einstein, *Ann. Phys.* (*Leipzig*) **22**, 180 (1907); **34**, 170 (1911).
[8] You may wonder why the electrons' contribution to the specific heat is negligible. We shall answer that question when we discuss applications of Fermi–Dirac statistics.

The total energy of N_A oscillating atoms should then be $3N_A h\nu/(e^{h\nu/kT} - 1)$ which, of course, reduces to $3N_A kT$ if $h\nu \ll kT$. Thus the fact that most solids do obey the Dulong–Petit law at room temperature could be explained by saying that for these solids, the characteristic frequency ν must be much smaller than kT/h. The solids for which the law was not obeyed were assumed to differ from the others only in the value of ν.

If Einstein's assumptions were correct, one should be able to fit the specific heat data for any solid simply by choosing ν correctly for that solid; the specific heat should be the same function of T/ν for *all* solids. This function can be easily computed from the above formula for the total energy; it has the interesting feature that it goes to zero as T goes to zero.

At about the same time that Einstein's paper appeared, Nernst began a series of measurements of specific heats of solids at low temperatures, to test his belief that molar specific heats of all solids should approach the same value (not necessarily zero) as T approaches zero. When he later compared his results with the curve predicted by Einstein's model, he found serious deviations at low temperatures; although the specific heat did approach zero as T went to zero, the temperature dependence was not that of Einstein's formula.

Nernst and Lindemann tried to fit the data by assuming that there are several characteristic frequencies instead of just one for a given solid, but such an empirical approach could not be very convincing. Obviously, if we have a few curves to fit, we can do if it we use a sufficient number of independent parameters, but no fundamental understanding is gained in this way. In 1912, P. Debye found a more general approach which was quite successful in explaining the data.[9] Einstein had assumed that the atoms were vibrating independently, but Debye considered the possibility of collective motions. He recognized that the normal modes of oscillation of the atoms in a crystal lattice can have *many* different characteristic frequencies, and that the problem was to decide, on other than empirical grounds, what the frequency *distribution* should be for a given solid. He then made the reasonable assumption that the problem is similar to the problem of electromagnetic oscillations in a cavity, so that the number of modes per unit frequency range is given by $dN/d\nu = A\nu^2$, where A is a constant. This result follows from Eq. (19), if one assumes that the photon gas in a cavity has the same modes of vibration as the phonon gas in a solid. But the phonon gas does differ from the photon gas, in that it has a limited number—$3N_A$—of modes of oscillation, and consequently there must be an upper limit on the frequency of a mode. We can write the constant A in terms of this upper limit ν_m as follows:

$$3N_A = \int_{\nu=0}^{\nu=\nu_m} dN = \int_0^{\nu_m} A\nu^2 \, d\nu = \frac{A\nu_m^3}{3}$$

[9] P. Debye, *Ann. Phys.* (*Leipzig*) **39**, 789 (1912). Debye later (1936) won the Nobel Prize in chemistry for his work on molecular structure.

therefore

$$A = \frac{9N_A}{v_m^3}$$

The energy of the dN modes whose frequency lies between v and $v + dv$ is therefore $dN\,\bar{\varepsilon}$, where $\bar{\varepsilon}$, the average energy of a mode whose frequency is v, is given by Eq. (14). And the total energy[10] is simply

$$E = \int_{v=0}^{v=v_m} \bar{\varepsilon}\,dN$$

$$= \int_0^{v_m} \frac{hv}{e^{hv/kT} - 1} \frac{9N_A}{v_m^3} v^2\,dv$$

We can simplify this expression by substituting the dimensionless variable $x = hv/kT$ into the integral, which then becomes

$$E = \frac{9N_A k^4 T^4}{h^3 v_m^3} \int_0^{hv_m/kT} \frac{x^3}{e^x - 1}\,dx$$

$$= 9N_A kT \left(\frac{T}{\Theta_D}\right)^3 \int_0^{\Theta_D/T} \frac{x^3\,dx}{e^x - 1} \tag{20}$$

where $\Theta_D \equiv hv_m/k$ is called the *Debye temperature* of a solid.

When the temperature T is much greater than the Debye temperature, the variable x is much smaller than 1 over the range of integration. In that case the integrand becomes approximately $x^3/(1 + x - 1) = x^2$, and the energy of the solid becomes $E = 3N_A kT$, which is just the classical expression leading to a specific heat of $3R$ per mole. So the solids which obey the Dulong–Petit law are those for which Θ_D is small, compared to the temperature at which the specific heat is measured.

At low temperatures, that is, when $T \ll \Theta_D$, the upper limit in the integral of Eq. (20) is very large, and the integral approaches the limit

$$\int_0^\infty \frac{x^3}{e^x - 1}\,dx = \frac{\pi^4}{15}$$

At these temperatures, E must then be proportional to T^4, and the specific heat $\partial E/\partial T$, should be proportional to T^3. Experiments on a wide variety of solids has verified this temperature dependence for the specific heat at low temperatures, and Fig. 3 shows that the Debye theory also agrees with experiment over a wide range of temperatures for several solids. In each case,

---

[10] We are neglecting the ground state, or "zero point" energy of $hv/2$ for each mode (see Problem 4). This energy, of course, makes no contribution to the specific heat.

Θ_D is determined from the experimental data, but it is impressive that this is the only parameter needed to fit all the points. The value of Θ_D obtained in this way is also in agreement with values of the maximum vibration frequency v_m obtained by independent methods, such as measurement of elastic constants. But the theory is not perfect; the weakness in the theory is in its assumption that the crystal is a continuous medium, like the vacuum in a cavity. As expected, the simple theory breaks down for anisotropic materials, where the spectrum of frequencies is much more complicated than the simple v^2 dependence; more careful measurements have shown that it also breaks down in other materials. The assumption of a v^2 dependence is a great oversimplifica-

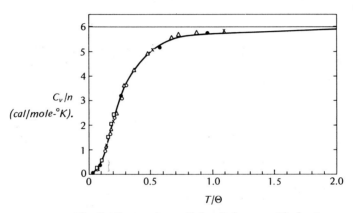

Fig. 3. *Comparison of the Debye specific heat curve and the observed specific heats of a number of simple substances. (●) Ag; (△) Al; (□) C (graphite); (○) Al₂O₃; (×) KCl. [From "The Modern Theory of Solids" by F. Seitz. Copyright McGraw-Hill, New York, 1940. Used by permission of McGraw-Hill Book Company.]*

tion, and a great deal of effort has been devoted to the determination of the actual "phonon spectrum" in many solids and to the development of "modified Debye theories" which are based on different sets of normal modes. Nevertheless, the simple Debye theory is a remarkably accurate and useful first approximation.

c. Liquid Helium and Superfluidity. Our first two examples have shown us how to use Bose–Einstein statistics as the basis for a calculation; however, they did not bring out some distinctive features of these statistics, because they are somewhat special applications in which the number of particles is not constant. In fact, it is not even necessary to use Bose–Einstein statistics to solve those problems, because we can apply classical Boltzmann statistics to the oscillators instead of using Bose–Einstein statistics for the photons or phonons; we have already (Section 3.1) seen how this is done for black-body radiation.

But application of Bose–Einstein statistics to a system of material particles, in which N is constant, reveals a startling new possibility, which had been suggested by Einstein in 1924. Einstein pointed out that at low temperatures, a gas of Bose–Einstein particles would undergo a "condensation" which is totally different from the ordinary gas–liquid condensation. In this condensation, a large fraction of particles would occupy the lowest energy state. In a large scale system in which the quantum number is zero, the correspondence principle would no longer apply as it does to most large systems, and thus the *mechanical* behavior of the system would no longer be correctly described by classical mechanics. Quantum effects might then be visible, and would not have to be deduced from indirect evidence.

To see why this condensation should be peculiar to Bose–Einstein statistics, consider a system containing N particles, with a set of nondegenerate energy levels ($g_s = 1$ for all s). The number of possible distributions becomes enormous for large N, but compare just two possibilities:

(a) There is one particle in each of the lowest N levels.

(b) There are $N - 1$ particles in the lowest level, and one particle in a higher level, with energy equal to that of the $N - 1$ excited particles in (a).

In Bose–Einstein statistics, both distributions are equally likely, each containing one arrangement. But classically, distribution (a) contains $N!$ arrangements while (b) contains only N arrangements, so that (a) occurs $(N - 1)!$ times as often as (b). Thus a distribution like (b), with a large fraction of the particles in the ground state, is always highly improbable in classical statistics; but we can see the possibility that, at low temperatures, such distributions may begin to make their presence felt in a Bose–Einstein gas.

Einstein's idea was intriguing, but apparently unrealistic, because at the very low temperature (about $3°K$) at which such a condensation should be expected, no known substance remains in gaseous form. The condensation had been deduced from the properties of an *ideal gas*, in which one assumes no interaction between the particles of the gas; but clearly, when a substance is held together in the liquid or solid state, the interaction between particles must be considerable. In 1938, however, Fritz London pointed out[11] that many peculiar properties of liquid ^4He could be explained if it is treated as a Bose–Einstein "gas," even though it is in the liquid state.

[11] F. London, *Phys. Rev.* **54**, 947 (1938). See also London's "Superfluids," Vol. II. Wiley, New York, 1954.

⁴He atoms, having total angular momentum of zero, obey Bose–Einstein statistics. The normal condensation into a liquid occurs at 4.2°K, at atmospheric pressure. The temperature can be reduced below that point by reducing the pressure. As the pressure is reduced, the liquid boils until the temperature reaches about 2.2°K. At this point, called the lambda (λ) point,[12] *boiling* suddenly ceases, although *evaporation* continues. Of course, boiling occurs in the first place because the liquid is not at a uniform temperature throughout, and bubbles form at the "hot" points. The cessation of boiling indicates a sharp increase in thermal conductivity; in fact, the thermal conductivity appears infinite, for all practical purposes, so that the liquid is always at a perfectly uniform temperature throughout.

Below the λ point, liquid ⁴He has another remarkable property: it can penetrate through the tiniest capillary, as if the viscosity were zero. But when the viscosity is measured by means of a torsion pendulum, by observing the drag on a set of parallel plates moving through the fluid, the viscosity is *not* zero, and it shows no discontinuity as a function of temperature at the λ point.

To account for these properties, London suggested that liquid ⁴He below the λ point (called liquid He II) consists of two interpenetrating fluids—a "normal" fluid and a "superfluid." The superfluid is that part of the liquid whose atoms are in the ground state; the normal fluid is the rest. The normal fluid causes the drag on a torsion pendulum; the superfluid seeps through capillaries. It appears that the atoms of the superfluid, being in a state of almost perfectly defined momentum, cannot be localized in space. Heating the liquid destroys superfluid; but the superfluid, which cannot be confined to one region of the liquid, is destroyed uniformly throughout the entire volume of the liquid whenever heat is applied to any point. Destruction of superfluid is equivalent to a rise in temperature; since this rise takes place everywhere (almost) simultaneously, the fluid appears to have infinite thermal conductivity.

In a moment we shall discuss further the reason for this strange behavior of atoms in the ground state. Right now let us look more carefully into the properties of an *ideal* Bose–Einstein gas at low temperatures, to see just how it is that so many atoms condense into the ground state, and why this condensation makes its presence felt at a well defined nonzero temperature. We begin with the Bose–Einstein distribution numbers

$$n_s = \frac{g_s}{e^\alpha e^{\varepsilon_s/kT} - 1} \tag{21}$$

---

[12] Called the λ point because the curve of specific heat versus temperature resembles the letter λ near this temperature.

The zero level of energy is arbitrary, so let us set the energy of the lowest level equal to zero; that is, we let $\varepsilon_1 = 0$. In this case, it is clear that $\alpha > 0$, for if α were zero, as in the cases when the number of particles is not fixed, the value of n_1 would be infinite, and if α were negative, n_1 would be negative, according to Eq. (21), and a negative value of n_1 makes no sense.

Now in dealing with a macroscopic sample, the usual procedure is to replace n_s by a continuous function $n(\varepsilon)$, as we did in parts a and b of this section, because in a macroscopic system the energy levels may be spaced so closely that their discrete nature is not observable. But we must be careful when we are dealing with low temperatures, because the total energy involved is so small that the discreteness of the levels may be important even in a macroscopic system. With this in mind, let us proceed to a continuum description and see what happens. We write, in place of Eq. (21),

$$n(\varepsilon) = \frac{g(\varepsilon)}{e^{\alpha}e^{\varepsilon/kT} - 1} \tag{22}$$

an equation similar to Eq. (15), except for the factor e^{α}.

We may find $g(\varepsilon)$ by the same basic method used for photons. In a cube of side a, the permissible *wavelengths* are the same for material particles as for photons, and there is the same connection between wavelength and *momentum* in both cases. Therefore the *momentum* components of a given particle are given by Eqs. (16). But if the particle has mass M, the momentum-*energy* relation at low energies is $\varepsilon = p^2/2M$ rather than $\varepsilon = pc$, and Eq. (17) is replaced by

$$\varepsilon = \frac{h^2(l_x^2 + l_y^2 + l_z^2)}{8Ma^2}$$

The integers for which the particle energy is ε or less obey the relation

$$l_x^2 + l_y^2 + l_z^2 \leqslant \frac{8Ma^2\varepsilon}{h^2}$$

and the points corresponding to these integers lie within a sphere of radius $(8Ma^2\varepsilon/h^2)^{1/2}$. The number G of states within this sphere is equal to $\frac{1}{8}$ of the volume of the sphere, or

$$G = \frac{1}{8} \cdot \frac{4\pi}{3} \left(\frac{8Ma^2\varepsilon}{h^2}\right)^{3/2} \tag{23}$$

and the density of states is

$$g(\varepsilon) = \frac{dG}{d\varepsilon} = \frac{3}{2} \cdot \frac{1}{8} \cdot \frac{4\pi}{3} \left(\frac{8Ma^2}{h^2}\right)^{3/2} \varepsilon^{1/2}$$

or, with $a^3 = V$,

$$g(\varepsilon) = 2\pi V \left(\frac{2M}{h^2}\right)^{3/2} \varepsilon^{1/2} \tag{24}$$

Now we may find a relation between N and α by using the fact that

$$N = \int_0^\infty n(\varepsilon)\, d\varepsilon = \int_0^\infty \frac{g(\varepsilon)\, d\varepsilon}{e^\alpha e^{\varepsilon/kT} - 1}$$

$$= 2\pi V \left(\frac{2M}{h^2}\right)^{3/2} \int_0^\infty \frac{\varepsilon^{1/2}\, d\varepsilon}{e^\alpha e^{\varepsilon/kT} - 1}$$

Let us make the substitutions $A = 2\pi V (2Mk/h^2)^{3/2}$ and $x = \varepsilon/kT$ to obtain

$$N = A T^{3/2} \int_0^\infty \frac{x^{1/2}\, dx}{e^{\alpha+x} - 1} \tag{25}$$

We may rewrite Eq. (25) as

$$N = A T^{3/2} \int_0^\infty \frac{x^{1/2}}{e^{\alpha+x}} (1 - e^{-\alpha-x})^{-1}\, dx$$

which may be expanded to give

$$N = A T^{3/2} \int_0^\infty x^{1/2} e^{-\alpha-x} (1 + e^{-(\alpha+x)} + e^{-2(\alpha+x)} + \cdots)\, dx$$

and integrated term by term. If one remembers the definition of the gamma function:

$$\Gamma(n) = \int_0^\infty x^{n-1} e^{-x}\, dx$$

the integral may be written

$$N = A T^{3/2} \Gamma\left(\frac{3}{2}\right) \left[e^{-\alpha} + \frac{e^{-2\alpha}}{2^{3/2}} + \frac{e^{-3\alpha}}{3^{3/2}} + \cdots \right]$$

or

$$N = A T^{3/2} \Gamma\left(\frac{3}{2}\right) f(\alpha) \tag{26}$$

where

$$f(\alpha) = \sum_{p=1}^\infty \frac{e^{-p\alpha}}{p^{3/2}}$$

The series sum $f(\alpha)$ is plotted versus α in Fig. 4. We may use Eq. (26) to. determine the value of α for a given temperature, as follows: We simply use the values of N, A, $\Gamma(\tfrac{3}{2})$, and T to compute the value of $f(\alpha)$, and we then read off the corresponding value of α from Fig. 4. (*Note*: $\Gamma(\tfrac{3}{2}) = \sqrt{\pi}/2 \approx 0.88623$.)

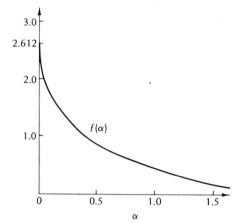

Fig. 4. *The function $f(\alpha)$ appearing in Eq. (26).*

But there is a serious difficulty as T approaches zero. Since N, A, and $\Gamma(\tfrac{3}{2})$ are all independent of T, we have, directly from Eq. (26),

$$f(\alpha) = \frac{N}{AT^{3/2}\Gamma(\tfrac{3}{2})} \to \infty \qquad \text{as} \quad T \to 0.$$

But $f(\alpha)$ is a mathematical function whose largest value, according to Fig. 4, is 2.612, at $\alpha = 0$,[13] so Eq. (26) *cannot possibly be correct* in the limit $T \to \infty$.

What went wrong? Remember the warning about the passage from Eq. (21) to Eq. (22). There is one obvious discrepancy between Eq. (21) and our later equations: We found that $g(\varepsilon)$ is proportional to $\sqrt{\varepsilon}$, so that $g(\varepsilon) = 0$ at $\varepsilon = 0$, but obviously a value of the degeneracy smaller than one does not make sense physically. In fact, we know that the degeneracy of the lowest level is $g_1 = 1$; for this level $l_x = l_y = l_z = 1$. Therefore this level is not represented in Eq. (22), because it has a weight $g(\varepsilon) = 0$ when we use the continuous form for the degeneracy g. We can correct this situation simply by writing n_1 separately in the expression for N:

$$N = n_1 + \int_0^\infty n(\varepsilon)\, d\varepsilon = \frac{1}{e^\alpha - 1} + \int_0^\infty \frac{g(\varepsilon)\, d\varepsilon}{e^\alpha e^{\varepsilon/kT} - 1}$$

[13] Remember that α can never be negative.

or

$$N = \frac{1}{e^\alpha - 1} + AT^{3/2}\Gamma(\tfrac{3}{2})f(\alpha) \tag{27}$$

an equation identical to Eq. (26) except for the additional term $1/(e^\alpha - 1) = n_1$ on the right-hand side. As $T \to 0$, the second term of Eq. (27) goes to zero, but the first term has no upper limit as $\alpha \to 0$, so there is always some positive value of α which satisfies the equation. Notice that α may now be expressed in terms of n_1, for if

$$n_1 = \frac{1}{e^\alpha - 1}$$

then

$$e^\alpha = 1 + \frac{1}{n_1}$$

and

$$\alpha \simeq \frac{1}{n_1} \qquad (n_1 \gg 1) \tag{28}$$

We would now like to determine the temperature T_c at which the system begins to "condense" into the ground state—that is, where n_1 becomes very large. For $T > T_c$ we assume that n_1 is negligible compared to N, because the particles are spread out over an enormous number of levels, and there is no loss of accuracy in using Eq. (26) instead of the more correct Eq. (27). We may then *define T_c* as *the temperature below which Eq. (26) can no longer be satisfied by any value of α*, and we may determine the value of T_c simply by setting $f(\alpha)$ in Eq. (27) equal to its maximum value, 2.612, and solving for T:

$$T = T_c = \left(\frac{N}{2.612A\Gamma(\tfrac{3}{2})}\right)^{2/3} \tag{29}$$

We may use Eq. (27) to study temperatures below T_c. Since $f(\alpha) = 2.612$ at $T = T_c$, we might conclude that $\alpha = 0$ at $T = T_c$, but this cannot be correct, because according to Eq. (28), n_1 becomes *infinite* at $\alpha = 0$. However, we are considering systems in which the total number of particles N is of the order of 10^{23}, so that α is certainly very close to zero when n_1 becomes at all comparable to N. For example, suppose that n_1 is "only" 10^{10}; then $\alpha \simeq 10^{-10}$, and $f(\alpha) \simeq f(0) = 2.612$ to much better than four-place accuracy. Yet in this case n_1 is certainly *negligible* compared to N, so that Eq. (26) is still valid! As n_1 increases, α becomes still smaller, so that whenever it is necessary to use Eq. (27) instead of Eq. (26) we may set $f(\alpha)$ equal to 2.612, even though α is not *quite* zero. With $f(\alpha) = 2.612$, Eq. (27) becomes

$$N = n_1 + 2.612AT^{3/2}\Gamma(\tfrac{3}{2}),$$

or, from Eq. (29),

$$N = n_1 + N\left(\frac{T}{T_c}\right)^{3/2}$$

so that

$$n_1 = N\left\{1 - \left(\frac{T}{T_c}\right)^{3/2}\right\} \tag{30}$$

Figure 5 shows the temperature dependence of n_1, as given by Eq. (30).

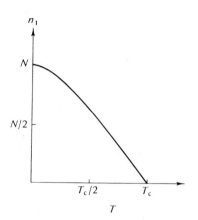

Fig. 5. *Temperature dependence of n_1 for an ideal Bose–Einstein gas. When $T > T_c$, n_1 is not necessarily zero, but it is so "small" (perhaps of order 10^{10} or less) that it does not show on this graph.*

If we use the density and particle mass of liquid ⁴He to compute the constant A/N of Eq. (29), we obtain for the critical temperature $T_c = 3.13$ K instead of the actual value of 2.2 K. But the analysis here applies only to an *ideal* Bose-Einstein gas, in which there is no interaction between particles, so we should not expect it to be *numerically* correct for liquid helium. Interaction between the helium atoms certainly affects the energy levels, so that $g(\varepsilon)$ is not given correctly by Eq. (24). However, the derivation does show how a condensation into the ground state can occur, and it shows that Bose-Einstein statistics must have some relevance to the situation.

There have been attempts to explain the properties of liquid He (II) without reference to Bose–Einstein statistics. But convincing evidence of the role of Bose–Einstein statistics is found in the behavior of liquid ^{3}He, an isotope which one would expect to behave very much like ^{4}He, and which fulfills this expectation except in its superfluid behavior. Liquid ^{3}He is a superfluid only at temperatures below 0.0025 K. The difference results from the fact that the atoms of ^{3}He have spin $\frac{1}{2}$ rather than spin zero, so they obey Fermi–Dirac rather than Bose–Einstein statistics. This makes condensation into the ground state impossible until the temperature becomes so low that ^{3}He atoms form *pairs* whose *total* spin is integral.[14]

Thus it seems that condensation into the ground state is a necessary condition for the occurrence of superfluid behavior. But it is not a *sufficient* condition. We have not yet answered the question of why such atoms behave as a superfluid, and the answer depends on the fact that there are interactions between the particles. An *ideal* Bose–Einstein gas, even though it condenses into the ground state, does *not* become a superfluid. To understand this, we must begin by considering why ordinary fluids are not superfluids.

A fluid flowing through a tube is slowed down because interactions between the fluid and the wall of the tube convert the fluid's translational energy into internal energy—that is, into random motion or heat. This occurs in an apparently continuous fashion, because in a normal fluid there are many possible internal motions which require very little energy to excite them. *Superfluid behavior requires that there be no low-energy states which can be easily excited*—that there be a gap between the ground state and the lowest state which can be easily excited by friction with the walls of the tube. Then if the fluid flows sufficiently slowly through the tube, the atoms cannot acquire enough energy from the walls to enable them to cross this energy gap; they stay in the ground state, and their translational motion continues unimpeded. (We still consider the atoms moving through the tube to be in the ground state, even though they have translational kinetic energy; we can consider the atoms to be at rest in a different frame of reference, and the tube to be moving past them.) As long as the speed of the fluid is below a certain "critical velocity," superfluid flow continues. If the speed exceeds the critical velocity, the fluid–tube interaction becomes able to excite turbulent states of the fluid, and the fluid is slowed down.

Precise determination of the nature of the excited states and the magnitude of the energy gap is difficult. It has been observed that a critical velocity exists for superfluid flow of liquid helium, but it is difficult to calculate theoretically what its value should be; oversimplified calculations can lead to a value of the order of 100 times too large. However, it is not too hard to show

[14] D. D. Osheroff, N. J. Gully, R. C. Richardson, and D. M. Lee, *Phys. Rev. Letters* **28**, 885 (1972). The magnetic properties indicate that the total spin is one rather than zero. See "The Helium Liquids," J. G. Armitage & I. E. Farquhar, eds., Academic Press, N.Y., 1975.

qualitatively that an energy gap exists, and that its existence depends upon an interaction between the helium atoms.

The interaction potential between two helium atoms has been calculated by quantum mechanics to be as shown in Fig. 6. Notice the sharp rise with decreasing r near 2.5 Å, as if a helium atom has a very hard "core" which repels other helium atoms. Now consider the ground state and first excited state of a collection of N helium atoms, taking the wavefunctions to be the unperturbed functions discussed previously—single-particle states for a particle in a box. The first excited state of the *system* is one for which all but one of the *particles* is in the ground state, and one particle is in the first excited state. (Here, as elsewhere in this section, we are considering the atoms to be particles; we are not considering internal energy levels of the atoms, but rather levels of the *particles* in the "box.") We state without proof that, because of the symmetry of the wavefunction for the system, there is for the first excited state a greater probability of finding two helium atoms within 2.5 Å of each other, where the potential energy becomes very large. Thus the energy of the first excited state is raised, relative to that of the ground state, as a result of the interaction and an energy gap is thereby created.

This discussion is necessarily incomplete and somewhat vague. Because the interaction exists, there are other possible excited states in addition to the single-particle states discussed. But further analysis of the nature of these states would lead us far beyond the scope of this book, into areas which are still imperfectly understood.

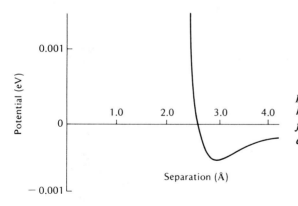

Fig. 6. *Interaction potential between two helium atoms as a function of their separation.*

We conclude this discussion with a description of the most spectacular of the phenomena attributable to superfluidity—the fountain effect. To achieve this effect, a tube is plugged with emery powder and cotton at the bottom, with the top left open, and then it is immersed in liquid HeII, as in Fig. 7. Heat is then supplied to the helium inside the tube, so that some of the He atoms are raised from the ground state; superfluid is "destroyed" inside the tube. Destruction of superfluid is equivalent to raising the temperature. In order to maintain uniformity of temperature and of superfluid fraction throughout the volume of the fluid, superfluid rushes in toward the heater. (Remember the observation of the high thermal conductivity of liquid HeII.) If the plug were not present in the bottom of the tube, movement of superfluid toward the heater would be accompanied by movement of normal fluid away from the heater, both movements tending to transfer heat from the heater and maintain a uniform temperature throughout the fluid, without any net transfer of *mass* in either direction. But the plug prevents the normal fluid, with its nonzero viscosity, from flowing out the bottom of the tube, so there *is* a transfer of mass into the heated region, and the liquid level builds up in the tube. If the tube has a narrow top, a fountain of liquid helium eventually shoots out the top of the tube.

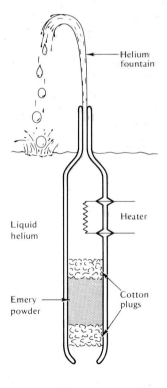

Fig. 7. Schematic illustration of the fountain effect. [From "Principles of Modern Physics" by R. B. Leighton. Copyright McGraw-Hill, New York, 1959. Used by permission of McGraw-Hill Book Company.]

11.4 APPLICATION OF FERMI–DIRAC STATISTICS: FREE ELECTRON THEORY OF METALS

The large electrical and thermal conductivity of metals indicates that many electrons—presumably the valence electrons—are not bound to individual atoms in a metal, but instead move freely throughout the volume of the metal. Apparently a valence electron can move in any direction with ease; as long as it remains inside the metal it is being pulled from all sides, and the forces in all directions tend to cancel each other. It is only when the electron reaches the surface of the metal that the positive charge of the atomic lattice pulls it back into the metal. In other words, the valence electrons in a metal behave like a gas confined within a box.

a. Electronic Specific Heat. The picture of metallic electrons as a gas is rather old, but it caused a puzzle for a long time, because this gas makes very little contribution to the specific heat of the metal. A classical gas, having an energy of $kT/2$ for each degree of freedom, would contribute a specific heat of $R/2$ per mole per degree of freedom; the electron gas, being like a monatomic gas, might therefore be expected to contribute $3R/2$ per mole. But the electronic contribution to the specific heat of a metal is known to be much smaller than this at room temperature, and it is not constant, but is proportional to the temperature.

Of course it should be no surprise to you that classical statistics fails to give the right answer, because an electron gas should obey Fermi–Dirac statistics. Using Fermi–Dirac statistics, we can easily show in a semiquantitative way that the electronic specific heat should vary linearly with temperature. To do this, we begin as we did with the other statistics, by investigating the parameter α. As before, α is a function of temperature which is determined by the condition that the total number of particles in the system is constant. In Fermi–Dirac statistics, it is convenient to write α as $-\mu/kT$, with μ another function of the temperature. This is merely a mathematical substitution; we shall see in a moment how it simplifies the analysis. The occupation index is now

$$\frac{n_s}{g_s} = \frac{1}{e^{(\varepsilon_s - \mu)/kT} + 1}$$

or, for a continuous distribution,

$$\frac{n(\varepsilon)}{g(\varepsilon)} = \frac{1}{e^{(\varepsilon - \mu)/kT} + 1} \tag{31}$$

and μ is seen to be the energy at which the occupation index is $\frac{1}{2}$. We can evaluate μ (called the chemical potential) for a given system at any temperature by applying the condition $N = \int_0^\infty n(\varepsilon)\, d\varepsilon$ and solving the equation for μ.

Fortunately we can solve many problems without determining the precise temperature dependence of μ, simply by considering the situation near $T = 0$ K. At $T = 0$ we should expect all particles to be in their lowest possible energy states; as only one particle can occupy a given state, the occupation index should be 1 for the N lowest states and 0 for all higher states. Equation (31) bears out this expectation; at $T = 0$, the exponent becomes $+\infty$ for $\varepsilon > \mu$ and $-\infty$ for $\varepsilon < \mu$, so that

$$n(\varepsilon) = \begin{cases} 1 & (\varepsilon < \mu) \\ 0 & (\varepsilon > \mu) \end{cases}$$

and at $T = 0$ K, μ must be equal to the energy of the Nth state. This energy is called the Fermi energy ε_f, a constant for a given metal.

Figure 8 shows the effect of raising the temperature. The curve remains symmetrical about the point $\varepsilon = \mu$, where the occupation index $n(\varepsilon)/g(\varepsilon) = \frac{1}{2}$. As T increases, μ becomes smaller, but as long as kT is much smaller than ε_f, μ remains very close to ε_f. For example, if $kT = 0.01\varepsilon_f$ and $g(\varepsilon)$ is constant, it is not too hard to calculate that μ differs from ε_f by less than one part in 10^{45}. (See Problem 6 for another example.)

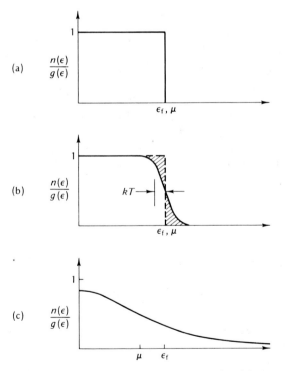

Fig. 8. The Fermi–Dirac occupation index; (a) at $T = 0°K$; (b) at small T, such that kT is of magnitude shown; (c) at very high T, where μ becomes noticeably smaller than ε_f.

If the temperature is raised so that $kT \gg \varepsilon_f$, then μ eventually becomes negative, so that only the "tail" of the Fermi–Dirac distribution appears in the positive energy levels, and the distribution merges into the Boltzmann distribution (as it should, for at such high temperatures, the density is so low that individual particles are distinguishable by their positions).

It happens that ε_f is several electron volts for metals, so that $kT \ll \varepsilon_f$ for any solid metal, and we have no further need to consider the temperature dependence of μ; we shall simply set μ equal to ε_f in the subsequent discussion.

The fact that $kT \ll \varepsilon_f$ immediately explains the linear temperature dependence of the electronic specific heat of metals. The occupation index is virtually 1 until $\varepsilon - \varepsilon_f$ is of order kT, and it falls from 1 to zero over an energy range of a few times kT. Thus only a small fraction of the electrons gain energy when the temperature is raised from $0°K$ to a temperature T; electrons are transferred from the shaded area below ε_f (Fig. 8) to the shaded area above ε_f. The *average* gain in energy per electron is the distance, on the energy axis, between the centroids of these areas; this distance is proportional to kT. The *number* of electrons which gain energy is proportional to the magnitude of the shaded area below ε_f; this quantity is also proportional to kT. The *total* energy gained by the electrons is proportional to the product of these two factors, that is, to $(kT)^2$. Since the specific heat is the temperature derivative of this gain in energy, the specific heat is proportional to T.

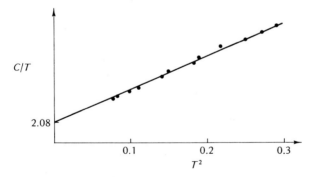

Fig. 9. *Specific heat of potassium at low T, plotted as C/T versus T^2. The fit to a straight line, $C/T = 2.08 + 2.57T^2$, shows that C is given by the sum of a linear term and a cubic term [W. Lien and N. E. Phillips, Phys. Rev. **133**, A1370 (1964)].*

Thus the specific heat of a metal at low temperatures is the sum of an electronic contribution proportional to T, and the lattice contribution, proportional to T^3, which we have already discussed (Section 11.3b). Figure 9 shows how well this is verified by experiment.

b. Calculation of ε_f for a Metal. The general conclusions of the preceding discussion depended primarily on the fact that the electrons in a metal form a Fermi gas, and did not depend on the actual form of the potential energy for each electron. But if we wish to calculate the Fermi energy ε_f for a metal we must determine $g(\varepsilon)$; this requires that we calculate the energy levels of the system, which we can only do if we know the potential energy.

The simplest assumption we can make is that the potential energy is constant inside the metal. This is somewhat unrealistic, for we know that the electrons are strongly attracted to each atomic core in the metal, so that the potential energy should become very negative at these points. But the simple assumption of constant potential energy turns out to give very good results, and is a good starting point in any case.

If the potential energy is constant, we have already solved the problem of finding $g(\varepsilon)$, because it is almost the same as in the liquid helium problem (Section 11.3c). The metal is a box within which the electrons are confined, and we require that the wavefunction be zero at the walls of the box. The possible wavelengths are therefore the same as those already found for He atoms (Sec. 11.3). To find G, the number of states below a given energy ε, we therefore use Eq. (23), replacing the helium mass M by the electron mass m_e, and multiplying by two, because there are two possible spin states for each set of wavenumbers l_x, l_y, l_z. Thus

$$G = \frac{\pi V}{3} \left(\frac{8 m_e \varepsilon}{h^2} \right)^{3/2} \tag{32}$$

When G = N, the total number of valence electrons, then $\varepsilon = \varepsilon_f$, and we may write

$$N = \frac{\pi V}{3} \left(\frac{8 m_e \varepsilon_f}{h^2} \right)^{3/2}$$

and solve for the Fermi energy:

$$\varepsilon_f = \frac{h^2}{8 m_e} \left(\frac{3N}{\pi V} \right)^{2/3} \tag{33}$$

The appropriate electron density N/V to use here is the number of *valence* electrons per unit volume; the other electrons are bound tightly to the atomic cores and are not part of the "gas."

EXAMPLE PROBLEM 2. Compute the Fermi energy of aluminum.

Solution. The atomic mass is 27, the density is 2.7 gm/cm^3, and the valence is 3, so the valence electron density is

$$\frac{N}{V} = \frac{3 \times 6.022 \times 10^{23}}{26.98 \text{ g}} \times \frac{2.702 \text{ g}}{cm^3} = 1.809 \times 10^{23} \text{ cm}^{-3}$$

and

$$\varepsilon_f = \frac{h^2}{8m_e} \left(\frac{3}{\pi} \frac{N}{V} \right)^{2/3}$$

$$= \frac{h^2 c^2}{8m_e c^2} \left(\frac{0.1809 \text{ Å}^{-3}}{1.047} \right)^{2/3}$$

$$= \frac{(1.240 \times 10^4)^2 \text{ eV}^2\text{-Å}^2}{8 \times 5.11 \times 10^5 \text{ eV}} \times (0.1727)^{2/3} \text{ Å}^{-2}$$

$$\varepsilon_f = 11.66 \text{ eV}$$

Aluminum *boils* at $T = 2330$ K, where kT is still only 0.19 eV. The Fermi energy is, of course, lower for monovalent metals, but it still corresponds to a very high temperature; Fermi energies for several monovalent metals are listed in Table 4, with the equivalent temperatures (ε_f/k).

Table 4

Fermi Energies and Equivalent Temperatures for Some Metals

Metal	ε_f (eV)	$\dfrac{\varepsilon_f}{k}$ (K)
Li	4.7	5.5×10^4
Na	3.1	3.7
K	2.1	2.4
Rb	1.8	2.1
Cs	1.5	1.8
Cu	7.0	8.2
Ag	5.5	6.4
Au	5.5	6.4

In Chapter 13, we shall see how it is possible to use positrons as a "probe" to measure the momentum distribution of electrons in a metal and thereby obtain a fairly direct test of the value of ε_f. The final result for ε_f has been found to be in agreement with Eq. (33), with a logical number of valence electrons, for every metal which has been tested. Thus, in spite of the crudeness of the underlying assumption that the potential energy is constant, this model is a useful approximation which is a good starting point for a study of electronic structure of metals.

c. Work Function and Contact Potential Difference. The Fermi energy is the *kinetic* energy of the electrons in the highest occupied states. We can relate this to the work function of the metal by a diagram like Fig. 10, showing the potential well in which the electrons reside and the filled states up to energy ε_f above the bottom of the well. If the well depth is W, the work function, being the energy needed to remove an electron from the metal, is obviously

$$e\phi = W - \varepsilon_f.$$

A diagram similar to Fig. 10 is useful in explaining the contact difference of potential between two metals. When a wire is connected between two metals, electrons can flow from one metal to the other until the Fermi levels[15]—the energies at the top of the filled states—are equal. Figure 11

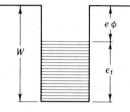

Fig. 10. *Relationship between well depth W, Fermi energy ε_f, and work function $e\phi$ for electrons in a metal. Horizontal lines indicate filled energy levels.*

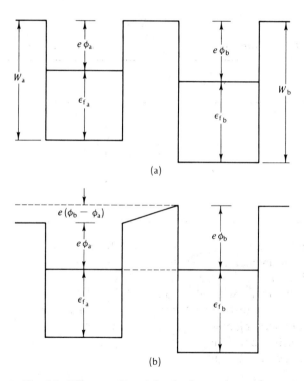

Fig. 11. *Effect on Fermi level when two metals are (a) separated (b) connected. Change in level results from change in the potential well, with negligible change in Fermi energy of either metal.*

[15] Note the distinction here between Fermi *level* and Fermi *energy*. The Fermi *energy* is the *kinetic* energy of each electron in the highest occupied states; the Fermi *level* is the *total* energy of each of these electrons, with respect to some outside reference energy.

shows what happens. Initially the metals are separated and uncharged, so that the potential is zero in the space between them. The metals have different well depths, different Fermi energies, and different work functions, so that there are electrons in metal A with higher *total* energy than that of any electron in metal B (at $T = 0$). When the metals are connected together, electrons flow from metal A to metal B in order to occupy the lower energy states available in B. Metal B becomes negatively charged, metal A becomes positively charged, and the potential energy is no longer zero in the region between the metals, so the potential well in B rises, relative to that in A, until the energy at the top of the filled levels in B equals that in A. The number of electrons transferred is minute[16] in comparison to the total number present, *so the values of ϕ and ε_f are unchanged in each metal*. The figure makes it clear that the difference in electrostatic potential between A and B becomes $\phi_b - \phi_a$.

Figure 10 helps us to understand the circumstances of a photoelectric experiment. If metals A and B are the two electrodes in a photocell, with $\phi_b > \phi_a$, the applied retarding potential V that is required to stop all the electrons is given by

$$eV = h\nu - e\phi_b$$

regardless of whether metal a or metal b is irradiated.[17]

It should also be noted that the contact potential difference cannot be measured with a conventional voltmeter. (Why not?) There is, however, a fairly direct way to measure it, by making a parallel-plate capacitor of the two metals, with a wire joining them, and measuring the current in the wire as the distance between the plates is varied. Changing the distance changes the capacitance, while the voltage across the capacitor remains constant (because it is the contact potential difference), so the charge on the capacitor must change, and a flow of current results. A known additional potential difference can then be inserted into the circuit, and varied until there is no longer a current flow when the capacitance is changed; at this point the voltage on the capacitor must be zero, so the additional potential difference must be just equal to the contact potential difference (and opposite in direction, of course.)

d. Emission of Electrons from a Metal. In the preceding discussion we ignored the electrons whose kinetic energy is greater than the Fermi energy. At any temperature above $0°$K there are always some electrons, in the tail of the distribution, with enough kinetic energy to go over the hill and escape

---

[16] See Problem 7 for verification of this statement.
[17] For further discussion of this point, see J. Rudnick and D. S. Tannhauser, *Am. J. Phys.* **44**, 796 (1976).

from the metal completely. If one measures the energy distribution of the electrons after they have escaped, one finds that it is a Boltzmann distribution. This seems odd at first glance, because the electrons in the metal obey Fermi–Dirac statistics. What causes them to switch over to Boltzmann statistics?

If you have followed us to this point, you will realize that no switch is involved. The tail of the Fermi-Dirac distribution is a simple exponential, just like the Boltzmann distribution; the energies of the escaping electrons exceed the Fermi energy by many times kT, so that $e^{(\varepsilon - \varepsilon_f)/kT} \gg 1$, and the occupation index is

$$\frac{1}{e^{(\varepsilon - \varepsilon_f)/kT} + 1} \to e^{-(\varepsilon - \varepsilon_f)/kT}$$

$$\to \text{constant} \times e^{-\varepsilon/kT}$$

which is the same as that for the Boltzmann distribution.

But in addition to measuring the energy of the escaping electrons, we can also measure the *number* of electrons which escape, as a function of temperature. The ability of Fermi–Dirac statistics to explain this temperature dependence, after classical calculations failed, was a great triumph for quantum theory. According to Boltzmann statistics, the current density of escaping electrons should be proportional to $T^{1/2}e^{-e\phi/kT}$, but experiments show that this current density is proportional to $T^2 e^{-e\phi/kT}$. The derivation of the latter temperature dependence is straightforward, if Fermi–Dirac statistics are used. Details are left as a problem (Problem 8).

Another interesting situation arises when an electric field is applied at the surface of a metal. The effect of such a field is to change the potential energy for an electron near the surface. Therefore, let us abandon our approximation that the electrons are in a *square* potential well, and let us examine more closely the shape of the potential near the metallic surface. Standard electrostatic theory tells us that an electron just *outside* a metal is *not* in a region of constant (zero) potential, but rather that the electron is attracted to the metal by a force which is equal to that of an "image charge" located inside the metal at the point where the mirror image of the electron would be. If the electron is at a distance x from the surface, the distance between the electron and its image is $2x$, so the electron feels a force of $e^2/4\pi\varepsilon_0(4x)^2$ (in mks units), and the corresponding potential energy is

$$V(x) = -\frac{e^2}{16\pi\varepsilon_0 x} \tag{34}$$

This potential would go to $-\infty$ as x goes to zero, but we know that the well has a finite depth W. Therefore we cut off the potential (34) at the point where it is equal to the well depth, that is, at $x = x_c$, where

$$-W = -\frac{e^2}{16\pi\varepsilon_0 \, x_c}$$

so that $V(x) = -W$ for $x \leqslant x_c$, and the resulting potential is as shown in Fig. 12a. This potential is still a bit unrealistic, because the metal is not really a continuous medium, but we can use this model to make rough calculations of the effect of applying an external electric field.

When a uniform external field is applied to the metal surface, in a direction toward the surface, so that it tends to pull electrons out of the metal, the potential becomes that of Fig. 12b. The potential energy never reaches

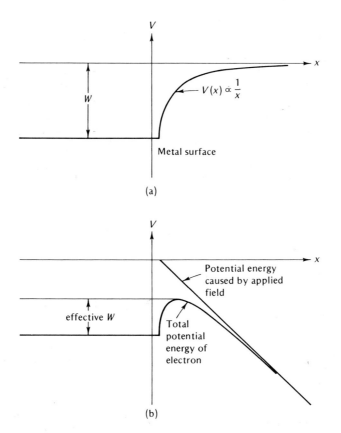

Fig. 12. (a) *Potential energy of an electron near the surface of a metal, including the effect of the image charge, but ignoring the effect of individual atoms.* (b) *Potential energy after a uniform electric field is applied to the surface; the linear potential of the applied field is added to the potential shown in* (a).

zero, but instead goes to a maximum and then decreases. Thus the *effective well depth*—the difference in energy between the bottom of the well and the maximum value of $V(x)$—is smaller than W, and the work function is correspondingly smaller than in the field-free situation.

At any given temperature the application of an electric field increases the current of electron emission, not only because of the decrease in the effective work function, but also because it is then possible for some electrons to *tunnel through* the barrier formed by the potential. This tunneling is possible because the barrier is no longer infinitely thick; as you can see from Fig. 12b, there is a region outside the metal where the potential energy of an electron is lower than it is inside. Calculation of the electric-field dependence of the electron emission current, taking account of these two effects, gives results which are in reasonable agreement with experiment.

e. Pauli Paramagnetism. Paramagnetic susceptibility is another property of metals which is not correctly accounted for by classical statistics. Consider a gas of N particles of spin $\frac{1}{2}$ and magnetic moment $\boldsymbol{\mu}$. If a magnetic field **B** is applied, a particle whose moment is parallel to **B** has a magnetic energy of $-\mu B$,[18] and a particle whose moment is antiparallel to **B** has a magnetic energy of $+\mu B$. According to classical statistics, the number of particles in each group is determined simply by the Boltzmann factor $e^{-\varepsilon/kT}$, so that if $\mu B \ll kT$, the number of particles whose moment is parallel to **B** is

$$n_1 \simeq \frac{N}{2} e^{+\mu B/kT}$$

$$\simeq \frac{N}{2}\left\{1 + \frac{\mu B}{kT}\right\} \tag{35}$$

and the number whose moment is antiparallel to B is

$$n_2 \simeq \frac{N}{2} e^{-\mu B/kT} \simeq \frac{N}{2}\left\{1 - \frac{\mu B}{kT}\right\} \tag{36}$$

We may use the numbers n_1 and n_2 to compute the paramagnetic susceptibility, as follows: The difference between n_1 and n_2, multiplied by μ, is the total magnetic moment of the N particles. The paramagnetic susceptibility χ is defined by the equation[19]

$$\mathbf{M} = \chi\mathbf{H} \tag{37}$$

where **M**, the magnetization, is the total magnetic moment divided by the volume. According to Eqs. (35) and (36), the magnitude of **M** must be

---

[18] Since the z component of μ has only one possible magnitude, we denote that magnitude simply by the symbol μ. For an electron, $\mu = e\hbar/2m$ (Section 7.3). As usual, we assume **B** to be in the z direction, with $B_z \equiv B$.

[19] In mks units, where $\mathbf{B} = (\mathbf{H} + \mathbf{M})/\epsilon_0 c^2$.

$$M = \frac{\mu(n_1 - n_2)}{V}$$

$$= \frac{\mu\left(\dfrac{N\mu B}{kT}\right)}{V} = \frac{N\mu^2 B}{kTV} \tag{38}$$

where V is the volume of the gas. We are concerned here with a situation in which χ is quite small, in which case we may write $\mathbf{B} \approx \mathbf{H}/\varepsilon_0 c^2$, and Eq. (37) becomes

$$\mathbf{M} = \chi\mathbf{B}\,\varepsilon_0 c^2$$

Comparison with Eq. (38) then yields

$$\chi = \frac{N\mu^2}{kTV\varepsilon_0 c^2} \tag{39}$$

Equation (39) says that χ is proportional to $1/T$, a result known as the *Curie law*, which is valid for many materials. But the measured paramagnetic susceptibilities of solid metals are *independent of T*, and are only about 1 percent as large as the values predicted by Eq. (39) for room temperature. Of course, a solid metal does not behave like a classical gas, but in this case the electrons in the metal do form a *Fermi* gas. We can quickly explain the temperature independence of χ by means of Fermi–Dirac statistics, as follows: Electrons with energy much less than ε_f cannot contribute to the total magnetic moment, because there are equal numbers of states with spin parallel to $\mathbf{B}$ and with spin antiparallel to $\mathbf{B}$, and all of the states are filled. The only electrons which can contribute to the susceptibility are those in the region where some states are filled and some are empty. The number of such electrons is proportional to kT; the T dependence of this number cancels the $1/T$ dependence expected from Eq. (39), leaving a susceptibility which is independent of T.

Pauli explained this point in 1927, and he derived the correct formula to replace Eq. (39), using the following line of reasoning: Since we expect the result to be independent of T, let us compute it at $T = 0$. We consider the electron gas to be *two* gases, one with spin parallel to $\mathbf{B}$ and density $n_2(\varepsilon)$ per unit energy interval, the other with spin antiparallel to $\mathbf{B}$ and with density $n_1(\varepsilon)$ per unit energy interval. We may plot both densities on the same graph, using the upward axis for $n_1(\varepsilon)$ and the downward axis for $n_2(\varepsilon)$ (see Fig. 13). Because the occupation index is 1 for $\varepsilon < \varepsilon_f$, each curve is the same as the curve of $g(\varepsilon)$, which is proportional to $\varepsilon^{1/2}$. Figure 13a represents the situation when $B = 0$; there are equal numbers of electrons in each gas. Figure 13b shows what happens when a field $\mathbf{B}$ is applied. The energy of each electron in gas 1 is reduced by an amount μB, and the energy of each electron in gas 2 is increased by the same amount. Electrons in the shaded part of gas 2 now find that lower energy states of opposite spin are available to them, so they flip their spins and join gas 1, filling the shaded area there. The number of elec-

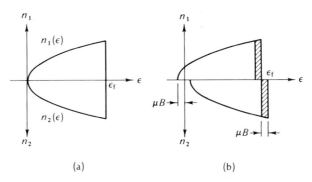

Fig. 13. *Shift in occupation density $n(\varepsilon)$ for an electron gas when a magnetic field is applied. See text for details.*

trons which flip is equal to the shaded area in either gas; if $\mu B \ll kT$, this number is

$$n_f = \tfrac{1}{2}\mu B\, n(\varepsilon_f) = \tfrac{1}{2}\mu B\, g(\varepsilon_f)$$

the factor of $\tfrac{1}{2}$ entering because the density of electrons in each gas is only one-half the total occupation density $n(\varepsilon)$.

Each electron's magnetic moment changes by 2μ when it flips, so the net resulting magnetic moment is $2\mu n_f = \mu^2 B\, g(\varepsilon_f)$, and the resulting magnetization is of magnitude

$$M = \frac{\mu^2 B g(\varepsilon_f)}{V}$$

The susceptibility is thus

$$\chi = \frac{\mu^2 g(\varepsilon_f)}{V\varepsilon_0 c^2}$$

The density of states is

$$g(\varepsilon_f) = \left.\frac{dG}{d\varepsilon}\right|_{\varepsilon=\varepsilon_f}$$

with G as given in Section 11.4(b), so that

$$\frac{g(\varepsilon_f)}{V} = \frac{\pi}{2}\left(\frac{8m_c}{h^2}\right)^{3/2}\varepsilon_f^{1/2}$$

and

$$\chi = \frac{\pi\mu^2}{2\varepsilon_0 c^2}\left(\frac{8m_c}{h^2}\right)^{3/2}\varepsilon_f^{1/2}$$

The susceptibility is often written in terms of the electron density N/V. Since, from Eq. (33),

$$\frac{N}{V} = \frac{\pi}{3}\left(\frac{8m_e\varepsilon_f}{h^2}\right)^{3/2}$$

we may write

$$\chi = \frac{\mu_0\,\pi\mu^2}{2}\left(\frac{8m_e}{h^2}\right)^{3/2}\varepsilon_f^{1/2}\left(\frac{\dfrac{N}{V}}{\left(\dfrac{\pi}{3}\right)\left(\dfrac{8m_e\varepsilon_f}{h^2}\right)^{3/2}}\right)$$

or

$$\chi = \frac{3\mu_0\,\mu^2 N}{2V\varepsilon_f}\qquad(\mu_0\equiv 1/\varepsilon_0 c^2)\tag{40}$$

While it is a vast improvement on Eq. (39), Eq. (40) yields values of χ which differ from experimental results by almost a factor of 2 even for alkali metals, for which the agreement is best. The error must lie in the calculation of $g(\varepsilon)$; we cannot expect to obtain $g(\varepsilon)$ accurately from an assumption that the potential energy of an electron is constant inside a metal. The actual potential energy is periodic, with a minimum at each lattice point, where a positively charged atomic core resides. In an alkali metal, the Fermi energy is so small that the wavelengths of the free electrons are all much greater than the lattice spacing; thus the periodic variation of the potential has less effect than in other metals and we should expect Eq. (40) to work best in this case.

f.   Electrical Conductivity.   One might expect the exclusion principle to have a great effect on the electrical conductivity of a metal, because acceleration of an electron requires the transition of an electron to another state, and many states are unavailable because they are already occupied. However, the effect is not drastic, as we show with the aid of Fig. 14.

When the field is applied, electrons at the "front" of the distribution can easily be accelerated, because they can transfer to empty states of larger p_x. This means that *all* the electrons can be accelerated, because the electrons with smaller p_x simply move into the states vacated by the electrons with larger p_x. The entire distribution can thus be rigidly accelerated by the field, without running afoul of the exclusion principle.

Eventually the whole distribution is displaced by an amount mv, where v is the "terminal velocity," determined by energy losses to the lattice as electrons are scattered by atoms in the lattice. These energy losses result from the recoil of the atomic cores; since these cores are much more massive than an electron, an electron loses only a small fraction of its energy in each scattering event. If the electric field is increased, the electrons will go to a larger average velocity, until the larger energy losses from scattering just compensate for the larger energy put in by the electric field.

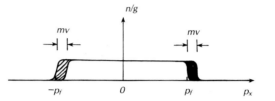

Fig. 14. *Occupation index n/g as a function of x component of momentum. In zero electric field, distribution extends from $-p_f$ to $+p_f$. When field is applied in x direction, electrons are accelerated into the blackened region, and accelerated out of the shaded region, shifting the whole distribution by the amount mv.*

When the field is turned off, these energy losses to the lattice cause the average velocity to return to zero, and the current ceases. Although the energy change in each scattering event is very small (relative to the Fermi energy), the momentum change is usually large (relative to the Fermi momentum p_f), because an electron can only be scattered if its final state, after scattering, was previously unoccupied. Thus electrons are scattered from the blackened region (Fig. 14) into the shaded region. For example, an electron with momentum $p_x = p_f + mv$ can lose energy by scattering to a state with $|p_x'| < p_x$. But the lowest-energy unoccupied state such that $|p_x'| < p_x$ is at approximately (from Fig. 14)

$$p_x' = -p_f + mv$$

Thus the *maximum* energy *loss* is

$$\Delta E_{max} = \frac{p_x^2 - p_x'^2}{2m}$$

$$= \frac{(p_f + mv)^2 - (p_f - mv)^2}{2m}$$

$$= 2p_f v$$

Since $E_f = p_f^2/2m$, we can write this result as

$$\frac{\Delta E_{max}}{E_f} = \frac{2p_f v}{p_f^2/2m}$$

$$= \frac{2mv}{p_f} \tag{41}$$

In most cases, $mv \ll p_f$, so $\Delta E_{max} \ll E_f$. But since the atomic recoil energy is small even for classical collisions, this restriction on the size of ΔE, which results from the exclusion principle, has no serious effect on the relaxation of a current when the electric field is turned off, except in special cases (e.g., superconductivity).

In a real metal, the scattering process results from imperfections in the lattice; we shall see in Chapter 12 that electrons can move through a perfect, static lattice with no energy loss from scattering. But defects, impurities, or vibrations of the lattice lead to scattering processes in which energy is lost. Thus all of these tend to increase the electrical resistivity. In Chapter 12 we shall discuss the connection between these ideas and the explanation of the conduction process in semiconductors and superconductors.

SUMMARY

In this chapter, armed with nothing but elementary quantum statistics, we have been able to explain an enormous variety of phenomena which had puzzled investigators for a long time. It is remarkable that we could do this even though we used only the simplest possible models for the structure of the systems which we have studied. By now we should be convinced of the validity of the quantum statistical formulas, and we are in a position to study more realistic models. In the next chapter, we shall see how various models of solids can be analyzed by straightforward application of the elements of quantum mechanics which we have already developed in Chapters 4-11.

PROBLEMS

1. Use the general formulas (3), (4), and (5) to compute the number of arrangements for each distribution in the four-particle example of Section 11.1, and verify that your results agree with Table 1.

2. Derive the Einstein expression for the specific heat in the limits of high T ($T \gg hv/k$) and low T ($T \ll hv/k$). Show that your low-T formula gives a specific heat of $0.27R$ at $T = hv/6k$.

3. Find the value of the specific heat at $T = \Theta_D/2$, according to the Debye theory. To do this, note that Eq. (20) may be written

$$E = 9R \frac{T^4}{\Theta_D^3} f\left(\frac{\Theta_D}{T}\right)$$

where

$$f\left(\frac{\Theta_D}{T}\right) = \int_0^{\Theta_D/T} \frac{x^3\,dx}{e^x - 1},$$

so that the specific heat may be written in terms of $f(\Theta_D/T)$ and $f'(\Theta_D/T)$. You may find $f(\Theta_D/T)$ and $f'(\Theta_D/T)$ by constructing a graph of the function $x^3/(e^x - 1)$ versus x.

4. The Debye theory, as given in this chapter, was worked out before it was known that a harmonic oscillator has a "zero-point energy" of $h\nu/2$. Thus Eq. (20) does not give the correct total energy of a solid in the limit $T \to 0$, because the average energy per oscillator goes to $h\nu/2$, rather than zero, at $T = 0$. Starting with this fact, use the Debye model to show that the zero-point energy of a solid should be $E_{T=0} = 9N_A k\Theta_D/8$.

5. Consider a system with many uniformly spaced energy levels, 0.01 eV apart, with the lowest level at zero energy, and with degeneracy $g_s = 2$ for each level. In this system are 9 particles, obeying Fermi–Dirac statistics, whose total energy is 0.17 eV.

    (a) Write down all of the possible distributions of the 9 particles into the various levels.

    (b) Find the probability of occurrence of each distribution.

    (c) Find the average number of particles in each level.

    (d) Find the value of the Fermi energy ε_f in eV for this system.

    (e) Compare the average distribution found in (c) with the Fermi–Dirac distribution, Eq. (31), and deduce an approximate "temperature" for the system (the value of T which fits best).

6. Consider a Fermi gas in which $g(\varepsilon)$ is a constant g, independent of ε. Find the order of magnitude of the difference between μ and ε_f, in units of ε_f—that is, find $(\mu - \varepsilon_f)/\varepsilon_f$—at temperature $T = 0.1\ \varepsilon_f/k$.
    Hint: At $T = 0°$K we may write the number of particles as

$$N = \int_0^\infty n(\varepsilon)\,d\varepsilon = \int_0^{\varepsilon_f} g(\varepsilon)\,d\varepsilon = g\varepsilon_f$$

In general, at temperature T, $N = \int_0^\infty n(\varepsilon)\,d\varepsilon = \int_0^{2\mu} n(\varepsilon)\,d\varepsilon + \int_{2\mu}^\infty n(\varepsilon)\,d\varepsilon$ and because of the symmetry of the curve of $n(\varepsilon)/g(\varepsilon)$, we know that

$$\int_0^{2\mu} n(\varepsilon)\,d\varepsilon = g\mu$$

Therefore

$$g\varepsilon_f = g\mu + \int_{2\mu}^\infty n(\varepsilon)\,d\varepsilon$$

Convince yourself that this is so; then complete the calculation.

7.   In the discussion of contact difference of potential (sec. 11.4c), it was stated that the "number of electrons transferred is minute in comparison to the total number present." Verify this statement by computing the charge transferred when two metals of area A and thickness d are brought together to form a parallel-plate capacitor of plate separation d. The plates are joined by a fine wire; assume a contact potential difference of 1.0 volt, $d = 0.1$ mm, and $A = 1.0$ cm^2. Compute the number of electrons transferred. Assuming that one plate is aluminum (see Example Problem 2), find the resulting fractional change in ε_f in that plate.

8.   Use the following procedure to find an expression for the "thermionic emission" current density, verifying that it has the temperature dependence given in sec. 11.4d:

(a)   Consider the electrons in a ring like the one illustrated in Fig. 3, chapter 1. All electrons in the ring are at a distance between r and $r + dr$ from the area element A on the surface of the metal, and the radius vector from A to any point in the ring makes an angle between θ and $\theta + d\theta$ with the normal to the surface. Show that the number of electrons in this ring with energy between ε and $\varepsilon + d\varepsilon$ is $n(\varepsilon) 2\pi r^2 \sin \theta\, dr\, d\theta\, d\varepsilon/V$, where V is the total volume of the metal and $n(\varepsilon)$ is the Fermi distribution function.

(b)   Multiply the result of (a) by the probability that a given electron is moving in the right direction to strike A, and then integrate on r from 0 to $v\, dt$ (where $v = \sqrt{2\varepsilon/m}$ is the speed of an electron whose energy is ε), to find the number of electrons striking A, in time dt, at an angle between θ and $\theta + d\theta$, with an energy between ε and $\varepsilon + d\varepsilon$. Your result should be

$$dN = \left(\frac{\varepsilon}{2m}\right)^{1/2} n(\varepsilon)\, d\varepsilon \sin \theta \cos \theta\, d\theta\, dA\, dt/V.$$

(c)   When an electron reaches A it can escape only if its x-component of velocity is sufficiently great; its other velocity components do not help it to escape. Thus an electron of a given energy ε can escape only if it strikes A at a sufficiently small angle θ. Show that the condition on θ is $\cos \theta \geqslant \sqrt{W/\varepsilon}$, where W is the well depth, and integrate the result of (b) over the allowed angles. Your result should be that the *current density* of electrons escaping from the metal with energy between ε and $\varepsilon + d\varepsilon$ is

$$j(\varepsilon) = \frac{e}{dA\, dt} \int_{\theta=0}^{\theta=\cos^{-1}\sqrt{W/\varepsilon}} dN = \frac{e\, n(\varepsilon)\, d\varepsilon}{4V} \left(\frac{2\varepsilon}{m}\right)^{1/2} \left(1 - \frac{W}{\varepsilon}\right)$$

(d)   Now insert the proper expression for $n(\varepsilon)$ and integrate on ε between the appropriate limits to find the total current. Notice that over the energy range considered $e^{(\varepsilon-\mu)/kT} \gg 1$. Your final result should be that the emission current density is

$$J = \frac{4\pi m e}{h^3} (kT)^2 \, e^{-e\phi/kT}$$

This is called the Richardson–Dushman equation.

9. The observed specific heat of metallic Ni at low temperatures is, in arbitrary units:

T:	2	4	6	8	10	12	14	16	°K
C_v:	0.4	0.8	1.3	1.8	2.4	3.0	3.8	4.8	

    The Debye temperature is 375 K.
    (a)   Deduce the value of the electronic specific heat, in these units, at 10 K.

    (b)   What value would you expect for the total specific heat, in these units, at 20 K?

10. (a)   Show that application of a uniform electric field of magnitude $\mathscr{E}$, directed into the surface of a metal, lowers the work function by an amount $e^{3/2}\mathscr{E}^{1/2}$ (cgs) or $e^{3/2}\mathscr{E}^{1/2}/(4\pi\varepsilon_0)^{1/2}$ (mks).

    (b)   Compute the amount by which the work function is lowered, in electron volts, if $\mathscr{E} = 10^6$ V/m.

11. Show that the average energy of the particles in a Fermi gas is equal to $3\varepsilon_f/5$ at $T=0$.

12. Show that the specific heat of a Bose–Einstein gas is proportional to $T^{3/2}$ for $T < T_c$.

13. At low temperatures, He^3 atoms injected into liquid He^4 form a mono-layer on the liquid surface. We can consider this layer to form a *two-dimensional* Fermi gas; we neglect any motion perpendicular to the surface.
    (a)   Show that the density of states $g(\varepsilon)$ is independent of ε for such a gas.

    (b)   Compute the value of ε_f if the surface density of He^3 atoms is 6×10^{14}/cm^2.

14. Consider a two-dimensional Bose–Einstein gas similar to the two-dimensional Fermi gas of Problem 13. Go through the steps used to deduce the condensation of a Bose–Einstein gas into the ground state (Section 11.3c), and show that there is *no* condensation for the two-dimensional Bose–Einstein gas.

15. A current of 200 amperes flows in a cylindrical aluminum wire whose radius is 1.0 mm. (At this current, the Joule heating equals the black-body radiation loss at a temperature of about 900 K. See Problem 4, Chapter 3.) Find the value of mv/p_f, where p_f is the Fermi momentum, and mv is the average momentum of the conduction (valence) electrons.

The Electronic
Structure of Solids

The assumptions of the previous chapter worked well enough that we can now use them as a starting point for a genuine theory of solids. Of course, in the short space available here, we cannot approach a complete survey of solid state physics, so our aim will be to study those aspects of the subject which form a good illustration of the quantum mechanical principles we have already developed. It is hoped that you will thereby acquire the basis for understanding more advanced or more detailed treatments.

We shall study the properties of those solids which have a regular crystalline lattice structure. We begin by modifying the free-electron theory of Chapter 11 to obtain instead a "nearly-free-electron" theory which takes account of the periodicity of the potential energy of an electron in a periodic lattice. The need for a modification of the free-electron theory is clear when one considers that the simple free-electron theory cannot even answer such an elementary question as: "Why are some solids conductors while others are insulators?"

429

Before the development of quantum mechanics, it was thought that the answer to this question was that electrons are tightly bound to individual atoms in an insulator, whereas they are free to move throughout the whole solid in a conductor. But it was not known why electrons should be tightly bound in one case and not in another; there was no way to correlate the conductivity with other properties, such as the cohesive energy. The quantum theory answers the question in a simple unambiguous way, when one takes account of the periodicity of the potential energy. According to the quantum theory, electrons are not all localized on individual atoms in *any* solid; whether the solid is a conductor or not depends on the energy levels available to the electrons.

We shall study two important consequences of the periodic nature of the potential energy: first, there is a discontinuity in the energy of an electron as a function of wavenumber—a so-called "energy gap"—which accounts for the difference between an insulator and a conductor; second, there is an anisotropy in the relation between the energy and momentum of an electron, so that the energy depends on the direction of **p** as well as on its magnitude.

12.1   ENERGY LEVELS FOR A SYSTEM OF N ATOMS

Before going into the details of the effect of the lattice on the energy levels, let us consider the electron energies in a solid in a qualitative way. From one point of view, the existence of an energy gap, or a band of forbidden energies for electrons in a solid, is a natural consequence of the fact that atomic energy levels are discrete. In a collection of N identical atoms there are N electron states for each state of a single atom. If the atoms are far apart, each of these N states has the same energy; there is N-fold degeneracy. We could say that there is an "energy gap" between one level and the next. When the atoms are close enough together to interact, this degeneracy is broken, and each level splits into a band containing many levels.[1] But when the interaction is sufficiently weak, the splitting is not large enough to mix the levels of one band with those of the next, and the gap remains.

To understand these points, consider what happens when two hydrogen atoms are brought together. Suppose that both are in the ground state. When the wavefunctions begin to overlap, we can no longer speak of an electron as

---

[1] The effect is analogous to the effect of coupling two identical simple harmonic oscillators. There are two normal modes, which are of the same frequency when the coupling is zero. A weak coupling shifts the frequency of each mode, so that one mode has a slightly higher frequency and the other mode has a slightly lower frequency than it had before. When there are N identical coupled oscillators, there are N modes, each with a slightly different frequency, in general.

belonging to one atom or another; instead we must write an antisymmetric wavefunction to describe the *system*. If the electron spins are parallel, the space part of the wavefunction must be antisymmetric with respect to interchange of the two electrons, and if the spins are antiparallel, the space part of the wavefunction is symmetric. We had previously (Section 8.2) gone through such a discussion in order to show that the triplet (parallel spin) level in a two-electron *atom* has lower energy than the singlet level, because the electrons are farther apart, on the average, when the space part of the wavefunction is antisymmetric. But now it happens that the argument goes the opposite way; the energy is lower when the electrons are closer together, because in that way the electrons can both be closer to the two protons.

We have seen (Section 9.3) that this effect causes the total electron spin in the H_2 molecule to be zero, yielding the antisymmetric spin state. Figure 9 of Chapter 9 shows the energy of this state and the three symmetric spin states as r varies. The three $S = 1$ states remain degenerate in the triplet level (curve A), but the $S = 0$ state (curve S) splits off as r decreases. In the presence of an external field, the triplet level would split further into three levels; such an external field could be provided by bringing a *third* atom near. There would then be four distinct energy levels instead of the single ground state level of the two hydrogen atoms. In general, there would be $2N$ 1s levels for N atoms.

A similar argument can be made for the energy levels in any set of N identical atoms. Each atomic level splits into a band as the atoms are brought near to one another. Figure 1 shows graphically the dependence of the

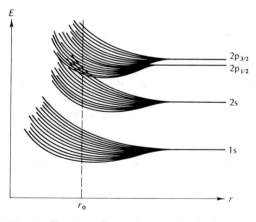

Fig. 1. *Energy of one-electron levels in a collection of N atoms, as a function of the nearest neighbor distance r.*

allowed energy levels on r. (We assume that the atoms are in a regular array, and r is now the distance between each atom and its nearest neighbor.)

The states are described by quantum numbers in much the same way as hydrogen-atom states. The 1s level splits into a band of $2N$ levels as r is decreased, and the 2s level does the same, because each level is $2N$-fold degenerate at large r. (At large r, an electron has a choice of two spin states and N atoms, for a total of $2N$ states.) The $2p_{3/2}$ and $2p_{1/2}$ levels are very closely spaced; these levels form a band which contains $6N$ states.

Now suppose that there is an equilibrium distance r_0 at which the atoms form a solid. Consider first the alkali metals. There are N valence electrons which occupy the $2N$ states in the highest occupied band (the "valence band"). Thus there are always many empty states with energy very close to the energy of the occupied states, and it is very easy to excite, or accelerate, an electron. (See Section 11.4f.) This means that the solid is a good conductor. By the same line of reasoning, we can see that any solid which has an *odd number* of valence electrons per atom has a half-filled valence band and is therefore a good conductor.[2] Any solid which has an *even* number of valence electrons per atom has just enough electrons to fill an integral number of bands, so it *could* be an insulator. The determining factor in this case is whether two bands overlap, as the 2s and 2p bands do in Fig. 2. If the highest occupied states are in a band which overlaps another band, the solid is a conductor, because, as in the alkali metal, it is easy to excite electrons into nearby unoccupied states. If the bands do not overlap, the solid is either a semiconductor or an insulator, depending on the magnitude of the gap between the highest occupied band and the next higher band.

12.2   TRAVELING ELECTRON WAVES IN A SOLID

If we were to pursue the line of inquiry outlined in the previous section, we would be faced with the task of calculating the energy levels for all conceivable configurations of N atoms as they are brought together. The configuration with the lowest ground-state energy would be the stable one, and the energies of the excited states (when the atoms are in this configuration) would determine the electronic properties of the solid. If we succeeded in this calculation, we could then find every property of the solid—its crystal structure, melting temperature, elastic constants, et cetera—from first principles. Given enough time and a large enough computer, we might be able to do this by successive approximations.

Unfortunately, no existing computer could complete such a program in our lifetimes, so let us choose another approach. We make two assumptions:

[2] There are exceptions. Solid fluorine, for example, is an insulator because it forms a *molecular* crystal rather than an atomic one. The atoms combine to form F_2 molecules, and these molecules form the solid. Each *molecular* energy level broadens into a band, and there are an *even* number of electrons per *molecule*, so that the highest band is just filled.

1.   The lattice structure is known.
2.   The electrons interact only with the lattice, not with one another.

Assumption 2 is successful because the electron–electron interactions tend to cancel one another, and one can imagine a single electron to move in a somewhat uniform field produced by the other electrons. This sounds like the argument made in Chapter 11 to justify the free-electron model. The difference is that we now wish to go one step further in the calculation; we wish to include the effect of discrete lattice points on a given electron, while we still average out the effects of the other electrons. At the outset, we do not know whether we will learn significantly more about solids by doing this; we simply have to try it and see how our results compare with experiment.

We shall make the further assumption that the valence electrons are "nearly free"; we consider each electron state to be primarily one of the plane-wave states of a particle in a box—the same states considered in Chapter 11—but now we permit each state to have a small admixture of other plane waves, as a result of the perturbation caused by the variation of potential energy throughout the lattice. It may seem strange that such an assumption would lead to any worthwhile results; the most obvious effect of the lattice potential is to cause the electron wavefunction to oscillate much more rapidly near each ionic site, where the potential energy becomes more negative, and it would require a very large number of plane waves to adequately reproduce such a wavefunction. But we are not interested so much in finding the correct *wavefunction* as in obtaining the correct *energy levels*, and we can do this by using a wavefunction which matches the true wavefunction over *most* of the crystal, even if it fails to reproduce the rapid oscillations near an atomic core.

How then does the treatment of this chapter differ from the original free-electron treatment, which had somewhat the same rationale? It turns out that even the first-order perturbation of the lattice on the electronic states is quite noticeable, especially on those plane-wave states whose wavelength is nearly equal to the wavelength at which the electron is diffracted by the lattice. In order to treat such diffraction, we need to discuss *motion* of electrons through the crystal, so we represent the electron states by *running* waves rather than the *standing* waves used in Chapter 11. This representation requires that we abandon the boundary condition that the wavefunction be equal to zero at the surface of the solid. We replace this condition by a condition which is equivalent to it, for all practical purposes; the condition, called the Born–von Karmann boundary condition, is that the wave be periodic in each space coordinate, with a period (wavelength) equal to the corresponding dimension of the solid.

It is easy to see that this new boundary condition could be appropriate for one coordinate, if the solid is in the shape of a ring (Fig. 2). It is more difficult to see how it can be appropriate for all three coordinates. But we must bear

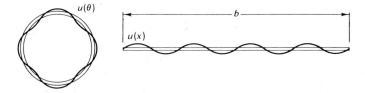

Fig. 2. *If a thin rod of length b is bent into a ring, the boundary disappears, and the wavefunction must have period b. This is equivalent to having the slope and magnitude of u equal at both ends of the rod, without requiring that u be zero there.*

two things in mind: (1) We do not expect the behavior of electrons deep inside a solid to be strongly dependent on conditions imposed at the surface; external boundary conditions are just an intermediate step toward finding the energy levels. (2) These boundary conditions give the same results as the previous boundary conditions, for any observable quantity such as the Fermi energy.

The truth of the latter assertion may not be completely obvious, because the set of possible wavelengths of an electron is different for the two kinds of boundary conditions. Under the Born–von Karmann condition, the possible wavelengths for an electron are equal to the dimension b divided by an integer l_x, whereas the possible wavelengths were formerly equal to $2b/l_x$. Thus the possible values of p_x differ by h/b instead of by $h/2b$, and the states are only *half as dense* in momentum space[3] (see Fig. 3). You might therefore think that

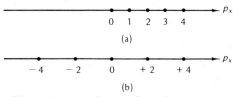

Fig. 3. *Points indicate independent states in terms of p_x (in units of $h/2b$) (a) for boundary conditions of Chapter 11 and (b) for Born–von Karmann boundary conditions. In each case there are five states with $|p_x| \leqslant 4(h/2b)$.*

the states would be occupied up to a greater value of p_x, and that the Fermi energy would be correspondingly greater. But there are now two states for each value of the integer l_x, because a running wave may travel in either direction, and each direction belongs to an independent state. Thus there are

---

[3] Momentum space, of course, is just a system in which the coordinate axes represent the momentum components of a particle rather than its position coordinates.

just as many states as before for which the *magnitude* of p_x is below any given value (see Fig. 3).

In three dimensions this is also true. The density of the allowed states in momentum space is only $(\frac{1}{2})^3 = \frac{1}{8}$ as great as before, but now we count the whole volume of the sphere, instead of the volume of the positive octant, in determining the number of states below a given energy, and we come out with the same answer we obtained previously.

Now we are ready to take account of the periodic potential of the lattice. The running electron waves are reflected from the atomic planes in the lattice, just as electrons are reflected in the Davisson–Germer experiment, or as X-rays are reflected in X-ray diffraction experiments. An electron is reflected if its wavelength λ satisfies the Bragg condition (Section 4.1 or Appendix B):

$$n\lambda = 2d \sin \theta \qquad (1)$$

where θ is the angle between the incident electron's direction and the reflecting plane (Fig. 4).

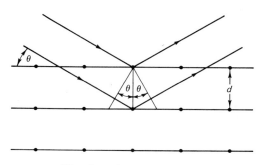

Fig. 4. *Electron reflection.*

Before discussing the complications of the problem in three dimensions, let us explore the consequences of electron reflection in one dimension. If a wave is incident on a set of equally spaced points in a straight line along its direction of propagation, it is reflected backwards if $n\lambda = 2a$, where a is the distance between consecutive points. (This is just the Bragg condition with $\theta = 90°$ and $d = a$.) In terms of the wavenumber $k = 2\pi/\lambda$, the condition for reflection is

$$k = \frac{n\pi}{a} \qquad (2)$$

We must remember that when this condition is satisfied, the electron *remains inside* the crystal, being reflected back and forth, so that its wavefunction must be a *standing wave*. For this value of k, therefore, the appropriate wavefunctions are not two independent waves running in opposite directions, but rather two independent *standing* waves, *each* of which is a superposition of *both* running waves:

$$u_1 = \sin kx = \frac{e^{ikx} - e^{-ikx}}{2i}$$

$$u_2 = \cos kx = \frac{e^{ikx} + e^{-ikx}}{2} \tag{3}$$

where $k = n\pi/a$.

It appears that we have been forced to return to our original treatment of the free electron model, except that both sine and cosine wavefunctions are now allowed. But there is a second very important difference: The wave-functions u_1 and u_2 *must correspond to different energies,* for reasons which become clear when we study Fig. 5. Near the atomic cores (ions) the potential

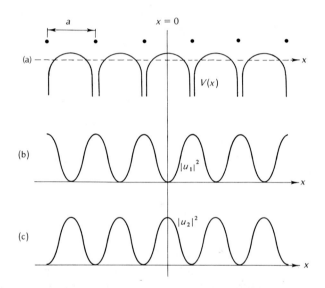

Fig. 5. *Potential energy of an electron in a row of atomic cores (ions). Dots indicate the positions of the ions. The broken line indicates the average level of the potential energy, averaged over all values of x. (b) Probability density for an electron whose wavefunction is $u_1 = \sin(\pi x/a)$. (c) Probability density for an electron whose wavefunction is $u_2 = \cos(\pi x/a)$.*

energy is lower than its average value, because the electron is attracted to the positively charged ions. The probability density $|u_1|^2$ is much larger than $|u_2|^2$ in the vicinity of the cores, whereas $|u_2|^2$ is the larger of the two in the region between the cores. Therefore the expectation value of the *potential energy* is smaller for an electron with wavefunction u_1 than for an electron with wavefunction u_2. The *kinetic* energy is the same for both electrons (because k is the same), and therefore the electron with wavefunction u_1 has smaller total energy.

Of course neither u_1 nor u_2 is precisely an allowed wavefunction—an eigenfunction of the exact Schrödinger equation—for an electron in the potential of Fig. 5. But if we can consider the fluctuation in potential at each core to be a small perturbation,[4] $v(x)$, on the *average* potential indicated by the broken line in the figure, we know that the first-order result is that the energy is shifted, relative to the unperturbed energy, by the amounts

$$\int_{-\infty}^{+\infty} u_1^* v(x) u_1 \, dx \qquad \text{or} \qquad \int_{-\infty}^{+\infty} u_2^* v(x) u_2 \, dx \qquad (4)$$

respectively. Thus the energy levels when $k = \pi/a$ are approximately

$$E_1 = \frac{\hbar^2 k^2}{2m} + \int_{-\infty}^{+\infty} u_1^* v(x) u_1 \, dx$$

and

$$E_2 = \frac{\hbar^2 k^2}{2m} + \int_{-\infty}^{+\infty} u_2^* v(x) u_2 \, dx$$

The perturbation $v(x)$ is positive where u_2 is large, and it is negative where u_1 is large, so that E_2 is larger than the free-particle energy $\hbar^2 k^2/2m$, and E_1 is smaller than the free-particle energy.

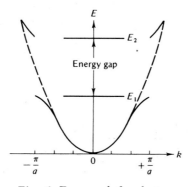

Fig. 6. *E versus k for electron wavefunctions for a periodic potential with period a.*

In general, the effect of the perturbing potential of the lattice is to produce a region of forbidden energies; energies in between the energy E_1 and the energy

---

[4] Obviously the perturbation is not *small,* but because of screening effects, it does extend over a small region, so one can obtain some insight by use of perturbation theory.

E_2 are not allowed for *any* value of k. The energy as a function of k varies somewhat as shown in Fig. 6. The broken line in this figure shows the parabolic k dependence expected for a truly free electron, for which the energy is simply $p^2/2m = \hbar^2 k^2/2m$. For an electron in a square well the E versus k curve is the same as for a free electron, except, of course, that the allowed states are located at discrete points along the broken line, because only certain values of k are permitted by the boundary conditions.

The energy in a periodic potential follows the broken line for *small* values of k, because an electron with sufficiently long wavelength is not sensitive to the rapid periodic variations in potential; such an electron sees only the average potential energy, as if it were in a square well. Near $k = \pm\pi/a$ the effect of the lattice becomes pronounced, and the E versus k curve deviates greatly from the free electron parabola, producing the gap between the energies E_1 and E_2.

> *Comments on the Perturbation Treatment.*   In view of the fact that the wavefunctions u_1 and u_2 are degenerate when the perturbation is zero, you may wonder why we could not use some linear combination of u_1 and u_2 as our wavefunction in expressions (4). Such a wavefunction would yield an energy between the values E_1 and E_2, because the expectation value of the perturbing potential $v(x)$ would lie between the two extremes given by the functions u_1 and u_2. However, to obtain an accurate result in perturbation theory, we should use a wavefunction which is as close as possible to the exact wavefunction of the system. We do this by using a wavefunction which is the *limit* of the exact wavefunction as the perturbation is slowly "turned off." In this particular case, we know that the exact wavefunction is either an odd function of x or an even function of x, because of the symmetry of the potential energy : $V(x) = V(-x)$. (See Section 5.4.) Therefore the limit of the exact wavefunction as the "perturbation" is turned off must be either the odd function u_1 or the even function u_2, *not* a linear combination of them which would have no definite parity. We conclude that u_1 and u_2 are the functions to use in order to find the best first-order approximation to the energy eigenvalues.

12.3   SOLUTIONS OF SCHRÖDINGER'S EQUATION FOR A PERIODIC POTENTIAL

We cannot fully understand the E versus k curve of Fig. 6 until we know more about the actual eigenfunctions for a periodic potential. As was mentioned in the previous section, these eigenfunctions are *not* the eigenfunctions e^{ikx} for a square well, but if the periodic part of the potential could be slowly turned off, the eigenfunctions would gradually reduce to the square-well eigenfunctions.

Let us pursue this idea further with the aid of perturbation theory. Suppose that $v(x)$ is a small periodic perturbation on a square-well potential which extends from $-b/2$ to $+b/2$. If $v(x)$ has period a, we may expand it in a Fourier series:

$$v(x) = \sum_{s=-\infty}^{\infty} v_s e^{2\pi six/a}$$

The first-order wavefunctions $u'_m(x)$ may be expanded in a series of eigenfunctions $u_n(x)$ of the unperturbed system:

$$u'_m(x) = \sum_{n=-\infty}^{\infty} a_{mn} u_n(x)$$

where $u_n(x) = e^{ik_n x} = e^{2\pi nix/b}$ is one of the unperturbed wavefunctions. (We have assumed periodic boundary conditions, so that $u_n(x)$ has period b.)

Now we make the standard assumption of perturbation theory, that

$$a_{mm} \approx 1 \qquad \text{and} \qquad a_{mn} \ll 1 \qquad (m \neq n)$$

so that the coefficients may be written[5]

$$a_{mn} = \frac{\int_{-\infty}^{\infty} u_n^* v(x) u_m \, dx}{E_m - E_n} \qquad (m \neq n) \tag{5}$$

where $E_m = \hbar^2 k_m / 2m$ and $E_n = \hbar^2 k_n^2 / 2m$ are the energies of the unperturbed states. Equation (5) may be expanded to

$$a_{mn} = \frac{2m \int_{-b/2}^{b/2} e^{-ik_n x} \left(\sum_s v_s e^{2\pi isx/a} \right) e^{ik_m x} \, dx}{\hbar^2 (k_m^2 - k_n^2)}$$

and integrated term-by-term. (The integral extends only from $-b/2$ to $+b/2$, because this is the location of the well. The width b must be an integral multiple of the period a of the perturbing potential.)

We may rewrite the integral as

$$\sum_{s=-\infty}^{\infty} \int_{-b/2}^{b/2} e^{i(k_m + (2\pi s/a) - k_n)x} \, dx$$

and since $k_n = 2\pi n/b$, and b is an integral multiple of a, the integral is of the form $e^{i(\text{integer})\pi}$, which has the same value in the upper and the lower limits, and thus vanishes, *unless*

$$k_m + \frac{2\pi s}{a} - k_n = 0$$

---

[5] See Chapter 8, Eq. (8). Here we replace E'_m by E_m, and E' by E_n, because the difference between the primed and the unprimed energies must be quite small, compared with $E_m - E_n$, if the approximations of perturbation theory are to be valid.

Thus a term appears in the first-order expansion of $u'_m(x)$ only if its k-value differs from k_m by the amount $2\pi s/a$, where s is an integer. If we carried the perturbation theory to higher and higher orders, by repeatedly using the perturbed wavefunction as the starting point for a new calculation, we would still find that the same set of k-values appear in the wavefunction, because integrals involving other k-values always go to zero. In other words, the wavefunction in a periodic potential of period a may always be written as a series of exponentials with wavenumber, k, $k \pm 2\pi/a$, $k \pm 4\pi/a$, ...; for a given value of k, the wavefunction is

$$u(x) = \sum_{s=-\infty}^{\infty} a_s(k)e^{i(k+2\pi s/a)x}$$

We have dropped the subscript m, on the wavefunction u and on the coefficients a_s, but we must understand that this is a *particular* wavefunction, for a *given* k value, and that the coefficients a_s depend on the value of k, in general; therefore we write them as $a_s(k)$.

If we factor out the e^{ikx} from each term of the series, we have

$$u(x) = e^{ikx} \sum_{s=-\infty}^{\infty} a_s(k)e^{2\pi isx/a}$$

or

$$u(x) = e^{ikx}u_k(x) \tag{6}$$

where

$$u_k(x) = \sum_{s=-\infty}^{\infty} a_s(k)e^{2\pi isx/a}$$

The function $u_k(x)$ is periodic, with a period a equal to the period of the potential. Equation (6) is a one-dimensional statement of *Bloch's theorem*, which says that the eigenfunctions of the Schrödinger equation for a periodic potential are always of the form

$$\boxed{u(\mathbf{r}) = u_k(\mathbf{r})e^{i\mathbf{k}\cdot\mathbf{r}}} \tag{7}$$

where $u_k(\mathbf{r})$ is periodic in x, y, and z, with the same period as the period of the potential in each coordinate.

Kronig–Penney Potential.[6]   In order to find the actual solutions of the Schrödinger equation for a solid, we would have to know the potential energy $v(x)$ for an electron in the field of the atomic cores. But even without knowing this, we gain some insight simply by solving the equation for a very simple potential which is a reasonable facsimile of the potential one might expect in

[6] R. Kronig and W. Penney, *Proc. Roy. Soc. (London)* **A130**, 499 (1930).

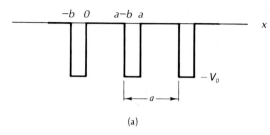

(a)

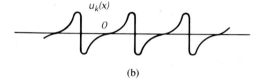

(b)

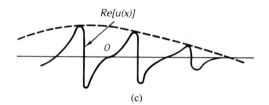

(c)

Fig. 7. *(a) Periodic square-well potential. (b) Periodic function $u_k(x)$ which is a factor in the eigenfunction, according to Bloch's theorem. (c) Real part of complete eigenfunction, with real part of the factor e^{ikx} shown as broken line.*

a solid. Such a potential, a periodic square well, is shown in Fig. 7. With the aid of Bloch's theorem, we shall now find the energy levels for this potential.

The potential is of the form

$$V(x) = \begin{cases} 0 & (0 < x < a - b) \\ -V_0 & (-b < x < 0) \\ V(x + a) & \text{(for all } x) \end{cases}$$

The Schrödinger equation is

$$\frac{d^2u}{dx^2} + \frac{2mE}{\hbar^2}\, u = 0 \qquad (0 < x < a - b)$$

$$\frac{d^2u}{dx^2} + \frac{2m(E + V_0)}{\hbar^2}\, u = 0 \qquad (-b < x < 0)$$

with the solution, as usual,

$$u(x) = \begin{cases} Ae^{i\alpha x} + Be^{-i\alpha x} & (0 < x < a - b) \quad (8) \\ Ce^{i\beta x} + De^{-i\beta x} & (-b < x < 0) \quad (9) \end{cases}$$

where $\alpha = (2mE/\hbar^2)^{1/2}$ and $\beta = (2m(E + V_0)/\hbar^2)^{1/2}$.

So far this is the standard solution which we examined in Chapter 5. But now we apply boundary conditions and Bloch's theorem. At $x = 0$, $u(x)$ and du/dx are continuous, so

$$A + B = C + D \qquad (10)$$

and

$$i\alpha A - i\alpha B = i\beta C - i\beta D \qquad (11)$$

According to Bloch's theorem, solutions (8) and (9) must be a product of a factor e^{ikx} and a function $u_k(x)$ which is periodic, with period a. Therefore we may multiply both sides of Eq. (8) and Eq. (9) by e^{-ikx}, to obtain

$$u_k(x) = \begin{cases} u(x)e^{-ikx} = Ae^{i(\alpha - k)x} + Be^{-i(\alpha + k)x} & (0 < x < a - b) \\ u(x)e^{-ikx} = Ce^{i(\beta - k)x} + De^{-i(\beta + k)x} & (-b < x < 0) \end{cases}$$

We may now apply boundary conditions to the function $u_k(x)$. We know that $u_k(x)$ has period a, thus

$$u_k(a - b) = u_k(-b)$$

and

$$\left.\frac{du_k}{dx}\right|_{x = a - b} = \left.\frac{du_k}{dx}\right|_{x = -b}$$

These conditions give us the equations

$$Ae^{i(\alpha - k)(a - b)} + Be^{-i(\alpha + k)(a - b)} = Ce^{-i(\beta - k)b} + De^{i(\beta + k)b} \qquad (12)$$

$$i(\alpha - k)Ae^{i(\alpha - k)(a - b)} - i(\alpha + k)Be^{-i(\alpha + k)(a - b)}$$
$$= i(\beta - k)Ce^{-i(\beta - k)b} - i(\beta + k)De^{i(\beta + k)b} \qquad (13)$$

Equations (10), (11), (12), and (13) are a set of linear, homogeneous equations in A, B, C, and D. If they are written in standard form (with zero on the right-hand side) it is clear that they are satisfied by nonzero values of A, B, C, and D only if the determinant of the coefficients is zero.

After considerable algebra, this determinant reduces to

$$\frac{-\beta^2 - \alpha^2}{2\alpha\beta} \sin \beta b \sin[\alpha(a - b)] + \cos \beta b \cos[\alpha(a - b)] = \cos ka \qquad (14)$$

Equation (14) is somewhat difficult to analyze as it stands, but we can simplify it by considering the case in which $b \ll a$. We must be careful how we do this; if we simply let $b \to 0$, we are left with nothing but a free-particle solution.

But the effect of the wells is not lost if we let the well depth go to infinity as the width goes to zero, in such a way that the product $V_0 b$ remains finite. Then $\beta^2 b$ remains finite, and $\beta b \rightarrow 0$, so that

$$\sin \beta b \approx \beta b$$
$$\cos \beta b \approx 1$$
$$a - b \approx a$$

and Eq (14) reduces to

$$-P \frac{\sin \alpha a}{\alpha a} + \cos \alpha a = \cos ka \tag{15}$$

where $P = \beta^2 ba/2$, a dimensionless number which expresses the strength of the periodic part of the potential. If $P = 0$,

$$\cos \alpha a = \cos ka$$

and therefore

$$\alpha = \left(\frac{2mE}{\hbar^2}\right)^{1/2} = k$$

which is just the result for a constant potential. Figure 8 shows the dependence of E on k for the case $P = 0.5$. Notice that the curve is parabolic from $ka = 0$ to beyond $ka = 2$, and is discontinuous at $ka = \pi$, as anticipated.[7]

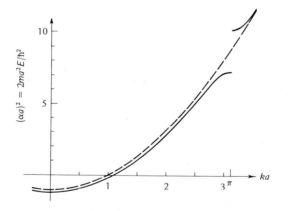

Fig. 8. *Dependence of the quantity $(\alpha a)^2$, or $2ma^2 E/\hbar^2$, on ka, as determined by Eq. (15), with $P = 0.5$. Broken line shows $(\alpha a)^2$ for a particle with a constant potential energy equal to the average value of $V(x)$ when $P = 0.5$.*

[7] Do not worry about the fact that the energy E becomes negative for small k; the potential energy is also negative in each well. The minimum value of E, at $k = 0$, is very close to the *average* value of $V(x)$, which is $-V_0 b/a$.

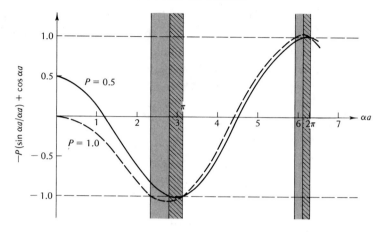

Fig. 9.   *Graph of the left-hand side of Eq. (15) versus αa, for two different values of P. When the curve is beyond the limits ±1, shown by the horizontal broken lines, the corresponding values of αa are not allowed. Shading indicates forbidden regions for αa in the two cases.*

Figure 9 shows how this discontinuity follows directly from Eq. (15). The only allowed values of energy correspond to αa such that

$$\left| -P \frac{\sin \alpha a}{\alpha a} + \cos \alpha a \right| \leq 1 \tag{16}$$

because $|\cos ka|$ cannot be greater than 1. Each region in which αa satisfies this condition is bounded by points where $\cos ka = +1$ or $\cos ka = -1$, so that $ka = n\pi$. For the example of Fig. 8, condition (16) is not satisfied when

$$2.79 < \alpha a < \pi$$

or, since $\alpha^2 = 2mE/\hbar^2$, the forbidden band of energies is

$$\frac{(2.79)^2 \hbar^2}{2ma^2} < E < \frac{\pi^2 \hbar^2}{2ma^2}$$

Effective Mass.   Apart from the fact that there are discontinuities where ka is an integral multiple of π, there is considerable difference in shape between the free electron parabola and the E versus k curves of Fig. 6 or 8. Let us investigate the effect of this difference by finding the acceleration of an electron in a crystal when an electric field $\mathscr{E}$ is applied.

The force on the electron is (in one dimension)

$$e\mathscr{E} = \frac{dp}{dt} = \hbar \frac{dk}{dt} \tag{17}$$

To define the acceleration of the electron we must deal with the electron's particlelike aspect, so we must assume the electron to be a wave packet, containing many different k-values. We assume that the k-values in the packet are reasonably close together, so that we can apply Eq. (17) unambiguously. The acceleration is then defined as the rate of change of the group velocity v_g for the wave packet. We know that

$$v_g = \frac{d\omega}{dk}$$

where $\omega = E/\hbar$; thus

$$v_g = \frac{1}{\hbar}\frac{dE}{dk} \tag{18}$$

Therefore the acceleration of the electron is

$$\frac{dv_g}{dt} = \frac{1}{\hbar}\frac{d^2E}{dk\,dt}$$

$$= \frac{1}{\hbar}\frac{d^2E}{dk^2}\frac{dk}{dt}$$

Now we may use Eq. (17) to eliminate dk/dt, obtaining

$$\frac{dv_g}{dt} = \frac{d^2E}{dk^2}\left(\frac{e\mathscr{E}}{\hbar^2}\right) \tag{19}$$

We see that the acceleration is proportional to the force. By analogy to Newton's law, we define an *effective mass m** as

$$m^* = \hbar^2\left(\frac{d^2E}{dk^2}\right)^{-1} \tag{20}$$

Notice that, for a free particle,

$$m^* = \frac{\hbar^2}{\dfrac{d^2}{dk^2}\left(\dfrac{\hbar^2k^2}{2m}\right)} = m$$

as one should expect. When the E versus k curve deviates from a parabola, then

$$m^* \neq m$$

There is nothing mysterious about this inequality of m^* and m. The force $e\mathscr{E}$ which appears in Eq. (17) is not the *total* force acting on the electrons, but only that *part* of the force which is due to the applied field $\mathscr{E}$. This part of the force would equal the electron's mass times its acceleration only if the lattice happened to exert zero force on the electron. This happens for low k, in the parabolic part of the E versus k curve, because the wave packet spreads out so far that the *average* force exerted by the lattice is zero. But when the packet's average k-value is near $k = n\pi/a$, Bragg reflection occurs, so it is obvious that the lattice exerts considerable force on the electron.

Notice from Fig. 8 that d^2E/dk^2 is negative when k is just below π/a, so that, according to Eq. (20), the effective mass is *negative* in this region. This means that the application of an external force in one direction causes the electron to acquire momentum in the opposite direction. This is not hard to understand, for when ka is just less than π, a force in the forward direction, which tends to increase k, results in Bragg reflection so that the electron goes in the opposite direction. Or to put it another way, when $k \approx \pi/a$, the wavefunction is primarily[8] a mixture of e^{ikx} and $e^{i(k-2\pi/a)x}$; as k increases, approaching $+\pi/a$, the proportion of $e^{i(k-2\pi/a)x}$ increases, but this part of the wavefunction has a negative wavenumber and thus corresponds to negative momentum. When $k = \pi/a$, the wave function is a pure standing wave, of the form $e^{ikx} \pm e^{-ikx}$; the electron's momentum is zero, and so its group velocity v_g must be zero. Thus, from Eq. (18), we see that $\partial E/\partial k$ must be zero, and Fig. 8 shows that the slope of E vs. k is zero at $ka = \pi$.

12.4   SUPERCONDUCTIVITY

The above discussion does not account for the phenomenon of electrical resistivity. In a perfect lattice, electrons would be continually accelerated when an electric field is present, so they would reach higher and higher velocities, and we would observe that the current increased with time although the voltage was constant. If the field were then turned off, the current would continue, because Bragg reflections do not remove energy from the electron gas. However, in a real lattice, the current ceases quickly, because of scattering of electrons from "imperfections"—vacancies, impurities, or atoms that are temporarily displaced from their equilibrium positions because of thermal agitation.

This scattering from imperfections also limits the current when the field is turned on; it is the mechanism of electrical resistivity, as discussed in Section 11.4f. We should expect the resistivity to increase with increasing temperature, as thermal agitation increases and the lattice becomes more "imperfect." When the lattice is cooled, this thermal contribution to the resistivity becomes very small in comparison to the contribution from impurities, and the resistivity reaches a lower limit, called the *residual resistivity*. Measurement of the residual

---

[8] The wave packet is composed of eigenfunctions of the Schrödinger equation. But each eigenfunction in a periodic potential is a superposition of plane waves of the form e^{ikx}. The proportion of the various plane waves in the eigenfunction may be estimated by perturbation theory, which tells us that the coefficients are proportional to $1/(E_m - E_n)$. [See Eq. (5).] When $k_m = \pi/a$, the corresponding unperturbed energy E_m is very close to that of $k_n = k_m - 2\pi/a = -\pi/a$; the corresponding coefficient a_{mn} becomes quite large, and the coefficients for other possible k-values are quite small in comparison.

resistivity is a way to test a material for impurities or for "frozen-in" defects which could be produced by irradiation or fatigue (the effect of repeated application of stresses).

However, Kamerlingh Onnes discovered in 1911 that, when mercury was cooled to a very low temperature, an entirely new phenomenon appeared. At a well-defined temperature T_c (now known to be 4.15 K for mercury), the metal became a superconductor, whose resistance was too small to measure. Onnes had measured the resistance of his sample to be 0.125 ohms when T was 4.3 K, and when T was reduced to 4.2 K he found 0.115 ohms; but when T was reduced further, he could only say that the resistance was less than 10^{-5} ohms (the limit of sensitivity of his apparatus), thereby justifying the use of the word "superconductor."

The most sensitive way to measure resistance in such a case is to cool a loop of wire in a magnetic field to a temperature $T < T_c$, and then induce a current by turning off the field. If the resistance of the loop is R and the inductance L, the current should decay exponentially with time, as $e^{-Rt/L}$. If the time t is sufficiently great, and $R \neq 0$, one should see a drop in the current. But careful measurements of "superconducting" loops at $T < T_c$ have never shown any change in the current over a period of time measured in months. The precision of the current measurement puts an upper limit on the value of R, but this upper limit seems to be simply a measure of the patience of the experimenter; we can assume that $R = 0$ in a superconductor. Dozens of materials have now been observed to become superconducting, at transition temperatures as high as $T_c = 20$ K or more. (At present the highest known transition temperature is for Nb_3Ge, at slightly over 23 K.)

No material ever fabricated has the absolute perfection needed to permit such a complete absence of resistivity on the basis of theories described so far in this book. The correct theory was not completed until 1957—almost half a century after the discovery of the phenomenon—by Bardeen, Cooper, and Schrieffer. The complete "BCS theory" is beyond the scope of this book, but we can describe its essential elements briefly. We begin by pointing out that the previously discussed theory of electrons in solids is a gross oversimplification, because it treats each electron as if it were the only one present, neglecting interactions between one electron and another. We have already seen (Section 9.1) that electron–electron interactions give observable correlation effects between electrons in atoms, and we should expect the same to be true for solids. The BCS theory is based on one such effect, the formation of so-called "Cooper pairs"— pairs of electrons in which the momentum is correlated, so that the average momentum of the two is equal to the overall average momentum of the conduction electron gas.

[9] For details of the theory, see "States of Matter" by D. L. Goodstein, Academic Press, New York, 371–411, (1975).

The electrons of a Cooper pair are bound together by an attractive force which results from the polarization of the lattice; one electron polarizes the lattice, and the other electron is attracted by the polarized lattice. At sufficiently low temperatures in many materials, this attractive force is larger than the Coulomb repulsion between the electrons, so the net force is attractive, and the electrons can be bound together with a small binding energy. (If it seems strange to you that like charges should attract each other, consider the H_2 molecule. There the two protons attract each other, in spite of their Coulomb repulsion, because of the two electrons in between them. In a solid, the electrons attract each other because of the lattice in between them.)

We saw in Chapter 6 that, in three dimensions, the presence of an attractive force is not always sufficient to produce a bound state for a single particle. But the presence of the other electrons in the Fermi gas changes the situation, as Cooper showed in 1956, so that *any* net attractive force, no matter how small, can produce a bound state. The correlation between the electrons of this bound pair imposes a constraint on the momentum states that each can occupy, so it is no longer correct to approximate these electron states as plane waves occupying the entire solid. As a consequence, the distribution of occupied states is altered, at $T = 0$, from that of the pure Fermi gas, becoming the distribution shown in Fig. 10. The distribution shown there becomes one grand quantum state, like the condensed state of a superfluid. Although the *kinetic* energies are not the lowest possible energies (the ones occupied in the normal ground state), the *overall* state, because of the negative *potential* energy resulting from pair binding, is the ground state of the system at $T = 0$. If the temperature is raised, eventually kT becomes comparable to the pair binding energy; thermal energy then excites pairs out of the "condensed" state, and the metal becomes "normal."

We can understand why the superconducting state has zero resistivity by referring to the discussion of Section 11.4f. Removal of the applied voltage from a *normal* conductor causes the current to decay, because electrons lose energy when they are scattered by the lattice. But a current in a superconductor cannot decay unless pairs are broken; an intact pair always has the same momentum and carries the same current, determined by the overall average velocity of the electron gas. Breaking a pair requires conversion of kinetic energy into potential energy, but we saw in Section 11.4f that the exclusion principle imposes an upper limit ΔE_{max} on the kinetic energy that can be lost in a scattering event. If ΔE_{max} is less than the energy Δ needed to break up a pair, then the pair must remain intact and the current persists unchanged. Eq. (41), Chapter 11, showed us that ΔE_{max} depends on the current; we had

$$\Delta E_{max} = (2mvp_f) = vp_f$$

and we know that v, the average velocity of the electron gas, is proportional to the current. The existence of superconductivity requires that

$$\Delta E_{max} = vp_f < \Delta$$

At a certain current called the "critical current," vp_f is equal to Δ; pairs then break up, and superconductivity disappears.

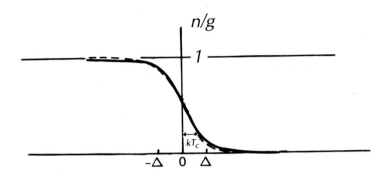

Fig. 10.   *Occupation index n/g vs. electron kinetic energy measured from the Fermi energy. Solid line: n/g for a superconductor at T = 0. Dashed line: n/g for a normal metal at T = T_c, where T_c is the "critical temperature" at which a superconductor becomes normal. (Adapted from "Introduction to Superconductivity," by M. Tinkham, McGraw-Hill, New York, 1975.)*

12.5   BRILLOUIN ZONES AND THE FERMI SURFACE

The one-dimensional discussions of Sections 12.2 and 12.3 have given us some idea of the principles which must be incorporated into the solution of a real problem in three dimensions. Let us first review our definitions for the plane wave states of a particle (Section 6.1). The kinetic energy is

$$\frac{p^2}{2m} = \frac{\hbar^2 k^2}{2m}$$

$$= \frac{\hbar^2}{2m}(k_x^2 + k_y^2 + k_z^2)$$

$$= \frac{(p_x^2 + p_y^2 + p_z^2)}{2m}$$

and the wavelength is $\lambda = 2\pi/k$. When the particle is confined within a rectangular "box," the possible states have equally spaced values of k_x, k_y, and k_z; that is, the states are represented by equally spaced points in "k-space."

For a nearly free electron in a rectangular lattice, the states may still be represented by equally spaced points in k-space, although in this case the wavefunction is not a pure plane wave. In this case the position of the state in k-space indicates the value of k for the plane wave which is the dominant wave in the wavefunction. If we plot the *total* energy of the electron versus *any one* of the wavenumbers k_x, k_y, or k_z, the curve is like that of Fig. 7; again the lattice potential causes an energy gap at the value of the wavenumber for which Bragg reflection occurs.

Now consider a simple cubic lattice[10] in which the spacing is a. Electrons are Bragg reflected, and thus there is an energy gap, when

$$k_x = \frac{n_x \pi}{a}, \qquad k_y = \frac{n_y \pi}{a}, \qquad \text{or} \qquad k_z = \frac{n_z \pi}{a}$$

where n_x, n_y, and n_z are integers. These conditions are easily seen to be equivalent to the Bragg condition, for reflection from the yz, xz, and xy planes, respectively. [See Eq. (1) and Fig. 4, with $d = a$.] For example, for the yz plane the Bragg condition is

$$\lambda n_x = 2a \sin \theta$$

But $\lambda = 2\pi/k$, so that

$$2\pi n_x = 2ka \sin \theta$$

or, since $\sin \theta = p_x/p = k_x/k$, we have

$$\frac{\pi n_x}{a} = k_x$$

The six planes $k_x = \pm \pi/a$, $k_y = \pm \pi/a$, and $k_z = \pm \pi/a$ enclose a cube in k-space. This cube, called the first Brillouin zone, contains all points for which k_x, k_y, and k_z are all smaller than the minimum value at which there is an energy gap. Within the zone, the energy is a continuous function of k, but across each of the zone boundaries—the cube faces for this lattice—the energy gap appears, so that a state represented by a point just outside the boundary has a much greater energy than a state represented by a point lying just across the boundary, inside the zone.

A two-dimensional view of this zone is shown in Fig. 11, which is a sort of "contour map" in the $k_x k_y$ plane, in which the lines connect points of

---

[10] An element rarely forms a *simple* cubic lattice, but it is the simplest lattice to consider in order to develop the principles. We can then apply these principles to more common types of lattice.

equal energy. If the electrons were completely free, the energy would be simply $E = \hbar^2 k^2/2m$, and a contour of constant energy would be a circle, on which k^2 is constant. (Or, in three dimensions, a *surface* of constant energy would be a *sphere*.) But the effect of the periodic lattice potential is to pull the constant-energy lines toward the zone boundary, so that there are bulges where the lines approach the boundary. Notice that the energy just inside the center of a zone boundary is E_4, whereas the energy just outside is E_6. (The contour for energy E_6 is partly inside and partly outside the zone.) Therefore the energy gap must be approximately $E_6 - E_4$ at this zone boundary.

These bulges in the constant energy contours may be easily understood by reference to Fig. 6. Notice in that figure that as k increases, approaching the value $k = \pi a$ at the zone boundary, the slope of the E versus k curve

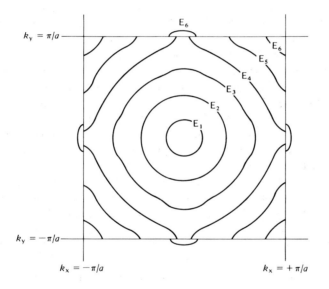

Fig. 11. *Two-dimensional contour map, showing loci of points of equal energy in the $k_x k_y$ plane for a square lattice of spacing a.*

decreases. This means that equal increments in energy correspond to progressively greater increments in k, as the boundary is approached. When this behavior is translated into two or three dimensions, it means that lines or surfaces of constant energy must be farther apart in the regions of k-space where they are close to a zone boundary.

One of the constant-energy surfaces is especially important, because it marks the boundary between the occupied and unoccupied states at 0 K.

This surface, whose energy is the Fermi energy, is called, naturally, the Fermi surface. To obtain some idea of the shape of this surface, we next count the number of states in the Brillouin zone.

For a simple cubic crystal containing N atoms, the first Brillouin zone contains $2N$ electronic states. This may be shown as follows: Let the entire crystal be in the shape of a cube, because the macroscopic shape of the crystal should not affect the conclusion. Then the length of an edge of this cube is $a(N)^{1/3}$, where a is the period of the lattice. With periodic boundary conditions, the possible *wavelengths* of an electron in this cube are $a(N)^{1/3}/n$, where n is a positive integer, and the possible values of k_x, k_y, or k_z are $2\pi n/a(N)^{1/3}$. In other words, *successive allowed values* of any component of k *differ by* $2\pi/a(N)^{1/3}$, and since the width of the Brillouin zone is $2\pi/a$, the number of possible values of *each component* of k is $(N)^{1/3}$ for states within the first zone. The number of different *sets* of allowed values of k_x, k_y, and k_z must therefore be the cube of $(N)^{1/3}$, or simply N; and the number of possible electron *states* is $2N$, because there are two spin states for each set of k values.

Suddenly we see a connection between the nearly-free-electron model and the discussion of energy bands with which we began this chapter. The $2N$ states in the first Brillouin zone are simply another representation of the $2N$ states in an energy band. In an alkali metal containing N atoms, there are only N valence electrons, so that the zone is only half occupied, and the occupied states are not very close to the zone boundary. Under these conditions, the free-electron-gas model is a good approximation, because the Fermi surface is very nearly spherical.

Like the alkali metals, the elements in group II of the periodic table are also conductors, but for a different reason. In these elements there are $2N$ valence electrons for N atoms, just enough to completely fill the first Brillouin zone. If the energy gap at the zone boundary were sufficiently great, all of the electrons would be confined to the first zone, and these elements would be insulators or semiconductors, because the total momentum of the electrons must be zero when the zone is filled. But the gap is not that great; the valence electrons spill out into the second zone, where they find states of lower energy than the energy of states in the corners of the first zone. In such a case the Fermi surface has a cross section something like the energy contour E_6 of Fig. 11. Both zones contain states of energy E_6, and an electron of such energy can be easily excited to states with slightly higher energy in either zone. There is no energy gap between *bands*, because there are some states within the first zone whose energy is equal to or higher than that of some states outside the first zone. Of course, there is still an energy gap *at the zone boundary*, but this gap only separates states *directly across the boundary from each other*.

Before discussing elements with a greater number of valence electrons, we must determine the location of other Bragg reflection planes. Even in a simple cubic lattice, there are many such planes at various angles, and each plane defines a boundary of a Brillouin zone. For example, Fig. 12 shows an edge-wise view of a set of diagonal planes (labeled [110]) and a set of planes (labeled [210]) which pass through atoms which are one "knight's move" apart (on the corners of a 2×1 rectangle). The distance between successive planes in each case is given by $d = a/(n_1^2 + n_2^2 + n_3^2)^{1/2}$, where a is the lattice spacing and the indices n_1, n_2, and n_3 determine the direction of each plane

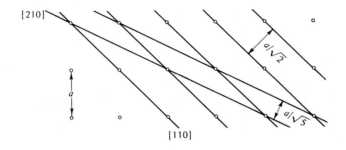

Fig. 12. *Bragg reflection planes in a cubic lattice.*

relative to the cubic axes. For a lattice with cubic symmetry, n_1, n_2, and n_3 are simply the smallest set of whole numbers which are proportional to the direction cosines of the line perpendicular to the plane.[11] The derivation of the expression for d is thus a straightforward exercise in trigonometry.

If we substitute this expression for d into the Bragg condition (1), we obtain

$$n\lambda = \frac{2n\pi}{k} = \frac{2a \sin \theta}{(n_1^2 + n_2^2 + n_3^2)^{1/2}}$$

and therefore

$$k \sin \theta = \frac{n\pi}{a} (n_1^2 + n_2^2 + n_3^2)^{1/2} \qquad (21)$$

where θ is, of course, the angle between the direction of the **k** vector and the plane defined by the numbers n_1, n_2, and n_3.

For a given set of the numbers n_1, n_2, n_3, and n, the locus of all points in

[11] The general definition of these indices, known as *Miller indices,* differs from the one given here, but the difference matters only when the lattice does not have cubic symmetry. The Miller indices of a plane are defined in general as the smallest set of whole numbers which are *inversely proportional* to the *intercepts* of the plane on the principal axes of the crystal. For example, for a [210] plane the intercepts are, respectively, 1, 2, and ∞, with reciprocals 1, $\frac{1}{2}$, and 0, which are in the ratio 2, 1, and 0.

k-space which satisfy Eq. (21) is a plane. For example, when n_1, n_2, and n_3 are 1, 0, and 0, respectively, with $n = 1$, Eq. (21) defines the plane

$$k \sin \theta = \frac{\pi}{a} \qquad \text{or} \qquad k_x = \frac{\pi}{a}$$

one of the boundaries of the first Brillouin zone. With indices 1, 1, 0, and $n = 1$, we obtain the plane $k \sin \theta = \pi \sqrt{2}/a$, where θ is now the angle between the **k** vector and the [110] plane. The plane thus defined is a [110] plane in k-space, and it is a boundary of the *second* Brillouin zone.[12] The second Brillouin zone is defined as the region of k-space which lies directly across a zone boundary from the first zone. Crossing another zone boundary brings one into the third zone (or back into the first zone), and so on. The second Brillouin zone is bounded on the inside by the first Brillouin zone and on the outside by twelve planes of the types [110], [101], [$\bar{1}$10], and so on. (The bar over the 1 indicates a minus sign.) The second Brillouin zone, like the first, contains $2N$ electron states.

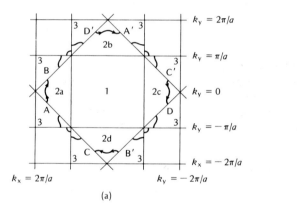

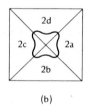

Fig. 13. (*a*) *Brillouin zones for a square lattice; zones identified by number. Curved lines show a possible Fermi surface under the influence of this zone structure; energy gaps at each zone boundary distort the Fermi surface from the circular shape it would have if there were no electron–lattice interaction. Arrows indicate motion of an electron along the surface in the second zone; Bragg reflections keep the electron confined in one zone, so that there are three possible orbits on this Fermi surface—one orbit in each of the second, third, and fourth zones. (b) Orbit in the second zone re-assembled to form a continuous curve; the segments 2a, 2b, 2c, and 2d have each been translated by an amount* $2\pi/a$ *in the directions needed in order to assemble them into the first zone's area.*

[12] That is, the direction cosines of the perpendicular to the plane are in the ratio 1:1:0. The perpendicular distance from the origin to the plane is $\pi\sqrt{2}/a$.

Electron Orbits.   Figure 13 shows the first, second, and third zones for a square lattice in two dimensions. Notice that all of these zones have the same area, and thus they contain the same number of states. The various apparently disconnected segments of each zone are connected in a sense by Bragg reflections. To illustrate this point, Fig. 13 shows a situation in which the Fermi surface passes through the second zone. An electron can be forced to move on this surface by application of a magnetic field, because the field changes the direction of the electron's momentum without changing its energy.[13] Thus, starting at point A, an electron moves in the direction of the arrow to point B. Here the **k**-vector is just right for Bragg reflection; this process reverses the component of **k** which is perpendicular to the plane of reflection, so that the electron goes to point B' in k-space. The magnetic field then moves the electron to point C, from which it is reflected to C', and so on, eventually returning to A again to begin another cycle. During this cycle, the electron always remains within the second zone; Bragg reflections prevent the electrons from reaching states of the same energy in other zones.

  Reduced-Zone Scheme.   To aid in the visualization of these orbits in k space, one can resort to a translation of the **k**-vectors in such a way that the orbit becomes closed. Figure 13b shows how this is accomplished. Translation of the four segments of the second zone in the manner indicated causes the second zone to occupy exactly the same part of k-space occupied by the first zone, and it makes the orbit a continuous path, as shown in Fig. 13b. The translation is legitimate, because in any experiment involving orbits of electrons we are concerned with changes in **k** rather than in the instantaneous **k**-vector itself. The discussion which follows will clarify this point.

  Mathematically this translation simply amounts to choosing a different way to decompose the wavefunction into a product of the type required by Bloch's theorem [Eq. (7)]. Let us write the one-dimensional Bloch function

$$u(x) = u_k(x)e^{ikx}$$

and multiply it by unity, in the form $e^{2\pi ix/a}e^{-2\pi ix/a}$, so that it becomes

$$u(x) = u_k(x)e^{2\pi ix/a}e^{ikx}e^{-2\pi ix/a} \qquad (22)$$

We know from Bloch's theorem that $u_k(x)$ has period a; therefore the product $u_k(x)e^{2\pi ix/a}$ also has period a, and we may write the function $u(x)$ in the Bloch form as

---

[13] Of course, the electron is free to change states (that is, move through k-space) only because there are many unoccupied states at the Fermi surface. Electrons whose energy is much less than ε_f are "locked in" and cannot move through k-space.

$$u(x) = u_{k'}(x)e^{ik'x}$$

where

$$u_{k'}(x) = u_k(x)e^{2\pi i x/a} \quad \text{and} \quad k' = k - \frac{2\pi}{a} \tag{23}$$

Thus as far as Bloch's theorem is concerned, we can characterize the wave-function by either the wavenumber k or the wavenumber k'. Notice that the translation of the second zone into the first zone as in Fig. 13b is equivalent to replacing the wavenumber k_x or the wavenumber k_y by a new number which differs from the previous number by $+2\pi/a$ or by $-2\pi/a$. The actual state of the electron has not been changed by this translation; only our representation of this state in k-space has changed, because we are incorporating one of the electron's Fourier components into the function $u_{k'}(x)$, rather than displaying this component in k-space.

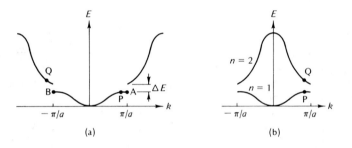

Fig. 14. (a) E versus k in one dimension, in the extended zone scheme. Notice that the points A and B represent the same state, which is an equal mixture of plane waves e^{ikx} and e^{-ikx} for this value of k. (b) E versus k for the same system, in the reduced zone scheme, in which k is only defined in the range $-\pi/a < k < +\pi/a$. States in the second zone of Fig. (a) are translated into the first zone and given a second quantum number n to denote the band of energy to which they belong; notice the new position of point Q.

In a similar manner, one can also translate the other zones—third, fourth, et cetera—into an area of the same size and shape as the first zone. Representation of all **k**-vectors in the first zone is called the "reduced zone scheme," in contrast to the "extended zone scheme" illustrated in Fig. 13a. When a magnetic field is applied, the k' of the reduced zone scheme changes in the same way as the k of the extended zone scheme, so that the reduced zone scheme can always be used in analyzing the effects of a magnetic field on the electron's momentum.[14]

Figure 14 shows E versus k for both the reduced and the extended zone schemes in one dimension. In the reduced zone scheme, E is no longer a single-valued function of k, so a second quantum number, the *band number n*, is introduced to distinguish the energy states from one another. The bottom curve, which was originally in the first zone, has band number 1; the next higher curve, which was translated from the second zone, has band number 2; and so on. The bands differ from one another in that part of the electron's wavefunction which has the period of the lattice; by our notation [Eqs. (22) and (23)], this part is $u_k(x)$ in the first band, $u_{k'}(x)$ in the second band, and so on.

It is possible to study the shape of the Fermi surface by measuring the frequencies of electron motions on orbits like the one shown in Fig. 13. As the electron moves through k-space, it also moves through real space, in the manner shown in Fig. 15. (We assume that the electron is not perfectly localized in

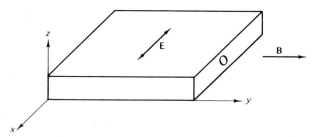

Fig. 15. *Magnetic field* **B** *applied in y direction causes electron to move in orbit in xz plane. As electron reaches surface of metal, it receives an impulse from the alternating electric field* **E** *in the x direction.*

either k-space or real space; the position of a wave packet can be defined in *both* spaces with an accuracy which is quite sufficient for the purpose of the measurement.) If an alternating electric field is applied to the metal, it does not affect electrons while they are deep within the metal—beyond the "skin depth." But if an electron's orbit brings it close to the surface, the electron receives a periodic impulse from the field. If the frequency of the field just equals that of the electron in its orbit, the effect of these impulses is cumulative,

[14] If we measure the *instantaneous* momentum of the electron, we are, of course, making a Fourier analysis of the total wavefunction of the electron, which includes the function $u_k(x)$ as well as the plane wave part e^{ikx}. In such a case the *extended* zone scheme comes closer to indicating the k-values which result from the measurement; the **k**-vectors are displayed, rather than concealed in the function $u_k(x)$. One can make such an instantaneous measurement by annihilating positrons with the electrons in the metal. See Section 13.4.

and energy is absorbed by the metal. By varying the frequency of the electric field and looking for this "cyclotron resonance," one can determine the frequencies of the electrons in their orbits, and from this knowledge one can test hypotheses regarding the shape of the Fermi surface.

EXAMPLE PROBLEM 1. Show that the frequency of revolution of a free electron in a circular orbit under the influence of a magnetic field of strength B is $v = 2.8 \times 10^{10} B$, if B is in teslas (T). (1 T $\equiv$ 1 Wb/m^2 = 10^4G). Use this result to determine a typical distance that an electron might travel between Bragg reflections in a lattice, when $B = 1$ T.

Solution.   Setting force equal to mass times acceleration, we have

$$qvB = mv^2/r$$

where r is the radius of the orbit. Therefore

$$v = qBr/m$$

and the frequency v is equal to v divided by the circumference of the orbit, or

$$v = qB/2\pi m$$

Remembering that $mc^2 = 5.1 \times 10^5$ *electron* volts, and $q = 1$ *electron*, we write

$$v = \frac{qc^2}{2\pi mc^2} B = \frac{9.0 \times 10^{16}}{6.3 \times 5.1 \times 10^5} B$$
$$= 2.8 \times 10^{10} B$$

Let us assume that the electron moves on a circular arc, through an angle of about 1 rad between Bragg reflections. Then the time between reflections is about $1/2\pi v$, or, when $B = 1$ T, the time $t \approx 5 \times 10^{-12}$ sec. To find the distance traveled by the electron in this time, we remember that the electron is on the Fermi surface, so its kinetic energy is of the order of 10 eV, and its velocity is

$$v = \left(\frac{2E}{m}\right)^{1/2} = c\left(\frac{2E}{mc^2}\right)^{1/2}$$
$$\approx c\left(\frac{20}{5 \times 10^5}\right)^{1/2} \approx 6 \times 10^{-3}c$$

Therefore the distance traveled is

$$s = vt \approx 6 \times 10^{-3} \times 3 \times 10^8 \frac{m}{sec} \times 5 \times 10^{-12} \text{ sec}$$
$$\approx 10^{-5} \text{ m} = 10^5 \text{ Å}$$

Thus the orbit is of considerable size in atomic terms; the electron passes on the order of 10^5 atoms between Bragg reflections. An orbit of this size is large enough to be affected by the size of the sample in which the electron moves, and size effects have been seen in small samples.

Brillouin Zones for More Common Lattice Structures of Cubic Symmetry.
Lattices with cubic symmetry generally form in either a face-centered cubic (fcc) or a body-centered cubic (bcc) structure (Fig. 16), rather than in the simple cubic structure which we have been discussing.

A moment's reflection should convince you that the bcc structure is equivalent to *two* interpenetrating simple cubic lattices, displaced from each other by one-half of a body diagonal. Thus for a given value of the cube edge a there are exactly twice as many atoms per unit volume as in a simple cubic lattice. The Brillouin zone structure also differs from that of a simple cubic lattice, because the conditions for Bragg reflection are not satisfied by the same set of **k**-vectors.

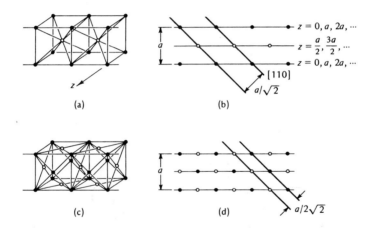

Fig. 16. (a) Two adjacent cubes, bcc structure, seen in perspective; (b) same structure, looking along the z axis, with points at different z values distinguished by closed or open circles; (c) fcc structure, perspective; (d) fcc, view along the z axis as in (b). (●)z = 0, a, 2a, . . .; (○)z = a/2, 3a/2, ...

In the bcc structure, adjacent planes of the [100] type are a distance of $a/2$ apart, rather than a distance of a. Thus the **k** vector required for Bragg reflection from these planes is just twice as great as in the simple cubic lattice; the planes $k_x = \pm\pi/a$, $k_y = \pm\pi/a$, and $k_z = \pm\pi/a$ do *not* constitute a zone boundary in k-space, although there are still zone boundaries at $k_x = 2\pi/a$, et cetera.[15] (The latter plane in k-space is called the [200] plane, to distinguish it from the [100] plane on which $k_x = \pi/a$.)

However, you can see from Fig. 16 that the spacing between adjacent [110] planes is still $a/\sqrt{2}$, as it is in the simple cubic lattice, because each atom at a "body center" lies in one of the [110] planes which was already occupied in the simple cubic lattice. Thus the [110] planes in k-space still define k-vectors for which Bragg reflection occurs, as in the simple cubic lattice; because the [100] planes no longer define a zone boundary, the [110] planes now define boundaries of the *first* Brillouin zone (rather than the second zone, as in the simple cubic structure). There are twelve such planes, which enclose a regular dodecahedron (Fig. 17) whose volume can be shown to equal $2(2\pi/a)^3$, just twice the volume of the cubic zone for the simple cubic lattice. As there are also twice as many atoms in a given volume for the bcc lattice, the first Brillouin zone still contains two states per atom.

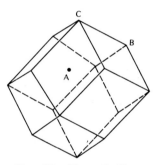

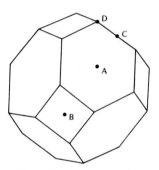

Fig. 17. *First Brillouin zone for the bcc lattice. Values of ka are $\pi\sqrt{2}$ at point A, $\pi\sqrt{3}$ at point B, and 2π at point C. [Adapted from "The Theory of the Properties of Metals and Alloys" by N. Mott and H. Jones. Dover Publications, Inc., New York, 1968. Reprinted through permission of the publishers.]*

Fig. 18. *First Brillouin zone for the fcc lattice. Values of ka are $\pi\sqrt{3}$ at point A, 2π at point B, $3\pi/\sqrt{2}$ at point C, and $\pi\sqrt{5}$ at point D. [Adapted from "The Theory of the Properties of Metals and Alloys" by N. Mott and H. Jones. Dover Publications, Inc., New York, 1968. Reprinted through permission of the publishers.]*

---

[15] We might say that the Bragg reflection for $k_x = \pi/a$ is destroyed by interference between waves from the planes at $x = 0,\ a,\ 2a,\ \dots$ and waves from the planes at $x = a/2, 3a/2, 5a/2, \dots$. When $k_x = \pi/a$, the wave reflected from the $x = 0$ plane is just 180° out of phase with the wave reflected from the $x = a/2$ plane, so the reflections cancel in pairs.

The fcc lattice has the same spacing of [100] planes as the bcc lattice, so again planes of this type do not form the boundaries of the first Brillouin zone. Nor do [110] planes form such a boundary in this case, because there are additional diagonal planes which make the spacing between diagonal planes equal to $a/2\sqrt{2}$, rather than $a/\sqrt{2}$ as in the bcc case (Fig. 16d). The smallest possible k-vectors for Bragg reflection are now perpendicular to the [111] planes, and these planes combine with the [200] planes in k-space to form the boundaries of the first Brillouin zone (Fig. 18). For a given value of a, this zone is twice as large as the zone for the bcc structure, but you should be able

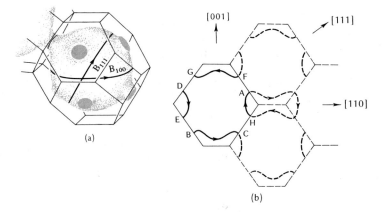

Fig. 19. (*a*) *Three-dimensional view of copper Fermi surface.* (*b*) *Cross section through center of copper Fermi surface, showing electron orbit through points ABCDEFGH. Electron reaching point A is Bragg reflected to B; moving to C, it is then Bragg reflected to D, etc. Notice that the orbit can be drawn as a closed orbit by reproducing the first zone as indicated by the dashed lines. This method of depicting the orbits is called the periodic zone scheme; its validity has the same basis as that of the reduced zone scheme. This particular orbit is called the "dog's bone," for obvious reasons.*

to see from Fig. 16d that the fcc structure has twice as many atoms per unit volume as the bcc structure, so again the first zone contains just two states per atom.

Figure 19 shows the Fermi surface and Brillouin zone for copper, which has the fcc structure. Although copper has a valence of one, so that the first zone is only half occupied, the Fermi surface touches the zone boundaries, forming a circular area of contact around the center of each [111] plane. Historically, this was the first Fermi surface to be determined experimentally,[16]

[16] A. B. Pippard, *Phil. Trans. Roy. Soc. (London)* **A250**, 325 (1957).

and a variety of electronic orbits on the surface have been identified. One of the more picturesque is the "dog's bone" orbit shown in Fig. 19b.

12.6   INSULATORS AND SEMICONDUCTORS

As mentioned in Section 12.5, the elements in group II (alkaline earths), with just enough electrons to fill the first Brillouin zone, actually contain occupied states in the second zone and unoccupied states in the first zone, so that these elements are conductors (although not such good conductors as the alkali metals or other metals of valence 1—for example, copper, silver, and gold). But the elements germanium and silicon of group IV are semiconductors, because in this case the four valence electrons can and do occupy all of the states in the first two Brillouin zones.

A pure semiconductor such as Si or Ge is a rather good insulator at temperatures near $0°K$, but the conductivity increases rapidly as T increases, because some electrons become thermally excited across the zone boundary (into the "conduction band"); these electrons then become capable of carrying a current, because their total momentum is no longer constrained to equal zero. The excitation of *one* electron actually produces *two* current carriers, because the empty state, or "hole," which the excited electron leaves in the lower energy band (the "valence band") permits the valence band as well as the conduction band to have nonzero momentum. In other words, there is an odd electron in each band whose momentum is not canceled out by the motion of another electron in that band.

A semiconductor which carries current primarily because of thermal excitation of electrons is called an *intrinsic* semiconductor. The only difference between an intrinsic semiconductor and an insulator is in the magnitude of the energy gap. For example, diamond is an excellent insulator, with a resistivity greater than about 10^{20} Ω-cm at $300°K$ (a factor of 10^{26} greater than that of a good conductor), whereas intrinsic silicon and germanium, with the same crystal structure and the same number of valence electrons per atom, have resistivities of 2.6×10^5 Ω-cm and 43 Ω-cm, respectively, at $300°K$. The difference is that the energy gap between the filled valence band and the empty conduction band is 5.33 eV for diamond, 1.14 eV for silicon and only 0.68 eV for germanium.[17]

The size of the energy gap has a tremendous effect on the number of electrons which can be excited into the conduction band at any given temperature. To understand this effect, we must return to the Fermi–Dirac distribution function [Chapter 11, Eq. (31)], which determines the occupation

---

[17] The next question which may occur to you is *why* the energy gaps are so much different in the three materials. This question can also be answered, but not in this text; it requires a detailed calculation of the energy eigenfunctions for electrons in the potentials of each of the three kinds of atom.

index $n(\varepsilon)/g(\varepsilon)$, for a semiconductor just as it does for a metal. (It is the density of states, $g(\varepsilon)$, which varies from one material to another, and which depends on the band structure, the existence of zone boundaries, et cetera.) For an intrinsic semiconductor or an insulator, the Fermi energy is determined by the condition that the number of conduction electrons must equal the number of holes. If $g(\varepsilon)$ had the same value in the region on both sides of the gap, the Fermi energy would have to be exactly in the middle of the energy gap, in order to equalize the number of electrons and holes.

For a rough calculation, even though $g(\varepsilon)$ is different on the two sides of the gap, one can still assume the Fermi energy to be in the middle of the gap, because the number of electrons and holes is affected far more by the exponentially varying occupation index than by the slowly varying density of states. (See Fig. 20.) With an energy gap of 5.33 eV, the bottom of the conduction band is about 2.67 eV above the Fermi energy. At such a high energy, the Fermi–Dirac occupation index becomes a simple exponential; its value at the bottom of the conduction band is roughly $e^{-(\varepsilon-\varepsilon_f)/kT} \simeq e^{-2.67/0.026} \simeq e^{-100} \simeq 10^{-43}$ at room temperature, so that the number of conduction electrons for diamond is minuscule. On the other hand, when the energy gap is only 0.68 eV, the lowest state in the conduction band is about 0.34 eV above the Fermi energy, and the occupation index for this state is thus about $e^{-0.34/0.026} \simeq 10^{-13}$. This is still a small number, but it permits the conduction electron density in intrinsic germanium to be $2.5 \times 10^{13}/cm^3$, not a negligible number.

Of course, it is not the intrinsic conductivity which is so important in semiconductors; it is the fact that the conductivity can be varied over a wide range and can be controlled by various external means. For example, the conductivity of a semiconductor can be greatly increased, temporarily, when it is exposed to electromagnetic radiation or to a flux of charged particles, because each photon or each charged particle traversing the semiconductor can excite a great many electrons into the conduction band. These electrons very quickly fall back into the valence band, but the temporary rise in conductivity can easily be arranged to produce a pulse in an electrical circuit, for a variety of purposes, from burglar alarms to nuclear radiation detectors. This is not the place to discuss the details of such devices, but it might be pointed out that semiconductor detectors have become the most accurate detectors by far for measuring the energies of the various kinds of radiation emitted by atomic nuclei. Their accuracy stems from the fact that a very small amount of energy is needed to excite electrons in a semiconductor, in comparison to the energy needed to ionize the particles in other types of nuclear radiation detector.

Doping.   The possibilities of semiconductors have been further enhanced by the practice of "doping"—deliberately adding impurities which increase

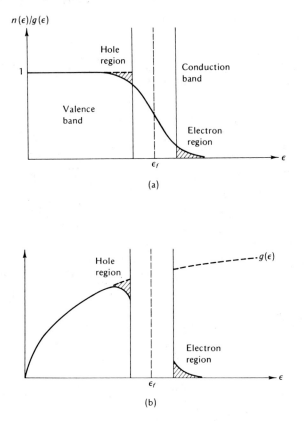

Fig. 20. (a) Occupation index and Fermi energy in relation to band gap for a semi conductor. The Fermi energy ε_f is in the middle of the gap, so that the number of holes will be equal to the number of conduction electrons. (b) Occupation number $n(\varepsilon)$ for a semiconductor (solid line), obtained by multiplying the somewhat idealized $g(\varepsilon)$ (dashed curve) by the occupation index shown in (a). The area of the electron region should equal the area of the hole region, for an intrinsic semiconductor. If ε_f were exactly at the center of the gap, the area of the electron region would be slightly greater than the area of the hole region, because $g(\varepsilon)$ increases with energy while the occupation index is symmetrical about ε_f. Placing ε_f only slightly closer to the valence band than to the conduction band is sufficient to equalize the two areas.

the conductivity of a specimen. Consider what happens if one adds an impurity of valence 3 to a semiconductor of valence 4. Each impurity atom, while taking up room in the lattice, contributes one fewer than its fair share of valence electrons, so that an empty state is left in the conduction band. Thus it is possible to produce material in which the current carriers are mostly *holes*, rather than electrons; these materials are called *p-type* semiconductors, because each hole behaves like a *positive* charge.

If one instead adds an impurity of valence 5 to a semiconductor of valence 4, the extra electron must go into the conduction band. In this case the current carriers are *negative*, so the material is called *n-type*.

At first sight, this distinction between n-type and p-type semiconductors might seem a bit pedantic, because in either case the flow of current involves the transfer of *electrons* from one point to another, and the electrons are, of course, always negative. But the two types of semiconductor behave quite differently, as we can illustrate by a discussion of the *Hall effect* (see Fig. 21).

If a magnetic field **B** is applied to a current-carrying sample, a voltage is induced across the sample, in a direction perpendicular to both the field and the current. This so-called Hall voltage is caused by the force exerted by **B** on the current carriers, which tends to deflect them toward the side of the sample. If the current carriers have *positive* charge q, then their velocity **v** is in the direction of the current I, so that $q(\mathbf{v} \times \mathbf{B})$ is in the direction shown in Fig. 21a, and the left-hand side of the sample becomes *positively charged*. Equilibrium is reached when the electric field of this charge distribution produces a force on the current carriers which is equal and opposite to the force exerted by the field **B**. On the other hand, if the current carriers are *negatively* charged, then **v** is *opposite in direction* to the current I, and q is *negative*; the force $q(\mathbf{v} \times \mathbf{B})$ is *still in the same direction*; the *negative* charges are deflected toward the left, and the left-hand side becomes *negatively* charged, as shown in Fig. 21b.

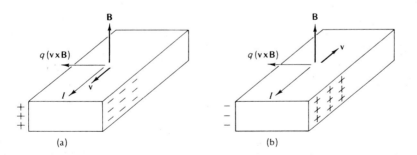

Fig. 21. (a) Hall effect for positive current carriers. (b) Hall effect for negative current carriers.

Thus the direction of the Hall voltage may be considered to be a direct indication of the sign of the current carriers. Hall voltages have been measured for many materials, and they indicate that the current carriers are indeed positive in what we have defined as p-type material, and they are negative in n-type material. This is really mysterious when you first think about it, because we know that the current carriers are actually electrons in both cases. The difference in sign between the Hall voltages in different materials was actually observed long before it was explained; it had been a famous unsolved problem in physics.

To explain the phenomenon, one needs to invoke the concept of *effective mass* (Section 12.3). We have been discussing the effect as if the charge carriers were *free* and could respond to external fields in the way that one would expect a free particle to respond. This approach gives the right answer for n-type material, because electrons in such material lie near the bottom of a band and therefore have positive effective mass. But in general one should not expect electrons in a lattice to respond to an *external* field in the same way that a free electron responds, because the lattice exerts considerable additional force on the electron, and this force changes when an external field is applied.

In particular we have seen that the electrons near the *top* of a band have *negative effective mass*, so that their response to an external field is just the opposite of the response of a free electron. The holes in p-type material lie near the top of a band, so that each hole represents the *absence* of a particle whose charge and effective mass are *both negative*. It should be clear that such a hole would behave in every respect like the *presence* of a particle of positive charge and positive effective mass—that is, like a *free* particle of positive charge.[18] However, we could equally well say that the *electrons* which conduct the current behave like *negative masses*; this too would yield the correct sign for the Hall voltage in p-type material. (The electrons which conduct the current are, of course, those which occupy states of *exactly opposite momentum* from the momentum of the unoccupied states; the total momentum of all the other electrons is zero. The electrons in these states have negative effective mass, because they lie near the top of the band just as the unoccupied states do.)

Of course, the difference between n-type and p-type material has implications far beyond those which we have discussed, for it is the basis for the transistor and a host of other "solid state" electronic circuit elements. However, a discussion of such elements would only divert us from the task of developing and illustrating the fundamental principles of modern physics;

---

[18] Notice the similarity of the argument here to the argument for the hole theory of the positron. The removal of an electron from the sea of negative energy states is like the removal of an electron from the valence band; in either case a negative charge and a negative (effective) mass is *subtracted*, which is equivalent to the *addition* of a positive charge with a positive mass.

to do justice to such a discussion would require an entire book, and there are many of them on the market.

It is hoped that this discussion of the energy and momentum of electrons in solids, abbreviated though it is, has been more than a mere recital of phenomena, and that it has been sufficient to illustrate the most important principles involved. Rather than immerse you in a number of rigorous proofs of these principles, or attempt to discuss the most general cases, we have felt it more helpful at this stage to try to develop a *feeling* for the principles by presenting a number of simple examples and inductive arguments. If we have succeeded, you should now have a better appreciation for the quantum theory, and you will be better able to follow and understand the motivation for the more general proofs and discussions found in more advanced texts.

PROBLEMS

1. Show that the energy gap at a zone boundary, in one dimension, is approximately

$$\Delta E = \frac{2}{a} \left| \int_0^a V(x)\, e^{2\pi i x/a}\, dx \right|$$

where $V(x)$ is the lattice potential, of period a, with the zero level chosen so that the average value of $V(x)$ is zero.

Hint: The wavefunction near the zone boundary may be written approximately as

$$u(x) = A_0 e^{ikx} + A_1 e^{ik_1 x} \tag{1}$$

where $k_1 = k - (2\pi/a)$. If this expression is substituted into the Schrödinger equation, you have

$$A_0 e^{ikx}[E - E_0 - V(x)] + A_1 e^{ik_1 x}[E - E_1 - V(x)] = 0 \tag{2}$$

where $E_0 = \hbar^2 k^2/2m$ and $E_1 = \hbar^2 k_1^2/2m$.

You can obtain two equations by multiplying Eq. (2) by either e^{-ikx} or $e^{-ik_1 x}$ and integrating on x from 0 to a. If you eliminate A_0 and A_1 from these two equations, and use the fact that $E_0 = E_1$ when $k = \pi/a$, you can deduce the desired result.

2. (a) Use the result of Problem 1 to show that the energy gap for the Kronig–Penney potential of Fig. 7 is given by first-order perturbation theory as $2V_0 b/a$, in the limit $b/a \to 0$ with $V_0 b/a$ remaining constant.

   (b) Rewrite the energy gap shown in Fig. 8 in terms of V_0, b, and a, and compare the resulting expression with your answer to part (a).

3. Repeat the calculation leading to Fig. 8, for the case $P = 1$, and deter-

mine the size of the energy gap in terms of V_0, b, and a. Compare with the result of Problem 2. Is there better or worse agreement between your result and the result of first-order perturbation theory for this value of P? Explain.

4. Show from first-order perturbation theory (Section 8.1) and the result of Problem 1 that the coefficient A_1 of Eq. (1) is given by

$$|A_1| = m\, \Delta E\, |A_0|/h^2\, |k^2 - k_1{}^2| \qquad (|k| < \pi/a)$$

with ΔE, k, and k_1 as defined or given in Problem 1. Why do we add the restriction that $|k| < \pi/a$?

5. By drawing additional Bragg reflection planes for Fig. 12, determine the location of the fourth Brillouin zone, and show that its area is equal to the area of each of the first three zones.

6. Show that the distance between successive occupied planes of the [111] type is $a/\sqrt{3}$ in a simple cubic lattice, but that it is only $a/2\sqrt{3}$ in a bcc lattice, so that in k-space in the [111] direction, the first energy gap is at $|\mathbf{k}| = \pi\sqrt{3}/a$ for a simple cubic lattice, but it is at $|\mathbf{k}| = 2\pi\sqrt{3}/a$ for a bcc lattice. (This latter fact is often expressed by saying that there is an energy gap at the [222] planes rather than at the [111] planes.)

7. Silicon, germanium, and diamond form in the so-called "diamond structure" illustrated in the figure. The structure may be described most simply as two interpenetrating fcc lattices, displaced from each other by $\frac{1}{4}$ of a body diagonal. Determine the Bragg reflection planes and the first two Brillouin zones for this structure. (You may check your answer by reference to Mott and Jones.[19])

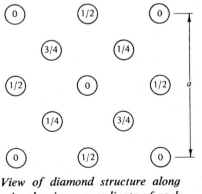

View of diamond structure along z axis, showing z coordinate of each atom in units of a.

[19] N. F. Mott and H. Jones, "The Theory of the Properties of Metals and Alloys," Dover, New York (1958).

8. Reassemble the third zone of Fig. 13a into a square, as was done for the second zone, and draw the appropriate part of the orbit of that figure in the reassembled third zone. (Notice that the orbit is perpendicular to the zone boundary at each point where it touches the boundary.)

9. Assume that the density of states $g(\varepsilon)$ for a semiconductor is the same as that for a metal (see Eq. (32), Chapter 11), and thereby show that the number of electrons in the conduction band of a semiconductor at temperature T is proportional to $(kT)^{3/2} e^{-E_g/2kt}$ when the energy gap $E_g \gg kT$.

10. The conductivity of a semiconductor depends not only on the density of conduction electrons, but also on how fast these electrons can travel under the influence of the lattice interactions which impede their motion. Thus the conductivity of a semiconductor may be written as $\sigma = n_e \mu_e + n_h \mu_h$, where n_e and n_h are the number of conduction electrons and number of holes, respectively, and μ_e and μ_h are the *mobilities* of electrons and holes, respectively. It has been shown that the mobility of electrons and holes in silicon is proportional to $T^{-3/2}$. Thus from the result of Problem 9, and on the assumption that the number of electrons is equal to the number of holes, we see that the conductivity is proportional to $e^{-E_g/2kT}$. The following values were recorded for the resistivity ρ of a relatively pure sample of silicon, at various temperatures:

$T(^\circ K)$:	600	700	800	900
ρ (ohm-cm):	10	1.2	0.4	0.14

Plot these data in such a way as to obtain a straight line and thus verify that the above arguments are correct. From the slope of the straight line determine the value of E_g. You might expect data taken at lower temperaures to deviate from this straight line. Why? (Data are from G. L. Pearson and J. Bardeen, *Phys. Rev.* **75**, 865 (1949); the derivation of the $T^{-3/2}$ dependence of the mobility is also in this paper.)

11. Given a two-dimensional rectangular lattice whose spacing is a in the x direction and b in the y direction, with $b = 2a/3$,

   (a) sketch energy E versus wave number k_x, and E versus k_y, on the same scale, and

   (b) sketch some surfaces of constant energy in the $k_x k_y$ plane; include a surface which does not approach any Brillouin zone boundary, and a surface which intersects the Brillouin zone boundaries.

12. Given the following energy–momentum relationship

$$E = Ak^2 + Bk^4$$

for an electron in a one-dimensional lattice, find for the electron

(a) the group velocity v_g, and

(b) the effective mass m^*.

Nuclear Radiation

The atomic nucleus rested undisturbed deep within the atom for millenia, while philosophers and scientists argued about the structure of matter. The nucleus might have remained undetected for a still longer time, if the naturally occurring nuclei were all stable. But unstable nuclei betray themselves by spontaneously emitting radiation, and some of this radiation found its way onto some film in the lab of the French physicist Henri Becquerel in 1896. The exposure of this film, as a result of the ionizing power of the radiation, touched off a flurry of activity by Becquerel, the Curies, Rutherford, and many others. Eventually this led to the use of these radiations as probes to explore the atom, and to Rutherford's discovery that the atom has a tiny nucleus.

471

Nuclear radiation has subsequently been used to study solids and liquids as well as the nucleus itself, so this topic forms a bridge between nuclear and solid-state physics. It is of further interest because of the interaction of such radiation with living matter, and the uses and hazards that result from such interactions.

13.1 EARLY WORK WITH RADIOACTIVITY

Classification and General Properties of Radiation. Early investigators classified the radiation into three types—alpha, beta, and gamma rays. Alpha rays, the least penetrating, could be stopped by a sheet of paper; beta rays were stopped by a millimeter of metal; and gamma rays would penetrate a centimeter or more of lead. It was not at first possible to observe the paths of individual particles, but the ionization produced by the particles was observed by the discharge of electrometers. By collimation and application of electric and magnetic fields, it was possible to make crude measurements of e/m for the rays. These led to the suspicion that the alpha particles might be helium ions, the betas, electrons, and the gammas, electromagnetic radiation (photons).

Rutherford finally proved that the alphas were indeed helium, by trapping them in a thin-walled glass bottle and subsequently observing the helium spectrum (1909).

Although measurements of e/m for the beta particles were eventually refined to the point where their e/m agreed with that of the electron to within 1.5 percent,[1] the most conclusive evidence for their identity with electrons was not found until 1948. The 1948 experiment was based on the exclusion principle: if a beta particle differs in any way from an atomic electron, it should be able to fall into an already filled K shell in an atom. Such a process would lead to emission of characteristic spectral lines where beta rays were stopped in an absorber. A careful search for these lines yielded a negative result[2]; an upper limit of 3 percent was established for the percentage of betas which fell into the K shell, a result consistent with the hypothesis that the betas are electrons.

The identification of gamma rays as photons was accepted when Rutherford and others succeeded in diffracting them with a crystal.

Exponential Decay Law. At first it was thought that radioactive materials were capable of radiating "perpetually," because no diminution of activity was seen over a period of time. (This should teach us a lesson about extrapolation.) Such a phenomenon would seem to conflict with the law of conservation of energy, but it was suggested by Lord Kelvin that some kind of energy we could not detect might be floating around in space, and that radio-

---

[1] C. Zahn and A. Spees, *Phys. Rev.* **53**, 365 (1938).
[2] M. Goldhaber and G. Scharff-Goldhaber, *Phys. Rev.* **73**, 1472L (1948).

active atoms were able to extract this energy and shoot it out in the form of alpha, beta, and gamma rays.[3]

The situation began to be clarified with the appearance of the 1903 paper of Rutherford and Soddy.[4] They had succeeded in separating the radio-active gas (now known as radon-220) which emanated from thorium. They found that the activity of the gas decreased by a factor of 2 every 52 sec, after it was isolated from the thorium. They concluded that a general property of radioactivity is the exponential decay law

$$I = I_0 e^{-\lambda t} \tag{1}$$

where I is the intensity of the radiation and λ is the "decay constant" characteristic of the material. It was soon pointed out by Schweidler that this law could be derived on the assumption that the decay process is statistical in nature—that is, each atom existing at any time t has the same probability of decay in the next short time interval dt. If this probability is $\lambda\, dt$, then the number N of radioactive atoms changes in a time interval dt by an amount

$$dN = -N\lambda\, dt$$

This differential equation can be integrated immediately to yield

$$N = N_0 e^{-\lambda t} \tag{1'}$$

where N_0 is, of course, the number of atoms present at time $t = 0$. Expression (1') is obviously identical in meaning to (1), because the intensity of the radiation is directly proportional to the number of radioactive atoms which are present. The reasoning here is precisely the same as that leading to Eq. (19), Chapter 10, dealing with atomic radiation.

Radioactive Series. In their 1903 paper, Rutherford and Soddy also proposed the transformation theory, which states that, in radiating, an atom changes its species. This idea, whose novelty is not so apparent to us now, provided the basis for great advances in understanding. Over the next ten years, Rutherford and co-workers explored the consequences of this idea, and were able to identify three *families* of naturally occurring radioactive elements. Each family has a long-lived "parent," which decays into another radioactive species; after a series of alpha and beta decays, the process ends with a stable element.

Because of the occurrence in these series of different species with the same chemical properties (same Z), but different atomic weight A, Soddy coined the term "isotope" in 1911. He then stated displacement laws which govern the development of a series: (1) in alpha decay, Z is reduced by 2 units and

---

[3] E. U. Condon, *Phys. Today*, 37, (Oct. 1962).
[4] E. Rutherford and F. Soddy, *Phil. Mag.* **5**, 576 (1903).

A by 4 units. (2) In beta decay, Z is increased by 1 unit while A is unchanged.[5] (Elements with the same A but different Z are called *isobars*). The value of A for each element in a given series has the same remainder after division by 4. The four possible series are:

Parent and half-life (yr)		A	End product
^{238}U	4.5×10^9	$4n + 2$	^{206}Pb
^{235}U	7.1×10^8	$4n + 3$	^{207}Pb
^{232}Th	1.4×10^{10}	$4n$	^{208}Pb
^{237}Np	2.2×10^6	$4n + 1$	^{209}Bi

where n is an integer. The last series is not found in nature, but it has been produced artificially.

The extremely long half-lives of the natural parents explain why no diminution of activity was observed in the early studies of radioactive rocks. Of course, since no known process on earth can produce these heavy elements, it is clear that only elements with half-lives comparable to the age of the earth remain on earth. If it is assumed that the heavy elements were produced in roughly equal abundance by some process which occurred before the earth was formed, the presence of three of these series and the absence of the fourth gives a limit on the age of the earth. Further applications are developed in the problems at the end of this chapter.

EXAMPLE PROBLEM 1. Estimate the age of the earth, on the assumption that ^{235}U and ^{238}U were equally abundant when the earth was formed. The relative abundance now is $N_{238}/N_{235} = 140$, and the respective decay constants are

$$\lambda_{235} = 1.0 \times 10^{-9}/\text{yr} \qquad \text{and} \qquad \lambda_{238} = 1.6 \times 10^{-10}/\text{yr} .$$

Solution. If equal quantities were present at $t = 0$, then

$$N_{238} = N_0 e^{-\lambda_{238}t} \qquad \text{and} \qquad N_{235} = N_0 e^{-\lambda_{235}t}$$

are the abundances at time t.

(The decay constant λ is found from the half-life $t_{1/2}$ as follows: $e^{-\lambda t_{1/2}} = 1/2$, so $\lambda t_{1/2} = \ln 2$. Notice that the *mean* lifetime $\tau = 1/\lambda = t_{1/2}/(\ln 2) = 1.44 \, t_{1/2}$; the proof that $\tau = 1/\lambda$ is mathematically identical to the derivation of mean free path in Chapter 1.)

---

[5] Obviously, these laws follow directly from the fact that α particles are helium nuclei and β particles are electrons.

We may construct the ratio

$$140 = \frac{N_{238}}{N_{235}}$$

$$= \frac{e^{-\lambda_{238}t}}{e^{-\lambda_{235}t}}$$

$$= e^{-\lambda_{238}t + \lambda_{235}t}$$

Taking the natural logarithm of both sides yields

$$4.9 = \lambda_{235}t - \lambda_{238}t$$

so that

$$t = \frac{4.9}{\lambda_{235} - \lambda_{238}}$$

$$= 5.8 \times 10^9 \quad \text{yr.}$$

Rutherford Scattering.   Early in this century it was found that individual alpha particles could be detected by observing the flashes of light produced by their impact on a zinc sulfide screen. At Rutherford's suggestion, this technique was used by Geiger and his student, Marsden, to look for large-angle scattering of alpha particles by thin foils. It was not thought that many large-angle deflections would be found, because according to the Thomson model of the atom, the positive charge fills most of the volume of the material, so that the deflection of a particle is made up of many small deflections. A large deflection would be produced by many small deflections added together, so the range of angles of deflection should, by the laws of statistics, be proportional to $\sqrt{t}$ in passage through a foil of thickness t.

When the experiment was tried, it was found that, although most of the deflections were through very small angles, there were some deflections of 90° or more, even with the thinnest foils that could be used. Rutherford has been quoted as saying: "It was quite the most incredible event that has ever happened to me in my life. It was almost as incredible as if you had fired a 15-in. shell at a piece of tissue paper and it came back and hit you."[6]

It is clear from this statement that Rutherford had a very keen insight into the nature of the problem, and, as we saw in Chapter 3, he soon produced the explanation that the positive charge is concentrated in a very small volume. But Rutherford was not satisfied with a qualitative explanation. In

---

[6] W. E. Burcham, "Nuclear Physics," p. 49. McGraw-Hill, New York, 1963.

order to provide a strict test of his hypothesis, he used it to derive a scattering law whose details could be carefully checked by further experiments.[7]

Rutherford's scattering law predicts the number of alpha particles which will be deflected through various angles in passing through a thin foil. The result depends on the thickness of the foil, the atomic number of the element in the foil, and the kinetic energy of the alpha particles. In order to derive the result, Rutherford first determined the path, and therefore the scattering angle θ, of a single alpha particle in terms of the "impact parameter" b (see Fig. 1). The impact parameter is the perpendicular distance between the

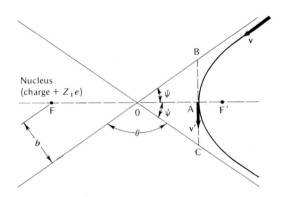

Fig. 1. *Hyperbolic path of an alpha particle in the field of a nucleus. Nucleus is at point F, the external focus of the hyperbola. Particle's initial direction is along line OB, and its final direction is along line OC; it is scattered through an angle θ.*

nucleus and the path which the alpha particle would follow if it were not repelled by the nucleus. After finding the relation between θ and b, Rutherford then assumed that an alpha particle would interact with only one nucleus as it passed through the foil, and that the impact parameter would be random. He was then able to calculate the distribution of scattering angles. Let us follow this calculation essentially as Rutherford gave it.[8]

Because the repulsive Coulomb force between the alpha particle and the nucleus obeys the inverse square law, the path of the alpha particle is a

---

[7] It may be appropriate to point out here that this prediction of experimentally verifiable consequences marks the difference between sound theory and idle speculation. The amateurs, who complain that physicists will not listen to their theories, usually do not make any physical prediction which could be tested by an experiment.

[8] E. Rutherford, *Phil. Mag.* **21**, Ser. 6, 669 (1911).

hyperbola. (This is proved in any intermediate mechanics text. The field of the atomic electrons may be ignored, for the alpha particle must approach quite close to the nucleus in order to be scattered through an appreciable angle). The nucleus, of charge $Z_1 e$, is located at F, the external focus of the hyperbola. We neglect the motion of the nucleus; if necessary, we can later correct for this motion by using the reduced mass of the alpha particle in the resulting formulas and converting the resulting angles from the center-of-mass coordinate system to one which is fixed in the laboratory. The alpha particle, of charge $Z_2 e$,[9] comes in with speed v and impact parameter b. We seek an expression for the scattering angle θ in terms of b and v. We begin by applying the laws of conservation of angular momentum and energy:

$$\text{Angular momentum:} \qquad vb = v'(\text{FA}) \qquad (2)$$

$$\text{Energy (mks units):} \qquad \tfrac{1}{2}mv^2 = \tfrac{1}{2}mv'^2 + \frac{Z_1 Z_2 e^2}{4\pi\varepsilon_0(\text{FA})} \qquad (3)$$

where v' is the velocity of the alpha particle when it is at point A, and non-relativistic mechanics is used because the alpha's energy is only a few Mev (0.1 percent of rest energy).

It is convenient to define a distance d which is the distance of closest approach of the alpha particle *in a head-on* collision. An expression for d can be found from (3) by putting $v' = 0$ and FA $= d$; in that case

$$d = \frac{Z_1 Z_2 e^2}{2\pi\varepsilon_0 mv^2} \qquad (4)$$

and Eq. (3) may be written in terms of d as

$$1 = \frac{v'^2}{v^2} + \frac{d}{\text{FA}}$$

But Eq. (2) tells us that $v'/v = b/\text{FA}$, so we may rewrite the above as

$$1 = \frac{b^2}{(\text{FA})^2} + \frac{d}{\text{FA}}$$

$$b^2 = (\text{FA})[(\text{FA}) - d] \qquad (5)$$

Now some straightforward geometry gives us the relation between b and θ. From the properties of the hyperbola we know that FO $=$ OB, so that

$$\text{FA} = \text{FO} + \text{OA}$$

$$= \text{FO} + \text{OB} \cos \psi$$

$$= \text{FO} + \text{FO} \cos \psi$$

[9] We write the charge as $Z_2 e$ for generality; the alpha particle charge is, of course, $2e$.

or

$$FA = FO(1 + \cos \psi)$$
$$= b \csc \psi (1 + \cos \psi)$$
$$= b \cot \left(\frac{\psi}{2}\right)$$

Substituting this value of FA into (5) yields

$$b^2 = \left[b \cot \frac{\psi}{2}\right]\left[b \cot \frac{\psi}{2} - d\right]$$

which may be solved for d to give

$$d = \frac{b\left[\cot^2 \frac{\psi}{2} - 1\right]}{\cot \frac{\psi}{2}} = 2b \cot \psi \qquad (6)$$

But the scattering angle is $\theta = \pi - 2\psi$, so

$$\frac{d}{2b} = \cot \psi = \tan\left(\frac{\pi}{2} - \psi\right) = \tan \frac{\theta}{2} \qquad (7)$$

Notice that as $b \to 0$, $\tan(\theta/2) \to \infty$, so that $\theta \to 180°$ and any scattering angle is therefore *possible* (in the center-of-mass coordinate system). But we must find the *relative probability* of the various scattering angles. This is done by assuming that the undeflected alpha particle path (line OB in Fig. 1) is equally likely to pass through any unit area; therefore the probability that a given particle's impact parameter lies between b and $b + db$ is proportional to the area $2\pi b \, db$ of a ring of radius b and thickness db (Fig. 2). This area is

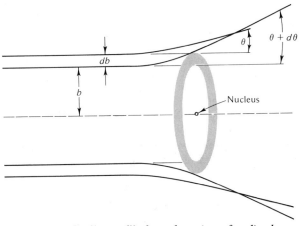

Fig. 2. *Particles "aimed" through a ring of radius b and area 2πb db centered on a nucleus are scattered into a cone of angles between θ and θ + dθ.*

then, by definition, equal to the differential cross section for scattering through the corresponding angle θ, as given by Eq. (7):

$$d\sigma = |2\pi b \, db|$$

but

$$b = \frac{d}{2} \cot \frac{\theta}{2}$$

and so

$$db = -\frac{d}{4} \csc^2 \frac{\theta}{2} \, d\theta$$

making

$$d\sigma = 2\pi \frac{d^2}{8} \cot \frac{\theta}{2} \csc^2 \frac{\theta}{2} \, d\theta$$

It is more convenient to express $d\sigma$ in terms of the differential solid angle $d\Omega = 2\pi \sin \theta \, d\theta$:

$$d\sigma = \frac{\dfrac{2\pi d^2}{8} \cot \dfrac{\theta}{2} \csc^2 \dfrac{\theta}{2} \, d\Omega}{2\pi \sin \theta}$$

$$= \frac{\dfrac{d^2}{8} \cot \dfrac{\theta}{2} \csc^2 \dfrac{\theta}{2} \, d\Omega}{2 \sin \dfrac{\theta}{2} \cos \dfrac{\theta}{2}}$$

with the final result that

$$d\sigma = (d^2/16) \csc^4(\theta/2) \, d\Omega$$

The differential cross section is the area around *each* nucleus which is effective for scattering alpha particles through an angle θ into a given amount of solid angle. In order to find the number of particles actually scattered into any particular solid angle, we must multiply the cross section by the number of nuclei per unit area in the target, and by the number of particles in the beam (on the assumption that all beam particles go through the target). This is a valid procedure as long as the foil is so thin that no alpha particle is scattered more than once as it passes through.[10] If the foil contains n nuclei

---

[10] It might seem that the alpha particle is scattered by every nucleus in the foil, since the Coulomb field goes to infinity. But the electron shells modify the field at distances of the order 10^{-8} cm from the nucleus; and even for closer approaches, the scattering angle may be negligible. For a 5-MeV alpha particle to be scattered through $1°$ or more by a nucleus with $Z = 50$, Eqs. (4) and (7) show that b must be 1.6×10^{-10} cm or less, making the cross section for scattering through $1°$ or more equal to πb^2 or about 10^{-19} cm^2. A typical foil 10^{-5} cm thick contains about 3×10^{17} nuclei per cm^2, so that about 3 percent of the total area is "covered," as far as scattering through $1°$ or more is concerned. Thus only 3 percent of the nuclei are scattered even once, and fewer than 0.1 percent are scattered more than once, through an angle of $1°$ or more.

per unit volume, and has thickness t, and if there are N_0 particles in the beam, the number of particles scattered is therefore

$$dN(\theta) = \frac{N_0\,ntd^2}{16}\,\csc^4\frac{\theta}{2}\,d\Omega \qquad (8)$$

Geiger and Marsden made careful checks of this formula over the years 1911–1913.[11] By this time their technique had been refined so that they could test the dependence of $dN(\theta)$ on each variable. The number of alphas detected by the detector at a certain point should be proportional to (a) $\csc^4(\theta/2)$, (b) the thickness of the scatterer, (c) the square of the Z of the scatterer, (d) the inverse square of the incident energy. (The latter two are contained in the d^2 factor in the formula.) Each point was verified precisely over a range of angles giving a variation of greater than a factor of 10^4 in $\csc^4(\theta/2)$. The Z of the scatterer was not known unambiguously at that time, but it was found that Z is roughly one-half the atomic mass A. This sort of experiment was used later by Chadwick (1920) to check Moseley's results on atomic numbers of elements (see Section 4.1).

13.2 PASSAGE OF RADIATION THROUGH MATTER

In addition to Rutherford scattering, particles may be subject to various other interactions during their passage through matter. Although most of these interactions are with electrons rather than with nuclei, the study of these interactions is important in nuclear physics, for it is only through such interactions that quantitative measurements of the properties of the radiation can be made.

The early classification of nuclear radiations as alpha, beta, or gamma rays was based on the different behavior of the radiations in passing through matter. It is still useful to retain three categories for discussing the interactions of these "rays" with electrons in matter, but to generalize the categories as follows:

1. Heavy charged particles—any charged particle of mass greater than the electron mass. The lightest of these is the muon, whose mass is about 200 electron masses.
2. Electrons (positive or negative).
3. Photons—gamma rays or X-rays.

Heavy Charged Particles. Except when rare interactions with nuclei occur, a heavy charged particle travels in an almost straight line through matter. It interacts with many electrons, losing a very small fraction of its

---

[11] H. Geiger and E. Marsden, *Phil. Mag.* **25**, Ser. 6, 604 (1913).

energy in each interaction, so that it may be considered to lose energy continuously, at a rate which is a continuous function of the kinetic energy of the particle.

If the energy loss is continuous, one can define a *range* which is reproducible; that is, a given kind of particle, with a given energy, always penetrates a given material for the same distance before stopping.[12] Knowledge of the energy loss rate and the relation between the initial energy and the range is useful in many ways. For example, if a particle track is observed in a bubble chamber or in a photographic emulsion, the density of bubbles or of exposed grains along the track is proportional to the rate of energy loss of the particle; this information is often used to identify the particle. Another application is simply to use a range measurement to determine the initial energy of particles whose identity is known.

For accurate work, one must determine the range–energy relation experimentally. But a rough theoretical analysis enables one to understand the general features of this relation. Let us therefore make an approximate calculation of the energy loss rate of a heavy charged particle. We assume that the path of the particle is a straight line, and we consider an interaction with an electron which lies at a distance b from this line. We also assume that the electron receives a *sudden* impulse, so that it does not move during the interaction, and we assume that the electron is free. This last assumption is reasonable if the electron acquires enough kinetic energy to free it from the atom to which it is bound; this is possible if b is sufficiently small. There is therefore an upper limit on the possible values of b; if b is too large the electron remains in a discrete state and does not contribute to the energy loss of the particle.

The momentum acquired by the electron in time dt is equal to $e\mathbf{E}\,dt$, or $e\mathbf{E}\,dx/v$, where v is the velocity of the incident heavy particle, and $\mathbf{E}$ is the electric field produced by the particle at the electron. The total momentum acquired by the electron is found by integrating this expression from $x = -\infty$ to $x = +\infty$. In doing this, we assume that the energy lost by the heavy particle is small in comparison with its initial energy, so that v can be considered constant. Symmetry then shows that the impulse on the electron is perpendicular to the path of the heavy particle, and the only component of $\mathbf{E}$ which need be considered is $E_\perp$ (see Fig. 3). The momentum acquired by the electron is therefore

$$|\Delta p| = \frac{1}{v} \int_{-\infty}^{+\infty} eE_\perp \, dx \tag{9}$$

---

[12] This statement might seem quite obvious, but it is not true if the particle loses a large fraction of its energy in each interaction (as do electrons and photons). Since the interactions occur at random, identical particles with the same initial energy could then penetrate to quite different distances in the same material. Even for heavy particles, small variations in range can be observed; these show that the energy loss is not really continuous but results from a finite number of interactions.

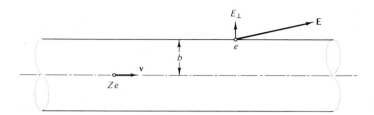

Fig. 3. *Electric field* **E** *and component* $E_\perp$ *produced at electron e by charge Ze. The infinitely long cylinder of radius b is used to evaluate* $\int_{-\infty}^{+\infty} E_\perp\, dx$, *by means of Gauss's law.*

This integral may be evaluated by using Gauss's law; the flux through the wall of an infinitely long cylinder of radius b, containing a charge Ze (the heavy particle's charge) is

$$\int_{-\infty}^{+\infty} E_\perp 2\pi b\, dx = \frac{Ze}{\varepsilon_0}$$

We conclude that $|\Delta p| = Ze^2/2\pi b\varepsilon_0 v$. The energy lost by the heavy particle is equal to that acquired by the electron, or

$$\Delta E = -\frac{(\Delta p)^2}{2m}$$

$$= \frac{-Z^2 e^4}{8\pi^2 mb^2 \varepsilon_0^2 v^2} \tag{10}$$

where ΔE is the change in energy of the heavy particle and m is the *electron* mass.

To find the total energy loss per centimeter, one multiplies Eq. (10) by the number of electrons per centimeter which are encountered at a distance between b and $b + db$. This product is then integrated over the allowed range of b. The upper limit on b has already been mentioned; the lower limit is that value of b at which the energy transfer given by Eq. (10) would be $2mv^2$, for it is easy to see that $2mv^2$ is the maximum energy that the electron can gain in an interaction with a particle of velocity v.[13] In traveling a distance dx the particle encounters $n2\pi b\, db\, dx$ electrons at a distance between b and

[13] Imagine the incident particle to be at rest. If the electron strikes it with velocity v, the maximum change in electron velocity is $2v$, which occurs when the electron bounces directly backwards. Therefore an electron at rest in the laboratory frame can acquire a maximum velocity of $2v$, or an energy of $m(2v)^2/2 = 2mv^2$.

$b + db$, if n is the number of electrons per unit volume in the material. The energy loss in distance dx is therefore

$$-dE = \int_{b_{min}}^{b_{max}} \frac{Z^2 e^4 n 2\pi b \; db \; dx}{8\pi^2 \varepsilon_0^2 \; mb^2 v^2}$$

$$= \frac{nZ^2 e^4 \; dx}{4\pi\varepsilon_0^2 \; mv^2} \ln \frac{b_{max}}{b_{min}} \tag{11}$$

The energy loss rate depends upon v explicitly through the factor of v^2 in the denominator, and implicitly through the factors b_{max} and b_{min}, which themselves depend on v. If the latter factors were independent of v, we could integrate Eq. (11) easily, for non-relativistic particles, because the energy E is proportional to v^2. Separation of variables would yield

$$-dE/dx = \text{constant}/E$$

If the particle has energy E_0 when $x = 0$, then by rearranging terms and integrating we obtain

$$\int_0^R dx = - \text{ constant} \times \int_{E_0}^0 E \; dE \tag{12}$$

because R, the range, must equal the value of x when E has been reduced to zero.

Integration of (12) indicates that R is proportional to E_0^2. This result cannot be precise because we have neglected the velocity dependence of b_{min} and b_{max} as well as the effect of the polarization of the medium on the electric field at the target electrons. However, experiment shows that it is surprisingly close to the truth; for protons in air, the range is proportional to the 1.8 power of E_0, over a large range of values of E_0. Numerically,

$$R = (E_0/9.3)^{1.8} \tag{13}$$

where E_0 is in MeV and R is in meters, for

$$3 \text{ MeV} \lesssim E_0 \lesssim 300 \text{ MeV}$$

For $E_0 \gtrsim 300$ MeV, the range increases more slowly with energy (see Fig. 4) because the *speed* increases more slowly with energy as the energy approaches the rest energy.

Using Eq. (13) or Fig. 4, we can calculate approximate ranges of other heavy particles as well as protons, because the energy loss rate does not depend on the mass M of the heavy incident particle. From Eq. (11) we know that

$$- \frac{dE}{dx} = Z^2 \, f_1(v)$$

where $f_1(v)$ expresses the dependence on speed, a dependence whose exact form is not essential to the argument. We can also write

$$E = M f_2(v)$$

where $f_2(v)$ is given by the relativistic expression for kinetic energy as

$$f_2(v) = c^2(1 - (v^2/c^2)^{-1/2} - c^2$$

although here again the exact form of $f_2(v)$ is not relevant to the argument. We can now integrate on x to find the range:

$$R = \int_0^R dx = - \frac{M}{Z^2} \int_{v=v_0}^{v=0} \frac{d(f_2(v))}{f_1(v)}$$

The right-hand side is in general another function f_3, which is a function of v_0, multiplied by the factor M/Z^2, so

$$R = (M/Z^2) f_3(v_0) \tag{14}$$

where $f_3(v_0)$ is *independent of the charge or the mass* of the incident particle.

We conclude that, for a given *speed,* the range is proportional to M/Z^2. For example, if a proton and a deuteron (deuterium nucleus) have the same initial speed, the deuteron will go twice as far as the proton. If a proton and an alpha particle have the same initial speed, they should have the same range according to (14), because M/Z^2 is the same for both. (However, the latter result is complicated by the fact that an alpha particle near the end of its range captures an electron and becomes a particle with an effective Z of only 1, which can penetrate a bit farther.) In general, given the kinetic energy of a particle of mass M and charge Ze, we must find the range of a proton with the same speed as this given particle, then multiply by $M/M_p Z^2$ (where M_p is the proton mass) to find the range of the given particle.

EXAMPLE PROBLEM 2. Find the range of a 10-MeV muon and the range of a 90-MeV muon.

Solution. The muon mass is approximately $M_p/9$, so the speed of a 10-MeV muon is equal to the speed of a 90-MeV proton, and the speed of a 90-MeV muon is equal to that of an 810-MeV proton. Eq. (13) yields $R = 59$ meters for 90-MeV protons, and Fig. 4 shows that $R \approx 2300$ meters for 810-MeV protons. Therefore, the *muon* ranges for these same speeds are

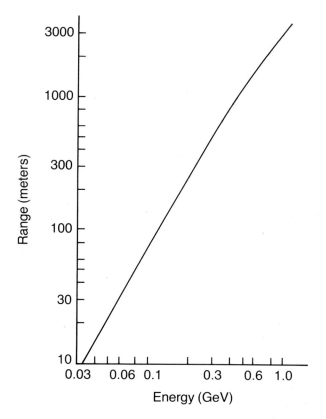

Fig. 4. *Range vs. kinetic energy for protons in air, up to 1 GeV. (1 GeV = 1000 MeV)*

$$R_\mu = \frac{59}{9} \approx 6.6 \text{ meters} \qquad \text{for 10-MeV muons}$$

$$R_\mu = \frac{2300}{9} \approx 260 \text{ meters} \qquad \text{for 90-MeV muons}$$

Notice that the range of a 90-MeV *muon* is several times the range of a 90-MeV *proton,* because of the muon's greater speed. Notice also that Eq. (13) is valid for the 90-MeV protons, but not for the 810-MeV protons, where it gives a range of about 3100 meters.

Ranges quoted in meters of air may be used to find ranges in other materials as well. There is an empirical rule that the range is proportional to $A^{1/2}\rho^{-1}$, where A is the average atomic mass of the absorber, and ρ its density. For example, the range of the 5.0 MeV alpha particle from ^{234}U is 3.50 cm of air, for which $A = A_{\text{air}} = 14.4$ and $\rho = \rho_{\text{air}} = 1.3 \times 10^{-3}$ g/cm^3. In water (or in human tissue, which is mostly water), we have $A_{water} = 6$ and $\rho_{water} = 1$ g/cm^3, so the range would be

$$R_{water} = (A_{water}/A_{air})^{1/2} \times (\rho_{air}/\rho_{water}) \times R_{air}$$

$$= (6/14.4)^{1/2} \times 1.3 \times 10^{-3} \times R_{air}$$

$$= 0.84 \times 10^{-3} \times R_{air} \simeq 3 \times 10^{-3} \, cm$$

or about half the thickness of this sheet of paper. We see that, in general, the range in water (or in a low-density solid) must be about 1/1000 of the range in air.

One should not expect the above treatment to work well for relativistic particles, for neither the electric field nor the polarization produced by such particles is spherically symmetrical, and one should use relativistic expressions for energy and momentum in the analysis of the energy transferred to the electron. But it happens that the *speed* of the particle is still the major factor determining the energy loss. Thus for relativistic particles, whose speed is almost independent of energy, the energy loss rate $-dE/dx$ has very little energy dependence. The rate actually reaches a minimum at a kinetic energy near the rest energy of the particle; beyond that it increases very slowly with increasing energy.

A significant consequence of this behavior of relativistic particles is that most of the tracks of high energy particles in a bubble chamber have the same density of bubbles, corresponding to what is called " minimum ioniza-tion." This minimum ionization provides a convenient way to calibrate the bubble densities observed in the chamber; one can then compute the energy loss rates of those other particles whose tracks have a greater bubble density, and thereby determine the energy of such particles. Of course such considera-tions also apply to other track-forming devices such as cloud chambers or photographic emulsions.

The *range* of relativistic particles is roughly proportional to their energy, since the greatest contribution to the range comes from the part of the path over which the energy loss rate is fairly constant. The complete range–energy relation shows the 1.8 power dependence of Eq. (13) at low energy, with a gradual transition to a linear dependence as the energy increases. The onset of this transition can be seen in Fig. 4, above 0.3 GeV.

Electrons. The passage of electrons through matter is more complicated than that of heavy particles. It is clearly incorrect to consider the electron to be undeviated from a straight line in a collision with an electron in the material. An electron can lose up to one-half of its energy in such a collision. (At first you might think that an electron could lose all of its energy, but you would be overlooking the indistinguishability of the electrons. Loss of all the energy to another electron is equivalent to the loss of no energy. If we arbitrarily consider the higher energy electron emerging from the collision to be the incident electron, we see that the maximum loss of energy is one-half.) Thus there is some probability that even a very high energy electron will be removed from a beam in a very short distance, and the concept of range becomes less exact. However, it is still possible to define a mean range for electrons of given energy, even though there are large deviations in the range of individual electrons. Just as for other charged particles, this range turns out to be proportional to E^2 at low energy (up to 30 keV) and to E at high energy (>1 MeV).

Absorption of electrons differs from that of heavy particles in one other important respect. Because electrons passing through matter are subject to larger accelerations, they also lose energy by emission of electromagnetic radiation, or bremsstrahlung (as we have seen in the discussion of X-ray production). For electrons up to about 2 MeV, the average fraction of energy converted into bremsstrahlung in a target of atomic number Z is about $0.0007ZE$, where E is in MeV.[14] Since the intensity of radiation is proportional to the square of the acceleration, which in turn is inversely proportional to the mass of the particle, we see that the bremsstrahlung energy loss by heavy particles is negligible.

A photon emitted in bremsstrahlung can have any energy up to the total energy of the incident electron. If the electron has enough energy, the emitted photon can have energy in excess of the 1.02 MeV required to produce a positron–electron pair. The positron produced by such a photon can then annihilate with another electron in the material to produce two more photons. In this way, if the initial electron energy is high, an "electron shower" can be produced. Such showers are often observed in cosmic rays; a single high energy electron can give rise to millions of electron–positron pairs.

Photons.   The absorption of X or gamma rays differs from that of charged particles because a photon is absorbed or scattered out of a beam in a single event, whereas a charged particle loses energy steadily while proceeding along its path. Photons lose energy principally by photoelectric absorption, Compton scattering, or, for photons of energy 1.02 MeV or more, production of positron–electron pairs. There are other possible interactions which are rarer and do not usually contribute much to the overall absorption. (But sometimes these interactions are of great interest, as we shall see in Section 13.4.)

The attenuation of a beam of photons resembles the removal of molecules from a beam, discussed in Chapter 1. The number removed from a beam in a distance dx is

[14] R. D. Evans, "The Atomic Nucleus," p. 617. McGraw-Hill, New York, 1955.

(15)

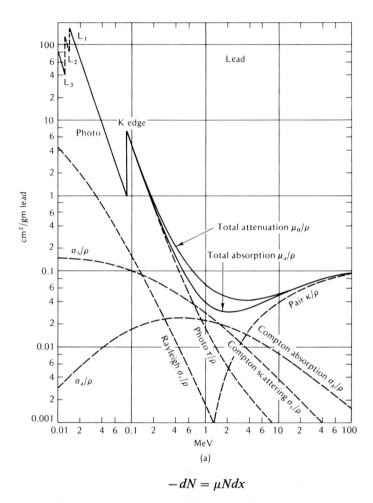

(a)

$$-dN = \mu N dx$$

where μ, the linear attenuation coefficient, may be written as the sum of three major contributions: σ, the Compton attenuation coefficient; τ, the photo-electric absorption coefficient; and κ, the pair-production absorption coefficient. The probability that a photon will traverse a distance x without interacting in *any* of these ways is thus

$$e^{-\mu x} = e^{-(\sigma + \tau + \kappa)x} = e^{-\sigma x} e^{-\tau x} e^{-\kappa x}$$

where, for example, the factor $e^{-\sigma x}$ is the probability that the photon will traverse distance x without a Compton interaction. The mean distance traveled by a photon before interacting is $1/\mu$, in analogy to the mean free path of a molecule (or to the mean lifetime $1/\lambda$ for radioactive decay).

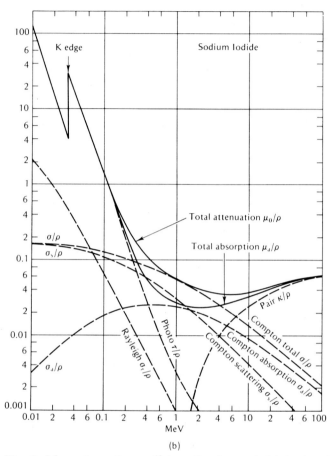

(b)

Fig. 5. *Mass attenuation coefficients for photons in (a) lead and (b) sodium iodide. Linear coefficients may be obtained by using* $\rho = 11.35$ *gm/cm³ for lead and* $\rho = 3.67$ *gm/cm³ for sodium iodide. [From "The Atomic Nucleus" Ch. 25, by R. D. Evans. Copyright McGraw-Hill, New York, 1955. Used by permission of McGraw-Hill Book Company.]*

Compton interactions differ from photoelectric effect or pair production because the photon is not absorbed but is only scattered, retaining some energy. Therefore, if one is concerned with the *energy given to the absorber* by a beam of photons, one writes σ as the sum of two terms, σ_a and σ_s, such that σ_a/σ is the average fraction of energy absorbed in Compton scattering, and σ_s/σ is the average fraction of energy retained by the photon. (It is necessary to refer to average values, because even in a beam of monoenergetic photons, the amount of energy given to the electron in a single event depends on the scattering angle, as we saw in Chapter 3). In a thickness dx, the energy absorbed from n photons is then $nhv(\sigma_a + \tau + \kappa)dx$, if each photon has energy hv.

Figure 5 shows how σ_a, σ_s, τ, and κ vary with photon energy in lead and in sodium iodide. (Sodium iodide is of interest because of its use in scintillation counters.) In this figure the "mass attenuation coefficients" are plotted; these are simply the linear attenuation coefficients divided by the density. The mass attenuation coefficient is more closely related to the atomic cross section for an interaction; we may compare the two kinds of coefficients as follows:

$$\text{linear attenuation coefficient}$$
$$= (\text{atomic cross section}) \times (\text{no. of atoms/cm}^3)$$

If we divide by density, we have

$$\text{mass attenuation coefficient}$$
$$= (\text{atomic cross section}) \times (\text{no. of atoms/gm})$$

But the number of atoms per *gram* is independent of the physical state (solid, liquid, or gas) of the absorber and depends only on constant properties of the material; it is equal to Avogadro's number divided by the atomic mass A of the atoms in the absorber.

A striking feature of Fig. 5 is that the Compton attenuation curves for the two absorbers are almost identical. The reason is that the atomic cross section for the Compton effect is simply Z times the electronic cross section, so that the mass attenuation coefficient varies from one element to another as Z/A, which is almost constant. Another general feature is that the Compton effect is the major effect at energies around 1 MeV, while the photoelectric effect predominates at lower energy and pair production at higher energy.[15]

13.3   POSITRON ANNIHILATION

At first thought, the behavior of positrons might not be expected to differ much from that of electrons in passing through matter, because the mass and the magnitude of the charge are the same for both particles. But matter contains electrons and not positrons; the annihilation of a positron by one of these electrons is a unique feature of positron interactions in matter. The gamma rays resulting from the annihilation carry away information about the state of the positron and the state of the electron with which it annihilates, so that the positron can be used as a "probe" to investigate properties of electrons in matter.

[15] Detailed discussions of these effects are found in R. Evans, "The Atomic Nucleus," Ch. 23–25. McGraw-Hill, New York, 1955, and in K. Siegbahn, "Alpha, Beta, and Gamma Ray Spectroscopy," Vol. 1. North-Holland, Amsterdam, 1965.

Angular Correlation of Annihilation Radiation. Perhaps the most important information carried by the annihilation radiation concerns the momentum of the annihilating pair. If the electron and positron are at rest when they are annihilated, momentum conservation requires that the two gamma rays be emitted in exactly opposite directions, and energy conservation requires that each have energy mc^2 (m = electron mass).[16] The momentum of each is thus equal to mc. But it was found[17] in 1950 that, although the angle between the gamma rays is very close to $180°$ for most annihilations of positrons in condensed matter, small deviations could be observed, and these were attributed to the motion of the centers of mass of the annihilating pairs.

The center-of-mass momentum of an annihilating pair (and hence the resultant momentum of the two gamma rays) may be resolved into components $p_{\parallel}$ and $p_{\perp}$, which are parallel and perpendicular, respectively, to the direction of emission of one gamma ray. In the cases of interest for the study of the condensed state, $p_{\parallel}$ and $p_{\perp}$ are both much smaller than $m_e c$, so that the momentum of each individual gamma ray is very close to $m_e c$. In that case the angle θ defined by the gamma ray directions (Fig. 6) is approximately equal to $p_{\perp}/m_e c$. The component $p_{\parallel}$ affects the energy of the gamma rays rather than the angle between them; this effect may be analyzed by application of the conservation laws, or it may simply be considered as a Doppler shift of the gamma ray frequencies. The analysis is left as an exercise. (See Problem 15.) Thus either the energy distribution or the angular correlation of the annihilation photons may give useful information about the electron–positron states that are being annihilated.

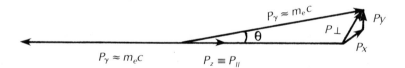

Fig. 6. *Relation between momentum and angular correlation of annihilation radiation. Magnitudes of p_x, p_y, and p_z exaggerated for clarity.*

Positron Lifetimes; Positronium. The time that a positron can survive in matter is determined by the annihilation *cross-section*, a quantity analogous to the scattering cross-section discussed in Section 6.7. In general, the probability that a single particle, striking a thin target at normal incidence, will interact with

---

[16] The total energy of the annihilating pair is $2mc^2$; this must be equally shared by the gamma rays if they are to have equal momenta.
[17] S. Debenedetti, C. E. Cowan, W. R. Konnecker, and H. Primakoff, *Phys. Rev.* **77**, 205 (1950).

that target in a given way is equal to the product of the cross-section σ for that interaction and the number of particles per unit area in the target.

The annihilation cross-section depends on the relative spin of the positron and the target electron. If the spins are *anti*parallel (total spin quantum number $S = 0$), annihilation occurs with the emission of two photons; Dirac showed that the cross section in this case is approximately

$$\sigma_{2\gamma} = 4\pi r_0^2 c/v \tag{16}$$

where r_0 (called the "classical radius of the electron") is equal to $e^2/4\pi\epsilon_0 m_e c^2$, or about 2.82×10^{-13} cm, and the positron speed is $v \ll c$.

By analogy to the discussion of the linear attenuation coefficient for gamma rays (Section 13.2), we relate the annihilation cross-section to the annihilation probability in distance dx by

$$\text{Probability} = \sigma_{2\gamma} n\, dx$$

where n is the density of electrons whose spins are antiparallel to the spin of the positron. (The parallel-spin electrons, which contribute very little to the annihilation probability, will be discussed in the following paragraphs.)

To find the positron lifetime, we need the annihilation probability per unit time, which we obtain simply by dividing the above probability by dt:

$$
\begin{aligned}
\text{Probability per unit time} \equiv \lambda &= n\sigma_{2\gamma}(dx/dt) \\
&= n\sigma_{2\gamma}v \\
&= 4\pi n r_0^2 c
\end{aligned}
\tag{17}
$$

Notice that λ depends only on the electron density n and constant factors. Unfortunately, it depends on the electron density in the presence of the positron, and this electron density is enhanced by the attractive force between positron and electron, so calculation of λ is difficult in most cases.

However, λ is easily calculated if the positron and electron form a bound state, positronium (Ps), a two-body system similar to the hydrogen atom (See Problem 7, Chapter 3). If the positron and electron spins are antiparallel (total spin $S = 0$), the state is called *para*positronium, and the electron density at the positron is known, because it is determined by the appropriate hydrogenlike wavefunction of table 1, Chapter 7. All we need do is set M, the mass of the "nucleus," equal to m_e, set r, the relative coordinate, equal to zero, and use the probability density $|u(0)|^2$ in place of the electron density n; we then obtain λ directly from Eq. (17). For Ps in the ground state, we have

$$|u(0)|^2 = |u_{100}(0)|^2 = \frac{1}{8\pi a_0^3}$$

with the result that

$$\lambda = r_0^2 c / 2a_0^3 = 8.04 \times 10^9 \, \text{s}^{-1} = \lambda_{\text{para}}$$

Thus if one could assemble a collection of para-Ps atoms in their ground states, the number would decay exponentially, with a mean lifetime of

$$\tau_{\text{para}} = 1/\lambda_{\text{para}} = 1.24 \times 10^{-10} \, \text{s}$$

These atoms would decay just as ordinary radioactive atoms do (Section 13.1), except that there would be no remaining matter after the decay—just gamma rays. A similar fate would eventually befall a collection of *ortho*-Ps atoms, in which the spins of electron and positron are parallel ($S = 1$), but in this case conservation of angular momentum requires the emission of three photons instead of two. This requirement permits ortho-Ps to survive longer than para-Ps; in general, for a transition involving photon emission, the square of the matrix element is multiplied by a factor of order α, the fine-structure constant, for each additional photon emitted. A detailed calculation[18] shows that the factor in this case is actually about $1/1110$, making the final result

$$\lambda_{\text{ortho}} = (7.0379 \pm 0.0012) \times 10^6 \, \text{s}^{-1}$$

(The uncertainty results from an estimate of the magnitude of neglected terms in a perturbation calculation.) The corresponding mean lifetime, for a collection of ortho-Ps atoms, is

$$\tau_{\text{ortho}} = 1/\lambda_{\text{ortho}} = 142.088 \pm 0.024 \, \text{nanosec}$$

The uncertainty is quoted here because, surprisingly, one can experimentally test this result to a high degree of accuracy, and thus obtain a good test of the basic theory of quantum electrodynamics. One can measure the lifetimes of positrons by using a positron-emitting isotope, ^{22}Na, which emits a 1.3-MeV gamma ray simultaneously with the positron (within a few picoseconds). The detection of this gamma ray, signaling that a positron has been created, starts an electronic "clock," which is stopped when one of the lower-energy gamma

---

[18] The original calculation by A. Ore and J. L. Powell (*Phys. Rev.* **75**, 1696 (1949)) was slightly corrected by W. E. Caswell, G. P. LePage, and J. Sapirstein, *Phys. Rev. Letters* **38**, 488 (1977).

rays[19] from the positron's annihilation is detected. A computer records the time interval registered by the clock for each detected gamma-ray pair.[20]

If the positrons are emitted into a gas, some of them will form ortho-Ps. These positrons will live much longer than the others because the parallel-spin electron shields them by repelling other electrons that could annihilate them by forming an $S = 0$ state. So the lifetime distribution consists of a number of short ($<< 100$ nanoseconds) lifetimes, plus an exponential "tail" produced by the decay of ortho-Ps atoms. The decay constant λ of this tail is the total annihilation probability per unit time, which includes the probability that the positron will be annihilated by "pickoff" with an outside electron, so

$$\lambda = \lambda_{ortho} + \lambda_{pickoff}$$

where $\lambda_{pickoff}$ is directly proportional to the electron density, and hence to the gas density ρ. Thus

$$\lambda \rightarrow \lambda_{ortho} \qquad \text{as} \qquad \rho \rightarrow 0$$

so that one can determine λ_{ortho} by plotting λ against gas density and extrapolating to zero density. Very careful measurements have established[21] that the limit at zero gas density is $\lambda = (7.045 \pm 0.006) \times 10^6$ s^{-1}, in excellent agreement with the above-quoted theoretical result for λ_{ortho}.

Positronium can also form in some liquids and solids. Although $\lambda_{pickoff}$ is much greater there, the Ps atom can survive long enough to react chemically like any ordinary atom. For example, ortho-Ps in water shows up as a "long-lived" component in the lifetime distribution, decaying exponentially with a mean lifetime of about 1.8 ns; this is easily distinguishable from the lifetime distribution of the rest of the positrons, whose mean lifetime is about 0.4 ns. Figure 7 shows the separation of the lifetime distribution into these two components, and it also shows what happens when a strong oxidizing agent is added to the water. The mean lifetime of the long-lived component is reduced, by a far greater factor than could be accounted for by any change in electron density. The explanation is that the oxidizing ion, in a collision with a Ps atom, captures the electron (as it

---

[19] The sum of the energies of the gamma rays is 1.02 MeV, or 0.51 MeV per photon in 2γ annihilation. In 3γ annihilation, the 1.02 MeV is divided among the photons in a way that depends upon the relative angles at which they are emitted; symmetrical emission gives each photon 0.34 MeV.

[20] Because each positron lives for a very short time, two positrons are rarely present at the same time, so the chance of measuring a spurious "lifetime" by matching the wrong gamma rays is very small. If no annihilation photon is detected within a reasonable time after the creation signal, the clock is reset and the event is not recorded.

[21] T. C. Griffith, *Advances in Atomic & Molecular Physics*, **15** (1979).

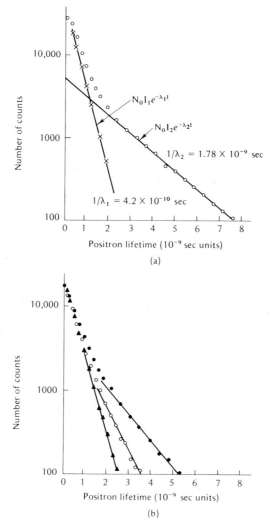

Fig. 7. *Lifetime distributions of positrons annihilating (a) in* H^2O, *and (b) in aqueous solutions of* $KMnO_4$ *in various concentrations. Curve (a) illustrates separation of lifetime distribution into two components.* (○) *Total counts;* (×) *counts remaining after subtraction of long-lived component.* [See Eq. (18).] *Curve (b) shows faster decay of long-lived component, because of oxidation of* Ps *by* MnO_4^- *ions, the oxidation rate increasing with* $KMnO_4$ *concentration.* (▲) *1/8M;* (○) *1/32M;* (●) *1/128M.*

tends to do when encountering any ordinary atom under these circumstances). The result is a bare positron, which is then annihilated more rapidly than the original ortho-Ps atom would be. Many experiments of this kind are now done by chemists in an effort to obtain unique information about the nature of chemical reactions in solutions.

The most obvious way to detect the presence of positronium would seem to be through observation of its optical spectrum, but success in this direction was not reported until 1975.[22] The positronium spectrum should be given by the same equation found for the hydrogen spectrum [Eq. (20), Chapter 3]; however, the reduced mass for positronium is $m_e/2$, making each frequency emitted by positronium very nearly one-half of the corresponding frequency in the hydrogen spectrum. The Ps Lyman-α line, the highest-frequency line of the Lyman series, should have a wavelength of 2430 Å; this was observed, and the fine structure of the line was also measured.

Positron Lifetimes in Solids; Defects. When a positron enters a solid, it reaches thermal equilibrium in a time of the order of picoseconds. Its wavelength then is of the order of 100 Å, which is many times the lattice spacing, so all positrons in a homogeneous solid should be expected to have the same environment and thus the same annihilation rate. This expectation is borne out in well-annealed metals, in which the positron lifetime distribution is a single exponential. But when defects are introduced (via fatigue, irradiation, or heat), the mean positron lifetime is increased, and careful measurements of the lifetime distribution show that it is no longer a single exponential—different positrons are seeing different environments.

Surprisingly, the positron mean lifetime can be noticeably lengthened by a concentraton of *single vacancies* as small as one in 10^6. (That is, one atom out of 10^6 is missing from its usual place in the lattice.) To explain such an effect from such a small concentration of vacancies, one must assume that positrons are *trapped* at vacancies, creating a second distinct population of positrons. This population sees a smaller electron density than that seen by the untrapped positrons, so it lives longer. Trapping is possible because the absence of the positively charged ion core leaves a net negative charge at the vacancy, creating a potential well that is often deep enough and wide enough to keep a positron in a bound state.

The overall positron lifetime distribution can become very complicated when a variety of defects are present, because positrons trapped in each kind of defect form a distinct population with its own particular decay constant λ. (For example, in aluminum it has been calculated that a positron in a single vacancy has a mean lifetime of 230 ps, and a positron in a cluster of eight vacancies has a mean lifetime of 400 ps.) But some situations can be analyzed on the basis of trapping into a single kind of defect where the positron decay constant is well defined.

---

[22] K. F. Canter, A. P. Mills, Jr., and S. Berko, *Phys. Rev. Lett.* **34**, 177 (1975).

When it is initially thermalized (defined as time $t = 0$), we assume the positron to be free (that is, bound to the whole solid, but free to diffuse through it), because the defects occupy such a small proportion of the whole solid. These free positrons are then either annihilated (annihilation rate λ_f) or trapped (trapping rate γ). After trapping, the annihilation rate is λ_t. Thus the number of free positrons (N_f) and trapped positrons (N_t) obeys the following equations:

$$dN_f/dt = -\lambda_f N_f - \gamma N_f$$
$$dN_t/dt = +\gamma N_f - \lambda_t N_t$$

With the conditions $N_f = N_0$ and $N_t = 0$ at time $t = 0$ (i.e., no positrons trapped until after they are thermalized), the solution is that the total number of positrons surviving at time t is

$$N = N_f + N_t = N_0(Ae^{-\lambda_1 t} + Be^{-\lambda_t t}) \tag{18}$$

where

$$A = 1 - \frac{\gamma}{\lambda_1 - \lambda_t} \quad , \quad B = \frac{\gamma}{\lambda_1 - \lambda_t} \quad , \text{ and } \quad \lambda_1 = \gamma + \lambda_f$$

Thus the population of positrons is the sum of two exponentials, with mean lifetimes of $\tau_1 = (\lambda_f + \gamma)^{-1}$, and $\tau_2 = \lambda_t^{-1}$, respectively.[23]

The positron technique has become increasingly useful as a nondestructive test for defects. It does not require the cutting off of an extremely thin sample for study (as does transmission electron microscopy), and it is possible to do studies on a single sample while defects are being produced or eliminated in that sample through fatiguing, heating, or annealing.

Angular Correlation of Annihilation Radiation from Solids. One of the first applications of positron annihilation was to study the momentum distribution of valence electrons in a metal. This is made possible by a combination of circumstances governing positron annihilation in metals:

1. The positron reaches thermal energy in about 1 ps, so its momentum is almost always negligibly small in comparison with the momentum of a typical electron.

2. A large fraction of the annihilations occurs with valence electrons, and this fraction is peaked more strongly near $\theta = 0$ (Fig. 6) than are the events from electrons in the atomic cores (because the core electrons generally have more momentum).

---

[23] It may seem strange to you that τ_1 is shorter than the mean lifetime λ_f^{-1} that is observed in the pure sample. The reciprocal of τ_1 is the *disappearance rate* of positrons from the free state only; this rate, $\lambda_f + \gamma$, is a *combination* of the annihilation rate and the trapping rate, making τ_1 shorter than it is in the pure sample, where only annihilation occurs. Many of the positrons still exist in the trapped state after disappearing from the free state.

3. The probability of annihilation with a given valence electron is almost independent of the electron's momentum.

These circumstances mean that the momentum distribution (angular correlation) of the gamma rays contains a component that is readily separable and that closely follows the momentum distribution of the valence electrons. This component has the expected shape (see Problem 11) and it cuts off at an angle θ_f corresponding to the Fermi momentum p_f of the valence electrons:

$$
\begin{aligned}
\theta_f = p_f/mc &= \frac{(2m\varepsilon_f)^{1/2}}{mc} \\
&= \frac{(2mc^2\varepsilon_f)^{1/2}}{mc^2} \\
&= (10^6\varepsilon_f)^{1/2}/(5 \times 10^5) \\
&= (\varepsilon_f)^{1/2} \times 2 \times 10^{-3}
\end{aligned}
$$

where ε_f, the Fermi energy, is in electron volts, and θ is in radians. If we apply the formula (Eq. (33), Chapter 11) for ε_f in terms of the valence electron density ρ, we find that $\theta_f = 1.195 \times 10^{-7}$ mrad cm $\times \rho^{1/3}$. Figure 8 shows the striking agreement between this equation and the experimental results; these results give the most direct verification of the validity of the Fermi gas model of electrons in metals. Angular correlation measurements with improved resolution have permitted exploration of the electron momentum distribution in sufficient detail to go beyond the Fermi gas model and to see Brillouin zone effects of the sort described in Section 12.5. These measurements have been greatly aided by the development of detection systems that record both the p_x and the p_y of the gamma-ray pair (Fig. 6). These systems have an efficiency that is independent of p_z, so the number of counts $N(p_x,p_y)$ for a given pair of values (p_x,p_y) is proportional to the number of occupied states along a line in momentum space characterized by those two values (i.e., parallel to the p_z axis).

If the Fermi surface marks the boundary between occupied and unoccupied states, with no holes inside the occupied region (in the extended-zone scheme, Section 12.5), then $N(p_x,p_y)$ is proportional to the length of the above-mentioned line inside the Fermi surface, and the set of such numbers, properly normalized, will map out the Fermi surface. Figure 9 shows how well this has been done for the case of copper. (Compare with Fig. 18, Chapter 12.) The advantage of this technique is that it can be applied to alloys as well as to pure metals (in contrast to other methods requiring observation of complete electron *orbits,* which can be broken up by the presence of "foreign" atoms in an alloy).

In contrast to the correlations obtained from metals, data from some other solids exhibit quite a different shape, whose explanation provides a good illustration of basic quantum mechanics. Figure 10 shows the angular correlation from a

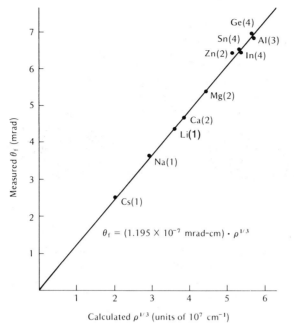

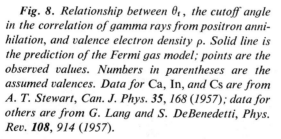

$$\theta_t = (1.195 \times 10^{-7} \ \text{mrad-cm}) \cdot \rho^{1/3}$$

Calculated $\rho^{1/3}$ (units of 10^7 cm^{-1})

Fig. 8. *Relationship between θ_t, the cutoff angle in the correlation of gamma rays from positron annihilation, and valence electron density ρ. Solid line is the prediction of the Fermi gas model; points are the observed values. Numbers in parentheses are the assumed valences. Data for* Ca, In, *and* Cs *are from A. T. Stewart, Can. J. Phys.* **35**, *168 (1957); data for others are from G. Lang and S. DeBenedetti, Phys. Rev.* **108**, *914 (1957).*

single crystal of ice at a temperature T of about 80 K. The sharp peak results from annihilation of Ps atoms, whose center-of-mass momentum is much smaller than the momentum of an electron in a crystal. The width of the peak suggests a maximum center-of-mass momentum of the order of $10^{-3} \ m_e c$. By the uncertainty principle, a Ps atom with such a momentum must have a wavefunction that spreads out over a distance of *at least* $\hbar/p \simeq 10$ Å, which is more than twice the lattice spacing in ice. So the lattice modifies the Ps wavefunction, giving it the character of a Bloch wave [Section 12.3, Eq. (7)], containing a factor whose period is equal to the lattice spacing a. By the Fourier Integral Theorem (Section 4.3), and the deBroglie relation (Section 4.2), this introduces momentum components of magnitude $p = h/a$ into the system. The annihilation process, giving the gamma rays a definite momentum, constitutes a momentum *measurement* in which the result $p = h/a$ must therefore appear in some fraction of the events.

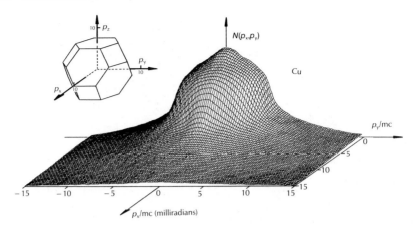

Fig. 9. *Two-dimensional angular correlation of annihilation radiation from a single crystal of copper. Number of counts as a function of p_x and p_y is shown in the vertical direction. Orientation of crystal is shown by the Brillouin zone in the inset. (From S. Berko, M. Haghgooie, and J. J. Mader, Physics Letters **63A**, 335 (1977).)*

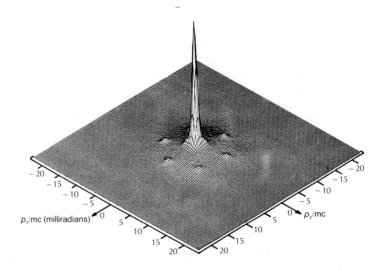

Fig. 10. *Two dimensional angular correlation of positron annihilation radiation from a single crystal of ice. Horizontal scales show $p/m_e c$ in units of five milliradians. (From R. N. West, J. Mayers, and P. A. Walters, J. Phys E **14**, 478 (1981).)*

The six smaller peaks in Fig. 10 (one hidden behind the main peak) demonstrate this effect, and their locations are in agreement with the known hexagonal crystal structure of ice at that temperature.

If the concept of a bound state that is spread out over a crystal lattice seems strange to you, remember that there is a precedent for such a thing. The Cooper pair of superconductivity theory (Section 12.4) also interacts with the lattice on a large scale (i.e., with many atoms).

Positrons in Medicine.   Gamma rays have long been used for medical diagnosis. A patient ingests a gamma emitter that has some chemical affinity for a specific organ (e.g., ^{131}I for the thyroid). If a tumor, which is more chemically active, is present, the gamma rays will be seen to emanate more strongly from the region of the tumor. Unfortunately, one cannot focus gamma rays to form an image, so to know where they are coming from, one must use lead collimating shields in front of the detector and many of the gamma rays are absorbed in the lead.

A more efficient alternative is to use a positron-emitting isotope in conjunction with two detectors, one on each side of the patient. Each detector is a so-called "scintillation camera," which can indicate, within a few millimeters, the x and y coordinates of a gamma ray interaction in the camera (the z axis being the direction perpendicular to the face of the camera). When an event is recorded in both cameras simultaneously, it is known that the positron annihilation occurred on the straight line joining the points recorded by the camera, because on the scale of this measurement the angle θ is negligible. A computer registering all of these events can build up a very good picture of the volume in which the source is concentrated.

This method, although it requires more equipment than the simple one-detector technique, is considerably safer, because, with no lead collimators needed, the same number of events can be recorded from a much weaker radioactive source.

13.4   *RECOILLESS RESONANT ABSORPTION OF GAMMA RAYS (MÖSSBAUER EFFECT)*

Photons can be absorbed by matter in another way which was not mentioned in Section 13.2. In this process a photon's energy is used to excite a nucleus to one of its higher energy levels. Because most of these energy levels are quite sharply defined, there is a very small probability that a random photon has precisely the energy required for all this so-called resonance, and it is impossible to depict the resulting gamma-ray absorption on a graph like those of Fig. 5. Nevertheless, the possibility of resonant absorption is of astonishing import, because through it one can compare gamma ray frequencies to an accuracy of one part in 10^{13}. This accuracy makes possible a wide variety of measurements—in nuclear physics, solid state physics, and chemistry—whose accomplishment was once considered to be an idle dream.

Resonance absorption of photons was known long ago on the atomic level, usually involving visible light. The "Fraunhofer lines," a series of dark lines in the spectrum of light from the sun, were observed by Fraunhofer in 1814; these result from selective absorption of certain wavelengths by gas in the sun's atmosphere. A "resonance" in the absorption occurs whenever the energy of a photon is exactly that required to cause a transition of an atom of gas from its ground state to an excited state. The excited atom subsequently returns to its ground state, usually[24] emitting a photon of the same frequency as the original photon, but in a random direction. The net effect is "resonant scattering" of the photon. This effect is easily observed in the laboratory; if light from a sodium flame is focused on a bulb containing sodium vapor, the bulb glows.

Recoil of the Nucleus. In principle, the same sort of effect should be possible with gamma radiation from nuclei. A gamma ray is emitted by a nucleus in a transition from one internal energy state of the nucleus to a lower energy state. If the gamma ray receives the full energy of the transition, a nucleus of the same species in the lower energy state should be able to absorb that gamma ray and go to the higher energy state. But conservation of momentum causes a problem. A nucleus recoils when it emits the gamma ray, so that some of the transition energy becomes kinetic energy of the recoiling nucleus instead of being given to the gamma ray. If this kinetic energy is greater than the natural linewidth of the radiation, the emitted gamma ray has insufficient energy to excite another nucleus (see Fig. 11). A short calculation will show why this effect is important in the case of nuclear radiation but could be ignored for atomic radiation.

We can assume the linewidths involved to be about the same in both cases; a typical mean life τ of about 10^{-7} sec leads to a linewidth $\Gamma = h/\tau$ of the order of 10^{-8} eV, according to the "uncertainty relation" for energy and time (Section 10.2). But the recoil energies are quite different simply because the emitted photons have such different energies. A free body of mass m, emitting a photon of energy $h\nu$, recoils with momentum $p = h\nu/c$, acquiring a kinetic energy equal to $p^2/2m$, or $(h\nu)^2/2mc^2$, Thus the energy of recoil is proportional to the square of the energy of the emitted photon. If we let $2mc^2$ have the typical value of the order of 10^{11} eV (corresponding to the mass of about 100 nucleons), and if the photon has energy 10^5 eV, which is quite typical for a nuclear gamma ray, then the recoil energy is

[24] If there is an energy level between the ground state and the excited state, the atom may return to the ground state via the intermediate level, emitting two lower-frequency photons. We have seen (Chapter 9) that such a sequence of events occurs in the operation of the laser and the maser.

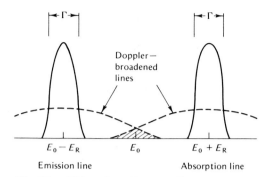

Fig. 11. *Effect of recoil energy E_R on spectrum of gamma ray energies. If nuclear transition energy is E_0, emission line appears at energy $E_0 - E_R$ because energy E_R is lost in recoil. Absorption requires energy $E_0 + E_R$, to provide energy E_R for recoil of absorbing nucleus. If natural linewidth Γ is much less than E_R, emission and absorption lines do not overlap and resonance absorption cannot occur. But if Doppler broadening of lines occurs (dotted lines), some overlap may occur (shaded area), and resonance is possible.*

$$\frac{(h\nu)^2}{2mc^2} = \frac{(10^5 \quad eV)^2}{10^{11} \quad eV}$$

$$= 0.1 \quad eV$$

On the other hand, if the photon's energy is only a few electron volts, as it is for atomic radiation, then the recoil energy is of order

$$\frac{10 \quad eV^2}{10^{11} \quad eV} = 10^{-10} \quad eV$$

In the latter case the photon's energy is decreased by far less than the natural linewidth as a result of the recoil, so that resonance absorption of the photon by another atom is still possible. But in the case of the nuclear gamma ray, the recoil eats up an energy which is 10^7 times as great as the natural line width.

Notice that the emitting and absorbing nuclei are out of resonance by *twice* the amount calculated above, because the second nucleus acquires the same amount of kinetic energy as the first nucleus, as a result of the recoil when the photon is *absorbed*. It is possible to compensate for the recoil energy by using the Doppler effect, which increases the frequency of a photon by an amount $\delta\nu$ approximately equal to $(v/c)\nu$ when the source and absorber approach one another with speed $v \ll c$. This increase compensates for the energy lost in recoiling, if

$$\frac{v}{c} = \frac{\delta v}{v} = \frac{h\,\delta v}{hv}$$

$$= \frac{\text{Energy lost by recoil}}{\text{Total transition energy}}$$

For the example cited here, v/c must equal $0.2 \text{ eV}/10^5$ eV, so

$$v = 0.2 \times 10^{-5} \times 3 \times 10^8 \quad \text{m/sec} = 600 \quad \text{m/sec}$$

Such a speed can be realized by using an ultracentrifuge, and resonant scattering was observed in this way by P. B. Moon in 1951.[25] Moon used a source of ^{198}Au which decays to an excited state of ^{198}Hg, and he observed resonant scattering of the 0.41-MeV gamma ray from the ^{198}Hg when the source moved toward the scatterer at a speed of about 700 m/sec.

Recoil of the Lattice. But such heroic measures are not absolutely necessary to achieve resonance. The analysis of the recoil energy has been carried out on the assumption that the nucleus is free; but the nucleus can be part of a crystal lattice. The lattice has two effects. First, lattice vibrations cause the nucleus to oscillate, so that Doppler shifts broaden the frequency range of the emitted gammas and the range of frequencies which may be absorbed. As indicated by Fig. 11, these shifts are large enough to make resonance possible in a small fraction of events. This resonant fraction might be expected to become smaller as the crystal is cooled and the Doppler broadening reduced, but in 1958, Mössbauer discovered that it does not always work that way. He found the second effect of the lattice, the Mössbauer effect,[26] which resulted in an increase rather than a decrease in resonant absorption when his source was cooled. The Mössbauer effect depends on the fact that, in a certain fraction of events, the *entire lattice* recoils as a rigid body when a γ ray is emitted or absorbed. We call such events "recoilless"; they are not strictly "recoilless," but the recoil *energy* $p_\gamma^2/2M$ becomes negligible in such events because M is the huge mass of the lattice. If the recoil energy is negligible in both emission and absorption, we have resonant absorption. As the lattice is cooled and becomes "more rigid," the fraction of "recoilless" events increases. This effect had already been seen in X-ray scattering, where the intensity of Bragg diffraction lines depends on the fraction of "recoilless" events. The temperature dependence of this intensity had been thoroughly investigated; the factor governing the effect was well known as the "Debye–

---

[25] P. B. Moon, *Proc. Phys. Soc. (London), Ser. A* **64**, 76 (1951).
[26] Mössbauer himself calls it "recoilless resonant absorption," but everybody else calls it the Mössbauer effect.

Waller factor.'' But curiously, nobody before Mössbauer had thought to look for similar effects involving radiation from nuclei.[27]

Although some features of the Mössbauer effect can be understood by using a classical approach, it may be less confusing simply to discuss the general features in terms of transitions of a quantum mechanical system. The system consists of the entire lattice containing the nuclei which emit or absorb the radiation. The states of this system are characterized by the phonons, which carry lattice vibrations, and by the energy levels of the nuclei. (We neglect the electron states, which are irrelevant to this discussion). When an emitted gamma ray carries energy and momentum away from the system, conservation of energy and conservation of momentum impose separate requirements on the state of the remaining *system*.[28]

In *all* events, conservation of *momentum* is eventually achieved by recoil of the entire lattice. Any shock wave triggered by the gamma emission is finally damped out, leaving the recoil momentum as translational momentum of the entire lattice. But the *energy* of the transition is shared by the gamma ray, the lattice vibrations, and translation of the lattice as a whole. (Translation of the nucleus within the solid does not occur; the nucleus is too tightly bound to the lattice). We have seen that the translational energy of the solid is negligible when its momentum equals the gamma ray momentum. Therefore the gamma ray receives less than the full transition energy only when a lattice vibration is excited by emission of the gamma ray; i.e., *only when a phonon is emitted simultaneously with the emission of the gamma ray.*

Doppler Broadening.   We have already seen in the theory of specific heats the lattice vibrations are quantized and cannot possess arbitrary amounts of energy. If the solid were an "Einstein solid," which vibrates at a single characteristic frequency ω_E, the smallest amount of energy which the lattice vibrations could acquire would be $\hbar\omega_E$. It has been shown[29] that the *average* amount of energy which the solid receives per event is equal to E_R, the free atom recoil energy. If $E_R \ll \hbar\omega_E$, we may neglect the probability that two or more phonons are emitted. The correct average energy is then obtained if a phonon of energy $h\omega_E$ is emitted in the fraction $E_R/\hbar\omega_E$ of events, and no phonon is emitted in other events. The energy spectrum of the emitted gamma rays then looks like Fig. 12. This analysis, of course, applies only when the temperature of the lattice is $0°K$; at higher temperatures there is also the

---

[27] Mössbauer was looking for something different when he began his experiments. But he knew what he had when he found it.

[28] It is confusing and erroneous to consider that energy and momentum are given to the nucleus (in its recoil) and that there is a subsequent transfer of energy and momentum to the lattice. The nucleus is always ''aware'' that it is in a lattice, and its behavior can only be described as part of a lattice. For a good discussion of this point, see L. Eyges, *Amer. J. Phys.* **33**, 790 (1965).

[29] H. J. Lipkin, *Ann. Phys. (New York)* **9**, 332 (1960).

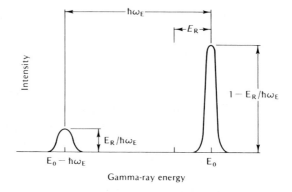

Fig. 12. *Spectrum of gamma ray energies from an "Einstein solid," if the free atom recoil energy E_R is much smaller than the phonon energy $\hbar\omega_E$.*

possibility of a gamma ray energy of $E_0 + \hbar\omega_E$, corresponding to *absorption* of a phonon as the gamma ray is emitted!

The spectrum of gamma rays from an actual solid (iridium) at nonzero temperatures is shown in Fig. 13. The source of the gamma rays is ^{191}Ir, the isotope originally used by Mössbauer. Because of the huge number of possible modes of vibration of the crystal, the phonon spectrum appears to be continuous. Nevertheless, a significant fraction of the total events appears in the sharp zero-phonon peak. The "Doppler broadening" seen in Fig. 11 now appears simply as the "phonon wing." Notice the small part of this wing which extends to positive energy shifts and results from absorption of phonons; it was this part which enabled resonant absorption to be observed at high temperatures before Mössbauer.

Our picture of gamma ray emission from a crystal is thus quite different from the free atom picture with which we started. Instead of a single line which is Doppler-broadened by the motion of the atom, we have a whole series of lines resulting from gamma decays accompanied by emission or absorption of phonons of various energies. Doppler broadening does not spread an individual line; it simply broadens the *overall distribution* by allowing more lines to appear in the spectrum (in the phonon wing). As the temperature is raised and more lines become possible, the intensity of the zero-phonon line is reduced, until finally the Mössbauer effect "disappears"; that is, the zero-phonon line becomes too weak to detect.

Since the zero-phonon line *itself* is *not* Doppler-broadened, it can appear

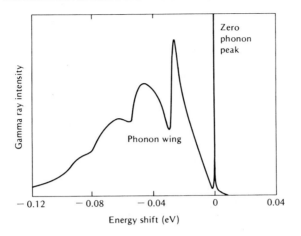

Fig. 13. *Spectrum of gamma ray energies from a real solid. In this case the zero-phonon line contains 5.7 percent of the area under the curve; it is too narrow and high to be shown accurately. [From G. K. Wertheim, "Mössbauer Effect: Principles and Applications," Academic Press, New York, 1964, based on calculations by W. M. Visscher, Ann. Phys. (New York) 9, 194 (1960).]*

with a width close to its natural linewidth.[30] For the 14.4-keV gamma ray of ^{57}Fe, the natural linewidth is about 10^{-8} eV, or less than 10^{-12} times the energy of the line. This is the exciting feature of the phenomenon, which allows the observation of small effects previously thought to be undetectable. Proof of the narrowness of the line is seen in Fig. 14, which shows the effect of moving the source at a small velocity relative to the absorber. A counter located behind the absorber detects the gamma rays which are *not* absorbed; therefore the counting rate is smallest when the source and absorber are at rest and resonant absorption occurs. When the source is moved, the Doppler shift destroys the resonance, so that the counting rate increases.

You can see from Fig. 14 that a source speed of slightly less than 0.02 cm/sec cuts the resonant absorption in half. This speed, being $6 \times 10^{-13}c$,

---

[30] The observed linewidth is often greater than the natural linewidth, because of effects of the lattice on the nuclear energy levels. Inhomogeneity of the crystal may result in different energy levels for different nuclei, effectively broadening the line, or internal electric fields may interact with the quadrupole moment of the nucleus to split the line. (Unresolved splitting appears as broadening.)

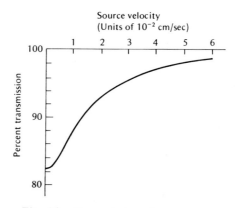

Fig. 14. *Transmission of gamma rays of ¹⁴Fe through ⁵⁷Fe absorber, as a function of source velocity. Transmitted intensity is shown as a percentage of transmission at large velocity. At $v = 0$, maximum resonant absorption occurs and transmission is minimum (82.5 percent in this case). [Data from R. V. Pound and G. Rebka, Jr., Phys. Rev. Lett. 3, 554 (1959).]*

produces a frequency shift of 6×10^{-13} of the original gamma ray frequency, just about what you would expect from the fractional linewidth quoted above.

The Gravitational Red Shift. The most fundamental application of this extreme precision in gamma ray energy has been a measurement of the so-called "gravitational red shift"—the effect of gravity on the frequency of a photon. According to the "principle of equivalence," which is the starting point for Einstein's general theory of relativity, a clock in a gravitational field should run slowly, by an amount proportional to the magnitude of the gravitational potential. Therefore a photon emitted as a result of a certain transition in such a potential should have a lower frequency than a photon resulting from the same transition at a higher gravitational potential (or in gravity-free space).

The effect, illustrated in Fig. 15, depends on the equivalence of gravitational and inertial mass and on the conservation of energy. In the cycle shown, an atom in the earth's gravitational field absorbs a photon of frequency v; then the atom falls a distance d, emits a photon, and is raised a distance d, ending in its original state. There are two net effects: (1) A photon at point A is replaced by a photon at point B, a distance d below A. (2) Since the falling

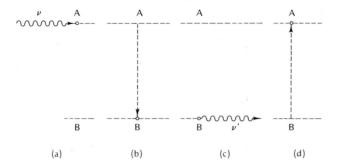

Fig. 15. *The "apparent weight of photons." (a) Atom at point A, in ground state, with mass m, absorbs a photon of frequency v; mass of atom becomes $m + (hv/c^2)$. (b) Atom falls to point B. (c) Atom emits a photon of frequency v' and returns to ground state. (d) Atom is raised back to point A. Atom was more massive during stage (b) than during stage (d); therefore net work was done during the cycle. This work could only have resulted from the photon energy; thus the photon absorbed at A was of higher energy (and higher frequency) than the photon emitted at B.*

atom is more massive[31] than the atom which is raised, a net amount of work is done, equal to gd times the difference in mass, hv/c^2. If energy is to be conserved, the energy of the photon emitted at B must be smaller than the energy of the photon absorbed at A, by the amount $hvgd/c^2$. Thus the characteristic frequency of the transition at B is not v, but rather v', such that

$$hv' = hv - \frac{hvgd}{c_2}$$

$$= hv\left(1 - \frac{gd}{c^2}\right)$$

The frequency v' can be compared with v by means of the Mössbauer effect. The most successful attempt to do this was by Pound and Rebka,[32] who used an ^{57}Fe source[33] and an ^{57}Fe absorber at top and bottom of a 22-m tower. The fractional frequency difference between source and

---

[31] Here is where the equivalence of inertial and gravitational mass comes in. The special theory of relativity simply says that the *inertial* mass increases when a photon is absorbed; now we say that this mass is acted on by gravity.

[32] R. V. Pound and G. Rebka, Jr., *Phys. Rev. Lett.* **4**, 337 (1960). See also R. V. Pound, *Sov. Phys.—Usp.* **3**, 875 (1961).

[33] The actual source used was ^{57}Co, which decays to the excited state of ^{57}Fe, from which the gamma ray is then emitted.

absorber lines should in this case be $gd/c^2 = 9.8 \times 22/(3.0 \times 10^8)^2 = 2.4 \times 10^{-15}$. This difference, which is far less than the fractional linewidth, was detectable because of the careful technique used by Pound and Rebka. Because the principle involved in this technique is of broad applicability, it is instructive to study it in some detail.

The source was oscillated upward and downward, and the number of transmitted gamma rays was counted when the source had its maximum upward velocity and when it had its maximum downward velocity. These velocities were chosen so that they corresponded to the steep parts of the transmission curve, giving the greatest possible change in counting rate for a given change in velocity (see Fig. 16). Because the characteristic frequency of the source differed from that of the absorber, the entire absorption curve was shifted, maximum absorption occurring when the source had some non-zero velocity. This velocity could be found indirectly by comparing the counting rates observed at equal positive and negative velocities provided by the oscillation of the source. If the exact shape of the curve of counting rate versus velocity were known, the counting rate difference could immediately be translated into a velocity and the shift in the absorption curve thereby determined. But a more reliable method was adopted; after the original counting rate difference was measured, a constant velocity of 6×10^{-4} cm/sec (a few times the expected shift) was superimposed on the oscillation of the source, and the counting rate difference measured again. This made possible

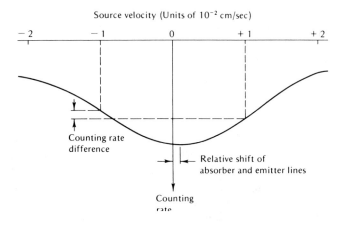

Fig. 16. *Effect of shift in absorption curve on gamma ray counting rates. Because of the shift, equal positive and negative velocities result in unequal counting rates. The effect on counting rates is greatest when the velocity corresponds to the steepest part of the curve.*

a calibration which showed the relation between the change in counting rate and the change in source velocity. Over such a small range of velocities, the counting rate curve could be assumed to be linear to a high degree of accuracy (especially in this region near the inflection point); the shift in the curve could therefore be calculated easily.

The shift in the absorption curve does not give the gravitational frequency shift directly, because the characteristic frequency of the source may differ from that of the absorber for other reasons. In the Pound–Rebka experiment, it was found that the characteristic frequency of all absorbers used differed from that of the source by several times the expected gravitational shift. This was apparently caused by slight differences in the lattice structures of source and absorbers, which caused different perturbations in the nuclear energy levels. It was, however, possible to isolate the gravitational shift by interchanging source and absorber in the tower and comparing the observed shifts. For example, in the first four days of counting, with the source at the top and the absorber at the bottom, the gamma ray frequency was too low for maximum absorption, by 15.5 ± 0.8 parts in 10^{15}, while with the source at the bottom and the absorber at the top, the gamma ray frequency was low by 19.7 ± 0.8 parts in 10^{15}. The mean shift of 17.6 parts in 10^{15} was therefore characteristic of the source–absorber combination, while the deviation of 2.1 parts in 10^{15} was the effect of the gravitational red shift and is quite close to the expected value. The reader may verify that the shift was in the right direction. After four months of counting, the observed gravitational shift equalled $(0.98 \pm 0.04)gd/c^2$, in excellent agreement with theory.[34]

Temperature Dependence of Gamma Ray Energy.   In the course of this experiment, Pound and Rebka discovered a complication which is of fundamental importance; they found that the frequency of the emission or absorption line is temperature dependent. They therefore were forced to attach thermocouples to both source and absorber and to correct the data for the effect of the observed temperature difference. (The data given above are the corrected data.) The significance of this temperature effect, as Pound and Rebka showed, is that it is evidence of time dilation, produced by the speed of the oscillating nucleus in the lattice. Let us examine this effect more closely.

A nucleus which is moving with mean-square velocity of $\overline{v^2}$ emits radiation whose frequency is low by the time dilation factor

$$\left(1 - \frac{\overline{v^2}}{c^2}\right)^{1/2} \qquad \text{or} \qquad 1 - \frac{\overline{v^2}}{2c^2} \qquad \text{for} \qquad \overline{v^2} \ll c^2$$

(This relativistic effect is sometimes called the "second-order Doppler effect"

[34] R. V. Pound, *Sov. Phys.—Usp.* **3**, 875 (1961).

because it depends on the square of the speed, or the "transverse Doppler effect" because it is observed even when the source is moving transversely to the line of sight.) The mean-square velocity may be found by using the fact that the mean kinetic energy of an oscillator is one-half of its total energy, so that $\frac{1}{2}M\overline{v^2} = \frac{1}{2}U$, where M is the mass of an atom in the lattice and U is the oscillation energy per atom. The fractional frequency shift should therefore be

$$\frac{\delta v}{v_0} = \frac{\overline{v^2}}{2c^2} = \frac{U}{2Mc^2}$$

The variation of frequency with temperature is then

$$\frac{\partial v}{\partial T} = \lim_{\delta T \to 0} \frac{\delta v}{\delta T}$$

$$= \frac{v_0}{2Mc^2} \frac{\partial U}{\partial T}$$

or

$$\frac{\partial v}{\partial T} = \frac{v_0 C_{\mathrm{L}}}{2Mc^2} \tag{19}$$

where v_0 is the unshifted frequency of the gamma ray and C_{L} is the lattice specific heat, which may be found from the Debye theory. Pound and Rebka measured the temperature dependence of the frequency shift for their absorbers and found excellent agreement with Eq. (19) over absorber temperatures ranging from 88 K to 320 K, on the assumption that the Debye temperature for their absorber was 420 K. Thus we have still another confirmation of special relativity, which has since been tested by others and is now the basis of an undergraduate laboratory experiment.[35]

Chemical Shift. Space permits only a brief survey of the many further applications of the Mössbauer effect. Before Mössbauer, effects of the environment on the nuclear energy levels were not often discussed at all, for it was generally assumed that such effects were so small that they would never be detected. One such effect is the chemical shift (sometimes called "isomer shift"), which causes gamma ray energies to vary with the chemical compound in which the nucleus resides. This effect is a result of the electrostatic interaction between the atomic electrons and the nucleus. If the distribution of atomic electrons changes as a result of a chemical change, the interaction with the nucleus changes and the nuclear energy levels shift slightly. Study of these shifts by means of the Mössbauer effect gives information not only about the electron distribution but also about the size of the

---

[35] J. W. Weinberg, B. L. Robinson, J. K. Major, and U. O. Herrmann, *Am. J. Phys.* **34**, 184 (1966).

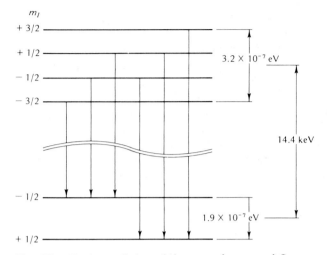

Fig. 17.   *Zeeman splitting of the ground state and first excited state of ^{57}Fe in solid iron, showing the six allowed transitions.*

nucleus, *in the excited state* as well as in the ground state. For example, it was found that the ^{57}Fe nucleus is *smaller* in its first excited state than it is in its ground state. Such information is obviously of great interest in nuclear theory.

Nuclear Zeeman Effect.   The Mössbauer effect has also been used to measure the magnetic fields at the nucleus, especially in ^{57}Fe; the magnetic field strength as well as the magnetic moments of the nucleus in the excited state and in the ground state can be found by observing the Zeeman splitting of the energy levels of each state. The magnetic energy which produces the Zeeman splitting is $-\mathbf{\mu} \cdot \mathbf{B}$, where $\mathbf{\mu}$ is the magnetic moment of the nucleus and $\mathbf{B}$ is the magnetic field at the nucleus. For a nucleus of spin I there are $2I + 1$ equally spaced levels, with quantum numbers $m_I = I, I - 1, \ldots, -I$, in exact analogy to the atomic case. Figure 17 shows the splitting for ^{57}Fe in a typical case; the excited state, of spin $\frac{3}{2}$, splits into four levels, while the ground state, of spin $\frac{1}{2}$, splits into two levels. Notice that the levels are reversed in the ground state relative to the excited state, with $m_I = -\frac{1}{2}$ lying above $m_I = +\frac{1}{2}$, indicating that the magnetic moments are in opposite directions in the two states. The allowed transitions are shown in the figure; the selection rule in this case is $\Delta m_I = 0$ or ± 1. Since the energies of these transitions differ from one another by considerably more than the linewidth, the six lines are easily observed.

13.5 BIOLOGICAL EFFECTS OF RADIATION

The effect of radiation on any given organism or cell depends not only on the energy of the radiation, but also on its nature (whether charged or uncharged particles, heavy or light particles, strongly or weakly interacting particles). For example, neutrinos (Section 15.2) interact with matter only through the so-called "weak interaction"; a beam of neutrinos millions of times as intense as any presently attainable laser beam would have no observable effect on any living thing, because less than one part in 10^{19} of the energy would be absorbed.

The most significant quantity for determining the biological effect of radiation is the *dose,* defined as the energy *absorbed* per unit mass of material. The unit of dose is the rad, defined by

$$1 \text{ rad} = 100 \text{ erg/g}$$

Human beings at sea level receive an average dose of 0.17 rads per year from all sources: cosmic rays bombarding earth from outer space; radioactivity in rocks, soil, buildings, the atmosphere, and the body itself; and medical and dental X rays. Cosmic radiation alone accounts for 0.03 rads per year at sea level, and it doubles with each 1.5 km increase in altitude. Thus the residents of Denver (altitude $\sim$1.6 km) receive an average yearly dose of over 0.20 rads. (Some of them receive considerably more than that because of the deposits of uranium in that area.) Persons flying on jet aircraft (altitude $\sim$11 km) receive a dose *rate* of over 2^7 times the sea-level rate, or about 4 rads per year. On a single trip of two hours duration, that would result in a total dose of only 0.001 additional rads. But a pilot who flies 1000 hours per year at that altitude receives an additional dose of about 0.5 rads; his total radiation exposure is therefore four times that of the average sea level resident.

Much of the radioactivity in buildings comes from decay of the radium that is a natural constituent of building materials; this releases radon gas (^{222}Rn), an alpha emitter with a half life of 3.83 days. Ironically, attempts to seal buildings tightly, to prevent energy losses, have led to buildup of radon gas to unacceptable levels.[36] Other radioactive gases that occur naturally in the atmosphere are ^{3}H and ^{14}C, both of which are produced by interactions of cosmic rays with atmospheric nitrogen. (see Problem 19.)

Let us now see how to determine dose rates for different types of radiation; then we will be able to discuss the factors that determine the biological effect of any given dose.

Gamma Irradiation. Most of the radiation dose that we normally receive

---

[36] "Indoor Air Pollution," *Science 80,* **30**, March–April 1980.

comes from gamma rays or X rays. To relate the dose rate to the intensity of radiation from these "rays," we use Eq. (15), which can be written in terms of intensity as

$$dI = \mu I \, dx$$

whose solution is

$$I = I_0 e^{-\mu x}$$

where x is the depth penetrated by the radiation in a medium whose absorption coefficient is μ, and I_0 is the intensity where $x = 0$.

If this radiation is incident upon a slab of thickness dx, density ρ and area A, then the total energy absorbed per unit time must be $A dI$, or $A \mu I dx$. The mass in which this energy is absorbed is $\rho A dx$, so the dose rate is

$$\text{Energy absorbed per sec per gm} = I \mu A dx / A \rho dx$$
$$= I \mu / \rho$$
$$= I \times \text{mass absorption coefficient}$$

which can be converted to rads/sec by using the factor 1 rad/(100 erg/g). For example, let us find the dose rate delivered to a slab of lead by a flux of 10^6 photons/cm^2-s, if each photon has an energy of 1.0 MeV. From Fig. 5, we see that the mass absorption coefficient at 1 MeV is approximately 0.04 cm^2/g. The intensity I is 10^6 MeV/cm^2-s, or 1.6 erg/cm^2-s, so the dose rate is

$$0.04 \text{ cm}^2/\text{g} \times 1.6 \text{ erg/cm}^2\text{-s} \times 1 \text{ rad}/(100 \text{ erg/g}) \simeq 6 \times 10^{-4} \text{ rad/s} .$$

This is the dose rate at the surface of the slab; the dose rate will be smaller inside the slab, because I is smaller there.

One can determine how great an energy *fluence* (energy per unit area) is required to deliver a dose of 1 rad to a thin slab of any given material, if one knows the absorption coefficient as a function of photon energy E_γ for that material. The result for water is of special interest, because organisms are mostly water. For photons in the energy range 0.05 MeV $< E_\gamma <$ 2.0 MeV, $\mu \approx 0.03 \text{ cm}^{-1}$, and therefore

$$I \approx I_0 e^{-0.03x} \quad , \ x \text{ in cm}$$

From an external source, tissue lying at a depth x within the body receives a dose in accordance with the intensity at that depth. At a depth of one cm, $I \approx 0.97 I_0$, so the dose is 97% of the dose received by the surface tissue. We can relate the surface dose to the incident fluence by noting that a one-cm cube of water has a mass of 1 g, and it must *absorb* 100 erg to receive a dose of 1 rad. Since it absorbs only 3% of the incident energy, the incident energy must be 100 erg/ 0.03, or about 3300 erg over the area of 1 cm², making the fluence about 3300 erg/cm² for a surface dose of 1 rad.

If there are N radioactive atoms whose decay constant is λ, it follows from Eq. (1) that λN atoms decay each second. The number of decays per second is called the *activity*, usually measured in Curies; 1 Curie = 3.7×10^{10} decays/s. For a gamma emitter, we can now relate the activity to the dose rate at a given distance. The energy flux (fluence per unit time) is the product of the activity and the energy emitted per decay, divided by the area $4\pi d^2$ over which this energy is distributed at a distance d. Since approximately 3300 erg/cm² yield 1 rad, we can find the number of rads per second by dividing the energy flux (in erg/cm²-s) by 3300. Thus the dose in rad/s, at a distance of d cm from a source of S Curies, emitting E_γ MeV per decay, is

$$\text{Dose rate} = \frac{(3.7 \times 10^{10} S)(1.6 \times 10^{-6} E_\gamma)}{(3300)(4\pi d^2)} \simeq 1.4\, SE_\gamma/d^2 \quad (20)$$

Particle Irradiation.   Irradiation by *uncharged* particles follows the same exponential law that is obeyed by photons. However, there is a large variation in μ. For neutrinos, μ is in the range 10^{-17} to 10^{-22} cm^{-1}, making the *neutrino* dose rate negligible even inside a nuclear reactor. For *neutrons*, μ is in the range 10^{-4} to 10^{+3} cm^{-1} in most solids; it can be greatly influenced by traces of elements such as boron and cadmium that strongly absorb neutrons. Furthermore, as we shall see below, neutrons can be quite deadly, not only because of the large μ, but also because they usually interact with nuclei, rather than electrons.

Charged particles, as discussed in Section 13.1, lose energy continuously. Therefore one can compute the dose received from a charged particle beam as long as one knows, from Fig. 4 or by some similar means, how much energy is lost by the particles in a given distance. For example, we deduce from Fig. 4 that a 0.2 GeV proton loses 0.1 GeV in passing through about 160 meters of air. A fluence of 1 proton of this energy per cm² would therefore deliver an energy of 0.1 GeV = 10^8 eV = 1.6×10^{-4} erg to a volume of air that is 1 cm × 1 cm × 160 m. Such a volume has a mass of about 20 g, so the dose in rads is $(1.6 \times 10^{-4}$ erg/20 g) × (1 rad/100 erg-g^{-1}), or about 0.8×10^{-7} rad. A proton whose energy is only 0.1 GeV delivers a greater average dose to the air along its subsequent path, because it loses 0.1 GeV in only 70 m instead of 160 m. This makes the dose 160/70 times the above dose, or about 1.8×10^{-7} rad *per proton*.

Ingestion of Radioisotopes; Biological Half-Life.   Radiation with a high penetrating power, such as neutrons or gamma rays, is equally dangerous whether the source is inside or outside the body. But swallowing, inhalation, or injection of a source does create the additional problem that the irradiation continues as long as the source remains in the body. A further problem with alpha or beta emitters is that an internal source can irradiate organs that would be too deep inside the body to be reached by such radiation from an external source.

If any element or isotope is ingested once and not replenished, the amount in the body decays exponentially, because its removal from the body proceeds at a rate that is proportional to the amount that is present at any given time. Thus the number of atoms of a *stable* isotope in the body would follow the equation

$$dN = -\lambda_b N dt$$

or

$$N = N_o e^{-\lambda_b t}$$

where λ_b is determined by the body's own removal processes for that element. We can define the *biological* half life t_b as the time required to remove half of this isotope from the body by biological processes. This must be

$$t_b = (ln\ 2)/\lambda_b$$

If the isotope is radioactive, then its disappearance from the body is hastened by the radioactive decay process, and

$$dN = (-\lambda_b N - \lambda N)dt$$

or

$$N = N_o e^{-(\lambda_b + \lambda)t}$$

where λ is the usual radioactive decay constant. Thus we can also define an *effective* half life as

$$t_{eff} = (ln\ 2)/(\lambda_b + \lambda) = (t_{1/2}^{-1} + t_b^{-1})^{-1}$$

where $t_{1/2}$ is the radioactive half life.

For example, iodine has a biological half life of 138 days, and the radioactive half life of ^{131}I is 8.05 days, so for ^{131}I the effective half life is

$$t_{eff} = (138^{-1} + 8.05^{-1})^{-1} = 7.61\ days$$

Effective half lives of some important isotopes are shown below:

Isotope	$t_{1/2}$	t_b	t_{eff}
^{3}H	12 yr	12 days	12 days
^{14}C	5730 yr	10 days	10 days
^{24}Na	0.63 days	11 days	0.60 days
^{90}Sr	28.1 yr	50 yr	18 yr
^{131}I	8.1 days	138 days	7.6 days
^{239}Pu	24,400 yr	200 yr	198 yr

Quality Factor.   The biological effects of radiation are not related in a simple way to the energy absorbed because this energy can do various things, from outright destruction of a cell to temporary impairment of one of the cell's functions. The end result depends on the type of radiation and the cell's environment as well as on the magnitude of the dose delivered; it can also depend on how rapidly the dose is delivered.

Breaking a bond in a molecule can result in temporary or permanent damage. A molecule usually suffers irreparable damage if its nucleus is transformed by a nuclear reaction. This is more likely with heavy particle radiation than with electrons or photons.

An attempt has been made to quantify this difference in radiation effects by defining a ''quality factor'' for each kind of radiation. Multiplying the dose by the quality factor (QF) gives a more accurate measure of the expected effect from any given dose; the product of the dose in rads and QF gives the equivalent dose in ''rems.''

Unlike a physical constant, QF is not a precise number, but rather an estimate based upon comparison studies of persons who have been exposed to large doses of various kinds of radiation. For electrons and photons, QF = 1. For heavy particles, QF ranges from 1 to 20, being approximately 10 for alpha and neutron irradiation.

The ''normal'' dose to humans consists almost entirely of QF = 1 radiation, so the annual average equivalent dose in rems is 0.17 rem, the same as the average dose in rads.

Effects on Humans.   Use of the rem as a unit permits one to make general observations concerning the effect of various doses. A whole-body dose of 500 rem is lethal in about 50% of the cases. A whole-body dose of over 600 rem is almost always lethal, because the bone marrow is destroyed; but persons receiving such doses can survive if an arm or a leg is shielded (so that bone marrow there can provide new blood cells). Doses of between 100 and 500 rem cause nausea, fatigue, possible hemorrhaging, and lowered resistance to infection, but recovery is possible. Doses of between 25 and 100 rem cause changes in blood cell counts, but the recipient usually feels little effect.

Smaller doses cause no measurable effects, but they do damage cells, and this can result in a decrease in life expectancy. Statistics on survivors of large radiation doses have shown that a whole-body dose of 100 rem eventually causes some form of cancer in about one person in 50. (This is the excess number of cancers in the exposed population, when it is compared with the number in the unexposed "control" population.) At *larger* doses, the observed cancer rate is proportional to the dose. In the absence of observations at smaller doses, one can extrapolate downward linearly to predict that a dose of one rem would lead to cancer in one person out of 5000.

The same sort of linear extrapolation would say that smoking just one cigarette per day would cause lung cancer in one person out of 2000. Apart from illustrating the lethal nature of cigarettes, this comparison shows the difficulty of this sort of extrapolation. When very low doses are involved (either of radiation or tobacco smoke), one might expect that damaged tissues *could* recover completely. But there is no way to test this possibility directly, because when the probability of disease is on the order of 0.001 or less, it becomes a practical impossibility to gather reliable statistics comparing an exposed population with a control. For radiation, it has been judged safest to use the linear extrapolation, because it is hard to envision a mechanism that would make a small dose more hazardous, per unit dose, than a large dose is.

Effects on Cells; Cancer Therapy.   It has been demonstrated that individual cells can recover from radiation damage, but that a second equal dose, administered before recovery is complete, can kill them. The mortality rate is strongly dependent on the timing of the second dose. It has also been shown that cells irradiated in the presence of oxygen are more sensitive to damage. Oxygen seems to inhibit recovery by removing electrons that are useful for recombination of molecules.

These facts have been useful in the search for improvements in radiation therapy for cancer. Dividing a dose into two parts, with the optimum time interval between application of the two, can kill more cancer cells than would be killed by giving the entire dose at once. The explanation is that cells are more susceptible to radiation in certain parts of the cell cycle. If the first dose catches a cell in a resistant phase, the second dose will hit it when it is more susceptible. However, dividing the dose into *many* small doses is less effective; a cell can recover from a series of small doses more quickly than from a single, massive dose, because the sudden large dose may break a chromosome in several places, making repair almost impossible. (When several pieces are present, they are far more likely to recombine the wrong way.)

Many tumors contain cells that are resistant to radiation because of an oxygen deficiency. One way to overcome this resistance is to use heavy particles for irradiation. The effect of these particles is less influenced by oxygen because they interact with nuclei and transmute elements. Use of such radiation also takes advantage of the greater effectiveness of a highly concentrated dose.

You should now have the *physical* basis for understanding radiation effects in living systems. Any further discussion would lead us too far into the realm of biology.

PROBLEMS

1. The human body contains about 0.35 percent potassium, by weight. Of the potassium, 0.012 percent is ^{40}K, a beta emitter with a half-life of 1.3×10^9 yr. Find the number of ^{40}K decays per minute in your body.

2. The presence of radioactive series gives information on the ages of individual rocks. In some cases a rock may have contained no lead when it was formed; this is indicated by the absence of ^{204}Pb. If the rock did contain a normal isotopic mixture of uranium, the decay of the ^{238}U and ^{235}U causes ^{206}Pb and ^{207}Pb to build up in the rock. Find an expression relating the age of the rock to the ratio of ^{206}Pb to ^{207}Pb in the rock, using the uranium abundance data in Example Problem 1.

3. In radioactive series decay, the amount of each species is increased by the decay of its parent and decreased by its own decay. If N_1 and N_2 are the number of atoms of parent and daughter, respectively, and λ_1 and λ_2 are the respective decay constants, then the change in N_2 in a time dt is $dN_2 = (\lambda_1 N_1 - \lambda_2 N_2)\,dt$. Starting with a pure sample, so that $N_1 = N_0$ and $N_2 = 0$ at $t = 0$, find N_2 as a function of time. Show that if the parent is sufficiently long-lived, the activity λN of each decay product eventually equals the activity of the parent.

4. ^{235}U decays to ^{231}Th, which in turn decays to ^{231}Pa. Suppose one starts at $t = 0$ with pure ^{235}U whose activity is $1\ \mu$Ci (3.7×10^4 decays/sec). Show that for $t \ll 35$ hours, the number of ^{231}Pa atoms present is proportional to t^2, and find the expected number of ^{231}Pa atoms present at $t = 100$ sec. (Mean life of ^{231}Th is 35 hr, of ^{231}Pa is 4.8×10^4 yr.)

5. One of the radioactive series begins as follows: ^{232}Th $\rightarrow$ ^{228}Ra $\rightarrow$ ^{228}Ac $\rightarrow$ ^{228}Th $\rightarrow$ ^{224}Ra $\rightarrow$ ^{220}Rn $\rightarrow$ ^{216}Po $\rightarrow$ ^{212}Bi. From ^{212}Bi there are two possibili-

ties: ^{212}Bi → ^{208}Tl → ^{208}Pb (33.7%) or ^{212}Bi → ^{212}Po → ^{208}Pb (66.3%). Suppose you came across a rock containing 1 gm of ^{232}Th; how many atoms of each species in the ^{232}Th series should it contain? (Assume equilibrium has been achieved; see Problem 3. Data on half lives are in Appendix D.)

6. Use Eq. (7) to verify that the impact parameter $b = 1.6 \times 10^{-10}$ cm for scattering a 5-MeV alpha particle through 1° by a nucleus of $Z = 50$, as stated in footnote 10. Also verify the other numbers given in that footnote.

7. A proton initially has kinetic energy of 93 MeV. Use Eq. (13) to find the energy it loses in penetrating (a) 50 m of air, and (b) 1 cm of aluminum.

8. Use Fig. 4 to estimate the energy lost by (a) a 200-MeV proton, (b) a 500-MeV proton, in the first 100 meters of its path in air. Compare with the result of using Eq. (13).

9. Using Eq. (13) and the reasoning of Example Problem 2, show that a particle of mass M_x, charge $Z_x e$, and kinetic energy E_x (in MeV) has a range R_x given by

$$R_x = (M_x/M_p Z_x^2)(M_p E_x/9.3\ M_x)^{1.8}$$

meters of air, where 3 MeV $< M_p E_x/M_x <$ 300 MeV, and M_p is the proton mass.

10. Find the range of a 50-MeV muon in air, using Fig. 4 and Eq. (14). Could you use the result of Problem 9 in this case?

11. In measuring the angular correlation of annihilation radiation, long, narrow slits have often been used to define the gamma ray directions. A given slit position then defines the projection of the angle θ onto the plane perpendicular to the slit, thereby defining one component of the momentum of the gamma ray pair (see figure). Use the Fermi gas model of valence electron states in a metal to show that the contribution of valence electrons to the number of gamma-ray pairs observed as a function of angle should be an inverted parabola. (Hint: The occupied valence electron states fill a sphere in momentum space.)

12. Consider a liquid containing positronium (Ps) atoms, and use the following definitions: N_p = Number of atoms of Ps present in the solution at any given time; N_b = Number of bare positrons present in the solution at any given time; λ_p = Probability per unit time that a positron in a Ps atom will be annihilated; λ_o = Probability per unit

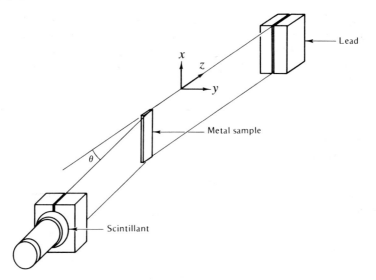

time that a Ps atom will be oxidized, that is, broken up, leaving a bare positron; λ_b = Probability per unit time that a bare positron will be annihilated.

(a) Write differential equations for dN_p/dt and dN_b/dt in terms of defined quantities.

(b) Solve these equations for N_p and N_b as functions of time, given the initial conditions

$$N_p = N_0 \quad \text{at} \quad t = 0$$

and

$$N_b = 0 \quad \text{at} \quad t = 0$$

(*Hint*: N_b is the sum of two exponentials.)

13. Compute the mean lifetime of parapositronium in the 2s state. Use Eq. (17) and the appropriate wavefunction from Table 1, Chapter 7.

14. (a) Show from Eq. (18) that $\lambda_f = (1 - B)\lambda_1 + B\lambda_t$.

(b) Positron lifetimes in a sample were observed to follow Eq. (18), with $1/\lambda_1 = 100$ ps, $1/\lambda_t = 400$ ps, and $B = 0.40$. After the sample was annealed, the positron lifetime distribution became a single exponential. What would you expect the mean lifetime of those positrons to be?

15. Find the energies of the two gamma rays emitted in positron annihilation if the positron is at rest and the electron has energy of 5 eV at the time of annihilation, and the electron's momentum is parallel to the direction of emission of one gamma ray. (*Hint*: The electron's *kinetic*

energy is negligible, but its *momentum* is not.) Suppose you wished to measure electron momentum distributions by measuring the energy distribution of the emitted gamma rays, rather than measuring their angular correlation. What energy resolution is needed to compete with an angular resolution of 1 mrad?

16. Figure 14 shows a case in which 17.5 percent of the gamma rays are resonantly absorbed when the absorption and emission lines fully overlap. Assume that the fraction of zero-phonon events is the same for emission and absorption, and compute this fraction for this case.

17. Suppose in a Mössbauer experiment that the source and absorber material has a Debye temperature of 420°K, and the absorber is at 300°K. Find the thermal frequency shift if the source is at 0, 100, and 200°K, and compare with gravitational shift. To check your answer, compare with the result given by R. Pound and G. Rebka, *Phys. Rev. Lett.* **4**, 275 (1960).

18. In studying hyperfine structure with the Mössbauer effect, one can simplify matters by using a *source* in which the line is unsplit; there are then just as many resonances as there are levels in the *absorption* spectrum. For Fe^{57}, this is accomplished by using a stainless steel source. Find the Fe^{57} absorber velocities at which resonance occurs (a) if the source is Fe^{57} in stainless steel (unsplit line), and (b) if the source is Fe^{57} in iron (split line).

19. ^{14}C is constantly produced in the atmosphere by reactions between protons and ^{14}N atoms. The half life of ^{14}C is 5730 yr. If one gram of carbon has an activity of 15 decays per minute, how many ^{14}C atoms does this gram contain?

20. Using the result of Problem 1, estimate the radiation dose (in rads) received by your body in one year from the ^{40}K that it contains. Assume that 0.4 MeV is absorbed as a result of each ^{40}K decay, and that the ^{40}K is uniformly distributed in your body.

21. A careless person leaves one millicurie of ^{60}Co on a table, unshielded, 5.0 meters away from a fellow worker. If this worker spends 40 hours at his workplace before the ^{60}Co is removed, how much whole-body radiation dose does he receive? ^{60}Co emits gamma rays totalling 2.5 MeV in each decay; neglect the reduction in gamma-ray intensity as the rays pass through the worker's body.

22. The π^- meson (pion) is now being used to irradiate tumors. One advantage over the conventional X-ray and gamma-ray treatments is that most of the particle's energy loss is near the end of its path (in contrast to the exponen-

tial attenuation of photons), and maximum energy can therefore be released in the tumor with minimum damage to the body on the way in. Use Eq. (13) to find the π^- energy needed to penetrate 10 cm of tissue, and find the energy released in the last cm of the path. (Assume π^- mass is $\frac{1}{7}$ the proton mass, and the tissue to be equivalent to water.) Compare the result with the energy losses resulting from irradiation with 1-MeV gamma rays.

23. A neutron bomb is designed to deliver a dose of 8000 rads at a distance of 850 m from the bomb explosion (F. M. Kaplan, "Enhanced Radiation Weapons," *Scientific American,* May 1978, pp. 44–51). Assuming that the absorption coefficient of air for these neutrons is 0.003 m^{-1}, find the dose delivered to an unshielded person who is 1.7 km from the point of the explosion. (The neutrons fly out in all directions, so you must use the inverse square law.)

Properties of the Nucleus

Just as the study of atomic structure led to important discoveries about the fundamental laws of nature (for example, quantum mechanics), the study of the structure of the atomic nucleus has led to further fundamental discoveries. But the structure of the nucleus is not so well understood as the electronic structure of the atom, for many reasons. The nuclear force is more complicated than the Coulomb force, and its short range makes it more difficult to study experimentally. Furthermore, the high density of nucleons in the nucleus makes it difficult to analyze the nucleus in terms of two-body interactions or closed shells, as was done for atomic electrons.

As a result of these and other difficulties, we do not have a single model of the nucleus which gives us the best understanding of all properties of all nuclei. Instead, there are several models, each of which works well for certain applications. Models which are successful in describing some properties of the nucleus are often completely incapable of handling other properties, and some models, which work well for nuclei whose Z or A values lie in certain ranges, can be quite inaccurate outside these ranges. We shall therefore discuss various properties of the nucleus in turn, introducing the appropriate models as we go along. Many of these properties have already been mentioned in other connections, but we include them here for completeness.

14.1   CHARGE

The charge on the nucleus is $+Ze$, where Z is the atomic number and e is the electronic charge. As we saw in Chapter 3, the atomic numbers were sorted out by Moseley's X-ray experiments and verified by Chadwick's alpha-particle scattering experiments. There remains the question of whether the nuclear charge is *exactly* Ze; that is, whether atoms are neutral. Some cosmological speculation has tried to explain the expansion of the universe by postulating that there is a small net charge on all atoms. However, very refined experiments have shown that the unit of nuclear charge—the proton charge—differs (in magnitude) from the electron charge by no more than 5×10^{-19} electron charges, a difference too small to account for the observed expansion, if indeed there is a difference at all.

14.2   RADIUS

Knowing the magnitude of the nuclear charge, we next would like to know how this charge is distributed. As a first approximation, we assume that the nucleus is spherical (a rather good approximation), and that it has a uniform charge distribution within a well-defined boundary (a not quite so good approximation); we can then attempt to find its radius.

In Rutherford's first experiments we have the earliest clue. In those experiments the nucleus was indistinguishable from a point charge. This means that the alpha particles used were not energetic enough to penetrate the nuclei in the foils; their entire path must have remained outside the nuclear charge distribution.

It is a simple matter to apply the law of conservation of energy in order to compute the closest approach d of an alpha particle to a nucleus in a head-on collision. At the distance of closest approach, the particle's energy is all potential energy, so we have, in mks units,

$$\text{Energy} = \frac{Z_1 Z_2 e^2}{4\pi\varepsilon_0 d}$$

where $Z_1 e = 2e$ is the charge of the alpha particle, and $Z_2 e$ is the charge of the scattering nucleus. In Rutherford's first experiments, the alpha particle energies were about 7 MeV, and the scatterers were gold and silver. If we write the energy in *electron* volts, one factor of e is canceled, and we have (since $1/4\pi\varepsilon_0 = 9 \times 10^9$ V-m/C),

$$d = \frac{Z_1 Z_2 e(9 \times 10^9 \text{ V-m/C})}{7 \times 10^6 \text{ V}}$$

With $e = 1.6 \times 10^{-19}$ C, $Z_1 = 2$, and $Z_2 = 79$, we find that

$$d \simeq 3 \times 10^{-14} \text{ m} = 3 \times 10^{-12} \text{ cm}$$

Later Rutherford used lighter elements as scatterers, and found the first positive effect of the size of the nucleus: deviations from his scattering law were observed at values of d of about 3×10^{-13} cm.[1]

In the years since Rutherford, many methods have been used to determine nuclear sizes. Various other types of scattering experiments have been done, using high energy electrons and neutrons as incident particles. Static effects of the nuclear size on atomic energy levels have also been studied. In order to analyze these experiments, we must clarify our mental image of the nucleus. We cannot think of the nucleus as being like a hard little marble; there is no sharp border at which incident particles receive a sudden impulse and bounce off. The nucleus can rather be thought of as a ball of electric charge, which can be penetrated by incident particles or by its own atomic electrons, and which exerts a specific nuclear force on *nucleons*—either neutrons or protons.

Isotope Shift. The penetration of atomic electrons into the nucleus provides an elegant way to determine the nuclear radius. If the nucleus were a point charge, the electronic energy levels of two isotopes should be equal, except for the slight difference caused by the difference in reduced mass, because the electron configurations are identical. But because the wavefunction of an atomic electron overlaps the nucleus, the electron spends part of its time inside the nucleus, where it feels a Coulomb force which is smaller than that given by the inverse-square law. Thus the average potential energy of the electron is smaller in magnitude than it would be if the nucleus were a point charge. Because nuclei of different isotopes have different radii, this effect varies from one isotope to another, causing an "isotope shift" in the energy levels. This isotope shift can easily be distinguished from the "shift" caused by the reduced mass, because the mass effect becomes negligible in the medium and heavy nuclei (why?), and it can be calculated in any case, whereas the effect of the nuclear radius *increases* with increasing nuclear size. (See Problems 1 and 2 for further development.)

[1] E. Rutherford, *Phil. Mag.* **37**, 537 (1919).

Muonic X-Rays. The isotope shift is very small for electron levels, but it can be used with great accuracy to determine nuclear radii if it is "magnified" by using muons instead of electrons. The negative muon differs from an electron only in its mass, which is 207 electron masses. If muons are stopped in matter, they are captured by atoms and they make transitions to lower energy states just as electrons do. But muon orbits are smaller than electron orbits by a factor of 207, so the muon spends a much greater part of its time inside the nucleus. For example, in the lowest orbit in lead, the muon spends half of its time inside the nucleus.

The muon in its lowest level is well within the electronic K shell; thus the electrons have little effect on the muon's energy, and it is easy to calculate the effect of the nuclear size on the muonic energy levels. The energies of the "muonic X-rays" emitted when the muon makes its transitions to lower energy levels can then be measured, and these energies compared with energies computed as a function of the nuclear radius, to determine which radius fits the data.

The nuclear size has a huge effect; therefore these energies provide a sensitive determination of the nuclear radius. For example, in lead the muon would release 16.4 MeV in going from the $2p_{3/2}$ state to the $1s$ state, if the nucleus were a point charge; but the observed X-ray energy for this transition is only 6.02 MeV.[2] The radii calculated from such X-ray energies are in agreement with results obtained by entirely different methods,[3] and they show that the nuclear radius R follows the general rule,

$$R = R_0 A^{1/3}$$

where $R_0 = 1.2 \times 10^{-13}$ cm $= 1.2$ fermis,[4] for values of A greater than about 10. This means that the volume of the nucleus is directly proportional to A, the number of nucleons, or that the *density* of a nucleus is independent of the number of nucleons in it. This rule is obeyed rather well, although there are significant deviations for nuclei containing "closed shells" of nucleons (Section 14.8).

Electron Scattering. The significance of a static measurement of the nuclear radius depends upon the validity of the assumptions we make about the distribution of charge in the nucleus. A shift in one energy level gives us only a single parameter, so it cannot tell us details about the way the charge is distributed within the nucleus. However, this charge distribution can be determined by scattering particles from nuclei. Electrons are ideal for mapping

---

[2] V. L. Fitch and J. Rainwater, *Phys. Rev.* **92**, 789 (1953).
[3] For an independent verification of the formula for R, see Section 15.1.
[4] It is convenient to introduce a unit called the fermi (fm), equal to 10^{-13} cm, when dealing with distances of the order of nuclear dimensions. The fermi is also known as the femtometer, because "femto-" is the standard prefix for 10^{-15}.

out the nuclear charge distribution in this way, because they are not affected by the specific nuclear force.

Electron scattering can resolve features whose dimensions are comparable to the de Broglie wavelength of the electron. To explore the nucleus with resolution of, say, 10^{-13} cm, one therefore needs an electron *momentum p* of at least $h/10^{-13}$ cm, or 1.24×10^9 eV/c. Such electrons are highly relativistic, so the electron *energy* is very close to pc, or about 1.2×10^9 eV = 1200 MeV.

Experiments have been performed at such energies, and they perhaps tell you more about the nucleus than you really want to know at this stage. But experiments at somewhat lower energy (a few hundred MeV) also give some information about the nuclear charge distribution.[5] These experiments may be explained by assuming the nuclear charge density to be distributed as

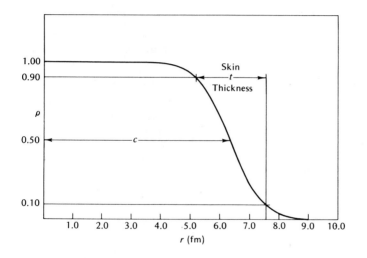

Fig. 1. *Parameters characterizing the nuclear charge distribution. Values shown are for the gold nucleus. [From R. Hofstadter, Ann. Rev. Nucl. Sci. 7, 296 (1957).]*

shown in Fig. 1. It was found that, except for very light nuclei ($A < 10$), the charge distribution can be specified quite well by use of the two parameters

$$c = c_0 A^{1/3} \quad \text{(the "half-density radius")}$$

and

$$t = 2.4 \quad \text{fm} \quad \text{(the "skin thickness")}$$

Later work[6] has indicated that the best value of c_0 is 1.08 fm. It is interesting

[5] R. Hofstadter, *Rev. Mod. Phys.* **58**, 214 (1956); *Ann. Rev. Nucl. Sci.* **7**, 231 (1957).
[6] R. D. Ehrlich *et al.*, *Phys. Rev. Lett.* **18**, 959 (1967).

to note that the skin thickness t is independent of A; this is true down to $A \approx 10$, where the nucleus is "all skin."

The half-density radius c is not identical to the radius R determined by the isotope shift or muonic X–rays, although R would be equal to c if the skin thickness t were zero. The energy shift that determines the value of R results from an integration of the charge density over the whole volume of the nucleus. If the nucleus were uniform, with a sharp cutoff, then R would be unambiguously defined; the charge density would be constant for $r < R$, and zero for $r > R$. This case may be used to construct a general definition of R by relating R to the root-mean-square radius of the distribution of charge, as follows:

The element of charge, for a spherically symmetric distribution, is $dq = 4\pi r^2 \rho dr$. Then, by definition, the mean square radius is

$$\overline{r^2} = \frac{\int_0^\infty r^2 (4\pi r^2 \rho)\, dr}{\int_0^\infty 4\pi r^2 \rho\, dr}$$

which becomes, for a charge distribution that cuts off at $r = R$,

$$\overline{r^2} = \frac{\int_0^R r^2 (4\pi r^2 \rho)\, dr}{\int_0^R 4\pi r^2 \rho\, dr}$$

or, since ρ is constant

$$\overline{r^2} = \frac{3R^2}{5}$$

Therefore

$$R = \sqrt{\tfrac{5}{3}\, \overline{r^2}}$$

For the general case, this equation is used to define R in quoting a value for the "nuclear radius" as derived from muonic X-rays or the isotope shift. For a real nucleus, whose skin thickness is not zero, R as defined in this way comes out to be somewhat larger than the half-density radius c.

Given the value of c and the value of t obtained from electron scattering, one can compute energy eigenvalues of the muon in the corresponding potential, and the values found are in agreement with the muonic X–ray data.

Some deviations from the general formula for c have been found, and confirmed by the various methods. For example, it is found that ^{48}Ca has a smaller radius than ^{40}Ca. This somewhat startling result is an indication of the shell structure of the nucleus. (See Sec. 14.8.)

14.3 ANGULAR MOMENTUM AND MAGNETIC DIPOLE MOMENT

You will recall from Chapter 7 that the existence of the magnetic moment of the nucleus could be inferred from the hyperfine structure of atomic spectral lines. This magnetic moment is associated with the total nuclear "spin" **I**, which is the vector sum of the spins of protons and neutrons in the nucleus *plus* the orbital angular momentum of these particles as they move about inside the nucleus. According to the principles developed in Chapter 7, the vector **I** must be quantized in the same way as any other angular momentum vector; it has a magnitude equal to $\hbar[I(I+1)]^{1/2}$, and the possible values of a component in any direction are $m_I\hbar = I\hbar, (I-1)\hbar, \ldots, -I\hbar$, where I is the quantum number for the nuclear "spin." The value of the *quantum number I* is often what is meant when one refers to the "spin" of a nucleus.

The number of levels produced by the hyperfine splitting was shown to be equal to $2I+1$ or $2J+1$, whichever is smaller. Therefore I can often be determined simply by counting the number of lines in the hyperfine structure. It has been found that I is always an odd multiple of $\frac{1}{2}$ for odd-A nuclei, and I is an integer for even-A nuclei. This was the earliest evidence that the nucleus contains Z protons and $A - Z$ neutrons, or a total of A particles, each of spin $\frac{1}{2}$, rather than A protons and $A - Z$ electrons. The latter combination would give the right charge and mass for each nucleus, but the nucleus would contain $2A - Z$ particles, each of spin $\frac{1}{2}$, so that in cases where A is even and Z is odd, I would be an *odd* multiple of $\frac{1}{2}$, in disagreement with the observations.

Once the spin of the ground state is known, the spins of excited states of the nucleus may be found by measuring angular correlations of the radiations from the nucleus. We shall discuss this point further in Chapter 15. This information is very valuable in testing predictions of various models of the nucleus.

Magnetic Moment and Magnetic Resonance. The magnetic dipole moment of the nucleus results from the intrinsic moments of the protons and neutrons plus the moment of the current loop formed by the orbiting protons. The intrinsic magnetic moment of the proton is parallel to its spin, as one might expect for a positively charged body. One would not necessarily expect the neutron to have any magnetic moment, but it happens that the neutron has a magnetic moment which is antiparallel to its spin. (This suggests that the neutron itself has internal structure consisting of positively and negatively charged regions. We shall return to this point in Chapter 16.) The resultant magnetic moment of the nucleus may therefore be either parallel or antiparallel to the nuclear spin; we may write it as

$$\boldsymbol{\mu} = g_N \left(\frac{e}{2M_p}\right) \mathbf{I}$$

where g_N is the *nuclear g factor*, analogous to the Lande g factor of atomic spectroscopy (Section 9.2), and M_p is the mass of the proton. The quantity $e\hbar/2M_p$ is called the *nuclear magneton* (in analogy to the Bohr magneton, $e\hbar/2m_e$), and denoted by the symbol μ_N.

It is possible to measure g_N by measuring the separation of an atomic beam into various components by an inhomogeneous magnetic field, as in the Stern–Gerlach experiment. But the separation between the components is very small when nuclear moments are involved, and much greater precision in measuring the magnetic moment is achieved by using the *resonance* phenomenon.

To measure a magnetic moment by a resonance method, one applies a uniform magnetic field **B** which causes energy differences to appear between states with different values of the m_I quantum number, because of differences in the interaction energy $-\mathbf{\mu} \cdot \mathbf{B}$. One can then induce transitions between these states by irradiating the system with radiation of frequency v such that hv equals the energy difference between two states. Various methods have been used to determine the frequency at which these transitions occur and thus to find g_N. One method, used by I. Rabi and his group, is illustrated in Fig. 2.

A beam of atoms passes through a region in which there is an upward magnetic field gradient, then through a constant-field region, and finally through a region in which there is a downward field gradient. Particles whose magnetic moment is unchanged in flight receive opposite impulses in the first and third regions, and are focused onto a detector. But if **μ** changes direction while the particle traverses region II, the particle receives the wrong impulse in region III and it does not strike the detector. The change in direction of **μ** in region II is caused by a horizontal magnetic field which varies as $\sin \omega t$.

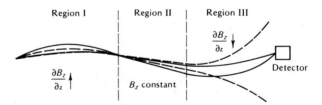

Fig. 2. *Possible paths of particles through an atomic beam apparatus. Solid lines show paths of particles whose z component of magnetic moment, μ_z, is unchanged during flight. Dashed lines show possible paths of particles whose μ_z is changed in region II. Notice that the solid lines are appropriate for particles with negative μ_z and with different initial velocities. The lower dashed line shows a particle whose μ_z changed sign in region II.*

If $\hbar\omega$, the "quantum" of this field, equals the energy difference between two orientations of $\boldsymbol{\mu}$, then transitions are induced and the direction of $\boldsymbol{\mu}$ changes.

The experiment is done by using different values of the *vertical* field **B** in region II, and counting the number of particles reaching the detector as a function of **B**. There is a striking decrease in this number when **B** is such that the quantum of interaction energy—the difference between two possible values of $\boldsymbol{\mu} \cdot \mathbf{B}$ —is equal to $\hbar\omega$, the quantum of energy of the oscillating *horizontal* field. The interaction energy $-\boldsymbol{\mu} \cdot \mathbf{B}$ is therefore precisely determined. This interaction energy is so small that, in fields of a few thousand gauss, the resonant frequency lies in the range of radio frequencies, which are easy to produce and control accurately.

Notice that the measured quantity is not $\boldsymbol{\mu} \cdot \mathbf{B}$ itself, but the *change* in $\boldsymbol{\mu} \cdot \mathbf{B}$. If **B** is along the z axis, one measures $(\Delta\mu_z)B_z$, or $(g_N eh/2M_p)(\Delta m_I)B_z$, and if $\Delta m_I = 1$, $\omega = g_N eB_z/2M_p$. The *maximum* value of μ_z, or $g_N e\hbar I/2M_p = g_N\mu_N I$, is what is customarily called "the magnetic moment," even though the magnitude of the actual magnetic moment of the nucleus is greater than this (by a factor of $(I(I + 1)/I^2)^{1/2}$).

Knowledge of the magnetic moment of the nucleus is not only important for theoretical reasons, as a test of varous models of the nucleus, but it also makes it possible to determine the magnetic field at the nucleus in various materials. As we have seen (Section 13.4), once the nuclear moment is known, the internal magnetic field can be determined quite accurately by using the Mössbauer effect to measure the hyperfine splitting of the *nuclear* energy levels.

14.4 ELECTRIC QUADRUPOLE MOMENT

It is usually convenient to consider the nucleus to be spherical, as we have done until now, and this is an extremely good approximation, but quite a few experimental results can only be explained by assuming that the nuclei involved are *not* spherical. In such cases, part of the *electrostatic* interaction energy between the nucleus and the surrounding electrons depends on the orientation of the nuclear spin. The simplest shape which can produce such an effect is that of an ellipsoid of revolution, whose axis of revolution is identical to the spin axis (see Fig. 3).

Such a shape is characterized by its "quadrupole moment," which is related to the electrostatic energy of the charge distribution in an external electric field, as follows: We observe that the energy of any charge distribution $\rho(x, y, z)$ in an external potential field $\phi(x, y, z)$ is given by $W = \int \rho\phi \, d\tau$, where $d\tau$ is the volume element $dx \, dy \, dz$, and the integral is over all space. If

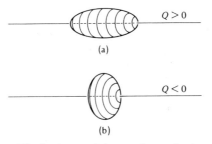

Fig. 3. *Assumed shapes of nonspheri- cal nuclei; (a) nucleus with positive "quadrupole moment" Q (prolate el- lipsoid) (b) nucleus with negative "quadrupole moment" Q (oblate ellipsoid).*

the distribution is concentrated in a small volume, one can expand ϕ in a Taylor series, using the center[7] of the charge distribution as origin, to obtain

$$W = \phi(0) \int \rho \, d\tau + \text{Dipole energy} + \text{Quadrupole energy} + \text{Higher order terms}$$

where

$$\text{Dipole energy} = \left(\frac{\partial \phi}{\partial x}\Big|_0 \int x\rho \, d\tau\right) + \left(\frac{\partial \phi}{\partial y}\Big|_0 \int y\rho \, d\tau\right) + \left(\frac{\partial \phi}{\partial z}\Big|_0 \int z\rho \, d\tau\right)$$

and

$$\text{Quadrupole energy} = \frac{1}{2}\left[\left(\frac{\partial^2 \phi}{\partial x^2}\Big|_0 \int x^2\rho \, d\tau\right) + \left(\frac{\partial^2 \phi}{\partial y^2}\Big|_0 \int y^2\rho \, d\tau\right)\right.$$
$$\left. + \left(\frac{\partial^2 \phi}{\partial z^2}\Big|_0 \int z^2\rho \, d\tau\right)\right] + \left(\frac{\partial^2 \phi}{\partial x \, \partial y}\Big|_0 \int xy\rho \, d\tau\right)$$
$$+ \left(\frac{\partial^2 \phi}{\partial y \, \partial z}\Big|_0 \int yz\rho \, d\tau\right) + \left(\frac{\partial^2 \phi}{\partial z \, \partial x}\Big|_0 \int zx\rho \, d\tau\right)$$

Because of the symmetry of the assumed shape of the nuclear charge dis- tribution, all of the dipole terms and the last three quadrupole terms are zero. Thus the energy may be written (as you may verify for yourself)

$$W = Ze\phi(0) + \frac{1}{2}\left[\frac{\partial^2 \phi}{\partial x^2}\Big|_0 Q_{11} + \frac{\partial^2 \phi}{\partial y^2}\Big|_0 Q_{22} + \frac{\partial^2 \phi}{\partial z^2}\Big|_0 Q_{33}\right]$$

---

[7] If the distribution is symmetric, like those in Fig. 3, the position of the center is obvious. In general, the choice of "center" is arbitrary, and the expansion is not unique, although the sum of *all* terms in the expansion must of course be independent of the choice of center.

where Q_{11}, Q_{22}, and Q_{33} are components of the *quadrupole moment tensor* defined by

$$Q_{ij} = \int x_i x_j \rho \, d\tau$$

with

$$x_1 = x, \qquad x_2 = y, \qquad \text{and} \qquad x_3 = z.$$

This expression may be further simplified. If the spin axis, the axis of revolution, is chosen as the z axis as usual, then $Q_{11} = Q_{22}$. We can then use the fact that the potential energy obeys Laplace's equation[8] in order to reduce the shape dependence of the energy to a single term; we define a quantity $Q = 2(Q_{33} - Q_{11})/e$, and we write the quadrupole energy as

$$W_Q = \frac{1}{2} Q_{11} \left(\frac{\partial^2 \phi}{\partial x^2} + \frac{\partial^2 \phi}{\partial y^2} \right) + \frac{1}{2} Q_{33} \frac{\partial^2 \phi}{\partial z^2}$$

or, since (Laplace's equation)

$$\frac{\partial^2 \phi}{\partial x^2} + \frac{\partial^2 \phi}{\partial y^2} + \frac{\partial^2 \phi}{\partial z^2} = 0$$

$$W_Q = \frac{1}{2} Q_{11} \left(-\frac{\partial^2 \phi}{\partial z^2} \right) + \frac{1}{2} Q_{33} \left(\frac{\partial^2 \phi}{\partial z^2} \right) = \frac{1}{2} \frac{\partial^2 \phi}{\partial z^2} (Q_{33} - Q_{11})$$

or

$$W_Q = \frac{1}{4} eQ \frac{\partial^2 \phi}{\partial z^2}.$$

Notice that $Q = 0$ for a spherical nucleus. The quantity Q is often called simply the "quadrupole moment" of the nucleus (a somewhat misleading designation, because the quadrupole moment is really defined as a tensor). The quantity $\partial^2 \phi/\partial z^2$ is the derivative of the z component of the electric field, and is often referred to as the "electric field gradient." In a *uniform* electric field there is no quadrupole energy.

A concrete example may give you a better feeling for the quadrupole energy. Compare the energy of a point charge q, at the origin, with the energy of two charges, each $q/2$, at $z = +a/2$ and at $z = -a/2$, respectively, in an external potential $\phi(z)$. The energy of charge q is simply $q\phi(0)$, whereas the total energy of the two charges is

---

[8] The potential is produced by the external electrons surrounding the nucleus. The charge density of these electrons within the nucleus may be neglected, and thus ϕ obeys Laplace's equation.

$$\frac{q}{2}\,\phi\!\left(\frac{a}{2}\right) + \frac{q}{2}\,\phi\left(\frac{-a}{2}\right) = q\left[\frac{\phi\!\left(\dfrac{a}{2}\right) + \phi\!\left(\dfrac{-a}{2}\right)}{2}\right]$$

Thus the energy of the two charges is proportional to the *average* potential at the points $z = a/2$ and $z = -a/2$. If E_z is constant, $\phi(z)$ has a constant slope and the average of $\phi(a/2)$ and $\phi(-a/2)$ is equal to $\phi(0)$; the energy of the "spread out" charge distribution—the two charges $q/2$—equals the energy of the point charge. There is no dipole energy, because the dipole moment is zero, and there is no quadrupole *energy*, although there is a quadrupole *moment*. But if the slope of $\phi(z)$ is not constant, the average potential $[\phi(a/2) + \phi(-a/2)]/2$ does not equal $\phi(0)$ (Fig. 4), so that the potential

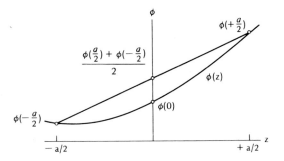

Fig. 4. *Graphical comparison of the average potential at $z = +a/2$ and $z = -a/2$ with the potential at $z = 0$, in a nonuniform electric field.*

energy of the two charges does not equal the energy of the single charge q. The difference between $q[\phi(a/2) + \phi(-a/2)]/2$ and $q\phi(0)$ is approximately the quadrupole energy. (If $\partial^2\phi/\partial z^2$ varies appreciably over the interval $-a/2 < z < +a/2$, higher order "poles" could also contribute to this energy.)

The quadrupole energy is observed in the splitting of energy levels which would otherwise be degenerate. It resembles the hyperfine splitting, in that nuclear states with different values of m_I are split; but the splitting is distinguishable from magnetic splitting, because the quadrupole interaction energy does not separate states which have the same absolute value of m_I.

The reason for this is that reversing the sign of m_I simply reverses the spin, without affecting the overall orientation of the nucleus (in its assumed shape), so that there is no change in the electrostatic interaction energy. Thus, for example, if $I = \frac{3}{2}$, there are only two energy levels, rather than four, and if $I = \frac{1}{2}$, no splitting occurs.

Measurements of nuclear quadrupole moments are important because of the information which they reveal about the structure of the nucleus. Systematic variation of the quadrupole moment with increasing number of nucleons is an indication of shell structure in the nucleus, and it also indicates that there are collective motions of nucleons in the filled shells. (See Section 14.8 and Fig. 10.)

The Mössbauer effect has been useful in studying quadrupole energies, and it provides a means of studying solids as well as nuclei. Attempts have been made to determine electric field gradients in solids and to determine the effect of impurities on these gradients. In these experiments, the observed energy, of course, depends on the product of the quadrupole moment and the field gradient, and if the quadrupole moment is unknown, as it often is, one cannot determine the field gradient separately. But by comparing the splitting of levels of the same nucleus in different environments, one can determine how the field gradient varies from one material to another, or how it depends on the concentration of impurities.

14.5   MASS AND BINDING ENERGY

The identification of radioactive isotopes by Soddy (1911) and the discovery of two isotopes of neon by J. J. Thomson (1913) cleared up the problem of fractional atomic weights and showed that the mass of each atom is close to an integral multiple of a basic unit, the atomic mass unit, which in turn is almost the mass of a nucleon. In 1961 the atomic mass unit (amu) was defined as 1/12 of the mass of the C^{12} atom; the energy equivalent is 931.481 MeV. (See Problem 7 for earlier definitions of atomic mass units.)

The atomic mass unit is about 0.8% smaller than the average mass of a nucleon. This means that nucleons are so tightly bound in a typical nucleus that an energy equivalent to 0.8% of their mass would be needed to free them. The energy needed to split up a nucleus completely into free protons and neutrons is called its *binding energy*. In practice, binding energies of *atoms* rather than of nuclei are determined, for it is difficult to keep electrons from congregating around a nucleus while you measure the mass. (However, the *binding energies* of the *electrons* can usually be ignored, although the electron *masses* cannot.) We therefore write the binding energy E_B, of an atom of mass M, containing Z protons and N neutrons, as the difference between the rest energy of the atom and the total rest energies of the Z hydrogen atoms and N neutrons into which it could be broken up. Thus

$$E_B = c^2(ZM_H + NM_n - M) \tag{1}$$

where M_H is the mass of the hydrogen atom and M_n the neutron mass. Notice that both M and ZM_H include the mass of Z electrons.

The measurement of nuclear masses was refined and developed by F. W. Aston right after World War I; his work won him the Nobel Prize for chemistry in 1922. The measurements were all based on the use of crossed electric and magnetic fields, just as were the measurements of e/m for the electron. The refinements introduced by Aston included the use of "matched doublets"; an example of a matched doublet is the methane ion $(^{12}C^1H_4)^+ -$ oxygen ion $(^{16}O)^+$ doublet. Since the methane ion and oxygen ion each contains 16 nucleons, the difference in mass of these ions is caused primarily by the difference in binding energy of the nuclei, which is less than 1 percent of the mass of either ion. It is clearly much easier to measure this difference directly than it would be to measure the masses separately and then subtract one from the other.

Numerous measurements of the mass difference between these ions have given values clustered around 36.4×10^{-3} amu. Within the accuracy of the measurements, this mass difference is in perfect agreement with the energy gained in nuclear reactions which can transform $^{12}C^1H_4$ into ^{16}O. A possible set of reactions is

$$
\begin{array}{lll}
^{12}_{6}C + ^{1}_{1}H \rightarrow ^{13}_{7}N & +1.94 & \text{MeV} \\
^{13}_{7}N \rightarrow (^{13}_{6}C)^- + \beta^+ & +1.20 & \text{MeV} \\
^{13}_{6}C + ^{1}_{1}H \rightarrow ^{14}_{7}N & +7.55 & \text{MeV} \\
^{14}_{7}N + ^{1}_{1}H \rightarrow ^{15}_{8}O & +7.29 & \text{MeV} \\
^{15}_{8}O \rightarrow (^{15}_{7}N)^- + \beta^+ & +1.74 & \text{MeV} \\
^{15}_{7}N + ^{1}_{1}H \rightarrow ^{16}_{8}O & +12.13 & \text{MeV}
\end{array}
$$

where the number of MeV shown in each case is the net amount of kinetic energy appearing as a result of the reaction (or decay).

The net effect of these reactions is

$$^{12}C + 4^1H \rightarrow ^{16}O + 2e^- + 2e^+ + 31.85 \quad \text{MeV (of kinetic energy)}$$

Annihilation of the two electron–positron pairs yields an additional 2.04 MeV, so that the total energy release is 33.89 MeV, which equals $c^2 \times 36.4 \times 10^{-3}$ amu. Thus, in the mass spectrometer, Einstein's relation $E = mc^2$ has been verified by a direct measurement of the change in inertia which results when energy is released in a collision (or collisions).

A complete table of nuclear masses may be constructed by starting with absolute measurements, then using doublets to relate them, and finally interpolating by observing energy release in nuclear reactions. Accuracy in determining the larger masses may be improved by using doublets which compare multiply charged heavy ions with singly charged light ions. For example, $(^{16}O)^+$ may be easily compared with $(^{48}Ti)^{+++}$, because their e/m ratios are very nearly equal.

The fact that atomic masses are close to integral multiples of the atomic mass unit indicates that *the binding energy per nucleon is almost independent of the size of the nucleus.* A plot of E_B/A versus A is shown in Fig. 5. There is

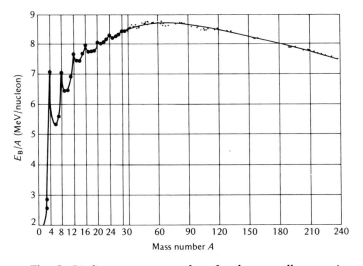

Fig. 5. *Binding energy per nucleon for the naturally occurring species. Note the magnification of the A scale below A =30. [From "The Atomic Nucleus" by R. D. Evans. Copyright McGraw-Hill, New York, 1955. Used by permission of McGraw-Hill Book Company.]*

a maximum around $A = 60$, with a slow decrease as A increases above 60. There are also secondary peaks at the "alpha particle nuclei" $^4_2$He, $^{12}_6$C, $^{16}_8$O, $^{20}_{10}$Ne, and $^{24}_{12}$Mg. (^{8}Be is not stable; it splits into two alpha particles in a time of the order of 10^{-15} sec.)

Fermi Gas Model. The secondary peaks as well as the general relationship of N to Z in stable nuclei can be understood from a simple energy level diagram, if we assume that each nucleon is in a single-particle state in a potential well formed by forces exerted by the other nucleons. This assumption, reminiscent of the treatment of electrons in a metal, forms the basis of the "Fermi gas model" of the nucleus. The energy levels of the protons are not the same as those of the neutrons, because of the Coulomb force, but in the light nuclei the levels are almost the same, because the *nuclear* force (that part which is not electromagnetic) is the same for protons and neutrons. Thus, since neutrons and protons have spin $\frac{1}{2}$ and obey the exclusion principle, two protons and two neutrons fill an energy level, producing the successive alpha particle nuclei. After a level is filled, the next nucleon must go into a higher level, so that the *average* binding energy is reduced, and there is a dip in the E_B/A curve.

The alpha particle effect disappears when the electrostatic repulsion of the protons becomes so great that a proton level is shifted upward to the next higher neutron level. (See Fig. 6.) Heavy nuclei thus contain more neutrons than protons, and all that remains of the alpha particle effect is a pairing effect,

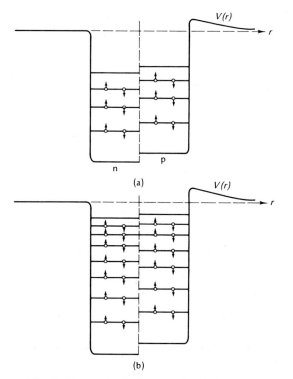

Fig. 6. *Comparison of energy levels for protons and neutrons. (a) Third proton level is higher than third neutron level, but not so high as fourth neutron level, so nucleus is stable with $N = Z = 6$. (b) Schematic representation of nucleus with large N and Z. Eventually, as N and Z increase, nth proton level becomes higher than $(n + 1)$th neutron level, and nucleus with $N = Z$ is no longer stable. (For N and Z even, this actually occurs at $Z = 22$. Difference between n and p levels is slightly exaggerated here).*

which makes nuclei with an even number of neutrons and an even number of protons more tightly bound than those with an odd number of each.

Semi-Empirical Mass Formula. We would like to have a theoretical formula which predicts the binding energies of all nuclei. Obviously it is not possible to derive such a formula from first principles; we do not know enough about the nuclear force, and even if we did, we could not solve such a complicated system as the nucleus. However it is possible, and quite useful, to construct a "semi-empirical" formula which matches the binding energies quite well after a few parameters have been correctly adjusted. The formula may be written in the form proposed by von Weizsäcker (1935) as

$$E_B = a_1 A - a_2 A^{2/3} - \frac{a_3(N - Z)^2}{A} - a_4 Z^2 A^{-1/3} \pm \delta(A) \qquad (2)$$

where a_1, a_2, a_3, and a_4 are adjustable parameters, and $\delta(A)$ is a function of A chosen to account for the pairing effect. If Eqs. (1) and (2) are combined, we have a formula for the *mass* of each nucleus:

$$M = ZM_H + NM_n - \frac{E_B}{c^2} \qquad (3)$$

Even an equation with this many adjustable parameters could not be made to fit the enormous number of known masses if each term were not based on a real physical property of the nucleus. Let us examine each term in Eq. (2) to see which property each accounts for.

$a_1 A$: This term and the next are based on the "liquid-drop model" of the nucleus, in which the nucleus is held together in the same manner as a liquid drop. You may wonder how useful this idea is, since we do not understand liquid drops very well, either. But the point here is that the short range of the nuclear force allows nucleons to interact with their nearest neighbors only, as do the molecules in a liquid. The number of nearest neighbors does not change as the total number of nucleons changes; hence the binding energy of each nucleon does not change, and the total binding energy is simply proportional to the number of nucleons present. (We must realize, of course, that these and the following considerations cannot be expected to hold for nuclei containing only a few nucleons.)

$-a_2 A^{2/3}$: The binding to nearest neighbors causes a "surface tension"; since nucleons on the surface of the nucleus have fewer neighbors than nucleons on the inside, we must add a correction to the first term. We *reduce* E_B by an amount proportional to the surface area and hence proportional to $A^{2/3}$.

$-a_3(N - Z)^2/A$: This term is based on the Fermi gas model of the nucleus. Because neutron and proton levels are almost equal, there is a tendency, if we neglect electrostatic forces (accounted for in the next term), for the binding energy to be greatest when $N = Z$. If we start with equal N and Z, and then convert neutrons to protons or vice versa, the converted nucleons must occupy higher energy levels. The quadratic variation with $N - Z$ is logical, for the energy is increased (and the *binding* energy *decreased*) by an amount proportional to the number converted times the increase in energy of each one. But this increase for each one is itself proportional to the fraction $(N - Z)/A$ converted; the difference between N and Z produces a difference in the " height " to which the respective levels are filled. The reasoning is similar to that which gave us the T^2 dependence for the energy of a free electron gas.

$-a_4 Z^2 A^{-1/3}$: The electrostatic energy of a charge distribution can be easily seen to be proportional to the square of the charge and inversely proportional to the radius, which for a nucleus goes as $A^{1/3}$ (see Problem 6.) This electrostatic energy results from a repulsive force, so it tends to make the nucleus fly apart. This means that it reduces the binding energy; therefore this term appears with a minus sign. It is interesting that a_4, evaluated simply by observing *masses* of nuclei, may be used to calculate the nuclear *radius*, if one assumes a uniform charge distribution in the nucleus. (See Problem 6 again.) The resulting value of R_0, 1.216 fm, is consistent with values obtained from muonic X-rays and from electron scattering.

$\pm \delta(A)$: This term allows for the pairing effect, which makes even–even nuclei (those with even Z and even N) more tightly bound than odd–odd nuclei. The plus sign is used for even–even nuclei and the minus sign for odd–odd nuclei. If A is odd, this term is zero.

More sophisticated " mass formulas " have been developed since 1935, to take account of other small effects which determine the energy of the nucleus; for example, there is an exchange energy, analogous to the exchange energy of atomic levels (Section 8.2). But Eq. (2) is sufficiently accurate to be quite useful. The best fit to the observed masses of nuclei is obtained when the parameters of Eq. (2) are chosen as follows (values are given in MeV)[9]:

a_1	a_2	a_3	a_4	$\delta(A)$
15.75	17.80	23.69	0.7100	$39 A^{-3/4}$

14.6   PARITY

We can see from the general form of the spherical harmonics (Section 6.2) that a state of odd l has odd parity; the wavefunction of such a state changes sign upon reflection of the coordinate system:

$$u(x, y, z) = -u(-x, -y, -z)$$

A state of even l has even parity.

It is difficult to write down wavefunctions for the nucleons in a nucleus, but we know that the nucleons have orbital motions which contribute to the

[9] A. E. S. Green, *Phys. Rev.* **95**, 1006 (1954).

nuclear spin. These orbital motions must follow the general rules for angular momentum, and thus they must be quantized, with a quantum number l. Since the parity of a state with a given value of l is determined by the angular momentum operator and not by any particular force law,[10] we expect each nucleus to have a definite parity which is determined by the sum of the l values of the nucleons.

Unlike spin, the parity of a nucleus is not a directly measurable quantity. However, it is possible to determine differences in parity between different states of nuclei. For example, if a reaction occurs in which emitted particles carry away one unit of *orbital* angular momentum, the parity of the nucleus must change. Such reactions can be identified in various ways, as we shall see later. Assignment of parity to most of the known nuclear states can then be made if one can somehow determine the parity of only a few states.

In order to make this parity assignment, consider the even–even nuclei. Every even–even nucleus is found to have a spin of 0 in its ground state. According to the Fermi gas model, the *spins* of the *nucleons* cancel in pairs. Thus if the *total* angular momentum of the nucleus is zero, the *orbital* angular momentum must also be zero, and the state must have even parity.

Assignment of parity to nuclear states may seem at first sight to be an unnecessary exercise, but it becomes significant in the analysis of nuclear reactions. In all nuclear reactions, it has been found that parity is conserved; that is, the total parity of the initial nuclei equals the total parity of the final nuclei. This conservation law prohibits certain nuclear reactions which could otherwise occur without violating any other law. The fact that parity is conserved indicates that the nuclear force itself has even parity; a Hamiltonian operator constructed from the nuclear force does not operate on a wavefunction in such a way as to change its parity. This is not true of all forces; we shall see in Section 15.2 that the interaction which causes beta decay does not conserve parity.

14.7 STABILITY

The construction of the mass formula has shown us that in any nucleus there is a delicate balance between the nuclear force holding the nucleons together and the Coulomb force attempting to push them apart. The effect of these forces, combined with the exclusion principle, is to produce the binding energy given by Eq. (2) for *any given combination* of Z protons and N neutrons. If the binding energy is positive for a certain pair of values of N and Z, that combination of protons and neutrons is stable against *complete* disintegration into its constituent nucleons. But the combination might still

[10] The force law does not enter into the angular part of the wave equation unless the force is noncentral. It happens that the nuclear force does have a noncentral part, so that the energy levels of nuclei are not pure eigenstates of angular momentum; but it is still true that each state has a definite parity, because the nuclear force has even parity.

be unstable with respect to transformation into some other *bound* configuration containing the same total number of nucleons.

Study of nuclear masses provides us with a systematic way to investigate the question of stability. A nucleus containing A nucleons is stable in *all* respects if its mass is smaller than the mass of *any other possible combination* of A nucleons. If, however, there is a combination with lower mass, the nucleus may spontaneously transform into that combination. The mass difference then appears as kinetic energy, as in the examples in Section 14.5. There is nothing mysterious about this so-called "transformation of mass into energy"; the energy is there all the time, in the form of potential energy. Measurement of the mass simply evaluates this energy before the transformation takes place.

In some cases, a nucleus may be stable even though there are nucleon combinations with smaller mass, because there is no way in which a transformation from one configuration to the other can take place. Let us now see what sort of transformations can occur (and why others cannot occur).

Beta Stability. A nucleus may transform into one of its isobars (a nucleus with the same A but different Z) if its mass is greater than that of the isobar. Since the transformation does not change the number of bound nucleons, all that is required is that Z be "adjusted." If Z is too low, it may be increased by one unit by β^- emission; a neutron in the nucleus is transformed into a proton plus an electron, and the electron is ejected. If Z is too high, it may be reduced by one unit by β^+ emission; a proton changes to a neutron plus a positron, and the positron is ejected. The nucleus may also reduce its charge by capturing one of its orbital electrons; the electron disappears, combining with a proton in the nucleus to form a neutron.

We might therefore suspect that only one of a set of isobars, the one of lowest mass, is stable, for the others could eventually transform into that one by successive beta decays. This is true for odd A. When A is odd, the $\delta(A)$ term in the mass formula is zero. A plot of the mass versus Z then yields a parabola, with the masses of actual atoms lying at the integral Z values on the parabola (see Fig. 7). The only stable isobar is the one whose Z value ($Z = Z_2$) places it nearest to the bottom of the parabola. The atom with $Z = Z_1$ can decay to that with $Z = Z_2$ by emitting an electron, while the one with $Z = Z_3$ can decay to $Z = Z_2$ by capturing an orbital electron, or possibly by emitting a positron.

But for even A there can be two or more stable isobars, because the $\delta(A)$ contribution to the mass causes the mass values to lie on two different parabolas (see Fig. 8). The mass for a given Z lies on the upper parabola if Z is odd, and on the lower one if Z is even. The value of δ is such that an odd–odd nucleus is always unstable,[11] decaying into an even–even nucleus either by

---

[11] The formula breaks down for low A values, and some light odd-odd nuclei are stable: $^2_1$H, $^6_3$Li, $^{10}_5$B, and $^{14}_7$N.

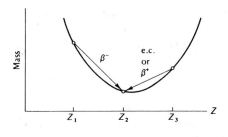

Fig. 7. *Dependence of mass on Z, for fixed odd A. Arrows indicate possible transitions, toward species of lowest mass.*

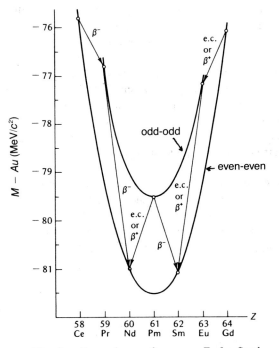

Fig. 8. *Dependence of mass on Z, for fixed even A. The vertical scale shows, in units of MeV/c², the so-called "mass excess" as listed in the Atomic Mass Table (Appendix D), for A = 146. The mass excess is simply the difference between the measured mass and the mass Au, where u is the atomic mass unit, equal to 931.50 MeV/c².*

beta-emission or electron capture. In the case illustrated, $_{60}$Nd is stable even though it is more massive than $_{62}$Sm, because it would have to decay to $_{62}$Sm by emitting two electrons simultaneously. Such *double* beta decay is extremely rare. (See Sec. 15.2.)

While β^- decays have been known since the discovery of radioactivity, β^+ decays were not proposed until 1933, because positron emitters do not occur in nature. The β decays in the long-lived radioactive families are all β^- decays, because emission of an alpha-particle from a heavy atom leaves it with a neutron excess rather than a proton excess. (The reason for the neutron excess is clear from Fig. 6 and the accompanying discussion.) The first positron decay observed was also the first observed case of induced or "artificial" radioactivity.[12] It was induced by the bombardment of ^{10}B with natural alpha rays. The reaction was

$$^{10}_{5}\text{B} + ^{4}_{2}\text{He} \rightarrow ^{1}_{0}\text{n} + ^{13}_{7}\text{N}$$

The $^{13}_{7}$N then decayed by β^+ emission to $^{13}_{6}$C; the half-life for this is about 10 min.

It sometimes happens that an atom may decay by electron capture but not by β^+ emission, although the converse is not true. Emission of a β^+ can occur only if the mass difference $M(Z) - M(Z - 1)$ is equal to two electron masses or more. To understand this, notice that when, for example, $^{13}_{7}$N emits a β^+, it becomes a ^{13}C *ion*, containing *seven* electrons, one more electron than the neutral ^{13}C atom contains. Adding the β^+ mass, we find that the total mass of the *products* is the mass of $^{13}_{6}$C plus *two* electron masses. In electron capture or β^- emission, this situation does not arise; when $^{14}_{6}$C goes to $^{14}_{7}$N plus a β^-, the ^{14}N has only 6 electrons. Its mass plus the mass of the β^- equals the mass of normal $^{14}_{7}$N.

Figure 8 shows that one species, $_{61}$Pm, can decay in several different ways. The probability that a given atom will decay in a time interval dt is the sum of the probabilities of the various types of decay. Thus for $_{61}$Pm, with decay constants λ_1 for β^- emission, λ_2 for β^+ emission, and λ_3 for electron capture, the number of atoms present at any given time t is given by

$$N = N_0 e^{-(\lambda_1 + \lambda_2 + \lambda_3)t}$$

and the half-life is $(\ln 2)/(\lambda_1 + \lambda_2 + \lambda_3)$. The value of each λ is independent of the others. For example, if all the electrons were removed from a $_{61}$Pm atom, so that λ_3 became zero, the total decay constant would be $\lambda = \lambda_1 + \lambda_2$ (the values of λ_1 and λ_2 being unchanged), and the half life would increase to $(\ln 2)/\lambda_1 + \lambda_2$. (See Problem 13).

One can measure a given decay constant λ_i by measuring the "partial activity" for that process, which is λ_i or $\lambda_i N_0 e^{-(\lambda_1 + \lambda_2 + \lambda_3)t}$. Notice that the partial

---

[12] I. Curie and F. Joliot, *C.R. Acad. Sci.* **198**, 254 (1934).

activity *decays* at a rate given by the *total* decay constant $\lambda_1 + \lambda_2 + \lambda_3$. One sometimes defines a "partial half life" *mathematically* as $(\ln 2)/\lambda_i$ for the *i*th process, but such a half life is never actually observed, because if several modes of decay are possible, one cannot restrict the decays to a single mode by any known technique. Thus the partial half life is always measured by determining the partial *activity*, i.e., by counting the number of decays of a known number of atoms via a specific mode. In doing this, it is necessary to accurately determine the detection efficiency for the mode in question.

Stability against Heavy-Particle Emission. Emission of a proton, a neutron, or an alpha particle may occur if the mass of the atom is greater than the mass of the emitted particle plus the mass of the atom which would remain after emission. Thus we could have

neutron emission if $\qquad M(Z, A) > M(Z, A - 1) + M_n \qquad$ (4a)

proton emission if $\qquad M(Z, A) > M(Z - 1, A - 1) + M_H \qquad$ (4b)

alpha particle emission if $\qquad M(Z, A) > M(Z - 2, A - 4) + M_{He} \qquad$ (4c)

where M_H and M_{He} are the masses of the hydrogen atom and the helium atom, respectively.[13]

We may, using Eq. (3), rewrite Eqs. (4a)–(4c) as

$$E_B(Z, A) < E_B(Z, A - 1) \qquad (4a')$$

$$E_B(Z, A) < E_B(Z - 1, A - 1) \qquad (4b')$$

$$E_B(Z, A) < E_B(Z - 2, A - 4) + E_B(He) \qquad (4c')$$

Thus neutron or proton emission cannot occur unless the binding energy decreases with increasing A. This does not occur for the species in the range of Fig. 5 (all naturally occurring species); you see there that E_B/A decreases at large A, but you may convince yourself by a brief numerical analysis that E_B continues to increase as A increases.

The situation is different for alpha decay, because of the presence of the binding energy of helium (about 28 MeV) in (4c'). In the vicinity of $A = 200$, the binding energy for beta-stable species increases by only about 6 MeV per nucleon, so that addition of four nucleons increases the binding energy by less than the binding energy of helium, making the heavier species unstable with respect to alpha decay. Study of the table of atomic species might even cause you to wonder why many more species are not listed as being alpha emitters. For example, ^{209}Bi is considered stable, although its mass exceeds that of ^{205}Tl plus the helium mass by 3.1 MeV; and there are other

---

[13] Masses of *atoms* rather than nuclei are used in all equations. In this case, the emitted particle is not a complete atom, but the remaining atom contains one or two extra electrons whose mass must be accounted for.

examples, all the way down to $^{142}_{58}$Ce, a "stable" species whose mass exceeds the sum of the masses of $^{138}_{56}$Ba and $^4_2$He by 1.4 MeV.

How do we account for this unexpected stability? The reason is found by examining the potential energy of an alpha particle in the field of a nucleus (Fig. 9). The potential is the sum of the Coulomb and nuclear potentials. An alpha particle with positive total energy E may still remain trapped inside the nucleus (at $r < R$). The alpha particle may eventually escape by tunneling through the barrier, but the penetration probability decreases rapidly with decreasing E. If E is too small, emission of alpha particles is too rare to be detected, and the species is considered stable. (We shall go into the relationship between energy and half-life in alpha decay in some detail in the next chapter.)

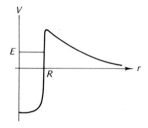

Fig. 9. *Potential energy and energy level of an alpha particle in a nucleus.*

Stability against fission may be analyzed in a somewhat similar way, even though the "ejected" particle in this case is almost as large as the nucleus which remains. Inspection of the nuclear mass table shows that *all* nuclei with $A \gtrsim 100$ are energetically unstable with respect to fission into two equal or nearly equal parts; but a barrier like that shown in Fig. 9 prevents fission from occurring *at a detectable rate* when A is much less than 230. The barrier is more effective than for alpha decay because the "particle" which must penetrate the barrier is more massive.

The fastest fission rate for a naturally occurring species is that for ^{238}U: 0.69×10^{-3} fissions per gram per second. This means that the partial half-life for fission is 10^{16} yr. Since the partial half-life for alpha decay is 4.5×10^9 yr, there are $10^{16}/(4.5 \times 10^9) = 2.2 \times 10^6$ alpha decays for each fission. The spontaneous fission rate of the famous isotope ^{235}U is much slower, with a partial half-life of 1.9×10^{17} yr.

Fission can also be *induced* by the absorption of a neutron by a heavy nucleus. In some cases, induced fission does not seem to be impeded by a barrier; the absorbed neutron provides the energy needed to surmount the barrier, and fission occurs immediately. This is best explained with an example: induced fission in ^{235}U and in ^{238}U. Although the *spontaneous* fission rate is higher for ^{238}U than for ^{235}U, *induced* fission goes more easily with ^{235}U.

Thermal neutrons induce fission in ^{235}U, while neutrons must have about 1 MeV of kinetic energy to induce fission in ^{238}U.

The difference in neutron energies needed to induce fission in ^{235}U and in ^{238}U can be understood in a simple way by reference to the mass formula. The key point is that the absorption of a neutron converts ^{235}U and ^{238}U into the "compound nuclei" ^{236}U and ^{239}U, respectively. (The term "compound nucleus" is often used to describe a nucleus which is formed when an incident particle is absorbed by a target nucleus. The idea of a compound nucleus will be very useful in our discussion of nuclear reactions.) If the incident neutron has thermal energy, the compound nucleus is formed with a mass equal to the total mass of the incident neutron and the target nucleus. This mass may be greater than the mass of the ground state of such a nucleus; if it is, the compound nucleus is formed in an excited state. The excitation energy of this excited state is the energy difference between this state and the ground state: $E_{exc} = c^2[M(Z, N - 1) + M_n - M(Z, N)]$, where the mass values $M(Z, N - 1)$ and $M(Z, N)$ are the ground state masses of target and compound nucleus, respectively, which are given by the mass formula. For U^{236} and U^{239}, we have the excitation energies:

$$E_{exc}^{236} = c^2[M(92, 143) + M_n - M(92, 144)]$$

$$E_{exc}^{239} = c^2[M(92, 146) + M_n - M(92, 147)]$$

Comparing the two, we see that, since the values of Z are all the same and the values of N nearly so, the major difference between the two values of E_{exc} should come from the $\delta(A)$ term in the mass formula. $M(92, 144)$ and $M(92, 146)$ each contain a term $-\delta$ because they are even–even, while the other M contain no δ term. This makes

$$E_{exc}^{236} - E_{exc}^{239} \approx 2\delta \approx 1 \text{ MeV}$$

(Check this from the data at the end of Section 14.5.) The excitation energy of ^{239}U is 1 MeV less than that of ^{236}U when they are both formed from thermal neutrons. Since we expect the barrier height to be about the same in both nuclei, we expect that fission will occur rapidly at about the same excitation energy in each. Thus we must provide an extra 1 MeV, in the form of kinetic energy of the incident neutron, to give the ^{239}U the same fission rate as the ^{236}U.

EXAMPLE PROBLEM 1. Use the semi-empirical mass formula, with constants given on page 542, to compute the total kinetic energy of the products in the induced fission reaction:

$$\tfrac{1}{0}n + \tfrac{235}{92}U \rightarrow \tfrac{95}{39}Y + \tfrac{139}{53}I + 2\tfrac{1}{0}n$$

Solution. The mass of ^{235}U equals 92 hydrogen masses plus 143 neutron masses minus the binding energy given by Eq. (2). Equating this, plus the incident neutron mass, to the product masses (which are found in the same way) plus the kinetic energy, we have

$$M_n + 92M_H + 143M_n - E(235, 92) = 39M_H + 56M_n - E_B(95, 39) + 53M_H$$
$$+ 86M_n - E_B(139, 53) + 2M_n$$
$$+ \text{kinetic energy}$$

The M_H and M_n terms balance, as they should, since no particles are created or annihilated. We find that

Kinetic energy

$$= E_B(95, 39) + E_B(139, 53) - E_B(235, 92)$$
$$= a_1(95 + 139 - 235) - a_2(95^{2/3} + 139^{2/3} - 235^{2/3})$$
$$- a_3\left(\frac{17^2}{95} + \frac{33^2}{139} - \frac{51^2}{235}\right) - a_4\left(\frac{39^2}{95^{1/3}} + \frac{53^2}{139^{1/3}} - \frac{92^2}{235^{1/3}}\right)$$
$$= 15.75 \cdot (-1) - 17.80 \cdot (9.57) - 23.69 \cdot (-0.28) - 0.7100 \cdot (-495.1)$$
$$= -15.75 - 170.35 + 6.64 + 351.52$$
$$= 172 \quad \text{MeV}$$

For a rough check on this answer, we may refer to Fig. 5. There we see that the binding energy *per nucleon* is about 8.5 MeV in the region near $A = 100$, and it is about 7.5 MeV for $A > 200$. Thus about 1 MeV per nucleon, or 235 MeV, should be released in the fission of ^{235}U. Notice that the 172 MeV calculated is not the full amount of energy which is eventually released, because the products ^{95}Y and ^{139}I are highly unstable, being extremely rich in neutrons. They decay by successive β^- emissions to $\tfrac{95}{42}Mo$ and $\tfrac{139}{57}La$, respectively, releasing additional energy. We see that the fission reaction is about 0.1% efficient; that is, about 0.1% of the initial rest mass appears as kinetic energy.

There are many other ways in which the fission of ^{235}U can go, but in each case two or three neutrons are released to induce other fissions, making a "chain reaction" possible. The release of neutrons is a consequence of the fact that the "neutron excess" $N - Z$ is much greater for heavy nuclei than for lighter nuclei. This fact may be understood from a simple energy-level diagram such as Fig. 6.

If each of the neutrons produced in a fission were to induce another fission, the fission rate would rapidly escalate. This occurs in a bomb when a "critical mass" is reached. When the mass is smaller than critical, the surface-to-volume ratio is so large that fewer than one neutron per fission is absorbed in the fissionable material. In a reactor, a stable situation is achieved by means of control rods which absorb just enough neutrons so that exactly one neutron per fission is captured to induce another fission. Thus the fission rate remains constant. To increase this rate, the control rods are partly removed; to reduce the rate, the rods are inserted more deeply.

The time required for a neutron to find another fissionable atom is of the order of microseconds. If only 1.01 neutrons were captured, on the average, within ten *micro*seconds of each fission, the fission rate would increase by a factor of $(1.01)^{1000}$, or more than 10^4, in ten *milli*seconds. Control of the reactor would be virtually impossible, because it would respond far too quickly to any movement of the control rods. Fortunately, some of the neutrons produced in a fission are "delayed"; they are emitted many seconds after the fission, from neutron-emitting nuclei that result from beta decays of the fission products. For example, one possible fission product is ^{87}Br, whose half life is 55 s. It decays to an excited state of ^{87}Kr, which immediately emits a neutron to become ^{86}Kr. Thus, when a control rod is withdrawn, the fission rate increases gradually as ^{87}Br and other fission products decay to neutron emitters, and there is time to make adjustments.

14.8 SHELL STRUCTURE

The major inaccuracy in the mass formula is caused by fluctuations attributable to shell structure in the nucleus. Just as atoms with closed electron shells are exceptionally stable, nuclei with certain "magic numbers" of neutrons or protons are more tightly bound than the average nucleus. These nuclei have a lower mass and a greater abundance than would otherwise be expected. They also have a very small quadrupole moment, indicating a nearly spherical shape (see Fig. 10). The magic numbers are 2, 8, 20, 28, 50, 82, and 126. Nuclei containing a magic number of nucleons of one type (neutron or proton) are very reluctant to add another nucleon of the same type. The additional nucleon, when added, is very loosely bound, like the valence electron in an alkali metal.

Shell effects in nuclei were known or suspected for a long time, but there were difficulties in explaining them theoretically. The shell effects in atomic structure are produced by the spin–orbit interaction, but it is hard to think of orbits in such a dense body as a nucleus. Wigner, however, pointed out that when all of the available nuclear states are filled, a nucleon can "collide" with another nucleon without changing the state of motion of either; in effect, one nucleon passes right through another one. Other difficulties which a theory must face are that there is no fixed center of force as there is in an atom, and that the exact shape of the potential is not known even for two nucleons.

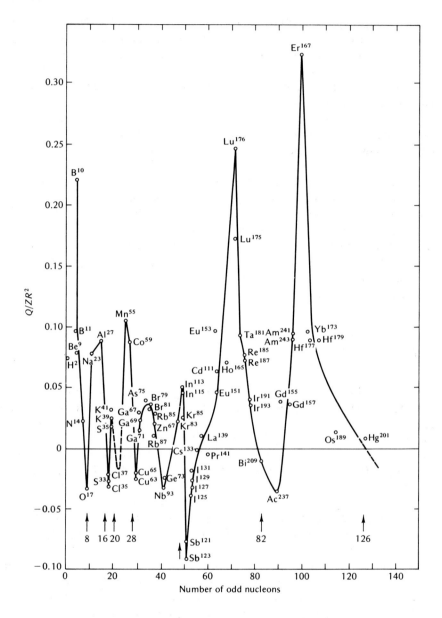

Fig. 10. *Reduced nuclear quadrupole moments as a function of the number of odd nucleons. The quantity Q/ZR^2 gives a measure of the nuclear deformation independent of the size of the nucleus. [From E. Segre, "Nuclei and Particles." W. A. Benjamin, New York, 1964.]*

The difficulties were overcome in 1949 by Maria Mayer and (independently) by Hans Jensen; they shared the 1963 Nobel Prize in physics with Wigner, whose ideas laid the groundwork for their success. They assumed that the nucleon moves in a potential well formed by all the other nucleons, and that the shape of the well must be somewhere between that of a square well and a harmonic oscillator. This in itself produces groups of energy levels separated by gaps which indicate "shell closure" at 2, 8, and 20 states. (See Fig. 11.) Unfortunately, with any reasonable potential, gaps also appear at "non-magic" numbers such as 40, while there is no gap at 50 or at 82.

In order to produce gaps at the correct magic numbers, Mayer and Jensen assumed that the nuclear force also includes a spin–orbit interaction, like that produced by electromagnetic forces for atomic electrons, but much stronger. Figure 11 shows the effect of the spin–orbit interaction on the levels produced by a potential intermediate between the harmonic oscillator and square-well potential. (The overall picture is not affected very much by the shape of the potential, within these limits.) The spin–orbit interaction splits each level of a given l into two levels, of total angular momentum $j = l + \frac{1}{2}$ and $j = l - \frac{1}{2}$, respectively.

By choosing the proper strength for the spin–orbit interaction, Mayer and Jensen were able to adjust this splitting to produce the correct magic numbers. The sign of the interaction also had to be chosen so that the state with $j = l - \frac{1}{2}$ has a higher energy than the state with $j = l + \frac{1}{2}$, for both neutrons and protons.

It may not seem to be much of a feat to produce the seven correct magic numbers, with so many parameters to manipulate. But measurements of the spins and parities of many nuclei make it clear that the shell model accurately describes real properties of the nucleus. For example, since a closed shell should have spherical symmetry, the angular momentum and parity of the nucleus should be determined by the nucleons outside the closed shells, and measurements confirm that this is so. If there is only one nucleon outside the closed neutron and proton shells, the nuclear spin is that of the next shell-model level above the closed shell: ^{17}O has spin $\frac{5}{2}$ and even parity because the ninth neutron is in a $d_{5/2}$ state. Further, if a nucleus lacks only one nucleon of having filled proton and neutron shells, its spin and parity are that of the unoccupied level, just as in the case of atomic electron shells: ^{15}N, with a full shell of neutrons and one proton less than a full shell, has spin $\frac{1}{2}$ and odd parity, because the empty proton state is a $p_{1/2}$ state.

These considerations apply also to the individual levels as well as shells: $^{33}_{16}S$ has spin $\frac{3}{2}$ and even parity, because it contains filled levels up to the $2s_{1/2}$ level, and the odd neutron is in the $d_{3/2}$ level. And a pair of like nucleons outside a filled level has zero total angular momentum, so that, for example, $^{35}_{17}Cl$, which, like ^{33}S, has filled levels up to the $2s_{1/2}$ level, also has spin $\frac{3}{2}$ and even parity; the

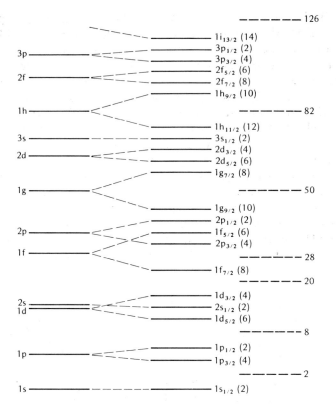

Fig. 11. *Shell-model energy levels. At left are the energy levels produced by a potential intermediate between the harmonic oscillator and square-well potentials. Levels are labeled by the radial quantum number and the spectroscopic symbol for the orbital angular momentum. At right are the levels after splitting by the spin-orbit force. The numbers at far right show the number of states below the various gaps between levels; notice the magic numbers. The order of the levels is correct as shown, for protons; above N = 50, the neutron levels differ a bit from those shown here. Energy differences between levels are not drawn to scale.*

odd proton is in the $1\,d_{3/2}$ level and the odd neutrons combine to have zero angular momentum.

An enormous number of shell model predictions are therefore possible.[14] You may wish to check the data of Appendix D to find your own examples, compar-

---

[14] For a five-page list of observed and predicted spins and parities, see the article by M. Mayer and H. Jensen, in " Alpha, Beta, and Gamma-Ray Spectroscopy," (K. Siegbahn, Ed.) Vol. I, p. 557, North-Holland, Amsterdam, 1965.

ing the observed spins and parities with the level assignments in Fig. 11. There are a few exceptional cases, such as ^{19}F, where the results do not agree, but even in these cases there is a low-lying excited state with the spin one would have expected from the shell model levels, and the observed ground-state spin agrees with the spin of a low-lying shell model level.

14.9 COLLECTIVE MOTIONS

In spite of the impressive success of the shell model, many problems remain in attempts to describe nuclear properties. The shells in nuclei are not nearly so "inert" as atomic shells, and it is not possible to explain all nuclear properties by considering only the nucleons in unfilled shells. Much recent work has been done in developing the theories of Bohr and Mottelson, which deal with effects of collective motions of the nucleons in the closed shells, in attempts to derive nuclear properties and to predict properties of excited states.

The presence of these collective motions may be deduced from the quadrupole moments of certain nuclei (Fig. 10), which are much larger than the moments which would result from uncorrelated motions of nucleons outside closed shells. These large quadrupole moments occur particularly in the region $150 < A < 190$, where there are many nucleons outside closed shells and the nuclei are consequently "deformed" into a nonspherical shape. One can assume that the nucleus as a whole takes on a shape like those of Fig. 3, and that the motions of the outer nucleons deform the closed shells as well.

Further evidence of collective motion in nuclei in this region is seen in the energy levels of the even–even nuclei, a few of which are shown in Fig. 12. Each of the lowest energy levels in one of these nuclei has an energy which may be expressed as $E = I(I + 1)\hbar^2/2\mathscr{I}$, which is simply the energy of a rotating body whose angular momentum is $\hbar(I(I + 1))^{1/2}$ and whose moment of inertia is $\mathscr{I}$. The quantum number I takes on only even values; thus these "rotational levels" have energies which are in the ratios $6:20:42:72$, corresponding to the values of $I(I + 1)$ for $I = 2, 4, 6$, and 8, respectively.

> The fact that the value of I is always even is of interest in itself, for it is one of the pieces of evidence which indicate the existence of a "pairing force" between nucleons in the nucleus. The pairing force causes nucleons to form pairs such that the members of a pair have equal and opposite momenta at all times. Incidentally we might notice that the spacing of these energy levels is further evidence for the validity of the quantum-mechanical formula for angular momentum eigenvalues; it would be difficult to explain these levels in any other way.

Of course the internal motions of nuclei are extremely complicated, and they do not rotate as *rigid* bodies even when they are in such rotational states. But the regularity in the energy of these levels is impressive. Great progress

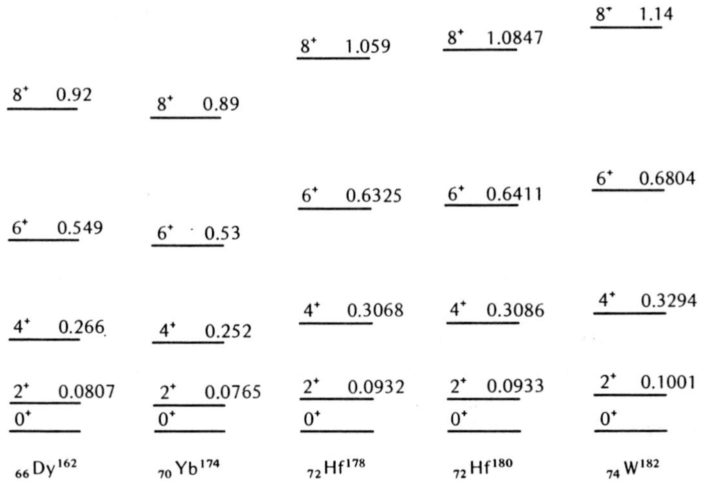

Fig. 12. *Rotational energy levels of some deformed nuclei showing spin and parity of each level. For each nucleus, the energy of each level is proportional to I(I + 1); thus energies are in the ratio 6:20: 42:72 for 2+, 4+, 6+, and 8+ levels, respectively.*

has been made in explaining other levels and other nuclear properties on the basis of collective motions.

14.10 PROPERTIES OF THE DEUTERON

Let us close our discussion of nuclear properties by focusing our attention on a single nucleus—the deuteron, the nucleus of ^{2}H. The deuteron, because it contains only two nucleons, occupies a unique position in nuclear physics, comparable to the position of the hydrogen atom in atomic spectroscopy. We might expect the deuteron to yield clues to the nature of the nuclear force, clues which are obscured when many nucleons interact.

The following properties of the deuteron are of special interest:

1. It has only one bound state, of binding energy 2.22 MeV.
2. The spin is 1.
3. The quadrupole moment is $+0.274 \times 10^{-26}$ cm^2.
4. The magnetic moment is $+0.85741$ nuclear magnetons.

In the first property, we see a striking difference between the deuteron and the hydrogen atom, which has many energy levels. The presence of only one bound state is a consequence of the short range of the nuclear force. We have seen that the number of possible states in a square well depends on the width of the well, while for the Coulomb potential, which extends to infinity, there are an infinite number of levels. In a way, the presence of only one bound deuteron state is unfortunate. If there were many levels, the energies of these levels could be used to determine the shape of the nuclear potential.

As it is, we may assume some arbitrary shape, such as a spherically symmetrical square well, for the potential, and then attempt to determine the depth of the well. In order to do this we must know, in addition to the binding energy, the radius of the well. This radius may be deduced from experiments in neutron–proton scattering to be about 2 fm. Calculation of the well depth is then a straightforward quantum-mechanical boundary value problem. The

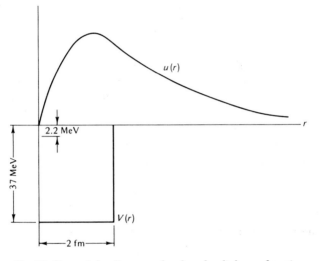

Fig. 13. *Potential well, energy level, and radial wavefunction for the deuteron.*

ground state of such a potential is an S state, so that the radial Schrödinger equation reduces to the same form as the one-dimensional square-well equation which we have previously solved (Chapter 5). Solution of the equation shows that the depth of the well is about 37 MeV, and that there are no excited bound states, in agreement with observation. (You can verify this; see Problem 17.) The situation is shown in Fig. 13. Notice that the wavefunction extends quite far beyond the limits of the well; each nucleon spends more than half the time beyond the range of the force exerted by the other nucleon (where the kinetic energy of the system is negative, so that it remains bound). It is clear that the nucleons are just barely bound, for the slope of the wavefunction is only slightly negative at the edge of the well.

Property 2 of the deuteron shows us another difference between the nuclear force and the electromagnetic force. In the hydrogen atom, the magnetic coupling between the proton spin and electron spin is so weak that it has little effect on the energy levels; the atom can have either spin 1 or spin 0, and the energy difference between the two spin states is minuscule—6×10^{-6} eV. But in the deuteron, the interaction between the spins of neutron and proton is so great that the spin 0 state is not even bound! The potential well shown in Fig. 13 therefore applies only to the spin 1 state; analysis of scattering of neutrons by protons shows that the well depth for the spin 0 state is only about 12 MeV, a depth insufficient to allow a bound state.

EXAMPLE PROBLEM 2. Make an order-of-magnitude calculation of the *magnetic* energy between the proton and neutron spins in the deuteron, to show that this energy cannot be responsible for the difference in well depths quoted above.

Solution. We shall base the calculation on the above-quoted energy of 6×10^{-6} eV for the spin-spin interaction in the hydrogen atom. The potential energy for a magnetic dipole pair is proportional to $1/r^3$, and r in the deuteron is smaller, by a factor of 3×10^4, than it is in the hydrogen atom. The energy is also proportional to the magnetic dipole moment of each particle, and the dipole moment of the electron is about 1000 times that of the neutron. The magnetic interaction energy in the deuteron is therefore about $(3 \times 10^4)^3 \times 10^{-3}$ times the energy in the hydrogen atom; this is only 200 keV, significantly smaller than the difference in the two well depths.

The strong spin-dependence of the nuclear force has other effects, one of which is the nonzero quadrupole moment. We have previously shown that the ground state of any central force is spherically symmetrical. A system in such a state could have no quadrupole moment. But, with its strong spin dependence, the nuclear force is noncentral, and the ground state of the deuteron is not spherically symmetrical.

The lack of spherical symmetry shows up in the magnetic moment as well as in the quadrupole moment. If the deuteron were in a pure S state, the deuteron spin would be simply a result of the alignment of the proton and neutron spins. In that case the magnetic moment would be the algebraic sum of the proton and neutron moments, or $2.79278 - 1.91314 = 0.87964$ nuclear magnetons. This is *close* to the actual value of the moment, so we may assume that the ground state is primarily the $^3S_1$ state,[15] but we must assume that there is also a small admixture of another state, which has to be the $^3D_1$ state. (Two other states, the $^3P_1$ and $^1P_1$, also have one unit of total angular momentum, but we rule these out because they have odd parity. Under the

---

[15] The labeling of states is as in atomic spectroscopy (Section 9.1). The superscript 3 indicates the triplet spin state (spins parallel); the S indicates zero orbital angular momentum, and the subscript 1 indicates that the total angular momentum quantum number is 1.

influence of the nuclear force, which conserves parity, states with odd parity cannot combine with states of even parity. And there is no combination of the two P states alone which would produce the observed magnetic moment.)

EXAMPLE PROBLEM 3. Evaluate the magnetic moment μ for a deuteron in the $^3S_1$ state and for one in the $^3D_1$ state, and find how these moments must be weighted to provide a mixture which gives the observed moment of the deuteron.

Solution.   We start with the general expression for μ:

$$\mu = \mu_{spin} + \mu_{orbital} \tag{5}$$

In both the $^3S_1$ and the $^3D_1$ states, neutron and proton *spins* are parallel, so that the internal *magnetic moments* of neutron and proton are in opposite directions, and their resultant is

$$\mu_{spin} = (2.793 - 1.913)\left(\frac{e}{2M_p}\right)\mathbf{S}$$

$$= 0.880\left(\frac{e}{2M_p}\right)\mathbf{S} \tag{6}$$

where **S** is the total spin vector. The orbital motion of the *proton* produces

$$\mu_{orbital} = \left(\frac{e}{2M_p}\right)\mathbf{L_p} = \left(\frac{e}{4M_p}\right)\mathbf{L} \tag{7}$$

because $\mathbf{L_p}$, the orbital angular momentum of the *proton*, equals one-half of the total orbital angular momentum **L**. Inserting Eqs. (6) and (7) into Eq. (5) yields

$$\mu = \left(\frac{e}{2M_p}\right)\left(0.880\ \mathbf{S} + \frac{\mathbf{L}}{2}\right)$$

or, since $\mathbf{I} = \mathbf{S} + \mathbf{L}$,

$$\mu = \left(\frac{e}{2M_p}\right)\left(0.380\ \mathbf{S} + \frac{\mathbf{I}}{2}\right) \tag{8}$$

We now apply the reasoning used in analyzing the Zeeman effect (Section 9.2). Because μ precesses rapidly around **I**, the average value of μ is a vector in the direction of **I**, whose magnitude is $\mu \cdot \mathbf{I}/|\mathbf{I}|$; the z component of the average μ is therefore[16]

---

[16] Be sure to notice the distinction between the vector **I** and the quantum number *I*.

$$(\mu_{av})_z = \frac{(\mathbf{\mu} \cdot \mathbf{I})I_z}{|\mathbf{I}|^2}$$

$$= \frac{(\mathbf{\mu} \cdot \mathbf{I})m_I \hbar}{I(I + 1)\hbar^2} \tag{9}$$

The measured magnetic moment is the maximum value of this $(\mu_{av})_z$, which we find by putting m_I equal to 1 in Eq. (9). Then, since the quantum number I is also 1, the *measured* moment is

$$\mu_m = \frac{(\mathbf{\mu} \cdot \mathbf{I})}{2\hbar}$$

or, using Eq. (8) for the *vector* $\mathbf{\mu}$,

$$\mu_m = \left(\frac{e}{4M_p\,\hbar}\right)(0.380\,\mathbf{S} \cdot \mathbf{I} + \tfrac{1}{2}\mathbf{I} \cdot \mathbf{I}) \tag{10}$$

To evaluate μ_m for the two states in question, we need $\mathbf{S} \cdot \mathbf{I}$, which we find by writing

$$\mathbf{L} \cdot \mathbf{L} = (\mathbf{I} - \mathbf{S}) \cdot (\mathbf{I} - \mathbf{S})$$
$$= \mathbf{I} \cdot \mathbf{I} + \mathbf{S} \cdot \mathbf{S} - 2\,\mathbf{S} \cdot \mathbf{I}$$

so that

$$\mathbf{S} \cdot \mathbf{I} = \tfrac{1}{2}(\mathbf{I} \cdot \mathbf{I} + \mathbf{S} \cdot \mathbf{S} - \mathbf{L} \cdot \mathbf{L})$$
$$= \tfrac{1}{2}\hbar^2[I(I + 1) + S(S + 1) - L(L + 1)]$$

For the $^3D_1$ state, $I = 1$, $S = 1$, and $L = 2$, so $\mathbf{S} \cdot \mathbf{I} = \tfrac{1}{2}\hbar^2(2 + 2 - 6) = -\hbar^2$. We also know that $\mathbf{I} \cdot \mathbf{I} = 2\hbar^2$, so we may substitute into Eq. (10) to obtain, for the $^3D_1$ state, $\mu_m = (e\hbar/4M_p)(-0.380 + 1.000) = 0.310$ nuclear magnetons. For the $^3S_1$ state we may deduce directly from Eq. (6) that $\mu_m = 0.880$ nuclear magnetons.

If the deuteron is a mixture of p parts of $^3D_1$ state and $(1 - p)$ parts of $^3S_1$ state, the observed moment should be

$$\mu_m = 0.310\,p + 0.880(1 - p)$$

Setting this expression equal to the measured value of 0.857, and using a little algebra, we find that $0.570p = 0.023$, or $p = 0.04$. Thus the deuteron state must be 4 percent $^3D_1$ and 96 percent $^3S_1$.

PROBLEMS

1. Assuming that the nucleus has a uniform charge distribution of radius R, show that the electrostatic potential at a point inside the nucleus, at a

distance r from the center, is given by $V(r) = Ze(3R^2 - r^2)/8\pi\varepsilon_0 R^3$, in mks units.

2. Use the result of Problem 1 to compute the electrostatic potential energy of a K-shell electron, using a hydrogen atom wavefunction (Table I of Chapter 7) for the electron. This is done by evaluating

$$W = \int_0^{+\infty} eV(r)P(r) \, dr,$$

where $P(r) = 4\pi r^2 |u(r)|^2$ is the radial probability density for the electron. (*Hint*: The integral must be evaluated in two steps, for $r < R$ and for $r > R$. Use the fact that R is much smaller than the Bohr radius a_0.). Compare W for the two isotopes $^{120}_{52}\text{Te}$ and $^{130}_{52}\text{Te}$.

3. Carry out the procedure outlined in Section 14.2 to find the effective radius of a nucleus with $A = 125$, using the values of c and t given in that section. You may approximate the charge distribution of Fig. 1 by straight lines in order to simplify the integration. Set your resulting value of R to $R_0 A^{1/3}$ and find R_0 for this case.

4. Show that the frequency which produces transitions in a magnetic resonance experiment is equal to the Larmor precession frequency caused by the torque exerted by the applied vertical field **B** on the magnetic moment. (Thus the oscillating horizontal field drives the system at a frequency which is equal to its natural frequency of oscillation, as in classical resonance.)

5. In one experiment done by Rabi's group, the transmission of a beam of ^7Li was sharply reduced when B was 3380 G in region II (Fig. 2). Find the frequency of the oscillating field, given that $\mu = 3.26$ nuclear magnetons for the ^7Li nucleus, and $I = \frac{3}{2}$. (Remember that the μ quoted here and in tables is always the maximum value of a component of μ along a given axis.) Why must you know I as well as μ to solve this problem?

6. Show that the energy of a uniform distribution of charge q with radius R is $\frac{3}{5}(q^2/4\pi\varepsilon_0 R)$. (*Hint*: Consider the work done in bringing a spherical shell of charge dq from infinity, when a charge q is already present.) Use this result to verify that the value $a_4 = 0.7100$ MeV in the mass formula leads to $R_0 = 1.216$ fm, as stated in Section 14.5.

7. On the 1961 scale of atomic mass units, one u is $\frac{1}{12}$ of the mass of ^{12}C. This scale replaces separate scales used previously in physics and chemistry; one u was $\frac{1}{16}$ of the mass of ^{16}O in the physical scale, whereas one u in the chemical scale was $\frac{1}{16}$ of the average mass of the normal isotopic mixture of

oxygen. Find the percent difference between the present scale and each of the older scales, using the table of atomic masses (Appendix D) and the fact that "normal" oxygen is 99.759% ^{16}O, 0.037% ^{17}O, and 0.204% ^{18}O.

8. Write the expression for Q_{ij} (Section 14.4) in a form suitable for point charges, by replacing the integral by a sum. Use this expression to compute the value of Q_{33} for two point charges, each of charge $q/2$, located at $z = +a/2$ and $z = -a/2$, respectively. Then compute the quadrupole energy of these charges in terms of $\partial^2\phi/\partial z^2$, and compare your result with the quadrupole energy of $(q/2)[\phi(a/2)+\phi(-a/2)] - q\phi(0)$ quoted in the text for such a pair of charges. (Assume that $\partial^2\phi/\partial z^2$ is constant for $-a/2 < z < +a/2$.)

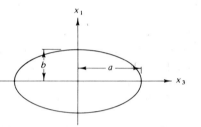

9. Consider an ellipsoid of revolution formed by rotating the ellipse shown about the x_3 axis. If this ellipsoid contains a uniform charge density ρ,

   (a) show that $Q_{33} = \frac{4}{15}\pi\rho b^2 a^3$;

   (b) show that $Q_{22} = Q_{11} = \frac{4}{15}\pi\rho b^4 a$;

   (c) show that the total charge is $q = \frac{4}{3}\pi\rho b^2 a$.

   (d) The quadrupole moment Q for $^{27}_{13}Al$ is quoted as $Q = -0.157 \times 10^{-24}$ cm^2. Use this value and the results of (a), (b) and (c) to compute the percent difference between the axis a and the axis b, if the ^{27}Al nucleus has this kind of shape. (*Hint:* Notice that the $x_1 x_2$ cross sections are circles, and the $x_2 x_3$ cross sections are ellipses.)

10. Why is it that there can be as many as nine or ten stable A values for a given Z, although there are, at most, three stable Z values for a given A? Explain with reference to the case of Xe, which has stable isotopes at $A = 124, 126, 128, 129, 130, 131, 132, 134$, and 136; draw qualitative mass versus Z diagrams for these values of A, showing how, as A changes, the Z-value of the minimum in the parabola shifts. (Notice that there are three stable species with $A = 124$ and three with $A = 136$.)

11. Use the mass formula to compute the Z-value of the beta-stable (lowest mass) species with $A = 27$ and with $A = 125$. (*Hint:* Find $\partial M/\partial Z$.) Comment on any discrepancies between your result and the observed facts.

12. Suppose that a certain species has partial half-lives of 1 yr for β^- decay, 2 yr for β^+ decay, and $\frac{1}{2}$ yr for electron capture. If there are N_0 atoms of this species present at $t = 0$, and each decay product is stable, find how many atoms of the original species and of each decay product are present at $t = 1$ yr.

13. Given that 43% of all ^{64}Cu decays are by electron capture, find the half life of a *bare nucleus* of ^{64}Cu. (See Appendix D for the half life of a ^{64}Cu *atom*.)

14. Using the masses given in Appendix D, and the stability rules of Section 14.7, construct the four radioactive series (Section 13.1). (One of these series is given in Problem 5, Chapter 13. Notice that a series splits into two branches when it reaches a species which is unstable with respect to both alpha and beta decay.)

15. Determine, from the masses given in Appendix D, the modes of decay which are energetically possible for each species with $A < 10$. Consider the possibilities of electron capture, beta decay (+ or −), and heavy-particle emission (n, p, or α).

16. We can use the shell model to predict nuclear magnetic moments as well as spins and parities. These predictions do not stand up so well as those on spin and parity, possibly because the structure of a nucleon is distorted in the nucleus so that the nucleon's *intrinsic* magnetic moment differs from the moment which it has as a free nucleon. However, the predictions for "closed shell ± 1" nuclei are good; for example, the magnetic moment of ^{17}O is -1.89, only 1 percent different from the intrinsic moment of the odd neutron. Compute the expected magnetic moment of ^{15}N, by computing the moment of a proton in the (unfilled) $p_{1/2}$ state; compare your answer with the observed value. (See Example Problem 3 for the technique of combining spin and orbital contributions to the magnetic moment of a nucleus.)

17. Find the depth of a square well which has a radius of 2 fm and which contains a single energy level at -2.2 MeV, in order to verify the accuracy of Fig. 13 (see Section 6.6).

18. Use the table of atomic masses to calculate the binding energy of the 50th neutron in various nuclei with $N = 50$, and show that this neutron is more tightly bound than the last neutron in comparable nuclei with $N \neq 50$. (Choose your comparisons carefully, in order to illustrate the shell effect rather than other effects; remember that the last neutron is always more tightly bound when N is even than when N is odd, so compare only nuclei of even N. Also remember that the binding energy depends on $N - Z$, so compare with nuclei which have the same value

of $N - Z$. For example, compare ^{86}Ki with ^{82}Se and ^{90}Sr; in all three of these, $N - Z = 14$, and N is even.)

19. Program a computer to use Eqs. (2) and (3), with values of the constants as given in Section 14.5, to make a table of atomic masses as a function of A and Z. Compare this table with Appendix D and identify shell effects which may cause the tables to differ. (You may prefer to write your program to give the mass *excess* rather than the mass, to facilitate comparison with Appendix D.)

20. Use Eq. (33), Chapter 11, to compute the Fermi energy of the neutrons in a nucleus of ^{27}Al.

21. Verify from the atomic mass table that a neutron-emitting excited state of ^{87}Kr is energetically possible as a result of the beta decay of ^{87}Br. (See Section 14.7.)

Nuclear Transformations

We now shift our attention from the static properties of nuclei to the study of nuclear transformations, including alpha, beta, and gamma decay as well as transformations caused by collisions. These processes are interesting in themselves; in addition, they provide information about excited states of nuclei which can be used to test various nuclear models.

15.1 THEORY OF ALPHA DECAY

The alpha decay process is relatively easy to understand, because the particles emitted exist in the nucleus before the decay. The mechanism of alpha decay was nevertheless a mystery before the development of the quantum mechanical theory of barrier penetration.

To analyze this process, we begin with the potential energy diagram for an alpha particle in the field of a nucleus (Fig. 1), which is the sum of the Coulomb and nuclear potentials.[1] One would think classically that an alpha

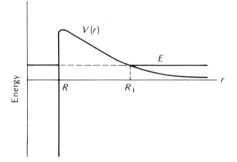

Fig. 1. *Energy level and potential energy of an alpha particle in the field of a nucleus. The bottom of the potential well is too deep to be shown on this scale.*

particle inside the nucleus could not escape unless its total energy E exceeded the maximum value of $V(r)$, for it would otherwise have to traverse a region in which its kinetic energy would be negative. But we know that alpha particles with energy of a few MeV do escape from heavy nuclei, and it is easy to calculate that the Coulomb potential energy of the alpha particle is, in MeV,

$$V(r) = 2.88 \frac{Z}{r} \tag{1}$$

if r is in fermis. Therefore, for a typical alpha particle, with $E = 4.6$ MeV and $Z = 80$, $V(r) = E$ at $r = 50$ fm, well beyond the range of the nuclear force; $V(r)$ must continue to increase as r decreases, until r approaches the nuclear radius ($\sim R$), so that $E < V(r)$ over a considerable range of values of r.

[1] See also Fig. 9, Chapter 14, and the accompanying discussion in Chapter 14.

The explanation of these facts appeared in 1928, when Gamow and (independently) Condon and Gurney applied the newly developed theory of barrier penetration to the problem. It is clear that this theory provides a mechanism whereby the alpha particle can escape. But it does more than simply account for this one fact; its success in accounting for the relation between the alpha particle energy and the half-life of the emitter[2] is equally important, constituting a major confirmation of the quantum theory. Let us see how this relation is derived.

If there is a preformed alpha particle knocking about inside the nucleus, the particle strikes the potential barrier each time it crosses the nuclear diameter, so if its velocity is v, it strikes the barrier $v/2R$ times per second. Each time it strikes the barrier the probability that it will escape is equal to the transmission coefficient T for that barrier. The decay probability per unit time is therefore

$$\lambda = \frac{v}{2R} T = \frac{T}{2R} \left(\frac{2E}{m}\right)^{1/2} \tag{2}$$

where m is the reduced mass of the alpha particle, equal to about 98 percent of the alpha particle mass when the A of the nucleus is 200.

To find T, we refer to Section 5.5, where we found that the penetration probability for a *square* barrier is

$$T = \left[1 + \frac{V_0^2 \sinh^2 \beta a}{4E(V_0 - E)}\right]^{-1} \qquad (E < V_0) \tag{3}$$

where a is the thickness of the barrier and V_0 is its height, and $\beta = (2m(V_0 - E)/\hbar^2)^{1/2}$. If $\beta a \gg 1$, we can write Eq. (3) as

$$T = \frac{16E}{V_0}\left(1 - \frac{E}{V_0}\right)e^{-2\beta a} \tag{4}$$

How can we apply this result to the alpha decay problem, where the barrier is not square? Let us concentrate on the exponential factor, which is produced entirely by the exponential decay of the wavefunction inside the barrier. In that region ($R < r < R_1$), the radial Schrödinger equation may be written

$$\frac{d^2 u(r)}{dr^2} = \beta^2 u(r)$$

with solution $u(r) = e^{-\beta r}$, so that the probability density, which is proportional to $|u(r)|^2$, goes as $e^{-2\beta r}$. If the barrier is not square, but $V(r)$ varies slowly [that is, much more slowly than $u(r)$], we can, for the purpose of computing the exponential part of $u(r)$, imagine the barrier to consist of a succession of

[2] As early as 1911, Geiger and Nuttal discovered a monotonic decrease in alpha-particle range (and hence energy) with increasing half-life of the alpha emitter.

square barriers of varying height, each of thickness Δr. Then $u(r)$ is proportional to the product of the successive attenuation factors $e^{-\beta_1 \Delta r}$, $e^{-\beta_2 \Delta r}$, $e^{-\beta_3 \Delta r}$, ..., where β_1 is the value of β in the first square barrier, β_2 the value in the second square barrier, etc. Since in our case the barrier extends from $r = R$ to $r = R_1$, the product of all the exponentials, in the limit $\Delta r \rightarrow 0$, becomes $e^{-\int_R^{R_1} \beta(r)\, dr}$. The transmission coefficient T is therefore proportional to the square of this factor, or to $e^{-2\int_R^{R_1} \beta(r)\, dr}$ (which clearly reduces to the exponential in Eq. (4), with $R_1 - R = a$, if β is constant).

The other factors in Eq. (4) result from matching the boundary conditions on the wavefunction at each side of the barrier; these factors also must be modified if the barrier is not square. But the energy dependence in these factors is slight in comparison to that in the exponential factor, so let us put aside that problem and examine the exponential.

The analysis is simplified by taking the logarithm of both sides of Eq. (2):

$$\ln \lambda = \ln T + \ln(2E/m)^{1/2} - \ln(2R)$$

$$= \ln T + \tfrac{1}{2} \ln E + \text{constant}$$

$$= -2 \int_R^{R_1} \beta(r)\, dr + \ln[f(E)] + \text{constant} \tag{5}$$

where the function $f(E)$ contains all the energy-dependent factors except the exponential.

It is clear that the most significant term in Eq. (5) is the integral, which may be evaluated as follows:

$$2 \int_R^{R_1} \beta\, dr = 2 \left(\frac{2m}{\hbar^2}\right)^{1/2} \int_R^{R_1} [V(r) - E]^{1/2}\, dr$$

$$= 2 \left(\frac{2mE}{\hbar^2}\right)^{1/2} \int_R^{R_1} \left(\frac{V(r)}{E} - 1\right)^{1/2} dr$$

or, since $E = V(R_1)$,

$$2 \int_R^{R_1} \beta\, dr = 2 \left(\frac{2mE}{\hbar^2}\right)^{1/2} \int_R^{R_1} \left(\frac{V(r)}{V(R_1)} - 1\right)^{1/2} dr$$

$$= 2 \left(\frac{2mE}{\hbar^2}\right)^{1/2} \int_R^{R_1} \left(\frac{R_1}{r} - 1\right)^{1/2} dr$$

The last integral is standard, being soluble by means of the substitution $r = R_1 \cos^2 \theta$. The integration then yields

$$\int_R^{R_1} \left(\frac{R_1}{r} - 1\right)^{1/2} dr = 2R_1 \left[\cos^{-1}\left(\frac{R}{R_1}\right)^{1/2} - \left(1 - \frac{R}{R_1}\right)^{1/2} \left(\frac{R}{R_1}\right)^{1/2}\right]$$

Having previously estimated R_1 to be about 50 fm, we know that R, which is close to the nuclear radius, is much smaller than R_1. We can therefore use the approximations $\cos^{-1}(R/R_1)^{1/2} \approx (\pi/2) - (R/R_1)^{1/2}$, and $1 - (R/R_1) \approx 1$. The result is

$$2 \int_R^{R_1} \beta \, dr \approx 2 \left(\frac{2mE}{\hbar^2} \right)^{1/2} \left[\frac{\pi R_1}{2} - 2(RR_1)^{1/2} \right] \tag{6}$$

Equation (6) is somewhat deceptive, for there is an energy dependence in R_1. To make this explicit, we substitute $E = V(R_1) = 2.88Z/R_1$ [from Eq. (1)]. Then

$$2 \int_R^{R_1} \beta \, dr \approx 2 \left(\frac{2m}{\hbar^2} \right)^{1/2} [1.44Z\pi E^{-1/2} - 2(2.88ZR)^{1/2}]$$

$$\approx 3.92ZE^{-1/2} - 2.94Z^{1/2}R^{1/2}$$

if E is in MeV and R in fermis.[3] Inserting this expression into Eq. (5), we find that the relation between λ and E is then

$$\ln \lambda = -3.92ZE^{-1/2} + 2.94Z^{1/2}R^{1/2} + \ln[f(E)] + \text{constant}$$

or, to the base 10,

$$\log_{10} \lambda = -1.70ZE^{1/2} + 1.28Z^{1/2}R^{1/2} + \log_{10}[f(E)] + \text{another constant} \tag{7}$$

In terms of the half-life $t_{1/2}$, we may write

$$\log_{10} t_{1/2} = -\log_{10} \lambda + \log_{10} 0.69$$

$$= 1.70ZE^{-1/2} - 1.28Z^{1/2}R^{1/2} - \log_{10}[f(E)] + \text{still another constant}$$

or

$$\log t_{1/2} + 1.28Z^{1/2}R^{1/2} = 1.70ZE^{-1/2} + \text{slowly varying terms} \tag{8}$$

Figure 2 shows the excellent agreement of the observed half-lives and energies with Eq. (8). The *position* of the solid line, but not its *slope*, was adjusted to fit the observed points. Inclusion of the term $1.28Z^{1/2}R^{1/2}$ in plotting the ordinates has a small effect; this term varies from 34.6 for Po^{212} to 37.5 for U^{238}, a change of only 2.9. But the half-lives range from *less than* $1 \mu sec$ to *over* 10^{10} *yr*; the fit of theory to experiment over this range of values is remarkable.

The value of R was computed, for each point in Fig. 2, from the expression $R = R_0 A^{1/3}$, with $R_0 = 1.5$ fm.[4] This value of R_0 is used because it gives the best fit to the data; it is higher than the 1.2 fm found in measurements of the nuclear radius, but it must be remembered that R here is not the radius of the

---

[3] We have allowed for the recoil of the nucleus by setting m equal to the *reduced* mass of the alpha particle in the field of a typical nucleus; this reduced mass is about 0.98 times the mass of the free alpha particle, because a typical alpha emitter has a mass of about 50 times the mass of the alpha particle. Notice also that Z here is the Z of the daughter nucleus.

[4] The value of R, like the value of Z, is, of course, the value for the *daughter* nucleus.

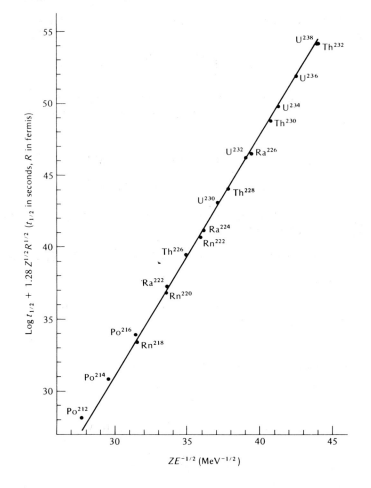

Fig. 2. *Comparison of Eq. (7) with experiment. Solid line has slope of 1.70, as required by the equation.*

nuclear charge distribution, but the radius at which the nuclear *force* cuts off. We might expect the nuclear force to be felt somewhat beyond the edge of the nuclear charge distribution.

Deviation of individual points from the line in Fig. 2 may result from any one of several possibilities:

1. The effect of the "slowly varying terms" in Eq. (8), which include the contribution of the alpha-particle velocity [from Eq. (3)], and possible effects of boundary conditions at $r = R$ and $r = R_1$ on the wavefunction.

2. Deviation of some nuclei from a spherical shape, so that the barrier varies with direction.

3. Variations in the probability of having a preformed alpha particle in the nucleus. This probability may vary from one nucleus to another because of shell effects.

4. Deviation of the nuclear radius from the value given by $R_0 A^{1/3}$. This can be expected for the "magic number" nuclei, which have a smaller radius than expected from the formula. For example, the point for Po^{212} is conspicuously high; in this case the daughter nucleus is $_{82}Pb^{208}$, which is doubly magic, having 126 neutrons and 82 protons, and which therefore has an exceptionally small radius for its value of A.

A few more features of Fig. 2 are of interest. It should be noticed that only even–even nuclei are shown. This is done to eliminate the effect of angular momentum. In alpha decay of an even–even nucleus to the ground state of the daughter (which is also even–even, of course), both parent and daughter have spin zero, so that the angular momentum of the emitted alpha particle must also be zero. In cases where the angular momentum of the alpha particle is not zero, the radial Schrödinger equation contains the centrifugal potential. The resultant increase in the size of the effective potential barrier can change the half-life by as much as a factor of 10. Decays involving a change in angular momentum of the nucleus are therefore said to be "hindered."

In many cases decay to an excited state or states of the daughter nucleus also occurs, resulting in the emission of lower energy alpha particles. Measurement of the energies of these alpha particles enables us to deduce the energies of the excited states of the daughter nucleus. This provided one of the earliest methods of finding energies of nuclear excited states.

Some of the nuclei plotted in Fig. 2 decay to excited states as well as to the ground state of the daughter. In such a case, the half-life used in the figure is of course the *partial* half-life for decay to the ground state, which is directly related to the relevant decay constant. Decays to excited states were not plotted here, because these decays involve a change in angular momentum.

15.2 THEORY OF BETA DECAY

The Neutrino. Beta decay differs from alpha decay in that the emitted particles have a continuous spectrum of energies (Fig. 5) rather than one or a few discrete energies. The maximum energy in the continuous spectrum corresponds to the mass difference between the parent and daughter atoms, so that energy is conserved when the emitted electron has the maximum possible energy. But when the electron's energy is less than the maximum value, it seems that some energy is lost somewhere.

Even before these facts were established, a test of energy conservation was made by placing a beta emitter in a heavy lead container in a calorimeter. If the emitted energy were in a form such that it could be absorbed by the lead container (for example, if it went off with a particle which stopped in the lead), this energy would appear as heat in the calorimeter. But comparison of the calorimeter results with measurements of the beta energy spectrum showed that the average energy left in the calorimeter, per beta decay, was just equal to the average energy of the beta particles themselves. No matter how the system was observed, some energy had disappeared.

After various unsatisfactory proposals had been made to account for these facts, Pauli in 1930 provided an explanation. He suggested that an extremely penetrating particle, the "neutrino",[5] was also emitted in beta decay, and that this particle carried away the missing energy. The presence of the neutrino also accounts for the continuous nature of the energy spectrum, for the momentum and energy can be shared by neutrino and electron in a way which depends on the relative angles at which they are emitted. Clearly, if no particle other than the electron were emitted, the electron and the recoiling nucleus would move in exactly opposite directions; the energy of each would be uniquely determined by energy and momentum conservation, and the spectrum would be discrete, as in alpha decay.[6] The beta decay reaction in a nucleus is therefore either

$$p \rightarrow n + e^+ + \nu_e \tag{9}$$

or

$$n \rightarrow p + e^- + \bar{\nu}_e \tag{10}$$

where ν_e stands for neutrino and $\bar{\nu}_e$ for antineutrino. The equations are written in this way, with two kinds of neutrinos, so that an antiparticle is always created when a particle is created. Since the neutrino was "invented" simply to account for the missing energy in beta decay, you may wonder why we need two kinds (antineutrino as well as neutrino), but we shall see in a moment that there is experimental evidence that the antineutrino is distinguishable from the neutrino.[7] Furthermore, if the neutrino is a real particle, it is entitled to have its antiparticle, just as the other particles have antiparticles.

[5] The neutrino received its fine Italian name from Fermi.

[6] It was suggested that the electrons might have a discrete energy when they are emitted from the nucleus, but that they give up varying amounts of energy to other electrons before their detection, so that there is a continuous spectrum of energies when they are finally observed. We can rule out this possibility on the basis of the calorimeter experiment; any energy lost by the electrons in this way would have to appear eventually as a rise in temperature of the system, but the temperature rise was just what one would expect if the spectrum of Fig. 5 is the *primary* spectrum.

[7] In Chapter 16 we shall see that there are *three* different *neutrinos* and *three antineutrinos—six* neutrinos in all. The subscript $_e$ is used to distinguish the neutrinos of decay (9) from the other two types (which are produced with muons and tau particles).

It is logical to assume that the neutrino has the following properties:

$$\text{charge} = 0; \quad \text{mass} = 0; \quad \text{spin} = \tfrac{1}{2}$$

The charge must be zero, for charge to be conserved in reactions (9) and (10). The mass must be zero (or near zero), because the maximum beta energy equals the total energy which can be released in the decay, so that the neutrino sometimes carries off no energy. Finally, half-integer spin is required by conservation of angular momentum, since the other particles in reactions (1) and (2) all have half-integer spin; the simplest assumption therefore is that the spin is $\tfrac{1}{2}$.

For a long time the neutrino remained simply a clever invention which accounted for the continuous nature of the beta-decay energy spectrum and which balanced the books for the law of conservation of energy. There is nothing particularly wrong with that, for the concept of energy itself is basically a mathematical one, expressing the fact that a certain function of the dynamical variables remains constant in all interactions. Whenever an interaction has been found for which the energy function, as then constituted, did not remain constant, it has simply been assumed that the energy function was not complete, and another term has been added. In this way, terms for the various kinds of potential energy—gravitational, electrostatic, magnetic, and so on, have been constructed.

We might simply say, therefore, that we have a new form of energy, "neutrino energy." But this form of energy is unique in one way. In all other cases, energy in one form, after being transformed into one of the other forms, could be transformed back into the original form, so that the conservation law is not a trivial definition. In other words, reactions, at least on the microscopic level, are reversible. But for a while, it seemed that this could not be demonstrated for neutrino reactions. Since the energy which was transformed into neutrino energy could not be transformed back into other forms, the preservation of the law of conservation of energy had the appearance of a hollow victory.

But the merit of the neutrino hypothesis is more apparent when we see that it is needed in order to save other conservation laws as well. Our deduction of the neutrino spin makes it clear that this spin is necessary in order that angular momentum be conserved in beta decay. Furthermore, experimental observations of the recoil of the nucleus in beta decay have shown that the neutrino is also needed for the conservation of linear momentum. Nuclei were found to recoil in a direction which was not exactly opposite to the direction of an emitted beta particle; rather, the recoiling nucleus and the emitted beta had a resultant momentum which had to be balanced by the momentum of a third unseen particle. This "missing momentum" and the missing energy were found in all cases to be consistent with the assumption that a particle of zero mass was emitted. In other words, within the experimental error (about 10 percent), $E = pc$, where E is the missing energy and p is the magnitude of the missing momentum.

By 1950, evidence of this sort had become quite convincing. But it was still rather vexing to have all this energy, momentum, and angular momentum simply disappear, with no way of recovering it, whether or not a neutrino was carrying it off Attempts were therefore made to detect the absorption of neutrinos. This is extremely difficult, because the cross section for inter-action of neutrinos with anything is minuscule. We can actually predict how small a typical cross section should be, by considering the decay of the free neutron. The free neutron decays via reaction (10) with a mean lifetime of about 10^3 sec, which means that the decay probability is $10^{-3} dt$ in a short time dt. The transition probability for the *inverse* process should be equal to this decay probability, if the wavefunctions of the particles involved are the same, because the matrix element[8] for the transition is then the same. The exact inverse of reaction (10) cannot be observed, because three particles must be brought together, but the same matrix element is also involved in the reaction

$$\bar{\nu}_e + p \;\rightarrow\; n + e^+ \tag{11}$$

The terms in the matrix element simply appear in a different order. Reaction (11) differs from the inverse of (10) only in the emission of a positron instead of the absorption of an electron. Therefore when the antineutrino has suffi-cient energy to create the necessary rest mass (about 1.8 MeV), and the anti-neutrino and proton wavefunctions overlap, reaction (11) should proceed with the same transition probability per unit time as reaction (10). But the antineutrino, traveling at the speed of light, has a wavelength of $\sim 10^{-10}$ cm when its energy is ~ 2 MeV, and it crosses that distance in $\sim 10^{-20}$ sec, so that a given antineutrino has only 10^{-20} sec during which it can interact with a given proton. The total interaction probability is therefore about $10^{-3} \times 10^{-20} = 10^{-23}$, whenever the antineutrino passes within 10^{-10} cm of a proton. In order to pass this close to a proton, the antineutrino must pass through an area of about 10^{-20} cm^2. The cross section for the interaction equals the area which the antineutrino must hit, multiplied by the probability of an interaction when it does hit that area, or 10^{-20} cm$^2 \times 10^{-23} = 10^{-43}$ cm^2. To comprehend the size of this number, consider that in solid lead, containing 2.7×10^{24} protons/cm^3, the mean free path of an antineutrino would be about $1/(2.7 \times 10^{24} \times 10^{-43}) \approx 4 \times 10^{18}$ cm, or about 4 light-years.

Nevertheless, in the 1950's,[9] reaction (11) was observed by Cowan and

---

[8] See the discussion of transition probabilities and matrix elements in Section 10.1.

[9] The experiment required a good part of the decade to be completed with all of its refinements. A preliminary description was published in 1953 (C. Cowan and F. Reines, *Phys. Rev.* **90**, 492); results from an improved setup appeared in 1956 (*Science,* **124**, 103); and measurement of the cross section was reported in 1959 (*Phys. Rev.* **113**, 273). In 1979, Reines gave a good description of the entire sequence of events (*Science* **203**, 11).

Reines, who utilized the large antineutrino flux ($\sim 10^{13}/\text{cm}^2$-sec) near a nuclear reactor. They observed the products of the reaction—the positron and the neutron—in a way which left little room for doubt that reaction (11) was actually occurring. The experiment was a masterpiece which must be explained in some detail to be fully appreciated.

The detector and the sequence of events after an antineutrino interaction are illustrated in Fig. 3. Actually there are two detectors, one above the other; S1, A1, and S2 act as one detector, while S2, A2, and S3 act as the other detector. In each detector, the antineutrino interaction takes place in the $CdCl_2$ solution (A1 or A2) creating a neutron and a positron in the solution; the gamma rays, which ultimately result from the presence of the neutron and positron, are detected in the liquid scintillator. The positron, of course, produces annihilation radiation, which is identified when one gamma ray of the correct energy is detected in each of the two scintillators of a given detector. The neutron, meanwhile, wanders through the solution until it is captured by a Cd nucleus (which is in the solution for that purpose, because of its large neutron absorption cross section). The resulting Cd isotope (^{114}Cd) is formed in an excited state, 9.1 MeV above its ground state; it then decays to the ground state by emission of gamma rays, which are detected in one or both scintillators. (Many gamma rays are usually emitted.) The connection between the neutron and the positron is established by the requirements that (1) both sets of scintillator pulses appear in the same two scintillators of a single detector, (2) the pulses resulting from the neutron capture follow the positron annihilation pulses within 25 μsec and (3) no pulses appear in the third scintillator (the one which belongs to the other detector) during this time interval. The last requirement eliminates false signals which might be produced when a single charged particle passes through both detectors.

When these requirements are met, the output of the photomultiplier tubes observing each scintillator is displayed on an oscilloscope, producing traces like those shown in Fig. 4c. The appearance of such a distinctive set of pulses is the signal that an antineutrino interaction has occurred; but before we can be sure of this, other possible origins of the pulses must be considered. For example, suppose two neutrons, coming directly from the reactor, are absorbed by Cd nuclei in A2, at times a few microseconds apart. Normally the resulting sets of pulses is distinguishable from those shown in 4c, because the pulses in the first pair are too large to be the result of a positron annihilation. But in many cases, the liquid scintillators do not absorb all of the gamma rays from the Cd decay, so it is possible for each pulse of the first pair to be small enough so that the pair passes as an annihilation pair. This is obviously a rare occurrence, but the neutrino interactions are also rare, so the following tests (among others) were made:

1. The number of *free* protons (that is, those not bound in nuclei) in the

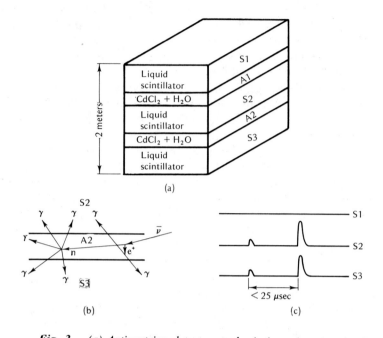

(a)

(b)

(c)

Fig. 3. (a) *Antineutrino detector stack. Antineutrinos interact in layers A1 and A2, containing $CdCl_2$ solution. (b) Enlarged view showing possible sequence of events following an antineutrino interaction. Positron and neutron are produced in A2. The positron annihilation gamma rays go in opposite directions, producing scintillations in S2 and S3. The neutron is captured by a Cd nucleus, causing the emission of gamma rays, most of which enter S2 and S3. (c) Oscilloscope traces showing outputs of photomultiplier tubes which view each scintillator. No pulses are seen in S1, since gammas stop before reaching there. The first pulse of each pair in S2 and S3 is produced by the annihilation gamma rays; the second, larger pulse is produced by the gamma rays from the Cd nucleus.*

solution was reduced by a factor of 2, by diluting the solution with heavy water. This did not affect the efficiency of the detector for detecting neutrons, but it cut the number of antineutrino "signals" by a factor of close to 2, as it should, since the rate of reaction (11) is proportional to the number of protons available to take part in the reaction. (The protons in heavy water are all bound in nuclei; hence much more antineutrino energy is needed to convert one of them into a *free* neutron, and antineutrinos of sufficiently high energy are relatively rare in the reactor output.)

2.   The amount of shielding between the reactor and the detector was greatly

increased. This reduced the number of neutrons reaching the detector from the reactor to one-tenth of its previous value, but it had no effect on the rate of events attributed to antineutrinos, whose number would be unaffected by any possible change in the amount of shielding. The rate was $1.74 \pm 0.12/h$ with the shield, and $1.69 \pm 0.17/h$ without it.

3. The concentration of $CdCl_2$ in A1 and A2 was changed, and the spectrum of time delays between the two pairs of pulses was observed to change. This indicated that the second pulse pair was truly correlated in time with the first, and that the delay between these pairs is the result of the time required for the neutron to find a Cd nucleus and be absorbed. The pulses identified as being produced by neutrons disappeared completely when the Cd was removed from the solution.

One other attempt to observe antineutrinos is of interest because it bears on the difference between the neutrino and the antineutrino. A large tank of CCl_4 was placed near a reactor, in an attempt to produce ^{37}Ar via the reaction

$$v_e + {}^{37}Cl \rightarrow e^- + {}^{37}Ar \tag{12}$$

Any ^{37}Ar which was produced could be swept away by bubbling He gas through the tank. The ^{37}Ar could then be separated from the He and detected[10] as it decayed back to ^{37}Cl. However, reaction (12) requires neutrinos, while the reactor produces *anti*neutrinos, from the negative beta decays of the fission products.[11] If antineutrinos differ from neutrinos, they cannot participate in reaction (12); hence no ^{37}Ar should be produced. The experiment was sensitive enough to detect 1/10 of the amount of ^{37}Ar which would have been produced if v_e and $\bar{v}_e$ were identical, but no ^{37}Ar was detected.

Other evidence that v_e's differ from $\bar{v}_e$'s comes from studies of *double* beta-decay—the simultaneous emission of two beta particles by a nucleus. In situations like the one illustrated in Chapter 14, Fig. 8, a species such as ^{146}Nd is energetically stable against ordinary beta decay but not against double beta decay; thus it cannot emit one beta followed by another beta, but it must emit both betas simultaneously in order to decay. This occurrence is obviously rare, but the degree of rarity depends on whether or not the v_e and the $\bar{v}_e$ are identical. If they are *not* identical, the double beta decay reaction must be

$$2n \rightarrow 2p + 2e^- + 2\bar{v}_e \tag{13}$$

while if the v and $\bar{v}$ are identical, the reaction could as well be written

---

[10] ^{37}A decays by K-electron capture. The decays can be detected by observing the characteristic gamma rays emitted as the vacancy in the K shell is filled.

[11] Fission products are always excessively rich in neutrons and thus undergo negative beta decay rather than positive beta decay. See Chapter 14, Example Problem 1.

$$2n \rightarrow 2p + 2e^- + \overline{v}_e + v_e$$

Since a particle and its antiparticle can be annihilated, the latter reaction could equally well be

$$2n \rightarrow 2p + 2e^- \tag{14}$$

It is not necessary for the neutrinos to be produced at all, if reaction (14) can occur. If this is allowed, reaction (14) is more probable than reaction (13), because fewer particles must be produced. But careful experiments have shown that double beta decay occurs far too rarely for reaction (14) to be allowed;[12] therefore we must assume that (14) is *not* allowed, and that it is therefore not legitimate to perform the steps leading from (13) to (14), because the v is *not* identical to the $\overline{v}$.

Fermi Theory of Beta Decay. The neutrino hypothesis made it possible to construct a theory of beta decay, the first requirement of which is that it account for the energy distribution of the emitted beta particles. Fermi's theory (1934) did just that, in a rather simple way. It turns out that the shape of this energy distribution depends mainly on statistical factors, as Fermi was able to show, even though the exact nature of the interaction which causes beta decay was not (and still is not) known. Knowledge of this general shape has provided work for several generations of physicists, for variations in the shape of the beta energy spectrum provide information about the properties of nuclear states. A study of the shape of the spectrum has also provided an upper limit on the possible mass of the neutrino.

The essence of the Fermi theory is that beta decay, in which a nucleus loses energy by emission of a neutrino and an electron, is similar in many ways to radiation of a photon by an excited atom. In both cases the emitted particle (or particles) is created at the time of emission. Therefore, as in the case of electromagnetic radiation, there is a perturbing potential which acts to convert the system from its initial state to the final state, and the transition probability per unit time for a transition to any *given state* can be found by the same method used for atomic radiation (Chapter 10). A given state is, of course, a state of the complete system, containing the final atom, the neutrino, and the electron. We may assume that the final atom is in a known, discrete state, but the state of the electron and the state of the neutrino are characterized by the momentum of each. We may make the latter states discrete by assuming that the system is in a large box, as we have done previously; then the total transition probability to any group of states for which the electron momentum lies in a given range is found by multiplying the transition probability per state, as given by perturbation theory, by the number of states in that group.

[12] By measurement of the excess concentration of ^{130}Xe (relative to other xenon isotopes) in tellurium minerals, the half-life for the double beta decay of ^{130}Te to ^{130}Xe has been deduced to be $10^{21.34 \pm 0.12}$ years. T. Kirsten, O. A. Schaeffer, E. Norton, and R. W. Stoenner, *Phys. Rev. Lett.* **20**, 1300 (1967). See also D. Bryman and C. Picciotto, *Rev. Mod. Phys.* **50**, 11 (1978).

We begin with Eq. 8, of Chapter 10, which gives the rate of change of the amplitude a_{nm} of the mth state in the expansion of the wavefunction of the system:

$$\dot{a}_{nm} = -\left(\frac{i}{\hbar}\right)e^{-i(E_n - E_m)t/\hbar}v_{mn} \tag{15}$$

where the system is assumed to have been in the nth state originally, and $v_{mn} = \int u_m^* V u_n \, d\tau$ is the matrix element of the perturbing potential V between the initial state, whose spatial wavefunction is u_n, and the mth state, whose spatial wavefunction is u_m.

We do not know the interaction potential V, but we shall assume that it is independent of time, and that it is "turned on" at time $t = 0$. The probability of a transition from state n to state m in time t is then equal to the probability that a measurement of the system at time t will show that it is in state m; this probability is just the absolute square of the coefficient a_{nm} which can be found by integrating Eq. (15) from $t = 0$ to $t = t$. We find that

$$|a_{nm}|^2 = \frac{4|v_{mn}|^2 \sin^2\left(\dfrac{t\,\Delta E}{2\hbar}\right)}{(\Delta E)^2} \tag{16}$$

The matrix element v_{mn} would seem to present a difficulty, because of our ignorance of V. However, we know that V must contain a constant factor g which determines the "strength" of the interaction, or how fast a reaction or decay process will proceed. For the time being, let us see how far we can go by simply letting $V = gO$, where O is an operator which converts a neutron to a proton (or vice versa) in the wavefunction of the nucleus.

The other ingredients of the matrix element are the wavefunctions of the initial and final states. The initial state wavefunction is simply the wavefunction of the parent nucleus, u_n; but the final state wavefunction is the product of the wavefunctions of the daughter nucleus, electron, and neutrino. For a first approximation, one can consider the electron and neutrino to be plane waves in a box of volume Ω, so that the final state wavefunction is

$$u_m = \frac{1}{\Omega}\, u_m\, e^{-i\mathbf{p}_e \cdot \mathbf{r}/\hbar} e^{-i\mathbf{p}_v \cdot \mathbf{r}/\hbar} \tag{17}$$

where $\mathbf{p}_e$ is the electron momentum and $\mathbf{p}_v$ is the neutrino momentum. Then, as in the case of atomic radiation, we find that the wavelength of each emitted particle is much greater than the diameter of the nucleus,[13] so that each exponential is approximately 1 at values of r for which the nuclear wavefunction u_m is not zero. The matrix element may therefore be written

$$v_{mn} = \frac{g}{\Omega}\int u_m^* O u_n \, d\tau$$

or

[13] Remember that the wavelength of a 1-MeV neutrino is of order 10^{-10} cm, or a few hundred nuclear diameters.

$$|v_{mn}|^2 = \left(\frac{g}{\Omega}\right)^2 \left|\int u_m^* O u_n \, d\tau\right|^2$$

$$= \left(\frac{g}{\Omega}\right)^2 |M|^2 \tag{18}$$

where M, the matrix element involving the initial and final states of the *nucleus*, may be ignored temporarily, because, being independent of the electron's momentum, it does not affect the energy spectrum of the emitted electrons.

Unfortunately, we have slightly oversimplified the situation. A plane wave is an acceptable wavefunction for the neutrino, but not for the electron, because the potential energy of an electron (or a positron) is not constant near the nucleus. The typical Coulomb potential energy of an electron and a positron, respectively, in the field of a spherical nucleus of radius R is shown in Fig. 4. The peak value of the Coulomb potential energy may be easily seen to be a few MeV, so that the effect of this potential cannot be neglected. But it is not too hard to calculate the actual potential energy for each nucleus, and from this to find the actual positron or electron wavefunction for each beta emitter. The value of the wavefunction at the origin can again be used as a first approximation in evaluating the matrix element. We shall not go into the details of this calculation, but simply state some results. The result of the calculation may be represented by multiplying the right-hand side of

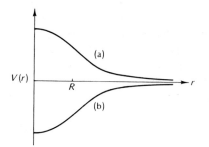

Fig. 4. *Coulomb potential energy of (a) a positron (b) an electron, in the field of a nucleus of radius R. (See also Chapter 14, Problem 1.)*

Eq. (18) by a factor $F(Z, p_e)$, called the nuclear Coulomb factor. The Coulomb effect reduces the amplitude of a positron wavefunction and increases the amplitude of an electron wavefunction, the effect being largest for low energy particles. Thus when all other factors are equal, positron emission is inhibited, relative to electron emission, and the inhibition is greatest for the particles of lowest energy. The *shape* of the positron energy spectrum therefore differs from the shape of the electron energy spectrum, in that there are relatively few low-energy positrons.[14] This effect is evident in Fig. 5.

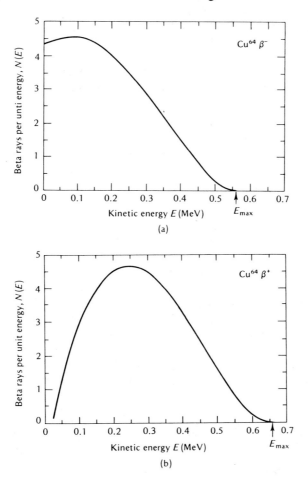

Fig. 5. *Energy spectrum of the (a) negative and (b) positive beta particles emitted by* ^{64}Cu, *showing the relative scarcity of low energy positrons.* [*From "The Atomic Nucleus" by R. D. Evans, Copyright McGraw-Hill, New York, 1955. Used by permission of McGraw-Hill Book Company.*]

[14] A pseudoclassical explanation of this phenomenon is given in some books. It is said that positrons leaving the nucleus are repelled, and pick up speed, and that this accounts for the dearth of low-energy positrons when the particles are observed. This provides a way to remember which way the effect goes, but the argument should not be taken seriously, for the Coulomb effect in beta decay is completely nonclassical. The potential barrier is so high that a positron must always tunnel through it to get out; the barrier actually inhibits the *creation* of the positron by reducing the magnitude of the positron wavefunction which appears in the matrix element.

Equations (16) and (18), with the factor F included, now give us the transition probability to the mth state:

$$|a_{nm}|^2 = \frac{g^2 F}{\Omega^2} |M|^2 \frac{4 \sin^2 \left(\frac{t \, \Delta E}{2\hbar}\right)}{(\Delta E)^2}$$

Of course, in practice, the mth state is not discrete but is part of a continuum, so we may write the probability for a transition to a group of states as

$$dP = |a_{nm}|^2 \, dn_e \, dn_v \tag{19}$$

if dn_e is the number of electron states and dn_v is the number of neutrino states in the group,[15] and the transition probability to each state (of the whole system) in this group is $|a_{nm}|^2$. The total transition probability P may now be found by integrating Eq. (19) over all the possible final states of the electron and the neutrino:

$$P = \frac{4g^2 F |M|^2}{\Omega^2} \iint \frac{\sin^2 \left(\frac{t \, \Delta E}{2\hbar}\right)}{(\Delta E)^2} \, dn_e \, dn_v \tag{20}$$

We assume that the momentum of the neutrino is independent of the momentum of the electron, because the recoil momentum of the atom can take care of momentum conservation, while energy conservation is automatically assured by the peaking of the function $\sin^2(t \, \Delta E/2\hbar)/(\Delta E)^2$ at $\Delta E = 0$. (Remember that ΔE is the difference in *total* energy between the initial and final states, so that energy is conserved when $\Delta E = 0$.) The distribution of electron momenta may then be found by differentiating Eq. (20) with respect to p_e. If we do this, letting $dn_v = (dn_v/dE_v) \cdot dE_v$, we obtain

$$\frac{dP}{dp_e} = \frac{dP}{dn_e} \frac{dn_e}{dp_e} = \frac{4g^2 F |M|^2}{\Omega^2} \left(\frac{dn_e}{dp_e}\right) \int \frac{\sin^2 \left(\frac{t \, \Delta E}{2\hbar}\right)}{(\Delta E)^2} \left(\frac{dn_v}{dE_v}\right) dE_v \tag{21}$$

To evaluate the integral, we write

$$\Delta E = E_f - E_i$$
$$= (E_f - E_i)_{\text{atom}} + E_e + E_v$$

or

$$E_v = \Delta E - E_e - (E_f - E_i)_{\text{atom}}$$

For a given value of E_e,

$$dE_v = d(\Delta E)$$

---

[15] For example, if there are four neutrino states and three electron states in the group, the number of ways in which these can be combined to form states of the whole system is $3 \times 4 = 12$, and, each of the 12 states having equal probability, the total probability is then 12 times the single state probability $|a_{nm}|^2$.

The integrand is close to zero everywhere except at $\Delta E = 0$, so we may remove the factor dn_ν/dE_ν from the integral, using only its value at $\Delta E = 0$. Then, since

$$\int_{-\infty}^{\infty} \frac{\sin^2 z}{z^2} \, dz = \pi$$

Eq. (21) finally becomes (with the aid of the substitution $z = -t \, \Delta E/2\hbar$),

$$\frac{dP}{dp_e} = \frac{4g^2 F |M|^2 \pi t}{2\hbar \Omega^2} \left(\frac{dn_e}{dp_e}\right)\left(\frac{dn_\nu}{dE_\nu}\right)_{\Delta E = 0} \tag{22}$$

We now see that statistical factors—the density-of-states factors dn_e/dp_e and dn_ν/dp_ν—do indeed determine the momentum distribution, and hence the energy distribution, of the emitted electrons. These factors can be found by using the same equation which gave us the number of states of a free electron in a metal. The number of states with momentum less than a given momentum p, for a particle in a box of volume Ω, may be written

$$n = \frac{\Omega p^3}{6\pi^2 \hbar^3} \tag{23}$$

This equation holds for both electrons and neutrinos, because it was derived simply by counting standing waves, and because the de Broglie wavelength formula $\lambda = h/p$ holds for all particles, regardless of rest mass. We can thus find dn_e/dp_e and dn_ν/dp_ν immediately, by differentiation of Eq. (23). But to find dn_ν/dE_ν, we must use the energy–momentum relation for neutrinos. Let us assume that $m_\nu = 0$, so that $E_\nu = p_\nu c$. Then

$$\begin{aligned}
\frac{dn_\nu}{dE_\nu} &= \frac{dn_\nu}{dp_\nu} \frac{dp_\nu}{dE_\nu} \\
&= \frac{\Omega p_\nu^2}{2\pi^2 \hbar^3} \frac{1}{c} \\
&= \frac{\Omega E_\nu^2}{2\pi^2 \hbar^3 c^3}
\end{aligned} \tag{24}$$

and

$$\frac{dn_e}{dp_e} = \frac{\Omega p_e^2}{2\pi^2 \hbar^3}$$

These are to be evaluated at $\Delta E = 0$, which simply means that $E_\nu = (E_i - E_f)_{atom} - E_e$, where $(E_i - E_f)_{atom}$, the difference between the initial and final energies of the beta emitting atom, is equal to the maximum observed electron energy, or E_m. Substitution of Eqs. (24) into Eq. (22), with $E_\nu = E_m - E_e$, yields

$$\frac{dP}{dp_e} = \frac{g^2 |M|^2 t F(Z, p_e)}{2c^3 \hbar^7 \pi^3} p_e^2 (E_m - E_e)^2 \tag{25}$$

Equation (25) is all we need for a comparison with experimental beta-decay spectra, but it appears to be in a somewhat complicated form. The relative probabilities of various electron *momenta* are given in terms of the momentum *and energy* of the electron. We could easily eliminate p_e by means of the relativistic energy–momentum relation, but this would not make a comparison with experiment any easier. The most convenient way to compare with experiment is to fit the experimental data to a straight line, and the proper straight line may be derived from Eq. (25) as follows: Divide both sides of Eq. (25) by p_e^2 and take the square root of each side; the result is that $[(dP/dp_e)/p_e^2 F(Z, p_e)]^{1/2}$ is proportional to $E_m - E_e$, so that a plot of $[(dP/dp_e)/p_e^2 F(Z, p_e)]^{1/2}$ versus E_e should yield a straight line.

Before the work of Fermi, various attempts were made to fit the beta decay spectrum to empirical equations. It is not hard to see why these attempts failed. In order to construct the quantity $[(dP/dp_e)/p_e^2 F(Z, p_e)]^{1/2}$ needed for a straight-line plot, one must divide the spectrum into equal *momentum* intervals, divide the number of counts in each interval by the square of the momentum and by the Coulomb factor $F(Z, p_e)$ for that momentum, and take the square root of the result. Then the numbers must be plotted against the

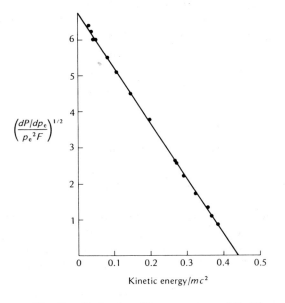

Fig. 6.   *Kurie plot of beta spectrum of* Pm^{147}.
[*From L. M. Langer, J. W. Motz, and H. C. Price, Jr., Phys. Rev. 77, 798 (1950).*]

electron *energy*. Figure 6 shows that this really works, but who could have guessed it? This way of displaying the data is called a Fermi plot or a Kurie plot (after F. N. D. Kurie, who first reported results in this way).[16]

The Kurie plot is useful in determining E_m, the energy difference between parent and daughter nuclei, for beta transitions. It also can be used to examine the assumptions made in deriving Eq. (25). There are cases in which the Kurie plot fails to follow a straight line, indicating that one of the basic assumptions was faulty. The main assumptions were:

1. that the neutrino rest mass is zero,

2. that the electron and positron wavefunctions can be approximated by using only the value of the wavefunction at the origin, and

3. that the interaction potential may be represented by the constant g times the simple operator O.

The first assumption is shown by the data to be very good. If the neutrino rest mass were not zero, the maximum observed beta energy would be less than $(E_i - E_f)_{atom}$, because some energy would always be used to create the rest mass of the neutrino. But if the neutrino mass is very small, the expression $E_v = p_v c$ is still valid when E_v is large, or when E_e is small. Thus if the Kurie plot is straight at small values of E_e, the extrapolation of the straight line intersects the E axis at $E_e = (E_i - E_f)_{atom}$. But if $m_v \neq 0$, the *data* do *not* follow the straight line as $E_e \to E_m$; a line through the data points would intersect the axis at $E_e = E_m \neq (E_i - E_f)_{atom}$. A careful search for such a deviation of the points has been made in the beta spectrum of ^3H (see Fig. 7). ^3H is particularly good for this search, because the value of E_m is only 18 keV, so that a small nonzero neutrino mass should have a greater effect than in other beta emitters. However, no effect could be seen. The accuracy of the experimental data places an upper limit of 200 eV on the neutrino rest energy.

It should be emphasized that no measurement can ever establish the value of any physical quantity to be *exactly* zero. Even if it is in fact zero, all that one can do experimentally is to put an upper limit on the value. The upper limit established in Fig. 7 for the neutrino mass, only 1/2000 of the electron's rest mass, would seem to be equivalent to zero for all practical purposes, and the elegant result shown there has stood unchallenged for many years. However, because there are so many neutrinos in the universe, a possible nonzero value even smaller than this would be of great interest to cosmological theorists. We shall see in Chapter 16 that for this and other reasons there is renewed interest in neutrino mass determinations, although there is still no clear evidence for a nonzero mass.

The second assumption, that we can evauate the electron and neutrino wavefunctions at the origin and use only these vaues in the matrix element,

[16] F. N. D. Kurie, J. R. Richardson, and H. C. Paxton, *Phys. Rev.* **48**, 167 (1935).

must be reconsidered when it yields a null result for the transition probability. This happens whenever $\int u_m^* O u_n \, d\tau = 0$. By analogy with the treatment of atomic radiation, we can see that this integral vanishes when the spin of the daughter nucleus differs from that of the parent, the functions u_m and u_n being orthogonal when they represent states of different angular momentum. (The operator O, of course, does not affect the form of the wavefunction.) We may therefore state the selection rule $\Delta I = 0$ for allowed transitions.

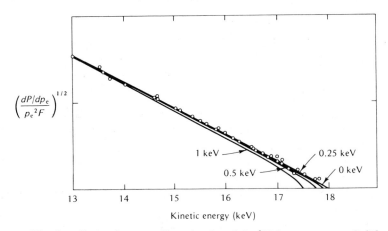

Fig. 7. *Kurie plot near the end point of the* ^{3}H *beta spectrum. Solid lines are the theoretical curves which the data should follow for neutrino rest energies of 0, 0.25, 0.5, and 1.0 keV, respectively. [From L. M. Langer and R. J. D. Moffat, Phys. Rev.* **88**, *689 (1952).]*

However, as in atomic radiation, forbidden transitions do occur, for which $\Delta I \neq 0$. The matrix elements for these transitions vanish only because of assumption (2); when the correct electron and neutrino wavefunctions are used, these matrix elements do not vanish. In order to simplify things, let us go back to the original assumption that both electron and neutrino wavefunctions are plane waves. But let us use the first *two* terms in the expansion of the exponential:

$$e^{i\mathbf{p}\cdot\mathbf{r}/\hbar} = 1 + i\mathbf{p}\cdot\mathbf{r}/\hbar + \cdots$$

The square of the matrix element then becomes

$$|v_{mn}|^2 = \frac{g^2 F}{\Omega^2} \left| \int u_m^* O u_n \, d\tau + i\left(\frac{\mathbf{p}_e + \mathbf{p}_\nu}{\hbar}\right) \cdot \int u_m^* \mathbf{r} O u_n \, d\tau \right|^2 \tag{26}$$

or

$$|v_{mn}|^2 = \frac{g^2 F}{\Omega} \left| \left(\frac{\mathbf{p}_e + \mathbf{p}_\nu}{\hbar}\right) \cdot \int u_m^* \mathbf{r} O u_n \, d\tau \right|^2 \tag{27}$$

for forbidden transitions, for which $\int u_m^* O u_n \, d\tau = 0$. This matrix element is nonzero when $\Delta I = \pm 1$ or 0. (Notice that the integral in Eq. (27) has the same form as the integral in the matrix element for dipole radiation; hence the selection rule for the angular momentum is the same.) Since Eq. (26) reduces to Eq. (27) when $\Delta I \neq 0$, the transition is governed by Eq. (27) only in cases when $\Delta I = \pm 1$. Such transitions are called *first-order forbidden*, or first forbidden.

The additional factor $(\mathbf{p}_e + \mathbf{p}_v)/\hbar$ in Eq. (27) changes the shape of the electron momentum spectrum, so that a normal Kurie plot does not yield a straight line. But a special Kurie plot can be devised to yield a straight line; this plot is made by introducing a "correction factor" to the quantity plotted on the ordinate. The correction factor may be understood by referring to Eq. (21). If the factor $(\mathbf{p}_e + \mathbf{p}_v)^2/\hbar^2$ is inserted into the integrand in Eq. (21), and the integration carried out, the effect is to multiply by the average value of $(\mathbf{p}_e + \mathbf{p}_v)^2$, the average being taken over all directions of $\mathbf{p}_e$ relative to $\mathbf{p}_v$. If these two directions are independent, then $|\mathbf{p}_e \cdot \mathbf{p}_v|_{av} = 0$, so that $(\mathbf{p}_e + \mathbf{p}_v)^2_{av} = \mathbf{p}_e^2 + \mathbf{p}_v^2$. Thus the momentum distribution of the electrons is given by an equation similar to Eq. (25),[17] but with the right-hand side multiplied by $p_e^2 + p_v^2$, or, in terms of the electron's momentum and energy, $p_e^2 + (E_m - E_e)^2/c^2$. Thus one must plot

$$\left[\frac{\dfrac{dP}{dp_e}}{\left[p_e^2 + \left(\dfrac{E_m - E_e}{c} \right)^2 \right] p_e^2 F(Z, p_e)} \right]^{1/2}$$

versus E_e in order to obtain a straight line. Surprisingly enough, this procedure works, even though, as we have seen, the electron wavefunction is not really a plane wave (see Fig. 8). However, its success is limited to elements of fairly low Z, for which the Coulomb effect is sufficiently small.

In cases where this procedure works, it provides a clear-cut identification of first-order forbidden transitions, which often makes it possible to determine the spin of one of the nuclear states involved. It is clear that the use of higher order terms in the wavefunction expansion would make it possible to devise special Kurie plots for second and higher order forbidden transitions, as well, and this has been done.

---

[17] The nuclear matrix element M is different, of course, but this difference does not affect the energy spectrum of the electrons.

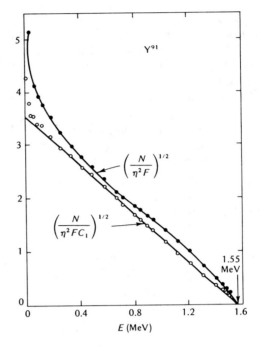

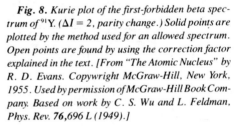

Fig. 8. *Kurie plot of the first-forbidden beta spectrum of ^{91}Y. ($\Delta I = 2$, parity change.) Solid points are plotted by the method used for an allowed spectrum. Open points are found by using the correction factor explained in the text. [From "The Atomic Nucleus" by R. D. Evans. Copywright McGraw-Hill, New York, 1955. Used by permission of McGraw-Hill Book Company. Based on work by C. S. Wu and L. Feldman, Phys. Rev. 76,696 L (1949).]*

Before leaving the subject of selection rules, we must consider assumption (3) of the above list, concerning the form of the interaction potential V. In 1936, Gamow and Teller pointed out that the selection rules are changed if V contains an operator which operates on the *spins* of the nucleons in the nucleus. There can then be a transition in which one unit of angular momentum is carried away by the electron and neutrino as *spin* angular momentum, the electron and neutrino spins being parallel. In this case a transition with $\Delta I = \pm 1$ is allowed, the distinguishing feature of an allowed transition being that the electron and neutrino carry away no *orbital* angular momentum, or in other words, that the nucleus may be regarded as a point source for these particles. Two sets of selection rules are therefore possible, one for "Fermi transitions," in which the electron and neutrino spins are antiparallel, and one for "Gamow–Teller (GT) transitions," in which electron and neutrino spins are parallel. The rules are listed in Table 1.

Table 1

Selection Rules in Beta Decay

	Fermi	Gamow–Teller (GT)
Allowed	$\Delta I = 0$	$\Delta I = \pm 1$ or 0, but not $I = 0$ to $I = 0$
First-forbidden	$\Delta I = \pm 1$ or 0	$\Delta I = \pm 2, \pm 1$, or 0
Second-forbidden	$\Delta I = \pm 2, \pm 1$, or 0	$\Delta I = \pm 3, \pm 2, \pm 1$, or 0

It should be noted that an allowed transition does not change the *parity* of the nucleus, because there is no change in the *orbital* angular momentum of the nucleus. The parity of the nucleus changes only in transitions which are forbidden to *odd* order.

In order to decide which set of selection rules is actually followed in beta decay, we must rely on experiment. We find that *both* sets are obeyed; for example, there are allowed transitions of the $I = 0$ to $I = 0$ type, which are allowed by the Fermi rules but forbidden by the GT rules, and there are allowed transitions with $\Delta I = \pm 1$, which are forbidden by Fermi rules. Thus we must conclude that the beta decay interaction potential contains at least two parts, one of which operates on the spin of the decaying nucleon and one of which does not. Because the magnitude of the matrix element decreases rapidly with increasing order of forbiddenness (see the following section), a transition between two given nuclear states will proceed by whichever interaction permits it to lowest order. For example, a transition from a 2^- state ($I = 2$, odd parity) to a 0^+ state ($I = 0$, even parity) will be a GT transition and first-forbidden, because as a Fermi transition it would be third-forbidden.

Half-Lives of Beta Emitters.   As we did for alpha decay, we can find a relationship between the half-life and the emitted energy for beta decay, although the result cannot be written so explicitly. The total transmission probability P

may be found by integrating Eq. (25) over all values of p_e, obtaining a result that is proportional to the time t. Dividing by t, we obtain the transition probability per unit time—that is, the decay constant λ, which equals $(\ln 2)/t_{1/2}$.

Without going into the details of the integration of Eq. (25), which depends on the knowledge of the nuclear Coulomb factor $F(Z,p_e)$, we can write the result as

$$1/t_{1/2} = g^2 \, |M|^2 f(Z,E_m)$$

where $f(Z,E_m)$, called the Fermi integral function, has been calculated for all values of Z and E_m, the maximum beta energy. Results of the calculation for β^- decay are plotted in Fig. 9. These results cannot be represented by a single formula, but one can see by inspection of the curves that f is roughly proportional to E_m^5 for small Z, if $E_m > 1$ MeV.

The product

$$f(Z,E_m)t_{1/2} = 1/g^2 \, |M|^2$$

is called the "ft value" in physicists' jargon. It can be determined experimentally for any given beta emitter by measuring E_m and $t_{1/2}$, then looking up $f(Z,E_m)$. (In this case, as in the Z dependence of alpha decay parameters, the Z is that of the *daughter* nucleus.) In a few simple cases, the nuclear matrix element M can be evaluated theoretically, so it is then possible to determine the value of the constant g. The result is $g = 1.4 \times 10^{-49}$ erg-cm³ for Fermi allowed transitions. For GT allowed transitions, it is slightly larger. Thus the two parts of the beta-decay interaction have roughly the same strength.

Values of ft range from 10^3 seconds to 10^{23} seconds, so it is convenient to quote the value of $\log_{10}ft$ instead (with $t_{1/2}$ always in seconds). Let us find this value for two allowed transitions:

1. *Decay of the neutron.* In this case, $E_m = 0.78$ MeV $= 1.53 \, m_e c^2$, so from Fig. 9, with $Z = 1$, we estimate $\log_{10}f$ to be about 0.3. From Appendix D, we see that $t_{1/2} = 660 \, s$, so $\log_{10}t_{1/2} \approx 2.8$, making $\log_{10}ft \approx 3.1$.
2. *Decay of ^{24}Na.* We can see from the values of I in Appendix D that decay to the ground state of ^{24}Mg is forbidden. But there is an excited state of ^{24}Mg, 4.12 MeV above the ground state, for which $\Delta I = 0$. ^{24}Na decays to this state with $E_m = 1.39$ MeV $= 2.73 \, m_e c^2$, and $t_{1/2} = 14.98 \, h = 5.39 \times 10^4 \, s$. From Fig. 9, $\log_{10}f(12, 2.73) \approx 1.4$, and $\log_{10}5.39 \times 10^4 \approx 4.7$, so $\log_{10}ft \approx 6.1$.

These two cases represent the extreme values of $\log_{10}ft$ for allowed transitions. Neutron decay proceeds rapidly (small ft), becauser it is a so-called favored transition, in which the matrix element M is as large as possible (corresponding to almost complete overlap between the initial and final state wavefunctions). We shall see in Section 15.4 other examples of favored transitions.

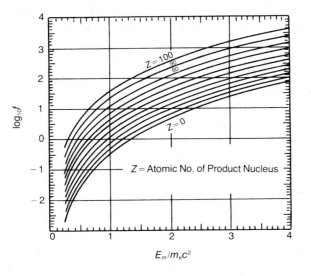

Fig. 9. *Fermi integral function f(Z,E_m) for β^- emission. (Adapted from E. Feenberg and G. Trigg, Rev. Mod. Phys. **22**, 399 (1950).)*

Values of $\log_{10}ft$ larger than 6.1 result from forbidden transitions. For a first-forbidden transition, the additional factor $(\mathbf{p}_e + \mathbf{p}_\nu)\cdot\mathbf{r}/\hbar$ in the matrix element is of the order of 1/100, making $|M|^2$ smaller by a factor of about 10^{-4}. Thus $\log_{10}ft$ values for first-forbidden transitions, as a group, tend to be about 4 more than the values for allowed transitions, and for each additional degree of forbiddenness, $\log_{10}ft$ increases by (approximately) another 4. For the fourth-forbidden decay of ^{115}In, $\log_{10}ft$ is 23.2. Thus the effect of degree of forbiddenness overshadows the variations in ft resulting from the wavefunctions in M, making it possible, in the absence of other information, to judge the degree of forbiddenness from the ft value alone, and hence to estimate the spin of a nuclear state.

An extreme example of a forbidden decay is that of ^{48}Ca, whose decay to ^{48}Sc is fifth forbidden; the half-life for this transition is so long that the decay has never been observed, and ^{48}Ca is listed as a stable species, even though it is more massive than ^{48}Sc (by 0.29 MeV/c^2). (For this transition, $Z = 21$, $E_m = 0.57\, m_e c^2$, so Fig. 9 gives $\log_{10}f \simeq -1.1$. We might expect $\log_{10}ft$ to be about 27, making $t_{1/2}$ of the order of 10^{28} seconds, or about 10^{21} years.

Conservation of Parity in Beta Decay. The beta decay interaction potential has another feature of great interest, whose existence was first suspected in 1956.

As proposed by Lee and Yang, the potential contains a factor which changes sign upon reflection of the coordinate system. This means that the result of a beta decay experiment may not be the same as the result of the mirror image of that experiment. This possibility was immediately tested by C. S. Wu, et al.,[18] in the beta decay of ^{60}Co. They found, by aligning the ^{60}Co nuclear spins in a strong magnetic field, that more electrons are emitted in the direction opposite to the nuclear spin than in the direction parallel to that spin. Since in a mirror image (Fig. 10) the spin direction is reversed while the electron direction is not, the mirror image of the beta decay of ^{60}Co shows more electrons being emitted in the direction of the spin than in the opposite direction. Thus in beta decay experiments, unlike other physics experiments, we can tell whether we are "looking at" the actual experiment or at its mirror image. This indicates that the "law of conservation of parity" is violated in beta decay. We shall return to this law and other relevant experiments in the next chapter.

Electron Capture.   As we have already seen (Section 14.7), when positron emission is possible, the nucleus may instead capture one of the electrons in the surrounding atomic shells. The reaction is

$$\text{p} + \text{e}^- \rightarrow \text{n} + \nu_e$$

The theory of electron capture differs very little from that of beta decay; the same interaction governs both processes. In fact, positron emission may be thought of as electron capture from an electron in the continuum of negative energy states. The selection rules of beta decay therefore apply also to electron capture, although, as we have seen in Chapter 14, electron capture may occur when positron emission is energetically impossible.

When both processes are possible, the ratio of electron captures to β^+ decays is determined by two factors: (1) the Coulomb potential between positron and nucleus, which inhibits β^+ decay, and (2) the density of bound electrons at the nucleus, which determines the electron capture probability. Both factors tend to favor electron capture more strongly in the heavier elements; the Coulomb potential becomes larger as Z increases, and the electron shells are pulled in closer to the nucleus, increasing the electron density at the nucleus. Thus there are no observable positron emitters among the heavy elements; potential positron emitters decay too quickly by electron capture.

Electron capture may involve electrons from any of the atomic shells— K, L, M, etc.— but obviously K capture is the most common event. After K capture occurs, the daughter atom is in an excited state, because of the

[18] C. S. Wu, E. Ambler, R. W. Hayward, D. D. Hoppes, and R. P. Hudson, *Phys. Rev.* **105**, 1413 (1957).

vacancy in the K shell. As this vacancy is filled, characteristic X-rays are emitted. For example, an L electron may fall into the K shell, with the emission of a K_α X-ray, and another X-ray may be emitted when the resulting vacancy in the L shell is filled. Instead of X-ray emission, Auger electron

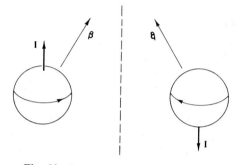

Fig. 10. *Beta decay and its mirror image. The vector* **I** *is not " reflected" in the same way as the rest of the figure, because* **I** *is not a real vector; it is constructed from the rotation of the nucleus. Thus* **I** *is called a pseudovector; if* **I** *points upward, its mirror image points downward. If an experiment shows more electrons emitted parallel to* **I** *than antiparallel to* **I***, the mirror image of the experiment shows the opposite result. Of course, in the mirror image, we do not actually see the nucleus rotating, but we can determine the direction of the current in a current loop, and see that it is reversed in the mirror image. The current in the loop produces a magnetic field which can be used to align the nuclear spins. The magnetic field* **B** *is related to the current in the same way that* **I** *is related to the rotation of the nucleus, so* **B***, like* **I***, is a pseudovector.*

emission is also possible. (This is the same process discussed in Section 9.1 in connection with atomic transitions.) The number of X-ray quanta emitted per vacancy in the K shell is called the K fluorescenece yield. This yield is nearly unity in the high Z elements and nearly zero in the low-Z elements. Since the neutrino is the only particle emitted in the primary reaction, the X-rays and Auger electrons provide the only means of detecting the electron capture process.

Electron capture has provided us with the only situation in which the half-life of a radioactive species is measurably affected by its environment. Chemical changes which alter the electronic structure of the atom can change the electron density at the nucleus, thereby changing the electron capture rate and thus the half-life. This was first demonstrated for ^{7}Be, which decays only by electron capture. The half-life of ^{7}Be is 0.07 percent longer in BeF$_2$ than in metallic Be.

15.3   GAMMA DECAY

The study of gamma decay is closely related to the study of beta decay, for most beta decays are followed by one or more gamma decays, and the study of these beta–gamma cascades yields valuable information about nuclear states.

The theory of gamma decay is simply electromagnetic theory. The only difficulty in the theory is in the determination of the wavefunctions of the nuclear states which produce the radiation. We shall not go into these problems, but simply state some general results. As we have seen before in the case of atomic radiation, the radiation may be classified by its electric or magnetic multipole nature, with selection rules for each type of multipole (see Table 2).

As always, the selection rules follow directly from the form of the matrix element governing the transition. (See Chapter 10). The maximum value of Δl equals the angular momentum carried by the gamma ray. An example illustrating the parity change rule may be helpful. For E1 transitions, the matrix element contains an integral of the form $\int u_m^* x u_k \, d\tau$. Since the variable x has odd parity, the matrix element vanishes unless the initial and final states have opposite parity. In general, all of the odd order "electric" transitions and all of the even order "magnetic" transitions result in a change in the parity of the nuclear state, while the other transitions do not.

Table 2 shows typical lifetimes calculated on the basis of the shell model for various types of radiation. Notice that radiation of a given multipole order is slower than that of the next lower order by a factor of about 10^4. This factor is of order $(R/\lambda)^2$, where R is the nuclear radius and λ is the wavelength of the radiation; the factor has the same origin as the similar factor for atomic radiation (Section 10.3a). Of course these calculations are based on a specific model, and on the assumption that the states involved are single particle states, so the lifetimes tabulated should not be expected to agree with measured lifetimes for all nuclei. For example, transitions between collective states (Section 14.9) are much faster than transitions between single-particle states, because many particles are cooperating in radiating the energy. Just as radiation from a charged oscillating dipole is proportional to the square of the charge, a transition involving n nucleons in a collective mode proceeds about n^2 times faster than a transition involving only one nucleon. It is found,

Table 2

Classification of Gamma Radiation

Transition and Symbol	Selection rules		Approximate mean lifetime[a] (sec)	
	$\Delta I \leqslant$	Parity change	1.0 MeV	0.1 MeV
Electric dipole E1	1	Yes	10^{-15}	10^{-12}
Magnetic dipole M1	1	No	10^{-13}	10^{-10}
Electric quadrupole E2	2	No	10^{-11}	10^{-6}
Magnetic quadrupole M2	2	Yes	10^{-9}	10^{-4}
Electric octupole E3	3	Yes	10^{-6}	10^{1}
Magnetic octupole M3	3	No	10^{-4}	10^{3}
Electric hexadecapole E4	4	No	10^{-1}	10^{8}
Magnetic hexadecapole M4	4	Yes	10^{1}	10^{10}

[a] From calculations by Blatt and Weisskopf, based on the shell model, for transitions between single-particle states.

for example, that the mean lifetime of the 0.1001 MeV state of ^{182}W (Fig. 12, Chapter 14) is only 2 nsec, whereas according to Table 2 a single-particle state radiating 0.1 MeV in an E2 transition should have a mean lifetime of 1 μsec, about 500 times as long.

Between any two given states, a transition normally proceeds by the fastest possible mechanism. For example, a transition from a 6^+ state to a 4^+ state could be E2, E4, E6, . . . or M3, M5, M7, . . . Of these, the electric quadrupole, E2, is the fastest, so the transition is E2. The next fastest transition rate, for M3, is slower by a factor of 10^7 (if the gamma ray energy is 1.0 MeV), so only one transition in 10^7 would be an M3 transition. A unique one-to-one correspondence can thus be made between the spin and parity change and the type of transition. This makes it possible to use the mean lifetime of a state as a rough guide to the type of transition which occurs from that state, so that one can often determine the spin and parity of an excited state simply by knowing its mean lifetime and the spin and parity of the state to which it decays.

In practice, the lifetime information is combined with other information in making these spin and parity determinations. Let us consider a well-known beta-gamma cascade, in the decay of $^{137}_{55}$Cs, to see how this works. The decay scheme is shown in Fig. 11. In $93\frac{1}{2}$ percent of the beta decays, the daughter $^{137}_{56}$Ba is left in an excited state, 0.66 MeV above the ground state, to which it decays with a half-life of 2.6 min. The ^{137}Ba ground state is known, from

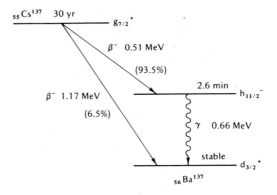

Fig. 11.   *Decay scheme of* ^{137}Cs, *showing spin,*
parity, and half-life of each level.

study of the hyperfine splitting, to have spin $\frac{3}{2}$, and the shell model
indicates that it should have even parity.[19] Atomic beam experiments have shown
that the ^{137}Cs ground state has a spin of $\frac{7}{2}$. Starting with this information, we can
deduce the parity of the ^{137}Cs ground state and the spin and parity of the ^{137}Ba
excited state, on the basis of the following data:

1. The partial half-life for the decay of ^{137}Cs to the ^{137}Ba ground state is 30
    yr/0.065 = 500 yr, which is quite long for a 1.17-MeV transition; $\log_{10} ft \approx 12$,
    indicating that the transition is second forbidden. (See Problem 4.)

2. The Kurie plot of the beta decays also indicates that this is a second-
    forbidden transition.[20] Since there is no parity change in a second-for-
    bidden transition, we deduce that the ^{137}Cs ground state parity is even. The
    spin change $\Delta I = \frac{7}{2} - \frac{3}{2} = 2$ is consistent with a second-forbidden transition.

3. The partial half-life for the 0.51 MeV transition is considerably smaller
    than for the 1.17-MeV transition, being only 30 yr/0.935 = 32 yr, while
    the energy is not so high as in the other transition. The ft value is thus
    considerably larger, and it indicates that this transition is first-forbidden.

4. The Kurie plot also indicates that the 0.51-MeV transition is first-forbidden.
    The indicated ΔI value is ± 2, and there is a parity change in a first-

---

[19] There is only one neutron lacking to form a closed shell at $N = 82$. The shell-model state of the
missing neutron must therefore be $d_{3/2}$ (Chapter 14, Fig. 11), with two units of orbital angular
momentum and hence even parity.

[20] At energies below 0.51 MeV, beta particles from the 0.51-MeV transition cannot be distinguished
from those resulting from the 1.17-MeV transition. Therefore only that part of the spectrum above
0.51 MeV can be used to determine the shape, but this is sufficient to establish that the transition is
second-forbidden.

forbidden transition,[21] so that the ^{137}Ba excited state must have odd parity and a spin of $\frac{7}{2} \pm 2 = \frac{11}{2}$ or $\frac{3}{2}$.

5. The half-life of the ^{137}Ba excited state, together with the fact that there is a parity change in the transition to the ground state, indicates (Table 2) that the transition is an M4 transition, making $\Delta I = 4$, so that the excited state spin is $\frac{11}{2}$ rather than $\frac{3}{2}$.

The nature of the excited state of ^{137}Ba is thus determined from overlapping pieces of information which provide several cross checks with complete consistency. To complete the story, we note that the shell model predicts the existence of an $h_{11/2}$ neutron state (a state of odd parity) in the outermost shell of ^{137}Ba.

Angular Correlation of Successive Radiations. The lifetime of a nuclear state is not the only indication of the multipole character of the radiation which it emits. We know from classical electromagnetic theory that there is a connection between the multipole nature and the angular distribution of the radiation. For example, the intensity of radiation from an oscillating electric dipole is proportional to $\sin^2 \theta$, where θ is the angle between the dipole axis and the direction of the emitted radiation. The electric dipole radiation from a nucleus should be distributed in the same way; the number of photons emitted into a solid angle of fixed magnitude centered on a given direction should be proportional to $\sin^2 \theta$; in this case, θ is the angle between the given direction and the axis of the "oscillating dipole" in the nucleus. This dipole axis must, of course, be defined in terms of the spin orientations of the initial and final states involved in the radiation.

In principle, the spins of the decaying nuclei could be aligned, and then the multipolarity of the radiation could be determined by measuring its angular distribution relative to the spin axis of the nuclei, but this is not usually feasible. Clearly, if the nuclear spins are not aligned but are distributed at random, the gamma rays are also emitted in random directions, regardless of the multipolarity of the radiation. However, information about the spin axis of a decaying nuclear state is available when that state has been produced by a *preceding* gamma decay. In this case, there is a correlation between the direction of the first gamma ray, associated with the production of the state, and the direction of the second gamma ray, associated with the decay of the state; these directions are related because of their common dependence on the spin direction of the emitting nucleus.

---

[21] According to Table 1, a transition with $\Delta I = \pm 1$ may also be first-forbidden. However, this choice of ΔI was ruled out because such a transition would be first-forbidden by both the Fermi and the GT selection rules; it would therefore be governed by two matrix elements rather than one, and the resulting Kurie plot would be more complex than the one observed. With $\Delta I = \pm 2$, the transition is pure GT, being first-forbidden only by the GT selection rules, and only one matrix element contributes appreciably to the transition probability.

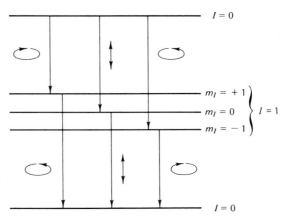

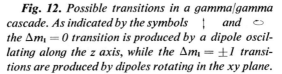

Fig. 12. *Possible transitions in a gamma/gamma cascade. As indicated by the symbols* ↕ *and* ⟲ *the Δm$_1$ = 0 transition is produced by a dipole oscillating along the z axis, while the Δm$_1$ = ±1 transitions are produced by dipoles rotating in the xy plane.*

The mathematics required to derive the actual angular correlation of the gamma rays corresponding to each multipole order is rather sophisticated. However, the principle involved is illustrated by the following simple (and somewhat artificial) situation (Fig. 12). An $I = 0$ state decays to an $I = 1$ state, which then decays to another $I = 0$ state. In each transition, Δm_I can be $+1$, 0, or -1. We need to know the angular distribution of the radiation in each case, and from this we must deduce the distribution of the angles between the two successive gamma rays.

We know from Chapter 10 (selection rules, Section 10.3) that the $\Delta m = 0$ transition radiates like a dipole oscillating along the z axis (matrix element $\int u^*_m z\, u_k\, d\tau$), while the $\Delta m = \pm 1$ transitions radiate like a dipole rotating in the xy plane (matrix elements $\int u^*_m (x \pm iy) u_k d\tau$). Therefore the intensity of the $\Delta m = 0$ radiation should be distributed as $\sin^2 \theta$, while the intensity of the $\Delta m = \pm 1$ radiations should be distributed as $1 + \cos^2\theta$.[22] In the absence of a magnetic field, the z direction is arbitrary, and the three magnetic substates of the $I = 1$ state are degenerate. We can then choose the z axis to be

---

[22] Consider a photon emitted in the yz plane. The direction of the photon makes an angle of θ with the z dipole, an angle of $\pi/2$ with an x dipole, and an angle of $\pi/2 - \theta$ with a y dipole. The respective intensities contributed by each dipole are therefore $\sin^2 \theta$, $\sin^2(\pi/2)$, and $\sin^2(\pi/2 - \theta)$. The dipole rotating in the xy plane is equivalent to the sum of an oscillating x dipole and an oscillating y dipole, so that it contributes an intensity proportional to the sum of $\sin^2(\pi/2)$ and $\sin^2(\pi/2 - \theta)$, or $1 + \cos^2 \theta$.

the direction of the first gamma ray, so that $\theta = 0$ for this ray. This ensures that $\Delta m = \pm 1$ in the first transition, for there are no gamma rays emitted at $\theta = 0$ when $\Delta m = 0$. In other words, in this case the nucleus can emit a gamma ray in a certain direction only if $\Delta m = \pm 1$ *with respect to that direction as z axis*. It is then clear from Fig. 12 that Δm must equal ± 1 in the second transition also. The angular distribution of the second gamma ray is then $1 + \cos^2 \theta$. Since θ is the angle between the direction of this second gamma ray and the z axis, and the z axis is the direction of the first gamma ray, the probability of a given angle θ *between the two gamma rays* is proportional to $1 + \cos^2 \theta$.

A complete theory of angular correlations has been worked out, giving the angular correlation corresponding to any possible combination of the angular momenta of the two gamma rays (the multipolarity of the radiation) of the spine of the three states involved in the cascade. Thus one can, by measuring the angular correlation, obtain information about the spins of these states. Information of a similar kind can be obtained by measuring the angular correlation between the *beta* particle and the subsequent gamma ray emitted in a beta–gamma cascade.

Implicit in the above discussion is the assumption that the intermediate state lives for a very short time, so that its spin orientation does not change appreciably during its lifetime. If the lifetime of the state is " long "—of the order of nanoseconds or longer—the nucleus has time to precess in an external field (for example in the magnetic field of the atomic electrons), between the time of emission of the two gamma rays. This, of course, perturbs the angular correlation between the gamma rays, but this can be turned to our advantage. We can influence the precession by changing the environment of the nucleus, and it is often possible to measure the precession rate by observing the dependence of the angular correlation on the time interval between the emission of the two gamma rays. In this way, magnetic moments of excited states of nuclei have been determined.

Internal Conversion.   Instead of emitting a gamma ray, an excited nucleus may lose its energy to an atomic electron whose wavefunction overlaps the nucleus. This process, called internal conversion, is a type of Auger transition. The internal conversion electrons, like all Auger electrons and unlike beta particles, have a definite energy, equal to the nuclear transition energy minus the binding energy of the electron which is emitted. When internal conversion occurs after beta decay, the spectrum of emitted electrons consists of the continuous beta spectrum plus a sharp peak from conversion electrons.

The total internal conversion coefficient is defined as the ratio of the probability of internal conversion to the probability of gamma emission from a given state. There are also separate internal conversion coefficients defining the probabilities of conversion of electrons in the various shells—K, L, M, etc.

In cases in which the excited state and the ground state each have $I = 0$, gamma decay is absolutely forbidden to all orders, but internal conversion can still occur, so the conversion coefficient is infinite. Existence of such cases proves that internal conversion is *not* an internal photoelectric effect, in which the gamma ray is first emitted and then absorbed by an electron. While such a two-step process is possible, its probability is generally negligible in comparison to the probability of direct transfer of the excitation energy from the nucleus to the electron.

Internal conversion thus provides an *additional* way for an excited nucleus to decay, without affecting the probability per unit time for gamma decay. (In other words, the decay constant for gamma decay would be unchanged if there were no atomic electrons around the nucleus to permit internal conversion.) There are still other possible ways for a nucleus to lose its excitation energy. For example, if the excitation energy of a state exceeds 1.02 MeV, "pair internal conversion" is possible. In this process the nucleus emits a positron–electron pair instead of the gamma ray. This process is equivalent to the internal conversion of an electron from the infinite sea of negative energy states, instead of an atomic electron.

15.4 NUCLEAR REACTIONS

Study of nuclear properties by observing decay of radioactive species is limited by the number of long-lived species available. Study of collisions between energetic particles and nuclei opens a much wider range of possibilities. Such collisions can excite the target nucleus, or can create new nuclear species, in the ground state or in excited states. This provides a systematic way to study excited states of nuclei and enables us to create new species for various purposes; we have now an enormous number of "artificial" radioactive species, which are man-made in nuclear reactions.

The possible results of a collision between an incident particle and a nucleus may be summarized as follows:

1. The incident particle may be elastically scattered, as a result of interaction with either the Coulomb force or the nuclear force exerted by the nucleus.

2. The incident particle may interact inelastically with the nucleus, either exciting it to a higher energy state or knocking one or more particles out of the nucleus. The emitted particles may be ones which were in the original nucleus, or they may be particles such as pions which are "materialized" by the energy of the collisions (see Section 16.1).

Two main physical pictures ("reaction mechanisms") which have been considered in studying these processes are the following:

a. *Compound nucleus formation*: The incident particle may interact with the nucleus as a whole, and become temporarily amalgamated with it. The resulting compound nucleus may then decay in a number of ways, as after induced fission (Section 14.7).

b. *Direct interaction mechanism*: Some reactions, especially at higher energies, may be best understood as direct interactions between the incident particle and one or more localized particles in the nucleus.

The probability of occurrence of each possible reaction may be conveniently expressed in terms of a *partial cross section*, which is related to the probability of occurrence of a *single* reaction in the way that the *total* cross section is related to the probability that some interaction will take place. A convenient unit for cross sections of nuclear reactions is the *barn*, equal to 10^{-24} cm^2.

While we have encountered the concept of a cross section before, let us now discuss the ways in which a cross section may be related to the number of events which are observed in an experiment. Basically, the probability that a single particle, striking at normal incidence, will interact with a target in a certain way is equal to the product of the cross section for that interaction and the number of target particles per unit area. If a *beam* of particles is incident on a target, there are two possibilities:

1. If the cross sectional area of the beam is smaller than that of the target, each beam particle has the interaction probability stated above, and to find the total number of events per second, we simply multiply the above probability by the total number of particles per second striking the target. Thus

$$\text{Number of events per second} = N_b \sigma n$$

where N_b is the number of particles per second in the beam, σ is the interaction cross section, and n is the number of target particles per unit area. Notice that it is not necessary to know the area of the beam, as long as the target intercepts the entire beam.

2. If the cross sectional area of the target is smaller than that of the beam,

$$\text{Number of events per second} = N_t \sigma n$$

where N_t is the total number of particles in the *target*, and n is the beam flux—the number of particles *per unit area* per second in the beam. We can understand this result by imagining a moving coordinate system in which the beam particles are at rest and the target moves toward them, so that the situation is the same as in (1) except that the roles of target and beam are reversed. In this case it is not necessary to know the area of the *target*, as long as one knows the total number of particles in the target.

In both cases it is assumed that the target is thin, so that there is no appreciable attenuation of the beam as it passes through the target, and each target particle has an equal chance of interacting with the beam particles. A thick target can, of course, be treated as a series of thin targets, and the number of events can be computed if the beam attenuation is known. If only one type of interaction occurs, the beam attenuation is easy to calculate; the probability that a beam particle will interact within a distance dx and be lost to the beam is $N\sigma\, dx$, where N is the number of target particles per unit area. (See Section 13.2.)

EXAMPLE PROBLEM 1. We wish to produce ^{64}Cu by bombarding natural copper (69% ^{63}Cu, 31% ^{65}Cu) with low-energy neutrons. The relevant cross-sections are $\sigma_{63} = 4.5$ barns for $^{63}Cu + n \rightarrow {}^{64}Cu$, and $\sigma_{65} = 2.3$ barns for $^{65}Cu + n \rightarrow {}^{66}Cu$. If the copper is 1.0 cm^2 in area and 0.01 cm thick, and the copper is in a reactor where the neutron flux is 10^{12}/cm^2-s, find the production rate of ^{64}Cu, and the ^{64}Cu activity after 12.8 hours in the reactor.

Solution. First let us determine whether or not the target is "thin," in the sense of the preceding discussion. The total probability P that a neutron will be absorbed in the thickness $dx = 0.01$ cm equals the probability for absorption by ^{63}Cu plus the probability for absorption by ^{65}Cu, or

$$P = (N_{63}\sigma_{63} + N_{65}\sigma_{65})dx$$

where

$$N_{63} = 0.69 \times (6.02 \times 10^{23}/64.4 \text{ g}) \times 8.9 \text{ g/cm}^3$$
$$= 5.74 \times 10^{22}/\text{cm}^3$$

and

$$N_{65} = 0.31 \times (6.02 \times 10^{23}/64.4 \text{ g}) \times 8.9 \text{ g/cm}^3$$
$$= 2.58 \times 10^{22}/\text{cm}^3$$

with the result that

$$P = 0.003$$

Thus the sample is "thin"; an attenuation of 0.3% can safely be neglected. Since the beam is obviously larger than the target, the total production rate R equals $N_t\sigma_{63}n$, where $N_t = N_{63} \times 0.01$ cm^3 is the *total* number of ^{63}Cu atoms in the target. Therefore

$$R = 5.74 \times 10^{20} \times 4.5 \times 10^{-24} \text{ cm}^2 \times 10^{12}/\text{cm}^2-\text{s}$$
$$= 2.6 \times 10^9 \text{ atoms/s}$$

This rate may be considered constant, because in any reasonable time interveral there is a negligible change in the number of ^{63}Cu atoms in the target. But the ^{64}Cu decays after it is produced, so we must take the decay as well as the production into account in determining how much ^{64}Cu activity is present at any given time. The situation is similar to that of production of a radioactive species by the decay of a very long-lived parent (Chapter 13, Problem 3); the production rate is constant and the decay rate is proportional to the number of daughter atoms present.

In our present problem, we may write the differential equation for N, the number of ^{64}Cu atoms present at any time t, as

$$dN/dt = R - \lambda N \, ,$$

where R is the production rate, 2.6×10^9/sec in this case, and λ is the decay constant. The solution to this equation is easily seen to be

$$N = \frac{R}{\lambda} (1 - e^{-\lambda t})$$

so that the activity at any time t is

$$\lambda N = R(1 - e^{-\lambda t})$$

From Appendix D, we see that $t_{1/2} = 12.8$ h, so when $t = 12.8$ h, $e^{-\lambda t} = \frac{1}{2}$ and we have

$$\lambda N = R(1 - \tfrac{1}{2}) = \frac{R}{2}$$

$$= 1.3 \times 10^9 \text{/sec} = 35 \text{ mCi}$$

The maximum possible activity which could be obtained for this piece of copper in the given neutron flux is simply equal to R. When the ^{64}Cu activity equals the production rate R, ^{64}Cu is decaying as fast as it is produced, so the number of ^{64}Cu atoms cannot increase further.

Resonances.   At certain values of incident energy, "resonances" occur and the cross section for a reaction becomes very large. These resonances may be detected by counting the number of ejected particles as a function of the energy of the incident particle. For example, Fig. 13a shows the number of alpha particles observed as a result of bombardment of ^{27}Al with protons of various energy. The reaction is

$$^{27}\text{Al} + \text{p} \rightarrow {}^{24}\text{Mg} + \alpha$$

which is written in condensed notation as $^{27}\text{Al}(\text{p},\alpha)^{24}\text{Mg}$, the first symbol in parentheses indicating the bombarding particle and the second indicating the ejected particle.

The resonances may be explained by assuming that the compound nucleus ^{28}Si is formed temporarily when the proton collides with the ^{27}Al nucleus. The ^{28}Si then decays rapidly by emitting an alpha particle, leaving ^{24}Mg—the "residual nucleus"—behind. The reaction only proceeds when the proton has just the right energy to form ^{28}Si in one of its excited states[23]; thus the resonances are observed at only those proton energies. If the energy of the proton is not a resonant energy, the compound nucleus cannot be formed, no alpha particles appear, and the proton is simply scattered.

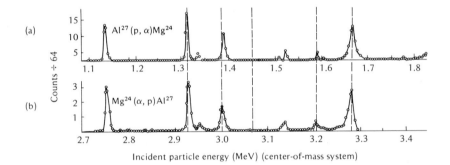

Fig. 13. (a) Number of alpha particles emitted at a fixed angle when ^{27}Al is bombarded with protons, plotted against the initial kinetic energy in the center-of-mass coordinate system. (b) Number of protons emitted at a fixed angle when ^{24}Mg is bombarded with alpha particles, plotted in the same way. [From S. G. Kaufmann, E. Goldberg, L. J. Koester, and F. P. Mooring, Phys. Rev. **88**, 673 (1952)]

Notice that Fig. 13 shows the initial kinetic energy in the center-of-mass coordinate system, which is a convenient system to use because in that system the compound nucleus is formed at rest. The rest energy of the compound nucleus formed at any resonance may therefore be found simply by adding the rest energies of the reactants to the kinetic energy at which the resonance appears. The *excitation* energy of the ^{28}Si level is then found by subtracting the ground-state rest energy of ^{28}Si. For example, the excitation energy corresponding to the resonance at 1.14 MeV is

---

[23] Only the *excited* states of ^{28}Si are involved, for the ground state is stable and cannot emit an alpha particle.

rest energy of ^{27}Al + rest energy of ^{1}H
+ 1.14 MeV − rest energy of ^{28}Si

or

$$(27u - 17.196) + (u + 7.289) + 1.14 \text{ MeV} - (28u - 21.490)$$
$$= +12.72 \text{ MeV}$$

In this calculation, u is the energy of one atomic mass unit, and it drops out of the final result, because no nucleons are created or destroyed in the reaction. Notice that the rest energies of the neutral *atoms,* as listed in the atomic mass table, are used in the calculation. (It might seem that this is incorrect, because the incident particle is a bare proton rather than a hydrogen atom; but the ^{28}Si which is formed is missing one electron also, so no error results from using atomic masses.)

Fig. 13b shows that the same arrangement of peaks, with the same spacing and the same relative sizes, results from the inverse reaction ^{24}Mg$(\alpha,p)^{27}$Al. This is a good indication that we are looking at the results of the same set of ^{28}Si energy levels. The only difference is that each peak in 13b appears at an initial kinetic energy which is about 1.6 MeV larger than the energy at the corresponding peak in 13a. This is simply because the rest energy of the reactants ^{4}He + ^{24}Mg is 1.6 MeV smaller than the rest energy of ^{1}H + ^{27}Al, so that ^{4}He and ^{24}Mg require an additional 1.6 MeV, in the form of kinetic energy, if they are to combine to form the same nuclear state formed by ^{1}H + ^{27}Al. The same difference is, of course, present in the decay; the 12.72-MeV excited state of ^{28}Si can decay into ^{4}He + ^{24}Mg with a total kinetic energy of 2.7 MeV, or into ^{1}H + ^{27}Al with a total kinetic energy of 1.1 MeV. Thus in the reaction ^{27}Al$(p,\alpha)^{24}$Mg, the kinetic energy of the reactants is 1.6 MeV less than that of the products. This kinetic energy difference is called the Q value of the reaction. When the Q value is positive, rest energy is converted to kinetic energy, and the reaction is called exoergic. When the Q value is negative, as in ^{24}Mg$(\alpha,p)^{27}$Al, kinetic energy is converted to rest energy, and the reaction is called endoergic. In writing the reaction, one may show the Q value explicitly, as we did in Chapter 14. Thus

$$^{27}\text{Al} + \text{p} \quad \rightarrow \quad ^{24}\text{Mg} + \alpha + 1.6 \text{ MeV}$$

Since all of these kinetic energies are given with respect to the center-of-mass (c.m.) coordinate system, and the energies are actually measured in the laboratory coordinate system, in which the target is at rest, we must know how to convert from one system to the other. This is not very difficult; we can easily derive a simple (nonrelativistic) formula which gives the total initial kinetic energy in the c.m. system in terms of the kinetic energy of the incident particle in the lab system, as follows (Fig. 14):

Let M_1 = mass of incident particle; M_2 = mass of target nucleus;

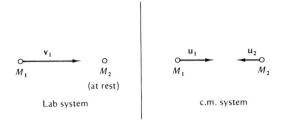

Fig. 14. *Velocities of incident particle and target nucleus, in laboratory and in center-of-mass coordinate systems.*

$\mathbf{v}_1$ = velocity of incident particle in lab system; $\mathbf{u}_1$ = velocity of incident particle in c.m. system; and $\mathbf{u}_2$ = velocity of target nucleus in c.m. system.

Then, since the total momentum in the c.m. system is zero,

$$M_1|u_1| = M_2|u_2| \tag{28}$$

But $|u_2|$ must equal the speed of the c.m. system relative to the lab system, so

$$|u_1| = |v_1| - |u_2| \tag{29}$$

Solving Eqs. (28) and (29) for $|u_1|$ and $|u_2|$, we obtain

$$|u_1| = \frac{M_2|v_1|}{(M_1 + M_2)}$$

$$|u_2| = \frac{M_1|v_1|}{(M_1 + M_2)}$$

The kinetic energy in the c.m. system is therefore

$$E_{\text{c.m.}} = \frac{M_1 u_1^2}{2} + \frac{M_2 u_2^2}{2}$$

$$= \frac{M_1 M_2^2 v_1^2}{2(M_1 + M_2)^2} + \frac{M_2 M_1^2 v_1^2}{2(M_1 + M_2)^2}$$

$$= \tfrac{1}{2}M_1 v_1^2 M_2 \left[\frac{M_2 + M_1}{(M_1 + M_2)^2} \right]$$

$$= \frac{E_1 M_2}{M_1 + M_2} \tag{30}$$

where E_1 is the kinetic energy of the incident particle in the laboratory system.

The compound nucleus lives for a very short time; its lifetime in any given state can often be estimated by determining the uncertainty in the energy of the state, which is indicated by the width ΔE of the appropriate peak in a curve like those of Fig. 13. The uncertainty relation $\Delta E \cdot \Delta t \approx \hbar$ then tells us Δt, the mean lifetime of the excited state. A typical width of 1 keV indicates a mean lifetime of $\sim 10^{-18}$ sec.

You may well wonder whether it makes much sense to speak of a compound nucleus at all, when it breaks up so quickly. But you must consider that the incident particle travels at a speed of $\sim 10^9$ cm/sec, so that it can cross the nucleus in less than 10^{-21} sec. Thus, even in the short time of 10^{-18} sec, the incident particle can become thoroughly mixed with the target nucleus, so that the compound nucleus forgets how it was formed, and we are able to obtain matched curves like those of Fig. 13, which would be hard to explain in any other way. Other examples of compound nucleus formation are abundant. One is ^{14}N, whose excited states may decay to $^{10}B + \alpha$, $^{12}C + d$, $^{13}C + p$, or $^{13}N + n$; the probability of each type of decay from a given excited state is independent of the method of production of that state.

Nuclear reactions may also be used to study the energy levels of the *residual* nucleus. To do this, one usually observes the energy and angle of emission of the *ejected* particle. The problem is then: Given the mass M_1 and kinetic energy E_1 of the incident particle, the mass M_2 of the target nucleus, and the mass M_3, kinetic energy E_3, and angle of emission θ of the ejected particle, deduce the mass M_4 of the residual nucleus. Since the residual nucleus is usually one whose ground state rest mass is known, one can then find the excitation energy of this nucleus, which is simply the difference between $M_4 c^2$ and the known ground state rest energy.

It is not too difficult to work this out relativistically. We can then find the nonrelativistic formula (which suffices for most purposes) by letting $E_1 = P_1^2/2M_1$ and $E_3 = P_3^2/2M_3$. Let us use units in which $c = 1$, and let $W = M + E$ be the total energy. We then have, from conservation of energy,

$$W_4 = W_1 + M_2 - W_3$$

But $W_4^2 = P_4^2 + M_4^2$, so that

$$P_4^2 + M_4^2 = (W_1 + M_2 - W_3)^2$$

and

$$M_4 = [(W_1 + M_2 - W_3)^2 - P_4^2]^{1/2} \qquad (31)$$

Here P_4 is unknown, but we may eliminate it by using momentum conservation. (See Fig. 15.)

$$P_4 \cos \phi = P_1 - P_3 \cos \theta$$
$$P_4 \sin \phi = P_3 \sin \theta$$

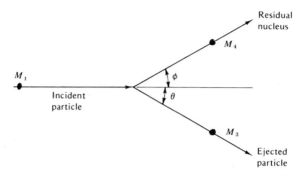

Fig. 15. *Angles involved in nuclear reactions with target nucleus at rest.*

By squaring both sides of each equation and adding, we obtain

$$P_4^2 = P_1^2 - 2P_1P_3 \cos \theta + P_3^2$$

so that Eq. (31) becomes

$$M_4 = [(W_1 + M_2 - W_3)^2 - P_1^2 - P_3^2 + 2P_1P_3 \cos \theta]^{1/2} \tag{32}$$

where the quantities on the right are all "given"; P_1 and P_3 may be found directly from the given quantities listed above.

So far the calculation is relativistically correct. For a nonrelativistic collision we now make the substitutions $P_1^2 = 2M_1E_1$ and $P_3^2\; 2M_3E_3$, which yield

$$M_4 = M_1 + M_2 - M_3$$
$$+ \; \frac{E_1(M_2 - M_3) - E_3(M_1 + M_2) + 2(M_1 M_3 E_1 E_3)^{1/2} \cos \theta}{M_1 + M_2 - M_3} \tag{33}$$

where we have neglected terms of order E^2, because $E \ll M$ for all particles.

EXAMPLE PROBLEM 2. Calculate the Q value of the reaction $^9\text{Be}(\alpha,\text{n})^{12}\text{C}$. If the alphas come from ^{210}Po, with an energy of 5.30 MeV, find the energy of the neutrons which emerge at angles of 0°, 90°, and 180° with the direction of the incident alpha particles, assuming that the residual nucleus ^{12}C is left in its first excited state, 4.43 MeV above the ground state.

Solution. From the atomic mass table (Appendix D) we find that the relevant masses are

$$M_1 = 4u + 2.42 \text{ MeV}/c^2 \qquad (^4\text{He})$$

$$M_2 = 9u + 11.35 \text{ MeV}/c^2 \qquad (^9\text{Be})$$

$$M_3 = u + 8.07 \text{ MeV}/c^2 \qquad (n)$$

$$M_4 = 12u + 4.43 \text{ MeV}/c^2 \qquad (^{12}\text{C})$$

where the first three are ground state masses and M_4 is the mass of the ^{12}C excited state. From these we find that $Q = 1.27$ MeV.

To find the neutron energy at each angle, we may simply insert the given quantities into Eq. (33) and solve for E_3. We may write Eq. (33) as

$$\frac{E_1(M_2 - M_3) - E_3(M_1 + M_2) + 2(M_1 M_3 E_1 E_3)^{1/2} \cos \theta}{M_1 + M_2 - M_3} = -Q \quad (34)$$

or, for $\theta = 0$,

$$\frac{5.30(8u) - E_3(13u) + 2(4u^2 \times 5.30E_3)^{1/2}}{12u} = -1.27$$

where we have neglected energies of a few MeV in comparison to uc^2, which is 931 MeV. This equation reduces to

$$13E_3 - 9.21(E_3)^{1/2} - 57.64 = 0$$

so that

$$E_3 = 6.20 \text{ MeV} \qquad \text{for } \theta = 0$$

With $\theta = 90°$, Eq. (34) becomes

$$13E_3 - 57.64 = 0$$

and

$$E_3 = 4.43 \text{ MeV} \qquad \text{for} \quad \theta = 90°$$

while with $\theta = 180°$

$$E_3 = 3.17 \text{ MeV} \qquad \text{for} \quad \theta = 180°$$

For a check on the answer for $\theta = 0°$, we may test to see if momentum is conserved by our value for E_3. The initial momentum is

$$P_1 = (2M_1E_1)^{1/2}$$
$$= (8u \times 5.30)^{1/2}$$
$$= (42.2u)^{1/2}$$
$$= 6.51(u)^{1/2}$$

The kinetic energy of the residual nucleus is $E_4 = 5.30 + Q - 6.20 = 0.37$ MeV, so that the final momentum is

$$P_3 + P_4 = (2M_3E_3)^{1/2} + (2M_4E_4)^{1/2}$$
$$= (2u \times 6.20)^{1/2} + (24u \times 0.37)^{1/2}$$
$$= (3.52 + 2.98)(u)^{1/2}$$

which agrees with P_1, within the error to be expected from rounding off in the calculation.

It may be instructive to look at the same collision in the c.m. system. In that system, the initial kinetic energy is, from Eq. (30), equal to

$$\frac{E_1M_2}{(M_1 + M_2)} = 5.30\left(\frac{9}{13}\right)$$
$$= 3.67 \quad \text{MeV}$$

The final kinetic energy is then $3.67 + Q = 4.94$ MeV. Since in this system the final particles have equal and opposite momenta, the energy is shared in inverse proportion to the mass. The neutron, with $\frac{1}{13}$ of the mass, then has $\frac{12}{13}$ of the energy, or 4.56 MeV. Notice that in the c.m. system, the neutron energy is independent of its direction of emission.

We must now take the vector sum of $\mathbf{v}_{c.m.}$, the velocity of the center of mass with respect to the lab, and $\mathbf{u}_3$, the neutron velocity in the c.m. system, in order to find the neutron velocity $\mathbf{v}_3$, and hence its energy, in the lab system. Let us do only the case in which the neutron emerges in the lab at a 90° angle from the initial alpha-particle direction (see Fig. 16). Then $\mathbf{v}_{c.m.}$ is pependicular to $\mathbf{v}_3$, so that

$$v_3^2 = u_3^2 - v_{c.m.}^2.$$

Fig. 16. Relation between velocity of ejected particle in lab system (**v**₃), *velocity of ejected particle in c.m. system* (**u**₃), *and velocity of c.m. system with respect to lab. The magnitudes of* **u**₃ *and* **u**₄ *are completely determined by conservation of momentum and energy.* (**u**₄ = *velocity of residual nucleus in c.m. system) The direction of* **u**₃ *is random, but in this case it has been chosen so that* **v**₃ *is perpendicular to* **v**_{c.m.}, *which is, of course, parallel to the incident particle direction.*

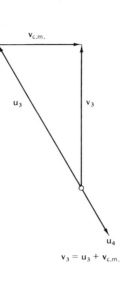

To find $\mathbf{v}_{c.m.}$, we use the fact that the initial momentum equals the product of $\mathbf{v}_{c.m.}$ and the total mass. Therefore

$$M_1 \mathbf{v}_1 = (M_1 + M_2) \mathbf{v}_{c.m.}$$

$$\mathbf{v}_{c.m.} = \frac{M_1 \mathbf{v}_1}{(M_1 + M_2)}$$

$$= \frac{4}{13} \mathbf{v}_1$$

But $\tfrac{1}{2} M_1 v_1^2 = 5.30$, and $M_1 \approx 4u$, so

$$|v_1| = \left(\frac{10.60}{M_1}\right)^{1/2} = \left(\frac{2.65}{u}\right)^{1/2}$$

and

$$|v_{c.m.}| = \frac{4}{13} \left(\frac{2.65}{u}\right)^{1/2}$$

The neutron energy is then

$$\tfrac{1}{2}M_3\,v_3^2 = \tfrac{1}{2}M_3\,u_3^2 - \tfrac{1}{2}M_3\,v_{\text{c.m.}}^2.$$

$$= 4.56 - 0.13 = 4.43 \quad \text{MeV}$$

in agreement with our previous result.

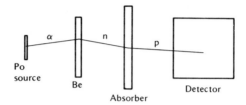

Fig. 17. *Experimental arrangement used in first observation of the neutron.*

The reaction analyzed in this example is of historical interest, for it was the one with which the neutron was discovered by Chadwick in 1932 (see Fig. 17). The neutrons, being uncharged, were not observed directly, and when the reaction was first seen it was thought that the particles ejected from the Be target were gamma rays, these being the only uncharged radiation then known. But Mme. Curie-Joliot and M. Joliot noticed that these "gamma rays" could in turn eject protons of energy up to 5.7 MeV from absorbers. In order to do this in a Compton-type collision, a gamma ray must have the unlikely energy of 55 MeV. Chadwick verified this observation, and then he found that the rays could also impart energies of up to 1.6 MeV to nitrogen atoms in the gas in his detector (an ionization chamber). This feat would require a gamma ray energy of 100 MeV. Thus the hypothesis that the rays were gammas began to appear more and more ridiculous. Not only was the required gamma ray energy unreasonably high, but the required energy varied with the method used to observe the rays. Chadwick then showed that the required energy did not vary in this way if it was assumed that the particle ejected by the Be had a mass approximately equal to the proton mass. Further experiments with boron in place of the beryllium enabled Chadwick to determine the mass—the neutron mass—to within a few tenths of a percent of the correct value. Chadwick's paper describing these investigations is well worth reading.[24]

Charge Independence of Nuclear Forces. We have assumed throughout our discussion of nuclear physics that the nuclear force is charge independent.

[24] J. Chadwick, *Proc. Roy. Soc.* **A136**, 692 (1932).

In other words, that part of the force between two nucleons which is not electromagnetic in nature is independent of the charge state of each nucleon; it is the same for proton–proton, proton–neutron, and neutron–neutron pairs. Much of the justification for this assumption comes from the systematic comparison of nuclear energy levels, made possible by study of nuclear reactions.

The easiest comparison to make is between "mirror nuclei." In a pair of mirror nuclei, the Z of one nucleus equals the N of the other, and vice versa. If the nuclear force is charge independent, the two nuclei should have the same shell structure, except that the proton shell of one nucleus has the

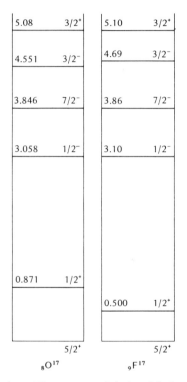

Fig. 18. Energy levels in one set of mirror nuclei, showing spin and parity of each state.

structure of the neutron shell of the other. The two nuclei should thus have the same set of energy levels. There are, of course, Coulomb energy differences caused by the difference in the number of protons, but these differences should not vary much from one level to another. Figure 18 shows the energy levels of the mirror nuclei $^{17}_{8}\text{O}$ and $^{17}_{9}\text{F}$. The energy levels correspond quite closely, and the spin and parity of each level is the same as the spin and parity of the corresponding level in the mirror nucleus.

There is other evidence of the great similarity between mirror nuclei. Allowed beta decay of a nucleus to its mirror proceeds very rapidly. The *ft* values for such transitions, called "favored" transitions, are smaller, by a factor of up to 1000, than the ft values for allowed transitions which are not favored. This is a result of the great similarity between the wavefunctions of parent and daughter in these cases. The matrix element $\int u_m^* O u_k \, d\tau$ has close to its maximum value, unity, because the daughter wavefunction u_m is almost identical to the product of the operator O and the parent wavefunction u_k. (You will recall that the operator O simply transforms a proton into a neutron or vice-versa; thus it transforms a nucleus into its mirror if $N = Z \pm 1$.)

Next let us look at the energy levels of a set of *three* isobars, such as $^{10}_{4}\text{Be}$, $^{10}_{5}\text{B}$,

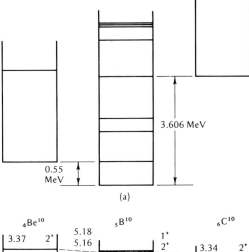

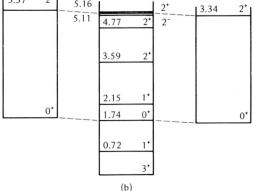

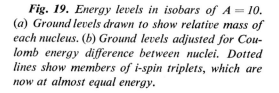

Fig. 19. *Energy levels in isobars of $A = 10$. (a) Ground levels drawn to show relative mass of each nucleus. (b) Ground levels adjusted for Coulomb energy difference between nuclei. Dotted lines show members of i-spin triplets, which are now at almost equal energy.*

and $^{10}_6$C (Fig. 19). There are two differences to notice here. First, if we subtract the Coulomb energy[25] from the energy of each level, the ground states of the three nuclei do not have equal energies; instead, the ground level of ^{10}B lies about 1.7 MeV below the ground levels of the other two, which are about equal. Second, there are many more levels in ^{10}B than there are in ^{10}Be or in ^{10}C. Both facts may be explained by the exclusion principle. Each of these nuclei contains filled levels of four protons and four neutrons, plus two additional nucleons. In ^{10}Be and in ^{10}C these nucleons are identical—both neutrons or both protons, respectively—while in ^{10}B they are different—one neutron and one proton. The two nucleons in ^{10}Be or ^{10}C are thus restricted by the exclusion principle; they cannot occupy a state in which all quantum numbers of the two are identical. There is no such restriction on the nucleons of ^{10}B, because they are not identical particles; they are free to form more states, and the result is the set of additional energy levels.

To explain the fact that the ground state of ^{10}B lies below the ground states of ^{10}B and ^{10}C, we recall from the analysis of the deuteron that the nuclear force is

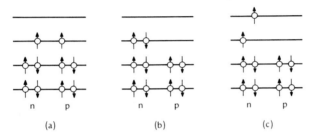

      (a)   (b)   (c)

Fig. 20. *Schematic representation of neutron and proton levels for A = 10. If two "odd" nucleons are not identical (a), they can occupy the lowest available level and have parallel spin. If odd nucleons are identical, they must either have antiparallel spin (b), or occupy different levels (c). In either case the total energy (excluding Coulomb energy) is greater than in (a).*

[25] If you wish to check the result by carrying out this subtraction yourself, you must be careful. First, you must begin with the *nuclear* masses, not atomic masses. Then you must allow for (1) the neutron–proton mass difference, 1.3 MeV, (2) the Coulomb energy of a spherical charge distribution, as it appears in the mass formula (Chapter 14), and (3) a negative correction to the Coulomb energy, amounting to about $0.46 Z^{4/3}$, which results from the requirement that the nuclear wavefunction be antisymmetric with respect to exchange of two protons. The last contribution is analogous to the electron exchange energy in atomic physics; the antisymmetry requirement increases the average distance between protons. This energy contribution was calculated by von Weizsäcker in 1935, but it is not usually included in the mass formula, because, for large Z, it is quite small in comparison to the Z^2 term. For further details see A. N. Cooper and E. M. Henley, *Phys. Rev.* **92**, 801 (1953).

stronger between nucleons of parallel spin. The lowest energy state for $A = 10$ must therefore be one in which the two odd nucleons have parallel spin *and* occupy the lowest available energy level. (See Fig. 20.) This is only possible in ^{10}B; thus there is no state in ^{10}B or in ^{10}C with an energy as low as that of the ground state of ^{10}B.

Nevertheless, we see that there is a correspondence between the energy levels in the three isobars. There are states of ^{10}B in which the odd neutron and odd proton do not have an identical set of quantum numbers, and these states correspond to states in ^{10}Be and ^{10}C. Two such sets of states are evident in Fig. 19.

I-Spin. We can say that each set of three levels mentioned above would be degenerate if the electromagnetic force could be "turned off." In that case the proton and neutron would be identical. This suggests that one consider neutron and proton to be simply two different states of a single particle, the nucleon. The nucleon then has an internal variable which has two possible values; when this variable has one value, the nucleon is a proton; when it has the other value, the nucleon is a neutron.

We have already encountered a variable, the spin, which has two possible values for its component along a given axis. Just as these two spin states have different energies when a magnetic field is turned on, the neutron and proton states of the nucleon have different energies when the electromagnetic force is "turned on." Thus it is useful to give this new internal variable a name that suggests this similarity to spin. We call it *isospin,* or simply *i-spin,* since this variable distinguishes the members of a set of *isobars.*[26] The variable itself has nothing to do with spin, or angular momentum, of course; the name simply reflects the *mathematical* similarity to ordinary spin.

We thus define the i-spin quantum number of the nucleon to be $T = \frac{1}{2}$, and we define an "axis" in "i-spin space" such that T_3, the "component" of the i-spin along this axis, is $+\frac{1}{2}$ if the nucleon is a proton and $-\frac{1}{2}$ if the nucleon is a neutron. The i-spins of nucleons are combined in the same way as ordinary spins are combined, so that the value of T_3 for a nucleus is always the sum of the values of T_3 for the individual nucleons in the nucleus; thus $T_3 = (Z - N)/2$. For example, $T_3 = -1$ for ^{10}Be, 0 for ^{10}B, and $+1$ for ^{10}C. The total i-spin, T, cannot be found so simply; it may vary from one state to another in a nucleus, but of course, $T \geq |T_3|$. T is found by looking at the level structure of a set of isobars. Continuing with the example of $A = 10$, we deduce that the levels in ^{10}B which have no counterpart in ^{10}Be or ^{10}C must be "i-spin singlets," with $T = 0$; obviously, if $T = 0$, there can be no state with $T_3 = \pm 1$, hence no ^{10}Be or ^{10}C state. The two sets of states with equal energies in ^{10}Be, ^{10}B, and ^{10}C are assumed to be members of an "i-spin triplet," with $T = 1$ and $T_3 = -1$, 0, and $+1$, respectively, just as in ordinary space a nucleus with a spin of 1 has three possible orientations.

---

[26] You may find i-spin referred to as "isotopic spin" in older texts (or by older professors).

Remember that this T assignment is simply a mathematical device to classify the states. A state with $T = 1$ has certain structural features which are the same for $N = Z$, for $N = Z + 2$, or for $N = Z - 2$; because of this similarity of structure all the members of a multiplet of states have the same spin and parity. Another look at Fig. 20 may help to clarify this. The states in 20b and 20c are shown with six neutrons and four protons, but the same distribution of the nucleons into the levels would be allowed if one or two of the neutrons were changed into protons. As shown, state (b) or state (c) has $T_3 = -1$; converting one neutron to a proton produces the $T_3 = 0$ state, and converting a second neutron to a proton produces the $T_3 = +1$ state. In each case $T = 1$; if the difference between proton and neutron is ignored, the arrangement of *nucleons* into the various levels, and thus the spin and parity, is the same for all three cases. On the other hand, the state shown in 20a is obviously an *i*-spin singlet, because the value of T_3 cannot be changed without changing the arrangement; if a neutron were converted to a proton, it would be necessary either to flip the spin or to put the neutron into a higher level to avoid violation of the exclusion principle.

Of course, we can recognize the *i*-spin triplets in Fig. 19 without knowing the detailed structure of each state, so we have a general classification scheme which does not depend on any specific model of the nuclear levels. But, you may wonder, what good is it? The important feature of *i*-spin seems to be that it is conserved in many nuclear reactions. The conservation of T_3 is not surprising, for that is equivalent to conservation of charge. But the *total i*-spin T is also conserved. For example, if we bombard ${}^{10}B$ nuclei with deuterons, both particles have $T = 0$, so that the T of the whole system is 0; inelastic scattering of the deuteron can then excite the ${}^{10}B$ to any state with $T = 0$, but not to the 1.74 or 5.16-MeV states, which have $T = 1$. Excitation of the ${}^{10}B$ to either of the latter two states is not forbidden by any conservation law except the "conservation of *i*-spin"; the T of the final state of the whole system would be 1 if either of these states were excited. The $T = 1$ states can be excited by a reaction such as ${}^{13}C(p,\alpha){}^{10}B$, since ${}^{13}C$ and a proton, each with $T = \frac{1}{2}$, can combine to form $T = 1$.

Conservation of *i*-spin is not a rigorous law, like the conservation of charge, conservation of energy, or conservation of angular momentum. It is violated in gamma and beta decay, which are governed by the electromagnetic and the beta decay interactions,[27] respectively. But the *nuclear* force does not seem to be able to alter the symmetry of a state in such a way as to change its *i*-spin.[28]

---

[27] The interaction which causes beta decay is distinct from the nuclear force, and is so weak that it has no effect on nuclear energy levels. It is usually called simply the "weak interaction." We shall have more to say about it in the next chapter.

[28] It should be observed that T is the magnitude of a vector—not a vector in real space, but a quantity that behaves like a vector—so T is conserved as a vector, not a scalar, and the rules of quantum-mechanical addition of vectors apply. Thus if two nuclei, each in a $T = 1$ state, were to collide and form a compound nucleus, the compound nucleus could be in a $T = 2$, $T = 1$, or $T = 0$ state, just as two atomic states each with angular momentum quantum number $j = 1$ can combine to form a state with j equal to 2, 1, or 0.

The concept of *i*-spin as a description of the internal structure, or symmetry, of a state becomes especially useful in the study of elementary particles, where it is impossible to visualize the internal structure even in the crude manner of Fig. 20, and we must simply resort to classification schemes. We shall see in the next chapter that these schemes lead to additional conservation "laws" whose origin is one of the great unsolved problems of physics.

PROBLEMS

1. $^{188}_{76}$Os is energetically unstable against alpha decay to $^{184}_{74}$W. Use the table of nuclear masses to find the kinetic energy which an emitted alpha particle would have; then use Fig. 2, extrapolating if necessary, to determine the expected half-life of this species in years.

2. In the experiment of Cowan and Reines, 10^{13} antineutrinos/cm^2-sec struck the detector, which contained 200 l. of H_2O. If the interaction cross section is 10^{-43} cm^2, how many interactions were there per hour in the detector?

3. Use Fig. 9 and Appendix D to find the value of $\log_{10}ft$ for the β^- decay of $^{35}_{16}$S to the ground state of $^{35}_{17}$Cl. Is this an allowed transition? If E_m for this decay were 1 MeV larger, all other factors being unchanged, what would the half life be?

4. For the decay of ^{137}Cs to the ground state of ^{137}Ba, determine $\log_{10}ft$ (from Fig. 9 and the information in Fig. 11) and $\log_{10}t_{1/2}$, to verify that $\log_{10}ft$ is about 12 for this transition, as stated in Section 15.3. Why should you use the 500-year *partial* half-life, rather than the 30-year half-life of ^{137}Cs, for $t_{1/2}$ in this calculation?

5. Compare the 2.6-min half-life of the 0.66-MeV excited state of ^{137}Ba with the half-life which would be expected on the basis of Table 2. (Notice that the transition rate for an M4 transition goes as the ninth power of the energy; you can use this fact to interpolate.)

6. The thermal neutron absorption cross section of ^{27}Al is 0.23 barns. If 1 g of ^{27}Al is placed in a thermal neutron flux of 10^{12}/cm^2-sec for several days, what will be the ^{28}Al activity at the instant the sample is removed? (^{28}Al half-life is 2.30 min.) What would the activity be if the ^{27}Al had only been in the neutron flux for 4.60 min?

7. A beam of thermal neutrons with an initial intensity of 10^6/cm^2-sec passes through a tantalum sheet 1.0 mm thick. The intensity of the beam emerging from the tantalum is 8.86×10^5 neutrons/cm^2-sec. What is the thermal neutron absorption cross section for tantalum?

8. The cross section for absorption of slow neutrons by ^{113}Cd is 27,000 barns. The density of cadmium is 8.64 g/cm^3, and about $\frac{1}{8}$ of it is ^{113}Cd. The cross

sections of the other isotopes of cadmium are negligible. Find the thickness of cadmium needed to absorb 99% of the slow neutrons striking it.

9. Investigations done by undergraduate students $\{Science\ \mathbf{215},\ 377,\ (1982)\}$ have shown that ^{59}Ni, which is produced in nuclear reactors by neutron irradiation of the ^{58}Ni in stainless steel, is a previously overlooked major contributor to the radioactivity remaining in a reactor long after it has been "decommissioned" (shut down). Given that the reactor flux is 10^{13} thermal neutrons per cm^2 per second, that the reactor has run for 30 years (10^9 s), that the thermal neutron absorption cross section of ^{58}Ni is 4.7 barns, and that ^{58}Ni constitutes about 5 atomic % of the stainless steel in the reactor, find

   (a) the ^{59}Ni activity per kg of stainless steel when the reactor is shut down,

   (b) the ^{59}Ni activity per kg of stainless steel 40,000 years after the reactor is shut down.

10. If the Q value of a reaction is negative, the incident particle must have a certain minimum energy, called the threshold energy, before the reaction can occur. What is the threshold energy (in lab coordinates, of course) for $^{13}C(p,d)^{12}C$? (d = deuteron.)

11. In the reaction $^{14}N(n,p)^{14}C$, a resonance occurs at an incident neutron energy of 3.1 MeV. Find the excitation energy of the compound nucleus at this resonance. At what incident alpha-particle energy might you expect to find a resonance in $^{11}B(\alpha,n)^{14}N$?

12. For the reaction of example Problem 2, find the energy of a neutron that emerges at an angle of 60° with the direction of the incident alpha particle

   (a) by direct application of Eq. (34)

   (b) by converting from the c.m. system, in which the neutron energy was shown to be 4.56 MeV.

13. Find the maximum energy of the neutron emerging from the reaction $^{10}B(\alpha,n)^{13}N$, assuming that the alpha particles have a kinetic energy of 5.30 MeV, the ^{10}B atoms are at rest, and the ^{13}N atoms are left in the ground state.

14. A possible reaction in a fusion reactor is $^3H(d,n)^4He$. (The reaction sustains itself by capture of the emerging neutron by a deuteron to form a new atom of ^3H.) Find the Q value of this reaction, and the kinetic energy of each emerging particle, if the ^3H atom and the deuteron have negligible kinetic energy.

15. In the reaction $^{12}C(\alpha,n)^{15}O$, suppose that the incident alpha particle has a kinetic energy of 16 MeV and the target ^{12}C nucleus is at rest.

(a) What is the total kinetic energy in the c.m. frame of reference before the reaction takes place?

(b) What is the total kinetic energy in the c.m. frame after the reaction?

(c) What is the kinetic energy of the neutron in the c.m. frame?

(d) What are the maximum and minimum kinetic energies with which the neutron could be observed in the lab? (*Hint*: Find the relative velocity of the c.m. frame and the lab.)

16. When ^{3}H is bombarded with protons, neutrons first appear at $0°$ from the direction of the bombarding protons when the proton energy is 1.019 MeV. The reaction is

$$^1H + {}^3H \rightarrow n + {}^3He$$

Find the Q-value of this reaction. Combine this value with the decay energy of tritium:

$$^3H \rightarrow {}^3He + 0.018 \text{ MeV}$$

to determine the neutron-hydrogen mass difference in MeV. (Use only the data given in this problem, not the nuclear mass table.)

17. In the reaction of Problem 16, what is the minimum proton energy for which neutrons can emerge at an angle of $90°$ from the incident beam direction? Draw a diagram showing the neutron velocity in lab and in c.m. system, to show why there is a difference between this "$90°$ threshold" and the $0°$ threshold.

18. Verify the statements about the required energy of the gamma ray in the experiments which led to the discovery of the neutron (Section 15.4). Using the data on recoil energies given there, 5.7 MeV for protons and 1.6 MeV for ^{14}N atoms, deduce the mass and energy of the neutron in those experiments.

19. Give an argument to show that T should equal T_3 for the ground state of any nucleus. What is the value of T for the ground state of ^{15}C? At what excitation energy might you expect to find the corresponding state (a member of the same i-spin multiplet) in ^{15}N? (Make a rough calculation only, using Appendix D and the Coulomb energy term in the mass formula of Section 14.5) Determine which, if any, of the following reactions could produce ^{15}N in this state: $^{14}C(d,n)^{15}N$, $^{12}C(\alpha p)^{15}N$, $^{18}O(p,\alpha)^{15}N$, $^{11}B + \alpha \rightarrow {}^{15}N$. Assume that the initial nuclei are all in the ground state and T is conserved.

16

Particles and Interactions

The study of the nuclear force leads us into the domain of the so-called elementary particles. Just as nuclear physics developed from the study of atomic structure, elementary particle physics has developed from the study of nuclear structure. But we are not yet finished with this process of finding structure at smaller and smaller levels. When we observe phenomena at the very limits of our experimental capabilities, things are not always what they seem, and the elementary particles have turned out to be not so elementary. They appear to have a number of internal quantum numbers, which are suggestive of a complex internal structure, and we have had difficulty in deciding which particles, if any, should be called elementary.

The year 1935 might be considered a turning point in the study of particles. Before 1935, "atomic" physics was concerned only with well-behaved objects—protons, neutrons, and electrons—which arranged themselves in orderly fashion to form nuclei, atoms, and molecules.

Of course, there was also the positron, but it was well accounted for by the Dirac equation; it was assumed that the proton and the neutron could also have antiparticles, but this was considered rather an academic assumption, as there seemed to be no way of experimentally observing these antiparticles.

In addition to these three particles and their antiparticles, the only elementary object known in 1935 was the photon.[1] The photon is an excitation of the electromagnetic field, and it can be produced or destroyed in great numbers. We might wonder whether a similar particle is associated with the nuclear force.

By analogy with electromagnetic fields, Yukawa in 1935 developed a field theory of the nuclear force, in which an object called a *meson* is the analog of the photon. The meson differs from the photon in a number of ways, in particular, in its possession of a nonzero rest mass. The achievement of Yukawa was remarkable, not only for the prediction that the meson exists, but also for the accurate prediction of the rest mass it possesses.

In the search for the meson, physicists came across a number of other particles whose existence had not even been suspected. First there was the *muon*, which was originally thought to be Yukawa's meson. Then, in the same year (1947) in which the Yukawa meson (now called π *meson* or *pion*) was discovered in cosmic radiation, other heavier mesons were discovered, as well as "strange particles" which decayed into nucleons.

The heavy particles that decay into nucleons are best described as excited states of the nucleon, rather than as totally new and different particles. The existence of an excited state of a nucleon, of course, implies the presence of some sort of internal structure that can be deformed in a high energy collision; this structure is now being explored by accelerating particles to higher and higher energies. Thus particle physics has reached a position analogous to that of atomic physics at the end of the nineteenth century; the excited states can be classified by means of a system of quantum numbers, certain selection rules governing the decay of these states have been learned, and some formulas have been developed that give the energies of these states to rather good accuracy. But there is still an enormous amount to learn about the actual structure of these states, or the true nature of the forces that govern the formation of these states.

All of the fundamental forces of nature—the nuclear force (or strong interaction), the electromagnetic force, the weak interaction(s) which cause beta decay, and even the gravitational force—may be involved in a unified approach to the problems of particle physics. We do not have the mathematical apparatus or the time to pursue an investigation of the details of these forces and fields, but we can, in this chapter, learn something about the rules that have been discovered which govern particle interactions under the influence of these forces.

---

[1] The neutrino was also "known," having been invented by Pauli in 1930 (Section 15.2), but it was still in the nature of a hypothesis rather than a particle.

16.1   MESON THEORY OF THE NUCLEAR FORCE

The creation of a photon by an isolated charged particle is a violation of the law of conservation of energy,[2] but we know that the energy of a system can only be defined by a measurement. If the measurement requires a time Δt, then the energy must be uncertain by an amount ΔE, where $\Delta E \, \Delta t > \hbar$; this relationship follows from the general properties of wave packets as applied to the photon (Chapter 4), or it can be deduced from an analysis of quantum mechanical transitions (Chapter 10).

Thus a charged particle, even when it is at rest, can emit a photon of energy ΔE, as long as the photon only lives for a time $\Delta t \approx \hbar/\Delta E$, so that energy is conserved within the uncertainty permitted by quantum theory. The force between two charged particles can be explained in terms of the existence of such transient photons, or *virtual* photons, which exist only long enough to transfer momentum from one charged particle to another. Because of the photon's zero rest mass, there is no lower limit on the value of ΔE, so that Δt can be as large as desired; thus a virtual photon can be sent from one particle to another, regardless of the distance between the particles. This fact accounts for the infinite range of the Coulomb force—in other words, for the fact that the Coulomb force obeys Gauss's law, because the lines of force stretch to infinity.

There is no law corresponding to Gauss's law for the nuclear force, because the nuclear force has a *finite* range. The nuclear potential falls off much more rapidly than $1/r$, so that it can be said to "cut off" at a fairly well-defined distance of order 10^{-13} cm from a nucleon. Nevertheless, Yukawa postulated that the nuclear force is carried by a particle, called a meson, which is analogous to the photon. The fact that the force is of short range was considered by Yukawa to be an indication that the meson has a *nonzero rest mass*.

An approximate value for this rest mass may be found by letting Δt be the time required for the meson to travel a distance equal to the range of the nuclear force. Because the meson can never transmit a force over a distance greater than this range, the meson, if it travels at a speed near the speed of light, must complete its journey from one nucleon to another in a time $\Delta t \simeq 10^{-13}$ cm/c. This value of Δt leads to a value of $E \simeq \hbar/\Delta t$ of order 10^8 eV or 100 MeV. Thus the meson could have a rest mass of the order of 100 MeV without requiring the energy of the system to exceed the amount permitted by the uncertainty relation. Conversely, if the meson had a much smaller rest mass, there is no reason why it could not transmit the nuclear force over a longer distance than that observed for the range of this force.

In order to deal with the problem in a more quantitative way, Yukawa introduced a potential-energy formula which fits the observed nuclear force

---

[2] Unless the particle is transformed into another particle in the process.

rather well; the formula gives the potential as

$$U(r) = \frac{e^{-Kr}}{r} g^2 \tag{1}$$

where K can be adjusted to provide the right "cutoff" distance for the force, and g is the "coupling constant" for the nuclear force (analogous to the charge q for the Coulomb force). The constant K can be related to the meson mass as follows:

Consider the wave equation for electromagnetic waves,

$$\nabla^2 V - \frac{1}{c^2} \frac{\partial^2 V}{\partial t^2} = 0 \tag{2}$$

which reduces to Laplace's equation,

$$\nabla^2 V = 0 \tag{3}$$

in the static case. There is no characteristic length here, because the force is of infinite range and the photon is massless. But if instead of the potential $V(r) = q^2/r$ of the Coulomb force, we work with the Yukawa potential $U(r)$ given by Eq. (1), it is easy to see, by applying the operator ∇^2 to $U(r)$ (for $r = 0$), that the corresponding static equation is

$$\nabla^2 U - K^2 U = 0$$

where the appearance of the constant K introduces the characteristic cutoff length. Thus we might expect that the corresponding *wave* equation would be obtained from Eq. (2) simply by adding the same term, $-K^2U$, which is added in the static case. Adding this term to Eq. (2) yields the wave equation

$$\nabla^2 U - \frac{1}{c^2} \frac{\partial^2 U}{\partial t^2} - K^2 U = 0 \tag{4}$$

called the Klein–Gordon equation, which is appropriate for a short range force which cuts off at $r \simeq 1/K$.

Let us assume that Eq. (4) has a wavelike solution, which we write in one dimension as

$$U(x) = e^{i(kx - \omega t)}$$

To test this assumption, we may insert this solution into Eq. (4), and we obtain

$$-k^2 + \frac{\omega^2}{c^2} - K^2 = 0 \tag{5}$$

At this point we can inject the meson into the theory. If the energy and momentum of the wave are carried by quanta, or mesons, then we expect that the usual relationships would apply between energy and frequency, and between momentum and wavenumber, namely $E = \hbar\omega$ and $p = \hbar k$. Then, in terms of E and p, Eq. (5) becomes

$$-\frac{p^2}{\hbar^2} + \frac{E^2}{\hbar^2 c^2} - K^2 = 0$$

or

$$E^2 = p^2 c^2 + \hbar^2 K^2 c^2 \tag{6}$$

If E and p are the energy and momentum of a particle, the mass of the particle follows directly from the relativistic energy–momentum relation

$$E^2 = p^2 c^2 + m^2 c^4 \tag{7}$$

Equations (6) and (7) are in agreement if and only if

$$m^2 = \frac{\hbar^2 K^2}{c^2} \qquad \text{or} \qquad m = \frac{\hbar K}{c}$$

To give the best agreement of Eq. (1) with experiment, k must be about 0.7×10^{13} cm^{-1}. The resulting energy is of the order of magnitude deduced previously:

$$mc^2 = \hbar c K$$
$$\simeq \hbar c (0.7 \times 10^{+13} \quad \text{cm}^{-1})$$
$$\simeq 140 \quad \text{MeV}$$

It should be remembered that this is only an approximate value for the rest energy, because the Yukawa potential (1) cannot be adjusted to fit the nuclear potential *exactly*. For example, the Yukawa potential does not account for the noncentral part of the nuclear force (see Section 14.9).

In 1947, Yukawa's meson, now called the π meson or pion, was discovered in cosmic radiation.[3] The pion normally has no independent existence outside the nucleon whose force it transmits. Each nucleon is surrounded by a cloud of virtual mesons, which it continually sends out in "search" of other nucleons, like Noah's doves seeking dry land; the meson is either absorbed by another nucleon within a radius of order 10^{-13} cm, or it "returns" to the original nucleon. But if there is a collision between a nucleon and another particle, the kinetic energy may be sufficient to materialize a meson from this cloud, so that it can travel far from the mother nucleon and be observed as a distinct entity. In a similar manner, one of the virtual photons discussed above can be converted to a real photon by a collision.

[3] C. Lattes, G. Occhialini, and C. Powell, *Nature* **160**, 453 (1947). The pion was called a π meson to distinguish it from the muon, which had been discovered previously and had been called a meson in the mistaken belief that it was Yukawa's meson. The original "meson" was then called a μ meson, later shortened to muon. We now have a large class of particles which carry the nuclear force and are called mesons, but the muon is *not* a meson, so we have refrained from calling it a μ meson. The *history* here is more confusing than the physics; everyone was confused from 1937 to 1947, because the muon had about the mass of Yukawa's meson, but it did not interact as it should.

The Yukawa theory was a crucial step in the development of the modern approach to the forces of nature. However, it turns out that the pion is not actually the analog to the photon as the *fundamental* carrier of the strong interaction, because the pion is a composite particle. The fundamental carrier of the strong interaction will appear in Section 16.8; it is the "gluon," which acts as the "glue" to bind together the "quarks" in nucleons as well as pions.

Nevertheless, to understand the nuclear force, we must begin by studying the properties of the pion.

16.2 PROPERTIES OF THE PION

a. Charge. The pion differs from the photon in another important respect; it can carry an electric charge. There are three "charge states" of the pion: π^+, π^0, and π^-, having $+1$, 0, and -1 units of charge, respectively. The charged pions each have a mass of 139.6 meV, slightly more than the mass of 135.0 MeV for the neutral pion, and in excellent agreement with Yukawa's prediction. The greater mass of the charged pions can be accounted for simply by the electrostatic energy of the charge.

The charged pions are involved in the transmission of a force between a proton and a neutron and vice versa; for example, the positive pion may be emitted by a proton and absorbed by a neutron, so that the proton becomes a neutron and the neutron becomes a proton.

b. *i*-Spin. We can describe the charge states of the pion very neatly by means of *i*-spin (see Section 15.4). The charged pion transfers one unit of *i*-spin from one nucleon to the other, so tnat the *i*-spin of each nucleon is "flipped," one going from $+\frac{1}{2}$ to $-\frac{1}{2}$, the other going from $-\frac{1}{2}$ to $+\frac{1}{2}$. Thus the charged pion must carry one unit of *i*-spin. The three pions from an *i-spin triplet*, with T_3, the third "component" of *i*-spin, being equal to $+1$, 0, and -1 for π^+, π^0, and π^-, respectively. Presumably, as in the case of nuclei which form *i*-spin triplets, the three pions would have the same mass if the electromagnetic force could be turned off.

The third component of *i*-spin, T_3, is conserved in reactions in which one or more pions are produced, or "materialized." For example, in the reaction

$$n + p \quad \rightarrow \quad n + n + \pi^+$$

the T_3 values are

$$-\tfrac{1}{2} + \tfrac{1}{2} \quad \rightarrow \quad -\tfrac{1}{2} - \tfrac{1}{2} + 1$$

and T_3 is conserved. Of course in pion production, as in nuclear reactions, conservation of T_3 is nothing more than conservation of charge. But it has been found that, as in nuclear physics, there are reactions which do *not* occur *only* because the *total i-*spin *T*—the magnitude of the *i*-spin " vector "—would not be conserved if the reaction took place. Examples of these will be discussed in Section 16.4.

Stating the situation in terms of the esoteric ''vector'' is, of course, simply a way of disguising our ignorance. We know that the forces involved in nuclear or particle collisions—the strong interactions—can only act in certain ways, and that there is some internal symmetry of the particles which is not upset by the force. We therefore invent a *mathematical* device which permits us to predict which reactions will occur, even though we do not know the *physical* nature of the restrictions which prevent certain reactions from occurring. We shall return to this point after we have discussed a few more particles.

c. Spin. The spin of the π^+ has been deduced to be zero, by analysis of the rates of the reaction

$$p + p \ \rightleftarrows \ \pi^+ + d$$

in the " forward " and " reverse " directions. The rate of the forward reaction is proportional to the density of final states, and this in turn depends on the spin of the pion. For example, a pion spin of 1 provides three times as many final states as a pion spin of zero, so that the forward rate would be three times as great for pion spin 1 as for pion spin zero. The rate of the *reverse* reaction, in the appropriate energy range, enables one to determine the matrix element; once this is done, one can easily distinguish among the various spin possibilities by measuring the rate of the forward reaction, and one finds that the π^+ has zero spin.

The spin of the π^0 can also be deduced to be zero, by consideration of the symmetry of the decay of the π^0 into two gamma rays. This decay, like the decay of para-positronium (Section 13.3), has the symmetry required of the decay of a zero-spin system. And it is a good thing that it does, because we have a general view in which the π^0 and π^+ are *simply different states of the same particle,* and hence must have the same spin, just as the neutron and proton must have the same spin. Thus we say in general that the pion has zero spin, regardless of its charge state.

d. Parity. Another very important (and somewhat confusing) property of the pion is its *intrinsic parity*. We have discussed the fact that a nucleus or an atom may have positive or negative parity (Chapter 14); this ''internal'' parity follows directly from the symmetry of the spherical harmonics whicn must be included in the wavefunction. But we have no way to write a wavefunction of this sort for the *internal* structure of the pion, so you may wonder how we can speak with any assurance of its possessing any *intrinsic* parity, or why we would even consider such a thing.

Of course we know that we can write a wavefunction for the *motion* of the pion, and that the parity of this wavefunction is determined by the l quantum number for the pion's orbital angular momentum; but the parity *could* also be determined by some internal parity quantum number, just as the total angular momentum of any particle is the sum of its orbital angular momentum and its spin angular momentum.

The idea of intrinsic parity is not some abstraction that a theorist concocted in order to make the subject more interesting; it was forced upon us by the results of various experiments. Consider the following reaction:

$$d + \pi^- \quad \rightarrow \quad n + n \tag{8}$$

which has been observed to occur after the π^- has been captured by a deuterium atom. After its capture, the pion often falls into the 1s state; this fact can be verified by observation of the photons emitted in the transitions which lead to the 1s state, (as one does in studying muonic atoms—Section 14.2). When the initial state of reaction (8) is the 1s state, its parity would be *even* (positive) if there were no contribution from intrinsic parity; however, the final state of the two neutrons can be demonstrated to have *odd* parity, as follows:

The final state must be one of four possibilities: $^3D_1$, $^3P_1$, $^3S_1$, or $^1P_1$, in the usual notation, in which the superscript 3 refers to the symmetric spin state, the superscript 1 to the antisymmetric spin state, and the subscript 1 to the quantum number for the total angular momentum. Of these, the only acceptable possibility is $^3P_1$, because the wavefunction must be antisymmetric in the exchange of the two neutrons. (See Section 9.1 for a discussion of the symmetry of two-particle wavefunctions.)[4] We cannot avoid this assignment by allowing the neutrons to have different *radial* quantum numbers, because in the center-of-mass frame of reference, the radial motions of the two neutrons must be identical. The $^3P_1$ state, whose orbital angular momentum has an odd quantum number, is a state of odd parity.

Therefore, if parity is to be conserved in reaction (8), the initial state must have odd parity; this is only possible if either the pion or the deuteron has odd intrinsic parity. We saw in Section 14.9 that the *deuteron* has *even* intrinsic parity, so we must say that *pion* has *odd* intrinsic parity. This parity assignment may seem to be arbitrary, and you may wonder why we do not simply abandon the law of conservation of parity for strong interactions, especially since we already know that the law does not hold for weak interactions (beta decay). But it happens that the definition of odd intrinsic parity for the pion is sufficient to ensure that parity is conserved in other strong

---

[4] The $^3D_1$ and $^3S_1$ states are symmetric in both space and spin; the $^1P_1$ state is antisymmetric in *both* space and spin, and thus symmetric overall.

interactions in which pions are involved, so that the law is a useful one—it describes how nature works.[5]

Do not waste your time trying to form a mental image of this odd intrinsic parity. It is a completely nonclassical attribute of the pion, and perhaps would not even be mentioned here if it were not for the fact that its existence is related to a famous puzzle which we shall discuss in Section 16.6.

e. Lifetime. The pion differs from the photon in one other striking way: It is not stable. An isolated neutral pion decays into two photons:

$$\pi^0 \quad \rightarrow \quad 2\gamma$$

the mean lifetime for this process is about 0.8×10^{-16} sec.[6]

This decay is an electromagnetic process, analogous to the decay of paraposi-tronium (Section 13.3). The lifetime of the π^0 is much shorter than that of para-Ps simply because so much more energy is released in its decay. As was mentioned in Section 15.3, the transition rate for gamma decay goes roughly as the cube of the energy emitted; the neutral pion has 133 times the energy of para-Ps, so we should expect its lifetime to be shorter than the para-Ps lifetime by a factor of the order of 100^3, as it is.

You may wonder why the pion would disappear in this manner, and the answer can only be "Why not?" M. Gell-Mann[7] has suggested that nature in general obeys the "totalitarian principle":

Everything that is not forbidden is compulsory.

Therefore any particle decays into particles of smaller rest mass, unless the decay would violate some conservation law. Apparently the neutral pion, which has no charge to be conserved, contains nothing else which cannot be conserved in the process of decay into two photons. (The odd intrinsic parity of the pion can be conserved by giving the photons the appropriate relative

---

[5] This definition is not the only workable one; since we can observe only *changes* in parity, not parity itself, there are other possibilities. For example, parity would still be conserved if we assumed that the intrinsic parity of the neutral pion and of the neutron were odd, and that the intrinsic parity of the proton and the charged pion were even. The intrinsic parity of the deuteron would then be odd, and both sides of Eq. (8) would have odd parity. But it seems more reasonable, and is more convenient, to define both nucleons to have even parity. See the discussion by R. P. Feynman, in "Symmetries in Elementary Particle Physics" (A. Zichichi, ed.). Academic Press, New York, 1965.

[6] This is the shortest lifetime ever measured by observation of the creation and decay of the particle as separate events, and it was duly listed as a "record," alongside the fastest mile run, tallest TV antenna, etc. in the first edition of "The Guinness Book of Records," published by the Guinness Brewing Company (presumably to settle arguments in pubs).

[7] Winner of the 1969 Nobel Prize in physics for classification of elementary particles (Section 16.5).

polarization.) Another decay mode which is not forbidden, and which is observed in about 1 percent of the decays, is

$$\pi^0 \quad \to \quad e^+ + e^- + \gamma$$

16.3 *LEPTONS AND THE WEAK INTERACTION*

The *charged* pion cannot decay into photons, because its charge must be conserved. But it can decay into a muon plus a neutrino, as follows:

$$\pi^+ \quad \to \quad \mu^+ + \nu_\mu$$

$$\pi^- \quad \to \quad \mu^- + \bar{\nu}_\mu$$

(9)

where ν_μ and $\bar{\nu}_\mu$ are neutrino and antineutrino, respectively. (We use the subscript μ to distinguish this neutrino from the neutrino that is emitted in beta decay. We shall see in a moment that they are indeed distinguishable experimentally.)

This decay takes much longer than the decay of the π^0; the mean lifetime of the charged pion is 2.6×10^{-8} sec, longer than that of the π^0 by a factor of more than 10^8. The reason for this seems to be that the process requires the creation of a neutrino or an antineutrino, and thus it can only occur via the same interaction that causes beta decay, which is the only interaction in which neutrinos participate.[8] This neutrino interaction is so weak that it is called the "weak interaction." In contrast, the force between nucleons, which holds a nucleus together, is often called the "strong interaction." We have seen that the weak interaction is so weak that it is of no consequence in determining the energy levels of nuclei[9]; it is only noticed because it causes particles to decay (slowly) into other particles, although once in a great while it can cause a neutrino to be absorbed (or scattered).

Conservation of Lepton Number. You might ask, "Why cannot the charged pion decay without the creation of a neutrino or antineutrino?" This is a reasonable question; to answer it, we must consider what the possible decay processes might be. One logical candidate is

$$\pi^\pm \quad \to \quad \mu^\pm + \gamma$$

(10a)

another is

$$\pi^\pm \quad \to \quad e^\pm + \gamma$$

(10b)

[8] Except for gravitation, which presumably acts on anything which carries energy.
[9] We shall see in Section 16.6 that the weak interaction causes a minute energy difference, of order 10^{-6} eV, between two states of the neutral K meson. This energy difference may be compared with the 5-MeV energy difference between π^+ and π^0, as another measure of the relative strengths of weak and electromagnetic interactions.

In these processes charge is conserved, and energy can certainly be conserved; but angular momentum cannot be conserved, because only one particle of half-integer spin takes part in the process.

We can rule out decays (10) in another way. In either of these decays, a particle is created but no corresponding antiparticle is created. The created particles, the muon and the electron, belong to a class of particles called leptons,[10] consisting of those particles that participate in the weak interaction but not in the strong interaction. It has been found that lepton number—the total number of leptons minus the total number of antileptons—is conserved in all interactions known to date (although there are theoretical grounds for believing that this conservation "law" could be violated on rare occasions, as we shall see). In pion decay (9), as well as in the ordinary beta decay process, lepton number is conserved. The law of lepton conservation is also obeyed in the decay of the muon, but with another complication.

Decay of the Muon. On the basis of the totalitarian principle, one might expect the muon to decay by the process

$$\mu^\pm \rightarrow e^\pm + \gamma \tag{11}$$

In this process, the number of leptons is conserved, charge is conserved, and energy, momentum, and angular momentum can all be conserved. But it just does not happen! If it did happen, then the electron's energy, in the muon's rest frame, would always have a fixed value, determined by conservation of energy and momentum. But it is found that muons at rest decay into electrons which have a broad spectrum of energies, suggestive of a decay into three particles rather than two.[11] When a muon decays, only an electron is detected, so we assume that the other two particles produced are neutrino and antineutrino, and that the decay is

or

$$\mu^+ \rightarrow e^+ + \nu_e + \bar{\nu}_\mu$$
$$\mu^- \rightarrow e^- + \bar{\nu}_e + \nu_\mu \tag{12}$$

The negative muon, being a lepton, is replaced by the ν_μ; the positive muon, an antilepton, is replaced by the antineutrino $\bar{\nu}_\mu$. Notice that these definitions of μ^+ as lepton and μ^- as antilepton are consistent with reactions (9), in each of which a lepton and an antilepton are created. In (12) we also see the creation of lepton and antilepton: either e^+ and ν_e, or e^- and $\bar{\nu}_e$.

The fact that (12) occurs, while (11) does not, is evidence that ν_μ and ν_e are not identical. If ν_μ were identical to ν_e, then $\bar{\nu}_\mu$ would be antiparticle to ν_e, and

[10] From the Greek λεπτον, a small coin, which is generalized to refer to anything which is small or light.
[11] Remember that the continuous spectrum of energies of the electron in beta decay led us to conclude that a third particle, the neutrino, was among the decay products.

the particle-antiparticle pair produced in Eq. (12) could just as well be replaced by a gamma ray.[12]

There is more direct evidence that ν_μ is not identical to ν_e. Consider the ν_μ produced in π^+ decay (9). If these were identical to ν_e, they could interact with neutrons via either

$$\nu + n \rightarrow e^- + p \tag{13a}$$

or

$$\nu + n \rightarrow \mu^- + p \tag{13b}$$

Pions of sufficiently high energy might be expected to produce neutrinos so energetic that they could participate almost equally well[13] in either of reactions (13). But when an attempt was made to observe reactions (13) by using neutrinos that resulted from decay of high energy pions, it was found that reaction (13b) occurs and (13a) does not.[14] On the other hand, we know that reaction (13a) occurs with the neutrinos produced in beta decay. The inescapable conclusion is that the neutrino of pion decay (9) is not the same as the neutrino of beta decay.

The Tau Particle, a Heavy Lepton. The discovery, in 1975, of a heavy lepton, now called the tau, demonstrates that small mass is not an essential characteristic of a lepton.[15] The tau, whose rest energy is 1784 MeV (almost twice that of the proton) behaves exactly like a lepton. All of its known decay modes appear to be into particles whose total lepton number is one.

Like the muon, the tau can decay into an electron plus a neutrino and an antineutrino. It can also decay into a *muon* plus a neutrino and an antineutrino. (As in the decay of the muon, the broad spectrum of energies of the observed decay product indicates that the unseen neutrino and antineutrino are emitted.) These decay modes of the τ *(the τ^-)* and the $\bar{\tau}$ (or τ^+) are written:

$$\tau^+ \rightarrow e^+ + \nu_e + \bar{\nu}_\tau$$
$$\tau^+ \rightarrow \mu^+ + \nu_\mu + \bar{\nu}_\tau$$
$$\tau^- \rightarrow e^- + \bar{\nu}_e + \nu_\tau$$
$$\tau^- \rightarrow \mu^- + \bar{\nu}_\mu + \nu_\tau$$

[12] A similar argument was used in Section 15.2, where the extreme rarity of double beta decay was cited to support the idea that ν and $\bar{\nu}$ are not identical.

[13] Reaction (13a) would be slightly favored over reaction (13b) because the smaller rest mass of the products would mean a greater kinetic energy and thus a greater density of states available to the *electron* and proton, compared with the *muon* and proton.

[14] The first such experiment was performed at Brookhaven in 1962, and reported by G. Danby *et al.*, *Phys. Rev. Lett.* **9**, 36 (1962). In this experiment, 15-GeV protons struck a target to produce pions; the pion beam was sent into a 13-m-thick iron wall which filtered out everything except the neutrinos resulting from the pion decay. The resulting neutrino interactions were observed photographically in a 10-ton spark chamber, and all events which appeared at all to resemble muon or electron production were carefully studied.

[15] M. L. Perl, *et al.*, *Phys. Rev. Lett.* **35**, 1489 (1975). See also "Heavy Leptons" by M. L. Perl and W. T. Kirk, in *Scientific American,* March, 1978, pp. 50–57.

Here, on the basis of experience with the muon, it is assumed that another distinct neutrino, the ν_τ, is involved. We believe this to be true because the decays

$$\tau \to \mu + \gamma \text{ and } \tau \to e + \gamma$$

are not observed. If ν_τ were identical to either ν_e or ν_μ, one of the above decays would occur, for the reason discussed previously in connection with the decay of the muon.

Because of its great mass, the tau has other decay modes in addition to the four shown above. For example, there is

$$\tau^- \to \pi^- + \nu_\tau$$

in which only one neutrino is needed to conserve lepton number.

So we now have six known leptons: e, μ, τ, and their associated neutrinos. Except for reactions (13), no differences have been observed between ν_e and ν_μ (or ν_τ). However, a small mass difference has not been ruled out, and this would tie in nicely with astrophysical observations (Section 16.6). Theorists are now constructing a unified model of quarks and leptons, in which there must be as many kinds of leptons as there are kinds of quarks (Section 16.8). This does not answer the question "Why are there different kinds of leptons?" But such a question may not be the sort that one can answer; it could be akin to asking a biologist "Why are there rabbits?"

16.4 DISCOVERY OF "STRANGE PARTICLES"

At about the same time as the discovery of the pion, the solution to one meson puzzle, physicists began to see the makings of another puzzle in their cloud chambers. Events such as the one shown in Fig. 1 were observed,[16,17] and they could only be explained as the decays of new particles, of masses different from the masses of any known "elementary particle" or nucleus.

The sudden appearance of two tracks at the inverted V of Fig. 1 was an indication that some neutral particle (which leaves no track, because it does not ionize the gas in the chamber) must have decayed into two charged particles. In some cases the charged particles were identified as π^+ and π^-; measurements of the curvature of their tracks provided a determination of the momentum of these pions, from which the momentum and energy of the decaying neutral particle could be deduced (see problem 6). It was found that the particle had a rest energy of about 500 MeV; this particle is now known as the K^0 meson, and the decay observed was therefore

$$K^0 \quad \to \quad \pi^+ + \pi^- \tag{14}$$

[16] L. LePrince-Ringuet and M. Lhertier, *C.R. Acad. Sci.* **219**, 618 (1944).
[17] G. D. Rochester and C. C. Butler, *Nature* **160**, 855 (1947).

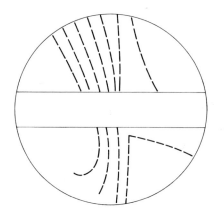

Fig. 1. *Decay of a neutral "strange particle." Cosmic rays moving from top to bottom strike the 3-in.-thick plate; one of them interacts to form the neutral particle, whose decay is observed below the plate, forming the characteristic Λ shape in the tracks of the decay products.*

In other cases, the decay products were found to be a π^- and a proton, and the decaying neutral particle was found to have a rest energy of about 1100 MeV; this particle is now known as the Λ particle, whose decay is via

$$\Lambda \ \rightarrow \ p + \pi^- \tag{15}$$

Discovery of the K^0 and the Λ was followed by discovery of other heavy particles which could be classified according to their ultimate decay products, as follows:

1. A *baryon* is a particle whose decay products ultimately include one proton (the proton being the lowest-mass baryon). Thus the Λ is a baryon. Baryon decays will be discussed further in Section 16.5.

2. A *meson* is a particle whose ultimate decay products are only leptons, antileptons, and/or photons. [The pions in decay (14) ultimately produce muons, neutrinos, and antineutrinos, via (9).]

Mesons and baryons are collectively known as "hadrons," a generic term for a particle that responds to the strong nuclear force.

Strangeness.   Masses and charges of some of the many hadrons discovered since 1944 are shown in Figs. 2 and 3. Most of these particles are called "strange particles." (The Λ_c^+ baryon and the D mesons are "charmed" rather than "strange"; we shall discuss charm in Section 16.8.) It must have seemed strange to discover these particles of unaccountable masses, just when it appeared that all of the elementary particles had been found that were needed in order to explain the structure of the known universe. But unexpected particles had been found before. What is it about these particles which makes them stranger than previously known particles (some of which are quite peculiar, as we have seen), so that the name "strange particles" has been retained?

The strange aspect of their behavior is in the speed with which they are produced, relative to the speed with which they decay. These particles are produced quite readily in encounters between nucleons, or of pions with nucleons; the production cross sections have been found to be of the order of 10^{-27} cm^2, indicative of a strong interaction. (Compare with the weak interaction cross section of order 10^{-40} cm^2, typical of neutrino interactions; see Section 15.2.) Yet the lifetimes of these particles are of order 10^{-10} sec, comparable to that of the charged pion, which decays only by the *weak* interaction. So the strange behavior is that the particles live too long, considering how quickly they are produced. For example, the inverse of decay (15) would be the reaction $\pi^- + p \to \Lambda$, which should require the same amount of time as the decay; that is, the wavefunctions of pion and proton must overlap for a time of order 10^{-10} sec, and this would indicate that the cross section for the process should be of order 10^{-40} cm^2.

There is no question of neutrino production in the decay of these particles, as there was in the case of the charged pions, which are also produced quickly and decay slowly. So we may ask, "Why are these particles forced to decay only by the weak interaction?" In particular, the K^0 meson lives a million times as long as the π^0 meson, although the K^0 is more massive than the π^0, and thus might be expected to decay into two photons even more rapidly than the π^0.

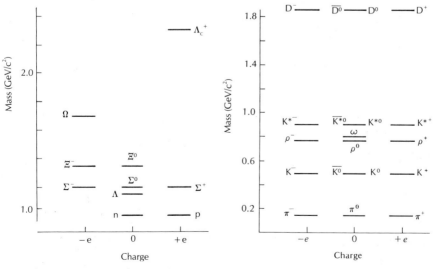

Fig. 2. *The masses of some "stable" baryons—those which do not decay by the strong interaction.*

Fig. 3. *Masses of some of the observed mesons, including some observed only as "resonances" (Section 16.7).*

In 1952, A. Pais showed the way to an explanation of this behavior by suggesting[18] that the strange particles must be produced *in pairs*. As accelerators were developed, and more interactions of strange particles could be seen, it was found that this prediction was essentially correct. A large variety of strange particles have now been discovered, as we see from Figs. 2 and 3 as well as from the table of "stable" particles.[19] All of these particles have been produced in reactions involving more than one strange particle, such as

$$\pi^- + p \quad \rightarrow \quad \Lambda + K^0 \tag{16a}$$

and

$$\pi^- + p \quad \rightarrow \quad \Sigma^0 + K^0 \tag{16b}$$

But how does the phenomenon suggested by Pais, called "associated production," explain the strange behavior of these particles? Consider the following analogy. Gamma rays of sufficiently high energy can create electron–positron *pairs* quite copiously, yet once they are separated, both e^+ and e^- are quite stable; i.e., neither of them *decays* into other particles. The reason for this stability is obvious. *Charge must be conserved*; e^+ and e^-, being the lightest charged particles, cannot decay into lighter particles.

By analogy, we can say that the Λ and the K^0 must possess some internal property, which has come to be called "strangeness," and we assume that strangeness is conserved when these particles are produced. That is, the strangeness of the Λ must be opposite to that of the K^0, just as the charge of the e^- is opposite to that of the e^+. The K^0, being the lightest strange particle, is "stable" once it is separated from the Λ. Of course the analogy is not perfect, because the K^0 does decay eventually, with a lifetime typical of the weak interactions. Thus we conclude that strangeness is not rigorously conserved, like charge; we must assume that strangeness is not conserved in weak interactions, although it is conserved in strong and electromagnetic interactions.

Universal Fermi Interaction.   We have seen numerous examples of decays which proceed via the weak interaction. One of the most fascinating, if not yet completely understood, features of these decays is that all of them are governed by two apparently universal constants. These decays can be classified according to whether or not strangeness is changed in the decay; it then appears that all of the decays of a given class are governed by the same coupling constant.

---

[18] A. Pais, *Phys. Rev.* **86**, 663 (1952).
[19] Appendix E. The particles in this Appendix are called "stable" because they do not decay by the strong interaction.

All of the decays in which strangeness does not change proceed at a rate which is determined by the same coupling constant g_1 which governs nuclear beta decay (Section 15.2). Thus, except for differences caused by the densities of final states, decay of the pion, the neutron, and the muon all proceed at the same rate.

The strangeness-changing decays, such as decay of the K mesons and the Λ particle, proceed at rates consistent with there being a second universal coupling constant g_2 which is approximately equal to $\frac{1}{4}g_1$.

16.5 CONSERVATION LAWS

Conservation of Baryon Number. Like the K^0, the Λ is also stable against decay by strong or electromagnetic interactions, although we can see from the list of particles in Appendix E that there are lighter particles that have the same strangeness as the Λ. The decay of the Λ into such lighter particles is not prohibited by conservation of strangeness or by any other law which we have previously discussed. But the Λ is not known to decay into anything which does not include a neutron or a proton. Apparently there is a different law at work here; the Λ must possess some property, in common with the proton and the neutron, which no lighter particles possess, and which is conserved in all known interactions. We call this property the baryon number, and we say that the nucleons and the Λ belong to a class of particles called baryons. We assign a baryon number of $+1$ to the neutron, the proton, and the Λ, and we assign baryon number -1 to each of the respective antiparticles. The "stable" baryons are shown in Fig. 2; each of these has baryon number $+1$, and each eventually becomes a proton after a series of decays.[20]

The conservation of baryon number is obeyed in weak and in strong interactions, involving nonstrange as well as strange particles. For example, the *anti*neutron *cannot* decay via

$$\overline{n} \rightarrow p + e^- + \overline{\nu}_e$$

although the neutron can. Conservation of baryon number requires the antineutron to decay by

$$\overline{n} \rightarrow \overline{p} + e^+ + \nu_e$$

---

[20] See also Appendix E. Notice that Appendix E shows only the baryons, not the antibaryons. Each baryon has an antiparticle; the properties of the antibaryon (charge, strangeness, T_3, etc.) are all opposite to those of the corresponding baryon.

The baryon number, of course, is also conserved in reactions in which particles are created. For example, in reaction (16a) the baryon number of the reactants is $+1$, and the baryon number of the products is also $+1$; the proton is simply replaced by the Λ as the carrier of baryon number.

Summary of Conservation Laws. We have now discussed a large enough number of conservation laws to confuse the reader thoroughly. It may be helpful to group these laws into three categories, so that similarities between the newer laws and the more familiar laws become apparent.

Category I involves quantities that are always conserved and have the same value in all frames of reference. Such a quantity may pertain to some immutable internal property of each particle that carries it. One example is electric charge. Other examples could be baryon number and lepton number. However, if some current theories prove to be correct, a baryon can, on very rare occasions, disappear, via, for example, $p \rightarrow e^+ + \pi^0$. In such a decay, neither baryon number nor lepton number would be conserved. (See Section 16.9.)

Category II involves quantities such as energy, momentum, and angular momentum, which are also conserved in all reactions, but which are not exclusively internal properties of particles, and which therefore have values which depend on the frame of reference.

Category III involves quantities such as parity and strangeness, which are not strictly conserved at all, but which are conserved in enough important situations to make the quantities useful. In this category, the laws in general seem to apply to strong interactions but not to weak interactions (with the effect of electromagnetic interactions being still open to question in some cases).

The conservation laws in category II are directly related to symmetry in nature. For example, conservation of momentum follows directly from the fact that physical laws are invariant with respect to translations through space.[21] The laws in category III may be said to involve a "broken symmetry"; for example, parity is conserved in strong interactions because the strong interaction potential is invariant with respect to mirror reflections. But the weak interaction potential does not possess this symmetry; there is a broken symmetry, and the conservation law is not obeyed. If baryon number changes via the decay of a baryon into a lepton, the law of conservation of baryon number would have to be put into category III. This would be the result of an "ultraweak" interaction whose potential lacks the symmetry needed for baryon conservation.

[21] For a good elementary discussion of invariance with respect to translation and its connection with physical laws, see "The Feynman Lectures in Physics," Vol. 1, Ch. 11. Addison-Wesley, Reading, Massachusetts, 1963. The laws of Category I can also be related to a "symmetry," which may or may not clarify the situation, depending on your sophistication. For example, conservation of charge follows from the invariance of electromagnetic interactions under a "gauge transformation."

Connection between Strangeness and *i*-Spin. Now that we have sorted out these laws, let us return to the problem of the strange particles. Refer to Fig. 2, in which the mass of each stable baryon is plotted versus its charge, and to Fig. 3, a similar plot for some of the mesons. When reactions involving all of these particles, mesons, and antiparticles are considered, it becomes necessary to place the concept of strangeness on a more quantitative basis.

In 1953, Gell-Mann and (independently) Nishijima developed a way of associating the concept of strangeness, which was then rather vague, with the more quantitative idea of *i*-spin. It turned out that the conservation of *i*-spin was sufficient to explain all of the laws involving production of strange particles, provided only that *i*-spin were assigned to these particles in a way which followed the method already developed for nuclear levels.

To see how this is done, notice that the particles of Fig. 2 fall into "multiplets," or sets of particles with roughly the same mass. As in nuclear physics, we assume that the various states (particles) of a multiplet would be identical if there were no electromagnetic interaction, and we assign an *i*-spin quantum number to these states according to their multiplicity, as we did with nuclear levels in Section 15.4. The states within a multiplet then differ from one another only in having different values of T_3, the quantum number for the "third component" of *i*-spin. For example, the Λ is a singlet; its total *i*-spin quantum number is $T = 0$, so that T_3 must also be zero. The Σ particles have $T = 1$, with $T_3 = +1$, 0, and -1 for Σ^+, Σ^0, and Σ^-, respectively. The Ξ have $T = \frac{1}{2}$, and the Ω^- is another singlet, with $T = 0$.

However, there is an important difference between these multiplets and the multiplets of nuclear levels. For all reactions involving only neutrons and protons, conservation of T_3 is identical with conservation of charge, as was pointed out in Section 15.4, because converting a neutron to a proton means adding one unit to T_3 as well as adding one unit of charge. This is *not* the case when strange particles are created, because T_3 can then change even though the charge is unchanged, and vice versa. For example, if a neutron were changed to a Λ, T_3 would change from $-\frac{1}{2}$ to zero, even though nothing else in the system changed. And if a proton were changed to a Ξ^0, the charge would change, although T_3 would not change.

The contribution of Gell-Mann and Nishijima was to suggest that T_3 is conserved in strong interactions, even in situations in which conservation of T_3 is *not* equivalent to conservation of charge. This new condition on particle interactions seems to be obeyed in all strong interactions. Thus the decay $\Lambda \rightarrow n + \pi^0$ is forbidden, because the T_3 values of the Λ and the π^0 are zero, whereas the neutron has $T_3 = -\frac{1}{2}$. And you can see for yourself by inspection of Fig. 2 that *all* of the decays of strange particles into nonstrange particles are forbidden if T_3 is conserved. Such decays must therefore proceed by the weak interaction. (This is also true for the "charmed" singlet Λ_c^+, with $T = T_3 = 0$, whose decay to a less-massive baryon is prohibited by T_3 and charge conservation. (See Section 16.8 for further discussion of the Λ_c^+.)

Similarly, the production reaction $\pi^- + p \rightarrow \Lambda$ is forbidden, because the reactants $\pi^- + p$ have a total T_3 equal to $-1 + \frac{1}{2} = -\frac{1}{2}$, while Λ has $T_3 = 0$. But the associated production reaction (16a) is allowed. This fact tells us that the K^0 must have $T_3 = -\frac{1}{2}$. The total i-spin for the K^0 must therefore be a half integer: $\frac{1}{2}$, $\frac{3}{2}$, or conceivably even greater. If $T = \frac{1}{2}$ for the K^0, then the K meson must exist in two different states: the K^0 with $T_3 = -\frac{1}{2}$, and the K^+ with $T_3 = +\frac{1}{2}$, in accordance with the general rule that a difference of one unit in T_3 accompanies a difference of one unit of charge, within a multiplet. If T were equal to $\frac{3}{2}$ for the K mesons, then there would be *four* of them, which would have to be K^{++}, K^+, K^0, and K^-. The question can only be resolved experimentally, and since no K^{++} has ever been observed, we conclude that $T = \frac{1}{2}$ for the K mesons.

We said before that the K^0 and the Λ must have opposite strangeness, and now we are in a better position to appreciate what that means. The strange property of these particles follows directly from the fact that the *center of charge* of the "strange" multiplets is displaced relative to the center of charge of the "normal" multiplet. If all of the baryon multiplets, for example, were centered at a charge of $\frac{1}{2}e$, then conservation of T_3 would be equivalent to conservation of charge, as it is in nuclear reactions which involve baryons only. But the Λ is at zero charge, $\frac{1}{2}$ unit *less* than the average charge of the np doublet, and the K^0–K^+ doublet is centered at a charge of $+\frac{1}{2}e$, which is $\frac{1}{2}$ unit *more* than the charge of the center of the "normal" meson triplet π^+, π^0, π^-. The displacement of the K meson doublet compensates for the displacement of the Λ, so that T_3 can be conserved as well as charge in reaction (16a).

The formulation of strangeness in terms of i-spin is a real advance, because it correctly accounts for particles that have more than one unit of strangeness. For example, the conjecture of Pais that strange particles must be produced in *pairs* is not sufficient to describe the production of the Ξ^0, because the reaction

$$\pi^- + p \quad \rightarrow \quad \Xi^0 + K^0 \tag{17}$$

does not occur. It is easy to see that reaction (17) is forbidden by T_3 conservation; in order to create a Ξ^0 from $\pi^- + p$, and still conserve T_3, we must create *two* K^0 mesons, via

$$\pi^- + p \quad \rightarrow \quad \Xi^0 + K^0 + K^0$$

The necessity for this is related to the fact that the Ξ^-–Ξ^0 doublet is displaced by a *full* unit of charge from the n–p doublet, rather than by one-half unit as is the Σ triplet or the Λ singlet.

We can describe the behavior of the Ξ particles by assigning to them *two units of strangeness*. We can define a strangeness quantum number S which is equal to twice the displacement of the center of a multiplet from the center of the normal multiplet—that is, from n–p for a baryon multiplet, or from the pion triplet for a meson multiplet. In symbols, we may write a general formula relating strangeness S, charge Q, i-spin T_3, and baryon number B as follows:

$$Q = (T_3 + \tfrac{1}{2}S + \tfrac{1}{2}B)e$$

where e is the charge of the proton. You may easily verify that this formula contains all of the relationships which we have discussed, concerning these quantities. Thus among the baryons, the Λ and the Σ have $S = -1$, the Ξ have $S = -2$, and the Ω^- has $S = -3$. The mesons K^0 and K^+ have $S = +1$. When these values are assigned, the conservation of strangeness is equivalent to conservation of T_3, in determining whether or not a strong reaction will occur. Since the normal particles all have $S = 0$, a particle of positive strangeness must always be produced with a particle or particles of negative strangeness, in reactions initiated with normal particles.

The strangeness of each particle is listed along with its other quantum numbers in the particle table, Appendix E. Each particle has its antiparticle, whose quantum numbers are all opposite to those of the particle. For example, there are three anti-Σ particles, each having strangeness $S = +1$; the anti-Σ of negative charge is the antiparticle to the Σ^+, and the anti-Σ of positive charge is the antiparticle to the Σ^-.

16.6   PROPERTIES OF THE K MESONS

The K mesons have a number of properties which are so fundamental and so intriguing that they deserve a section to themselves. Some of these properties led to the discovery of the fact that parity is not conserved in weak interactions, and all of the properties of the K mesons must be studied in the light of that fact.

The "$\theta - \tau$ Puzzle."   If you have looked up the early references to the K mesons, you may have had trouble understanding the language, because the K^+ is referred to as the "θ^+" or as the "τ^+" particle. It was believed that these were two different particles, whose decay modes are

$$\tau^+ \quad \rightarrow \quad \pi^+ + \pi^+ + \pi^- \tag{18a}$$

or

$$\tau^+ \quad \rightarrow \quad \pi^+ + \pi^0 + \pi^0 \tag{18b}$$

and

$$\theta^+ \quad \rightarrow \quad \pi^+ + \pi^0 \tag{19}$$

This older "τ" particle is not to be confused with the recently discovered τ *lepton*; it is now known that the old "τ" is really the K, so the letter τ has become available for use in designating the heavy lepton.

The belief that θ and τ are different particles was based on the law of conservation of parity. The analysis which led to this conclusion is a bit involved, but it can be understood in the following way. Let us assume that (1) parity is conserved, (2) the "τ" and the "θ" *are* the same particle. We can then show that these assumptions lead to a contradiction. We observe that the τ (and therefore the θ) must have zero spin, because the π^- emitted in decay (18a) has an isotropic angular distribution with respect to the relative momentum of the two π^+ mesons. (An isotropic angular distribution implies a spherically symmetric state, hence zero angular momentum.) Thus the three pions resulting from the decay of the τ^+ form a state of odd parity, because each pion has odd intrinsic parity, and there is no contribution of orbital angular momentum to the parity. If, as we assumed, the θ is identical to the τ, then the two-pion state resulting from the θ^+ decay must have even parity, because the zero spin of the initial state requires that the final state have zero orbital angular momentum. So we see that, if parity is conserved in both decays, the τ^+ must have odd parity and the θ^+ must have even parity.

Thus one of the two assumptions was wrong. Either it was wrong to say that θ^+ and τ^+ are the same particle, or it was wrong to say that parity is conserved. Since nobody questioned the law of conservation of parity, it was concluded that the θ and the τ were different particles.

This conclusion became more and more suspect as the masses and lifetimes of these particles were more accurately determined. It was found that τ and θ have equal masses *and* equal lifetimes (within the experimental accuracy of a few percent). These results led Lee and Yang to question the conservation of parity in weak interactions, and they decided to survey the experiments done in the field of beta decay. They found that no experiment had ever indicated that parity *is* conserved in beta decay, because *nobody had ever tested the law*. They then suggested the experiment on ^{60}Co decay discussed in Section 15.2, as well as another experiment on pion decay. Both experiments showed that parity is not conserved in the weak interaction, which governs these processes.[22] It then became apparent that τ and θ must be the *same particle*, which is now called the K^+ meson, and which has all of the decay modes (18a), (18b), and (19), plus a few others which are listed in Appendix E.

Symmetry Operations: Charge Conjugation, Parity, and Time Reversal.
The overthrow of parity conservation was somewhat jarring to theorists who believe in the symmetry of nature. A force which is not invariant with respect to mirror reflections seems to be out of place in the orderly structure of physics.

---

[22] The pion result was reported by R. L. Garwin, L. M. Lederman, and M. Weinrich, *Phys. Rev.* **105**, 1415 (1957).

It had been thought that all laws must be invariant with respect to *any one* of the following basic operations.

1. Charge conjugation, represented by an operator C which changes each particle into its antiparticle.

2. Mirror reflection, represented by an operator P (for parity) which changes the sign of one space coordinate in every equation. (The P operation may also be called space inversion, or a change in sign of all three space coordinates. This is equivalent to changing the sign of one coordinate and then rotating by 180° about that coordinate axis. Invariance with respect to rotation is as well established as anything can be in physics—it is the basis for the conservation of angular momentum—so that both definitions of P are equivalent.)

3. Time reversal, represented by an operator T which changes the sign of the time coordinate in every equation (so that velocity and angular velocity are reversed in direction).

After the K meson showed that at least one law is not invariant with respect to P, theorists quickly recovered, pointing out that laws should still be invariant with respect to simultaneous application of *all three* operations *CPT*. In other words, if we reverse time, *and* reflect the coordinates, *and* replace each particle by its antiparticle, then we must find all laws to be unchanged. Or if we take a motion picture of an experiment done with "matter," run the projector backwards, and view the projected images in a mirror, what we see should be consistent with all laws which are obeyed by "antimatter." The theorem stating this, called the *CPT* theorem (or Luders–Pauli theorem) can be "proven" on the basis of such fundamental assumptions about space and time, that if this theorem fails, a great deal of theoretical physics must be rewritten.

For a time it appeared that there was also invariance under CP and T *separately*; that is, the results of an experiment with particles should be the same as the results of the mirror image experiment with antiparticles (with no need to run the movie backwards). For example, in the beta decay of ^{60}Co, there are more electrons emitted in a direction opposite to the nuclear spin than there are emitted in a direction parallel to the nuclear spin. In the mirror image decay, there would be more electrons emitted parallel to the nuclear spin (Section 15.2, Fig. 10); therefore, we expect that in the decay of *anti-*^{60}Co the *positrons* would be emitted more often parallel to the nuclear spin.

Later experiments have shown that even the product CP is not strictly conserved, and thus if *CPT* invariance is correct, T invariance must also fail. The violation of CP invariance is quite small, but it is real.[23]

---

[23] J. H. Christenson *et al., Phys. Rev. Lett.,* **13**, 138 (1964). You will be in a position to understand this result after you study the K^0 meson.

The K^0 Meson.   If the K mesons and the heavy baryons are "strange," the *neutral* K meson seems positively weird in its behavior. Yet this behavior is an inevitable consequence of the principles which we have outlined here[24]; the behavior springs from the fact that there is an anti-K^0, or $\overline{K^0}$, which is distinct from the K^0 and which does not interact with matter in the same way as the K^0.

It is not completely obvious that there should be such a thing as an anti-meson, so let us review briefly the concept of antiparticles in general. The fact that a particle such as an electron has an antiparticle may be deduced from the existence of negative energy states in the solutions of the Dirac equation (Section 7.5). The Dirac equation itself does not describe spinless particles such as mesons. But the CPT theorem, which follows from the symmetry between the negative energy states and the positive energy states, shows that, whatever forces there are in nature, the same interactions must be possible between antiparticles as there are between particles. Thus we can take any interaction involving particles and transform it into an interaction involving antiparticles. In the process of transforming these interactions, we must change the sign of the charge of each particle (or meson)—hence the name "charge conjugation." For example, the decay

$$\pi^- \quad \rightarrow \quad \mu^- + \bar{v}_\mu$$
$$\hookrightarrow \quad e^- + \bar{v}_e + v_\mu$$

is transformed by charge conjugation into

$$\pi^+ \quad \rightarrow \quad \mu^+ + v_\mu$$
$$\hookrightarrow \quad e^+ + v_e + \bar{v}_\mu$$

so that the π^+ plays the role of "antimeson" to the π^-; every reaction involving a π^- must have a conjugate reaction involving the π^+, in which particles are replaced by antiparticles.

Now consider the reaction

$$\pi^- + p \quad \rightarrow \quad \Sigma^- + K^+ \tag{20}$$

There must be a conjugate reaction

$$\pi^+ + \bar{p} \quad \rightarrow \quad \Sigma^+ + K^- \tag{21}$$

so that the K^- is the antimeson to the K^+. To satisfy the conservation laws in reaction (21), the K^- must have $S = -1$ and $T_3 = -\frac{1}{2}$, the opposite of the

---

[24] Some say that God created the K^0 meson for the primary purpose of providing an illustration of the fundamental principles of quantum mechanics in their purest form. If the K^0 did not exist, it might be necessary to invent it.

strangeness and *i*-spin "component" of the K^+, just as any antiparticle has quantum numbers which are the opposite of the quantum numbers for the corresponding particle. Thus there exist K^+, K^0, and K^-, just as there exist π^+, π^0, and π^-, But the K mesons do *not* form an *i*-spin triplet as the π mesons do, because the *i*-spin of the K meson would then be 1, and we have seen that an *i*-spin of $\frac{1}{2}$ is required for the K meson as a result of reactions such as reaction (16). The only way to reconcile all the facts is to assume that the K must form *two doublets*: the K^+–K^0 with $T = \frac{1}{2}$ and $S = +1$, and the $\overline{K^0}$–K^-, with $T = \frac{1}{2}$ and $S = -1$.

Thus we are forced to conclude that the K^0 has an antiparticle which is a completely different entity, with different strangeness. Conversely, the fact that the pions have $T = 1$ indicates that there is only one neutral pion; the π^0 is its own antimeson. In other words, π^0 is identical to $\bar{\pi}^0$, so that the operation of charge conjugation transforms π^0 into itself. As a consequence, an isolated π^0 can decay into two gamma rays, which an isolated K^0 cannot do.

The $\overline{K^0}$ meson is easily distinguishable from the K^0 meson by its behavior in strong interactions. For example, the $\overline{K^0}$, because its strangeness is -1, can easily interact with a nucleon to produce a strange baryon, in a reaction such as

$$\overline{K^0} + \begin{pmatrix} n \\ p \end{pmatrix} \quad \rightarrow \quad \begin{pmatrix} \Lambda \\ \Sigma \end{pmatrix} + \pi \tag{22}$$

because $S = -1$ for the Λ or the Σ. The K^0, with $S = +1$, cannot do this, because there are no baryons with $S = +1$. (There are *antibaryons* with $S = +1$, but production of an antibaryon from a nucleon would violate the conservation of baryon number.) However, the K^0 can easily produce a K^+ via the "charge exchange" reaction

$$K^0 + p \quad \rightarrow \quad K^+ + n$$

whereas the $\overline{K^0}$ produces the K^- via the analogous reaction

$$\overline{K^0} + n \quad \rightarrow \quad K^- + p$$

The Two Lifetimes of the K^0 Meson. When we analyze the decay of the K^0, we see the most remarkable consequences of the existence of the $\overline{K^0}$; it is likewise remarkable that these consequences were predicted by Gell-Mann and Pais in 1955 before most of them were observed.[25] Consider the decay

$$K^0 \quad \rightarrow \quad \pi^+ + \pi^- \tag{23}$$

[25] M. Gell-Mann and A. Pais, *Phys. Rev.* **97**, 1387 (1955). It is fascinating to read this paper in the light of current knowledge; the authors seemed to be almost at the point of deducing the nonconservation of parity, but they did not take that next step in their reasoning.

Applying the *CPT* operator to (23), we obtain the time reversed, mirror reflected process

$$\pi^+ + \pi^- \quad \rightarrow \quad \overline{K^0} \tag{24}$$

which yields the anti-K meson. Since both (23) and (24) are possible, and the two-pion states are the same in both, it must also be possible to combine the processes (23) and (24) into a single process in which a K^0 becomes a $\overline{K^0}$, via

$$K^0 \quad \rightarrow \quad \pi^+ + \pi^- \quad \rightarrow \quad \overline{K^0}$$

in which a K^0 spontaneously transforms itself into its own antiparticle! The intermediate pions do not have to appear explicitly; they can be "virtual" pions, with the reaction being simply

$$K^0 \quad \rightarrow \quad \overline{K^0} \tag{25}$$

Gell-Mann and Pais made a brilliant extrapolation of Eqs. (23) through (25), showing that decay into two pions cannot be the only decay mode of the K^0, although it was the only one that had been observed up until that time. They showed that the K^0 must be a *superposition* of equal parts of *two other mesons*, the K_1 and the K_2, and that only the K_1 can decay into two pions. In other words, they said that decay into two pions can only account for one-half of all K^0 decays. Later observations confirmed the accuracy of this deduction.

To understand the K_1 and K_2, we shall rephrase the argument of Gell–Mann and Pais to bring it into line with current knowledge and notation, except that for convenience we shall assume *CP* invariance. (The results are only slightly affected by the small violations of *CP* invariance which have been discovered.) We observe that the final state of reaction (23) is an eigenstate of the *CP* operator (because application of *C* changes π^+ into π^- and π^- into π^+, and application of *P* interchanges the positions of the π^+ and the π^-). Using the notation of Dirac, we write

$$CP|\pi^+\pi^-\rangle = |\pi^+\pi^-\rangle$$

where the symbol $|\pi^+\pi^-\rangle$ indicates a *state* in which there exists a π^+ and a π^-. But the initial state $|K^0\rangle$, which decays into $|\pi^+\pi^-\rangle$, is not an eigenstate of *CP*, because

$$CP|K^0\rangle = |\overline{K^0}\rangle \quad \text{and} \quad CP|\overline{K^0}\rangle = |K^0\rangle \tag{26}$$

How can the state $|K^0\rangle$, which is *not* an eigenstate of *CP*, decay into the state $|\pi^+\pi^-\rangle$, which *is* an eigenstate of *CP*, if *CP* is conserved? The answer can only be that the state $|K^0\rangle$ must be a *superposition of states*, each of which is an eigenfunction of *CP*. The decay process is then equivalent to a *measurement of CP*; after the measurement, in accordance with the postulates of quantum mechanics (Section 5.6 and Section 6.3), the wavefunction of the system is an eigenfunction of the *CP* operator. The situation is analogous to the measurement of electron momentum by positron annihilation; the annihilation process Fourier-analyzes the electron–positron wavefunction into its momentum components.

To see what sort of a superposition the state $|K^0\rangle$ is, let us construct two normalized, independent linear combinations of $|K^0\rangle$ and $|\overline{K^0}\rangle$ as follows:

$$|K_1\rangle = \frac{|K^0\rangle + |\overline{K^0}\rangle}{\sqrt{2}} \tag{27}$$

$$|K_2\rangle = \frac{|K^0\rangle - |\overline{K^0}\rangle}{\sqrt{2}} \tag{28}$$

You may easily verify by application of the CP operator to Eqs. (27) and (28), using Eqs. (26), that

$$CP|K_1\rangle = |K_1\rangle \quad \text{and} \quad CP|K_2\rangle = -|K_2\rangle$$

Thus $|K_1\rangle$ and $K_2\rangle$ are eigenstates of CP, with eigenvalues $+1$ and -1, respectively.

We can solve Eqs. (27) and (28) for $|K^0\rangle$ and $|\overline{K^0}\rangle$, obtaining

$$|K^0\rangle = \frac{|K_1\rangle + |K_2\rangle}{\sqrt{2}} \tag{29}$$

and

$$|\overline{K^0}\rangle = \frac{|K_1\rangle - |K_2\rangle}{\sqrt{2}} \tag{30}$$

The factor $1/\sqrt{2}$ in Eqs. (27)–(30) is, of course, a normalization factor.

According to postulate 3, Section 6.3, the probability of obtaining a given eigenvalue as a result of a measurement is equal to the square of the coefficient of the corresponding eigenfunction in the expansion of the wavefunction. In this case, since the coefficients of $|K_1\rangle$ and $|K_2\rangle$ are either $+1/\sqrt{2}$ or $-1/\sqrt{2}$ in the expansions (29) and (30), a pure K^0 or a pure $\overline{K^0}$ must yield $CP = +1$ in half of its decays, and $CP = -1$ in the other half of its decays. This was the prediction of Gell-Mann and Pais; they said that one-half of all K^0's decay into a two-pion state (either $|\pi^+\pi^-\rangle$ or $|\pi^0\pi^0\rangle$), whose CP eigenvalue is $+1$. The other half decay more slowly, into states such as $|\pi^+\pi^-\pi^0\rangle$, whose CP eigenvalue is -1.[26]

To understand the full implications of these facts, let us "follow" a beam of K^0 mesons from the point where they are produced. Initially the beam is a

---

[26] It is now known that the long-lived component also decays into two pions in 0.16 percent of the decays, in violation of CP invariance.[23] The long-lived component is now called K_L, and the short-lived component K_s, in recognition of the violation of CP invariance upon which the definition of K_1 and K_2 is based. See V. L. Fitch, *Rev. Mod. Phys.* **53**, 367 (1981). Fitch and J. W. Cronin were awarded the 1980 Nobel Prize for this discovery of "CP violation."

pure K^0 beam, being produced by a reaction such as $K^+ + n \rightarrow K^0 + p$ or $\pi^- + p \rightarrow K^0 + \Lambda$, strong interactions in which strangeness is conserved, so that no $\overline{K^0}$ component can be produced. But as the K^0 travels, the part with CP equal to $+1$ decays exponentially, with a mean lifetime of 0.86×10^{-10} sec, into two pions; thus, after a few tenths of a nanosecond, half of the original beam has decayed, and the remainder is almost purely K_2, with CP equal to -1. But according to Eq. (28) a K_2 is a mixture of K^0 and $\overline{K^0}$, and the $\overline{K^0}$ component can participate in reactions such as (22), a fact which has been verified by experiment. Thus we see how the K^0 is transformed into the $\overline{K^0}$.

But this is not the end of the possible transformations; the K_2 can be transformed into the K_1. Let us follow our beam of neutral K a bit farther, and let the K_2 beam pass through an absorber. The $\overline{K^0}$ component of this beam is absorbed more strongly than the K^0 component, simply because it can participate in reactions such as (22). Therefore when the beam emerges from the absorber, it no longer contains equal parts of K^0 and $\overline{K^0}$, and thus it cannot be a pure K_2 beam; i.e., it contains a K_1 component again! This *regeneration* of the K_1 has been well verified by experiment; decay into two pions is again observed after a K_2 beam emerges from an absorber (see Fig. 4).

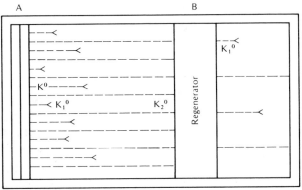

Fig. 4. *Schematic diagram showing regeneration of K_1 component from a beam of K_2 mesons. The symbol indicates the K_1 mode of decay, into two pions. Pions generate K^0 in target A, but only K_2 live long enough to reach the regenerator. The $\overline{K^0}$ component of the K_2 is absorbed in the regenerator, leaving K^0 with a regenerated K_1 component, so that 2π decays reappear.*

The entire sequence of events may be described graphically by means of vectors (see Fig. 5). The vector representing K^0 is orthogonal to that represent-

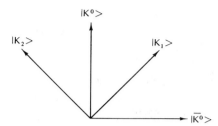

Fig. 5. *Vector diagram showing relationships among* $|K^0\rangle$, $|\overline{K}^0\rangle$, $|K_1\rangle$, *and* $|K_2\rangle$.

ing $\overline{K}^0$, and the K^0–$\overline{K}^0$ pair is at a $45°$ angle to the pair representing K_1 and K_2. Thus the K^0 has components K_1 and K_2 of equal amplitude; removal of the K_1 component leaves behind a K_2. The K_2, in turn, has components K^0 and $\overline{K}^0$ of equal amplitude, and removal of the $\overline{K}^0$ component from the K_2 leaves the K^0, which has a K_1 component. Even if only part of the $\overline{K}^0$ component were removed from a K_2, the resultant vector would no longer be parallel to the K_2 axis, and so it would contain some mixture of K_1.

Mass Difference between the K_1 and the K_2. Notice that the K^0 could be changed into a $\overline{K}^0$ by reversing the direction of the K_2 component; that is, by changing its *phase* by $180°$. This sort of transformation is also observed experimentally, and could happen even if the K^0 were completely stable. It is a result of the time dependent part of the wavefunction, $e^{-iEt/\hbar}$, and of the fact that K_1 and K_2 *have different masses*, so that their wavefunctions oscillate with different frequencies and thus develop a phase difference. If we include the time-dependent part, the wavefunction of a particle which is initially a K^0 would be written

$$|\psi\rangle = \frac{|K_1\rangle e^{-i\omega_1 t}e^{-t/2\tau_1} + |K_2\rangle e^{-i\omega_2 t}e^{-t/2\tau_2}}{\sqrt{2}} \tag{31}$$

where $\hbar\omega_1$ is the rest energy of the K_1, $\hbar\omega_2$ is the rest energy of the K_2, τ_1 is the mean lifetime for K_1 decay, and τ_2 is the mean lifetime for K_2 decay. (The mean lifetime is multiplied by 2 in the denominator of each exponent, because it is the *square* of the amplitude which determines the mean lifetime of the state.)

We can use Eq. (31) to find the fraction of $\overline{K}^0$ present in $|\psi\rangle$ as a function of time; this is a number that can be checked experimentally and used to determine the mass difference between the K_1 and the K_2. The probability that a particle, which is initially a K^0, will react as a $\overline{K^0}$ at any given time is given by the square of the "component" of $|\psi\rangle$ along the $|\overline{K^0}\rangle$ "axis," to use the terminology of vectors. We can find this component as in ordinary vector algebra, simply by taking the scalar product of $|\psi\rangle$ and the vector $|\overline{K^0}\rangle$ as given by Eq. (30). Using the usual technique of multiplying corresponding components together and adding, we obtain for this scalar product

$$\frac{e^{-i\omega_1 t}e^{-t/2\tau_1} - e^{-i\omega_2 t}e^{-t/2\tau_2}}{2}$$

whose absolute square is (as you may verify)

$$\frac{1}{4}\left\{e^{-t/\tau_1} + e^{-t/\tau_2} - 2e^{-t/2\tau_1}e^{-t/2\tau_2}\cos\frac{\Delta mc^2 t}{\hbar}\right\} \tag{32}$$

where Δm is, of course, the mass difference between the K_1 and the K_2.

Notice that the $\overline{K^0}$ fraction is zero at $t = 0$, and that it increases with time; at times large compared to τ_1 but small compared to τ_2, the $\overline{K^0}$ fraction would be $\frac{1}{4}$ if Δm were equal to zero. This fraction may be understood very simply; half of the original K^0 state decays, and half of the remaining state is $\overline{K^0}$. But the cosine term causes the number of $\overline{K^0}$ in the beam to oscillate, rather than to increase steadily at small t. The oscillation is analogous to the oscillation of energy between two coupled oscillators; energy flows back and forth at the "beat frequency," which is equal to the difference between the frequencies of the two normal modes of oscillation.

Measurements have established the existence of an oscillation in the number of $\overline{K^0}$ as a function of time, and have made it possible to determine the $K_1 - K_2$ mass difference to a remarkable degree of accuracy. The difference is $(4.0 \pm 0.2) \times 10^{-6}$ eV, which is about one part in 10^{14} of the total mass of either particle! This difference is significant because it is the only energy shift so far measured which is known to result from a weak interaction; the weak interaction must be responsible for this shift, because the only difference between K_1 and K_2 is in their behavior in weak interactions. This explains the enormous amount of effort which has been devoted to determining the size of this tiny effect.

Neutrino Oscillations.   It has been proposed that neutrinos may exhibit the same sort of oscillation that has been observed for the K^0 meson. If this is true, it would have far-reaching implications.

The theory can be set up in a manner similar to that of the K^0. For simplicity, let us consider ν_e and ν_μ only. If these neutrinos do not have zero mass, then they could be superpositions of two other neutrinos, ν_1 and ν_2, which are eigenstates of mass as are the K_1 and K_2. That is, ν_1 and ν_2 could each have a well-defined nonzero mass, and the states consisting of ν_e and ν_μ could be written, for example, as

$$|\nu_e\rangle = (|\nu_1\rangle + |\nu_2\rangle)/\sqrt{2}$$

$$|\nu_\mu\rangle = (|\nu_1\rangle - |\nu_2\rangle)/\sqrt{2}$$

in exact analogy to equations (29) and (30). The states $|\nu_e\rangle$ and $|\nu_\mu\rangle$ are the ones produced in beta decay and pion decay, respectively; they are eigenstates of "electron number" and "muon number," which are separately conserved in such decays (Section 16.3). But because each of these is a superposition of $|\nu_1\rangle$ and $|\nu_2\rangle$, which can have *different masses* m_1 and m_2, it can be shown that each can spontaneously change into the other (like K_1 and K_2). We write the general state as

$$|\psi\rangle = \frac{|\nu_1\rangle\, e^{-i\omega_1 t} \pm |\nu_2\rangle\, e^{-i\omega_2 t}}{\sqrt{2}}$$

in analogy to Eq. (31), where the plus sign makes $|\psi\rangle = |\nu_e\rangle$ at $t = 0$, the minus sign makes $|\psi\rangle = |\nu_\mu\rangle$ at $t = 0$, $\omega_1 = m_1 c^2/\hbar$, and $\omega_2 = m_2 c^2/\hbar$.

Then, if $|\psi\rangle = |\nu_e\rangle$ at time $t = 0$, some later time t, such that $e^{-i\omega_2 t}$ equals $-e^{-i\omega_1 2}$, the second term in $|\psi\rangle$ will be opposite in sign to the first term, making $|\psi\rangle$ the state $|\nu_\mu\rangle$ instead of the state $|\nu_e\rangle$. The time required is given by

$$\omega_2 t = \omega_1 t + \pi$$

or

$$t = \pi/(\omega_2 - \omega_1) = \pi\hbar/c^2(m_2 - m_1)$$

One complete oscillation returns the state to its original configuration; the period is $T = 2t = h/c^2(m_2 - m_1) = 4.14 \times 10^{-15}\ \text{eV-}s/\Delta E_r$ where ΔE_r is the difference in rest energy between ν_1 and ν_2.

The existence of these "neutrino oscillations" is somewhat speculative at present, because there is no direct way to observe them. Furthermore, a non-zero rest mass for the neutrinos is necessary (but not sufficient) to produce the oscillations, and the neutrino mass, if any, is so small that it has so far been impossible to detect it.

However, astrophysical observations and theory give indirect support for neutrino oscillations. It is now possible, using reaction (12) of chapter 15, to detect neutrinos from beta decays in the sun. Theories of the sun's structure and composition are now sufficiently detailed to give us a value for the expected neutrino flux from the sun, and the cross section, as a function of neutrino energy, is known, so it is possible to predict how many solar neutrinos should be observed in a given detector in a given time. But experiments to date have detected only about one-third of the expected number of neutrinos. Neutrino oscillations provide an explanation for this, as follows:

The period of oscillation, if it exists, is likely to be of the order of nanoseconds, during which time the neutrino travels a distance of order one meter or less. Each neutrino thus undergoes many oscillations on its way out of the sun, and neutrinos created at different points arrive here with different phases. If the oscillation involves three possible observable states (ν_e, ν_μ, and ν_τ), neutrinos in the detector will, on the average, have only a probability of ⅓ of being in the ν_e state to which the detector can respond, and the result is explained.

Of course there are alternative explanations, but none of them seem palatable. For example, our model of the sun could be wrong, because the sun might be in a non-equilibrium state—it could be about to blow up! So the solution to this problem may be of more than academic interest.

16.7 RESONANCES

All of the particles that we have discussed so far have been those which are produced by a strong intereaction or an electromagnetic interaction, in a time of the order of 10^{-23} sec in some cases, but which, for various reasons, do not decay so rapidly, so that they live long enough to travel a measurable distance in a detector. Now you should be wondering whether other particles might be produced that would decay by a strong interaction, in a time of the order of 10^{-23} sec. If such particles were produced, how would we know it, since they invariably would decay before traveling more than a few nuclear diameters? And if there were some means of detecting these particles, would we be justified in calling them particles and placing them on a par with particles whose tracks can be seen? Are such particles any less real than the particles which we have already discussed?

The answers to these questions are not so difficult as you might think. Such particles *have* been produced, in great profusion. There are heavy mesons with $S = 0$, such as the η (550 MeV), which decays into three pions, the ρ (750 MeV), which decays into two pions, the ω (780 MeV), which decays into two or three pions (usually three), the ϕ (1020 MeV), which decays into two K's or into three pions, and a host of others. There are also mesons with $S = 1$, which are heavy enough to decay into a K plus a pion, and heavy baryons which decay into a baryon plus pion. In each case, the particle is able to decay in a way which conserves strangeness, so that the decay is quite rapid, occurring in a time of order 10^{-20} to 10^{-23} sec.

How do we know that such particles are produced, when they never leave a measurable track? The answer is best shown by an example. Consider the ρ^- meson, which may be produced by the reaction

$$\pi^- + p \quad \rightarrow \quad \rho^- + p$$

After it is produced, the ρ^- decays via

$$\rho^- \quad \rightarrow \quad \pi^- + \pi^0$$

The *net* reaction is simply

$$\pi^- + p \quad \rightarrow \quad \pi^- + \pi^0 + p$$

The observations simply show that a π^- interacts in a bubble chamber, producing a π^-, a π^0, and a proton emerging from the point of interaction. But *analysis of the energies* of the emerging particles makes it possible to deduce that the interaction *must go in two stages*, with the ρ^- as an intermediate state in the process. The reasoning goes as follows:

If the ρ^- were *not* formed, the total momentum would be shared by *three* products when the reaction occurs; there would therefore be a smoothly varying distribution of energies for all three particles, the distribution being determined solely by the density of states (as in beta decay, Section 15.2). But suppose that the ρ^- *is* formed. The emerging proton's energy is then uniquely determined by conservation of energy and momentum, because the initial reaction creates only two products instead of three. Furthermore, the ρ^- energy is also uniquely determined, and when it decays, the energy of each pion is again determined solely by conservation of energy and momentum. Thus if the energy distribution of any one of the emerging particles is plotted, a peak is found in the distribution at the energy which is dictated by the production of a ρ^- as an intermediate state. (See Problem 4.) Measurement of the energy of this peak enables one to solve for the rest energy of the ρ^- meson.

Our confidence in the reality of these short-lived mesons is reinforced by the observation of *different decay modes*; analysis of the energies of the emerging particles in various reactions shows that the *same state* is involved in *different situations*, just as the same compound nucleus can be produced in various nuclear reactions and can decay via various modes, independently of the mode of production. (See Section 15.4.)

Figure 6 illustrates another meson "resonance," the ω meson. The analysis is slightly more complicated in this case, because there are more pions involved, but the basic principle is the same. The figure shows the results of interactions between protons and antiprotons, for those cases in which *five* pions emerge:

$$\bar{p} + p \quad \rightarrow \quad 2\pi^+ + 2\pi^- + \pi^0$$

For each event, one can group the pions *three at a time*, and one can compute the expression

$$(\textstyle\sum E_i)^2 - c^2(\textstyle\sum p_i)^2$$

summed over the three particles. If the three pions came from a single parent, $\sum E_i$ must have been the energy of this parent, and $\sum p_i$ must have been the parent's momentum. Therefore

$$(\textstyle\sum E_i)^2 = E_{\text{parent}}^2 \quad \text{and} \quad (\textstyle\sum p_i)^2 = p_{\text{parent}}^2$$

and thus

$$E^2_{\text{parent}} - c^2 p^2_{\text{parent}} = (Mc^2)^2 \qquad (33)$$

where M must be the rest mass of the parent. (M is called the "effective mass.")

If the values of M computed by using Eq. (33) follow the distribution which can be expected from the density-of-states factor, then it can be said that there was no parent which decayed into three mesons; the mesons must have been created independently. The results obtained by Maglic *et al.* are plotted in Fig. 6. You can see that there was no parent which decayed into a *charged* triplet of pions ($\pi^+\pi^+\pi^-$, $\pi^+\pi^+\pi^0$, etc.), but that the *uncharged* triplet $\pi^+\pi^-\pi^0$ shows a definite peak, corresponding to the ω meson, a neutral meson whose rest energy is about 780 MeV.

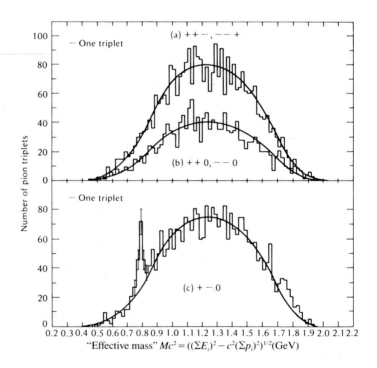

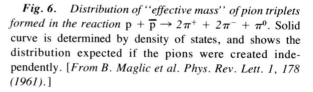

Fig. 6.   *Distribution of "effective mass" of pion triplets formed in the reaction* p $+$ $\bar{\text{p}}$ $\rightarrow$ $2\pi^+$ $+$ $2\pi^-$ $+$ π^0. *Solid curve is determined by density of states, and shows the distribution expected if the pions were created independently.* [*From B. Maglic et al. Phys. Rev. Lett.* **1**, *178 (1961).*]

Mean Lifetimes of Resonant States.   Although there is no way to observe the creation and decay of these so-called "resonant states" separately, one can still obtain a good idea of their mean lifetimes by using the uncertainty relation $\delta E \, \delta t \geqslant \hbar$, just as one does in estimating the lifetimes of short-lived nuclear states (Section 15.4). The value of δE, the uncertainty in the energy of the state, is given by the "width" of the resonance—the width of the peak in a curve like those of Fig. 6. From the figure we see that the ω meson appears to have a width of order 100 MeV, so that its lifetime must be of order $\hbar/100$ MeV, or 10^{-23} sec.[27]

16.8   STRUCTURE OF "ELEMENTARY" PARTICLES

The total of all of the resonances and antiparticles discovered to date is in the hundreds, and we can hardly feel comfortable in calling all of these things "elementary particles." We would like to be able to reduce nature to a few elementary constituents, as we thought we had done when we knew only the proton, neutron, and electron. But where do we begin to weed out the particles which are not so elementary?

Clearly we cannot say that a meson like the K^0 is any more elementary than the η meson; the K^0 lives much longer than the η only because it possesses a property, strangeness, which inhibits its decay. We cannot even say that the neutron is more fundamental than the Λ; in the center of a "neutron star," the Fermi energy of the neutrons might be so great that the Λ is a stable particle![28]

Such observations lead us to suppose that the baryons and mesons (which are collectively called hadrons) could be composites. For example, a neutral pion could be composed of a particle and an antiparticle; its decay into two gamma rays (Section 16.2e) would be similar to the decay of singlet positronium, and it could have excited states which would appear as resonances.[29]

If particles are composite, we can try to gain some insight into their structure by studying symmetries. The method can be based on the study of nuclei, where i-spin is introduced to classify levels according to their symmetry with respect to transformation of neutrons into protons, and vice versa.

The classification scheme for nuclei is really rather simple, because we have two fundamental "building blocks," the proton and the neutron, from which all

[27] Figure 6 does not give a true picture of the width, because the energy resolution is too coarse. More refined measurements indicate a width of about 12 MeV for this meson, making its mean lifetime of the order of 10^{-22} sec.

[28] A neutron star is believed to consist of dense nuclear matter. If the Fermi energy of the neutrons in such a star is greater than the mass difference between n and Λ, a neutron would decay into a Λ, which would be permitted to occupy a lower energy state.

[29] See "Are Mesons Elementary Particles?", E. Fermi and C. N. Yang, *Phys. Rev.*, **76**, 1739 (1949)—written only two years after the discovery of the K meson.

nuclei are constructed. But when we study particles we have several questions to answer before the solution to the problem begins to take shape:

1. Are the particles indeed composite, in the sense that there are a small number of building blocks which can be used to construct all particles?
2. What are the building blocks, if they exist?
3. How do we introduce the element of strangeness into the construction of a particle?

The best way to answer question 1 is to attempt to answer questions 2 and 3. We select various possible building blocks to see if we can use them to construct all particles—the strange ones as well as the nonstrange ones.

The Sakata Model. The first thing one might try is to add one strange particle to the basic np doublet, to see if all particles may be built up by combining particles from this triplet. Sakata did this, using a basic triplet consisting of n, p, and Λ; the Λ is a logical choice for the strange particle because it is an i-spin singlet.

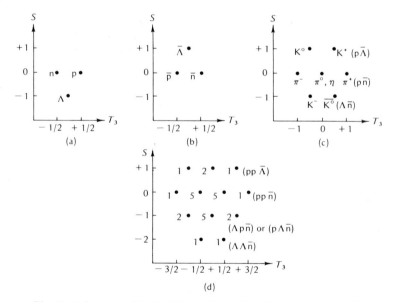

Fig. 7. *Sakata model.* (a) *The basic triplet,* (b) *its antitriplet,* (c) *the meson octet formed by combining one member of the triplet with one member of the antitriplet in all possible ways,* (d) *the possible baryons which result from combining two members of the triplet with one of the antitriplet; numbers indicate the degeneracy of each combination. There are a total of 27 different permutations, a few of which are indicated beside the states.*

One can easily use the Sakata model to construct the better known mesons. There are nine ways to combine one of the three basic particles—n, p, or Λ—with one of the basic antiparticles—n̄, p̄, or Λ̄; and these nine combinations can be related directly to the observed mesons. In Fig. 7, we plot the strangeness of each particle versus its i-spin, and on the assumption that these properties are additive, we see that each baryon–antibaryon combination gives the right strangeness and i-spin to be one of the observed mesons. Of course, the masses of these mesons are nowhere near the masses of the "constituent" particle plus antiparticle, but we must assume that the difference results from a huge binding energy.

But when we try to use the Sakata model to build up the baryons, it just does not work. In order to make a combination for which the baryon number is $B = 1$, one must assume that each heavy baryon consists of two basic *particles* plus one basic *antiparticle*. There are $3 \times 3 \times 3 = 27$ ways to do this; each permutation contributes a new state. These 27 states form in twelve different combinations of T_3 and S, as shown in Fig. 7d. But the combinations do not agree with those for the observed particles; for example, there is no i-spin triplet with $S = +1$, as shown in the top row of Fig. 7d. This point will become clearer as we examine a model that does work.

The Quark.   The successful model was developed by Gell-Mann and, independently, by George Zweig. It has the same symmetry as the Sakata model, but not the same basic particles; it does not identify the basic triplet with n, p, Λ, or any other previously known particles. Rather, it is based on three hypothetical particles, hitherto unknown and unsuspected, whose properties are dictated by the necessity of combining them to form the then-known "elementary" particles. Gell-Mann called these particles "quarks"; Zweig's name, "aces," has not survived.[30]

To produce strange particles, a strange quark s was needed, with strangeness $S = -1$. The proton and neutron then consist solely of two kinds of quark, called u (or "up" quark) and d (or "down" quark). Because there was no need to make these hypothetical quarks individually have the same properties as any known particles, they could be given noninteger values of charge number and baryon number. It turned out that all of the known baryons could be constructed from three-quark combinations of u, d, and s, provided that the baryon number of each quark is $+\frac{1}{3}$, and the charges are $+2e/3$, $-e/3$, and $-e/3$, respectively. For example, the proton has two u and one d; the neutron, two d and one u. Every meson is a combination of a quark and an antiquark; for example, K^+ has a u and a s̄ (an anti-s).

Fig. 8 shows the values of S and T_3 obtained by combining the three quark "flavors" (up, down, and strange) in all possible ways. Notice that, although there are only ten distinct combinations of S and T_3, there are 27 *states* (which

---

[30] Gell-Mann's reference for the word "quark" was "Finnegan's Wake," by James Joyce, but Joyce may have picked up the word from the German language, in which quark literally means cottage cheese. In colloquial German, quark means nonsense.

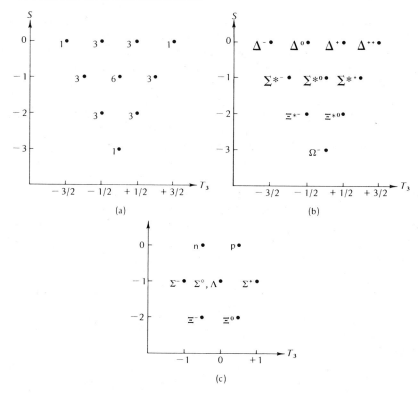

Fig. 8.   *The states and degeneracies which result from combining three quarks; (b) and (c) show how the known particles and resonances group themselves in agreement with these T_3 and S assignments.*

means 27 possible particles), because each of the $3 \times 3 \times 3$ permutations of the three flavors is a distinct state (as in the Sakata model, above). For example, the state $S = 0$, $T_3 = +\frac{1}{2}$ consists of two up quarks and one down quark, but it is triply degenerate, because uud, udu, and duu are distinct (just as the two-particle spin states $(+)(-)$ and $(-)(+)$ are distinct). One forms three linearly independent combinations of these states to obtain entities with the proper transformation properties to represent known particles. One of these particles is the proton; the other two are resonances that have been observed.

From Fig. 8a you can see that the up quark must have T_3 equal to $+\frac{1}{2}$, the down quark must have $T_3 = -\frac{1}{2}$, and the strange quark must have $T_3 = 0$. (This is sometimes expressed by calling s the "sideways" quark.) Using these facts, you can easily verify that each of the points in that figure has the degeneracy shown there, and that the total number of states is indeed 27. (See Problem 7.)

The transformation properties lead to a grouping of these 27 states into "supermultiplets," just as angular momentum and i-spin groups states into multiplets. These supermultiplets are the decuplet shown in Fig. 8b, two octets with the symmetry displayed in Fig. 8c, and one singlet with $T_3 = 0$ and $S = -1$.[31] This grouping is not arbitrary (for example, one could *not* have two septets and three singlets instead of two octets and one singlet, even though the required degeneracy could be maintained that way); it is determined by the possibility of using raising and lowering operators to transform one state into another, as was done in Section 7.3 for the angular momentum states. (See Problem 8 for the method as it applies to the supermultiplets.) The particles in the decuplet all have spin $\frac{3}{2}$ and even parity; the particles in the baryon octet of Fig. 8c all have spin $\frac{1}{2}$ and even parity.

The quark model also reproduces the success of the Sakata model in describing the observed *mesons*. A quark-antiquark pair is just as good as a baryon-antibaryon pair for this purpose, providing us with the basic meson octet (Fig. 7c).

If the quark model did nothing more than to provide a rationale for the existence of these supermultiplets, it would be worthy of notice. But when we investigate the *masses* of the particles in the various groups, we see the possibility of making further predictions from the theory. You will recall that members of an *i*-spin multiplet are supposed to have the same mass if the electromagnetic interactions could be turned off, and it is easy to believe that the small mass differences which are observed could result from electromagnetic interactions. We can extend this idea, saying that members of a *supermultiplet* would all have the same mass if some *additional interaction* could be turned off. This additional interaction must be a strong interaction; it produces a mass difference of several hundred MeV between members of the same supermultiplet. Since we have difficulty even accounting properly for the small mass differences produced by the electromagnetic force, you may wonder how we could hope to make any quantitative prediction about the mass differences produced by this strong interaction. But it turns out, almost miraculously, that this interaction has effects which can be accounted for very simply. To do this we study the symmetry of Figs. 7 and 8.

Notice that all of the diagrams of Figs. 7 and 8 have a threefold rotational symmetry; each pattern is invariant with respect to rotations through 120°. It is not hard to convince yourself that this should be true of any pattern which is built up from basic triplets with the symmetry of Fig. 7a or 7b. We can describe this symmetry in a *purely formal* way by inventing two more so-called "spin" variables—*u*-spin and *v*-spin—such that a 120° rotation causes U_3—the "third component" of *u*-spin—to replace T_3 as the variable plotted horizontally. In this terminology, for example, p and Σ^+ form a *u*-spin doublet, just as p and n form an *i*-spin doublet; we can say that p has $T_3 = +\frac{1}{2}$

---

[31] In order to make these diagrams symmetric about the origin, it is customary to plot the *hypercharge* $Y \equiv S + B$, rather than S, on the vertical axis. The point $Y = 0$, $T_3 = 0$ lies at the center of all groups.

and $U_3 = +\frac{1}{2}$, while Σ^+ has $T_3 = +1$ and $U_3 = -\frac{1}{2}$. And the fundamental triplet uds consists of the i-spin doublet ud, the u-spin doublet ds, and the v-spin doublet su.

The fact that members of a u-spin multiplet all have the same charge suggests that members of such a multiplet would be identical if the strong interaction could be turned off, just as members of an i-spin multiplet would presumably be identical if the electromagnetic interaction could be turned off. This mean that the mass difference between members of a u-spin multiplet can be attributed entirely to the strong interaction.[32]

In the decuplet, the Ξ^{*-} is about 150 MeV more massive than the Σ^{*-}, and the latter is about 150 MeV more massive than the Δ^-. These three belong to a u-spin quadruplet with the Ω^-, which had not yet been discovered when the quark theory was proposed. Gell-Mann, in using the quark theory to predict the existence of the Ω^- (which would be the first particle ever observed with $S = -3$), also predicted its mass by assuming that the mass depends linearly on U_3, because of the masses already observed. This would make the Ω^- about 150 MeV more massive than the Ξ^{*-}, or about 1680 MeV.

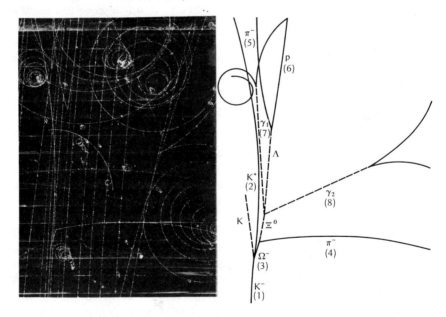

Fig. 9. *Trace and line diagram of event showing decay of Ω^- [From V. E. Barnes et al., Phys. Rev. Lett. 12, 204 (1964). Photograph provided through the courtesy of N. Samios and the Brookhaven National Laboratory.]*

[32] This expectation is based on absence of *interference* between the two interactions, so that turning on one interaction does not change the effect of the other interaction. The assumption is not necessarily correct; you could imagine that the strong interaction could affect the arrangement of quarks enough to change the electromagnetic interaction.

With its strangeness of -3, the Ω^-, unlike the other members of the decuplet, was expected to be "stable," because there is no way that strangeness can be conserved in its decay. Therefore there was a search for bubble chamber events that might contain the complicated series of tracks resulting from a cascade of decays to $S = -2$, $S = +1$, and finally $S = 0$. The search was successful and the result is shown in Fig. 9; the chain of events depicted is:

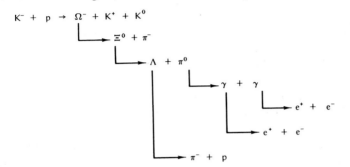

The mass of the Ω^- has been found to be within a few MeV of the predicted value.

Quark Searches. The predictions based upon quarks or quark-related symmetry are impressive, but one still wonders whether the quark is "real"—whether individual quarks can be "seen," as one sees individual tracks of other particles in bubble chambers. However, we are now accustomed to accepting indirect evidence for such things; we have come a long way since people wondered whether atoms are real.

Quarks have been observed individually *inside* the nucleon, in the same way that the nucleus was first observed inside the atom. As in Rutherford's experiment (Section 13.1), the inner structure of a scattering target—in this case the proton or neutron—is deduced from the angular distribution of the scattered particles; the occurrence of a small number of large-angle events shows that the nucleon is mostly empty space, containing pointlike charged scattering centers, whose properties agree with those of the quarks.

The word "parton" has been used to describe constituents found in scattering experiments; these are now identified as quarks and "gluons" (see below). Partons were not evident in earlier scattering experiments that determined the nuclear radius (Section 14.2), because the wavelength of the incident radiation was too long to resolve them. The first evidence for partons appeared in 1970, when electrons of 20-GeV energy were scattered from protons and neutrons in the Stanford Linear Accelerator Center. The results have been confirmed by subsequent experiments in which *protons* were scattered by protons.[33]

[33] M. Jacob and P. Landshoff discuss a number of these experiments in "The Inner Structure of the Proton," *Sci. Amer.* **242**, #3, 66 (1980).

However, even such high energies have been unable to separate an identifiable quark from a nucleon. There is additional evidence that such a separation is extremely rare. If a quark had ever been separated from a nucleon in the past (perhaps by a collision with an extremely energetic cosmic ray), it would be unlikely to recombine with two other quarks to form a normal nucleon. Once such a free quark slowed down, it would be captured by an atom and become tightly bound to the nucleus of that atom, where it would be unable to find two partner quarks. Such a quark must be a stable particle, because there is nothing into which it can decay without violating the law of conservation of electric charge; an atom containing such a quark would be forever fractionally charged, and would have been detectable by Millikan if it were present in one of his oil drops (Section 1.3).[34] Refinements of the oil-drop experiment, using niobium spheres instead of oil drops as the charge carriers, have now yielded several fractionally charged spheres, but these results have not been duplicated in other laboratories, and thus doubt remains.[35]

The difficulty in separating a quark from a nucleon, and the failure of quark searches to provide conclusive evidence for a separated quark, have led many theorists to postulate "confinement rules" that would prevent a quark from *ever* escaping from a nucleon or other hadron. These rules are based on the concept of quark "color."

Quark "Colors." The spins of the baryons are easily explained by assigning a spin of $\frac{1}{2}$ to the quark; this should make it a fermion, governed by Fermi-Dirac statistics and the exclusion principle. This leads to a problem when one considers the baryons containing three quarks of the same "flavor," such as the Δ^{++} with three "up" quarks. Such a baryon should have a much greater mass than it does, if one of its three quarks were forced by the exclusion principle into a higher energy state.

Instead of simply postulating that the exclusion principle does not apply in this situation, it is more fruitful to preserve the principle by assuming that a new quantum number exists, and to explore the consequences of this assumption. The new quantum "number" is called "color," and it is assumed to have three different "values"—red, green, and blue. It is further assumed that every baryon contains one quark of each color. The confinement of quarks appears as a byproduct of this picture, when one postulates that quarks can only appear in "colorless" combinations—red plus green plus blue in a baryon, or a color and its anticolor in a meson.

Color is believed to be the strong interaction's "charge," the source of the force. Quantum chromodynamics (QCD), based upon analogy with quantum electrodynamics (QED), has been very successful in predicting masses of ob-

---

[34] Millikan did indeed find a charge of 0.3 e on one oil drop, but he dismissed this result because it was not reproducible. R. A. Millikan, *Phil. Mag.* **19**, 209 (1910).

[35] For a discussion of this and other quark searches, see L. W. Jones, *Rev. Mod. Phys.* **49**, 717 (1977). Fractionally-charged niobium was reported by G. S. LaRue, W. M. Fairbank, and A. F. Hebard, *Phys. Rev. Lett.* **38**, 1011 (1977).

served particles as energy levels of quark combinations, using the concept of color as a sort of charge. In QCD, the force between quarks is carried by "gluons," which are the analog to photons in QED. (Thus the gluon, rather than the meson, is the carrier of the nuclear force.) The gluon, unlike the photon, carries a "charge"—that is, a *color* charge—so that emission or absorption of a gluon can change a quark of one color into a quark of another color. Each gluon is characterized by two colors; there are eight varieties of gluon in all, forming another octet.

The color of any given quark in a baryon is constantly changing as gluons are absorbed and emitted. For example, at a given instant the d quark in the proton may be blue, and the two u quarks red and green; at another instant, the d quark could be green and the u quarks red and blue. The overall properties of the proton are unaffected by the location of the color charge on the individual quarks.

16.9 UNIFIED THEORIES

We have seen that quantum chromodynamics and quantum electrodynamics can now be applied to quark interactions. But that is not the whole story, because quarks participate in weak interactions (and gravitational interactions) as well as electrodynamic and strong interactions. The ultimate goal of theorists is to produce a unified theory that includes all four forces in a description of particle interactions.

One early step toward unification was the development of a weak-interaction theory that explains this interaction in a manner analogous to the meson theory of the strong interaction, or the subsequent gluon theory. The particles that participate in a weak interaction are, in this theory, coupled by particles called intermediate vector bosons ($W^\pm$ and Z^0), just as quarks are coupled by gluons. (Generically, these bosons, the gluon, and the photon are called "gauge" particles.)

With the involvement of the W, weak decays are believed to proceed in two stages. For example, a positive muon can decay by emitting a W^+, leaving a $\bar{\nu}_\mu$:

$$\mu^+ \rightarrow W^+ + \bar{\nu}_\mu \tag{34}$$

This is a very short-lived "virtual" W^+, because the mass of the μ^+ is far too small for energy to be conserved in this process. The W^+ decays into a positron plus an electronic neutrino in this case, so the net result is the well-known decay mode of the muon:

$$\mu^+ \rightarrow e^+ + \nu_e + \bar{\nu}_\mu$$

The mass of the W has been predicted to be about 82 GeV/c^2. This large mass is consistent with the short range of the weak interaction (which in turn is associated with the long lifetimes of beta emitters, or with the small neutrino cross sections. The first observations of W bosons have yielded masses of 81 ± 5 GeV/c^2, in excellent agreement with the theory. These observations have been associated with quark transformations in colliding beams of protons and antiprotons. For example a u quark in the proton can interact with the $\bar{\text{d}}$ quark in the antiproton to produce a W^+; the complete reaction would be:

$$p + \bar{p} \to u\bar{d} + X$$
$$\phantom{p + \bar{p} \to u\bar{d}}\,\llcorner\!\!\!\blacktriangleright W^+ \to e^+ + \nu_e \tag{35}$$

where X stands for the other quarks and antiquarks in the p$\bar{\text{p}}$ pair, plus any gluons and quark-antiquark pairs that may be materialized in the collision. These other quarks and antiquarks form hadrons, while the u and $\bar{\text{d}}$ are annihilated, becoming the W^+—a process analogous to other particle-antiparticle annihilations in which photons are created. The W^+ can decay as in (35), leaving the e^+ and the ν_e; other decay modes are also possible, but this is the one that was first observed.

The Z^0 boson gives rise to a "neutral current" responsible for *elastic scattering* of neutrinos, according to an "electroweak" theory unifying weak interactions and electromagnetism.[36] The Z^0 has also been produced in proton-antiproton collisions, and its mass has been found to be 95 ± 2.5 GeV/c^2, in good agreement with the theoretical prediction of about 94 GeV/c^2.[37] Elastic scattering of neutrinos has also been observed experimentally.

The connection between the electroweak field and the *strong* interaction first began to emerge when the quark model gained recognition. George Zweig used the word "aces" for what we now call quarks, because he believed that there should be four of them, in order to produce symmetry with the four leptons known at that time. The idea that such symmetry exists has now been almost universally accepted.

Impetus for this acceptance was provided by the discovery of evidence for the fourth quark in 1974, by groups at Brookhaven and Stanford independently. This quark has a property analogous to strangeness, called "charm." It was unknown previously because of its large mass; replacement of a strange quark by a charmed quark in any strange hadron produces a new "charmed" hadron whose mass is roughly 1 GeV greater than that of the original strange hadron.

Inclusion of the charmed quark c gives us $4^3 = 64$ permutations of three quarks to form a baryon. Just as the 27 charmless permutations form the decuplet shown in Fig. 8a, now the 64 permutations form the 20-plet shown in Fig. 10. Degeneracies are not shown in Fig. 10, but you can work them out for yourself. For examples, the point $s = 0$, $u = 1$, $d = 1$, $c = 1$ is sixfold degenerate (like the central $S = -1$ point in Fig. 8a). One of the six baryons with that quark combination is the Λ_c^+, to be discussed below.

The existence of the charmed quark was first inferred from the occurrence of a resonance at an effective mass of about 3.1 GeV in the products of collisions

[36] See H. Georgi and S. L. Glashow, *Phys. Today* **33**, #9, 30 (Sept. 1980). Glashow shared the 1979 Nobel Prize with A. Salam and S. Weinberg for development of "electroweak" theory.
[37] G. Arnison, et al., *Physics Letters* **B** (1983).

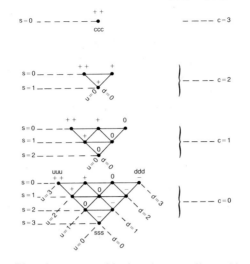

Fig. 10. *Baryon 20-plet, showing all possible combinations of three quarks of u, d, s, or c type. The c = 0 grouping is identical to the decuplet shown in Fig. 8b. Lines indicate groups of particles with a given number of u, d, or s quarks; e.g., particles on the s = 2 lines have two s quarks (and therefore strangeness S = −2).*

between particles and antiparticles. This resonance, called J/ψ (called J at Brookhaven and ψ at Stanford) was a new meson which decayed too quickly to be a strange meson, but not quickly enough to be a nonstrange meson; the narrowness of the resonance ($\Delta E \sim 0.1$ MeV) indicated a lifetime of about 10^{-20} s. It was concluded that this was a bound state of c and $\bar{c}$.

To see the difference as well as the parallel between charm and strangeness, compare the charmed baryon Λ_c^+ with the strange baryon Λ. (See Fig. 2 and Appendix E.) Each is an *i*-spin singlet whose strong decay is forbidden by T_3 conservation; thus each has a relatively long mean lifetime, considering the energy available for the decay. If Λ_c^+ were "strange," we would assign it a strangeness of +1, because it is displaced $\frac{1}{2}$ unit, along the charge axis (Fig. 2), from the center of the np doublet. (But we know that it cannot be a strange particle, because the s quark has strangeness of −1, not +1, and a baryon cannot contain an antiquark.) As a *charmed* particle, it has a *charm* of +1 (equal to the charm of the c quark) instead. Notice too that Λ_c^+ is much more massive than Λ, reflecting the greater mass of the c quark relative to the s quark. Thus all of the differences (and similarities) between Λ_c^+ and Λ result from the replacement of the s quark by the c quark.

Although the existence of the c quark permits a large number of additional hadrons to exist, there are definite constraints on their properties and behavior, which convince us that the quark model is useful and correct. To illustrate, we consider the D meson, which is the charmed analogue of the K meson. (See Fig. 3 and Appendix E.) There is a D^+D^0 doublet with charm of $+1$, and a $\overline{D^0}D^-$ doublet with charm of -1, paralleling the K^+K^0 and $\overline{K^0}K^-$ doublets with $S = +1$ and $S = -1$, respectively. D mesons could be plotted on Fig. 7c in place of K mesons, if the ordinate were charm instead of S. Each D can decay into a K by conversion of the c quark into the s quark (or $\bar{c}$ into $\bar{s}$) with emission of a charged virtual W boson, as follows: (quark constituents of each particle shown in parentheses)

$$D^+ (\bar{d}c) \to \overline{K^0} (\bar{d}s) + W^+ \qquad\qquad D^0 (\bar{u}c) \to K^- (\bar{u}s) + W^+$$

$$\overline{D^0} (u\bar{c}) \to K^+ (u\bar{s}) + W^- \qquad\qquad D^- (d\bar{c}) \to K^0 (d\bar{s}) + W^-$$

The W boson can then decay into leptons, as it does in neutron decay, or into quarks and antiquarks. In the latter case, the quarks can be rearranged somewhat, as follows:

$$D^+ (\bar{d}c) \to \bar{d}s + W^+$$
$$\qquad\qquad \hookrightarrow\ u\bar{u}\bar{d} \left.\begin{array}{c}\\\end{array}\right\} \to K^- (\bar{u}s) + \pi^+ (u\bar{d}) + \pi^+ (u\bar{d})$$

Without the quark model, you might think that the above decay would be no more likely than the decay $D^+ \to K^+ + \pi^- + \pi^+$. However, the latter decay is *not* observed, because it requires the transformation of c into $\bar{s}$, for which there is no mechanism. (Transformation of c into $\bar{s}$ requires emission of a charge of $e/3$, but there is no boson with such a charge.)

Unfortunately, as soon as the fourth quark vindicated Zweig's idea that the number of quarks should equal the number of leptons, the tau lepton was found (Section 16.3). But the quark-lepton symmetry was too good an idea to simply discard; we only need another pair of quarks to preserve it. One more quark has now been found, in a manner similar to the way that the c quark was found; it is called the "bottom," or b quark, and its charge is $-e/3$, like that of the d and the s. To complete the pair, we should have a "top" quark t, with a charge of $+2e/3$ like that of u and c. The t quark is believed to be extremely massive ($20\,\mathrm{GeV}/c^2$ or more), and accelerators are now being prepared to reach the energy needed to produce it.

The t and b quarks have been assigned the properties of "truth" and "beauty," respectively. We can now summarize the properties of the quarks as follows:

Charge	Quark	Property
	u (up)	
$+2e/3$	c (charmed)	Charm $= +1$
	t (top)	Truth $= +1$
	d (down)	
$-e/3$	s (strange)	Strangeness $= -1$
	b (bottom)	Beauty $= -1$

Each quark has an antiquark with opposite properties; e.g., $\bar{b}$ has a charge of $+e/3$ and beauty of $+1$.

The parallel between quarks and leptons is shown strikingly by the following table (constructed by S. Glashow in analogy to the periodic table of the elements), which shows how quark and lepton properties repeat periodically, just as properties of chemical elements repeat.

Periodic Table of Quarks and Leptons

charge:	-1	$-\frac{1}{3}$	0	$+\frac{2}{3}$
particle:	e^-	d	ν_e	u
	μ^-	s	ν_μ	c
	τ^-	b	ν_τ	t

The last two entries have not been observed at this writing, but their imminent discovery seems highly probable. At present there is no theoretical reason to expect a fourth row to appear in this table, but there is also no reason to rule it out.

The possibility of constructing such a table suggests that there could be an underlying reason for it that is similar to the reason for the periodicity in atomic properties. That is, might not quarks and leptons have an internal structure that is revealed here, just as the structure of atoms is revealed in the periodic table of the elements? This conjecture is currently receiving a great deal of attention.

Decay of the Proton.   The symmetry between quarks and leptons (if it exists) implies that there is some relationship between the two—that we might even consider a lepton to be a "colorless quark" (albeit with a different electric charge as well). If that is a correct view, one might wonder whether it is possible for a quark to be transformed into a lepton. Such a process must certainly be rare, because it would cause a proton to disappear, to be replaced by a positron and a neutral pion. If the proton had a mean lifetime as "short" as 10^{17} years, we would not exist; we would be killed by the radioactivity of our own bodies.

Surprisingly, it is not only possible to conceive of a proton decay on a time scale much longer than this, but it is also possible to calculate an order of magnitude for the proton's expected lifetime, simply by comparing the strengths and the ranges of the weak, strong, and electromagnetic interactions. These forces vary differently with distance r from the source, and it happens that at $r \approx 10^{-29}$ cm, all of these forces are about equally strong. The energy needed to probe matter at such distances—to produce wavelengths so short—is so enormous that even the 80-GeV mass of the W boson is negligible in comparison; W bosons and gluons, as well as photons, are relatively "massless" at such a range.

If two quarks approach each other to within such a distance, they are in a domain where the interaction between them is not significantly stronger than that between two leptons. The unified theory says that, in such a case, a quark can become a lepton.[38] The mechanism for this is a hypothetical X particle, an extremely massive gauge particle which can transform a quark into a lepton. In emitting an X particle, a u changes into a $\bar{u}$. The X particle must be very short lived, being able to exist only because of the uncertainty relation for energy and time. It is normally reabsorbed by the $\bar{u}$, recreating the original u. But if a d quark is sufficiently close by, it can absorb the X particle and become an e^+. In this process a charge of $+4e/3$ is transferred from the u to the d; this therefore is the X particle's charge.

When this process occurs inside a proton, the proton disappears, leaving the e^+, the $\bar{u}$, and the proton's other u. The u and the $\bar{u}$ combine to form a π^0, so the observed result is

$$p \rightarrow e^+ + \pi^0 \ .$$

Here a baryon is replaced by an antilepton. Neither baryon number nor lepton number is conserved, but the *difference* between baryon number and lepton number is conserved.

The only way to tell whether the X particle really exists is to look for its effect. If we "see" a proton decay, the X particle presumably is what caused it. Several searches for proton decay are now underway. If the lifetime estimate of 10^{31} years is correct, there should be one proton decay per year in a collection of 10^{31} protons. To see this decay, large numbers of detectors (photomultiplier tubes) have been placed throughout huge volumes of water or other material located deep underground in mines or tunnels. If 1000 tons of water can be monitored efficiently, roughly 100 proton decays per year would be expected. The main difficulty in the experiment is in unambiguously identifying these events as proton decays rather than, for example, events involving neutrino capture (since no mine is deep enough to shield out neutrinos). If the proton lifetime is 10^{33} years or more, the neutrino-induced background may be too large to permit a definitive result.

The importance of this search can be gauged from the list of thirteen massive experiments (either underway or being constructed) published in the June, 1981 issue of the Scientific American ("The Decay of the Proton," by S. Weinberg, page 244). If the proton is found to decay, it will be a superb triumph for present elementary-particle theories. But it will surely not mean that nature has no more surprises in store for us.

---

[38] See "A Unified Theory of Elementary Particles and Forces," H. Georgi; *Scientific American* **244**, #4, 48 (1981).

PROBLEMS

1. A "direct" measurement of the π^0 lifetime was performed in the following way[39]: K^+ mesons were stopped in a photographic emulsion, where some of them decayed into $\pi^+ + \pi^0$. Some of the π^0 decayed via $\pi^0 \rightarrow e^+ + e^- + \gamma$. When such a sequence of events was studied, the intersection of the K^+ track with the π^+ track gave the point of *origin* of the π^0, and the intersection of the e^+ and e^- tracks gave the point of *decay* of the π^0. The distance between these points could then be used to compute the π^0 lifetime; the mean lifetime was found (somewhat inaccurately) to be 1.9×10^{-16} sec.

    (a) Show that the π^0 must have a speed of 0.835 c when it is produced in the K^+ decay.

    (b) How far does such a π^0 travel in a time equal to the mean lifetime given above? (Be sure to account for the effect of time dilation.)

2. The π^+ and the π^- have precisely the same mass, whereas the Σ^+ and Σ^- masses are unequal. Explain the difference between the two situations. Would you expect that a small mass difference might be found between the K^+ and the K^-?

3. The cross section for the interaction $K^0 + n \rightarrow \Lambda + \pi^0$ is zero, and the cross section for the interaction $\overline{K^0} + n \rightarrow \Lambda + \pi^0$ is σ_n. Consider a pure K^0, of kinetic energy 125 MeV, produced in a strong interaction. What is the cross section, in terms of σ_n, for this particle to interact with a neutron and produce $\Lambda + \pi^0$, after the particle has traveled 2.0 cm from its point of production? Compute your answer on the basis of (a) zero mass difference between K_1 and K_2 (b) the mass difference reported in Section 16.6. Make the approximation that the K_2 lifetime τ_2 is infinite.

4. (a) Determine the minimum kinetic energy that a π^- meson must have in order to produce a ρ^- in the reaction

    $$\pi^- + p \rightarrow \rho^- + p$$

    with a stationary proton target. (See Section 16.7)

    (b) Suppose that the incident π^- in the above reaction has a kinetic energy of 1400 MeV, and that the proton is observed to recoil in the forward direction (the direction of the incident π^-). What is the kinetic energy of the recoiling proton?

---

[39] Glasser, N. Seeman, and B. Stiller, *Phys. Rev.* **123**, 1015 (1961).

(c)   Suppose that the reaction were

$$\pi^- + p \to \pi^- + \pi^0 + p$$

without production of a ρ^-, and that the proton recoils in the forward direction, as in part (b). What are the maximum and the minimum kinetic energies which the proton can have?

5.   Which of the following processes (a) cannot occur as *strong* interactions or decays; (b) cannot occur at all? State your reason in each case.

$$p + p \ \to \Xi^0 + K^+ + K^+$$
$$n + p \ \to \Sigma^+ + \Lambda$$
$$p + \bar{p} \ \to n + \Lambda + K^0$$
$$p + \bar{p} \ \to \Lambda + \bar{\Lambda} + \pi^0$$
$$\pi^- + p \ \to \Xi^- + K^+ + \overline{K^0}$$
$$\Lambda \ \to \pi^- + p$$
$$\Lambda \ \to K^- + p$$
$$\Lambda + d \ \to \Sigma^0 + d \quad (d = \text{deuteron})$$
$$\pi^0 \ \to e^+ + e^- + \nu$$

6.   Pages 671 and 672 show tracks resulting from K^- interactions in a hydrogen bubble chamber. The chamber was in a magnetic field, so that each track is curved, with a radius of curvature that is proportional to the momentum of the particle that made the track. On the scale shown, for the magnetic field used, the relation between momentum p and radius of curvature R is

$$p = (2.55 \pm 0.1)R$$

where p is in MeV/c and R is in cm.

(a)   Refer to the line drawings which show the geometry of the event of interest in each photograph. Why do the tracks coming from vertex A in each event indicate that the K^- was at rest at the time of the interaction? (All tracks are in the plane of the picture.)

(b)   From the conservation laws and the identities of the other three particles marked in each line drawing, determine the charge, baryon number, and strangeness of particle X in each event. (In each case the K^- interacts with a proton in the chamber.)

(c)   For each event determine the mass of X by measuring the momentum of the pion and applying conservation of energy and momentum to the reaction at A. (Remember that the K^- and the proton were both at rest initially. You can measure the radius of curvature of a track by drawing a chord of length y from one point on the track to another, and measuring the distance x from the track to the midpoint of the chord; it is easy to show that $R = y^2/8x$, if $y \gg x$.) Identify particle X, for each event.

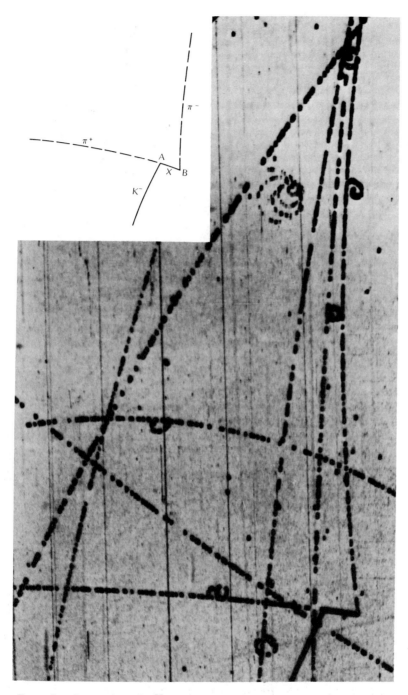

Event 1. *Interaction of a* K⁻ *meson in a hydrogen bubble chamber. The* K⁻ *stops at point A, and interacts with a proton to produce a* π⁺ *and the particle labeled* X. *Particle* X *decays at point B into a* π⁻ *and a neutral particle or particles. [Photographs for Events 1 and 2 provided through the courtesy of H. Whiteside, Univ. of Maryland and CERN, Geneva. First published in Amer. J. Phys.* **34***, 1005 (1966).]*

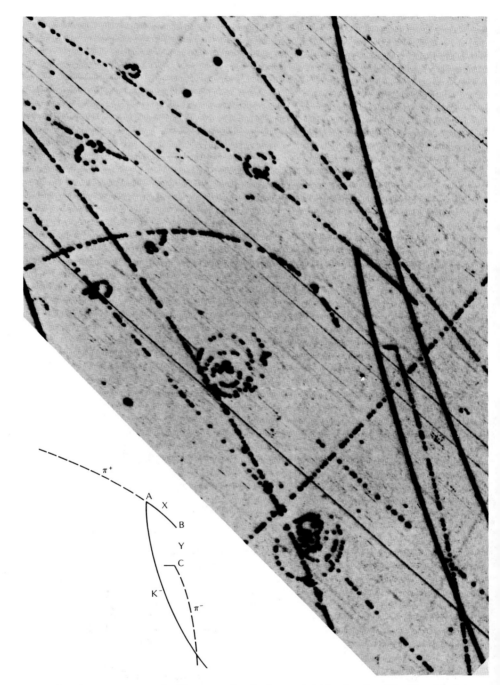

Event 2. *Interaction of a K⁻ meson in a hydrogen bubble chamber. The K⁻ stops at point A, where it interacts with a proton to produce a π⁺ and the particle labeled X. Particle X interacts with a proton at point B to produce two or more neutral particles. One of these, labeled Y, decays at point C into a π⁻ and a proton.*

(d) For event 1, estimate the time of flight of particle X by assuming it travels with constant speed from its point of creation at A to its point of decay at B. (The speed is, of course, known from your calculations of part (c). Is this time consistent with the mean lifetime of X, according to your identification in part (c)? (Figure is twice actual size.)

(e) When particle X decays at point B (event 1) a neutral particle is emitted as well as the π^-; the neutral particle of course leaves no track. Measure the momentum of the π^-, and use conservation of energy and momentum to determine the energy and momentum of the neutral particle. From this information determine the mass of the particle and identify it.

(f) In event 2, particle X interacts with a proton at point B, and only neutral particles are emitted. One of these neutral particles decays at point C into a π^- and a proton. Identify this particle by measuring the momentum of the π^- and using conservation of energy and momentum. (*Hint*: You know the *direction*, but not the magnitude, of the proton's momentum and the neutral particle's momentum.)

7. Write the three-quark combination that corresponds to each point in Fig. 8a, and verify that the degeneracy shown for each of those points is correct.

8. There are nine possible quark–antiquark pairs, but it was stated in Section 16.8 that these pairs form an *octet*. Thus one possible state is left over; this state obviously must be a singlet. We would like to understand why the states are grouped in this way, and to prove that one of the states must be a singlet.

We can easily determine the locations of quark-antiquark pairs on the perimeter of the octet (see figure). But the center can be formed in three ways, from the states $u\bar{u}$, $d\bar{d}$, and $s\bar{s}$. One can form various sets of orthogonal linear combinations of these states, and our task is to show that two orthogonal combinations belong to the octet, and that the third combination which is orthogonal to both of the first two is a singlet.

$$
\begin{array}{ccc}
\underset{\circ}{d\bar{s}} & & \underset{\circ}{u\bar{s}} \\
\\
\underset{\circ}{d\bar{u}} & \underset{\circ}{u\bar{u}}\ \ \underset{}{\circ\ d\bar{d}} & \underset{\circ}{u\bar{d}} \\
& \underset{}{s\bar{s}} & \\
\\
\underset{s\bar{u}}{\circ} & \underset{s\bar{d}}{\circ} &
\end{array}
$$

To understand the meaning of these statements, we must consider the transformation properties of these states. In Section 7.3 we saw that the members of an angular momentum multiplet could be transformed into one another by the raising and lowering operators $J_+ = J_x + iJ_y$ and $J_- = J_x - iJ_y$. Thus J_+ (J_-) transforms a state into a new state whose quantum number m_j is increased (decreased) by one. When $m_j = j$, the state cannot be raised further, and application of J_+ to this state therefore yields zero. Thus we could determine how many states were in a given multiplet simply by repeated application of the operators J_+ and J_-.

We can do the same thing with the "supermultiplets," except that we now have *three* "spin" variables—i-spin, u-spin, and v-spin. These variables all have exactly the same mathematical properties as ordinary angular momentum, so we can again construct raising and lowering operators: T_+, T_-, U_+, U_-, V_+, and V_-; T_+ increases the T_3 eigenvalue by 1, U_+ increases the U_3 eigenvalue by 1, and so forth.

The application of these operators to a composite state follows the usual rule for operating on a product; for example, application of T_+ to a pair of neutrons ($T_3 = -1$) yields

$$T_+nn = n(T_+n) + (T_+n)n$$
$$= np + pn$$

the symmetric state with T_3 equal to zero, because T_+ applied to n yields p. We can perform the same operations to generate the members of the octet, simply by applying rules based on operations on u, d, s, and their antiparticles. The rules, which follow from the i-spin values of these particles, are

$$T_-u = d,$$
$$T_+u = 0,$$
$$T_-d = 0,$$
$$T_+d = u,$$
$$T_-s = 0,$$
$$T_+s = 0,$$

$$T_+\bar{u} = -\bar{d}$$
$$T_-\bar{u} = 0$$
$$T_+\bar{d} = 0$$
$$T_-\bar{d} = -\bar{d}$$
$$T_+\bar{s} = 0$$
$$T_-\bar{s} = 0$$

(a)   Write out analogous rules for U_+, U_-, V_+, and V_-. For example $U_+s = d$, because the value of U_3 is $-\frac{1}{2}$ for s and $+\frac{1}{2}$ for d; ds is a

u-spin doublet. (See Fig. 7a, with the substitutions p→u, n→d, and Λ→s.)

(b)   Show that the normalized superposition of composite states

$$\frac{1}{\sqrt{3}} \ (u\bar{u} + d\bar{d} - 2s\bar{s})$$

must be a singlet in i-spin, in u-spin, and in v-spin. That is, show that application of each of the operators T_+, T_-, U_+, U_-, V_+, or V_- yields zero. Thus this state cannot belong to the octet; it must be a singlet in all three "spin" variables. But there are two other ways to combine these three quark-antiquark pairs to form states which are orthogonal to the above state; these other two combinations must be the members of the octet.

(c)   Show that the superposition

$$\frac{1}{\sqrt{2}} \ (d\bar{d} - s\bar{s})$$

belongs to the i-spin triplet and has $T_3 = 0$. (*Hint*: Apply the T_+ operator to the state with $T_3 = -1$.)

(d)   Show that the state of part (c) is orthogonal to the singlet state of part (b). (*Hint*: The states $u\bar{u}$, $d\bar{d}$, and $s\bar{s}$ may be considered to be orthogonal "unit vectors," just as the states of the K^0 and the $\overline{K^0}$ were treated as unit vectors in Section 16.6. Then the states of parts (b) and (c) are just linear combinations of these vectors, and the "angle" between them is found by taking the scalar product according to the usual rule of vector algebra.)

(e)   Show that the superposition

$$\frac{1}{\sqrt{6}} \ (u\bar{u} + d\bar{d} - 2s\bar{s})$$

is an i-spin singlet and is orthogonal to the states given in parts (b) and (c). Also show that this state is *not* a u-spin singlet.

(f)   Show that the superposition

$$\frac{1}{\sqrt{2}} \ (d\bar{d} - s\bar{s})$$

belongs to the u-spin triplet and has $U_3 = 0$. Show that this state may be written as a superposition of the states of part (c) and part (e).

This last result shows that a particle that belongs to the i-spin triplet does *not* belong to the u-spin triplet. For example, the π^0, the central member of the pion triplet in the meson octet, does *not* form a u-spin triplet with the K^0 and the $\overline{K^0}$. The significance of this point is seen in the masses of the particles. If the mass is linearly dependent on U_3 (as in the decuplet), and if $K^0 - \pi^0 - \overline{K^0}$ formed a u-spin triplet, the π^0 mass would be the average of the masses of the K^0 and the $\overline{K^0}$, which it clearly is not.

Since the same transformation rules apply to all octets, the Σ^0, which is a member of the i-spin triplet in the baryon octet, is not a member of the u-spin triplet. But we can say something about the masses of the baryons, because there is a linear combination of the Σ^0 and the Λ which *is* a member of the u-spin triplet; the result of part (f) shows that this combination is

$$\tfrac{1}{2}\Sigma^0 + \tfrac{1}{2}\sqrt{3}\,\Lambda$$

The mass of this *combination* should be equal to the average mass of the other two members of the triplet, the n and the Ξ^0. Taking the *squares* of the *amplitudes* in the above expression, we find that this mass is $\tfrac{1}{4}$ of the Σ^0 mass plus $\tfrac{3}{4}$ of the Λ mass. Thus we should find that

$$\tfrac{1}{4}m(\Sigma^0) + \tfrac{3}{4}m(\Lambda) = \tfrac{1}{2}m(\mathrm{n}) + \tfrac{1}{2}m(\Xi^0)$$

This formula, the *Gell-Mann–Okubo mass formula*, agrees with observed masses to within 7 MeV, an amount comparable to the electromagnetic mass differences which are not accounted for in the theory.

9. If the proton can decay, the neutron can also decay without conserving baryon number. Suppose the u in a neutron emits an X particle, which is then absorbed by one of the d quarks. What is the final state (a) in terms of quarks and other particles, (b) in terms of observed particles only?

10. Verify the statement that we would be killed by the radioactivity of our own bodies if the proton had a mean lifetime of 10^{17} years or less. (Assume that all of the decaying proton's rest energy is absorbed by the body. See Section 13.5.)

Appendix A

PROBABILITY AND STATISTICS

A knowledge of probability has always been essential in experimental physics, in order that the experimenter can judge the significance of his results. With the advent of quantum mechanics, the theoretician must also be concerned with probabilities, because quantum-mechanical laws can only be formulated as statements of probabilities. Thus the language of probability and statistics is impossible to avoid in a book of this sort. Unfortunately, many students arrive at this stage of their studies with an almost total ignorance of this subject, and for the benefit of these students we present here a brief survey of the fundamental language of this discipline.

Like any other branch of mathematics, the study of probability is based on axioms and definitions which are accepted without proof. As physicists, however, we wish to apply these axioms to the real world, so we shall attempt to make these axioms plausible on physical grounds as we introduce them.

We begin by assigning a number which expresses a likelihood or probability that a certain event will occur. We *define* a probability of 1 to mean that the event is certain to occur, and we define a probability of 0 to mean that the event is certain *not* to occur. Numbers between 0 and 1 are assigned to events which may or may not occur. In order to assign these numbers, we may begin with the assumption that we can identify events which are equally likely. Two events are equally likely if, in our ignorance, we have no reason to expect one to occur rather than the other. To understand the use of the word "ignorance," consider the throw of a single die. We say that a one and a six are equally likely results (events) because we have not examined that particular die, and we have no reason to expect one face of a cube to appear more often than any other. But an inspection of the die might change our evaluation of the probabilities; the die might be "loaded."

The identification of equally likely events permits us to define probabilities in certain specific situations, as follows: Given n mutually exclusive, exhaustive, and equally likely events, we say that the probability of occurrence of any one of these events is $1/n$. In this definition, "mutually exclusive" events are those such that the occurrence of one such event precludes the occurrence of any of the others; exhaustive events are such that one of the events *must* occur.

In applying this definition, we will obtain meaningful results only if each of the events is indeed equally likely. If we have misidentified the events (as in the case of the loaded die), then the probabilities which we calculate will not be useful. We would still have a consistent mathematical structure, but there would be nothing of value to physics in it. For example, consider the probability of obtaining one head and one tail in two flips of a coin. We can enumerate three events which are mutually exclusive and exhaustive : (1) two heads, (2) one head and one tail, and (3) two tails. If these events were equally likely, the probability of each would be 1/3. But the events would be equally likely only if there were no reason to expect one of the events more often than another. This is not true in this situation, because event 2 can occur in two possible ways, whereas event 1 or event 3 can each occur in only one way. We can distinguish *four* equally likely events by taking account of the order in which the head or the tail appears: (1) HH, (2) HT, (3) TH, (4) TT. We have no reason to expect one rather than another of these four, so each has a probability of 1/4.

We have chosen to define fractional probabilities in this *a priori* way, by considering the symmetry of various situations before any trials are made. But it is also possible to define fractional probabilities in an *a posteriori* way, by reference to actual events. If the same initial conditions occur n times, and a given outcome occurs m times, then, in the limits $m \to \infty$ and $n \to \infty$, the ratio m/n can be defined as the probability of that particular outcome, whenever the identical initial conditions recur. We shall not work with this definition directly, but we can use it to illustrate the additive nature of probability; if $p(A + B)$ is the probability that either event A or event B occurs, and the events are mutually exclusive, then

$$p(A + B) = p(A) + p(B)$$

where $p(A)$ is the probability of occurrence of event A, and $p(B)$ is the probability of occurrence of event B. We can see that this follows directly from the *a posteriori* definition of probability; it must be considered a *postulate* when one uses the *a priori* definition, so that the two definitions will lead to the same results. Then, according to either definition, the probability of obtaining one head and one tail in two flips of a coin should be the probability of HT plus the probability of TH, or 1/2.

Compound probability. Let us now consider events which may not be mutually exclusive, but which may occur at the same time or in sequence. Let AB be the event consisting of the occurrence of both events A and B. It is possible that the probability of B is affected by the occurrence of A, so we write $p_A(B)$ as the probability that B occurs, given that A occurs. Then it follows that

$$p(AB) = p(A)p_A(B)$$

If $p(B)$ is not affected by the occurrence of A, then $p_A(B) = p(B)$, and

$$p(AB) = p(A)p(B) \tag{1}$$

for independent events.

Binomial Distribution. Consider n occasions (trials) on which event A can occur, with the probability p of occurence each time. The probability that A will *not* occur on a given trial is $q = 1 - p$. We wish to find the probability that A will occur *exactly* r times in n trials; we denote this probability as $P_r(n, p)$. There are many different possibilities which must be considered in computing this probability. For example, A could occur on each of the first r trials, and on none of the remaining $n - r$ trials; by a simple extension of Eq. (1), we see that the probability of this set of occurrences is $p^r q^{n-r}$. There is an equal probability that event A will occur on any other *specific set* of r trials. Thus the total probability that A will occur on exactly r trials is equal to the product of $p^r q^{n-r}$ times the number of distinct sets of r trials in a total of n trials. Since the number of distinct sets of r objects which can be chosen from n objects is equal to $n!/r!(n - r)!$ (footnote 1, Chapter 11), we have

$$P_r(n, p) = p^r q^{n-r} \frac{n!}{r!(n - r)!} \tag{2}$$

This expression is identical to the $(r + 1)$th term in the expansion of $(q + p)^n$ in powers of p:

$$(p + q)^n = q^n + nq^{n-1}p + \frac{n(n - 1)}{2} q^{n-2}p^2 + \cdots$$

$$+ \frac{n(n - 1) \cdots (n - r + 1)}{r!} q^{n-r}p^r + \cdots + p^n$$

Therefore we may write a generating function $(q + ps)^n$ such that the coefficient of s^r in the expansion of this function is equal to the probability that event A will occur r times in n trials.

Poisson Distribution. The binomial distribution is useful for predicting results of coin-tossing experiments, but as it stands it has limited usefulness in physics. There are situations to which the binomial distribution could be applied, but usually p is very small and n is very large, so that the formula is inconvenient to use. For example, one could count gamma rays from a rare radioactive decay, using a counter whose resolving time is 10^{-6} sec. This means that each microsecond constitutes a "trial" in which there are only two possible outcomes—either a gamma ray is counted or it is not. But the mean number of gamma rays of the type required might be only one per second. Thus in each second there are 10^6 trials, with a probability p equal to 10^{-6} for the event to occur in any given trial. Clearly, the use of the binomial distribution to analyze such an experiment presents computational difficulties.

However, we can modify the binomial distribution to a form appropriate for this situation. We write the generating function, with $q = 1 - p$, as

$$(1 - p + ps)^n = \left(1 - \frac{m}{n} + \frac{ms}{n}\right)^n$$

$$= \left(1 - \frac{m(1 - s)}{n}\right)^n$$

and we take the limit as $n \to \infty$ and $p \to 0$ in such a way that the product np remains equal to the constant m. Using the definition of the exponential,

$$e^x = \lim_{n \to \infty} \left(1 + \frac{x}{n}\right)^n$$

we obtain as a limit for the generating function

$$e^{-m(1-s)} = e^{-m}e^{ms}$$

$$= e^{-m}\left(1 + ms + \frac{m^2s^2}{2!} + \frac{m^3s^3}{3!} + \cdots\right) \tag{3}$$

Since series (3) is still the generating function for the distribution, the coefficient of any given power of s is still equal to the probability of that given number of occurrences of the event. That is, the probability of zero occurrences is e^{-m}, the probability of one occurrence is me^{-m}, and so forth, where m is the mean, or expected number of occurrences. In general, the probability of r occurrences is

$$P_r(m) = \frac{m^r}{r!} e^{-m}$$

A distribution obeying such a law is called a Poisson distribution.

Figure 1 compares the Poisson distribution with two different binomial distributions; for all three distributions shown, the mean value of r is $m = 3$. One of the binomial distributions shows the probabilities of obtaining various numbers of heads in six tosses of a coin; for this distribution, $n = 6$, $p = 1/2$, and of course $m = 3$. The second binomial distribution shows the probabilities of obtaining various numbers of events when $n = 18$ and $p = 1/6$; the event in this case might be the appearance of a 1 on the throw of a single die. You can see how the second distribution more closely resembles the Poisson distribution. You might try some coin-tossing or dice-throwing to verify the accuracy of these figures.

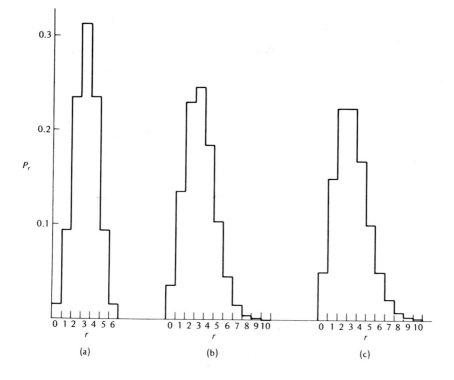

Fig. 1. *Probability P_r of r occurrences of an event in n trials if probability of occurrence is p on each trial. (a) Binomial distribution, $n = 6$, $p = 1/2$. (b) Binomial distribution, $n = 18$, $p = 1/6$. (c) Poisson distribution, limiting case of binomial distribution, with $n \to \infty$, $p \to 0$, but $np = 3$ as in (a) and (b).*

Normal Distribution. The various probabilies are easy to compute for the Poisson distribution when m is small, and one could, in principle, compute them for any m, but for large m there is a shortcut, whose nature is illustrated in Fig. 2. Figure 2 shows the Poisson distribution for $m = 100$, and we see that the distribution may be approximated rather well by the solid curve, which is the curve of the expression

$$P(x) = \frac{1}{10\sqrt{2\pi}} e^{-(x-100)^2/200}$$

This expression is a special case of the general form

$$\frac{1}{\sigma\sqrt{2\pi}} e^{-(x-m)^2/2\sigma^2} \tag{4}$$

A function of this form is called a Gaussian, or a normal error function; a probability distribution which fits this function is called a normal distribution. Thus the probability distribution shown in Fig. 2 is very nearly a normal distribution, with $m = 100$ and $\sigma = 10$. Now we ask the questions inspired by the fit shown in Fig. 2:

(1)   Under what conditions does a Poisson distribution or a binomial distribution closely resemble a normal distribution?

(2)   In general, how does one determine the value of σ which yields the best fit of a normal distribution to a Poisson or a binomial distribution?

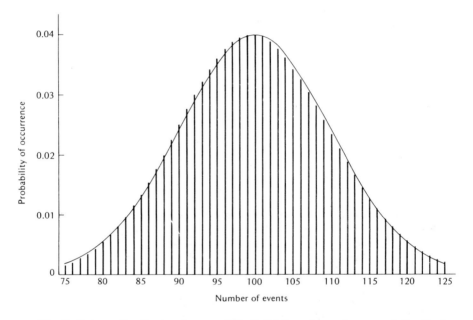

Fig. 2. *Poisson distribution for $m = 100$. Solid line shows the normal distribution which most closely fits this Poisson distribution, with $\sigma = 10$.*

The answer to question 1 is suggested by a comparison of Figs. 1 and 2. The Poisson distribution is much more symmetrical in Fig. 2, where m is larger, and it fits the normal distribution much better as a result. Thus we may attempt to answer question 2 by matching expression (4) to the limit of the Poisson distribution as $m \to \infty$. Because the Poisson distribution itself is only a limiting case of the binomial distribution, we return to Eq. (2), which gives the probability of r occurrences of an event in n trials for the binomial distribution. As we did in obtaining the Poisson distribution, we let $n \to \infty$, but now we keep p fixed, so that np also goes to infinity. We are concerned with the values of P_r for values of r near $r = np$, so we also let $r \to \infty$ and $n - r \to \infty$. This permits us to use Stirling's formula for the factorial:

$$r! \to (2\pi)^{1/2} r^{r+\frac{1}{2}} e^{-r} \qquad \text{as } r \to \infty \tag{5}$$

Substituting from expression (5) into Eq. (2), and canceling and rearranging terms, we obtain

$$P_r(n, p) \to \left(\frac{n}{2\pi r(n - r)}\right)^{1/2} \left(\frac{np}{r}\right)^r \left(\frac{nq}{n - r}\right)^{n-r} \tag{6}$$

We are interested in values of r near $r = np$, where $P_r(n, p)$ should have its maximum value, so we may write the right-hand side of expression (6) in terms of the deviation $\delta = r - np$, to obtain

$$P_r(n, p) \to \left(\frac{n}{2\pi(np + \delta)(nq - \delta)}\right)^{1/2} \frac{1}{\left(1 + \dfrac{\delta}{np}\right)^{\delta + np} \left(1 - \dfrac{\delta}{nq}\right)^{nq - \delta}}$$

We see that the first factor on the right-hand side reduces to $(1/2\pi npq)^{1/2}$ in the limits $np \to \infty$, $nq \to \infty$. The second factor is easiest to evaluate by taking its logarithm, which is

$$-(np + \delta) \ln\left(1 + \frac{\delta}{np}\right) - (nq - \delta) \ln\left(1 - \frac{\delta}{nq}\right)$$

Using the expansion

$$\ln(1 + x) = x - \frac{x^2}{2} + \frac{x^3}{3} - \cdots$$

we find that this expression becomes

$$-(np + \delta)\left(\frac{\delta}{np} - \frac{\delta^2}{2n^2 p^2} + \cdots\right) + (nq - \delta)\left(\frac{\delta}{nq} + \frac{\delta^2}{2n^2 q^2} + \cdots\right)$$

which reduces to

$$-\frac{\delta^2}{2np} - \frac{\delta^2}{2nq}$$

to terms of order δ^2, and since $p + q = 1$, this becomes $-\delta^2/2npq$. Thus we find that

$$P_r(n, p) \to \left(\frac{1}{2\pi npq}\right)^{1/2} e^{-\delta^2/2npq} \qquad (np \to \infty, \delta = r - np)$$

an expression of Gaussian form, with $\sigma^2 = npq$.

Thus the binomial distribution does approach the normal distribution as the mean value $m = np$ becomes large. And since the Poisson distribution is just a limiting case of the binomial distribution, the Poisson also approaches the normal distribution as m becomes large. Figure 2 shows how closely this approximation holds for a Poisson distribution when $m = 100$. Notice that the value of σ is equal to $m^{1/2}$ in this case, because when $q = 1$ (as it does in the Poisson distribution), $\sigma = (npq)^{1/2} = (np)^{1/2} = m^{1/2}$.

The quantity σ is called the standard deviation; it determines the width of the Gaussian curve, and thus it is a measure of the amount by which r is likely to deviate from the mean value m. The total probability of obtaining a value of r such that $m - \sigma < r < m + \sigma$ is equal to 0.68; this may be proven by analysis of the Gaussian curve, or by inspection of tabulated values. For example, for the Gaussian curve of Fig. 2, the sum of the ordinates from $r = 90$ to $r = 110$, inclusive, is equal to 0.706. If we do not include the end points ($r = 90$ and $r = 110$), the sum is equal to 0.658. In the limit as $m \to \infty$, this "end effect" disappears; the probability that r lies between $m + \sigma$ and $m - \sigma$ is a sum which approaches the integral

$$\int_{m-\sigma}^{m+\sigma} \frac{1}{\sigma\sqrt{2\pi}} e^{-(m-r)^2/2\sigma^2} dr = 0.68$$

Similarly, the probability that r will be within two standard deviations of the mean is given by

$$\int_{m-2\sigma}^{m+2\sigma} \frac{1}{\sigma\sqrt{2\pi}} e^{-(r-m)^2/2\sigma^2} dr = 0.95$$

and the probability that r will be within three standard deviations of the mean is equal to 0.997. These probabilities, and those for many intermediate points, are tabulated in the *Handbook of Chemistry and Physics* (Chemical Rubber Publishing Co., Cleveland, Ohio), under the heading "Normal Curve of Error." The tabulation is done for a standard curve

$$\phi(x) = \frac{1}{\sqrt{2\pi}} e^{-x^2/2}$$

for which $\sigma = 1$ and $m = 0$; to obtain results for other values of σ and m, one simply expresses $r - m$ as some multiple of σ. For example, the ordinate for $r = 110$, with $m = 100$ and $\sigma = 10$, is found at $x = 1$, because $(r - m)/\sigma$ is equal to 1. Of course the tabulated ordinate must be divided by the value of σ, because the general curve contains a factor of $1/\sigma$.

As an exercise, let us use the binomial formula to compute the probability of 8 occurrences of an event in 16 trials, when $p = 1/2$, and let us compare this result to the result that one would obtain from the normal curve of error, to test the accuracy of the "normal approximation" for this small value of n. According to the binomial formula (2)

$$P_8(16, \tfrac{1}{2}) = (\tfrac{1}{2})^8(\tfrac{1}{2})^8 \frac{16!}{8!\,8!}$$

$$= (\tfrac{1}{2})^{16}\left(\frac{16 \cdot 15 \cdot 14 \cdot 13 \cdot 12 \cdot 11 \cdot 10 \cdot 9}{8 \cdot 7 \cdot 6 \cdot 5 \cdot 4 \cdot 3 \cdot 2}\right)$$

$$= (\tfrac{1}{2})^{15}(15 \cdot 13 \cdot 11 \cdot 3)$$

$$\simeq 0.196$$

To compare with the tabulated value, we look up the ordinate for $x = 0$ (because r in this case is equal to m), and we divide by 2, because $\sigma = \sqrt{npq} = \sqrt{16 \cdot \tfrac{1}{2} \cdot \tfrac{1}{2}} = 2$. The result is 0.199, which differs by less than 2 percent from the accurate value computed above.

Obviously, this is a rather small value of n, so we should not expect the normal approximation to give extremely good results, and the fit is poorer at the "tail" of the distribution. For example, at $r = 14$, we have $P_{14}(16, \tfrac{1}{2}) = 16 \cdot 15(\tfrac{1}{2})^{17} = 0.0019$, whereas the normal curve gives a probability of about 0.0022 (about 15 percent too large) at $x = (r - m)/\sigma = 3$. If we wish to have accuracy of order 10 per cent for values of r out to three standard deviations from the mean, we cannot use the normal approximation unless m is greater than 100. (Notice the deviations of the normal curve from the Poisson distribution in Fig. 2.) Nevertheless, the normal distribution provides a useful rule of thumb for determining the range of values of r which are likely to occur—that is, the values of r which an experimenter should accept before he begins to suspect that his equipment is faulty.

As an example of the use of the normal distribution in an experiment, consider the number of gamma rays recorded by a detector in successive one-minute intervals. In twelve intervals, the detector counts 424, 398, 414, 396, 395, 410, 520, 394, 391, 414, 362, 422. The mean value is slightly over 400/min, and the standard deviation should be the square root of this, or about 20. We do indeed find that ten of the twelve results are within two standard deviations of the mean. But the result of 520 is more than five standard deviations above the mean. According to the normal distribution, the probability that r will be five or more standard deviations from the mean is less than 10^{-6}. (This is the *total* of *all* values of P_r for $r > m + 5\sigma$ or $r < m - 5\sigma$.) Of course an event with a probability of less than one in a million can occur, so we might hesitate to throw away this number, for fear of biasing the data. But no matter how well run any laboratory is, there is a probability of greater than 10^{-6} that something might go wrong with the apparatus or the experimental conditions. So further investigation into a possible mishap is warranted. In this case, the investigation happens to show that, during the minute when 520 gammas were counted, a graduate student carried a hot radioactive source past the detector! Therefore we cheerfully discard the result of 520.

Appendix B

DERIVATION OF THE BRAGG SCATTERING LAW

We wish to derive the law (Section 4.1)

$$2d \sin \theta = n\lambda \tag{1}$$

giving the angle θ at which radiation is strongly reflected from a crystal lattice, where d is the distance between consecutive parallel planes of atoms in the crystal, θ is the angle which the incident and the reflected radiation both make with these planes (specular reflection), n is an integer, and λ is the wavelength of the radiation.

For simplicity, we shall derive the law only for the case of a simple orthorhombic (rectangular) lattice (Fig. 1). We assume that one crystal plane coincides with the plane of the figure, and that the direction of the incident radiation lies in this plane. Contrary to the more usual treatment, we do not

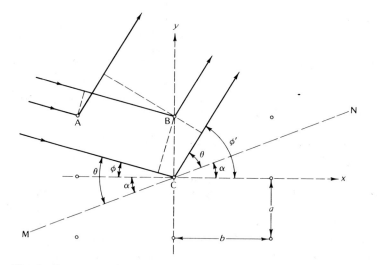

Fig. 1. *Scattering of rays from atoms* A, B, *and* C *in a rectangular array;* MN *is constructed so that the angle of incidence* θ *is equal to the angle of reflection.*

assume that the angle of scattering is equal to the angle of incidence with respect to any particular atomic plane in the crystal; rather, we *prove* that this is the case, by applying the condition that there must be constructive interference between rays scattered into a given direction if reflection is to occur.[1]

Consider the rays scattered from points A, B, and C. The ray scattered from A is in phase with the ray scattered from B if

$$b(\cos \phi - \cos \phi') = k\lambda \qquad (2)$$

where k is an integer. When this condition is satisfied, *all* rays scattered in this direction from atoms in the same horizontal row are in phase.

Similarly, the ray scattered from B is in phase with the ray scattered from C if

$$a(\sin \phi + \sin \phi') = l\lambda \qquad (3)$$

where l is another integer, and a, the distance between B and C, is in general not equal to b. Radiation scattered from all atoms is then in phase if and only if it is incident at angle ϕ and scattered at angle ϕ' such that ϕ and ϕ' satisfy Eqs. (2) and (3). (Equations (2) and (3) are the standard equations for diffraction by reflection from a grating.) Of course, there are many different sets of values of ϕ and ϕ' which satisfy Eqs. (2) and (3); there is a different set for each pair of values of the integers k and l.

For *any* pair of values of ϕ and ϕ', we can always *construct* a plane which makes equal angles with the incident and reflected rays. Fig. 1 shows such a plane, MN, which makes an angle θ with the direction of the incident radiation, and an equal angle with the direction of the scattered (reflected) radiation. Plane MN makes an angle α with the horizontal, such that

$$\theta = \phi + \alpha = \phi' - \alpha \qquad (4)$$

Our task now is to show that

1. the plane MN, constructed to go through atom C, also contains many other atoms; that is, it is one of the crystal planes illustrated in Fig. 2; and
2. the angle θ obeys Eq. (1), if d is the distance between any two successive atomic planes parallel to the plane MN.

We prove point 1 simply by dividing Eq. (2) by Eq. (3); after we use Eq. (4) to eliminate ϕ and ϕ', the result is

$$\tan \alpha = \frac{ka}{lb}$$

Thus the plane MN passes through points whose vertical displacements from C are integral multiples of a, and whose horizontal displacements from C are integral multiples of b. Atoms of the lattice are located at all such points.

---

[1] This derivation is based upon the treatment given by L. R. B. Elton and D. F. Jackson. *Amer. J. Phys.* **34**, 1036 (1966).

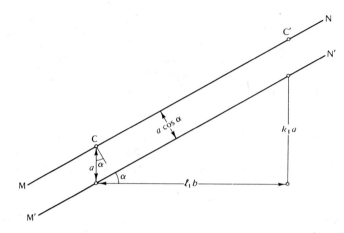

Fig. 2. *Parallel atomic planes for specular reflection. Other parallel atomic planes lie between MN and M'N'.*

To prove point 2, we let n be the largest integer which is a common factor of k and l, so that

$$k = nk_1$$

and

$$l = nl_1$$

where k_1 and l_1 are also integers. Equations (2) and (3) may now be written (with the help of Eq. (4))

$$2b \sin \theta \sin \alpha = nk_1\lambda \tag{5}$$

$$2a \sin \theta \cos \alpha = nl_1\lambda \tag{6}$$

It is very simple to put both of these equations into the *form* of Eq. (1); we simply let

$$d = \frac{b \sin \alpha}{k_1} = \frac{a \cos \alpha}{l_1} \tag{7}$$

but it remains to be shown that d here is actually the distance between plane MN and the next parallel atomic plane. We can show this by the following geometrical constructions and arguments:

(a) If the origin of coordinates is at C, with x and y axes as in Fig. 1, the *nearest neighbor* to atom C in plane MN is the atom located at C' (Fig. 2), whose (x, y) coordinates are (l_1b, k_1a).[2] *Proof*: Because the structure is peri-

[2] Throughout these arguments we refer only to atoms lying in the plane of the paper (Figs. 1 and 2).

odic, all atoms in MN are located at coordinates that are integral multiples of the coordinates of the nearest atom. Let us *assume* that the nearest atom is *not* the one at $(l_1 b, k_1 a)$. Then it must follow that $l_1 = n' l_2$ and $k_1 = n' k_2$, and k_1 and l_1 have a common divisor n'. But by definition, k_1 and l_1 have no common divisor. Thus the assumption leads to a contradiction, and the atom at $(l_1 b, k_1 a)$ *is* the nearest atom.

(b)   We construct a second plane $M'N'$, which is parallel to MN and which passes through the first atom directly beneath C. We consider the parallelogram bounded by MN, $M'N'$, and the vertical planes through C and C'. The number of atoms that lie inside this parallelogram is $l_1 - 1$. *Proof*: One and only one atom of each vertical plane between C and C' lies within the parallelogram, because the vertical atomic spacing is equal to the vertical distance a between MN and $M'N'$, and no atoms in these planes lie *on* MN or $M'N'$. (The latter point follows from (a).) The number of vertical planes must equal $l_1 - 1$, because they divide the distance from C to C' into l_1 equal parts. Therefore there are $l_1 - 1$ atoms in this region.

(c)   Through each of the atoms inside the parallelogram we can construct another plane parallel to MN. Each such plane contains only one atom within the parallelogram. *Proof*: Because of the symmetry of the lattice, the distance between nearest neighbours in each plane is equal to the distance between C and C'. Therefore the nearest neighbor (in the same plane) for each atom inside the parallelogram must lie outside the parallelogram, and so only one atom in the plane lies inside the parallelogram.

(d)   From (b) and (c) it follows that there are $l_1 - 1$ planes parallel to MN inside the parallelogram. These planes divide the parallelogram into l_1 smaller parallelograms; all of these smaller parallelograms are of equal width, because of the symmetry of the lattice. Therefore the perpendicular distance between these planes is equal to the perpendicular distance between MN and $M'N'$, divided by l_1, or $(a \cos \alpha)/l_1$.

Thus we have shown that point 2 is correct, and Eq. (1) is obeyed, with d equal to the distance between successive atomic planes. By making use of the relation

$$\tan \alpha = \frac{k_1 a}{l_1 b}$$

we could also show that Eq. (7) may be written

$$d = \frac{ba}{(k_1^2 a^2 + l_1^2 b^2)^{1/2}} \qquad (8)$$

Appendix C

SOLUTION OF THE RADIAL EQUATION FOR THE HYDROGEN ATOM OR HYDROGENLIKE ION

The equation is Eq. (5), Chapter 7, which in cgs units becomes

$$\frac{d^2\,(rR)}{d^2r} = -\frac{2m_r}{\hbar^2}\left[E_n + \frac{Ze^2}{r} - \frac{l(l+1)\hbar^2}{2m_r r^2}\right](rR) \tag{1}$$

where the subscript n has been introduced to distinguish between the different possible energy levels.

We wish to find solutions of Eq. (1) for bound states, for which E_n is negative. Following previous notation, we write $\alpha_n = (-2m_r E_n/\hbar^2)^{1/2}$; since E_n is negative, α_n is a real number. We also combine the various constants into a single constant C: $C = 2m_r Ze^2/\hbar^2$. With these substitutions, Eq. (1) becomes

$$\frac{d^2(rR)}{d^2r} = \left[\alpha_n^2 - \frac{C}{r} + \frac{l(l+1)}{r^2}\right](rR) \tag{2}$$

A useful strategy in solving an equation of this sort is to begin by determining the form of the solution at large values of r. Clearly the equation approaches

$$\frac{d^2(rR)}{d^2r} = \alpha_n^2(rR)$$

as $r \to \infty$, so the solution should be proportional to the solution of this equation, namely $e^{-\alpha_n r}$, for large r. We may show that any function of the form $rR = r^q e^{-\alpha_n r}$ satisfies Eq. (2) in the limit $r \to \infty$, for we may write

$$rR = r^q e^{-\alpha_n r}$$

$$\frac{d(rR)}{dr} = -\alpha_n r^q e^{-\alpha_n r} + qr^{q-1}e^{-\alpha_n r} \xrightarrow[r \to \infty]{} -\alpha_n r^q e^{-\alpha_n r}$$

so that

$$\frac{d^2(rR)}{dr^2} \xrightarrow[r \to \infty]{} \alpha_n^2 r^q e^{-\alpha_n r} = \alpha_n^2(rR)$$

which is just the limit of Eq. (2) as $r \to \infty$.

690

Therefore it seems reasonable to *try* the product of $e^{-\alpha_n r}$ and a *power series* in r, as a solution for *all* values of r. Therefore we write

$$rR(r) = g(r)e^{-\alpha_n r} \tag{3}$$

where $g(r)$ is a power series whose form we must now determine. We can obtain a differential equation for $g(r)$ by substituting expression (3) into Eq. (2). After we divide out the common factor $e^{-\alpha_n r}$, the result is

$$\frac{d^2 g}{dr^2} - 2\alpha_n \frac{dg}{dr} + \frac{Cg}{r} - \frac{l(l+1)g}{r^2} = 0 \tag{4}$$

We now substitute the power series

$$
\begin{aligned}
g(r) &= r^s \sum_{q=0}^{\infty} b_q r^q \\
&= b_0 r^s + b_1 r^{s+1} + \cdots + b_q r^{s+q} + \cdots
\end{aligned} \tag{5}
$$

into Eq. (4), and we attempt to find relationships among the constant coefficients b_q in order to determine $g(r)$. We obtain the following series:

$$
\begin{aligned}
\frac{d^2 g}{dr^2} &= s(s-1)b_0 r^{s-2} + (s+1)sb_1 r^{s-1} + \cdots \\
&\quad + (s+q)(s+q-1)b_q r^{s+q-2} + \cdots
\end{aligned}
$$

$$-2\alpha_n \frac{dg}{dr} = -2\alpha_n[sb_0 r^{s-1} + (s+1)b_1 r^s + \cdots + (s+q)b_q r^{s+q-1} + \cdots]$$

$$\frac{Cg}{r} = C[b_0 r^{s-1} + b_1 r^s + b_2 r^{s+1} + \cdots + b_q r^{s+q-1} + \cdots]$$

$$-\frac{l(l+1)g}{r^2} = -l(l+1)[b_0 r^{s-2} + b_1 r^{s-1} + \cdots + b_q r^{s+q-2} + \cdots]$$

If the sum of these four series is to be zero *for all* r, as demanded by Eq. (4), then the sum of the coefficients of each power of r must equal zero. The lowest power, the $(s-2)$ power, appears only twice, in the first series and in the last; the sum of the coefficients of r^{s-2} is

$$s(s-1)b_0 - l(l+1)b_0$$

so if we rule out the possibility $b_0 = 0$ ($b_0 \neq 0$ by definition, because b_0 is the coefficient of the smallest power of r appearing in the series), then we have

$$s(s-1) - l(l+1) = 0$$

This equation is called the indicial equation; it determines the value of s, the lowest power of r appearing in the series. There are two solutions: $s = -l$ or $s = l+1$. But we know that l is positive, and s must be positive, because $R(r)$ must be finite at $r = 0$. Therefore $s = l+1$.

From the coefficients of r^{s+q-1} we can now obtain a relation between the coefficients b_q and b_{q+1}. (To find the coefficient of r^{s+q-1} in the first and the fourth series, simply replace q by $q + 1$ everywhere in the general term shown. This converts the coefficient b_q into b_{q+1}; it produces the next term in the series.) We obtain

$$(s + q + 1)(s + q)b_{q+1} - 2\alpha_n(s + q)b_q + Cb_q - l(l + 1)b_{q+1} = 0$$

which, with the substitution $s = l + 1$, reduces to

$$b_{q+1} = b_q \left[\frac{2\alpha_n(l + 1 + q) - C}{(l + q + 2)(l + q + 1) - l(l + 1)} \right]$$

$$= \left[\frac{2\alpha_n(l + q + 1) - C}{ql + (q + 2)(l + q + 1)} \right] b_q \tag{6}$$

The relation (6) enables us to write out a series solution to Eq. (4), and therefore to solve Eq. (2), for any value of α_n. But the solution that we write for an arbitrary value of α_n may not be acceptable as a wavefunction. The solution must go to zero as r goes to infinity, because there is negligible probability of finding the particle at large r, far beyond the point where its kinetic energy becomes negative (see Chapter 7, Fig. 2.)

It is easy to show that any infinite series in powers of r, whose coefficients obey the relation (6), cannot go to zero as $r \to \infty$. The ratio of successive coefficients in the series approaches $2\alpha_n/q$ as $q \to \infty$; this is the same as the ratio of the coefficients in the series for $e^{+2\alpha_n r}$, and there is a theorem which states that two series for which this ratio is the same have the same behavior as $r \to \infty$. Therefore, since $e^{2\alpha_n r}$ goes to infinity as $r \to \infty$, the infinite series solution to Eq. (4) also goes to infinity as $r \to \infty$.

There is nothing strange about this behavior. The infinite series solution corresponds to an arbitrary value of α_n, or E_n; unless E_n has precisely the right value, the solution *should* blow up, in general, as we showed in our attempts to sketch solutions for the square-well and other potentials.

How do we find a value of E_n which prevents this blowup? We arrange things so that we have a *finite* series solution rather than an infinite series. This is possible if one of the coefficients, b_{q+1}, is zero, for then all succeeding coefficients are also automatically zero. From Eq. (6) we see that b_{q+1} becomes zero (with $b_q \neq 0$) when

$$2\alpha_n(l + q + 1) - C = 0 \tag{7}$$

This provides the condition on α_n which we are seeking;

$$\alpha_n = \frac{C}{2(l + q + 1)}$$

or, from the definitions of α_n and C:

$$\left(\frac{-2m_r E_n}{\hbar^2}\right)^{1/2} = \frac{2m_r Z e^2}{2\hbar^2(l + q + 1)}$$

or

$$E_n = \frac{-m_r Z^2 e^4}{2\hbar^2 n^2} \tag{8}$$

where we have defined a new quantum number, the radial quantum number $n = l + 1 + q$. If $q = 0$, $n = l + 1$; the possible values of n are therefore $l + 1$, $l + 2$, $l + 3$, ... When $q = 0$, $b_1 = 0$, so there is just one term in the series for $g(r)$, or $R(r)$; when $q = 1$, there are two terms. In general the quantum number n is equal to $l + t$, where t is the number of terms in the radial wavefunction, $R(r)$.

Equation (8), which is identical to Bohr's formula, gives the energy levels for any one-electron system: hydrogen atom, singly ionized helium, doubly ionized lithium, and so on, when one uses the appropriate Z value for the nuclear charge.

We may use Eq. (6) to write down some of the radial wavefunctions, now that we know the values of α_n. When $n = l + 1$, the radial function contains just one term [and we do not need Eq. (6)]; according Eqs. (3) and (5).

$$rR(r) = g(r)e^{-\alpha_n r} = b_0 r^s e^{-\alpha_n r}$$

or

$$R(r) = b_0 r^l e^{-\alpha_n r} \qquad (n = l + 1)$$

where b_0 is still arbitrary and may be chosen to normalize the wavefunction.

We could now substitute for the value of α_n, but we defer that for a moment until we look at a few more solutions. When $n = l + 2$, there are two terms, so we know that only b_0 and b_1 are nonzero. From Eq. (6) we have (first with $q = 0$ and then with $q = 1$):

$$b_1 = \frac{2\alpha_n(l + 1) - C}{2(l + 1)} b_0 \tag{9}$$

and

$$b_2 = \frac{2\alpha_n(l + 2) - C}{l + 3(l + 2)} b_1 = 0$$

or

$$2\alpha_n = \frac{C}{l + 2}$$

so that Eq. (9) becomes

$$b_1 = -\frac{C}{2(l+1)(l+2)} b_0$$

Therefore we may write the radial function $rR(r) = (b_0 r^s + b_1 r^{s+1})e^{-\alpha_n r}$ in terms of α_n, l, and C as

$$R(r) = b_0 r^l \left[1 - \frac{Cr}{2(l+1)(l+2)} \right] e^{-\alpha_n r} \qquad (n = l + 2)$$

where again b_0 is determined by normalization. You can now easily work out any other case for yourself.

In tabulating these functions, it is convenient to write the constants in terms of the first Bohr radius for hydrogen, except that we use the reduced mass m_r in the formula [Chapter 3, (Eq. 16)] instead of the electron mass; This "reduced Bohr radius," which physically is the distance between the proton and electron in the ground state of the Bohr hydrogen atom, is $a_0' = \hbar^2/m_r e^2$ in cgs units; in terms of a_0', we have $C = 2Z/a_0'$ and $\alpha_n = C/2n = Z/na_0'$. The first few radial functions are shown in Table 1.

Table 1

$n = l + 1$	$n = l + 2$
$R_{10} = b_0 e^{-Zr/a_0'}$	$R_{20} = b_0 \left(1 - \frac{Zr}{2a_0'}\right) e^{-Zr/2a_0'}$
$R_{21} = b_0 r e^{-Zr/2a_0'}$	$R_{31} = b_0 r \left(1 - \frac{Zr}{6a_0'}\right) e^{-Zr/3a_0'}$
$R_{32} = b_0 r^2 e^{-Zr/3a_0'}$	$R_{42} = b_0 r^2 \left(1 - \frac{Zr}{12a_0'}\right) e^{-Zr/4a_0'}$

Appendix D

TABLE OF ATOMIC SPECIES

This table gives, for each atomic species with well-measured characteristics,
1. the mass excess (see explanation below) of the *neutral atom*, in MeV/c²,
2. the spin and parity I^P of the nucleus,
3. the magnetic moment μ of the nucleus, in nuclear magnetons,
4. the % abundance (as atomic percentage of all isotopes of that element) if the species occurs naturally, and/or the half life if the species is radioactive.

The mass excess is the difference between the mass of the neutral atom and the mass of A atomic mass units (u), for a species of mass number A. To find the mass of the neutral atom, add Au to the mass excess. For example, the mass of ⁴He, in units of MeV/c^2, is $2.4249 + Au$, or $2.4249 + 4 \times (931.5016 \pm 0.0026)$, since $uc^2 = 931.5016 \pm 0.0026$. This means that the rest *energy* of the ⁴He atom is $2.4249 + 4 \times (931.5016 \pm 0.0026)$ MeV, or 3728.43 ± 0.01 MeV.

Data are given for the ground state only, except that in rare cases, where it is not known which of two states is the ground state, values for both states are given side by side. Some long-lived excited states, or *isomeric* states, have half lives that are long enough to make them useful; for example, an excited state of ⁹⁹Tc, with $t_{1/2} = 6$ hours, is used as a tracer in medical studies. For information on such states consult Nuclear Data Tables (Academic Press), from which the *half lives and I^P values* listed here were obtained. Each issue contains a "Cumulated Index to A-Chains", which lists the most recent reference for each value of A. Values listed in this table are taken from the references listed in Volume 34 (1981).

Mass values listed here are from A. H. Wapstra and K. Bos, Atomic Mass Table, Atomic Data and Nuclear Tables **19**, 185 (1977).

Magnetic moments are taken from American Institute of Physics Handbook, 3d Edition (1973); they were originally compiled by G. H. Fuller and V. W. Cohen, Nuclear Data **A5**, 6 (1968).

A	Element (symbol)	Z	Mass excess (MeV/c²)	I^P	μ (nucl. magnetons)	% Abundance or Half-life
1	n	0	8.0714	$\frac{1}{2}+$	−1.91314	10.6 m
	H	**1**	7.2890	$\frac{1}{2}+$	+2.792846	**99.985%**

A	Element (symbol)	Z	Mass excess (MeV/c^2)	I^P	μ (nucl. magnetons)	% Abundance or Half-life
2	H	**1**	13.1358	1^+	+0.857406	**0.015%**
3	H	1	14.9499	$\frac{1}{2}^+$	+2.97886	12.3 y
	He	**2**	14.9313	$\frac{1}{2}^+$	−2.12755	**0.00013%**
4	He	2	2.4249			**99.99987%**
5	He	2	11.39	$\frac{3}{2}^-$		
	Li	3	11.68	$\frac{3}{2}^-$		
6	He	2	17.5940	0^+		0.808 s
	Li	**3**	14.0873	1^+	+0.822010	**7.42%**
	Be	4	18.375			
7	He	2	26.111			
	Li	**3**	14.9082	$\frac{3}{2}^-$	+3.256424	**92.58%**
	Be	4	15.7701	$\frac{3}{2}^-$		
	B	5	27.94			53.3 d
8	He	2	31.609	0^+		122 ms
	Li	3	20.9469	2^+	+1.6534	0.84 s
	Be	4	4.9418	0^+		≈ 2 × 10^{-16} s
	B	5	22.9219	2^+		770 ms
	C	6	35.085			
9	Li	3	24.9548	$\frac{3}{2}^-$	+3.4359	178 ms
	Be	**4**	11.3480	$\frac{3}{2}^-$	−1.1778	**100%**
	B	5	12.4161			≈ 8 × 10^{-19} s
	C	6	28.9121			127 ms
10	Be	4	12.6076	0^+		1.6 × 10^6 y
	B	**5**	12.0517	3^+	+1.80065	**19.78%**
	C	6	15.7029	0^+		19.26 s
11	Li	3	40.94			8.5 ms
	Be	4	20.176	$\frac{1}{2}^+$		13.81 s
	B	**5**	8.6679	$\frac{3}{2}^-$	+2.688637	**80.22%**
	C	6	10.6500	$\frac{3}{2}^-$	−0.964	20.40 m
	N	7	25.23			
12	Be	4	25.030	0^+		
	B	5	13.3695	1^+	+1.00306	20.20 ms
	C	**6**	0.0 (standard)	0^+		**98.89%**
	N	7	17.3380	1^+	+0.4571	11.00 ms
	O	8	32.07	0^+		

A	Element (symbol)	Z	Mass excess (MeV/c^2)	I^P	μ (nucl. magnetons)	% Abundance or Half-life
13	B	5	16.562		+3.1771	17.36 ms
	C	**6**	3.1250	$\frac{1}{2}^-$	+0.7020	**1.11%**
	N	7	5.3456	$\frac{1}{2}^-$	−0.32212	9.961 m
	O	8	23.105	$\frac{3}{2}^-$		8.90 ms
14	B	5	23.657			16.1 ms
	C	6	3.0199	0^+		5730 y
	N	7	2.8634	1^+	+0.40376	**99.63%**
	O	8	8.0083	0^+		70.60 s
15	C	6	9.8732	$\frac{1}{2}^+$		2.449 s
	N	7	0.1015	$\frac{1}{2}^-$	−0.2831	**0.37%**
	O	8	2.8554	$\frac{1}{2}^-$	±0.7189	122.24 s
16	C	6	13.693	0^+		747 ms
	N	7	5.6816	2^-		7.13 s
	O	8	−4.7370	0^+		**99.759%**
	F	9	10.692	$(1)^-$		$\approx 10^{-19}$ s
	Ne	10	24.11	(0^+)		
17	N	7	7.870	$\frac{1}{2}^-$		4.169 s
	O	8	−0.8099	$\frac{5}{2}^+$	−1.89370	**0.037%**
	F	9	1.9517	$\frac{5}{2}^+$	+4.722	64.50 s
	Ne	10	16.478	$\frac{1}{2}^-$		109.0 ms
18	N	7	13.274			0.63 s
	O	8	−0.7830	0^+		**0.204%**
	F	9	0.8725	1^+		109.77 m
	Ne	10	5.319	0^+		1.672 s
19	N	7	15.60			
	O	8	3.3314	$\frac{5}{2}^+$		26.91 s
	F	**9**	−1.4874	$\frac{1}{2}^+$	+2.628866	**100%**
	Ne	10	1.7509	$\frac{1}{2}^+$	−1.887	17.22 s
	Na	11	12.930			
20	O	8	3.799	0^+		13.57 s
	F	9	−0.0171	2^+	+2.094	11.0 s
	Ne	**10**	−7.0430	0^+	$<2 \times 10^{-4}$	**90.92%**
	Na	11	6.844	2^+	+0.3694	446 ms
	Mg	12	17.568			

A	Element (symbol)	Z	Mass excess (MeV/c²)	I^P	μ (nucl. magnetons)	% Abundance or Half-life
21	O	8	8.120			
	F	9	−0.047	$\frac{5}{2}^+$		4.32 s
	Ne	**10**	−5.7331	$\frac{3}{2}^+$	−0.66176	**0.257%**
	Na	11	−2.1858	$\frac{3}{2}^+$	+2.39	22.48 s
	Mg	12	10.912			122 ms
22	O	8	9.490			
	F	9	2.826	3^+		4.23 s
	Ne	**10**	−8.0261	0^+	≈0	**8.82%**
	Na	11	−5.1840	3^+	+1.746	2.602 y
	Mg	12	−0.3941			3.86 s
23	F	9	3.35			2.23 s
	Ne	10	−5.1551	$\frac{5}{2}^+$	−1.08	37.2 s
	Na	**11**	−9.5296	$\frac{3}{2}^+$	+2.21751	**100%**
	Mg	12	−5.4706	$\frac{3}{2}^+$		11.32 s
	Al	13	6.768			470 ms
24	Ne	10	−5.949	0^+		3.38 m
	Na	11	−8.4175	4^+	+1.690	15.020 h
	Mg	**12**	−13.9306	0^+	≈0	**78.70%**
	Al	13	−0.052	4^+		2.07 s
	Si	14	10.74			
25	Ne	10	−2.15			600 ms
	Na	11	−9.357	$\frac{5}{2}^+$		60 s
	Mg	**12**	−13.1908	$\frac{5}{2}^+$	−0.85512	**10.13%**
	Al	13	−8.9129	$\frac{5}{2}^+$		7.18 s
	Si	14	3.824			220 ms
26	Na	11	−6.888			1.07 s
	Mg	**12**	−16.2124	0^+	≈0	**11.17%**
	Al	13	−12.2076	5^+		7.2×10^5 y
	Si	14	−7.1431	0^+		2.21 s
27	Na	11	−5.63			302 ms
	Mg	12	−14.5850	$\frac{1}{2}^+$		9.46 m
	Al	**13**	−17.1943	$\frac{5}{2}^+$	+3.64140	**100%**
	Si	14	−12.3853	$\frac{5}{2}^+$		4.16 s
28	Na	11	−1.13			30.5 ms
	Mg	12	−15.0164	0^+		20.90 h
	Al	13	−16.8482	3^+		2.241 m
	Si	**14**	−21.4912	0^+		**92.21%**
	P	15	−7.1595	$(2,3)^+$		270 ms
	S	16	4.19			

A	Element (symbol)	Z	Mass excess (MeV/c²)	I^P	μ (nucl. magnetons)	% Abundance or Half-life
29	Na	11	2.66			43 ms
	Mg	12	−10.75			1.4 s
	Al	13	−18.212	$\frac{5}{2}^+$		6.6 m
	Si	**14**	−21.8937	$\frac{1}{2}^+$	−0.55525	**4.70%**
	P	15	−16.9493	$\frac{1}{2}^+$		4.14 s
	S	16	−3.16	$\frac{5}{2}^+$		187 ms
30	Na	11	8.38			53.3 ms
	Al	13	−15.89			3.6 s
	Si	**14**	−24.4317	0^+	**3.09%**	**3.09%**
	P	15	−20.2045	1^+		2.498 m
	S	16	−14.0620	0^+		1.24 s
31	Al	13	−15.10			644 ms
	Si	14	−22.9487	$\frac{3}{2}^+$		2.62 h
	P	**15**	−24.4395	$\frac{1}{2}^+$	+1.13166	**100%**
	S	16	−19.0441	$\frac{1}{2}^+$		2.58 s
	Cl	17	−7.07			
32	Si	14	−24.092	0^+		330 y
	P	15	−24.3047	1^+	−0.2523	14.26 d
	S	**16**	−26.0151	0^+	≈0	**95.0%**
	Cl	17	−13.329	2^+		298 ms
	Ar	18	−2.21			
33	Si	14	−20.57			6.1 s
	P	15	−26.3369	$\frac{1}{2}^+$		25.3 d
	S	**16**	−26.5859	$\frac{3}{2}^+$	+0.64327	**0.76%**
	Cl	17	−21.0030	$\frac{3}{2}^+$		2.511 s
	Ar	18	−9.385	$\frac{1}{2}^+$		173 ms
34	P	15	−24.55	1^+		12.4 s
	S	**16**	−29.9312	0^+		**4.22%**
	Cl	17	−24.4383	0^+		1.526 s
	Ar	18	−18.3792			845 ms
35	P	15	−24.94			47 s
	S	16	−28.8463	$\frac{3}{2}^+$	±1.00	87.5 d
	Cl	**17**	−29.0137	$\frac{3}{2}^+$	+0.82183	**75.53%**
	Ar	18	−23.0489	$\frac{3}{2}^+$	+0.632	1.775 s
	K	19	−11.169			
36	S	16	−30.6659	0^+		
	Cl	17	−29.5218	2^+	+1.28538	3.01×10^5 y
	Ar	**18**	−30.2313	0^+	≈0	**0.337%**
	K	19	−17.426			342 ms
	Ca	20	−6.65			

A	Element (symbol)	Z	Mass excess (MeV/c²)	I^P	μ (nucl. magnetons)	% Abundance or Half-life
37	S	16	−26.908	$\frac{7}{2}^-$		5.05 m
	Cl	**17**	−31.761	$\frac{3}{2}^+$	+0.6841	**24.47%**
	Ar	18	−30.9479	$\frac{3}{2}^+$	+0.95	35.04 d
	K	19	−24.7994	$\frac{3}{2}^+$	+0.204	1.23 s
	Ca	20	−13.164			175 ms
38	S	16	−26.862	0^+		2.83 h
	Cl	17	−29.7980	2^-		37.24 m
	Ar	**18**	−34.7150	0^+		**0.063%**
	K	19	−28.8020	3^+	+1.373	7.64 m
	Ca	20	−22.060	0^+		450 ms
39	Cl	17	−29.803	$\frac{3}{2}^+$		55.6 m
	Ar	18	−33.241	$\frac{7}{2}^-$	−1.3	269 y
	K	**19**	−33.8062	$\frac{3}{2}^+$	+0.3914	**93.10%**
	Ca	20	−27.282	$\frac{3}{2}^+$		860 ms
40	Cl	17	−27.54			1.35 m
	Ar	**18**	−35.0402	0^+	≈0	**99.60%**
	K	**19**	−33.5352	4	−1.2981	1.28×10^9y; **0.0118%**
	Ca	**20**	−34.8468	0^+	≈0	**96.97%**
	Sc	21	−20.527	4^-		182 ms
	Ti	22	−9.04			
41	Cl	17	−27.40			34 s
	Ar	18	−33.0677	$\frac{7}{2}^-$		1.83 h
	K	**19**	−35.5597	$\frac{3}{2}^+$	+0.21483	**6.88%**
	Ca	20	−35.1385	$\frac{7}{2}^-$	−1.59460	1.03×10^5 y
	Sc	21	−28.6435	$\frac{7}{2}^-$		596 ms
	Ti	22	−15.78			80 ms
42	Ar	18	−34.42	0^+		33 y
	K	19	−35.0228	2^-	−1.141	12.36 h
	Ca	**20**	−38.5439	0^+		**0.64%**
	Sc	21	−32.1207	0^+		681 ms
	Ti	22	−25.122			199 ms
43	Cl	17	−23.14			
	Ar	18	−31.98			5.4 m
	K	19	−36.588	$\frac{3}{2}^+$	±0.163	22.3 h
	Ca	**20**	−38.4054	$\frac{7}{2}^-$	−1.3172	**0.145%**
	Sc	21	−36.1850	$\frac{7}{2}^-$	+4.62	3.89 h
	Ti	22	−29.324	$\frac{7}{2}^-$		510 ms

A	Element (symbol)	Z	Mass excess (MeV/c^2)	I^P	μ (nucl. magnetons)	% Abundance or Half-life
44	Ar	18	−32.271			11.9 m
	K	19	−35.807	2$^-$		22.1 m
	Ca	**20**	−41.4660	0$^+$		**2.06%**
	Sc	21	−37.8107	2$^+$	+2.56	3.93 h
	Ti	22	−37.5462	0$^+$		47 y
45	Ar	18	−29.730			21 s
	K	19	−36.611	$\frac{3}{2}^+$	±0.173	20 m
	Ca	20	−40.8096	$\frac{7}{2}^-$		163 d
	Sc	**21**	−41.0665	$\frac{7}{2}^-$	+4.75626	**100%**
	Ti	22	−39.0040	$\frac{7}{2}^-$	±0.095	3.08 h
	V	23	−31.879			
	Cr	24	−19.460	$(\frac{7}{2}^-)$		50 ms
46	Ar	18	−29.730	0$^+$		8 s
	K	19	−35.420	(2^-)		107 s
	Ca	**20**	−43.1382	0$^+$		**0.0033%**
	Sc	21	−41.7556	4$^+$	+3.04	83.83 d
	Ti	**22**	−44.1227	0$^+$		**7.93%**
	V	23	−37.0708	0$^+$		422.33 ms
	Cr	24	−29.461	0$^+$		260 ms
47	K	19	−35.698	$\frac{1}{2}^+$		17.5 s
	Ca	20	−42.3429	$\frac{7}{2}^-$		4.536 d
	Sc	21	−44.3305	$\frac{7}{2}^-$	+5.34	3.351 d
	Ti	**22**	−44.9310	$\frac{5}{2}^-$	−0.7883	**7.28%**
	V	23	−42.0011	$\frac{3}{2}^-$		32.6 m
	Cr	24	−34.618	$\frac{3}{2}^-$		460 ms
48	K	19	−32.220	(2^-)		6.9 s
	Ca	**20**	−44.216	0$^+$		**0.18%**; > 2 × 10^{6}y
	Sc	21	−44.498	6$^+$		43.7 h
	Ti	**22**	−48.4877	0$^+$		**73.94%**
	V	23	−44.4728	4$^+$		15.971 d
	Cr	24	−42.818	0$^+$		22.96 h
49	Ca	20	−41.286	$\frac{3}{2}^-$		8.716 m
	Sc	21	−46.555	$\frac{7}{2}^-$		57.4 m
	Ti	**22**	−48.5587	$\frac{7}{2}^-$	−1.1036	**5.51%**
	V	23	−47.9569	$\frac{7}{2}^-$	±4.46	330 d
	Cr	24	−45.3290	$\frac{5}{2}^-$	±0.48	42.09 m
	Mn	25	−37.613	$(\frac{5}{2}^-)$		
	Fe	26	−24.470	$(\frac{7}{2}^-)$		75 ms

A	Element (symbol)	Z	Mass excess (MeV/c^2)	I^P	μ (nucl. magnetons)	% Abundance or Half-life
50	Ca	20	−39.572	0^+		13.9 s
	Sc	21	−44.539	5^+		1.708 m
	Ti	**22**	−51.4321	0^+		**5.34%**
	V	**23**	−49.2193	6^+	+3.347	**0.24%**; $>4 \times 10^{16}$y
	Cr	**24**	−50.2580	0^+		**4.31%**
	Mn	25	−42.6257	0^+		283.2 ms
51	Sc	21	−43.220	$(\frac{7}{2}-)$		12.4 s
	Ti	22	−49.7330	$\frac{3}{2}-$		5.76 m
	V	**23**	−52.1991	$\frac{7}{2}-$	+5.148	**99.76%**
	Cr	24	−51.4478	$\frac{7}{2}-$	±0.94	27.704 d
	Mn	25	−48.2398	$\frac{5}{2}-$	±3.57	46.2 m
	Fe	26	−40.228	$(\frac{5}{2}-)$		270 ms
52	Ti	22	−49.469	0^+		1.7 m
	V	23	−51.4389	3^+		3.75 m
	Cr	**24**	−55.4153	0^+		**83.76%**
	Mn	25	−50.7042	6^1	±3.05	5.591 d
	Fe	26	−48.332	0^+		8.275 h
53	Ti	22	−46.890	$(\frac{3}{2})^-$		32.7 s
	V	23	−51.863	$\frac{7}{2}-$		1.61 m
	Cr	**24**	−55.2837	$\frac{3}{2}-$	−0.47434	**9.55%**
	Mn	25	−54.6874	$\frac{7}{2}-$	±5.01	3.7×10^6 y
	Fe	26	−50.9442	$\frac{7}{2}-$		8.51 m
	Co	27	−42.640	$(\frac{7}{2}-)$		262 ms
	Ni	28	−29.410	$(\frac{7}{2}-)$		45 ms
54	V	23	−49.930			49.8 s
	Cr	**24**	−56.9313	0^+		**2.38%**
	Mn	25	−55.5543	3^+	±3.29	312.5 d
	Fe	**26**	−56.2514	0^+		**5.82%**
	Co	27	−48.0096	0^+		193.23 ms
55	Cr	24	−55.1063	$\frac{3}{2}-$		3.55 m
	Mn	**25**	−57.7100	$\frac{5}{2}-$	+3.444	**100%**
	Fe	26	−57.4786	$\frac{3}{2}-$		2.7 y
	Co	27	−54.0239	$\frac{7}{2}-$	±4.3	17.54 h
	Ni	28	−45.334	$(\frac{7}{2}-)$		5 s

A	Element (symbol)	Z	Mass excess (MeV/c²)	I^P	μ (nucl. magnetons)	% Abundance or Half-life
56	Cr	24	−55.265	0^+		5.94 m
	Mn	25	−56.9088	3^+	+3.218	2.5785 h
	Fe	**26**	−60.6041	0^+	+1.1	**91.66%**
	Co	27	−56.0367	4^+	±3.83	78.76 d
	Ni	28	−53.902	0^+		6.10 d
57	Mn	25	−57.487	$\frac{5}{2}^-$		1.61 m
	Fe	**26**	−60.1790	$\frac{1}{2}^-$	+0.0902	**2.19%**
	Co	27	−59.3424	$\frac{7}{2}^-$	±4.62	270.9 d
	Ni	28	−56.099	$\frac{3}{2}^-$		36.08 h
	Zn	30	−32.630	$(\frac{7}{2}^-)$		40 ms
58	Mn	25	−56.210	3^+		65.3 s
	Fe	**26**	−62.1518	0^+		**0.33%**
	Co	27	−59.8440	2^+	+4.03	70.78 d
	Ni	28	−60.2243	0^+		**67.88%**
	Cu	29	−51.6617	1^+		3.204 s
59	Mn	25	−55.478			
	Fe	26	−60.6614	$\frac{3}{2}^-$		45.1 d
	Co	**27**	−62.2264	$\frac{7}{2}^-$	+4.62	**100%**
	Ni	28	−61.1529	$\frac{3}{2}^-$		7.5×10^4 y
	Cu	29	−58.3433	$\frac{3}{2}^-$		82.0 s
60	Fe	26	−61.437	0^+		1×10^5 y
	Co	27	−61.6466	5^+	±3.78	5.26 y
	Ni	**28**	−64.4702	0^+		**26.23%**
	Cu	29	−58.3433	2^+	+1.22	23.2 m
	Zn	30	−54.184	0^+		2.38 m
61	Fe	26	−59.010	$(\frac{3}{2})^-$		5.98 m
	Co	27	−62.8970	$\frac{7}{2}^-$		1.650 h
	Ni	**28**	−64.2191	$\frac{3}{2}^-$	−0.74868	**1.19%**
	Cu	29	−61.9807	$\frac{3}{2}^-$	+2.13	3.408 h
	Zn	30	−56.580	$\frac{3}{2}^-$		89.1 s
62	Fe	26	−58.930	0^+		68 s
	Co	27	−61.504	2^+		1.50 m
	Ni	**28**	−66.7454	0^+		**3.66%**
	Cu	29	−62.796	1^+	−0.38	9.74 m
	Zn	30	−61.169	0^+		9.26 h

A	Element (symbol)	Z	Mass excess (MeV/c^2)	I^P	μ (nucl. magnetons)	% Abundance or Half-life
63	Co	27	−61.850	$(\frac{7}{2})^-$		27.4 s
	Ni	28	−65.5126	$\frac{1}{2}^-$		100.1 y
	Cu	**29**	−65.5785	$\frac{3}{2}^-$	+2.223	**69.09%**
	Zn	30	−62.2111	$\frac{3}{2}^-$	−0.282	38.1 m
	Ga	31	−56.690	$\frac{3}{2}^-, \frac{5}{2}^-$		32.4 s
64	Co	27	−59.791	1^+		300 ms
	Ni	**28**	−67.0979	0^+		**1.08%**
	Cu	29	−65.4230	1^+	±0.216	12.701 h
	Zn	**30**	−66.0012	0^+	≈0	**48.89%**
	Ga	31	−58.836	0^+		2.630 m
	Ge	32	−54.430	0^+		63.7 s
65	Ni	28	−65.1245	$\frac{5}{2}^-$		2.520 h
	Cu	**29**	−67.2615	$\frac{3}{2}^-$	+2.382	**30.91%**
	Zn	30	−65.9096	$\frac{5}{2}^-$	+0.7692	244.1 d
	Ga	31	−62.6538	$\frac{3}{2}^-$		15.2 m
	Ge	32	−56.410	$\frac{3}{2}^-, \frac{5}{2}^-$		30.9 s
66	Ni	28	−66.021	0^+		54.6 h
	Cu	29	−66.2567	1^+	±0.283	5.10 m
	Zn	**30**	−68.8983	0^+	≈0	**27.81%**
	Ga	31	−63.7233	0^+		9.40 h
	Ge	32	−61.621	0^+		2.27 h
67	Ni	28	−63.470			18 s
	Cu	29	−67.305	$\frac{3}{2}^-$		61.88 h
	Zn	**30**	−67.8796	$\frac{5}{2}^-$	+0.8754	**4.11%**
	Ga	31	−66.8785	$\frac{3}{2}^-$	+1.850	78.26 h
	Ge	32	−62.450	$\frac{1}{2}^-, \frac{3}{2}^-$		18.7 m
68	Cu	29	−65.390	1^+		31 s
	Zn	**30**	−70.0063	0^+	≈0	**18.57%**
	Ga	31	−67.0852	1^+	±0.0117	68.1 m
	Ge	32	−66.972	0^+		287 d
69	Cu	29	−65.940	$(\frac{3}{2})^-$		3.0 m
	Zn	30	−68.4170	$\frac{1}{2}^-$		55.8 m
	Ga	**31**	−69.3215	$\frac{3}{2}^-$	+2.01602	**60.4%**
	Ge	32	−67.096	$\frac{5}{2}^-$		39.05 h
	As	33	−63.120	$(\frac{5}{2}^-)$		15.2 m
	Se	34	−56.300			27.3 s

A	Element (symbol)	Z	Mass excess (MeV/c²)	I^P	μ (nucl. magnetons)	% Abundance or Half-life
70	Cu	29	−63.390	1^+		4.5 s
	Zn	**30**	−69.5599	0^+		**0.62%**
	Ga	31	−68.9052	1^+		21.15 m
	Ge	**32**	−70.5614	0^+		**20.52%**
	As	33	−64.339	$4(^+)$		52.6 m
	Se	34	−61.590	0^+		41.0 m
71	Zn	30	−67.324	$(\frac{1}{2})^-$		2.45 m
	Ga	**31**	−70.1415	$\frac{3}{2}^-$	+2.56161	**39.6%**
	Ge	32	−69.9058	$\frac{1}{2}^-$	+0.546	11.8 d
	As	33	−67.893	$\frac{5}{2}^-$		64.8 d
72	Zn	30	−68.134	0^+		46.5 h
	Ga	31	−68.5910	3^-	−0.13220	14.10 h
	Ge	**32**	−72.5826	0^+		**27.43%**
	As	33	−68.232	2^-		26.0 h
	Se	34	−67.894	0^+		8.40 d
73	Zn	30	−65.030			23.5 s
	Ga	31	−69.730	$(\frac{3}{2})^-$		4.91 h
	Ge	**32**	−71.2935	$\frac{9}{2}^+$	−0.8792	**7.76%**
	As	33	−70.949	$\frac{3}{2}^-$		80.30 d
	Se	34	−68.209	$(\frac{7}{2})^+$		7.1 h
	Br	35	−63.670			3.3 m
	Kr	36	−56.980			25.9 s
74	Zn	30	−65.670	0^+		95 s
	Ga	31	−68.020	$(4)^-$		8.1 m
	Ge	**32**	−73.4221	0^+		**36.54%**
	As	33	−70.8597	2^-		17.78 d
	Se	**34**	−72.2127	0^+		**0.87%**
	Br	35	−65.295	$(0^-, 1^-)$		25.3 m
	Kr	36	−62.020	0^+		11.50 m
75	Ga	31	−68.560	$\frac{3}{2}^-$		2.10 m
	Ge	32	−71.8561	$\frac{1}{2}^-$	±0.51	82.78 m
	As	**33**	−73.0339	$\frac{3}{2}^-$	+1.439	**100%**
	Se	34	−72.1690	$\frac{5}{2}^+$		119.770 d
	Br	35	−69.159	$\frac{3}{2}^-$		97 m
	Rb	37	−57.510			4.3 m

A	Element (symbol)	Z	Mass excess (MeV/c²)	I^P	μ (nucl. magnetons)	% Abundance or Half-life
76	Zn	30	−62.550	0^+		5.7 s
	Ga	31	−66.440	(3^-)		27.1 s
	Ge	**32**	−73.2135	0^+		**7.76%**
	As	33	−72.2906	2^-	−0.905	26.32 h
	Se	**34**	−75.2592	0^+		**9.02%**
	Br	35	−70.303	1^-	±0.548	16.2 h
	Kr	36	−69.100	0^+		14.8 h
	Rb	37	−60.610			36.8 s
77	Ge	32	−71.2143	$\frac{7}{2}^+$		11.30 h
	As	33	−73.9157	$\frac{3}{2}^-$		38.83 h
	Se	**34**	−74.6061	$\frac{1}{2}^-$	+0.534	**7.58%**
	Br	35	−73.2415	$\frac{3}{2}^-$		57.036 h
	Kr	36	−70.236	$\frac{5}{2}^+$		74.4 m
	Rb	37	−65.110	$\frac{3}{2}^-$		3.70 m
	Sr	38	−57.960	$(\geqslant\frac{5}{2})$		9.0 s
78	Zn	30	−58.080	0^+		1.47 s
	Ga	31	−63.680	(3)		5.09 s
	Ge	32	−71.760	0^+		88 m
	As	33	−72.740	(2^-)		90.7 m
	Se	**34**	−77.0315	0^+	−1.02	**23.52%**
	Br	35	−73.458	1^+		6.46 m
	Kr	**36**	−74.150	0^+		0.35%
	Rb	37	−67.090	$0(^+)$		17.66 m
79	Ga	31	−62.810			3.00 s
	Ge	32	−69.570	$(\frac{1}{2})^-$		42 s
	As	33	−73.720	$\frac{3}{2}^-$		9.01 m
	Se	34	−75.9206	$\frac{7}{2}^-$		≤65000 y
	Br	**35**	−76.0700	$\frac{3}{2}^-$	+2.106	**50.54%**
	Kr	36	−74.439	$(\frac{1}{2})^-$		35.04 m
	Rb	37	−70.860			22.9 m
80	Ge	32	−69.430	0^+		24.5 s
	As	33	−72.060	$1(^+)$		16.5 s
	Se	**34**	−77.7613	0^+		**49.82%**
	Br	35	−75.8910	1^+	±0.514	17.4 m
	Kr	**36**	−77.897	0^+		**2.27%**
	Rb	37	−72.190	1^+		34 s
	Sr	38	−70.39	0^+		100 m

A	Element (symbol)	Z	Mass excess (MeV/c^2)	I^P	μ (nucl. magnetons)	% Abundance or Half-life
81	As	33	−72.64			33 s
	Se	34	−76.3910	$(\frac{1}{2})^-$		18.5 m
	Br	**35**	−77.967	$\frac{3}{2}^-$	+2.270	**49.46%**
	Kr	36	−77.707	$\frac{7}{2}^+$		2.1×10^5 y
	Rb	37	−75.445	$\frac{3}{2}^-$	+2.05	4.58 h
	Sr	38	−71.460	$(\frac{1}{2}^-)$		25.5 m
82	Se	34	−77.586	0^+		**9.19%** 1.4×10^{20} y
	Br	35	−77.498	5^-	±1.626	35.30 h
	Kr	**36**	−80.591	0^+		**11.56%**
	Rb	37	−76.213	1^+		1.25 m
	Sr	38	−75.999	0^+		25.0 d
83	As	33	−69.95			14.1 s
	Se	34	−75.41	$(\frac{9}{2})^+$		22.5 m
	Br	35	−79.025	$(\frac{3}{2})^-$		2.39 h
	Kr	**36**	−79.9846	$\frac{9}{2}^+$	−0.97	**11.55%**
	Rb	37	−78.987	$\frac{5}{2}^-$	+1.4	86.2 d
	Sr	38	−76.737	$\frac{7}{2}^+$		32.4 h
84	Se	34	−75.942	0^+		3.2 m
	Br	35	−77.759	2^-		31.80 m
	Kr	**36**	−82.4319	0^+		**56.90%**
	Rb	37	−79.752	2^-	−1.32	32.87 d
	Sr	**38**	−80.641	0^+		**0.56%**
	Y	39	−73.692	(5^-)		40 m
	Zr	40	−71.44	0^+		5.05 m
85	Br	35	−78.67	$(\frac{3}{2})^-$		2.90 m
	Kr	36	−81.4718	$\frac{9}{2}^+$	±1.005	10.72 y
	Rb	**37**	−82.1588	$\frac{5}{2}^-$	+1.3524	**72.15%**
	Sr	38	−81.095	$\frac{9}{2}^+$		64.84 d
	Y	39	−77.835	$(\frac{1}{2})^-$		2.68 h
86	Br	35	−75.96	(2^-)		55.0 s
	Kr	**36**	−83.263	0^+		**17.37%**
	Rb	37	−82.7377	2^-	−1.691	18.66 d
	Sr	**38**	−84.5127	0^+		**9.86%**
	Y	39	−79.239	4^-		14.74 h
	Zr	40	−77.94	0^+		16.5 h

A	Element (symbol)	Z	Mass excess (MeV/c²)	I^P	μ (nucl. magnetons)	% Abundance or Half-life
87	Kr	36	−80.707	$\frac{5}{2}+$		76.31 m
	Rb	37	−84.5957	$\frac{3}{2}-$	+2.7500	**27.85%**;
						4.80×10^{10}y
	Sr	**38**	−84.8689	$\frac{9}{2}+$	−1.093	**7.02%**
	Y	39	−83.0072	$\frac{1}{2}-$		80.3 h
	Zr	40	−79.43	$(\frac{9}{2}+)$		104.0 m
88	Kr	36	−79.689	0^+		2.84 h
	Rb	37	−82.602	2^-	±0.51	17.8 m
	Sr	**38**	−87.9106	0^+		**82.56%**
	Y	39	−84.298	4^-		106.60 d
	Zr	40	−83.621	0^+		83.4 d
	Nb	41	−76.42	(8^+)		14.3 m
	Mo	42	−72.92	0^+		8.2 m
89	Kr	36	−76.79	$(\frac{5}{2}+)$		3.07 m
	Rb	37	−81.717	$(\frac{3}{2}-, \frac{5}{2}-)$		15.4 m
	Sr	38	−86.203	$\frac{5}{2}+$		50.55 d
	Y	**39**	−87.6953	$\frac{1}{2}-$	−0.1373	**100%**
	Zr	40	−84.8595	$\frac{9}{2}+$		78.43 h
	Nb	41	−80.621	$(\frac{1}{2})^-$		66 m
90	Kr	36	−75.18	0^+		32.32 s
	Rb	37	−79.57	(1^-)		153 s
	Sr	38	−85.9347	0^+		28.6 y
	Y	39	−86.4807	2^-	−1.63	64.1 h
	Zr	**40**	−88.7646	0^+		**51.46%**
	Nb	41	−82.654	8^+		14.60 h
	Mo	42	−80.167	0^+		5.67 h
91	Kr	36	−71.77	$(\frac{5}{2}+)$		8.57 s
	Rb	37	−77.97	$\frac{3}{2}-$		58.4 s
	Sr	38	−83.666	$(\frac{5}{2})^+$		9.52 h
	Y	39	−86.3495	$\frac{1}{2}-$	±0.164	58.51 d
	Zr	**40**	−87.8925	$\frac{5}{2}+$	−1.303	**11.23%**
	Nb	41	−86.6369	$(\frac{9}{2})^+$		$\sim 1 \times 10^4$ y
	Mo	42	−82.199	$\frac{9}{2}+$		15.49 m
	Tc	43	−75.98	$(\frac{9}{2}+)$		3.14 m

A	Element (symbol)	Z	Mass excess (MeV/c²)	I^P	μ (nucl. magnetons)	% Abundance or Half-life
92	Kr	36	−69.15	0^+		1.85 s
	Rb	37	−75.12	$0^{(-)}$		4.50 s
	Sr	38	−82.892	0^+		2.71 h
	Y	39	−84.822	2^-		3.54 h
	Zr	**40**	−88.4561	0^+		**17.11%**
	Nb	41	−86.4481	$(7)^+$		3.5×10^7 y
	Mo	**42**	−86.807	0^+		**15.84%**
	Tc	43	−78.936	(8^+)		4.4 m
93	Rb	37	−72.92			5.8 s
	Sr	38	−80.28	$(\frac{5}{2}^+)$		7.6 m
	Y	39	−84.227	$\frac{1}{2}^-$		10.1 h
	Zr	40	−87.1167	$\frac{5}{2}^+$		1.53×10^6 y
	Nb	**41**	−87.2090	$\frac{9}{2}^+$	+6.167	**100%**
	Mo	42	−86.803	$\frac{5}{2}^+$		3.5×10^3 y
	Tc	43	−83.610	$\frac{9}{2}^+$		2.75 h
94	Sr	38	−78.96	0^+		78 s
	Y	39	−82.382	2^-		19.1 m
	Zr	**40**	−87.2639	0^+		**17.40%**
	Nb	41	−86.3671	6^+		2.03×10^4 y
	Mo	**42**	−88.4123	0^+		**9.04%**
	Tc	43	−84.156	$(7)^+$		293 m
95	Rb	37	−66.55			0.36 s
	Sr	38	−75.14			26 s
	Y	39	−81.233	$(\frac{1}{2})^-$		10.7 m
	Zr	40	−85.6634	$(\frac{5}{2})^+$		63.98 d
	Nb	41	−86.7865	$(\frac{9}{2})^+$		35.15 d
	Mo	**42**	−87.7121	$\frac{5}{2}^+$	−0.9133	**15.72%**
	Tc	43	−86.013	$(\frac{9}{2})^+$		20 h
	Ru	44	−83.452	$(\frac{5}{2})^+$		1.63 h
	Rh	45	−78.34			4.75 m
96	Sr	38	−73.07			
	Y	39	−78.43			2.3 m
	Zr	**40**	−85.4447	0^+		**2.80%**
	Nb	41	−85.608	(6^+)		23.35 h
	Mo	**42**	−88.7949	0^+		**16.53%**
	Tc	43	−85.821	7^+		4.28 d
	Ru	**44**	−86.075	0^+		**5.51%**
	Rh	45	−79.633	$(6^+, 7^+)$; $(1^+, 2^+, 3^+)$		9.25m; 1.55 m

A	Element (symbol)	Z	Mass excess (MeV/c²)	I^P	μ (nucl. magnetons)	% Abundance or Half-life
97	Y	39	−76.28			1.11 s
	Zr	40	−82.9542	$\frac{1}{2}^+$		17 h
	Nb	41	−85.6116	$(\frac{9}{2})^+$		72.1 m
	Mo	**42**	−87.5445	$\frac{5}{2}^+$	−0.9325	**9.46%**
	Tc	43	−87.224	$(\frac{9}{2})^+$		2.6×10^6 y
	Ru	44	−86.07	$(\frac{5}{2}^+)$		2.9 d
	Rh	45	−82.56			40 m
98	Zr	40	−81.292	0^+		31 s
	Nb	41	−83.530	$(4,5); (1^+)$		51.5 m; 2.8 s
	Mo	**42**	−88.1154	0^+		**23.78%**
	Tc	43	−86.434	$(7, 6^+)$		4.2×10^6 y
	Ru	**44**	−88.226	0^+		**1.87%**
	Rh	45	−83.168	(3^+)		9.05 m
	Pd	46	−81.27	0^+		18 m
99	Y	39	−71.5			0.8 s
	Zr	40	−77.89			2.4 s
	Nb	41	−82.346	$(\frac{9}{2})^+$		14.3 s
	Mo	42	−85.9695	$\frac{1}{2}^+$		66.02 h
	Tc	43	−87.3262	$\frac{9}{2}^+$	+5.68	2.13×10^5 y
	Ru	**44**	−87.7198	$\frac{5}{2}^+$	−0.63	**12.72%**
	Rh	45	−85.517	$(\frac{1}{2}^-)$		16 d
	Pd	46	−82.112	$(\frac{5}{2}^+)$		21.4 m
100	Zr	40	−76.6	0^+		7.1 s
	Nb	41	−79.96			1.5 s; 3.1 s
	Mo	**42**	−86.189	0^+		**9.63%**
	Tc	43	−86.0188	1^+		15.8 s
	Ru	**44**	−89.2216	0^+		**12.62%**
	Rh	45	−85.592	1^-		20.8 h
	Pd	46	−85.23	0^+		3.63 d
	Ag	47	−77.93			2.3 m
101	Nb	41	−78.95			7.1 s
	Mo	42	−83.516	$\frac{1}{2}^+$		14.6 m
	Tc	43	−86.327	$\frac{9}{2}^+$		14.2 m
	Ru	**44**	−87.9516	$\frac{5}{2}^+$	−0.69	**17.07%**
	Rh	45	−87.41	$\frac{1}{2}^-$		3.3 y
	Pd	46	−85.428	$\frac{5}{2}^+$		8.47 h

A	Element (symbol)	Z	Mass excess (MeV/c²)	I^P	μ (nucl. magnetons)	% Abundance or Half-life
102	Mo	42	-83.562	0^+		11.1 m
	Ru	**44**	-89.1005	(0^+)		**31.61%**
	Rh	45	-86.777	$(5^+, 6^+)$		$\approx$2.9 y
	Pd	**46**	-87.925	0^+		**0.96%**
	Ag	47	-82.33	5^+		12.9 m
103	Tc	43	-84.91			54.2 s
	Ru	44	-87.2614	$(\frac{3}{2})^+$		39.35 s
	Rh	**45**	-88.024	$\frac{1}{2}^-$	-0.0883	**100%**
	Pd	46	-87.478	$\frac{5}{2}^+$		16.96 d
	Ag	47	-84.80	$\frac{7}{2}^+$	$+4.4$	65.7 m
	Cd	48	-80.6	$(\frac{5}{2}^+)$		7.3 m
104	Mo	42	-80.5	0^+		1.3 m
	Tc	43	-82.7			18.2 m
	Ru	**44**	-88.099	0^+		**18.58%**
	Rh	45	-86.952	1^+		42.3 s
	Pd	**46**	-89.400	0^+		**10.97%**
	Ag	47	-85.15	5^+	$+4.0$	69.2 m
	Cd	48	-83.57	0^+		57.7 m
105	Tc	43	-82.54			7.7 m
	Ru	44	-85.938	$\frac{3}{2}^+$		4.44 h
	Rh	45	-87.855	$\frac{7}{2}^+$		35.36 h
	Pd	**46**	-88.422	$\frac{5}{2}^+$	-0.642	**22.23%**
	Ag	47	-87.075	$\frac{1}{2}^-$	±0.101	41.29 d
	Cd	48	-84.336	$\frac{5}{2}^+$	-0.74	55.5 m
106	Ru	44	-86.333	0^+		371.63 d
	Rh	45	-86.372	1^+		29.80 s
	Pd	**46**	-89.913	0^+		**27.33%**
	Ag	47	-86.929	1^+	$+2.9$	24.0 m
	Cd	**48**	-87.131	0^+		**1.22%**
	In	49	-80.586	$(7)^+$		6.2 m
107	Ru	44	-83.71			4.2 m
	Rh	45	-86.86	$(\frac{5}{2}, \frac{7}{2})^+$		21.7 m
	Pd	46	-88.371	$(\frac{5}{2}^+)$		6.5×10^6 y
	Ag	**47**	-88.404	$\frac{1}{2}^-$	-0.1135	**51.82%**
	Cd	48	-86.987	$\frac{5}{2}^+$	-0.6144	6.49 h
	In	49	-83.5	$(\frac{9}{2}^+)$		32.7 m

A	Element (symbol)	Z	Mass excess (MeV/c^2)	I^P	μ (nucl. magnetons)	% Abundance or Half-life
108	Ru	44	−83.82	0$^+$		4.5 m
	Rh	45	−85.02			16.8 s; 5.9 m
	Pd	**46**	−89.523	0$^+$		**26.71%**
	Ag	47	−87.602	1$^+$	+2.80	2.41 m
	Cd	**48**	−87.602	0$^+$		**0.88%**
	In	49	−84.10			58 m; 40 m
	Sn	50	−81.90	0$^+$		10.5 m
109	Pd	46	−87.606	$\frac{5}{2}^+$		13.46 h
	Ag	47	−88.722	$\frac{1}{2}^-$	−0.1305	**48.18%**
	Cd	48	−88.540	$\frac{5}{2}^+$	−0.8270	464 d
	In	49	−86.524	$\frac{9}{2}^+$	+5.53	4.2 h
110	Rh	45	−82.93			3.0 s
	Pd	**46**	−88.335	0$^+$		**11.81%**
	Ag	**47**	−87.456	1$^+$	+2.85	24.6 s
	Cd	**48**	−90.349	0$^+$		**12.39%**
	In	49	−86.409	7$^+$		4.9 h
	Sn	50	−85.834	0$^+$		4.0 h
111	Pd	46	−86.03	$\frac{5}{2}^+$		23.4 m
	Ag	47	−88.226	$\frac{1}{2}^-$	−0.145	7.45 d
	Cd	**48**	−89.254	$\frac{1}{2}^+$	−0.5943	**12.75%**
	In	49	−88.405	$\frac{9}{2}^+$	+5.53	2.83 d
	Sn	50	−85.941	$\frac{7}{2}^+$		35.3 m
	Te	52	−73.47	$(\frac{5}{2}^+)$		75 s
112	Pd	46	−86.326	0$^+$		21.045 h
	Ag	47	−86.62	2$^-$	±0.054	3.14 h
	Cd	**48**	−90.5779	0$^+$		**24.07%**
	In	49	−88.000	1$^+$	+2.81	14.4 m
	Sn	**50**	−88.658	0$^+$		**0.96%**
	Sb	51	−81.74	3$^+$		51.4 s
113	Ag	47	−87.04	$\frac{1}{2}^-$	+0.159	5.37 h
	Cd	**48**	−89.0503	$\frac{1}{2}^+$	−0.6217	**12.26%**; 9.3 × 10^{15} y
	In	49	−89.372	$\frac{9}{2}^+$	+5.523	**4.28%**
	Sn	50	−88.332	$\frac{1}{2}^+$	±0.88	115.09 d
	Sb	51	−84.443	$\frac{5}{2}^+$		6.67 m

A	Element (symbol)	Z	Mass excess (MeV/c²)	I^P	μ (nucl. magnetons)	% Abundance or Half-life
114	Ag	47	−85.16	1^+		4.52 s
	Cd	**48**	−90.0196	0^+		**28.86%**
	In	49	−88.576	1^+		71.9 s
	Sn	**50**	−90.560	0^+		**0.66%**
	Sb	51	−84.87	$(3)^+$		3.45 m
	Te	52	−82.1	0^+		16 m
115	Ag	47	−84.91	$(\frac{1}{2}^-)$		20.0 m
	Cd	48	−88.093	$\frac{1}{2}^+$	−0.6477	53.46 h
	In	**49**	−89.541	$\frac{9}{2}^+$	5.534	**95.72%**;
						4.4×10^{14} y
	Sn	**50**	−90.0351	$\frac{1}{2}^+$	−0.918	**0.35%**
	Sb	51	−87.005	$\frac{5}{2}^+$	+3.46	32.1 m
	Te	52	−82.42	$\frac{7}{2}^+$		5.8 m
116	Ag	47	−82.90			2.68 m
	Cd	**48**	−88.7176	0^+		**7.58%**
	In	49	−88.253	1^+		14.10 s
	Sn	**50**	−91.5261	0^+		**14.30%**
	Sb	51	−86.930	3^+		15.8 m
	Te	52	−85.37	0^+		2.50 h
	I	53	−77.61	1^+		2.91 s
	Xe	54	−73.27	0^+		56 s
117	Ag	47	−82.24	$(\frac{1}{2}^-)$		72.8 s
	Cd	48	−86.416	$\frac{1}{2}^+$		2.49 h
	In	49	−88.944	$\frac{9}{2}^+$		43.8 m
	Sn	**50**	−90.3989	$\frac{1}{2}^+$	−1.00	**7.61%**
	Sb	51	−88.654	$\frac{5}{2}^+$	+2.67	2.80 h
	Te	52	−85.164	$\frac{1}{2}^+$		62 m
	I	53	−80.85	$(\frac{5}{2}^+)$		2.3 m
118	Cd	48	−86.707	0^+		50.3 m
	In	49	−87.45	1^+		5.0 s
	Sn	**50**	−91.6536	0^+		**24.03%**
	Sb	51	−87.967	1^+	±2.5	3.6 m
	Te	52	−87.671	0^+		6.00 d
	I	53	−81.37			13.7 m
	Xe	54	−78.07	0^+		6 m

A	Element (symbol)	Z	Mass excess (MeV/c²)	I^P	μ (nucl. magnetons)	% Abundance or Half-life
119	Cd	48	−84.23	$(\frac{1}{2}^+)$		2.69 m
	In	49	−87.73	$\frac{9}{2}^+$		2.4 m
	Sn	**50**	−90.0667	$\frac{1}{2}^+$	+1.046	**8.58%**
	Sb	51	−89.483	$\frac{5}{2}^+$	+3.45	38.1 h
	Te	52	−87.189	$\frac{1}{2}^+$	±0.25	16.05 h
	I	53	−83.820	$(\frac{5}{2}^+)$		19.1 m
	Xe	54	−78.83	$(\frac{7}{2}^+)$		5.8 m
120	Cd	48	−83.981	0^+		50.80 s
	In	49	−85.7	1^+		3.08 s
	Sn	**50**	−91.1018	0^+		**32.85%**
	Sb	51	−88.421	1^+	±2.3	15.89 m
	Te	**52**	−89.404	0^+		**0.089%**
	I	53	−84.0	2^-		81.0 m
	Xe	54	−82.05	0^+		40 m
	Cs	55	−73.64	0^+		60.2 s
121	In	49	−85.842	$\frac{9}{2}^+$		23.1 s
	Sn	50	−89.2018	$\frac{3}{2}^+$	±0.70	27.06 h
	Sb	**51**	−89.5884	$\frac{5}{2}^+$	+3.359	**57.25%**
	Te	53	−88.508	$\frac{1}{2}^+$		16.78 d
	I	54	−86.14	$\frac{5}{2}^+$		2.12 h
	Xe	55	−82.35	$(\frac{5}{2}^+)$		40.1 m
	Ba	56	−70.57	$\frac{3}{2}^+$		125.6 s
122	In	49	−85.842			10.0 s; 1.5 s
	Sn	**50**	−89.946	0^+		**4.72%**
	Sb	51	−88.3233	2^-	−1.90	2.70 d
	Te	**52**	−90.304	0^+		**2.46%**
	I	53	−86.16	1^+		3.62 m
	Xe	54	−85.16	0^+		20.1 h
123	In	49	−83.44	$(\frac{9}{2})^+$		5.98 s
	Sn	50	−87.821	$\frac{11}{2}^+$		129.2 d
	Sb	**51**	−89.2175	$\frac{7}{2}^+$	+2.547	**42.75%**
	Te	**52**	−89.1655	$\frac{1}{2}^+$	−0.7359	**0.87%**; >1 × 10¹³ y
	I	53	−87.97	$\frac{5}{2}^+$		13.2 h
	Xe	54	−85.29	$(\frac{1}{2})^+$		2.08 h

A	Element (symbol)	Z	Mass excess (MeV/c²)	I^P	μ (nucl. magnetons)	% Abundance or Half-life
124	In	49	−81.10			3.21 s
	Sn	**50**	−88.240	0^+		**5.94%**
	Sb	51	−87.6134	3^-		60.20 d
	Te	**52**	−90.5183	0^+		**4.61%**
	I	53	−87.361	2^-		4.18 d
	Xe	**54**	−87.54	0^+		**0.096%**
	Cs	55	−81.53	(1^+)		26.5 s
125	In	49	−80.5	$(\frac{9}{2})^+$		2.33 s
	Sn	50	−85.902	$\frac{11}{2}^-$		9.64 d
	Sb	51	−88.252	$\frac{7}{2}^+$	±2.6	2.73 y
	Te	**52**	−89.019	$\frac{1}{2}^+$	−0.8871	**6.99%**
	I	53	−88.841	$\frac{5}{2}^+$	+3.0	60.14 d
	Xe	54	−87.11	$(\frac{1}{2})^+$		16.9 h
	Cs	55	−84.04	$\frac{1}{2}^+$	+1.41	45 m
	Ba	56	−79.46			3.5 m
126	In	49	−77.9			1.53 s
	Sn	50	−86.024	0^+		$\approx 10^6$ y
	Sb	51	−86.402	$(8^-, 7^-)$		12.4 d
	Te	**52**	−90.066	0^+		**18.71%**
	I	53	−87.911	2^-		13.02 d
	Xe	**54**	−89.162	0^+		**0.090%**
	Cs	55	−84.33	1^+		1.64 m
	Ba	56	−82.56	0^+		96.5 m
127	In	49	−77.17			2.07 s
	Sn	50	−83.6	$(\frac{11}{2}^-)$		2.10 h
	Sb	51	−86.704	$\frac{7}{2}^+$		3.85 d
	Te	52	−88.285	$\frac{3}{2}^+$		9.35 h
	I	**53**	−88.980	$\frac{5}{2}^+$	+2.808	**100%**
	Xe	54	−88.316	$(\frac{1}{2}^+)$		36.41 d
	Cs	55	−86.206	$\frac{1}{2}^+$	+1.46	6.25 h
	Ba	56	−82.76			18 m
128	In	49	−74.34			3.7 s
	Sn	50	−83.44	0^+		60.0 m
	Sb	51	−84.73			9.01 h
	Te	**52**	−88.9923	0^+		**31.79%**
	I	53	−87.734	1^+		24.99 m
	Xe	**54**	−89.8612	0^+		**1.92%**
	Cs	55	−85.935	1^+		3.9 m
	Ba	56	−85.482	0^+		24.3 d

A	Element (symbol)	Z	Mass excess (MeV/c²)	I^P	μ (nucl. magnetons)	% Abundance or Half-life
129	In	49	−73.12			
	Sn	50	−80.64			7.5 m
	Sb	51	−84.63	$\frac{7}{2}^+$		4.32 h
	Te	52	−87.007	$\frac{3}{2}^+$		69.6 m
	I	53	−88.505	$\frac{7}{2}^+$	+2.617	1.57×10^7 y
	Xe	**54**	−88.6985	$\frac{1}{2}^+$	−0.7768	**26.44%**
	Cs	55	−87.563	$\frac{1}{2}^+$	(+)1.479	32.06 h
	B	56	−85.116	$\frac{1}{2}^+$		2.20 h
130	Sn	50	−80.38	0^+		3.72 m
	Sb	51	−82.38	(8^-)		40 m
	Te	**52**	−87.348	0^+		**34.48%**; 2.51×10^{21}y
	I	53	−86.897	5^+		12.36 h
	Xe	**54**	−89.8811	0^+		**4.08%**
	Cs	55	−86.863	1^+		29.9 m
	Ba	**56**	−87.303	0^+		**0.101%**
131	Te	52	−85.201	$\frac{3}{2}^+$		25.0 m
	I	53	−87.451	$\frac{7}{2}^+$	+2.74	8.04 d
	Xe	**54**	−88.421	$\frac{3}{2}^+$	+0.6908	**21.18%**
	Cs	55	−88.066	$\frac{5}{2}^+$	+3.54	9.69 d
	Ba	56	−86.726	$\frac{1}{2}^+$		11.8 d
	La	57	−83.77	$\frac{3}{2}^+$		59 m
132	Sn	50	−76.39	0^+		40 s
	Sb	51	−79.61	(8^-)		4.2 m
	Te	52	−85.213	0^+		78.2 h
	I	53	−85.706	4^+	±3.08	2.30 h
	Xe	**54**	−89.286	0^+		**26.89%**
	Cs	55	−87.175	$(2)^-$	+2.22	6.475 d
	Ba	**56**	−88.453	0^+		**0.097%**
	La	57	−83.74	2^-		4.8 h
	Ce	58	−82.34	0^+		4.2 h
133	Sb	51	−78.98			≈2.7 m
	Te	52	−82.93	$(\frac{3}{2}^+)$		12.45 m
	I	53	−85.902	$\frac{7}{2}^+$	+2.84	20.8 m
	Xe	54	−87.662	$\frac{3}{2}^+$		5.29 d
	Cs	**55**	−88.089	$\frac{7}{2}^+$	+2.578	**100%**
	Ba	56	−87.569	$\frac{1}{2}^+$		10.5 y

A	Element (symbol)	Z	Mass excess (MeV/c²)	I^P	μ (nucl. magnetons)	% Abundance or Half-life
134	Te	52	−82.67	0^+		41.8 m
	I	53	−83.97			52.6 m
	Xe	**54**	−88.125	0^+		**10.44%**
	Cs	55	−86.909	$4^{(+)}$	+2.990	2.062 y
	Ba	**56**	−88.968	0^+		**2.42%**
	La	57	−85.268	1^+		6.67 m
u	Ce	58	−84.77	0^+		72.0 h
135	Te	52	−77.6			18 s
	I	53	−83.796	$\frac{7}{2}^+$		6.61 h
	Xe	54	−86.506	$\frac{3}{2}^+$		9.083 h
	Cs	55	−87.665	$\frac{7}{2}^+$	+2.729	2.3×10^6 y
	Ba	**56**	−87.870	$\frac{3}{2}^+$	+0.8365	**6.59%**
	La	57	−86.67	$\frac{5}{2}^+$		19.5 h
	Ce	58	−84.55	$\frac{1}{2}^{(+)}$		17.6 h
	Pr	59	−80.99	$\frac{3}{2}^{(+)}$		≈22 m
136	I	53	−79.43	(2^-)		84 s
	Xe	**54**	−86.425	0^+		**8.87%**
	Cs	55	−86.358	5^+	+3.70	13.16 d
	Ba	**56**	−88.906	0^+		**7.81%**
	La	57	−86.04	1^+		9.87 m
	Ce	**58**	−86.50	0^+		**0.193%**
	Pr	59	−81.40	2^+		13.1 m
	Nd	60	−79.19	0^+		50.65 m
137	I	53	−76.72	$(\frac{7}{2}^+)$		24.7 s
	Xe	54	−82.215	$(\frac{7}{2}^-)$		3.83 m
	Cs	55	−86.560	$\frac{7}{2}^+$	+2.838	30.17 y
	Ba	**56**	−87.733	$\frac{3}{2}^+$	+0.9357	**11.32%**
	La	57	−87.13	$\frac{7}{2}^+$		6×10^4 y
	Ce	58	−85.91	$\frac{3}{2}^+$		9.0 h
	Pr	59	−83.21	$(\frac{5}{2}^+)$		76.6 m
138	Xe	54	−80.03	0^+		14.13 m
	Cs	55	−82.77	$3^{(-)}$	±0.5	32.2 m
	Ba	**56**	−88.273	0^+		**71.66%**
	La	**57**	−86.966	5^+	+3.707	**0.089%;** 1.35×10^{11} y
	Ce	**58**	−87.565	0^+		**0.250%**
	Pr	59	−83.128	1		1.45 m
	Nd	60	−82.03	0^+		5.04 h

A	Element (symbol)	Z	Mass excess (MeV/c^2)	I^P	μ (nucl. magnetons)	% Abundance or Half-life
139	Xe	54	−75.75			39.68 s
	Cs	55	−80.63	$\frac{7}{2}$		9.27 m
	Ba	56	−84.925	$\frac{7}{2}^-$		84.6 m
	La	**57**	−87.231	$\frac{7}{2}^+$	+2.778	**99.991%**
	Ce	58	−86.966	$\frac{3}{2}^+$		137.66 d
	Pr	59	−84.854	$\frac{5}{2}^+$		4.41 h
	Nd	60	−82.05	$\frac{3}{2}^+$		29.7 m
	Pm	61	−77.5	$(\frac{5}{2})^+$		4.15 m
	Sm	62	−72.3	$(\frac{1}{2}^+)$		2.57
140	Xe	54	−73.18	0^+		13.60 s
	Cs	55	−77.24	1^-		63.7 s
	Ba	56	−83.285	0^+		12.746 d
	La	57	−84.320	3^-		40.272 h
	Ce	**58**	−88.081	0^+		**88.48%**
	Pr	59	−84.693	1^+		3.39 m
	Nd	60	−84.22	0^+		3.37 d
	Pm	61	−78.18	1^+		9.2 s
141	Xe	54	−69.0			1.72 s
	Cs	55	−75.0	$(\frac{7}{2}^+)$		24.94 s
	Ba	56	−79.98	$(\frac{3}{2}^-)$		18.27 m
	La	57	−83.008	$(\frac{7}{2}^+)$		3.93 h
	Ce	58	−85.438	$\frac{7}{2}^-$	+0.9	32.50 d
	Pr	**59**	−86.018	$\frac{5}{2}^+$	+4.3	**100%**
	Nd	60	−84.203	$\frac{3}{2}^+$		2.49 h
	Pm	61	−80.47	$\frac{5}{2}^+$		20.90 m
	Sm	62	−75.91	$\frac{1}{2}^+$		10.2 m
	Eu	63	−69.88	$(\frac{5}{2}^+)$		40.0 s
142	Xe	54	−66.05	0^+		1.22 s
	Cs	55	−70.95			1.80 s
	Ba	56	−77.82	0^+		10.6 m
	La	57	−80.018	2^-		92.5 m
	Ce	**58**	−84.535	0^+		**11.07%**; >5 × 10^{16} y
	Pr	59	−83.790	2^-	±0.25	19.13 h
	Nd	**60**	−85.949	0^+		**27.11%**
	Pm	61	−81.06	1^+		40.5 s
	Sm	62	−78.978	0^+		72.49 m

A	Element (symbol)	Z	Mass excess (MeV/c²)	I^P	μ (nucl. magnetons)	% Abundance or Half-life
143	La	57	−78.31			14.23 m
	Ce	58	−81.610	$\frac{3}{2}-$		33.0 h
	Pr	59	−83.065	$\frac{7}{2}+$		13.58 d
	Nd	**60**	−84.000	$\frac{7}{2}-$		**12.17%**
	Pm	61	−82.959	$\frac{5}{2}+$		265 d
	Sm	62	−79.511	$\frac{3}{2}+$		8.83 m
	Eu	63	−74.41	$\frac{5}{2}+$		2.63 m
144	Ce	58	−80.431	0^+		284.9 d
	Pr	59	−80.750	0^-		17.28 m
	Nd	**60**	−83.746	0^+		**23.85%**; 2.4×10^{15} y
	Pm	61	−81.416	5^-		363 d
	Sm	**62**	−81.964	0^+		**3.09%**
	Eu	63	−75.636	1^+		10.2 s
145	Ce	58	−77.12			2.98 m
	Pr	59	−79.625	$(\frac{7}{2})^+$		5.98 h
	Nd	**60**	−81.430	$\frac{7}{2}-$	−0.66	**8.30%**; $>1 \times 10^{17}$ y
	Pm	61	−81.270	$\frac{5}{2}+$		17.7 y
	Sm	62	−80.656	$\frac{7}{2}-$		340 d
	Eu	63	−77.936	$\frac{5}{2}+$		5.93 d
146	Ce	58	−75.76	0^+		14.2 m
	Pr	59	−76.84			24.07 m
	Nd	**60**	−80.923	0^+		**17.22%**
	Pm	61	−79.442	$2^-, 3, 4^-$		2020 d
	Sm	62	−80.984	0^+		1.03×10^8 y
	Eu	63	−77.111	4^-		4.61 d
	Gd	64	−75.91	0^+		48.3 d
147	Pr	59	−75.44	$(\frac{5}{2}+)$		13.6 m
	Nd	60	−78.144	$\frac{5}{2}-$	±0.59	10.98 d
	Pm	61	−79.040	$\frac{7}{2}+$	+2.7	2.6234 y
	Sm	**62**	−79.265	$\frac{7}{2}-$	−0.813	**14.97%**; 1.06×10^{11} y
	Eu	63	−77.535	$\frac{5}{2}+$		24 d
	Gd	64	−75.207	$\frac{7}{2}-$		38.1 h

A	Element (symbol)	Z	Mass excess (MeV/c²)	I^P	μ (nucl. magnetons)	% Abundance or Half-life
148	Pr	59	−72.6	(3)		2.30 m
	Nd	**60**	−77.407	0^+		**5.73%**
	Pm	61	−76.87	1^-	+2.0	5.37 d
	Sm	**62**	−79.335	0^+		**11.24%**; 8 × 10¹⁵ y
	Eu	63	−76.235	5^-		54.5 d
	Gd	64	−76.268	0^+		93 y
	Tb	65	−70.64	2^-		60 m
149	Pr	59	−71.37	$(\frac{5}{2}^+, \frac{7}{2}^+)$		2.5 m
	Nd	60	−74.374	$\frac{5}{2}^-$		1.73 h
	Pm	61	−76.063	$\frac{7}{2}^+$		53.08 h
	Sm	**62**	−77.135	$\frac{7}{2}^-$	−0.67	**13.83%**; >1 × 10¹⁶ y
	Eu	63	−76.439	$\frac{5}{2}^+$		93.1 d
	Gd	64	−75.131	$\frac{7}{2}^-$		9.4 d
	Tb	65	−71.434	$(\frac{3}{2}^+, \frac{5}{2}^+)$		4.15 h
150	Nd	**60**	−73.682	0^+		**5.62%**; >5 × 10¹⁷ y
	Pm	61	−73.55	(1^-)		2.68 h
	Sm	**62**	−77.049	0^+		**7.44%**
	Eu	63	−74.756	$0(^-)$		12.62 h
	Gd	64	−75.765	0^+		1.79 × 10⁶ y
	Tb	65	−71.098	(2^-)		3.27 h
	Dy	66	−69.14	0^+		7.17 m
151	Nd	60	−70.945	$(\frac{3}{2}^+)$		12.44 m
	Pm	61	−73.386	$\frac{5}{2}^+$	±1.6	28.40 h
	Sm	62	−74.574	$\frac{5}{2}^-$		90 y
	Eu	**63**	−74.650	$\frac{5}{2}^+$	+3.464	**47.82%**
	Gd	64	−74.168	$\frac{7}{2}^-$		120 d
	Tb	65	−71.608	$\frac{1}{2}^+$		17.6 h
	Dy	66	−68.601	$\frac{7}{2}^-$		16.9 m
152	Nd	60	−70.146	0^+		11.4 m
	Pm	61	−71.29	1^+		4.1 m
	Sm	**62**	−74.761	0^+		**26.72%**
	Eu	63	−72.884	3^-	±1.924	13.33 y
	Gd	**64**	−74.703	0^+		**0.20%**; 1.08 × 10¹⁴ y
	Tb	65	−70.853	2^-		17.5 h
	Dy	66	−70.116	0^+		2.38 h
	Ho	67	−63.71	(3^+)		2.35 m
	Er	68	−60.41	0^+		10.1 s

A	Element (symbol)	Z	Mass excess (MeV/c²)	I^P	μ (nucl. magnetons)	% Abundance or Half-life
153	Pm	61	−70.76	$(\frac{5}{2})^-$		5.4 m
	Sm	62	−72.557	$\frac{3}{2}^+$	−0.022	46.7 m
	Eu	**63**	−73.363	$\frac{5}{2}^+$	+1.530	**52.18%**
	Gd	64	−73.119	$\frac{3}{2}^-$		242 d
	Tb	65	−71.329	$\frac{5}{2}(+)$		2.34 d
	Dy	66	−69.155	$\frac{7}{2}(-)$		6.5 h
	Ho	67	−64.954			9.3 m
154	Pm	61	−68.45	(0,1)		1.7 m
	Sm	**62**	−72.454	0^+		**22.71%**
	Eu	63	−71.726	3^-	±2.000	8.8 y
	Gd	**64**	−73.704	0^+		**2.15%**
	Tb	65	−70.24	$0(-)$		21.4 h
	Dy	66	−70.392	0^+		~1 × 10⁷ y
	Ho	67	−64.635	1		11.8 m
155	Sm	62	−70.196	$\frac{3}{2}^-$		22.1 m
	Eu	63	−71.825	$\frac{5}{2}^+$		4.96 y
	Gd	**64**	−72.071	$\frac{3}{2}^-$	−0.254	**14.73%**
	Tb	65	−71.256	$\frac{3}{2}^+$		5.32 d
	Dy	66	−69.157	$\frac{3}{2}^-$		10.0 h
	Ho	67	−66.055	$\frac{5}{2}$		48 m
	Er	67	−62.057			5.3 m
156	Sm	62	−69.368	0^+		9.4 h
	Eu	63	−70.083	0^+		15.19 d
	Gd	**64**	−72.536	0^+		**20.47%**
	Tb	65	−70.098	$3(-)$		5.34 d
	Dy	**66**	−70.527	0^+		**0.052%**
						>1 × 10¹⁸ y
	Tm	69	−56.94			80 s
157	Sm	62	−66.86			83 s
	Eu	63	−69.465	$(\frac{5}{2}^+)$		15.15 h
	Gd	**64**	−70.825	$\frac{3}{2}^-$	−0.39	**15.68%**
	Tb	65	−70.767	$\frac{3}{2}^+$		150 y
	Dy	66	−69.425	$\frac{3}{2}^-$		8.1 h
	Ho	67	−66.89	$\frac{7}{2}(-)$		12.6 m
158	Eu	63	−67.24	$(1-)$		45.9 m
	Gd	**64**	−70.691	0^+		**24.87%**
	Tb	65	−69.475	3^-	±1.74	150 y
	Dy	**66**	−70.410	0^+		**0.090%**
	Ho	67	−66.433	5^+		11.3 m
	Er	68	−65.03	0^+		2.25 h

A	Element (symbol)	Z	Mass excess (MeV/c²)	I^P	μ (nucl. magnetons)	% Abundance or Half-life
159	Eu	63	−65.93	$(\frac{5}{2}^+)$		18.1 m
	Gd	64	−68.562	$\frac{3}{2}^-$		18.56 h
	Tb	**65**	−69.536	$\frac{3}{2}^+$	±1.99	**100%**
	Dy	66	−69.171	$\frac{3}{2}^-$		144.4 d
	Ho	67	−67.318	$\frac{7}{2}^-$		33 m
	Er	68	−64.39	$\frac{3}{2}^-$		36 m
160	Gd	**64**	−67.943	0^+		**21.90%**
	Tb	65	−67.840	3^-	±1.68	72.3 d
	Dy	**66**	−69.674	0^+		**2.29%**
	Ho	67	−66.388	5^+		25.6 m
	Er	68	−66.052	0^+		28.58 h
	Tm	69	−60.13	1		9.2 m
161	Gd	64	−65.507	$\frac{5}{2}^-$		3.7 m
	Tb	65	−67.466	$\frac{3}{2}^+$		6.91 d
	Dy	**66**	−68.056	$\frac{5}{2}^+$	−0.46	**18.88%**
	Ho	67	−67.203	$\frac{7}{2}^-$		2.5 h
	Er	68	−65.197	$\frac{3}{2}$		3.24 h
162	Gd	64	−64.36	0^+		9 m
	Tb	65	−65.76	$(1)^-$		7.7 m
	Dy	**66**	−68.181	0^+		**25.53%**
	Ho	67	−66.047	1^+		15 m
	Er	**68**	−66.335	0^+		**0.136%**
	Tm	69	−61.54	1^-		21.7 m
163	Tb	65	−64.68	$\frac{3}{2}^+$		19.5 m
	Dy	**66**	−66.382	$\frac{5}{2}^-$	+0.64	**24.97%**
	Ho	67	−66.379	$(\frac{7}{2})^-$		33 y
	Er	68	−65.168	$\frac{5}{2}^-$	+1.1	75.0 m
	Tm	69	−62.77	$\frac{1}{2}^+$	+0.08	1.81 h
164	Tb	65	−62.11	(5^+)		3.0 m
	Dy	**66**	−65.967	0^+		**28.18%**
	Ho	67	−64.937	1^+		29.0 m
	Er	**68**	−65.940	0^+		**1.56%**
	Tm	69	−61.978	1^+		2.0 m
	Yb	70	−60.88	0^+		75.8 m
165	Dy	66	−63.611	$\frac{7}{2}^+$		2.334 h
	Ho	**67**	−64.896	$\frac{7}{2}^-$	+4.12	**100%**
	Er	68	−64.518	$\frac{5}{2}^-$	±0.65	10.36 h
	Tm	69	−62.924	$\frac{1}{2}^+$		30.06 h
	Yb	70	−60.161	$(\frac{5}{2}^-)$		9.9 m

A	Element (symbol)	Z	Mass excess (MeV/c^2)	I^P	μ (nucl. magnetons)	% Abundance or Half-life
166	Dy	66	-62.583	0^+		81.6 h
	Ho	67	-63.067	0^-		26.80 h
	Er	**68**	-64.921	0^+		**33.41%**
	Tm	69	-61.874	2^+	±0.047	7.70 h
	Yb	70	-61.582	0^+		56.7 h
	Lu	71	-56.1	(6^-)		2.65 m
167	Ho	67	-62.316	$(\frac{7}{2}^-)$		3.1 h
	Er	**68**	-63.286	$\frac{7}{2}^+$	-0.564	**22.94%**
	Tm	69	-62.537	$\frac{1}{2}^+$		9.24 d
	Yb	70	-60.583	$\frac{5}{2}^-$		17.5 m
	Lu	71	-57.45	$\frac{7}{2}^+$		51.5 m
168	Ho	67	-60.27	3^+		3.0 m
	Er	**68**	-62.985	0^+		**27.07%**
	Tm	69	-61.306	3^+		93.1 d
	Yb	**70**	-61.565	0^+		**0.135%**
	Lu	71	-57.1	$(6)^-$		5.3 m
	Hf	72	-55.10	0^+		25.95 m
169	Ho	67	-58.793	$(\frac{7}{2}^-)$		4.6 m
	Er	68	-60.917	$\frac{1}{2}^-$	$+0.513$	9.3 d
	Tm	**69**	-61.269	$\frac{1}{2}^+$	-0.232	**100%**
	Yb	70	-60.361	$\frac{7}{2}^-$		30.7 d
	Lu	71	-57.881	$\frac{7}{2}^+$		34 h
	Hf	72	-54.53	$(\frac{5}{2})^-$		3.25 m
170	Ho	67	-56.1			2.8 m
	Er	**68**	-60.104	0^+		**14.88%**
	Tm	69	-59.791	1^-	±0.246	128.6 d
	Yb	**70**	-60.759	0^+		**3.03%**
	Lu	71	-57.319	0^+		2.00 d
	Hf	72	-56.12	0^+		12.2 h
171	Er	68	-57.714	$\frac{5}{2}$	±0.70	7.52 h
	Tm	69	-59.205	$\frac{1}{2}$	±0.229	1.92 y
	Yb	**70**	-59.302	$\frac{1}{2}^-$	±0.4919	**14.31%**
	Lu	71	-57.821	$\frac{7}{2}$		8.22 d
172	Er	68	-56.491	0^+		49.3 h
	Tm	69	-57.38	2^-		63.6 h
	Yb	**70**	-59.250	0^+		**21.82%**
	Lu	71	-56.726	(4^-)		6.70 d

A	Element (symbol)	Z	Mass excess (MeV/c²)	I^P	μ (nucl. magnetons)	% Abundance or Half-life
173	Er	68	−53.73	$(\frac{7}{2})^-$		1.4 m
	Tm	69	−56.226	$(\frac{1}{2}^+)$		8.24 h
	Yb	**70**	−57.546	$\frac{5}{2}^-$	−0.6776	**16.13%**
	Lu	71	−56.871	$\frac{7}{2}^+$		1.37 y
	Hf	72	−55.27	$\frac{1}{2}^-$		24.0 h
174	Tm	69	−53.85	(4^-)		5.4 m
	Yb	**70**	−56.940	0^+		**31.84%**
	Lu	71	−55.562	(0^-)		3.31 y
	Hf	**72**	−55.830	0^+		**0.18%**; 2.0×10^{15} y
	Ta	73	−51.98	(4^-)		1.2 h
	W	74	−50.08	0^+		29 m
175	Tm	69	−52.29	$(\frac{1}{2}^+)$		15.2 m
	Yb	70	−54.691	$\frac{7}{2}^-$		4.19 d
	Lu	**71**	−55.159	$\frac{7}{2}^+$	+2.23	**97.41%**
	Hf	72	−54.548	$\frac{5}{2}^-$		70 d
176	Yb	**70**	−53.490	0^+		**12.73%**
	Lu	71	−53.381	7^-	+3.18	3.60×10^{10} y
	Hf	**72**	−54.567	0^+		**5.20%**
	Ta	73	−51.47	1^-		8.08 h
	W	74	−50.57	0^+		2.3 h
177	Yb	70	−50.986	$\frac{9}{2}^+$		1.9 h
	Lu	71	−52.382	$\frac{7}{2}^+$	+2.24	6.71 d
	Hf	**72**	−52.879	$\frac{7}{2}^-$	+0.61	**18.50%**
	Ta	73	−51.721	$\frac{7}{2}^+$		56.6 h
178	Yb	70	−49.66	0^+		74 m
	Lu	71	−50.30	$1(^+)$		28.4 m
	Hf	**72**	−52.434	0^+		**27.14%**
	Ta	73	−50.52	1^+		9.31 m
	W	74	−50.43	0^+		21.7 d
	Re	75	−45.77	(3)		13.2 m
179	Lu	71	−49.11	$(\frac{7}{2}^+)$		4.59 h
	Hf	**72**	−50.462	$\frac{9}{2}^+$	−0.47	**13.75%**
	Ta	73	−50.347	$(\frac{7}{2}^+)$		664.9 d
	W	74	−49.283	$(\frac{7}{2}^-)$		37.5 m
	Re	75	−46.59	$(\frac{5}{2}^+)$		19.7 m

A	Element (symbol)	Z	Mass excess (MeV/c²)	I^P	μ (nucl. magnetons)	% Abundance or Half-life
180	Lu	71	−46.68	$(3^-, 4^-)$		5.7 m
	Hf	**72**	−49.779	0^+		**35.24%**
	Ta	**73**	−48.914	(8^+)		**0.0123%**;
						$>1.0 \times 10^{13}$ y
	W	**74**	−49.624	0^+		**0.14%**;
						6×10^{14} y
	Re	75	−45.829	$(1)^-$		2.43 m
181	Hf	72	−47.403	$(\frac{1}{2}^-)$		42.4 d
	Ta	**73**	−48.425	$\frac{7}{2}^+$	+2.36	**99.988%**
	W	74	−48.237	$\frac{9}{2}(^+)$		121.2 d
182	Hf	72	−45.99	0^+		9×10^6 y
	Ta	73	−46.417	3^-		115.0 d
	W	**74**	−48.228	0^+		**26.41%**
	Re	75	−45.43	2^+		12.7 h
183	Hf	72	−43.269	$(\frac{3}{2}^-)$		64 m
	Ta	73	−45.279	$\frac{7}{2}^+$		5.1 d
	W	**74**	−46.347	$\frac{1}{2}^-$	+0.117	**14.40%**
	Re	75	−45.791	$(\frac{5}{2})^+$		70.0 d
184	Hf	72	−41.48	0^+		4.12 h
	Ta	73	−42.821	(5^-)		8.7 h
	W	**74**	−45.687	0^+		**30.64%**;
						$>3 \times 10^{17}$ y
	Re	75	−44.191	3^-		38.0 d
	Os	**76**	−44.233	0^+		**0.018%**;
						$>1 \times 10^{17}$ y
	Ir	77	−39.51	5		3.02 h
	Pt	78	−37.21	0^+		17.3 m
185	Ta	73	−41.36	$(\frac{7}{2}^+)$		49 m
	W	74	−43.370	$\frac{3}{2}^-$		75.1 d
	Re	**75**	−43.802	$\frac{5}{2}^+$	+3.172	**37.07%**
	Os	76	−42.787	$\frac{1}{2}^-$		93.6 d
	Ir	77	−40.29	$\frac{5}{2}(^-)$		14.0 h
186	Ta	73	−38.60	(3^-)		10.5 m
	W	**74**	−42.498	0^+		**28.41%**
	Re	75	−41.910	$1(^-)$	+1.73	90.64 h
	Os	**76**	−42.987	0^+		**1.59%**;
						2×10^{15} y
	Ir	77	−39.156	(5)		15.8 h
	Pt	78	−37.83	0^+		2.0 h

A	Element (symbol)	Z	Mass excess (MeV/c²)	I^P	μ (nucl. magnetons)	% Abundance or Half-life
187	W	74	−39.893	$\frac{3}{2}^-$		23.9 h
	Re	75	−41.205	$\frac{5}{2}^+$		**62.93%;**
						5×10^{10} y
	Os	76	−41.208	$\frac{1}{2}^-$	+0.0643	**1.64%**
	Ir	77	−39.71	$(\frac{3}{2}^+)$		10.5 h
188	W	74	−38.657	0^+		69.4 d
	Re	75	−39.006	1^-	+1.78	16.98 h
	Os	76	−41.125	0^+		**13.3%**
	Ir	77	−38.323	$(2)^-$		41.5 h
	Pt	78	−37.788	0^+		10.2 d
189	W	74	−35.47			11.5 m
	Re	75	−37.97	$(\frac{5}{2}^+)$		24.3 m
	Os	76	−38.978	$\frac{3}{2}^-$	+0.6566	**16.1%**
	Ir	77	−38.48	$(\frac{3}{2}^+)$		13.3 d
	Pt	78	−36.57	$(\frac{3}{2}^-)$		10.87 h
190	W	74	−34.22			
	Re	75	−35.52	$2,3,4^-$		3.1 m
	Os	76	−38.699	0^+		**26.4%**
	Ir	77	−36.70	$4,5$		12.1 d
	Pt	78	−37.318	0^+		**0.0127%;**
						6×10^{11} y
	Au	79	−32.876	$1(^-)$	±0.066	42.0 m
	Hg	80	−30.96	0^+		20.0 m
	Tl	81	−24.16			
191	Re	75	−34.343	$(\frac{3}{2}^+, \frac{1}{2}^+)$		9.8 m
	Os	76	−36.388	$\frac{9}{2}^-$		15.4 d
	Ir	77	−36.698	$\frac{3}{2}^+$		**37.3%**
	Pt	78	−35.698	$\frac{3}{2}^-$		2.9 d
	Au	79	−33.870	$\frac{3}{2}^+$	±0.137	3.18 h
	Hg	80	−30.480	$(\frac{3}{2}^-)$		49 m
	Tl	81	−25.67			10 m
192	Os	76	−35.875	0^+		**41.0%**
	Ir	77	−34.826			74.02 d
	Pt	78	−36.283	0^+		**0.78%**
	Au	79	−32.768	$1(^-)$	±0.00785	5.03 h
	Hg	80	−31.97	0^+		4.9 h
	Tl	81	−25.59	(2^-)		9.5 m

A	Element (symbol)	Z	Mass excess (MeV/c²)	I^P	μ (nucl. magnetons)	% Abundance or Half-life
193	Os	76	−33.387	$\frac{3}{2}-$		30.5 h
	Ir	**77**	−34.519	$\frac{3}{2}+$	+0.158	**62.7%**
	Pt	78	−34.458	$(\frac{1}{2}-)$		50 y
	Au	79	−33.36	$\frac{3}{2}+$	±0.139	17.65 h
	Hg	80	−31.02	$\frac{3}{2}-$	−0.62	3.80 h
	Tl	81	−27.02	$\frac{1}{2}(+)$		21.6 m
194	Os	76	−32.417	$0+$		6.0 y
	Ir	77	−32.514	$1-$		19.15 h
	Pt	**78**	−34.765	$0+$		**32.9%**
	Au	79	−32.256	$1-$	±0.074	39.5 h
	Hg	80	−32.206	$0+$		260 y
	Tl	81	−26.81	$2-$		33.0 m
195	Os	76	−26.69			6.5 m
	Ir	77	−31.692	$(\frac{3}{2}+)$		2.5 h
	Pt	**78**	−32.802	$\frac{1}{2}-$	+0.6060	**33.8%**
	Au	79	−32.572	$\frac{3}{2}+$	±0.147	183 d
	Hg	80	−31.05	$\frac{1}{2}-$	+0.538	9.9 h
	Tl	81	−27.85	$\frac{1}{2}+$		1.16 h
	Pb	82	−23.55	$(\frac{3}{2}-)$		17 m
	Bi	83	−17.68			170 s
196	Ir	77	−29.44	$(0-)$		52 s
	Pt	**78**	−32.652	$0+$		**25.3%**
	Au	79	−31.162	$2-$		6.183 d
	Hg	**80**	−31.846	$0+$		**0.146%**
	Tl	81	−27.35	$2-$		1.84 h
	Pb	82	−25.15	$0+$		37 m
	Bi	83	−17.76			
	Po	84	−13.21	$0+$		
197	Ir	77	−28.43	$(\frac{3}{2}+)$		5.8 m
	Pt	78	−30.431	$\frac{1}{2}-$		18.3 h
	Au	**79**	−31.150	$\frac{3}{2}+$	+0.14486	**100%**
	Hg	80	−30.735	$\frac{1}{2}-$	+0.524	64.14 h
	Tl	81	−28.33	$\frac{1}{2}+$		2.84 h
	Pb	82	−24.63			

A	Element (symbol)	Z	Mass excess (MeV/c²)	I^P	μ (nucl. magnetons)	% Abundance or Half-life
198	Ir	77	−25.52			8 s
	Pt	**78**	−29.921	0^+		**7.21%**
	Au	79	−29.591	2^-	+0.590	2.696 d
	Hg	**80**	−30.964	0^+		**10.02%**
	Tl	81	−27.50	2^-	±0.002	5.3 h
	Pb	82	−25.90	0^+		2.40 h
	Bi	83	−19.30	(7^+)		11.85 m
	Po	84	−15.07	0^+		1.76 m
199	Pt	78	−27.42	$(\frac{5}{2}^-)$		30.8 m
	Au	79	−29.104	$\frac{3}{2}^+$	+0.270	3.139 d
	Hg	**80**	−29.557	$\frac{1}{2}^-$	+0.5027	**16.84%**
	Tl	81	−28.08	$\frac{1}{2}^+$	+1.59	7.42 h
	Pb	82	−25.28	$\frac{5}{2}^-$		90 m
	Bi	83	−20.61	$\frac{9}{2}^-$		27 m
	Po	84	−15.05	$(\frac{3}{2}^-)$		5.2 m
200	Pt	78	−26.60	0^+		11.5 h
	Au	79	−27.30	$1^{(-)}$		48.4 m
	Hg	**80**	−29.514	0^+		**23.13%**
	Tl	81	−27.060	2^-	±0.15	26.1 h
	Pb	82	−26.16	0^+		21.5 h
	Bi	83	−20.46	7^+		36.4 m
	Po	84	−16.74	0^+		11.5 m
201	Pt	78	−23.74	$(\frac{5}{2}^-)$		2.5 m
	Au	79	−26.40	$(\frac{3}{2}^+)$		26 m
	Hg	**80**	−27.672	$\frac{3}{2}^-$	−0.5567	**13.22%**
	Tl	81	−27.185	$\frac{1}{2}^+$	+1.60	73.1 h
	Pb	82	−25.327	$\frac{5}{2}^-$		9.4 h
	Bi	83	−21.41	$\frac{9}{2}^-$		108 m
	Po	84	−16.41	$\frac{3}{2}^-$		15.3 m
202	Au	79	−23.86	$1^-, 0^-$		28 s
	Hg	**80**	−27.356	0^+		**29.80%**
	Tl	81	−25.988	2^-		12.23 d
	Pb	82	−25.942	0^+		$\sim 3 \times 10^5$ y
	Bi	83	−21.04	5^+		1.72 h
	Po	84	−17.78	0^+		44.7 m
	At	85	−10.52	(5^+)		181 s
	Rn	86	−5.88	0^+		13 s

A	Element (symbol)	Z	Mass excess (MeV/c²)	I^P	μ (nucl. magnetons)	% Abundance or Half-life
203	Au	79	−22.98	$(\frac{3}{2}+)$		53 s
	Hg	80	−25.277	$\frac{5}{2}-$	+0.84	46.60 d
	Tl	**81**	−25.769	$\frac{1}{2}+$	+1.6115	**29.50%**
	Pb	82	−24.794	$\frac{5}{2}-$		52.1 h
	Bi	83	−21.60	$\frac{9}{2}-$	+4.59	11.76 h
	Po	84	−17.36	$\frac{5}{2}-$		36.7 m
	At	85	−11.97	$\frac{9}{2}-$		7.37 m
204	Au	79	−20.20	(2^-)		40 s
	Hg	**80**	−24.703	0^+		**6.85%**
	Tl	81	−24.353	2^-	±0.089	3.78 y
	Pb	**82**	−25.117	0^+		**1.48%**; $\geqslant 1.4 \times 10^{17}$ y
	Bi	83	−20.82	6^+	+4.25	11.22 h
	Po	84	−18.25	0^+		3.53 h
	At	85	−11.97			9.3 m
	Rn	86	−7.77	0^+		75 s
205	Hg	80	−22.299	$\frac{1}{2}-$		5.2 m
	Tl	**81**	−23.837	$\frac{1}{2}+$	+1.6274	**70.50%**
	Pb	82	−23.777	$\frac{5}{2}-$		1.43×10^7 y
	Bi	83	−21.070	$\frac{9}{2}-$	+5.5	15.31 d
	Po	84	−17.576	$\frac{5}{2}-$	≈+0.26	1.80 h
	At	85	−12.96	$(\frac{9}{2}-)$		26.2 m
	Rn	86	−7.60	$(\frac{5}{2}-)$		2.83 m
206	Hg	80	−20.955	0^+		8.15 m
	Tl	81	−22.269	0^-		4.20 m
	Pb	**82**	−23.795	0^+		**23.6%**
	Bi	83	−20.033	6^+	+4.56	6.243 d
	Po	84	−18.190	0^+		8.8 d
	At	85	−12.73	5^+		29.4 m
	Rn	86	−8.97	0^+		5.67 m
207	Tl	81	−21.041	$\frac{1}{2}+$		4.77 m
	Pb	**82**	−22.463	$\frac{1}{2}-$	+0.5895	**22.6%**
	Bi	83	−20.058	$\frac{9}{2}-$		38 y
	Po	84	−17.150	$\frac{5}{2}-$	~+0.27	350 m
	At	85	−13.310	$\frac{9}{2}-$		1.80 h
	Rn	86	−8.69	$\frac{5}{2}-$		9.3 m
	Fr	87	−2.65			14.8 s

A	Element (symbol)	Z	Mass excess (MeV/c²)	I^P	μ (nucl. magnetons)	% Abundance or Half-life
208	Tl	81	−16.768	(5^+)		3.07 m
	Pb	**82**	−21.759	0^+		**52.3%**
	Bi	83	−18.879	$(5)^+$		3.68×10^5 y
	Po	84	−17.475	0^+		2.898 y
	At	85	−12.64			1.63 h
	Rn	86	−9.56	0^+		24.35 m
209	Tl	81	−13.650	$(\frac{1}{2}^+)$		2.20 m
	Pb	82	−17.624	$\frac{9}{2}^+$		3.253 h
	Bi	**83**	−18.268	$\frac{9}{2}^-$	+4.080	**100%**
	Po	84	−16.373	$\frac{1}{2}^-$	+0.76	102 y
	At	85	−12.888	$\frac{9}{2}^-$		5.41 h
	Rn	86	−8.994	$\frac{5}{2}^-$		28.5 m
210	Tl	81	−9.251	$(4^+, 5^+)$		130 m
	Pb	82	−14.738	0^+		22.3 y
	Bi	83	−14.801	1^-	±0.0442	5.012 d
	Po	84	−15.963	0^+		138.378 d
	At	85	−11.976	(5^+)		81 h
	Rn	86	−9.608	0^+		2.5 h
211	Pb	82	−10.4199	$(\frac{9}{2}^+)$		36.1 m
	Bi	83	−11.865	$\frac{9}{2}^-$		2.14 m
	Po	84	−12.444	$\frac{9}{2}^+$		516 ms
	At	85	−11.653	$\frac{9}{2}^-$		7.214 h
	Rn	86	−8.761	$\frac{1}{2}^-$		14.6 h
	Fr	87	−4.22	$\frac{9}{2}^-$		3.10 m
	Ra	88	0.78	$(\frac{5}{2}^-)$		13 s
212	Pb	82	−7.562	0^+		10.64 h
	Bi	83	−8.135	$1(^-)$		60.55 m
	Po	84	−10.381	0^+		298 ns
	At	85	−8.625	(1^-)		314 ms
	Rn	86	−8.666	0^+		24 m
213	Pb	82	−3.140			10.2 m
	Bi	83	−5.243	$\frac{9}{2}(^-)$		45.59 m
	Po	84	−6.663	$\frac{9}{2}^+$		4.2 μs
	At	85	−6.589	$\frac{9}{2}^-$		110 ns
	Rn	86	−5.706	$(\frac{9}{2}^+)$		25.0 ms
	Fr	87	−3.556	$\frac{9}{2}^-$		34.6 s
	Ra	88	0.29	$(\frac{1}{2}^-)$		2.74 m

A	Element (symbol)	Z	Mass excess (MeV/c^2)	I^P	μ (nucl. magnetons)	% Abundance or Half-life
214	Pb	82	−0.1853	0$^+$		26.8 m
	Bi	83	−1.209	(1$^-$)		19.9 m
	Po	84	−4.479	0$^+$		164.3 μs
	At	85	−3.389			
	Rn	86	−4.328			0.27 μs
	Fr	87	−0.965	(1$^-$)		5.0 ms
	Ra	88	0.09	0$^+$		2.46 s
215	Bi	83	1.71			7.4 m
	Po	84	−0.5405	($\frac{9}{2}^+$)		1.780 ms
	At	85	−1.262	($\frac{9}{2}^-$)		0.10 ms
	Rn	86	−1.179	($\frac{9}{2}^+$)		2.30 μs
	Fr	87	0.309	($\frac{9}{2}$)$^-$		0.09 μs
	Ra	88	2.531	($\frac{9}{2}$)$^+$		1.59 ms
	Ac	89	5.95	($\frac{9}{2}$)$^-$		0.17 s
	Th	90	10.87	($\frac{1}{2}^-$)		1.2 s
216	Po	84	1.769	0$^+$		150 ms
	At	85	2.237	(1)$^-$		300 μs
	Rn	86	0.245	0$^+$		45 μs
	Fr	87	2.975			700 ns
	Ra	88	3.285	0$^+$		182 ns
217	Po	84	5.960			<10 s
	At	85	4.382	$\frac{9}{2}(^-)$		32.3 ms
	Rn	86	3.649	$\frac{9}{2}^+$		540 μs
	Fr	87	4.307	$\frac{9}{2}^-$		22 μs
	Ra	88	5.881	($\frac{9}{2}^+$)		1.6 μs
	Ac	89	8.701	$\frac{9}{2}^-$		111 ns
	Th	90	12.141			252 μs
218	Po	84	8.3546	0$^+$		3.05 m
	At	85	8.099			~2 s
	Rn	86	5.212	0$^+$		35 ms
	Fr	87	7.050			0.7 ms
	Ra	88	6.644	0$^+$		14 μs
	Ac	89	10.837			0.27 μs
	Th	90	12.362	0$^+$		109 ns

A	Element (symbol)	Z	Mass excess (MeV/c²)	I^P	μ (nucl. magnetons)	% Abundance or Half-life
219	At	85	10.53			0.9 m
	Rn	86	8.8307	$(\frac{5}{2}^+)$		3.96 s
	Fr	87	8.617	$(\frac{9}{2}^-)$		21 ms
	Ra	88	9.377			10 ms
	Ac	89	11.56	$(\frac{9}{2}^-)$		7 μs
	Th	90	14.47			1.05 μs
220	Rn	86	10.599	0^+		55.6 s
	Fr	87	11.470			27.4 s
	Ra	88	10.263	0^+		23 ms
	Ac	89	13.747			26.2 ms
	Th	90	14.663	0^+		9.7 μs
221	Rn	86	14.38	$(\frac{7}{2}^+, \frac{9}{2}^+)$		25 m
	Fr	87	13.265	$\frac{5}{2}(^-)$		4.9 m
	Ra	88	12.957			28 s
	Ac	89	14.518			52 ms
	Th	90	16.934			1.68 ms
222	Rn	86	16.3700	0^+		3.8235 d
	Fr	87	16.338	(2)		14.4 m
	Ra	88	14.312	0^+		38.0 s
	Ac	89	16.617			4.2 s
	Th	90	17.197	0^+		2.8 ms
	Pa	91	21.959			5.7 ms
223	Fr	87	18.3823	$(\frac{3}{2})$		21.8 m
	Ra	88	17.2348	$\frac{1}{2}(^+)$		11.434 d
	Ac	89	17.825	$(\frac{5}{2}^-)$		2.2 m
	Th	90	19.256			0.66 s
	Pa	91	22.33			6.5 ms
224	Ra	88	18.813	0^+		3.64 d
	Ac	89	20.219	$0(^-), 1$		2.9 h
	Th	90	19.993	0^+		1.04 s
	Pa	91	23.798			950 ms
225	Ra	88	21.9873	$(\frac{3}{2}^+)$		14.8 d
	Ac	89	21.626	$(\frac{3}{2})$		10.0 d
	Th	90	22.303	$(\frac{3}{2}^+)$		8.0 m
	Pa	91	24.32			1.8 s

A	Element (symbol)	Z	Mass excess (MeV/c²)	I^P	μ (nucl. magnetons)	% Abundance or Half-life
226	Fr	87	27.46			48 s
	Ra	88	23.6657	0^+		1600 y
	Ac	89	24.3010	(1^-)		29 h
	Th	90	23.189	0^+		30.9 m
	Pa	91	26.029			1.8 m
	U	92	27.186	0^+		0.5 s
227	Fr	87	29.58			2.4 m
	Ra	88	27.185	$(\frac{3}{2}^+)$		42.2 m
	Ac	89	25.8500	$\frac{3}{2}(^-)$	$+1.1$	21.773 y
	Th	90	25.8063	$(\frac{3}{2}^+)$		18.718 d
	Pa	91	26.832	$(\frac{5}{2}^-)$		38.3 m
228	Ra	88	28.941	0^+		5.75 y
	Ac	89	28.895	(3^+)		6.13 h
	Th	90	26.758	0^+		1.91313 y
	Pa	91	28.870	(3^+)		22 h
	U	92	29.221	0^+		9.1 m
229	Ac	89	30.72	$(\frac{3}{2}^+)$		62.7 m
	Th	90	29.5809	$(\frac{5}{2}^+)$	$+0.38$	7340 y
	Pa	91	29.887	$(\frac{5}{2})$		1.4 d
	U	92	31.201	$(\frac{3}{2}^+)$		58 m
	Np	93	33.758			4.0 m
230	Th	90	30.8613	0^+		7.7×10^4 y
	Pa	91	32.1655	(2^-)		17.4 d
	U	92	31.607	0^+		20.8 d
	Np	93	35.232			4.6 m
231	Ac	89	35.91	$\frac{3}{2}^+$		7.5 m
	Th	90	33.8122	$\frac{5}{2}(^+)$		25.52 h
	Pa	91	33.4231	$\frac{3}{2}^-$	± 1.98	32760 y
	U	92	33.78	$(\frac{5}{2})$		4.2 d
	Np	93	35.626	$(\frac{5}{2})$		48.8 m
232	Th	**90**	35.4472	0^+		**100%** 14.05×10^9 y
	Pa	91	35.934	(2^-)		1.31 d
	U	92	34.597	0^+		72 y
	Np	93	37.29	(4^+)		14.7 m
	Pu	94	38.362	0^+		34.1 m

A	Element (symbol)	Z	Mass excess (MeV/c²)	I^P	μ (nucl. magnetons)	% Abundance or Half-life
233	Th	90	38.7323	$(\frac{1}{2}^+)$		22.3 m
	Pa	91	37.4871	$\frac{3}{2}^-$	+3.4	27.0 d
	U	92	36.9147	$\frac{5}{2}^+$	+0.54	1.592×10^5y
	Np	93	38.01	$(\frac{5}{2}^+)$		36.2 m
	Pu	94	40.042			20.9 m
234	Th	90	40.612	0^+		24.10 d
	Pa	91	40.349	(4^+)		6.70 h
	U	**92**	38.1426	0^+		**0.0057%**; 2.445×10^5y
	Np	93	39.951	(0^+)		4.4 d
	Pu	94	40.342	0^+		8.8 h
235	Pa	91	42.32	$(\frac{3}{2}^-)$		24.1 m
	U	**92**	40.9164	$\frac{7}{2}^-$	-0.35	**0.72%**; 7.038×10^8y
	Np	93	41.0395	$\frac{5}{2}^+$		396.2 d
	Pu	94	42.16	$(\frac{3}{2}^+)$		25.3 m
236	Pa	91	45.54	(1^-)		9.1 m
	U	92	42.4420	0^+		2.3416×10^7y
	Np	93	43.426	(6^-)		1.15×10^5 y
	Pu	94	42.889	0^+		2.851 y
237	Pa	91	47.64	$(\frac{1}{2}^+)$		8.7 m
	U	92	45.3887	$\frac{1}{2}^+$		6.75 d
	Np	93	44.8693	$\frac{5}{2}^+$	+3.3	2.14×10^6 y
	Pu	94	45.087	$\frac{7}{2}^-$		45.3 d
238	Pa	91	51.27	(3^-)		2.3 m
	U	**92**	47.3070	0^+		**99.27%**; 4.468×10^9y
	Np	93	47.4526	2^+		2.117 d
	Pu	94	46.1608	0^+		87.74 y
	Am	95	48.417	1^+		98 m
	Cm	96	49.398	0^+		2.4 h
239	U	92	50.5722	$\frac{5}{2}^+$		23.50 m
	Np	93	49.3064	$\frac{5}{2}^+$		2.355 d
	Pu	94	48.5851	$\frac{1}{2}^+$	+0.200	24110 y
	Am	95	49.389	$(\frac{5}{2}^-)$		11.9 h

A	Element (symbol)	Z	Mass excess (MeV/c²)	I^P	μ (nucl. magnetons)	% Abundance or Half-life
240	U	92	52.712	0^+		14.1 h
	Np	93	52.21	(5^+)		65 m
	Pu	94	50.1228	0^+		6537 y
	Am	95	51.443	(3^-)		50.8 m
	Cm	96	51.712	0^+		27 d
241	Np	93	54.31	$(\frac{5}{2}^+)$		16.0 m
	Pu	94	52.9530	$\frac{5}{2}^+$	-0.73	14.4 y
	Am	95	52.9322	$\frac{5}{2}^-$	$+1.59$	432.2 y
	Cm	96	53.696	$\frac{1}{2}^+$		32.8 d
242	Pu	94	54.7150	0^+		3.763×10^5 y
	Am	95	55.4627	1^-	±0.382	16.02 h
	Cm	96	54.8015	0^+		162.8 d
	Cf	98	59.332	0^+		3.68 m
243	Pu	94	57.7525	$\frac{7}{2}^+$		4.956 h
	Am	95	57.1701	$\frac{5}{2}^-$	$+1.4$	7380 y
	Cm	96	57.1774	$\frac{5}{2}^+$		28.5 y
	Bk	97	58.685	$(\frac{3}{2}^-)$		4.5 h
244	Pu	94	59.803	0^+		8.26×10^7 y
	Am	95	59.8786	(6^-)		10.1 h
	Cm	96	58.4496	0^+		18.11 y
	Bk	97	60.646	(4^-)		4.35 h
	Cf	98	61.465	0^+		19.4 m
245	Pu	94	63.157	$(\frac{9}{2}^-)$		10.5 h
	Am	95	61.8973	$(\frac{5}{2}^+)$		2.05 h
	Cm	96	61.0013	$\frac{7}{2}^+$		8500 y
	Bk	97	61.811	$\frac{3}{2}^-$		4.94 d
	Cf	98	63.377			43.6 m
246	Pu	94	65.29	0^+		10.85 d
	Am	95	64.92	(7^-)		39 m
	Cm	96	62.6160	0^+		4730 y
	Bk	97	64.02	$2(^-)$		1.80 d
	Cf	98	64.0962	0^+		35.7 h
247	Cm	96	65.530	$\frac{9}{2}^-$		1.56×10^7 y
	Bk	97	65.484	$(\frac{3}{2}^-)$		1380 y
	Cf	98	66.15	$(\frac{7}{2}^+)$		3.11 h
	Es	99	68.55			5.0 m

A	Element (symbol)	Z	Mass excess (MeV/c²)	I^P	μ (nucl. magnetons)	% Abundance or Half-life
248	Cm	96	67.389	0^+		3.40×10^5 y
	Bk	97	67.99	$(6^+, 8^-)$		>9 y
	Cf	98	67.243	0^+		333.5 d
	Es	99	70.22	$(2^-, 0^+)$		27 m
	Fm	100	71.891	0^+		36 s
249	Cm	96	70.748	$\frac{1}{2}(^+)$		64.15 m
	Bk	97	69.8480	$\frac{7}{2}^+$		320 d
	Cf	98	69.7216	$\frac{9}{2}^-$		351 y
	Es	99	71.116	$\frac{7}{2}(^+)$		1.70 h
250	Cm	96	72.986	0^+		~7400 y
	Bk	97	72.950	2^-		3.217 h
	Cf	98	71.1698	0^+		13.08 y
	Es	99	73.17	(6^+)		8.6 h
	Fm	100	74.069	0^+		30 m
251	Cf	98	74.130	$\frac{1}{2}^+$		898 y
	Es	99	74.503	$(\frac{3}{2}^-)$		33 h
	Fm	100	76.0	$(\frac{9}{2}^-)$		5.30 h
252	Cf	98	76.031	0^+		2.638 y
	Es	99	77.15	$(5^-, 4^+)$		471.7 d
	Fm	100	76.822	0^+		25.39 h
	No	102	82.862	0^+		2.30 s
253	Cf	98	79.299	$(\frac{7}{2}^+)$		17.81 d
	Es	99	79.0125	$\frac{7}{2}^+$		20.47 d
	Fm	100	79.346	$\frac{1}{2}^+$		3.00 d
254	Cf	98	81.342	0^+		60.5 d
	Es	99	81.992	(7^+)		275.7 d
	Fm	100	80.899	0^+		3.240 h
	Md	101	83.39			10 m
	No	102	84.729	0^+		55 s
255	Es	99	84.08	$(\frac{7}{2}^+)$		39.8 d
	Fm	100	83.796	$\frac{7}{2}^+$		20.07 h
	Md	101	84.88	$(\frac{7}{2}^-)$		27 m
	No	102	86.87	$(\frac{1}{2}^+)$		3.1 m
256	Fm	100	85.481	0^+		157.6 m
	Md	101	87.42			76 m
	No	102	87.801	0^+		3.3 s

A	Element (symbol)	Z	Mass excess (MeV/c^2)	I^P	μ (nucl. magnetons)	% Abundance or Half-life
257	Fm	100	88.588	$(\frac{9}{2}+)$		100.5 d
	Md	101	89.04	$(\frac{7}{2}-)$		5.2 h
	No	102	90.223	$(\frac{7}{2}+)$		25 s
258	Md	101	91.82	(8^-)		55 d
	No	102	91.52	0^+		1.2 ms
	Lr	103	94.82			4.3 s
	Rf	104	96.55			0.1 s
259	No	102	94.026	$(\frac{9}{2}+)$		60 m
	Lr	103	96.00			5.4 s
	Rf	104	98.50			3.1 s
260	Lr	103	98.14			180 s
	Rf	104	99.23	0^+		$\leqslant$80 ms
		105	103.65			
261	Rf	104	101.25			65 s
		105	104.46			1.8 s
262		105	106.04			34 s
263		106	110.31			0.8 s

Appendix E[a]
"STABLE" PARTICLES

Class	Symbol	Spin, parity	T	T_3	Strangeness	Mass (MeV/c²)	Mean life (sec)	Common decay modes (percent)	
Photon	γ	1				0			
Lepton	ν_e	$\frac{1}{2}$				<0.00006			
	ν_μ	$\frac{1}{2}$				<0.57			
	$e^\pm$	$\frac{1}{2}$				0.5110034 ± 0.0000014	$>10^{29}$		
	$\mu^\pm$	$\frac{1}{2}$				105.65946 ± 0.00024	$(2.19712 \pm .000077) \times 10^6$	$e\nu\nu$	(100)
	τ^-	$\frac{1}{2}$				1782 ± 4	$(4.6 \pm 1.9) \times 10^{-13}$	$\mu\nu\nu$	(18)
								$e\nu\nu$	(17)
								hadrons + neutrals	(65)
Meson	$\pi^\pm$	0^-	1	± 1	0	139.5669 ± 0.0012	$(2.6030 \pm .0023) \times 10^{-8}$	$\mu\nu$	(100)
	π^0	0^-	1	0	0	134.9626 ± 0.0039	$(0.828 \pm .057) \times 10^{-16}$	$\gamma\gamma$	(98.8)
								$\gamma e^+ e^-$	(1.2)
	$K^\pm$	0^-	$\frac{1}{2}$	$\pm\frac{1}{2}$	± 1	493.669 ± 0.015	$(1.2371 \pm .0026) \times 10^{-8}$	$\mu\nu$	(63.5)
								$\pi\pi^0$	(21.2)
								$\pi\pi\pi$	(5.6)
								$\pi\pi^0\pi^0$	(1.7)
								$\mu\pi^0\nu$	(3.2)
								$e\pi^0\nu$	(4.8)
	K^0	0^-	$\frac{1}{2}$	$-\frac{1}{2}$	$+1$	497.67 ± 0.13	50% K_S, 50% K_L		
	$K_S(K_1)$						$(0.8923 \pm .0022) \times 10^{-10}$	$\pi^+\pi^-$	(68.6)
								$\pi^0\pi^0$	(31.4)
	$K_L(K_2)$						$(5.183 \pm .040) \times 10^{-8}$	$\pi^+\pi^-\pi^0$	(21.5)
								$\pi^+\pi^-\pi^0$	(12.4)
								$\pi\mu\nu$	(27.0)
								$\pi e\nu$	(38.8)
								$\pi^+\pi^-$	(0.20)
	$D^\pm$	0^-	$\frac{1}{2}$	$\pm\frac{1}{2}$	0 (Charm = ± 1)	1868.3 ± 0.9	$(2.5 \begin{smallmatrix} +3.5 \\ -1.5 \end{smallmatrix}) \times 10^{-13}$		
	D^0	0^-	$\frac{1}{2}$	$-\frac{1}{2}$	0 (Charm = + 1)	1863.1 ± 0.9	$(2.3 \begin{smallmatrix} +0.8 \\ -0.5 \end{smallmatrix}) \times 10^{-13}$		

Appendix E (continued)

Class	Symbol	Spin, parity	I-spin T	T_3	Strangeness	Mass (MeV/c²)	Mean life (sec)	Common decay modes (percent)
Baryon	p	$\frac{1}{2}$	$\frac{1}{2}$	$+\frac{1}{2}$	0	938.2796 ± .0027	$>10^{37}$	$pe^-\nu$ (100)
	n	$\frac{1}{2}$	$\frac{1}{2}$	$-\frac{1}{2}$	0	939.5731 ± .0027	917 ± 14	
	Λ	$\frac{1}{2}$	0	0	−1	1115.60 ± 0.05	$(2.632 ± .020) \times 10^{-10}$	$p\pi^-$ (64.2) $n\pi^0$ (35.8)
	Σ^+	$\frac{1}{2}$	1	+1	−1	1189.36 ± 0.06	$(0.800 ± .004) \times 10^{-10}$	$p\pi^0$ (51.6) $n\pi^+$ (48.4)
	Σ^0	$\frac{1}{2}$	1	0	−1	1192.46 ± 0.08	$(5.8 ± 1.3) \times 10^{-20}$	$\Lambda\gamma$ (100)
	Σ^-	$\frac{1}{2}$	1	−1	−1	1197.34 ± 0.05	$(1.482 ± .011) \times 10^{-10}$	$n\pi^-$ (100)
	Ξ^0	$\frac{1}{2}$	$\frac{1}{2}$	$+\frac{1}{2}$	−2	1314.9 ± 0.6	$(2.90 ± 0.1) \times 10^{-10}$	$\Lambda\pi^0$ (100)
	Ξ^-	$\frac{1}{2}$	$\frac{1}{2}$	$-\frac{1}{2}$	−2	1321.32 ± 0.13	$(1.641 ± .016) \times 10^{-10}$	$\Lambda\pi^-$ (100)
	Ω^-	$\frac{3}{2}$	0	0	−3	1672.22 ± 0.31	$(0.82 ± 0.03) \times 10^{-10}$	ΛK^- (68.6) $\Xi^0\pi^-$ (23.4) $\Xi^-\pi^0$ (8.0)
	Λ_c^+	$\frac{1}{2}$	0	0	0 (Charm = +1)	2273 ± 6	$\sim 7 \times 10^{-13}$	

[a] Data obtained from Particle Data Group, Review of Particle Properties, Rev. Mod. Phys. **52**, S18 (1980). (This is a compilation of results from various other papers whose references are there listed.) Mean lifetime of τ lepton from G. J. Feldman, et al., Phys. Rev. Lett. **48**, 66 (1982), and of D⁰ meson from N. Ushida, et al., Phys. Rev. Lett. **48**, 844 (1982).

Appendix F

TABLE OF PHYSICAL CONSTANTS

Values are taken from Particle Data Group, Rev. Mod. Phys. **52**, S33 (1980). Numbers in parentheses are standard deviations in the last digits of the quoted value, on the basis of consistency of fits of all constants to all relevant experiments.

Quantity	*Symbol*	*Value and Units*
Speed of light in vacuo	c	$2.99792458(1.2) \times 10^8$ m/s
Avogadro's number	N_A	$6.022045(31) \times 10^{23}$ mole^{-1}
Electron charge	e	$1.6021892(46) \times 10^{-19}$ C
Planck's constant	h	$4.135701(11) \times 10^{-15}$ eV-s
Fine-structure constant	α	$1/137.03604(11)$
Atomic mass unit	u	$931.5016(26)$ MeV/c^2
Electron mass	m_e	$0.5110034(14)$ MeV/c^2
Proton mass	M_p	$938.2796(27)$ MeV/c^2
		$= 1836.15152(70)\, m_e$
		$= 1.007276470(11)\, u$
Classical electron radius	r_e	$2.8179380(70)$ fm (1 fm $= 10^{-15}$ m)
First Bohr radius	a_o	$52.917706(44)$ pm (1 pm $= 10^{-12}$ m)
Compton wavelength of electron	λ_c	$2.4263089(40)$ pm
Rydberg constant for infinite-mass nucleus	R_∞	$10973731.77(83)$ m^{-1}
Bohr magneton	μ_B	$0.57883785(95) \times 10^{-10}$ MeV/T
Electron magnetic moment	μ_e	$1.00115965241(20)\, \mu_B$
Nuclear magneton	μ_N	$3.1524515(53) \times 10^{-14}$ MeV/T
Proton magnetic moment	μ_p	$2.7928456(11)\, \mu_N$
		$= 0.001521032209(16)\, \mu_B$

$$\hbar = \frac{h}{2\pi}$$

Boltzmann's constant	k	$8.61735(28) \times 10^{-5}$ eV/K
Gravitational constant	G	$6.6720(41) \times 10^{-11}$ m^3-kg^{-1}-s^{-2}

Conversion Factors

1 GeV $= 10^3$ MeV $= 1.6021892(46) \times 10^{-10}$ J

Wavelength of 1-eV photon $= 1.2398520(32)\,\mu$m

Temperature for 1 eV per particle $= 11604.50(36)$ K (from $E = kT$)

1 year (sidereal) $= 365.256$ days $= 3.1558 \times 10^7$ seconds

Answers to Selected Problems

Chapter One

1. 3.4Å
3. 0.50 mg at 5 mm; 1.25×10^{-3} mg at 10 mm
7. 70 cm^{-3}
9. $4.2 \times 10^4 \text{ cm/s}$
11. Graph shows that $\ln n$ changes by one unit in about 5.9×10^{-2} mm; this yields $N_A = 6.2 \times 10^{23}$ mole^{-1}
12. $a = 9.56 \times 10^{-4}$ mm; smaller.
13. $1.8 \times 10^{-19} \text{C}$

Chapter Two

3. Frame has a speed of 0.77 c relative to earth.
5. Speed for light moving with water is $(\frac{c}{n} + v)(1 + \frac{vc}{nc^2})^{-1} \approx \frac{c}{n} + v - \frac{v}{n^2}$ to terms of first order in v.
7. $1 - \beta = 1.8 \times 10^{-6}$
9. 9.4 m
13. Neutrino: 546 keV; electron: 237 keV; proton: 0.16 keV
15. $E = (55/12) Mc^2$; $cp = -(23/12) Mc^2$; mass $= 4.163 M$

Chapter Three

1. 5780 K
5. 2430 Å
7. (a) Orbit radii are the same as in hydrogen
   (b) $E_n = -6.80/n^2$
9. ^{1}H: 6564.7 Å; ^{2}H: 6562.9 Å
   (These are wavelengths in vacuum; Fig. 4 shows wavelengths in air.)
11. $\approx 10^{-8}$s
13. $\approx 10^{-5}$ times its original kinetic energy

742

Chapter Four

1. 3.55×10^5 eV
3.   Transition from M shell to K shell; $(9\nu_\beta/8\nu_0)^{\frac{1}{2}}$
     Ca: 18.23   Ti: 20.14   V: 21.11   Cr: 22.16   Mn: 23.10
5.   (a) 0.0005%; -1.2×10^{-5} eV   (b) 19,700%; -99.5 MeV
7.   (a) 12.8 eV   (b) 9.1×10^4 eV
13. (a) 1.3×10^6 m/s; 5.5 Å   (b) 2.99×10^8 m/s; 0.0023 Å
     (c) 3.1×10^7 m/s; 1.3×10^{-4} Å   (d) 3.3×10^{-6} m/s; 1.4×10^{-17} Å
19. $\phi(k) = \frac{2 \sin kb}{k \sqrt{2\pi}}$
21. (a) 10^{-5} rad   (b) $\sim 10^{-3}$
23. $\sim 10^{-16}$ m

Chapter Five

1.   0.04
5.   No
7. $ka = 1.68$
9. 4.9×10^{-4}
11. (a) $1.026 \, V(a)$ (b) $0.794 \, V(a)$
13. (b) $\delta x = 0.59 \, a$; $\delta p_x = 1.81 \, \hbar/a$ (c) $x = \pm \, a/2$
15. $h\nu/4$; $mh\nu/2$
17. $u_1 = $ constant $\times \, x \, e^{-ax^2/2}$ $u_2 = $ constant $\times \, (1 - 2ax^2)e^{-ax^2/2}$

Chapter Six

1.   10.7 eV, 17.8 eV, 23.2 eV; third level doubly degenerate
3. $\hbar^2/2$; Yes
5. $\frac{1}{4}(1 + \cos \alpha)^2$, $\frac{1}{2}\sin^2\alpha$, $\frac{1}{4}(1 - \cos \alpha)^2$
9. $\sqrt{\frac{105}{32\pi}} \sin^2 \theta \cos \theta \, e^{2i\phi}$
11. $E = [(q + \frac{1}{2})\omega_x + (r + \frac{1}{2})\omega_y + (s + \frac{1}{2})\omega_z]\hbar$, where q, r, and s are posi-
     tive integers or zero. $E_1 = (7/6)\hbar\omega_z$, degeneracy 1; $E_2 = (11/6)\hbar\omega_z$, degen-
     eracy 2; $E_3 = (13/6)\hbar\omega_z$, degeneracy 1; $E_4 = (5/2)\hbar\omega_z$, degeneracy 3.
     No, because the Hamiltonian does not commute with the L^2 operator.
13. $u_{200} = [A(Y_{2,2} + Y_{2,-2}) + BY_{2,0} + CY_{0,0}](a^3/\pi)^{1/4}e^{-ar/2}$
     where $A = 2ar^2/\sqrt{15}$, $B = 4ar^2/\sqrt{90}$, and $C = \sqrt{2}(ar^2 - 3)/3$
     Probabilities: $P_{2,2} = P_{2,-2} = 1/4$; $P_{2,0} = 1/6$; $P_{0,0} = 1/3$
     $\langle L^2 \rangle = 4\hbar^2$; $\langle L_z^2 \rangle = 2\hbar^2$; No.

15. 4.2 eV

17. $j_3(x) = \left(\frac{15}{x^4} - \frac{6}{x^2}\right) \sin x - \left(\frac{15}{x^3} - \frac{1}{x}\right) \cos x$

19. $\sigma = 42.8 \, \pi a^2$; $d\sigma/d\Omega = \sigma/4\pi = 10.7 \, a^2$

Chapter Seven

3. u_{100}: $3 \, a_0'/2Z$ u_{210}: $5 \, a_0'/Z$ u_{320}: $21 \, a_0'/2Z$

5. Approximately 10^{-15} and 2×10^{-26}, respectively.
   Higher ℓ means more circular orbit, less chance of penetrating nucleus.

9. (a) $S_z' = (S_x + S_z)/\sqrt{2}$

   (b) $a\left(\begin{smallmatrix}1\\0\end{smallmatrix}\right) + b\left(\begin{smallmatrix}0\\1\end{smallmatrix}\right)$, where $a = \frac{1}{2}\sqrt{2-\sqrt{2}}$ and $b = \sqrt{1-a^2}$

   (c) 0.854

11. $\left\{Y_{1,1}(-)_s + \sqrt{2}\, Y_{1,0}(+)_s\right\}/\sqrt{3}$; $J^2 = 15\,\hbar^2/4$
    $\left\{\sqrt{2}\, Y_{1,1}(-)_s - Y_{1,0}(+)_s\right\}/\sqrt{3}$; $J^2 = 5\,\hbar^2/4$

13. $3d_{5/2}$: -2.235×10^{-6} eV $2p_{3/2}$: -1.1315×10^{-5} eV

$\left.\begin{matrix}3d_{3/2}\\3p_{3/2}\end{matrix}\right\} -6.705 \times 10^{-6}$ eV $\left.\begin{matrix}2p_{1/2}\\2s_{1/2}\end{matrix}\right\} -5.6573 \times 10^{-5}$ eV

$\left.\begin{matrix}3p_{1/2}\\3s_{1/2}\end{matrix}\right\} -2.0115 \times 10^{-5}$ eV

In Angstroms: 6564.664, 6564.679, 6564.726, 6564.522, and 6564.569, for transitions $3d_{5/2} \rightarrow 2p_{3/2}$, $3d_{3/2} \rightarrow 2p_{3/2}$, $3s_{1/2} \rightarrow 2p_{3/2}$, $3d_{3/2} \rightarrow 2p_{1/2}$, and $3p_{1/2} \rightarrow 2s_{1/2}$, respectively.

Chapter Eight

1. $\Delta E_1 = 0.297\,\delta$; $a_{13} = 2\,ma^2\delta/\hbar^2\pi^2$ $a_{15} = 2\,ma^2\delta/27\hbar^2\pi^2$

3. $\Delta E_1 = 0.296\,\delta$ if $\delta \simeq 0.05\,E_1$

5. Probability of $A^+B^+ = 0.5$, of $A^+C^+ = 0.146$, of $B^+C^+ = 0.146$
   $0.5 > 0.146 + 0.146$, so inequality is violated.

7. Li^+: -196.46 eV, or 0.80% high
   Be^{2+}: -369.86 eV, or 0.44% high
   B^{3+}: -597.66 eV, or 0.30% high
   C^{4+}: -879.86 eV, or 0.23% high

9. $E_g = -14.40$ eV; $E_{predicted} = -12.86$ eV.   No.

11. 118

Chapter Nine

1. 0.002 eV

3. 20

5. Principal series; s states at -4.34, -1.73, -0.94, and -0.59; p states at -2.73 (splitting 0.007), -1.28, -0.745, and -0.49.

7. $(6.61/K) \times 10^6$ Å

9. 0.479, 0.240, 0.160, 0.120, 0.096, and 0.080 mm

11. 1.284 Å. Increases to 1.303 Å. Increases to 1.285 Å.

13. $1.00 \pm (0.00348 + 0.00232\,n)$ Hz, where n is a positive integer or zero.

Chapter Ten

1. Zero

3. $\phi\omega = \dfrac{iE}{\sqrt{8\pi}} \left\{ \dfrac{1 - e^{-i(\omega_0 - \omega)t'}}{\omega_0 - \omega} + \dfrac{1 - e^{-i(\omega_0 + \omega)t'}}{\omega_0 + \omega} \right\}$

   $|\phi(\omega)|^2 = \dfrac{E^2}{2\pi} \dfrac{\sin(\omega_0 - \omega)t'/2}{(\omega_0 - \omega)^2}$ for $\omega \simeq \omega_0$

9. All transitions are allowed; $7.26 \times 10^{-7} \left(\dfrac{1-\epsilon}{1+\epsilon} \right)^4$ s

11. (a) 0.13 T (b) 1.18×10^{10} Hz (c) population in level 3 is 10% higher than population in level 4.

Chapter Eleven

3. $2.476\,R$

5. (a) Starting with n_0, distributions are 2, 2, 2, 2, 0, 1, and 2, 2, 2, 1, 2.
   (b) Probability is 0.50 for each distribution
   (c) 2, 2, 2, 1.5, 1, 0.5   (d) 0.04 eV   (e) 105 K

7. 1.4×10^8; one part in 1.9×10^{13}.

9. (a) 2.0   (b) 7.2

13. (a) $g(\epsilon) = 4\,Ma^2\pi/h^2$ (b) 2.61×10^{-4} eV

15. $\sim 10^{-9}$

Chapter Twelve

3. $2.21\,V_0 b/a$. Worse agreement with $P = 1$, because "perturbation" is stronger

7. Bragg reflection planes are [111], [220], [311], and [400]. Second zone has shape of Fig. 17, but is bounded by [220] planes rather than [110] planes.

Chapter Thirteen

1. 5.1×10^5 min^{-1}

3. $N_2 = (\lambda_1 N_0/(\lambda_2 - \lambda_1))(e^{-\lambda_1 t} - e^{-\lambda_2 t})$

5. ^{232}Th: 2.60×10^{21} ^{228}Ra: 1.062×10^{12} ^{228}Ac: 1.292×10^8
   ^{228}Th: 3.53×10^{11} ^{224}Ra: 1.84×10^9 ^{220}Rn: 3.25×10^5
   ^{216}Po: 8.8×10^2 ^{212}Pb: 2.24×10^8 ^{212}Po: < 1
   ^{212}Bi: 2.13×10^7 ^{208}Tl: 3.63×10^5

7. (a) 54 MeV   (b) 13.2 MeV

13. 0.995 ns
15. $m_e c^2 \pm 1.1$ keV; 0.25 keV
17. $\delta\nu/\nu = 4.3 \times 10^{-13}, 3.7 \times 10^{-13}$, and 2.1×10^{-13}, for iron
19. 6.5×10^{10}
21. 2.0 millirad
23. 160 rad

Chapter Fourteen

3. $R = 6.286$ fm; $R_0 = 1.257$ fm
5. 5.60×10^6 Hz
7. Physical scale: 1 old amu $= (1 - 0.00032)$ new amu
   Chemical scale: 1 old amu $= (1 - 0.00004)$ new amu
9. (d) a is 4% smaller than b
11. $A = 27: Z = 13; A = 125: Z = 53$
13. 22.3 h
15. β^- possible for n, ^{3}H, ^{6}He, ^{8}He, ^{8}Li, ^{9}Li
   β^+ possible for ^{6}Be, ^{8}B, ^{9}B, ^{9}C
   ec possible when β^+ is possible, and also for ^{7}Be
   n emission for ^{5}He; p emission for ^{5}Li; α emission for ^{8}Be

Chapter Fifteen

1. 3×10^{23} y
3. 4.95; yes; 110 minutes
5. Half-life is expected to be about 5 minutes, which is the correct order of magnitude.
7. 22 barns
9. (a) 0.20 Ci   (b) 0.14 Ci
11. 13.7 MeV; 3.7 MeV
13. 6.00 MeV
15. (a) 12 MeV   (b) 3.5 MeV   (c) 3.3 MeV   (d) 5.3 MeV and 1.7 MeV
17. 1.146 MeV
19. $T = 3/2$; 11.6 MeV; ^{14}C$(d,n)^{15}$N and ^{18}O$(p,\alpha)^{15}$N

Chapter Sixteen

5. (a) Numbers 2, 5, and 6: S not conserved; number 3: B not conserved; number 7: energy not conserved; number 9: lepton number not conserved.

   (b) numbers 3, 7, and 9, for reasons given in part (a).
9. (a) $d\bar{u}e^+$ (b) $e^+\pi^-$

Index

A

Aberration, stellar, 29–30
Absorption
   of nuclear radiation, 480—489, 515–516
   "recoilless" resonant, 501–503
   of X-rays, 132
Absorption edge, 132
Aces, 657, 664
Actinides, 310
Alkali atoms, 314–316
Alkali halides, 331–333
Alkali metals, 314–316, 431–432
Alpha decay, 473–474, 566–571, 618
Alpha particles, 472
   range of, 484
   scattering of, 77, 475–480, 521, 526–527
Angular correlation
   of annihilation radiation, 491, 497–501
   in gamma decay, 597–599
Angular momentum, 190–211, 229, 234
   addition of vectors, 251–258, 317
   in Bohr theory, 77
   of hydrogen-atom states, 242–246, 251–258,
      274
   measurement of, 199–211
   of nucleus, 531–533, 555–556
   operators for, 193, 250–258
   orbital, 193, 248–260, 308
   spin, 248–260, 292–293, 308
   in three-dimensional harmonic oscillator,
      211–214
   uncertainty relations for, 200, 207–211
Annihilation
   neutrino-antineutrino, 578
   positron-electron, 273
   proton-antiproton, 664
   quark-antiquark, 664
Antimatter, 60, 239, 271–273
Antiparticle, 60, 578, 641, 657
Antisymmetric wavefunction, 289–294, 296–297,
   303, 323–324, 334
Associated Legendre function, 196, 361
Associated production, 636

Atmospheres, law of, 7
Atom
   Bohr model of, 75–82, 101–103
   electronic structure of, 102–103, 304–311
   hydrogen, 75–82, 239–278
   one-electron, 75–82, 239–278, 690–694
   Rutherford model of, 77
   size of, 23, 79
   Thomson model of, 77
Atomic beam, 207–211, 532–533, 561
Atomic clock, 369
Atomic mass
   table, 695–737
   unit, 335, 538, 561–562, 605
Atomic number, 101–103, 480
Atomic species (table), 695–737
Atomic spectroscopy, 313–331
Attenuation coefficient, 487–490
Auger effect, 321, 593, 599
Autoionization, 321
Avogadro's number, 3
   determination of, 16–23, 73

B

Balmer series, 76, 81, 93, 264
Band, spectral, 342
Band theory of solids, 431–432, 438–446
Barn, definition of, 601
Barrier penetration, 164–170, 566–571
Baryon, 739 (table)
   definition of, 637
   number, 637–638, 668
Beauty, 666
Bell inequality, 296, 311
Beta decay, 571–594
   favored transitions in, 590–591, 614
   Fermi theory of, 578–589
   Fermi transitions in, 588–589
   forbidden transitions in, 586–589, 591,
     595–597

Gamow–Teller transitions in, 588–589
inverse, 574
stability against, 544–547
theory of, 571–594
Beta ray (*see also* Electron), 472
Binding energy
molecular, 336–337, 342–343
nuclear, 537–542, 547, 557
Binomial distribution, 679–685
Black-body radiation, 64–73, 93, 355–357, 394–397
Bloch function, 440, 499
Bloch's theorem, 440–442, 455
Body-centered cubic lattice, 459–460
Bohr atom, 75–82, 101–103
Bohr magneton, 249, 265, 278
Bohr radius, 79, 246, 278, 694
Boltzmann constant, 7
Boltzmann distribution, 7, 368, 382–394, 418
Boltzmann factor, 7, 69, 93, 356, 370, 380, 382
Boltzmann statistics, 382–394
Bose–Einstein condensation, 401, 406
Bose–Einstein statistics, 382–392, 394–410
Boson, intermediate vector, 663–664
Bottom quark, 666
Boundary conditions, 141–145, 153, 185, 215
Born–von Karmann, 433–435, 439
Boyle's law, 2
Brackett series, 81, 94
Bragg condition, 98, 435, 450
Bragg reflection, 97–99, 450–454, 468, 686–689
Bremsstrahlung, 100, 487
Brillouin zones, 449–462, 467–469, 498, 500

C

Cathode rays, 12–14, 78
Cavity radiation, *see* Black-body radiation
Central force field, 190
Centrifugal potential, 216, 218–223
Chain reaction, 550–551
Charge conjugation, 642–644
Charged particles
absorption of, 480–487
energy loss of, 480–497
range of, 483–487
Charm, 634, 664-667
Chemical shift of nuclear energy levels, 512–513
Collective model of nucleus, 537, 555–556, 594
Commuting operators, 206–207, 268–271
angular momentum and, 206–207, 252–254, 268–271
Complementarity, 127–130
Compound nucleus, 601, 604–607

Compound probability, 678
Compton effect, 103–106, 132–133, 487–490
inverse, 132
Compton wavelength of electron, 105
Conduction band, 463–463
Conductivity
electrical, 410, 423–425, 462–465, 469
thermal, 402, 410
Conjugate variables, 126
Conservation
of energy
in beta decay, 571–574
in relativity theory, 54
in transitions, 351–355
of momentum
in beta decay, 572–574
in Mössbauer effect, 502–505
in relativity theory, 54
Contact potential difference, 415–417, 427
Continuity
equation, 147
of wavefunction, 141, 149, 151
Contraction, Lorentz, 38–39, 45–48
Cooper pair, 447–449, 501
Copper, Fermi surface of, 461, 498, 500
Correspondence principle, 82–89, 93, 184
harmonic oscillator and, 179
Cosmic rays, 132, 514, 634
Coulomb potential, 190, 566
Coupling
jj, 318–321
LS, 318–321
Covalent binding, 332–335
Covariant vectors, 50, 53
Cross section
differential, 224, 232–234, 236–237, 479
neutron absorption, 575, 602, 618
in nuclear reactions, 601–603
positron annihilation, 491–492
Rutherford scattering, 479
total, 224–225, 230–234, 236, 237
Crystal lattice, 450, 459–461, 468, 499, 501
Curie (unit), 516
Curie law, 421
Cyclotron resonance, 457–458

D

Dalton's law, 3
Davisson–Germer experiment, 108–111, 133, 435
DeBroglie wavelength, 107–111, 133, 137–138, 499

DeBroglie wave, 106–111, 133–134
Debye temperature, 399, 428
Decay constant (radioactivity), 474
Decuplet, 659–661
Defects in solids, 496–497
Degeneracy, 161, 212, 283, 286–287, 381, 673
Delta function, 123–124, 146, 185
Density of states, 395, 403, 422, 463, 469, 583
Deuterium, 82
Deuteron, 556–560, 615–616
Diamond structure, 468
Diffraction
   of electrons, 108–111
   of heavy particles, 111
   of light, 124–127, 130
   and uncertainty principle, 124–127
   of X-rays, 23, 96–99
Dirac delta function, 123–124
Dirac equation, 239, 266–273, 278, 622, 644
Displacement law, Wien, 65–66, 92
Dissociation energy, 336–337, 342–343
Doping in semiconductors, 463–465
Doppler effect, 60–61, 132
   transverse, 511–512
Doppler broadening of spectral line, 505–508
Dose, radiation, 514–519
Dulong–Petit law, 397–398

E

Earth, age of, 474
Effective mass, 444–446, 466, 470
Effective potential
   in hydrogen atom, 243–244
   in square well, 219
Eigenfunctions, 175
   of angular momentum operators, 195–207,
      213–214, 251–258, 274–277, 330
   for harmonic oscillator, 181, 185, 212–214
   for hydrogen atom, 246
   for one-dimensional square well, 150–164
   for one-electron system, 685
   parity of, 161–164
   for three-dimensional square well, 217–223
Eigenvalues, *see also* Eigenfunctions
   of angular momentum, 195–200, 251–258, 287
   of Dirac operators, 278
Einstein heat capacity, 397–398
Einstein photoelectric equation, 74–75
Einstein theory of relativity, 27–60
Einstein theory of spontaneous transitions,
   355–357

Electric dipole transitions, 266, 349–354,
   359–363
   in gamma decay, 594–595
Electric quadrupole moment, 533–537, 551–552,
   556, 558
Electric quadrupole transitions, 363–364, 377
   in gamma decay, 594–595
Electron, 12-16
   capture in nuclei, 544–547, 563, 592–594
   charge of, 13, 18–23, 73
   discovery of, 12–16
   energy loss by, 423–425, 486–487
   energy spectrum in beta decay, 571, 580–581
   "free", 410–427
   magnetic moment of, 249, 421–422
   "nearly free", 433
   orbits in a metal, 455–459, 498
   passage through matter, 100, 486–487
   radiation by, 100
   recoil, 104–106, 132
   rest energy of, 738
   scattering from nuclei, 528–530
   shower, 487
   spin of, 248–260, 292–293, 304, 334
   thermionic emission of, 417–420, 428
Electronic-vibrational-rotational spectra, 340–342
Electronic structure of solids, 429, 470
Electroweak theory, 664
Elementary particles, 621–676
   table, 738–739
Elements, periodic table of, 304–311
Elliptical orbits, 85–90
Energy gap
   in solid, 430, 438, 443, 450–451, 462–464,
      467, 469
   in superfluid, 408
Energy levels
   of hydrogen atom, 80
   molecular, 331–343
   nuclear, 556, 594–600
   for one-dimensional square well, 153–160
   of perturbed system, 281–287
   positive, 321–322
   for three-dimensional square well, 217–223
Energy operator, 173, 212, 270, 278
Equipartition of energy, 66, 71
Equivalent electrons, 342
Ether, 28
Exchange energy, 293–294, 318
Exclusion principle, 271, 292–293, 322–326, 332,
   335, 342, 423–425, 472
Expectation value, 170–176, 185, 202, 204, 234,

273, 294, 436
of momentum, 174, 185
Exponential decay, 357–359, 472–473
Extended-zone scheme, 456–457, 498

F

Face-centered cubic lattice, 459–461
Factoring of operator, 178
Faraday (unit), 12
Fermi–Dirac statistics, 382–392, 462–463
    applications of, 410–426
Fermi energy, 412, 464, 498
    calculation of, 414–415, 426
Fermi level, 416
Fermi surface, 449–462, 498, 500
Fine structure
    in alkali-metal spectra, 86–87, 315–316
    in hydrogen spectrum, 247–266, 277–278
Fine-structure constant, 81, 259
Fission, nuclear, 548–551
Fitzgerald–Lorentz contraction, 38–39, 45–48
Fizeau's experiment, 33
Forbidden region, 151–153, 163, 184
Forbidden transitions, 359, 363–364, 366
    in beta decay, 586–589, 591, 595–597
Fountain effect, 409–410
Four-force, 54
Four-vectors, 43, 50–58
Fourier integral, 117–118, 499
Fourier series, 113–117, 280
Fourier transform, 118, 121–124, 126, 134–135,
    376
Franck–Hertz experiment, 90–92
Free electron theory of metals, 410–428, 433, 436
Fresnel dragging coefficient, 30, 32–34
Fusion, nuclear, 619–620

G

Galilean transformation, 48, 51
Gamma decay, 594–600
    angular correlation in, 597–599
    mean lifetime for, 595
Gamma rays, 60, 472, 487–490, 501–503,
    515–516
Gauge particle, 663, 668
Gaussian wavefunction, 121–123, 182
Gay–Lussac's law, 3
Geiger–Marsden experiment, 77, 475–480
Gell-Mann–Okubo mass formula, 676
Gluon, 626, 661, 667
Grating, diffraction, 24, 126–127
Gravitational red shift, 508–511
Group velocity, 118–124, 445, 470
Gyromagnetic ratio, 330

H

Hadron, 634, 655
Half-life, 618–619
    of atomic state, 358–359
    of beta emitter, 546–547, 589–591, 593–594,
        596–597
    biological, 516–518
    and energy in alpha decay, 569–571, 618
    of radioactive species (table), 695–737
Hall effect, 465–466
Hamiltonian operator, 173, 212, 270, 278
    for harmonic oscillator, 178, 212
    for perturbed system, 280, 346
Harmonic oscillator, 176–182, 185, 362
    anisotropic, 235–236
    energy levels of, 181
    perturbed, 311
    three-dimensional, 211–216, 235–236
Heat capacity
    Debye theory, 398–400, 425–428
    Einstein theory, 397–398, 425
    Electronic, 411–415, 428
Heisenberg microscope, 124–126
Heisenberg uncertainty principle, 123–131, 136
Helium
    atom, 296–305
    ionized, 82
    liquid, 400–410
Hidden variables, 130–131
Hoek's experiment, 32
Hole
    in Dirac theory, 271–273, 466
    in semiconductor, 465–466, 469
Hydrogen, 190, 239–278, 376–377
    Bohr theory of, 75–82
    energy levels of, 80, 242–246, 693
    isotopes of, 82
    molecules, 332–334
    ortho and para, 293
    radial equation, 242, 690–694
    spectrum of, 76
    wavefunctions, 246, 297, 301, 690–694
Hypercharge, 659
Hyperfine structure, 265–266, 523, 536

I

i-spin, 616–617, 626, 657, 673–676
    and strangeness, 639–641
Ice, positron annihilation in, 499–500
Identical particles, 288–296
Impact parameter, 476
Indeterminacy, see Uncertainty principle
Indistinguishability, 288–289, 380, 382
Induced transitions, 346–357

Inertial frame of reference, 27
Insulators, 429–430, 462–463
Internal conversion, 599–600
Intrinsic semiconductor, 462–467, 469
Ionic bond, 331–332, 335
Ionization energy or potential, 102, 298–301,
    305–309, 312
Isobar, 474, 544, 615–617
Isobaric spin, *see i*-spin
Isomer shift, 512–513
Isotope, 82, 471
    shift, 527–528
Isotopic spin, *see i*-spin

J

J/ψ resonance, 665
jj coupling, 318–321

K

K⁰ meson, two lifetimes of, 645-650
K₁ meson
    regeneration of, 648–649
    and K₂ meson, mass difference, 649–650
K X-ray, 99–103, 593
Kinetic energy, relativistic, 56–59, 260
Kinetic theory of gases, 2–12
Klein–Gordon equation, 624
Kronig–Penney potential, 440–444, 467
Kurie plot, 584, 588, 596

L

L X-ray, 103, 593
Lagrange multipliers, 388
Lamb shift, 266
Lambda particle, 634, 636, 639–641, 655–657,
    665
Lambda point, 402, 406
Landé *g* factor, 330
Landé interval rule, 320, 342
Lanthanides, 310
Larmor precession, 208–211, 328, 561
Laser, 367–376
Legendre polynomial, 196, 227
    associated, 196, 361
Lepton, 630–633, 738 (table)
Lepton number, conservation of, 630–631, 638,
    668
Line spectra, 75–81
Liquid-drop model of nucleus, 541
Lithium, ionized, 82, 301, 307, 312
Local realism, 294–296

Lorentz contraction, 38–39, 45–48
Lorentz transformation, 40–50, 52, 107, 267
LS coupling, 318–321
Lyman series, 76, 81, 94
    alpha line, positronium, 496

M

Magic number, 551–555, 571
Magnetic dipole transition, 266, 364–365
    in gamma decay, 595
Magnetic moment
    of atom, 207–211
    of deuteron, 556, 559
    of electron, 249
    of neutron, 558, 559
    of nucleus, 265, 531, 563, 695–737 (table)
    of proton, 247, 265, 558, 741
Maser, 367–376
    ammonia, 368–370
    solid state, 370–373
Mass
    atomic, unit of, 335
    conversion into kinetic energy, 58, 61, 538
    defect, 58
    effective, 444–446, 470
    invariant, 53
    reduced, 80–82, 685
    relativistic, 53
Mass spectrometer, 538–539
Matrix
    element, 281, 353, 359–366, 377, 493,
        579–589, 594, 598, 614
    representation, 274–277
Maxwell velocity distribution, 6, 380
Maxwell–Boltzmann distribution, *see* Boltzmann
    distribution
Maxwell's equations, 28, 345
Mean free path, 3–6
Mean lifetime, 474
    of atomic state, 358–359
    of nuclear state, 595
    of pion, 629
Mercury, 92, 94
Meson, 622–626, 634, 738 (table)
    D, 634
    eta, 652
    K, 633–637, 640–650, 655, 657
    omega, 654
    pi, 58, 523–524, 622–630
    phi, 652
    rho, 652–653, 669
    virtual, 623–626
Metals
    band theory of, 431–432, 438–446

free electron theory of, 410–427
Michelson–Morley experiment, 34–36
Miller indices, 453
Millikan oil-drop experiment, 18–23, 75, 662
Mirror nuclei, 613–614
Mobility of charge carriers, 469
Molecule, 3, 10, 331–343
    alkali halide, 331–333
    ammonia, 368–370
    hydrogen, 332–334
Momentum
    four vector, 53
    operator, 172–173, 175
    relativistic, 54, 134
Morse potential, 336, 343
Moseley's law, 100–103, 131
Mössbauer effect, 501–513, 523, 533, 537
Multiplicity, 319
Multiple radiation, 363–364
Muon, 58, 93, 131
    decay of, 631–632
    range in matter, 484–486, 521
Muonic atom, 93
Muonic X-ray, 527–528
Muon number, 631–632, 651
Muon-type neutrino, 572, 631–632, 651–652

N

Natural linewidth, 351–354, 502–503, 507
Neutrino, 58, 516, 571, 668
    anti-, 60–61, 572–578, 630
    cross-section for interaction of, 514, 574, 635
    detection of, 574–577
    mass of, 573, 585–586
    muon type, 572, 631–632, 651–652
    oscillations, 650–652
    scattering of, 644
    solar, 651–652
    tau type, 572, 632–633, 651–652
Neutron, 9
    absorption of, 516, 548–551, 575–576, 618
    bomb, 524
    capture, 548–551, 575–576
    delayed emission of, 551, 564
    diffraction, 111
    discovery of, 612, 620
    existence in nucleus, 531
Noble gases, 308–310, 312
Normal distribution, 681–685
Normalization, 146–147, 153, 164, 185, 213, 258, 281
Nuclear Coulomb factor, 580, 584
Nuclear fission, 548, 551
Nuclear force
    charge independence of, 612–616

coupling constant, 624, 636–637
meson theory of, 622–626
range of, 525, 541
Nuclear fusion, 619–620
Nuclear magnetic resonance, 531
Nuclear magneton, 265, 278, 532, 558
Nuclear models, 539–542
Nuclear reactions, 600–620
Nuclear reactor, 551, 619
Nucleon, 613, 622
Nucleus
    angular momentum of, 531–533, 555–556
    binding energy of, 537–542, 547
    charge of, 526
    collective motions in, 537, 555–556, 594
    compound, 601, 604–607
    deformed, 555–556
    electric quadrupole moment of, 533–537, 551–552, 562
    electrostatic energy of, 533–536, 542, 560–561
    Fermi gas model of, 539–542, 564
    magnetic dipole moment of, 265, 531, 563, 695–737 (table)
    mass of, 537–542, 562–564
    parity of, 542–543, 563, 589, 594
    properties of, 525–564
    radius of, 526–530, 561, 571
    recoil of, 502
    skin thickness of, 529
    spin of, 531–533, 536–537, 555–556, 616
    stability of, 543–551, 562–563

O

Occupation index, 383–384, 412, 449, 464
Octet, 659, 673
Omega meson, 654
Omega particle, 639, 660–661
Operators
    for angular momentum, 193
    commutation of, 206–207
    momentum, 172–173, 175
    in quantum theory, 170–176
Orbit
    Bohr, 78
    dog's bone, 462
    elliptical, 85–89, 309
    on Fermi surface, 455
    in metal, 457–458
Ortho-hydrogen, 293
Ortho-positronium, 493
Orthogonality
    of particle states, 648–650, 673–674
    of wavefunctions, 204, 281, 301, 347, 362
Oscillator
    in Debye theory, 398–400

in Planck theory, 65, 396–397
simple harmonic, 176–182, 211–216, 235–236

P

Pair production, 272, 487
Pairing force, 555
Para-hydrogen, 293
Para-positronium, 492
Paramagnetic susceptibility, 420–423
Paramagnetism, 420–423
Parity, 642–643
non-conservation in beta decay, 543, 591–592,
642
in K-meson decay, 641–642
of nucleus, 542–543, 563
of pion, 627–629
of wavefunctions, 161–164, 217, 283–285,
361–362
Partial half life, 547, 563, 596
Partial waves, 225–234, 236–237
Particle in a box, 188–190
Parton, 661
Paschen–Back effect, 331
Paschen series, 76, 81
Pauli exclusion principle, *see* Exclusion principle
Pauli spin matrices, 276
Periodic table
of elements, 304–311
of quarks and leptons, 667
Perturbation of degenerate eigenfunctions, 283,
286–287, 438
Perturbation theory, 250, 297, 303, 311, 334, 433,
437–438, 468
time dependent, 346–355, 376–377, 578
time independent, 280–287, 297–301, 311, 348
Pfund series, 81
Phase shift, 226–233, 237
Phase velocity, 119
Phonons, 397, 505–506
Photoelectric effect, 73–75, 93, 487–490
inverse, 100
Photons, 73–75, 93, 103–106, 738
absorption of, 487–490
attenutation coefficient, 489–490
virtual, 623
Physical constants (table), 740
Pion, 58, 523–524, 622–630, 652–654, 669–670
Planck's constant, 63, 139
Planck's radiation law, 68–73, 92
Poisson distribution, 5, 679–685
Polarization, 15, 353, 377
Population inversion, 367–374
Positron, 273, 466, 490–501, 521–523, 538
annihilation in condensed matter, 415, 457,
494–501

annihilation in gases, 494
creation, 272, 487, 600
emission, 544, 546, 592
energy spectrum in beta decay, 580–581
lifetimes, 491–496
three-photon annihilation of, 493
trapping in defects, 496–497
Positronium, 93, 491–496, 521–522, 655
chemistry, 494–496
excited states of, 93, 496
Lyman alpha line, observation of, 496
Postulates
of Bohr, 77
of quantum theory, 175, 203, 280
of special relativity, 39
Potential barrier, 164, 170
in alpha decay, 567–571
Potential energy
centrifugal, 216, 218–223, 243
Coulomb, 190, 243, 349
Morse, 336
one-dimensional square well, 150–164,
182–184
periodic, 436–446
simple harmonic oscillator, 173, 211, 215
spherically symmetric, 190–191, 217–237
three-dimensional square well, 217–223
Yukawa, 622–626
Potential step, 142–145
Principal quantum number, 243
Principle of relativity, 39
Probability, 184, 236, 274, 285, 677–685
amplitude, 141
compound, 678
current, 146–150, 168–169
definition, 677
density, 122, 141, 146–150, 184, 197, 235,
245, 293, 436, 567
distributions, 679–685
expansion of wavefunction and, 202
in observation of waves, 111, 130
Proper time, 52
Proton, 60
decay of, 667–668
i-spin, 616
magnetic moment, 247, 265, 558, 741
mass, 739
range in matter, 483–486, 516, 521
Pumping
of laser, 374
of maser, 370

Q

Q value, 605, 608, 610, 619–620
Quadrupole moment, 533–537, 551–552, 558

Quadrupole radiation, 363–364, 377, 595
Quantization
of angular momentum, 78, 86
of energy, 80, 153–160, 217–223
Quantum electrodynamics, 662
Quantum numbers
F, 265
I, 265
J and m_J, 265, 318–320
j and m_j, 252–258
K, 337–338
L, 318–320
l and m, 195–196
n (principal), 243
S strangeness, 634–637, 641, 657
total spin, 293, 303
s and m_s, 251
T and T_3, 616–617
U and U_3, 659, 674–676
V and V_3, 674
vibrational, 336–337
Quantum statistics, 379–426
Quark, 626, 633, 657–661
colors, 662–663
flavors, 657
searches, 661–662

R

Rad, 514
Radial equation, 214–217, 261
Radial quantum number, 243
Radiation
atomic, 75–82, 100–103
nuclear, 471–524
Radioactive series, 473–475, 520–521
Radioactivity, discovery of, 471
Radius
atomic, 23, 79
Bohr, 79, 246, 278
nuclear, 526–530, 561, 571
Raman effect, 343
Ramsauer–Townsend effect, 237
Range
of charged particles, 483–487, 523–524
of strong (nuclear) interaction, 570
of weak interaction, 663
Range-energy relation, 483–485
Rare earth, 310
Rayleigh–Jeans law, 66–68, 397
Reaction, nuclear, 600–620
Reactor, nuclear, 551
Reduced mass, 80–82, 94, 241–242, 261, 334
Reduced zone scheme, 455–459
Reflection coefficient, 144–145, 166–170

Regeneration of K_1 meson, 648–649, 669
Relativistic addition of velocities, 48–50, 59–60
Relativistic correction to hydrogen atom energy
levels, 260–264
Relativistic dynamics, 50–58, 134, 607–608, 625
Relativity, special theory of, 27–60
Rem, 518
Residual nucleus, 607–609
Residual resistivity, 446
Resonance
in nuclear reactions, 603–612, 619
nuclear magnetic, 531
in particle interactions, 652–655
Rest energy, 55
Rest mass, 55
Rho meson, 652–653, 669
Richardson–Dushman equation, 428
Rotation of coordinates, 43, 201–203, 234–235
Rotational quantum number, 337–338
Rotational spectra
molecular, 334, 337–338, 342–343
nuclear, 556
Ruby maser, 372
Russell–Saunders coupling, *see LS* coupling
Rutherford scattering, 77, 475–480, 526–527
Rydberg constant, 76
Rydberg formula for hydrogen spectrum, 76,
78–82

S

Sakata model, 656–657
Scattering
of electron in metal, 423–425, 448, 455, 498
neutrino, 644
from "perfectly rigid" sphere, 232–234
from spherically symmetric potential, 223–234,
236–237
X-ray, 98, 103–106
Schrödinger equation
construction of, 138–141
for hydrogen atom, 242–246
for hydrogen molecule, 332–334
one-dimensional, 137–185
radial part, 214–217, 690–694
relativistic, 260–264
three-dimensional, 187–237
time dependent, 141
time-independent, 139, 150, 211, 280
two-particle, 240–242, 290–292
Scintillation camera, 501
Selection rules, 359–366
in beta decay, 587–589
for electric dipole transition, 353, 359–363
for electric quadrupole transition, 363–364
in gamma decay, 513, 594–598

on, 264–266, 314, 361, 377
on *m*, 359–360, 377
for magnetic dipole radiation, 364–366
for one-electron system, 86
on parity change, 361–362, 589
for simple harmonic oscillator, 362
Semiconductor, 425
  intrinsic, 462–467, 469
  n-type and *p*-type, 465
Semi-empirical mass formula, 541, 562, 620
Separation of variables, 189, 192–194
Series solution of Schrödinger equation, 677–685
Shell model of nucleus, 551–555, 594, 596
Shell structure
  atomic, 102–103, 304–311
  nuclear, 530, 551–555
Shielding number, 102–103
Sideways quark, 658
Sigma particle, 636, 639–641, 669–670
Simple cubic lattice, 452
Simple harmonic oscillator, *see* Harmonic
  oscillator
Simultaneity, 40, 47, 59
Single-valuedness of wavefunction, 194–195, 235
Singlet state, 293–296, 659, 673–675
Sodium
  atom, 314–315
  D lines, 315
Sommerfeld theory of hydrogen atom, 86–87,
  260, 263, 274
Specific heat
  Debye, 398–400
  Einstein, 397–398, 425
  of free-electron gas, 411–415, 428
Spectroscopic notation, 245–246, 314, 319
Spectroscopy
  atomic, 314–331
  molecular, 331–343
  nuclear, 594–600
Speed of light, 28, 740
Spherical Bessel functions, 218–221, 226, 236
Spherical coordinates, 191
Spherical Hankel functions, 221
Spherical harmonics, 196–198, 201–204,
  213–214, 234–235, 255
Spherical Neumann functions, 218–221, 226
Spherically symmetric potential, 190–191
Spin
  of deuteron, 556
  of electron, 248–260, 292–293, 304, 312
  of nucleus, 265
  of pion, 627
  of proton, 294–296
  relativity and, 266–273
Spin matrices, 276

Spin operator, 250–258, 311–312
Spin-orbit energy, 249–259, 258–259, 287,
  304–305, 315–319
Spin-orbit interaction, 249–250, 303–304,
  316–319, 551, 553
Spin-spin interaction, 316–319, 558
Spontaneous transitions, 355–357
Square-well potential
  with barrier inside, 162–164, 183
  one-dimensional, 150–164, 182–184
  periodic, 440–444
  three-dimensional, 217–223, 557
Standard deviation, 123, 173, 182, 684–685
Standing wave, 215, 394, 433, 435
Stark effect, 283
Stefan–Boltzmann law, 73, 92–93
Step potential, 142–145
Stepping operator, 178–180, 253–254, 674–675
Stern–Gerlach experiment, 207–211, 248, 532
Stimulated emission, 367–376
Stokes's law, 18, 21–22
Strange particles, 633–637
Strange quark, 657
Strangeness and *i*-spin, 634–637, 639–641, 657
Strong interaction, 622
Sudden approximation, 378
Superconductivity, 425, 446–449
Superfluid, 400–410
  critical temperature of, 402, 406
Supermultiplet, 658–659
Superposition
  of eigenfunctions, 202, 213–215, 220–221,
    236, 255–258, 276–277, 347
  of waves, 113–118
Susceptibility, magnetic, 420–423
Symmetry operations, 642–643
Symmetry of wavefunctions, 288–292, 297, 303,
  323–324

T

Tau lepton, 632–633, 641
Tau particle, 641–642
Tesla (unit), 458
Thermionic emission, 417–420, 428
Theta particle, 641–642
Thomas precession, 250
Thomson cathode-ray experiments, 12–14, 78
Thomson model of atom, 77
Time dilation (dilatation), 45–48, 59, 61
Time reversal invariance, 642–643
Top quark, 666
Totalitarian principle, 629
Transition elements, 310
Transition probability, 85, 349–357, 376–377, 578

Transitions, 78, 81–89, 345–366, 376–377
 forbidden, 359, 363–364, 366, 586–589,
  595–597
 induced, 346–357
 spontaneous, 346, 355–359
Transmission coefficient
 in alpha decay, 566–571
 for potential step, 141–145, 182
 for square barrier, 166–170, 185, 566
Triplet state, 293–294, 675–676
Tritium, 82, 585
Trouton–Noble experiment, 36–37
Truth, 666
Tunneling, *see* Barrier penetration

U

u-spin, 659, 674–676
Ultraviolet catastrophe, 67
Ultraweak interaction, 638
Uncertainty principle, 123–131, 136
 and angular momentum, 207–211
 and barrier penetration, 169–170
 and harmonic oscillator, 181–182
 and square well, 151–153, 182
 and trajectory of particle, 128–130
 and zero–point energy, 127
Uncertainty relation for energy and time, 351,
  354–355, 377, 502, 623, 655
Universal Fermi interaction, 636–637
Up quark, 657–658

V

v-spin, 674
Valence band, 432, 462–463
Variation method, 301–303
Vector
 covariant, 50, 53
 four-dimensional, 43, 50–58
Vibrational quantum number, 336–337
Vibrational spectra of molecules, 332, 334–338,
  342–343
Vibrational-rotational spectra of molecules,
  338–341, 343

W

W boson, 663–664, 667
Wave packet, 112–124, 135–136, 445, 457
 Gaussian, 121–123
Wavefunction, *see also* Eigenfunctions
 continuity of, 141, 149, 151
 and probability, 146–150
 single-valuedness of, 194–195, 235
 two-particle, 240–242, 290–292, 312
Wave-particle "duality", 95, 106, 111, 127–130
Weak interaction, 630–633, 635
Wien's law, 65–66, 92
Work function, 74, 415–417, 428
World line, 53

X

X particle, 668, 676
X-rays, 23, 95–106
 absorption edge, 132
 attenuation of, 487–490
 diffraction of, 23, 96–99, 435
 discovery of, 96
 muonic, 528
 production of, 99–103
 scattering of, 98, 103–106
Xi particle, 639–641, 670

Y

Yukawa potential, 622–626

Z

Z^0 boson, 663–664
Zeeman effect, 15, 326–331, 342, 377
 anomalous, 326, 331
 nuclear, 513
 quantum theory of, 326–331, 342
 in solid-state maser, 370–372
Zero-point energy, 127
 in harmonic oscillator, 127, 181
 in molecule, 127, 343
 in solid, 399